MODERN SYNTHETIC REACTIONS

Second edition

Herbert O. House
Georgia Institute of Technology

W. A. BENJAMIN, INC.
Menlo Park, California · Reading, Massachusetts
London · Amsterdam · Don Mills, Ontario · Sydney
1972

THE ORGANIC CHEMISTRY
MONOGRAPH SERIES

Organic Reaction Mechanisms
Second Edition
Ronald Breslow
Columbia University

Modern Synthetic Reactions
Second Edition
Herbert O. House
Georgia Institute of Technology

Introduction to Stereochemistry
Kurt Mislow
Princeton University

ISBN 0-8053-4501-9
CDEFGHIJKL-MA-798765

EDITOR'S FOREWORD TO
MODERN SYNTHETIC REACTIONS

It is now more than six years since the Organic Chemistry Monograph Series was launched with the publication of the first edition of *Modern Synthetic Reactions,* by Herbert House. It was followed by seven other books covering a wide range of material, and the continued success of many of them testifies to the great need for the brief monographs written at the level intended to enable the reader to follow a course in elementary organic chemistry. *Modern Synthetic Reactions* has been one of the most successful members of the series, and it has found wide use not only as a supplement and reference work but also as a very important text in courses in organic synthesis. The fact that it has been in such a great demand for six years after its publication indicates both the need for such a texbook and the high quality of the book itself.

However, synthetic organic chemistry is one of the most rapidly advancing branches of the field, with new methods appearing almost daily. The second edition of *Modern Synthetic Reactions,* while retaining to a great extent the successful format which proved so popular, expands and brings up to date the material it covers. It should therefore be even more successful as a teaching text as well as a book for reference and individual study.

It is our intention to maintain the entire series up to date with periodic new editions. *Organic Reaction Mechanisms,* for example, is another book of the series, which already appeared in a second edition in 1969. We continue to welcome any suggestions concerning other useful topics which might be included in the series, or any desirable changes in the books which have already been published in the Organic Chemistry Monograph Series.

Ronald Breslow

PREFACE TO THE SECOND EDITION

During the period since the first edition of this book was published, the development of synthetic organic chemistry has been characterized by introduction and use of a variety of new synthetic methods and improved reaction procedures. Although the objectives of this book and the general topics discussed—oxidations, reductions, formation of new carbon-carbon bonds—remain unchanged from the first edition, the substantial increase in the size of the second edition reflects mainly the new methodology which has been developed. Noteworthy examples of new or improved synthetic reactions which have been added to the second edition are homogeneous catalytic hydrogenation, dehydrogenation with quinones, reactions with alkylboranes, electrochemical reductions, reductions with hydrazine and diimide, oxidations with singlet oxygen, oxidations with lead tetraacetate, mercuric acetate, and other metal salts, stereospecific conversion of alcohols to alkyl halides, and reactions with sulfur ylids and related substances.

However, other synthetic reactions of importance, such as the Diels-Alder and related cycloaddition reactions, reactions of organometallic compounds, acylations at oxygen and nitrogen atoms, and elimination reactions, have not been included in this second edition, because an adequate discussion of any of these would make the present volume inconveniently large. Since each of the four topics has been reviewed extensively, a bibliography of selected reviews for each of these four synthetic methods has been included as an appendix.

Among the many people from whom I sought advice and suggestions about the content of this edition, I am indebted to Drs. E. C. Ashby, J. E. Baldwin, Drury S. Caine, Ernest E. Eliel, Alfred Hassner, Andrew S. Kende, Daniel J. Pasto, Thomas A. Spencer, and E. A. Sullivan for reading portions of the manuscript and offering helpful comments. I am particularly grateful to Drs. Robert E. Ireland, William S. Johnson, James A. Marshall, Norman A. Rabjohn, and Hugh W. Thompson for reading the entire manuscript and offering many useful suggestions. The task of checking the large number of references in this edition was accomplished by Miss Deborah Wiener and by my daughters, Margaret and Mary Ann. I am indebted to my wife, Jean, and to Mrs. Ann Edwards for substantial aid in preparing the subject index and, finally, my gratitude for typing the manuscript is extended to Miss Wiener and to Miss Laurie Walter.

H. O. H.

Atlanta, Georgia
November 1971

PREFACE TO THE FIRST EDITION

This book was written for advanced undergraduate and beginning graduate students of organic chemistry as a survey of certain reactions that currently enjoy widespread application to the synthesis of organic compounds in the laboratory. It was not my intention either to survey all of the important synthetic methods or to offer an exhaustive survey of those reactions which have been included. The Diels-Alder reaction, reactions of organometallic compounds, acylations at oxygen and nitrogen atoms, and elimination reactions leading to the formation of double and triple bonds are examples of important synthetic reactions which have received little, if any, mention. Nevertheless, I believe students will find that the assortment of reactions included have wide applicability to the practical solution of synthetic problems, and I can only hope that someone else, engaged in the training of students, will be sufficiently indignant about my omission of certain reactions to write a second volume which includes them.

The chapters have been written to survey reactions as they are currently used in the organic laboratory; accordingly, many of the references included refer to recent detailed descriptions of the applications of these reactions. In many cases examples have been taken from *Organic Syntheses* because of the detailed descriptions of experimental procedure which these preparations offer. In general, I have chosen not to trace the historical development of reactions and experimental procedures. However, these references may usually be found in the articles cited. Also, students of organic chemistry today are already burdened with the necessity to assimilate an immense body of factual knowledge, and I believe they should not also be required to concern themselves with which chemist should be allotted priority for discovering a particular reaction or procedure.

Discussions of the mechanisms of the various reactions included have been limited for the most part to statements and expressions of opinion, with leading references. Except for discussions of the stereochemical consequences of the various reactions, I have not reviewed the evidence that has led to the reaction mechanisms proposed since this is not the purpose of this book. In reading the mechanisms suggested, the student will do well to bear in mind that the evidence on which these proposals have been made is often tenuous and the mechanisms proposed may be shown to be partially or totally incorrect as additional evidence is accumulated.

Since the task of citing references to all of the synthetically important modifications of a reaction is impossible for one person to accomplish, I have asked a number of people in this country to offer suggestions of other material that might be included. I am indebted to Drs. W. G. Dauben, C. H. DePuy, E. L. Eliel, D. S. Heywood, K. B. Wiberg, and Mr. P. Starcher for offering suggestions relevant to certain portions of the manuscript, and I am particularly grateful to Drs. A. W. Burgstahler, R. K. Hill, R. E. Ireland, W. S. Johnson, G. Poos, and Mr. G. Mitchell for reading the entire manuscript and offering many helpful suggestions. The burden of preparing the manuscript and index as well as locating references, checking references, and proofreading the manuscript in various stages of preparation was lightened considerably by my five collaborators: Messrs. Joseph Ciabattoni, Theodore W. Craig, David B. Ledlie, George M. Rubottom, and Barry M. Trost, all

of whom are currently graduate students at the Massachusetts Institute of Technology. Finally, I am most grateful to Miss G. C. Martin, who typed the entire manuscript.

Herbert O. House

Cambridge, Massachusetts
February 24, 1964

CONTENTS

1

CATALYTIC HYDROGENATION
AND DEHYDROGENATION

Of the many reactions available for the reduction of organic compounds, catalytic hydrogenation—the reaction of a compound with hydrogen in the presence of a catalyst—offers the advantages of widespread applicability and experimental simplicity to a unique degree.[1] Catalytic hydrogenation is usually effected in the laboratory by stirring or shaking a solution of the compound to be reduced with a heterogeneous catalyst under an atmosphere of hydrogen gas. The progress of the reduction may be followed readily by measuring the uptake of hydrogen, and the crude reduction product is usually isolated simply by filtration of the catalyst followed by evaporation of the solvent.

CATALYSTS, SOLVENTS, AND EQUIPMENT

The selection of the solvent, reaction temperature, and hydrogen pressure for a given hydrogenation is dependent on the catalyst chosen. It is convenient to divide the common hydrogenation catalysts into the group of certain noble metal catalysts as well as active grades of Raney nickel, which are generally used with low (1 to 4 atm or 0 to 60 psi) hydrogen pressures and at relatively low (0 to 100°) temperatures; and the group of less active catalysts, which are normally used at higher (100 to 300 atm or 1500 to 4500 psi) hydrogen pressures and may require higher (25 to 300°) reaction temperatures.

The noble metal catalysts frequently used for low-pressure hydrogenations contain platinum, palladium, ruthenium, or rhodium. Whereas reductions over platinum frequently employ finely divided metallic platinum obtained by the reduction of a platinum compound[2] such as platinum oxide (PtO_2)[2a] in the hydro-

---

1. (a) H. Adkins and R. L. Shriner in H. Gilman, ed., *Organic Chemistry, An Advanced Treatise,* Vol. 1, 2d ed., Wiley, New York, 1943, pp. 779–832. (b) G. Schiller in E. Müller, ed., *Methoden der organischen Chemie* (*Houben-Weyl*), Vol. 4, Part 2, Georg Thieme Verlag, Stuttgart, Germany, 1955, pp. 283–332. (c) K. Wimmer, *ibid.,* pp. 163–192. (d) V. I. Komarewsky, C. H. Riesz, and F. L. Morritz in A. Weissberger, ed., *Technique of Organic Chemistry,* Vol. 2, 2d ed., Wiley-Interscience, New York, 1956, pp. 94–164. (e) F. J. McQuillin in A. Weissberger, ed., *Technique of Organic Chemistry,* Vol. 2, 3rd ed., Wiley-Interscience, New York, 1963, pp. 497–580. (f) R. L. Augustine, *Catalytic Hydrogenation,* Marcel Dekker, New York, 1965.
2. (a) R. Adams, V. Voorhees, and R. L. Shriner, *Org. Syn.,* **Coll. Vol. 1,** 463 (1944). (b) For the use of sodium borohydride to reduce platinum salts and other metal salts forming reactive hydrogenation catalysts, see H. C. Brown and C. A. Brown, *Tetrahedron,* **Suppl. 8(I),** 149 (1966); C. A. Brown and H. C. Brown, *J. Org. Chem.,* **31,** 3989 (1966); C. A. Brown, *J. Am. Chem. Soc.,* **91,** 5901 (1969); *Chem. Commun.,* **No. 3,** 139 (1970). (c) A very reactive platinum catalyst has been obtained by reducing chloroplatinic acid with triethylsilane in ethanol. C. Eaborn, B. C. Pant, E. R. A. Peeling, and S. C. Taylor, *J. Chem. Soc.,* C, 2823 (1969).

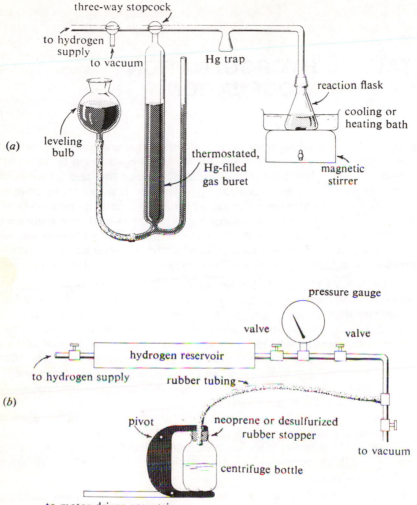

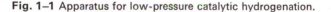

Fig. 1–1 Apparatus for low-pressure catalytic hydrogenation.

genation apparatus, the palladium,[3] ruthenium,[4] and rhodium[4] catalysts are usually deposits of the metal on the surface of an inert support such as carbon, alumina, barium sulfate, calcium carbonate, or strontium carbonate. Catalysts formed by

---

3. (a) E. R. Alexander and A. C. Cope, *Org. Syn., * **Coll. Vol. 3**, 385 (1955). (b) R. Mozingo, *ibid.,* **Coll. Vol. 3**, 685 (1955). (c) Preparations for nonpyrophoric supported palladium, rhodium, and ruthenium catalysts are given by W. M. Pearlman, *Tetrahedron Letters,* **No. 17**, 1663 (1967). (d) An osmium catalyst supported on alumina has been found useful for the selective reduction of α,β-unsaturated aldehydes to allylic alcohols. P. N. Rylander and D. R. Steele, *ibid.,* **No. 20**, 1579 (1969).

depositing platinum on an inert support are also used.[4] The activity of a catalyst on an inert support is normally diminished as the support is changed from carbon to barium sulfate to calcium or strontium carbonate. Also, the activity of a given catalyst is generally increased by changing from a neutral, nonpolar solvent to a polar, acidic solvent. Solvents frequently employed for low-pressure hydrogenation include ethyl acetate, ethanol, water, acetic acid, and acetic acid plus perchloric acid. Obviously, the acidic solvents cannot be employed with catalysts that are supported on metal carbonates. Diagrams of typical laboratory equipment used for low-pressure catalytic hydrogenations are provided in Figure 1-1a and b. The apparatus illustrated in Figure 1-1a utilizes a hydrogen pressure of 1 atm, and the hydrogen uptake is measured as the change in volume of the hydrogen gas in the system. Some workers prefer a modification of this apparatus in which the suspension of the catalyst in the reaction solution is shaken rather than stirred. Apparatus of this type is often used to obtain a quantitative measure of the number of reducible functions in a compound of unknown structure. Figure 1-1b represents apparatus[5] usually used with 2 to 4 atm hydrogen pressure, the hydrogen uptake being measured by observing the pressure change within the system. The following equations provide examples of typical laboratory hydrogenations that have been effected over noble metal catalysts.

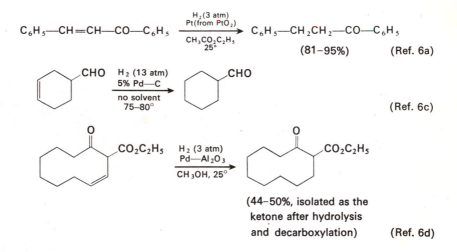

4. (a) Palladium, platinum, rhodium, and ruthenium catalysts are commercially available from Engelhard Industries, Inc., Chemical Division, 113 Astor St., Newark 14, New Jersey. (b) For a discussion of the effect of supports on the activity of platinum catalysts, see E. B. Maxted and J. S. Elkins, *J. Chem. Soc.*, 3630 (1963).
5. R. Adams and V. Voorhees, *Org. Syn.*, **Coll. Vol. 1**, 61 (1944).
6. (a) R. Adams, J. W. Kern, and R. L. Shriner, *Org. Syn.*, **Col. Vol. 1**, 101 (1944). See also (b) E. C. Horning, J. Koo, M. S. Fish, and G. N. Walker, *ibid.*, **Coll. Vol. 4**, 408 (1963). (c) H. E. Hennis and W. B. Trapp, *J. Org. Chem.*, **26**, 4678 (1961). (d) R. D. Burpitt and J. G. Thweatt, *Org. Syn.*, **48**, 56 (1968). (e) P. G. Gassman and K. T. Mansfield, *ibid.*, **49**, 1 (1969). (f) J. A. Marshall, N. Cohen, and A. R. Hochstetler, *J. Am. Chem. Soc.*, **88**, 3408 (1966).

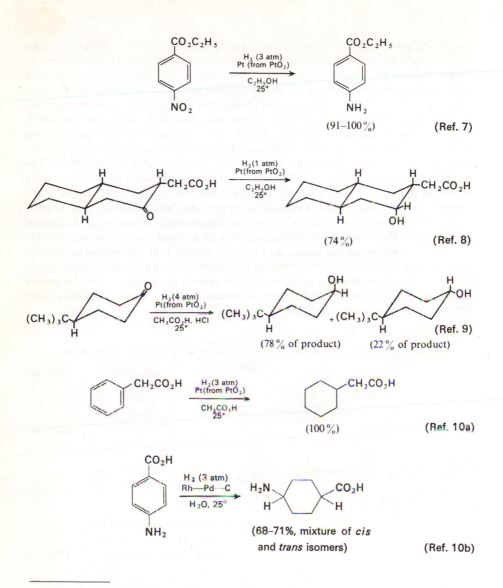

7. (a) R. Adams and F. L. Cohen, *Org. Syn.*, **Coll. Vol. I**, 240 (1944). (b) G. D. Mendenhall and P. A. S. Smith, *ibid.*, **46**, 85 (1966).

8. W. S. Johnson and co-workers, *J. Am. Chem. Soc.*, **83**, 606 (1961).

9. (a) E. L. Eliel and R. S. Ro, *J. Am. Chem. Soc.*, **79**, 5992 (1957). See also (b) R. J. Wicker, *J. Chem. Soc.*, 2165 (1956). (c) For a discussion of reaction conditions useful in the reduction of ketones, see E. Breitner, E. Roginski, and P. N. Rylander, *J. Org. Chem.*, **24**, 1855 (1959).   (d) The rate of ketone reduction over a platinum catalyst is enhanced by the use of trifluoroacetic acid rather than acetic acid as a solvent. P. E. Peterson and C. Casey, *ibid.*, **29**, 2325 (1964).

10. (a) R. Adams and J. R. Marshall, *J. Am. Chem. Soc.*, **50**, 1970 (1928). (b) W. M. Pearlman, *Org. Syn.*, **49**, 75 (1969).

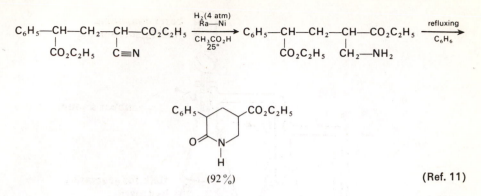

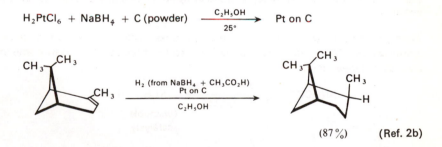

(92%) (Ref. 11)

A recent modification of the usual low-pressure catalytic hydrogenation technique utilizes the reaction of various metal salts with either sodium borohydride or a trialkyl- or triarylsilane to generate the hydrogenation catalyst.[2b,c] Salts of rhodium, platinum, and palladium are reduced to the finely divided free metals, which may be used directly as catalysts or, preferably, adsorbed on carbon prior to use.[2b] Reaction of excess sodium borohydride with an acid added to the

reaction mixture provides a source of hydrogen, permitting the direct hydrogenation of easily reducible functions (nitro groups and unhindered olefins). This procedure, illustrated in the preceding equations, offers the advantage of simplicity and can be combined with a device to automatically regulate the rate of hydrogen generation.[2b,12a] However, it suffers from the facts that the final hydrogenated product must be separated from a number of other components in the reaction mixture and that any functional groups present which are reduced by sodium borohydride (see Chapter 2) will also be altered by the reduction procedure. This difficulty can be overcome by use of a separate vessel for the generation of hydrogen (see Fig. 1-2)[2b,12a] A hydrogenation catalyst, thought to be nickel boride,[12b–d] has similarly

---

11. (a) R. K. Hill, C. E. Glassick and L. J. Fliedner, *J. Am. Chem. Soc.*, **81**, 737 (1959). See also (b) H. J. Dauben, Jr., H. J. Ringold, R. H. Wade, and A. G. Anderson, Jr., *ibid.*, **73**, 2359 (1951).
12. (a) This automatic device, called a Brown[2] Hydrogenator, is marketed by Delmar Scientific Laboratories, Inc., Maywood, Illinois. (b) R. Paul, P. Buisson, and N. Joseph, *Ind. Eng. Chem.*, **44**, 1006 (1952). (c) H. C. Brown and C. A. Brown, *J. Am. Chem. Soc.*, **85**, 1003, 1005 (1963). (d) C. A. Brown, *J. Org. Chem.*, **35**, 1900 (1970).

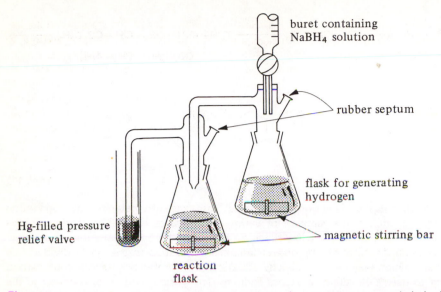

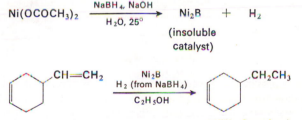

Fig. 1–2 Apparatus for generation and use of hydrogen in a low-pressure catalytic hydrogenation.

been prepared by the reduction of various nickel salts with sodium or potassium borohydride. This catalyst is reported[12b,d] to be more reactive than commercially available Raney nickel.

$$\text{Ni(OCOCH}_3)_2 \xrightarrow[\text{H}_2\text{O, 25}^\circ]{\text{NaBH}_4, \text{NaOH}} \underset{\substack{\text{(insoluble}\\\text{catalyst)}}}{\text{Ni}_2\text{B}} + \text{H}_2$$

$$\text{[cyclohexenyl]CH=CH}_2 \xrightarrow[\text{C}_2\text{H}_5\text{OH}]{\substack{\text{Ni}_2\text{B}\\ \text{H}_2 \text{ (from NaBH}_4)}} \text{[cyclohexenyl]CH}_2\text{CH}_3$$

(98% of product)   (Ref. 12d)

For hydrogenations at relatively high hydrogen pressures (100 to 300 atm), the common catalysts are Raney nickel, copper chromite, and ruthenium[4] supported on carbon or alumina. Raney nickel (abbreviated Ra-Ni)[13] is a porous form of metallic nickel, obtained by reaction of a nickel-aluminum alloy with aqueous sodium hydroxide.[14] Copper chromite (abbreviated CuCr$_2$O$_4$)[15] is a mixture of copper and

13. (a) R. Schröter in *Newer Methods of Preparative Organic Chemistry*, Wiley-Interscience, New York, 1948, pp. 61–101. (b) Raney nickel catalyst of ordinary activity, designated W-2, can be purchased or prepared as described by R. Mozingo, *Org. Syn.*, **Coll. Vol. 3**, 181 (1955) ; more active grades of Raney nickel, designated W-3 to W-7, which are effective at lower hydrogen pressures, may also be prepared. See (c) H. R. Billica and H. Adkins, *ibid.*, **Coll. Vol. 3**, 176 (1955), and (d) X. A. Dominguez, I. C. Lopez, and R. Franco, *J. Org. Chem.*, **26**, 1625 (1961). (e) Nickel catalysts are commercially available from the Raney Catalyst Co., Inc., Chattanooga, Tennessee, and from Girdler Catalysts, Chemical Products Division, Chemetron Corp., Louisville, Kentucky.

chromium oxides. Depending on the physical properties of the compound to be reduced, the high-pressure hydrogenations may be performed with or without a solvent. The solvents usually employed with these catalysts are ethanol, water, cyclohexane, or methylcyclohexane. Strongly acidic solvents or reactants cannot be used with nickel or copper chromite because these catalysts will dissolve. Figure 1-3 illustrates a typical autoclave used in the laboratory for high-pressure hydrogenations. The uptake of hydrogen in this apparatus is followed by observing the change in pressure. The total volume of liquid placed in such an apparatus should never exceed one-half the total volume of the autoclave, in order to allow room for the liquid to expand as it is heated. Typical examples of high-pressure laboratory hydrogenations are described by the following equations.

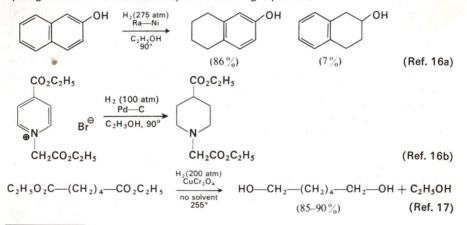

$$C_2H_5O_2C—(CH_2)_4—CO_2C_2H_5 \xrightarrow[\substack{no\ solvent \\ 255°}]{\substack{H_2(200\ atm) \\ CuCr_2O_4}} HO—CH_2—(CH_2)_4—CH_2—OH + C_2H_5OH$$

(85–90%)   (Ref. 17)

14. The reduction of organic compounds by reaction with the Raney alloy and aqueous alkali has also been used: for examples, see (a) D. Papa, E. Schwenk, and B. Whitman, *J. Org. Chem.,* **7**, 587 (1942). (b) D. Papa, E. Schwenk, and H. F. Ginsberg, *ibid.,* **16**, 253 (1951). (c) E. Schwenk, D. Papa, H. Hankin, and H. Ginsberg, *Org. Syn.,* **Coll. Vol. 3,** 742 (1955). (d) B. Staskun and T. van Es, *J. Chem. Soc.,* C, 531 (1966). This procedure may involve, at least in part, a low-pressure catalytic hydrogenation in which the hydrogen liberated from reaction of the aluminum with the aqueous alkali is absorbed on the surface of the newly formed Raney nickel catalyst; however, certain of the reductions performed by this procedure resemble more closely the subsequently discussed dissolving metal reductions (see Chapter 3).
15. (a) C. Grundman in *Newer Methods of Preparative Organic Chemistry*, Wiley-Interscience, New York, 1948, pp. 103–123. (b) W. A. Lazier and H. R. Arnold, *Org. Syn.* **Coll. Vol. 2,** 142 (1943). (c) This catalyst is commercially available from Girdler Catalysts, Chemical Products Division, Chemetron Corp., Louisville, Kentucky.
16. (a) H. Adkins and G. Krsek, *J. Am. Chem. Soc.,* **70**, 412 (1948). If the reaction mixture is alkaline during the hydrogenation, the alcohol rather than the phenol becomes the major product [G. Stork, *ibid.,* **69**, 576 (1947)]. (b) H. U. Daeniker and C. A. Grob, *Org. Syn.,* **44**, 86 (1964).
17. (a) W. A. Lazier, J. W. Hill, and W. J. Amend, *Org. Syn.,* **Coll. Vol. 2,** 325 (1943). (b) For a discussion and review of this reaction, see H. Adkins, *Org. Reactions,* **8**, 1 (1954). (c) Various rhenium oxide catalysts have been recommended for the reduction of carboxylic acids. See H. S. Broadbent and T. G. Selin, *J. Org. Chem.,* **28**, 2343 (1963); H. S. Broadbent and W. J. Bartley, *J. Org. Chem.,* **28**, 2345 (1963); H. S. Broadbent and D. W. Seegmiller, *J. Org. Chem.,* **28**, 2347 (1963). (d) The hydrogenation of cyclic imides and anhydrides over a platinum catalyst has been studied by A. J. McAlees and R. McCrindle, *J. Chem. Soc.* C, 2425 (1969).

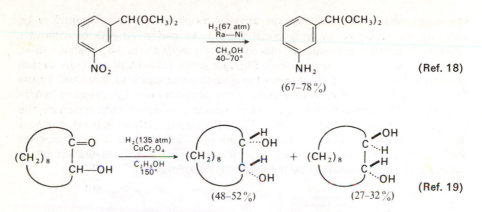

(Ref. 18)

(67–78%)

(Ref. 19)

(48–52%)   (27–32%)

The rate of a given hydrogenation may be increased by increasing the hydrogen pressure,[20] by increasing the amount of catalyst employed,[6a] or by increasing the temperature. However, many compounds can be reduced either by low-pressure or by high-pressure hydrogenation. The selection of the reaction conditions in such cases is usually determined by the quantity of material to be reduced. The type of low-pressure hydrogenation apparatus illustrated in Figure 1-1a is normally useful for reduction of quantities ranging from 10 mg to 25 g, whereas the apparatus shown in Figure 1-1b is most often used with 10- to 100-g quantities. The high-pressure equipment (Figure 1-3) available in most organic laboratories permits the reduction of 10- to 1000-g quantities.

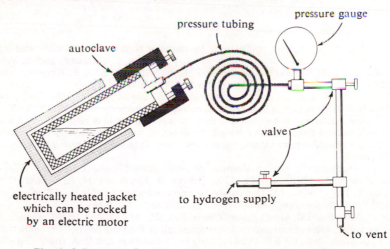

pressure gauge

pressure tubing

autoclave

valve

electrically heated jacket
which can be rocked
by an electric motor

to hydrogen supply

to vent

Fig. 1–3 Apparatus for high-pressure catalytic hydrogenation.

18. (a) R. N. Icke, C. E. Redemann, B. B. Wisegarver, and G. A. Alles, *Org. Syn.,* **Coll. Vol. 3,** 59 (1955). (b) K. Dimroth, A. Berndt, H. Perst, and C. Reichardt, *ibid.,* **49,** 116 (1969).
19. A. T. Blomquist and A. Goldstein, *Org. Syn.,* **Coll. Vol. 4,** 216 (1963).
20. For example, see C. F. H. Allen and J. VanAllan, *Org. Syn.,* **Coll. Vol. 3,** 63 (1955).

REDUCTION OF FUNCTIONAL GROUPS

From the equations previously used to illustrate reaction conditions, it is apparent that a variety of functional groups can be reduced by catalytic hydrogenation. In Table 1-1 the common functional groups are listed in approximate order of decreasing ease of catalytic hydrogenation. This order should not be considered inflexible since special structural features in the compound being reduced or changes in catalyst or reaction conditions sometimes result in interchanging the ease of reduction of functional groups having similar reactivity. However, it is almost always possible to selectively reduce functional groups found

Table 1–1 Approximate order of reactivity of functional groups in catalytic hydrogenation

Functional group	Reduction product(s)	Comments
R—CO—Cl	R—CHO	Most easily reduced
R—NO$_2$	R—NH$_2$	
R—C≡C—R	(H, H / C=C / R, R olefin)	
R—CHO	R—CH$_2$OH	With Pt catalyst, reduction is accelerated by ferrous ion
R—CH=CH—R	R—CH$_2$CH$_2$—R	Ease of reduction is decreased by the presence of additional substituents
R—CO—R	R—CHOH—R	
C$_6$H$_5$CH$_2$OR	C$_6$H$_5$CH$_3$ + ROH	
R—C≡N	R—CH$_2$NH$_2$	
(naphthalene)	(tetralin)	Also partial reduction of other polycyclic aromatic systems
R—CO—O—R′	R—CH$_2$OH + R′—OH	Pt and Pd catalysts fail to effect these reductions
R—CO—NH—R	R—CH$_2$NH—R	
(benzene)	(cyclohexane)	Least easily reduced
R—CO$_2{}^{\ominus}$Na$^{\oplus}$		Inert

near the top of Table 1-1 in the presence of functional groups listed at the bottom of the table, and it is rarely possible to selectively reduce the functional groups listed at the bottom in the presence of the more reactive functions listed at the top. For example, the reduction of an unsaturated ester or ketone to an unsaturated alcohol is normally accomplished with metal hydride reducing agents (Chapter 2) rather than by catalytic hydrogenation,[3d] whereas reduction of an unsaturated ester or ketone to a saturated ester or ketone is readily achieved by catalytic hydrogenation.[6] The platinum, palladium, and nickel catalysts are most often used for the

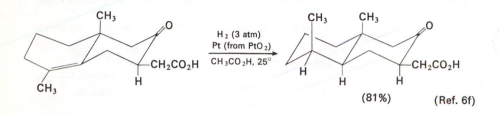

(81%)   (Ref. 6f)

reduction of isolated carbon-carbon double bonds; the platinum, copper chromite, and ruthenium catalysts are most often used to reduce carbonyl groups,[9c] and the platinum, nickel and rhodium catalysts are most often used to reduce aromatic, systems.

It will be noted in Table 1-1 that catalytic hydrogenation may serve either to saturate a multiple bond (e.g., C=C or C=O) or to cleave certain types of single bonds. The cleavage of a single bond by catalytic hydrogenation, termed hydrogenolysis, is normally found with allylic or benzylic amines or alcohols having the

structural units $-\overset{|}{C}=\overset{|}{C}-\overset{|}{C}-O-$ or $-\overset{|}{C}=\overset{|}{C}-\overset{|}{C}-N-$

and with compounds containing C—halogen and C—S single bonds. Hydrogenolysis is also observed with N—N, N—O, and O—O single bonds as well as with C—C single bonds of small rings and the C—O or C—N bonds of three-membered ring heterocycles.[21]

$$(CH_3)_2C\!\!-\!\!CH_2 \xrightarrow[\text{dioxane}]{\substack{H_2 \text{ (4–5 atm)} \\ Ra\text{—}Ni}} (CH_3)_3C\!\!-\!\!NH_2$$

with N and ·H below the ring.

(75–82%)

(Ref. 21c)

21. (a) J. Newham, *Chem. Rev.,* **63**, 123 (1963). (b) For a discussion of the mechanism of cyclopropane hydrogenolysis, see W. J. Irwin and F. J. McQuillin, *Tetrahedron Letters,* **No. 18**, 2195 (1968). (c) K. N. Campbell, A. H. Sommers, and B. K. Campbell, *Org. Syn.,* **Coll. Vol. 3**, 148 (1955). (d) For discussion of the stereochemistry of hydrogenolysis of epoxides and aziridines, see A. Suzuki, M. Miki, and N. Itoh, *Tetrahedron,* **23**, 3621 (1967); S. Mitsui and Y. Sugi, *Tetrahedron Letters,* **No. 16**, 1287, 1291 (1969).

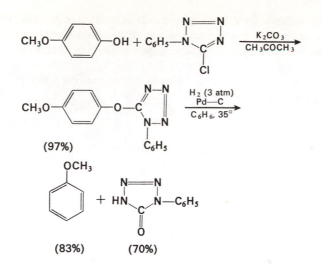

(97%)

(83%)   (70%)   (Ref. 23b)

The occurrence of hydrogenolysis may be considered either desirable or objectionable, depending on the goal of a given reaction. The hydrogenation of the aryl ester **[1]** to benzyl alcohol is difficult to realize because of further reaction

$$C_6H_5CO_2C_2H_5 \xrightarrow[\substack{C_2H_5OH \\ 200-250°}]{\substack{H_2(200 \text{ atm}) \\ CuCr_2O_4}} [C_6H_5CH_2OH] \xrightarrow{\text{hydrogenolysis}} C_6H_5CH_3$$
$$\text{[1]} \qquad\qquad + \qquad\qquad\qquad +$$
$$\qquad\qquad\qquad C_2H_5OH \qquad\qquad\qquad H_2O$$

to form toluene, and it requires the use of excess catalyst and lower reaction temperatures.[22] However, the comparable reduction of the keto acid salt **[2]** offers a useful procedure for the removal of the ketone function. The ready hydrogenolysis of benzyl ethers and amines[24] has prompted the extensive use of the benzyl group as a protecting group that can be removed under mild, nonhydrolytic conditions.

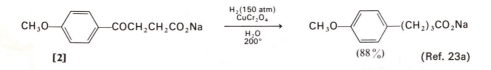

[2]   (88%)   (Ref. 23a)

22. R. Mozingo and K. Folkers, *J. Am. Chem. Soc.,* **70**, 229 (1948).
23. (a) H. O. House and R. J. McCaully, *J. Org. Chem.,* **24**, 725 (1959); see also L. F. Fieser and W. H. Daudt, *J. Am. Chem. Soc.,* **63**, 782 (1941). (b) W. J. Musliner and J. W. Gates, Jr., *ibid.,* **88**, 4271 (1966); for the hydrogenolysis of aryl *p*-toluenesulfonic esters to form aromatic hydrocarbons, see G. W. Kenner and M. A. Murray, *J. Chem. Soc.,* S 178 (1949).
24. (a) W. H. Hartung and R. Simonoff, *Org. Reactions,* **7**, 263 (1953). (b) R. E. Bowman, *J. Chem. Soc.,* 325 (1950).

The last step in Bowman's ketone synthesis is illustrated by the hydrogenolysis of the benzyl ester **[3]**; the hydrogenolysis of the benzyl urethane **[4]** illustrates the use of the carbobenzyloxy blocking group in peptide synthesis.

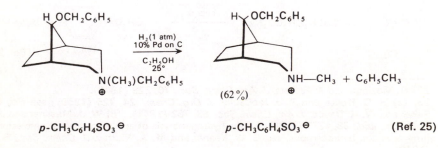

$$CH_3(CH_2)_6CO{\Large\diagdown} \atop CH_3(CH_2)_7{\Large\diagup} C(CO_2CH_2C_6H_5)_2$$

[3]

$$\xrightarrow[\substack{C_2H_5OH \\ CH_3CO_2C_2H_5 \\ 25-30°}]{\substack{H_2(1\ atm) \\ 10\%\ Pd\ on\ C}}$$

$$\left[CH_3(CH_2)_6CO{\Large\diagdown} \atop CH_3(CH_2)_7{\Large\diagup} C(COOH)_2 \atop +\ 2\ \ C_6H_5CH_3 \right]$$

$$\xrightarrow{heat} CH_3(CH_2)_6CO—CH_2(CH_2)_7CH_3\ +\ 2\ CO_2$$

$$(91\%)$$

(Ref. 24b)

$$C_6H_5CH_2O—CO—NH \atop HO_2C—CH_2CH_2—\overset{|}{CH}—CONHCH_2CO_2C_2H_5$$

[4]

$$\xrightarrow[C_2H_5OH,\ CH_3CO_2H]{\substack{H_2(1\ atm) \\ Pt(from\ PtO_2)}}$$

$$\left[HO—CO—NH \atop HO_2C—CH_2CH_2—\overset{|}{CH}—CONHCH_2CO_2C_2H_5 \atop +\ C_6H_5CH_3 \right]$$

$$\longrightarrow CO_2\ +\ {}^{\ominus}O_2CCH_2CH_2—\underset{\overset{|}{\overset{\oplus}{N}H_3}}{CH}—CONHCH_2CO_2C_2H_5$$

$$(80\%)$$

(Ref. 24a)

The relative ease of cleavage of benzyl groups, namely,

$$C_6H_5CH_2—\overset{|}{\underset{|}{N}}{}^{\oplus}— > C_6H_5CH_2—O— > C_6H_5CH_2—\overset{|}{N}—,$$

is illustrated in the following equation.

$$\xrightarrow[\substack{C_2H_5OH \\ 25°}]{\substack{H_2(1\ atm) \\ 10\%\ Pd\ on\ C}}$$

$$(62\%)$$

$$p\text{-}CH_3C_6H_4SO_3{}^{\ominus} \qquad\qquad p\text{-}CH_3C_6H_4SO_3{}^{\ominus}$$

(Ref. 25)

25. H. O. House, P. P. Wickham, and H. C. Müller, *J. Am. Chem. Soc.*, **84**, 3139 (1962).

The hydrogenolysis of allyl alcohols, ethers, and esters is less often synthetically useful because the rate of cleavage is often similar to the rate of reduction of the carbon-carbon double bond, leading to mixtures of products.

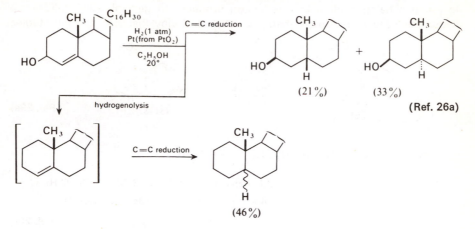

(21%)   (33%)

(Ref. 26a)

(46%)

Either palladium-on-carbon or copper chromite catalyst is usually employed for the hydrogenolysis of benzyl groups, since with these catalysts competing reduction of the aromatic ring occurs only very slowly. Rhodium and ruthenium catalysts have been found most effective when one wishes to minimize hydrogenolysis[27,28] during the reduction of an aromatic ring, or an olefinic double bond. The following examples are illustrative.

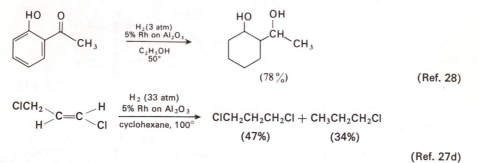

(78%)   (Ref. 28)

$CICH_2CH_2CH_2Cl + CH_3CH_2CH_2Cl$

(47%)   (34%)

(Ref. 27d)

26. (a) M. C. Dart and H. B. Henbest, *J. Chem. Soc.*, 3563 (1960) : These authors found that the presence of alkaline inorganic salts in the reaction mixture decreased the amount of hydrogenolysis and produced more saturated alcohol where the hydrogen has added *cis* to the hydroxyl function. (b) The use of alkali with platinum and palladium catalysts to minimize the hydrogenolysis of benzyl alcohols has been reported by W. Theilacker and H. G. Drössler, *Chem. Ber.*, **87**, 1676 (1954).
27. (a) A. W. Burgstahler and Z. J. Bithos, *Org. Syn.*, **42**, 62 (1962). (b) J. H. Stocker, *J. Org. Chem.*, **27**, 2288 (1962). (c) A. E. Barkdoll, D. C. England, H. W. Gray, W. Kirk, Jr., and G. M. Whitman, *J. Am. Chem. Soc.*, **75**, 1156 (1953). (d) G. E. Ham and W. P. Coker, *J. Org. Chem.*, **29**, 194 (1964). (e) A. I. Meyers, W. Beverung, and G. Garcia-Munoz, *J. Org. Chem.*, **29**, 3427 (1964).
28. I. A. Kaye and R. S. Matthews, *J. Org. Chem.*, **28**, 325 (1963).

Although the ready hydrogenolysis of carbon-halogen bonds offers a useful method for removing halogen from a molecule (e.g., reduction of [5]), the ease of this cleavage often prevents the use of catalytic hydrogenation when one desires to retain a halogen atom in a reduction product. In general, alkyl halides are hydrogenolyzed more slowly than aryl, allyl, and vinyl halides. Several studies

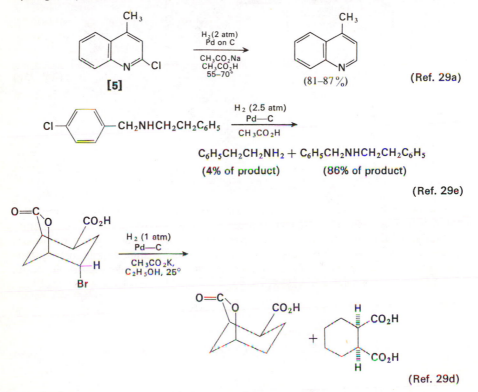

(Ref. 29a)

(Ref. 29e)

(Ref. 29d)

have indicated that the presence of bases in the reaction mixture not only serves to remove the hydrogen halide but also enhances the rate of carbon-halogen bond cleavage.[29b,c,d] The Rosenmund reduction of acid chlorides (such as [6]) to aldehydes illustrates the ease of carbon-halogen cleavage. In this synthesis, a mixture called a catalyst poison is usually added to reduce the activity of the hydrogenation catalyst.

29. (a) F. W. Neumann, N. B. Sommer, C. E. Kaslow, and R. L. Shriner, *Org. Syn.,* **Coll. Vol. 3**, 519 (1955). (b) For the use of hydrogen and Raney nickel or a palladium-on-carbon catalyst with potassium hydroxide for the hydrogenolysis of aryl, vinyl, and alkyl halides, see H. Kämmerer, L. Horner, and H. Beck, *Chem. Ber.,* **91**, 1376 (1958) ; C. K. Alden and D. I. Davies, *J. Chem. Soc.,* C, 700 (1968). (c) M. G. Reinecke, *J. Org. Chem.,* **29**, 299 (1964). (d) D. A. Denton, F. J. McQuillin, and P. L. Simpson, *J. Chem. Soc.,* 5535 (1964). (e) M. Freifelder, *J. Org. Chem.,* **31**, 3875 (1966). (f) For examples of the rearrangement of bromohydrins to form ketones during catalytic hydrogenolysis over a palladium-on-strontium carbonate catalyst, see R. L. Clarke and S. J. Daum, *ibid.,* **30**, 3786 (1965).

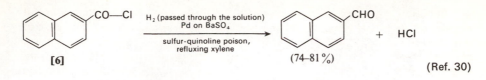

(Ref. 30)

A variety of substances are poisons for hydrogenation catalysts,[31] the most commonly encountered being mercury, divalent sulfur compounds, and, to a lesser degree, amines. These poisons are believed to function by being preferentially bonded to the catalyst surface, preventing bonding between the catalyst surface and the molecule to be reduced. It is sometimes possible to remove small amounts of impurities which are catalyst poisons by stirring a solution of the starting material with a portion of Raney nickel catalyst. The resulting mixture is filtered to remove the Raney nickel catalyst containing adsorbed impurities, and the filtrate is mixed with a fresh portion of a catalyst and subjected to the usual hydrogenation procedure. Although carbon sulfur bonds are liable to hydrogenolysis, the bond cleavage cannot be performed in the manner of a normal catalytic hydrogenation because both the reactants and the products are powerful catalyst poisons. However, the problem has been circumvented by allowing divalent sulfur compounds to react with the hydrogen adsorbed on a large (tenfold) excess of Raney nickel.[32] This process, Raney nickel desulfurization, is illustrated in the following equations. The reaction has been suggested to involve the removal of sulfur to form free radical intermediates, particularly in cases where the nickel catalyst used is deficient in adsorbed hydrogen. Whereas this procedure offers an extremely mild method for

$$n\text{-}C_4H_9\!-\!\underset{S}{\text{(thiophene)}}\!-\!COCH(CH_2)_4CO_2H$$
$$\overset{OH}{|}$$

$$\xrightarrow[\text{H}_2\text{O, reflux}]{\text{Ra—Ni, Na}_2\text{CO}_3} \quad \xrightarrow{\text{HCl}}$$

$$CH_3(CH_2)_7CH\!-\!CH(CH_2)_4CO_2H$$
$$\underset{OH}{|} \quad \underset{OH}{|}$$

(50%, mixture of *erythro* and *threo* isomers)   (Ref. 32c)

30. (a) E. B. Hershberg and J. Cason, *Org. Syn., Coll. Vol. 3*, 627 (1955). (b) See also E. Mosettig and R. Mozingo, *Org. Reactions,* **4**, 362 (1948). (c) The use of thiourea as a poison for the Rosenmund reduction has been recommended: see C. Weygand and W. Meusel, *Ber.,* **76**, 503 (1943). (d) In at least one instance a successful Rosenmund reduction was effected at 30–35° in the absence of a catalyst poison: see W. S. Johnson, D. G. Martin, R. Pappo, S. D. Darling, and R. A. Clement, *Proc. Chem. Soc.,* 58 (1957).
31. For examples, see L. Horner, H. Reuter, and E. Herrmann, *Justus Liebigs Ann. Chem.* **660**, 1 (1962).
32. (a) G. R. Pettit and E. E. van Tamelen, *Org. Reactions,* **12**, 356 (1962). (b) H. Hauptmann and W. F. Walter, *Chem. Rev.,* **62**, 347 (1962). (c) J. F. McGhie, W. A. Ross, D. H. Laney, and J. M. Barker, *J. Chem. Soc.,* C, 1 (1968); for a review of syntheses involving the desulfurization and reduction of thiophene intermediates, see Y. L. Gol'dfarb, S. Z. Taits, and L. I. Belen'kii, *Tetrahedron,* **19**, 1851 (1963).

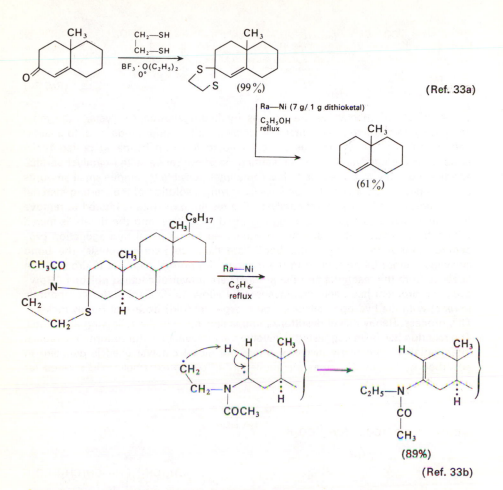

(Ref. 33a)

(Ref. 33b)

the reduction of an aldehyde or ketone to a methyl or methylene group, it becomes much less attractive for large-scale reductions because of the quantities of Raney nickel that must be employed.[34]

The reduction of nitriles to primary amines is complicated not only by the equilibria illustrated in the accompanying equations, which lead to the production of secondary amines as by-products, but also by the fact that the amine products

33. (a) F. Sondheimer and D. Rosenthal, *J. Am. Chem. Soc.,* **80**, 3995 (1958); for a similar reaction where a number of double bond isomers were obtained, see V. Permutti, N. Danieli, and Y. Mazur, *Tetrahedron,* **24**, 5425 (1968). (b) N. S. Crossley, C. Djerassi, and M. A Kielczewski, *J. Chem. Soc.,* 6253 (1965); E. L. Eliel and S. Krishnamurthy, *J. Org. Chem.,* **30**, 848 (1965). (c) For a brief review of the formation of ketals and thioketals, see J. F. W. Keana in C. Djerassi, ed., *Steroid Reactions,* Holden-Day, San Francisco, 1963, pp. 1–33. 34. Alternative methods for the reduction of dithioketals include the reaction with hydrazine and base reported by V. Georgian, R. Harrisson, and N. Gubisch, *J. Am. Chem. Soc.,* **81**, 5834 (1959); also reduction with sodium and liquid ammonia has been used. For an example, see R. E. Ireland, T. I. Wrigley, and W. G. Young, *ibid.,* **80**, 4604 (1958).

are catalyst poisons. With noble metal catalysts, these difficulties may be avoided by performing the hydrogenation in an acidic solvent or in acetic anhydride, a practice which serves to remove the primary amine from these equilibria as its salt or as its acetamide derivative. Either of these procedures, both of which have also been used for the reduction of oximes, converts the amine to a substance that is

$$R—C≡N \xrightarrow[\text{catalyst}]{H_2} R—CH=NH \xrightarrow[\text{catalyst}]{H_2} R—CH_2—NH_2$$

$$R—CH_2—NH_2 + R—CH=NH \rightleftharpoons R—CH_2NH—\underset{\underset{NH_2}{|}}{CH}—R \rightleftharpoons R—CH_2—N=CH—R + NH_3$$

$$R—CH_2—N=CH—R \xrightarrow[\text{catalyst}]{H_2} R—CH_2—NH—CH_2—R$$

no longer a catalyst poison. A similar principle is involved in the hydrogenation of pyridinium salts to tetrahydropyridines over a palladium catalyst. The reduction of azides is sufficiently rapid that no special precautions are needed to avoid catalyst poisoning. For the high-pressure hydrogenation of nitriles over Raney nickel,

$$N≡C—(CH_2)_3—\underset{\underset{HON}{\|}}{C}—COOH \xrightarrow[\substack{CH_3COOH \\ 25°}]{\substack{H_2(3\ atm) \\ Pt(from\ PtO_2)}} H_2N—(CH_2)_4—\underset{\underset{NH_2}{|}}{CH}—COOH$$

(43% as hydrochloride)   (Ref. 35a)

$$ClCH_2CO_2C(CH_3)_3 + NaN_3 \xrightarrow[\text{reflux}]{\text{acetone—}H_2O}$$

$$N_3—CH_2—CO_2—C(CH_3)_3 \xrightarrow[\text{CH}_3\text{OH, 25°}]{\substack{H_2\ (1\ atm) \\ Pd—C}} H_2N—CH_2CO_2C(CH_3)_3$$

(92%)

(75–82% isolated
as phosphite salt)

(Ref. 35d)

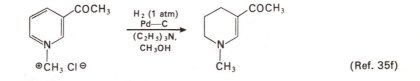

(Ref. 35f)

35. (a) A. F. Ferris, G. S. Johnson, F. E. Gould, and H. K. Latourette, *J. Org. Chem.*, **25**, 492 (1960). (b) The use of a rhodium catalyst for the reduction of nitriles to amines has been recommended: see M. Freifelder, *J. Am. Chem. Soc.*, **82**, 2386 (1960). (c) For the use of a palladium-on-carbon catalyst with oximes, see W. H. Hartung, J. H. R. Beaujon, and G. Cocolas, *Org. Syn.*, **40**, 24 (1960). (d) A. T. Moore and H. N. Rydon, *Org. Syn.*, **45**, 47 (1965). (e) A palladium-on-carbon catalyst has been used at elevated temperatures and high pressure (100 atm) to reduce β-amino-α,β-unsaturated esters to the saturated amino esters; R. L. Augustine, R. F. Bellina, and A. J. Gustavsen, *J. Org. Chem.*, **33**, 1287 (1968). (f) E. Wenkert and co-workers, *ibid.*, **33**, 747 (1968).

$$C_6H_5CH_2-C\equiv N \xrightarrow[\substack{\text{excess NH}_3 \\ \text{no solvent} \\ 120-130°}]{\substack{H_2 \text{ (130 atm)} \\ \text{Ra—Ni}}} C_6H_5CH_2CH_2-NH_2$$

$$(83-87\%) \qquad \text{(Ref. 36)}$$

where an acidic solvent cannot be used, an excess of ammonia is added to the reaction mixture to displace the equilibrium responsible for secondary-amine formation.

Similar reaction conditions have been used for the conversion of a ketone to an amine. In this process, called reductive alkylation,[37] a mixture of a ketone (e.g., [7]) and ammonia or a primary or secondary amine (e.g., [8]) is hydrogenated over a Raney nickel or platinum catalyst. The product is presumably formed by preferential reduction of an intermediate imine (e.g., [9]).

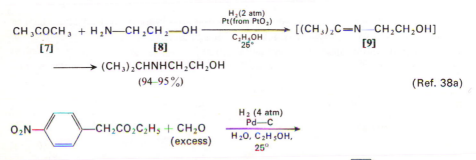

$$CH_3COCH_3 + H_2N-CH_2CH_2-OH \xrightarrow[\substack{C_2H_5OH \\ 26°}]{\substack{H_2 \text{(2 atm)} \\ \text{Pt(from PtO}_2)}} [(CH_3)_2C=N-CH_2CH_2OH]$$

[7]   [8]   [9]

$$\longrightarrow (CH_3)_2CHNHCH_2CH_2OH$$

$$(94-95\%) \qquad \text{(Ref. 38a)}$$

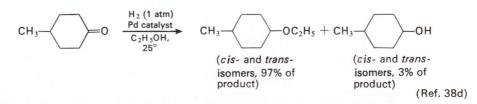

(67—77%)   (Ref. 38c)

An analogous conversion of cyclohexanone derivatives to ethers has been observed during hydrogenations in ethanol solution. The reaction is particularly favorable when a palladium catalyst is used.

CH$_3$—⬡=O $\xrightarrow[\substack{C_2H_5OH, \\ 25°}]{\substack{H_2 \text{ (1 atm)} \\ \text{Pd catalyst}}}$ CH$_3$—⬡—OC$_2$H$_5$ + CH$_3$—⬡—OH

(*cis-* and *trans-*
isomers, 97% of
product)

(*cis-* and *trans-*
isomers, 3% of
product)

(Ref. 38d)

36. J. C. Robinson, Jr., and H. R. Snyder, *Org. Syn.,* **Coll. Vol. 3**, 720 (1955).
37. W. S. Emerson, *Org. Reactions,* **4**, 174 (1948).
38. (a) E. M. Hancock and A. C. Cope, *Org. Syn.,* **Coll. Vol. 3**, 501 (1955). (b) See also D. M. Balcom and C. R. Noller, *ibid.,* **Coll. Vol. 4**, 603 (1963). (c) M. G. Romanelli and E. I. Becker, *ibid.,* **47**, 69 (1967). (d) S. Nishimura, T. Itaya, and M. Shiota, *Chem. Commun.,* **No. 9**, 422 (1967).

MECHANISM AND STEREOCHEMISTRY

The early studies by Linstead and co-workers[39] of the hydrogenation of phenanthrene and diphenic acid derivatives (e.g., [10]) over platinum led to the concept that the less hindered side of an unsaturated molecule is adsorbed on the catalyst surface. The adsorption was thought to be followed by the simultaneous transfer

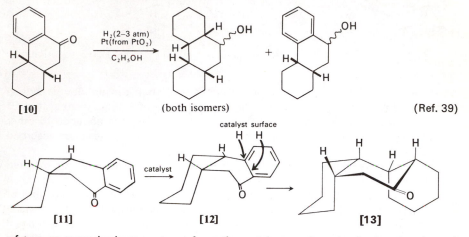

[10]   (both isomers)   (Ref. 39)

catalyst surface

[11]   [12]   [13]

of two or more hydrogen atoms from the catalyst to the adsorbed molecule and subsequent desorption of the reduced molecule, as illustrated in the sequence [11] → [12] → [13]. This concept has led to the useful generalization that catalytic hydrogenation of a multiple bond results in the *cis* addition of two hydrogen atoms from the less hindered side of the multiple bond. Two of the many examples of the utility of this generalization are illustrated below: the hydrogenation of disubstituted acetylenes (e.g., [14]) to *cis*-olefins over a palladium catalyst partially poisoned with lead acetate (Lindlar catalyst)[40] or quinoline,[41] and the reduction

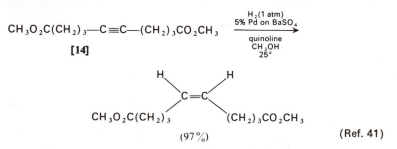

$$CH_3O_2C(CH_2)_3 - C \equiv C - (CH_2)_3CO_2CH_3$$

[14]

$$\xrightarrow[\substack{quinoline \\ CH_3OH \\ 25°}]{\substack{H_2(1\ atm) \\ 5\%\ Pd\ on\ BaSO_4}}$$

$$CH_3O_2C(CH_2)_3 \diagdown C = C \diagup (CH_2)_3CO_2CH_3$$

(97%)   (Ref. 41)

39. (a) R. P. Linstead, W. E. Doering, S. B. Davis, P. Levine, and R. R. Whetstone, *J. Am. Chem. Soc.,* **64**, 1985, 1991, 2003, 2007, 2009, 2014, 2022 (1942). (b) See also L. F. Fieser and M. Fieser, *Steroids,* Reinhold, New York, 1959, pp. 271–274.
40. (a) H. Lindlar, *Helv. Chim. Acta,* **35**, 446 (1952). (b) H. Lindlar and R. Dubuis, *Org. Syn.,* **46**, 89 (1966).
41. (a) D. J. Cram and N. L. Allinger, *J. Am. Chem. Soc.,* **78**, 2518 (1956). (b) N. A. Dobson, G. Eglinton, M. Krishnamurti, R. A. Raphael, and R. G. Willis, *Tetrahedron,* **16**, 16 (1961). (c) E. N. Marvell and J. Tashiro, *J. Org. Chem.,* **30**, 3991 (1965).

of the ketone [15] from the less hindered side. Terminal acetylenes are hydrogenated to olefins more rapidly than are disubstituted acetylenes:[41b] the relative rates

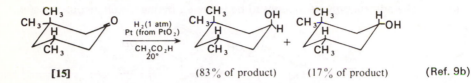

[15]   (83% of product)   (17% of product)   (Ref. 9b)

of reduction of carbon-carbon triple bonds and carbon-carbon double bonds become more nearly equal when the triple bond is part of a conjugated system.[41c] The very great selectivity for the reduction of acetylenes in preference to other functional groups is attributable to the strong tendency of acetylenes to be adsorbed on the catalyst surface.[1f]

Although the foregoing examples appear to be consistent with a concerted *cis* addition of two hydrogen atoms to a multiple bond, studies of the isomerization of olefins over hydrogenation catalysts, of hydrogen-deuterium exchange, and of the stereochemistry of catalytic hydrogenation have established that the transfer of hydrogen atoms to an absorbed molecule must occur in a stepwise manner.[42] The commonly accepted mechanism for this process, postulated by Horiuti and Polanyi,[43] involves the series of accompanying equilibria which implicate both π-bonded intermediates, such as [16] and [17] (or comparable intermediates where both carbon atoms are bonded to the catalyst surface), and a half-hydrogenated intermediate [18]. At relatively high hydrogen pressures the adsorption step (to form 16) is rate limiting, while the subsequent step 16 →18 is believed to be rate limiting at low hydrogen pressures.[42] More recently, evidence has been obtained for yet another mechanism, which involves the formation of a π-allyl intermediate [19] from the catalyst and the compound being reduced.[44] The

42. (a) R. L. Burwell, Jr., *Chem. Rev., 57*, 895 (1957); *Accts. Chem. Res., 2*, 289 (1969); E. I. Klabunovskii, *Russ. Chem. Rev., 35*, 546 (1966); S. Siegel, *Advan. Catalysis, 16*, 123 (1966). (b) J. F. Sauvage, R. H. Baker, and A. S. Hussey, *J. Am., Chem. Soc. 83*, 3874 (1961). (c) G. V. Smith and R. L. Burwell, Jr., *ibid., 84*, 925 (1962). (d) S. Siegel and B. Dmuchovsky, *ibid., 84*, 3132 (1962). (e) S. Siegel, G. V. Smith, B. Dmuchovsky, D. Dubbell, and W. Halpern, *ibid., 84*, 3136 (1962). (f) F. J. McQuillin, W. O. Ord, and P. L. Simpson, *J. Chem. Soc.*, 5996 (1963); I. Jardine and F. J. McQuillin, *ibid.*, C, 458 (1966). (g) S. Siegel, M. Dunkel, G. V. Smith, W. Halpern, and J. Cozort, *J. Org. Chem., 31*, 2802 (1966). (h) A. S. Hussey, G. W. Keulks, G. P. Nowack, and R. H. Baker, *ibid., 33*, 610 (1968); A. S. Hussey, T. A. Schenach, and R. H. Baker, *ibid., 33*, 3258 (1968); A. S. Hussey and G. P. Nowack, *ibid., 34*, 439 (1969). (i) G. V. Smith and J. R. Swoap, *ibid., 31*, 3904 (1966); G. V. Smith and J. A. Roth, *J. Am. Chem. Soc., 88*, 3879 (1966). (j) H. A. Quinn, M. A. McKervey, W. R. Jackson, and J. J. Rooney, *ibid., 92*, 2922 (1970).
43. I. Horiuti and M. Polanyi, *Trans. Faraday Soc., 30*, 1164 (1934).
44. (a) F. G. Gault, J. J. Rooney, and C. Kemball, *J. Catalysis, 1*, 255 (1962). (b) W. R. Moore, *J. Am. Chem. Soc., 84*, 3788 (1962). (c) J. F. Harrod and A. J. Chalk, *ibid., 88*, 3491 (1966). (d) J. J. Rooney, *Chemistry in Britain*, **June 1966**, 242; R. L. Burwell, Jr., *Chem. Eng. News*, **Aug. 22, 1966**, 56.

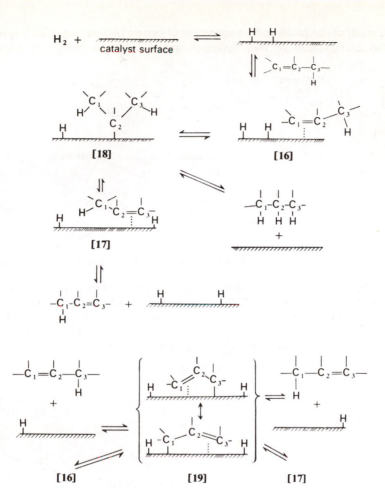

intervention of such intermediates serves to explain the facts that olefins are isomerized on hydrogenation catalysts: that reaction of an olefin with deuterium and a catalyst leads to a mixture of reduction products, including molecules that contain more than and fewer than two deuterium atoms per molecule: and that

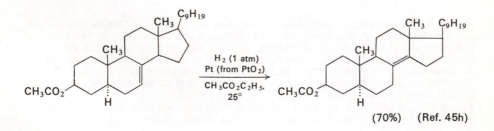

(70%)   (Ref. 45h)

the net result of the catalytic hydrogenation of certain olefins (e.g., [20]) is not the expected *cis* addition of two hydrogen atoms. As the foregoing discussion

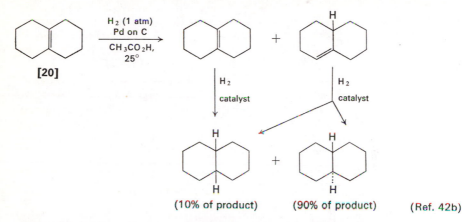

[20]

(10% of product)   (90% of product)   (Ref. 42b)

would indicate, the *cis* addition of hydrogen from the less hindered side of a double bond without the complication of accompanying double bond isomerization has been found to be favored by the use of relatively high hydrogen pressures[42g,45] and relatively small quantities of catalyst.[46] Furthermore, platinum and rhodium catalysts have been found distinctly superior to palladium catalysts[42,43,45] if

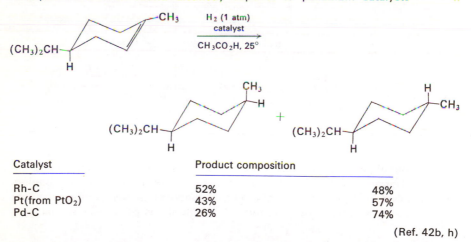

Catalyst	Product composition	
Rh-C	52%	48%
Pt(from PtO$_2$)	43%	57%
Pd-C	26%	74%

(Ref. 42b, h)

45. (a) S. Siegel and G. V. Smith, *J. Am. Chem. Soc.,* **82**, 6082, 6087 (1960). (b) W. Herz and R. H. Mirrington, *J. Org. Chem.,* **30**, 3198 (1965). (c) W. Cocker, P. V. R. Shannon, and P. A. Staniland, *J. Chem. Soc.,* C, 41 (1966). (d) H. C. Yao and P. H. Emmett, *J. Am. Chem. Soc.,* **83**, 799 (1961). (e) G. C. Bond, G. Webb, P. B. Wells, and J. M. Winterbottom, *J. Chem. Soc.,* 3218 (1965). (f) L. Crombie and P. A. Jenkins, *Chem. Commun.,* **No. 8**, 394 (1969). (g) W. S. Johnson and K. E. Harding, *J. Org. Chem.,* **32**, 478 (1967). (h) D. H. R. Barton and J. D. Cox, *J. Chem. Soc.,* 1354 (1948).
46. (a) H. O. House, R. G. Carlson, H. Muller, A. W. Notles, and C. D. Slater, *J. Am. Chem. Soc.,* **84**, 2614 (1962). (b) R. L. Augustine, *J. Org. Chem.,* **28**, 152 (1963).

addition of hydrogen to an unsaturated molecule without the competing olefin isomerization is desired. The rate of hydrogenation of multiple bonds over a platinum catalyst is also influenced by the acidity of the reaction solution; usually the use of an acid solution (e.g., acetic acid as a solvent) enhances the hydrogenation rate, and in some cases the rate is also enhanced by the presence of base.[45d]

Recent studies[42i,45e] suggest that the catalytic hydrogenation of carbon-carbon double bonds conjugated with phenyl groups and vinyl groups (i.e., 1,3-dienes) is facilitated by the simultaneous bonding of both the double bond and the adjacent unsaturated group to the catalyst surface. A similar phenomenon is postulated in the subsequently discussed hydrogenolysis reactions. When several nonconjugated functional groups are present, it would appear that hydrogenation of the most easily reduced group is completed before the reduction of less reactive functions occurs. The phenomenom has been explained as the result of a competition among molecules containing unsaturated functions for a limited number of coordination sites on the catalyst surface.[45f]

One possible course for the hydrogenolysis of benzyl and allyl groups can be pictured as involving an intermediate π-benzyl or π-allyl system **[21]**, as illustrated.

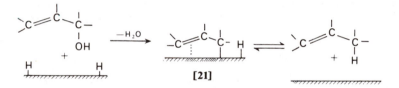

If such an intermediate **[21]** is formed with retention of configuration, the hydrogenolysis at an asymmetric center would occur with retention of configuration, as has been reported[47] for the alcohol **[22]** and similar structures over Raney nickel. However, the hydrogenolysis of other bonds over Raney nickel has been observed[47b,c] to occur with racemization or inversion of configuration, and the hydrogenolysis of epoxides,[21d] benzyl-amine derivatives,[21d,47g,i] and benzyl

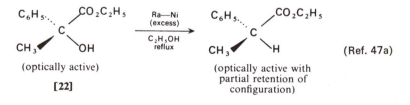

(optically active)

[22]

(optically active with partial retention of configuration)

(Ref. 47a)

47. (a) W. A. Bonner, J. A. Zderic, and G. A. Casaletto, *J. Am. Chem. Soc.,* **74**, 5086 (1952). (b) W. A. Bonner and J. A. Zderic, *ibid.,* **78**, 3218 (1965). (c) W. A. Bonner and R. A. Grimm, *J. Org. Chem.,* **32**, 3022, 3470 (1967). (d) S. Mitsui and S. Imaizumi, *Bull. Chem. Soc. Japan,* **34**, 774 (1961). (e) A. M. Khan, F. J. McQuillin, and I. Jardine, *Tetrahedron Letters,* **No. 24**, 2649 (1966); *J. Chem. Soc.,* C, 136 (1967). (f) E. W. Garbisch, Jr., L. Schreader, and J. J. Frankel, *J. Am. Chem. Soc.,* **89**, 4233 (1967). (g) C. O. Murchu, *Tetrahedron Letters,* **No. 38**, 3231 (1969). (h) S. Mitsui, Y. Kudo, and M. Kobayashi, *Tetrahedron,* **25**, 1921 (1969). (i) H. Dahn, J. A. Garbarino, and C. O'Murchu, *Helv. Chim. Acta,* **53**, 1370; (1970) H. Dahn and C. O'Murchu, *ibid.,* **53**, 1379 (1970).

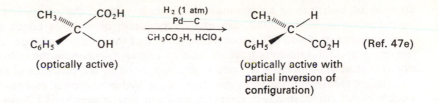

(Ref. 47e)

(optically active)

(optically active with
partial inversion of
configuration)

alcohol derivatives over palladium catalysts has been found to occur with inversion
of configuration.[47d–h] Both the rate and the stereochemical outcome of hydrogenolysis reaction may be influenced not only by the metal catalyst used but also
by the nature of the leaving groups,[47c,e] by the reaction solvent and amount of
hydrogen retained on the catalyst,[47c] and by the acidity or basicity of the reaction
mixture.[21d,47e]

Generally the rate of hydrogenolysis is increased as the bond being cleaved
becomes a better leaving group (e.g., rates: $OH < OCOCH_3 < OCOCF_3$) and as the
reaction mixture is made more acidic. This effect of acidity is presumably attributable
to the protonation of the nitrogen or oxygen function (e.g., $-OH + H^{\oplus} \rightarrow - {}^{\oplus}OH_2$)
to form a better leaving group. The accompanying equation illustrates the stereochemical changes which may result from altering the leaving group and the catalyst.

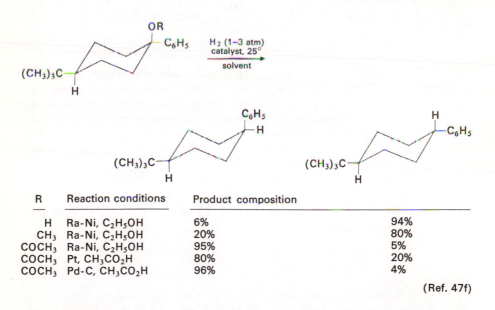

R	Reaction conditions	Product composition	
H	Ra-Ni, C_2H_5OH	6%	94%
CH_3	Ra-Ni, C_2H_5OH	20%	80%
$COCH_3$	Ra-Ni, C_2H_5OH	95%	5%
$COCH_3$	Pt, CH_3CO_2H	80%	20%
$COCH_3$	Pd-C, CH_3CO_2H	96%	4%

(Ref. 47f)

Under other conditions studied,[47f,] the stereochemistry of the hydrocarbon products
was comparable irrespective of the stereochemistry of the starting alcohol. Although
it is difficult to formulate a simple set of mechanisms which will accommodate all
the factors which appear to influence the stereochemistry of catalytic hydrogenolysis, the following scheme seems to account for a number of observations. If the
π-allyl or π-benzyl intermediates **[23a]** (formed with retention of configuration)

and **[23b]** (formed with inversion of configuration) are considered to arise by the processes indicated in the following equations, the backside displacement at the C—X bond to form the π-benzyl intermediate **[23b]** with inversion of configuration would be more favorable as the group X becomes a better leaving group. This inversion process forming **[23b]** is evidently favored for palladium catalysts, whereas the retention pathway forming **[23a]** is often favored with nickel and platinum catalysts. Even the hydrogenolysis of C—O bonds over Raney nickel may occur without retention of configuration since the illustrated equilibrium between **[21]** and the product may isomerize the initially formed reduction product to a more stable isomer. Such a result has been observed with the two cholestane derivatives **[24]** and **[25]**, both of which gave the same hydrocarbon, 3 β-phenyl-cholestane, with Raney nickel in boiling ethanol. However, the product **[26]** was produced from the alcohol **[24]** by hydrogenolysis, with retention of configuration

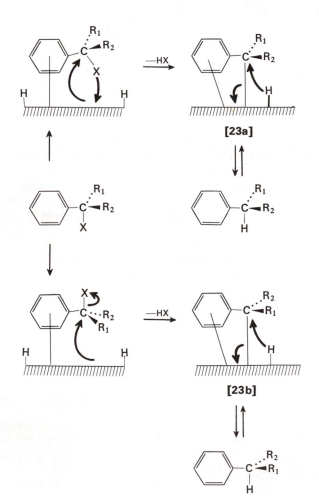

at lower temperatures. This hydrocarbon [26] was isomerized to the more stable isomer by Raney nickel in refluxing ethanol.[48] As would be expected from the previous discussion, hydrogenolysis of the 3 β-phenyl alcohol [25] over a palladium catalyst under nonequilibrating conditions results in inversion of configuration to form the less stable 3 α-phenyl hydrocarbon [26].[47e]

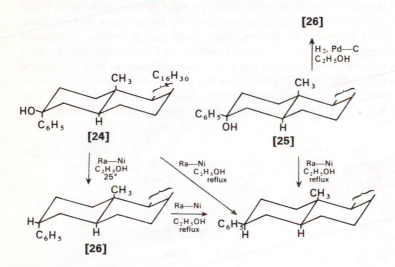

(Ref. 47e, 48)

The presence of acids or bases in the reaction mixtures employed for catalytic hydrogenation has been reported to alter the rate and steric course of many reductions.[26,47,49] For reductions of cyclohexanone derivatives, the modified von Auwers-Skita rule[9b,50] has led to the expectation that hydrogenation in neutral or basic media will produce predominantly equatorial alcohols, whereas axial alcohols are the major products from reduction in acidic solution. However, there is reason[9b] to believe that much of the earlier work on which this rule is based suffered from both inadequate product analysis and isomerization of initially formed products. As a result, it is unwise to consider this rule a reliable criterion for stereochemical assignments.

Both the ease and the stereochemical course of hydrogenations of α,β-unsaturated ketones are influenced by the nature of the solvent and by the presence of acid or base in the reaction mixture.[46b,49-52] For example, reduction of the

48. E. W. Garbisch, Jr., *J. Org. Chem.,* **27**, 3363 (1962).
49. F. J. McQuillin and W. O. Ord, *J. Chem. Soc.,* 2902, 3169 (1959).
50. J. H. Brewster, *J. Am. Chem. Soc.,* **76**, 6361 (1954).
51. (a) H. A. Weidlich, *Chemie,* **58**, 30 (1945). (b) A. L. Wilds, J. A. Johnson, Jr., and R. E. Sutton, *J. Am. Chem. Soc.,* **72**, 5524 (1950). (c) H. I. Hadler, *Experientia,* **11**, 175 (1955). (d) H. J. E. Loewenthal, *Tetrahedron,* **6**, 269 (1959).
52. (a) R. L. Augustine, *J. Org. Chem.,* **23**, 1853 (1958). (b) R. L. Augustine and A. D. Broom, *ibid.,* **25**, 802 (1960). (c) S. Nishimura, M. Shimahara, and M. Shiota, *ibid.,* **31**, 2394 (1966). (d) R. L. Augustine, D. C. Migliorini, R. E. Foscante, C. S. Sodano, and M. J.

unsaturated ketone [27] gave the product compositions indicated in the following equations. It must be noted that this example is clearly contradictory to the earlier

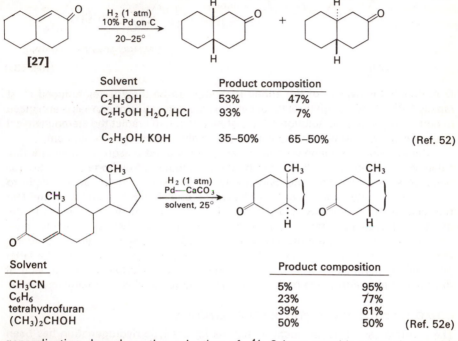

Solvent	Product composition	
C_2H_5OH	53%	47%
$C_2H_5OH\ H_2O, HCl$	93%	7%
C_2H_5OH, KOH	35–50%	65–50%	(Ref. 52)

Solvent	Product composition	
CH_3CN	5%	95%
C_6H_6	23%	77%
tetrahydrofuran	39%	61%
$(CH_3)_2CHOH$	50%	50%	(Ref. 52e)

generalization, based on the reduction of Δ^4, 3-keto steroids to products with cis-fused A and B rings,[51] that hydrogenation in neutral or alkaline medium favors the production of a cis-fused system. The interpretation of the results from hydrogenation of conjugated systems is further complicated by the possible addition of hydrogen at the ends of the conjugated system, as in the reduction of the hindered ketone [28], and by the possibility that the conjugated system may isomerize to a nonconjugated isomer prior to hydrogenation.[53] The increased amount of trans-fused product obtained from hydrogenation of the enone [27] in basic solution has been suggested to arise from hydrogenation of the relatively planar enolate anion present in the solution;[52d] in acidic solution (pH<2), the

Sisbarro, ibid., 34, 1075 (1969). (e) M. G. Combe, H. B. Henbest, and W. R. Jackson, J. Chem. Soc., C, 2467 (1967); H. B. Henbest, W. R. Jackson, and I. Malunowicz, ibid., C, 2469 (1967). (f) F. J. McQuillin, W. O. Ord, and P. L. Simpson, ibid., 5996 (1963). (g) I. Jardine, R. W. Howsam and F. J. McQuillin, ibid., C, 260 (1969). (h) For a review, see L. Velluz, J. Valls, and G. Nominé, Angew. Chem., Intern. Ed. Engl., 4, 181 (1965). (i) For the selective reduction of the γ,δ-double bond in conjugated dienones over a palladium-on-strontium carbonate catalyst, see E. Farkas, J. M. Owen, M. Debono, R. M. Molloy, and M. M. Marsh, Tetrahedron Letters, No. 10, 1023 (1966). (j) K. Mori, K. Abe, M. Washida, S. Nishimura, and M. Shiota, J. Org. Chem., 36, 231 (1971).
53. (a) R. C. Fuson, J. Corse, and C. H. McKeever, J. Am. Chem. Soc., 62, 3250 (1940). (b) H. Simon and O. Berngruber, Tetrahedron, 26, 161, 1401 (1970).

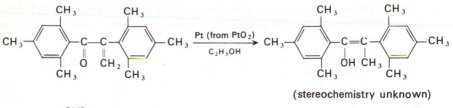

(stereochemistry unknown)

[28]   (Ref. 53a)

O-protonated derivative of the enone is thought to be the species reduced most rapidly.[50,52a,b,d] Although the ease of catalytic hydrogenation of enones is enhanced in both acidic and basic solution,[52d] it is often difficult to predict the stereochemical effect which will result from changes in the acidity of the reaction medium.

Consequently, in planning a synthesis there is good reason to anticipate that catalytic hydrogenation will proceed by the *cis* addition of hydrogen from the less hindered side of the molecule, particularly if the hydrogenation can be made to occur rapidly under mild conditions. However, one should be mindful of the fact that changes in the acidity and polarity of the reaction medium, changes in the quantity or type of catalyst employed, or changes in other, presumably inert, functional groups in the molecule being reduced[52e,54] may well alter the stereochemical course of the reaction sufficiently to be preparatively useful. The latter possibility is particularly worthy of exploration if steric hindrance to approach of the catalyst from either side of the molecule being reduced is approximately equal.

HOMOGENEOUS CATALYTIC HYDROGENATION

The possible use of homogeneous catalysts for catalytic hydrogenation has been the subject of a number of recent investigations.[55] The soluble tris-(triphenyl-phosphine)rhodium chloride catalyst[55-57] [29] is of special interest since it

54. For examples, see (a) T. G. Halsall, W. J. Rodewald, and D. Willis, *Proc. Chem. Soc.,* 231 (1958). (b) C. H. DePuy and P. R. Story, *J. Am. Chem. Soc.,* **82,** 627 (1960). (c) C. L. Arcus, J. M. J. Page, and A. J. Reid, *J. Chem. Soc.,* 1213 (1963). (d) S. Mitsui, Y. Senda, and H. Saito, *Bull. Chem. Soc. Japan,* **39,** 694 (1966). (e) M. Balasubramanian and N. Padma, *Tetrahedron,* **24,** 5395 (1968). M. Balasubramanian and A. D'Souza, *ibid.,* **24,** 5399 (1968).
55. (a) J. Tsuji, *Adv. Org. Chem.,* **6,** 109 (1969). (b) J. A. Osborn, *Endeavor,* **26,** 144 (1967). (c) R. Cramer, *Accts. Chem. Res.,* **1,** 186 (1968). (d) R. F. Heck, *ibid.,* **2,** 10 (1969).
56. (a) J. A. Osborn, F. H. Jardine, J. F. Young, and G. Wilkinson, *J. Chem. Soc.,* A, 1711 (1966). (b) F. H. Jardine, J. A. Osborn, and G. Wilkinson, *ibid.,* A, 1574 (1967). (c) F. H. Jardine and G. Wilkinson, *ibid.,* C, 270 (1967). (d) S. Montelatici, A. van der Ent, J. A. Osborn, and G. Wilkinson, *ibid.,* A, 1054 (1968). (e) D. Evans, G. Yagupsky, and G. Wilkinson, *ibid.,* A, 2660 (1968). G. Yagupsky, C. K. Brown, and G. Wilkinson, *ibid.,* A, 1392 (1970). (f) C. O'Connor and G. Wilkinson, *ibid.,* A, 2665 (1968); *Tetrahedron Letters,* **No. 18,** 1375 (1969); for the preparation and use of the soluble ruthenium catalyst [(C₆H₅)₃P]₃RuCl to reduce terminal alkenes see P. S. Hallman, D. Evans, J. A. Osborn, and G. Wilkinson, *Chem. Commun.,* **No. 7,** 305 (1967); P. S. Hallman, B. R. McGarvey, and G. Wilkinson, *J. Chem. Soc.,* A, 3143 (1968). (g) For the preparation and properties of soluble ruthenium catalysts, see D. Rose, J. D. Gilbert, R. P. Richardson, and G. Wilkinson,

reacts with hydrogen to form a metal hydride which rapidly transfers two hydrogen atoms to a carbon-carbon double bond. The preparation of this catalyst [29] and its function as a hydrogenation catalyst are illustrated in the following equations.[56a-k]

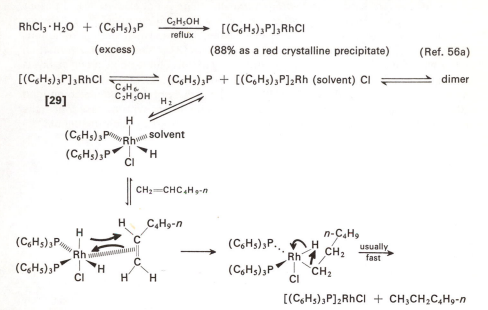

$$RhCl_3 \cdot H_2O + (C_6H_5)_3P \xrightarrow[\text{reflux}]{C_2H_5OH} [(C_6H_5)_3P]_3RhCl$$

(excess)   (88% as a red crystalline precipitate)   (Ref. 56a)

$[(C_6H_5)_3P]_2RhCl + CH_3CH_2C_4H_9\text{-}n$

The indicated rapid successive (or possibly simultaneous) transfer of two hydrogen atoms to the olefin is consistent with the ability of this catalyst to convert unhindered olefins

$$n\text{-}C_8H_{17}CH{=}CH_2 \xrightarrow[C_6H_6]{\substack{D_2 \, (ca. \, 1 \, atm) \\ [(C_6H_5)_3P]_3RhCl}} n\text{-}C_8H_{17}\underset{\underset{D}{|}}{CH}{-}CH_2D$$

(Ref. 57b)

ibid., A, 2610 (1969) ; I. Ogata, R. Iwata, and Y. Ikeda, *Tetrahedron Letters,* **No. 34,** 3011 (1970). (h) H. van Bekkum, F. van Rantwijk, and T. van de Putte, *ibid.,* **No. 1,** 1 (1969) (i) G. V. Smith and R. J. Shuford, *ibid.,* **No. 7,** 525 (1970). (j) L. Horner, H. Siegel, and H. Büthe, *Angew. Chem. Intern. Ed. Engl.,* **7,** 942 (1968). (k) P. Abley and F. J. McQuillin, *Chem. Commun.,* **No. 9,** 477 (1969) ; **No. 24,** 1503 (1969). For other soluble catalysts, see (l) I. Jardine and F. J. McQuillin, *ibid.,* **No. 9,** 477 (1969) ; **No. 11,** 626 (1970) ; (m) P. Legzdins, G. L. Rempel, and G. Wilkinson, *ibid.,* **No. 14,** 825 (1969). (n) For the use of iridium catalysts and alcohols to reduce acetylenes and ketones, see Y. M. H. Haddad, H. B. Henbest, J. Husbands, and T. R. B. Mitchell, *Proc. Chem. Soc.,* 361 (1964) J. Trocha-Grimshaw and H. B. Henbest, *Chem. Commun.,* **No. 11,** 544 (1967) ; **No. 13,** 757 (1968) ; P. A. Browne and D. N. Kirk, *J. Chem. Soc.,* C, 1653 (1969).
57. (a) A. S. Hussey and Y. Takeuchi, *J. Am. Chem. Soc.,* **91,** 672 (1969) ; *J. Org. Chem.,* **35,** 643 (1970). (b) J. R. Morandi and H. B. Jensen, *ibid.,* **34,** 1889 (1969). (c) R. L. Augustine and J. F. Van Peppen, *Ann. N. Y. Acad. Sci.,* **158,** 482 (1969) ; *Chem. Commun.,* **No. 8,** 495, 497 (1970) ; **No. 9,** 571 (1970). (d) T. R. B. Mitchell, *J. Chem. Soc.,* B, 823 (1970). (e) G. E. Hartwell and P. W. Clark, *Chem. Commun.,* **No. 17,** 1115 (1970). (f) M. Yagupsky, C. K. Brown, G. Yagupsky, and G. Wilkinson, *J. Chem. Soc.,* A, 937 (1970) ; M. Yagupsky and G. Wilkinson, *ibid.,* A, 941 (1970).

to dideuterated alkanes[56a,57b] without the extensive hydrogen-deuterium scrambling often seen with heterogeneous catalysts. Investigations of hydrogenation stereochemistry and investigations with deuterium labeling[56i,57] have indicated that the second of the two successive hydrogen transfer steps is not always fast. In cases where the second hydrogen transfer is slow, the reaction proceeds by a stepwise formation of an olefin π-complex and an alkyl metal hydride intermediate analogous to the previously discussed half-hydrogenated state [21]. The reversible formation of this alkyl metal hydride accounts for the ability of the soluble noble metal catalysts to isomerize olefins,[55,57,58] a reaction that is especially prevalent if the initial catalyst is not pretreated with hydrogen.[57c] Although the activity of this rhodium compound [29] to hydrogenate olefins is enhanced by pretreatment with oxygen,[56h,57c] the ability of the catalyst to isomerize olefins is also enhanced by such treatment.[57c]

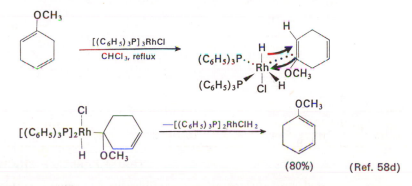

(80%)   (Ref. 58d)

Although very few comparisons of product stereochemistry have been made, it would appear that similar stereochemical results may be expected from the heterogeneous and homogeneous catalysts.[57] The principal advantage offered by the homogeneous rhodium catalyst is its degree of selectivity for reducing only unhindered carbon-carbon double bonds.[59] The following equations illustrate the ability of the catalyst to reduce unhindered carbon-carbon double bonds even in the presence of functions such as keto, nitrile, nitro or sulfide groups. Similar catalysts containing more basic phosphine ligands [e.g., $C_6H_5P(CH_3)_2$] have been used in the presence of catalytic amounts of water to reduce ketones to alcohols.[59f] An apparently related catalyst, prepared from the *tris*-pyridine complex of rhodium trichloride and sodium borohydride, is effective in catalyzing the hydrogenation of azo groups, imino groups, and nitro groups.[56l]

58. (a) J. F. Harrod and A. J. Chalk, *J. Am. Chem. Soc.,* **88,** 3491 (1966). (b) R. Cramer and R. V. Lindsey, Jr., *ibid.,* **88,** 3534 (1966). (c) J. F. Biellmann and M. J. Jung, *ibid.,* **90,** 1673 (1968). (d) A. J. Birch and G. S. R. Subba Rao, *Tetrahedron Letters,* **No. 35,** 3797 (1968).
59. (a) M. Brown and L. W. Piszkiewicz, *J. Org. Chem.,* **32,** 2013 (1967). (b) R. E. Harmon, J. L. Parsons, D. W. Cooke, S. K. Gupta, and J. Schoolenberg, *ibid.,* **34,** 3684 (1969). (c) A. J. Birch and K. A. M. Walker, *J. Chem. Soc.,* C, 1894 (1966); *Tetrahedron Letters,* **No. 20,** 1935 (1967). (d) J. J. Sims, V. K. Honwad, and L. H. Selman, *ibid.,* **No. 2,** 87 (1969). (e) C. Djerassi and J. Gutzwiller, *J. Am. Chem. Soc.,* **88,** 4537 (1966). (f) R. R. Schrock and J. A. Osborn, *Chem. Commun.,* **No. 9,** 567 (1970).

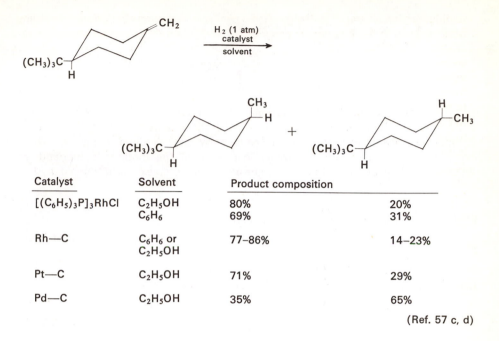

Catalyst	Solvent	Product composition	
[(C₆H₅)₃P]₃RhCl	C₂H₅OH	80%	20%
	C₆H₆	69%	31%
Rh—C	C₆H₆ or C₂H₅OH	77–86%	14–23%
Pt—C	C₂H₅OH	71%	29%
Pd—C	C₂H₅OH	35%	65%

(Ref. 57 c, d)

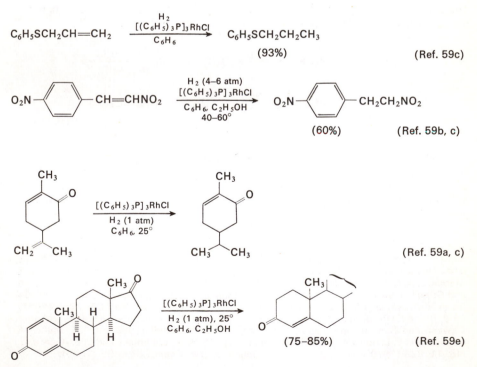

$$C_6H_5SCH_2CH{=}CH_2 \xrightarrow[C_6H_6]{\substack{H_2 \\ [(C_6H_5)_3P]_3RhCl}} C_6H_5SCH_2CH_2CH_3$$

(93%)   (Ref. 59c)

$$O_2N{-}\langle\rangle{-}CH{=}CHNO_2 \xrightarrow[\substack{C_6H_6, C_2H_5OH \\ 40-60°}]{\substack{H_2 (4-6 \text{ atm}) \\ [(C_6H_5)_3P]_3RhCl}} O_2N{-}\langle\rangle{-}CH_2CH_2NO_2$$

(60%)   (Ref. 59b, c)

(Ref. 59a, c)

(75–85%)   (Ref. 59e)

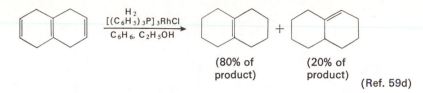

(80% of
product)

(20% of
product)

(Ref. 59d)

Although the unhindered carbon-carbon double bonds of unsaturated aldehydes may be reduced in reasonable yield, this reaction is complicated by a competing decarbonylation reaction caused by the rhodium complex. This competing reaction to form a carbonyl derivative of the rhodium compound is illustrated by reaction of the phosphine-rhodium complex with formaldehyde.

$$CH_3CH\!\!=\!\!CHCHO \quad \xrightarrow[\substack{C_6H_6,\ 25°}]{\substack{H_2\ (2\ atm)\\ [(C_6H_5)_3P]_3RhCl}} \quad CH_3CH_2CH_2CHO$$

(61%)   (Ref. 56c)

$$C_6H_5CH\!\!=\!\!CHCHO \quad \xrightarrow[\substack{C_2H_5OH,\ 60°}]{\substack{H_2\ (6\ atm)\\ [(C_6H_5)_3P]_3RhCl}} \quad C_6H_5CH_2CH_2CHO \ + \ C_6H_5CH_2CH_3$$

(60% of product)   (40% of product)

(Ref. 59b)

The corresponding hydride derivative of this rhodium carbonyl complex has been used as a hydrogenation catalyst of rather low activity.[56e,f,57f]

$$RhCl_3 \ + \ (C_6H_5)_3P \quad \xrightarrow[\text{reflux}]{C_2H_5OH} \quad [(C_6H_5)_3P]_3RhCl \quad \xrightarrow[\substack{C_2H_5OH,\\ H_2O}]{H_2CO}$$

$$[(C_6H_5)_3P]_2Rh(CO)Cl \quad \xrightarrow[C_2H_5OH]{NaBH_4} \quad [(C_6H_5)_3P]_2Rh(CO)H$$

(72%)   (Ref. 56e, f)

The ability of the rhodium catalyst to abstract carbon monoxide from aldehydes and from acid halides serves as a useful synthetic procedure, as does the reverse process, the hydroformylation reaction of an olefin with carbon monoxide and hydrogen to form an aldehyde.[55,56a,e,60] The following equations illustrate synthetic applications of the decarbonylation procedure.

---

60. (a) C. K. Brown and G. Wilkinson, *Tetrahedron Letters,* **No. 22,** 1725 (1969) ; D. Evans, J. A. Osborn, and G. Wilkinson, *J. Chem. Soc.,* A, 3133 (1968). (b) M. C. Baird, J. T. Mague, J. A. Osborn, and G. Wilkinson, *ibid.,* A, 1347 (1967). (c) M. C. Baird, C. J. Nyman, and G. Wilkinson, *ibid.,* A, 348 (1968). (d) J. Tsuji and K. Ohno, *J. Am. Chem. Soc.,* **90,** 94, 99 (1968). (e) J. Blum, E. Oppenheimer, and E. D. Bergmann, *ibid.,* **89,** 2338 (1967). (f) J. Blum, H. Rosenman, and E. D. Bergmann, *J. Org. Chem.,* **33,** 1928 (1968). (g) D. J. Dawson and R. E. Ireland, *Tetrahedron Letters,* **No. 15,** 1899 (1968). (h) E. Müller, A. Segnitz, and E. Lange, *ibid.,* **No. 14,** 1129 (1969). (i) H. M. Walborsky and L. E. Allen, *ibid.,* **No. 11,** 823 (1970). (j) J. Blum and G. Scharf. *J. Org. Chem.,* **35,** 1895 (1970).

$$CH_3(CH_2)_5—CHO + [(C_6H_5)_3P]_3RhCl \xrightarrow[40°]{C_6H_6} [(C_6H_5)_3P]_2\overset{H}{\underset{Cl}{Rh}}—CO(CH_2)_5CH_3 \longrightarrow$$

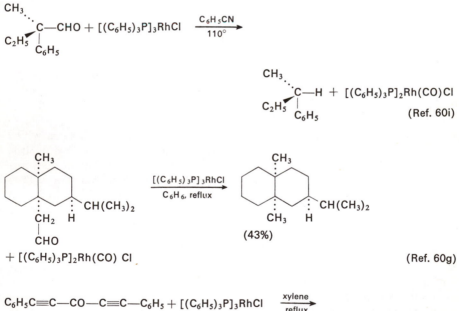

$$[(C_6H_5)_3P]_2Rh\overset{H}{\underset{Cl}{\overset{(CH_2)_5CH_3}{\diagdown}}}\underset{C=O}{} \longrightarrow [(C_6H_5)_3P]_2Rh\overset{Cl}{\underset{C=O}{\diagdown}} +$$

$$CH_3(CH_2)_4CH_3 + CH_3(CH_2)_3CH=CH_2 + CH_3(CH_2)_2CH=CHCH_3$$

(80% of the product)   (10% of the product)   (10% of the product)

(Ref. 56a, 60b, c)

It will be noted from the following example that the decarbonylation process is stereospecific, proceeding with retention of configuration.

$$\underset{C_2H_5}{\overset{CH_3}{\diagdown}}\underset{C_6H_5}{\overset{|}{C}}—CHO + [(C_6H_5)_3P]_3RhCl \xrightarrow[110°]{C_6H_5CN}$$

$$\underset{C_2H_5}{\overset{CH_3}{\diagdown}}\underset{C_6H_5}{\overset{|}{C}}—H + [(C_6H_5)_3P]_2Rh(CO)Cl$$

(Ref. 60i)

$$\xrightarrow[C_6H_6, \text{ reflux}]{[(C_6H_5)_3P]_3RhCl}$$

(43%)

+ [(C_6H_5)_3P]_2Rh(CO) Cl

(Ref. 60g)

$$C_6H_5C\equiv C—CO—C\equiv C—C_6H_5 + [(C_6H_5)_3P]_3RhCl \xrightarrow[\text{reflux}]{\text{xylene}}$$

$$C_6H_5C\equiv C—C\equiv CC_6H_5 + [(C_6H_5)_3P]_2Rh(CO)Cl + (C_6H_5)_3P$$

(78%)   (Ref. 60h)

Carboxylic acid chlorides may be decarboxylated at high temperatures with only a catalytic amount of the rhodium complex, as shown in the following examples. A similar process results in the desulfonylation of arylsulfonyl chlorides and bromides.[60j]

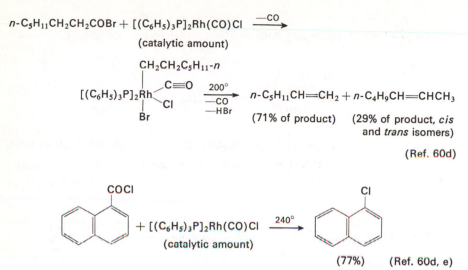

$$n\text{-}C_5H_{11}CH_2CH_2COBr + [(C_6H_5)_3P]_2Rh(CO)Cl \xrightarrow{-CO}$$

(catalytic amount)

$$n\text{-}C_5H_{11}CH=CH_2 + n\text{-}C_4H_9CH=CHCH_3$$

(71% of product)   (29% of product, *cis* and *trans* isomers)

(Ref. 60d)

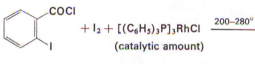

+ [(C_6H_5)_3P]_2Rh(CO)Cl $\xrightarrow{240°}$

(catalytic amount)

(77%)   (Ref. 60d, e)

+ I_2 + [(C_6H_5)_3P]_3RhCl $\xrightarrow{200–280°}$

(catalytic amount)

(82%)   (Ref. 60f)

DEHYDROGENATION REACTIONS

In contrast to the previous discussion is a group of reaction procedures which serve not to add hydrogen to an organic molecule but rather to remove it. It is appropriate to discuss these methods here, since one of the common dehydrogenation procedures is in essence the reverse of catalytic hydrogenation.[61] The most common application of this reaction has been the removal of two hydrogen atoms from a cyclohexadiene derivative or the removal of four hydrogen atoms from a cyclohexene derivative to form the corresponding aromatic compound. As a general rule, the reaction conditions required for dehydrogenation become milder the fewer the number of hydrogen atoms that must be removed or transferred to form an aromatic system. The reaction conditions which accomplish these changes normally do not affect saturated six-membered rings. Even if unsaturation is present in rings not containing six members, the formation of dehydrogenated products in satisfactory yield is only possible if a stable aromatic system (e.g., azulene) can be formed. Presumably, in other cases, any dehydrogenation products formed are less stable than the starting material and are further transformed by the reaction procedures employed.

---

61. (a) P. A. Plattner in *Newer Methods of Preparative Organic Chemistry*, Wiley-Interscience, New York, 1948, pp. 21–59. (b) L. F. Fieser, *Experiments in Organic Chemistry*, 2d ed., D. C. Heath, New York, 1941, pp. 454–464.

The common dehydrogenating agents include sulfur, selenium, a palladium catalyst (usually palladium on carbon), a platinum catalyst, and several quinones.[62] The hydrogenation catalysts are mixed with the compound to be dehydrogenated in a relatively high-boiling solvent such as cumene (b.p. 153°), p-cymene (b.p. 176°), decalin (b.p. ca. 185°), quinoline (b.p. 238°), naphthalene (b.p. 218°), or nitrobenzene (b.p. 211°), and the mixture is refluxed with stirring. Some workers pass a slow stream of nitrogen or carbon dioxide through the reaction mixture both to stir the mixture and to sweep the hydrogen from the system as it is liberated. If carbon dioxide is used, the gas leaving the reaction vessel may be passed through aqueous potassium hydroxide to remove the carbon dioxide, and the remaining hydrogen may be collected in a gas buret. In this way, the progress of the dehydrogenation may be followed. This dehydrogenation method appears to follow a mechanism which is the reverse of the previously discussed catalytic hydrogenation.

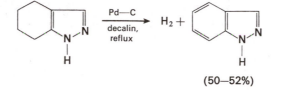

(50—52%) (Ref. 63a)

In other words, the partially unsaturated substrate is adsorbed on the catalyst surface, and it transfers hydrogen atoms to the catalyst; then the dehydrogenated product is desorbed from the catalyst. Since both hydrogen and a catalyst are present in these reaction mixtures, dehydrogenation of a compound containing easily reduced functions (e.g., [30]) may lead both to dehydrogenation and concurrent hydrogenation (i.e., disproportionation) or hydrogenolysis.

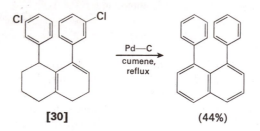

[30]　　　　　　　　　　(44%)　　　　　　　　(Ref. 64a)

62. (a) L. M. Jackman, *Adv. Org. Chem.,* **2**, 329 (1960). (b) D. Walker and J. D. Hiebert, *Chem. Rev.,* **67**, 153 (1967). (c) For studies of other easily reduced quinones, see K. Wallenfels, D. Hofmann, and R. Kern, *Tetrahedron,* **21**, 2231 (1965); K. Wallenfels, G. Bachmann, D. Hofmann, and R. Kern, *ibid.,* **21**, 2239 (1965); A. S. Hay, *Tetrahedron Letters,* **No. 47**, 4241 (1965).
63. (a) C. Ainsworth, *Org. Syn.,* **Coll. Col. 4**, 536 (1963). (b) For a discussion of the use of Raney nickel catalyst in high-boiling solvents to dehydrogenate unsaturated alcohols, see S. K. Banerjee, D. Chakravarti, R. N. Chakravarti, and M. N. Mitra, *Tetrahedron,* **24**, 6459 (1968).
64. (a) H. O. House and R. W. Bashe, *J. Org. Chem.,* **32**, 784 (1967). (b) *Ibid.,* **30**, 2942 (1965). (c) J. Blum and S. Biger, *Tetrahedron Letters,* **No. 21**, 1825 (1970).

Soluble hydrogenation catalysts such as *tris*-triphenylphosphine rhodium chloride have also been used for dehydrogenation.[64c] Presumably, these dehydrogenations proceed by a mechanism which is the reverse of the previously discussed reaction path for hydrogenation in the presence of these soluble catalysts.

Dehydrogenations on a preparative scale are often accomplished more economically by heating a mixture of the substrate with sulfur to 200–250° or with selenium to 250–300°. These reactions apparently involve the abstraction of hydrogen atoms from allylic or benzylic positions followed by attack of these stabilized radicals on sulfur or selenium. Subsequent elimination of hydrogen sulfide or hydrogen selenide results in the introduction of a new carbon-carbon double bond.

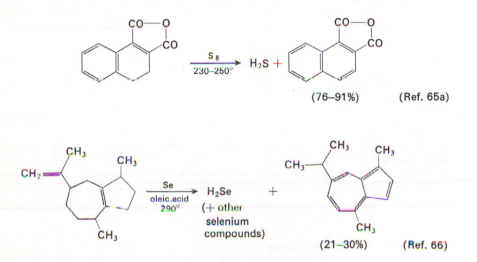

Aside from the possibility that functional groups may be reduced by the conditions of the above dehydrogenation procedures, other restrictions also limit the functional groups which may be present. Secondary and tertiary alcohols are usually dehydrated during the course of these dehydrogenation procedures. In fact, cyclohexanol derivatives can normally be substituted for the corresponding cyclohexenes as substrates for dehydrogenation. Although ketone functions often survive dehydrogenation reactions, if the carbonyl function is part of a six-membered ring (e.g., [31]), a phenol is the expected product. Carboxylic acids and their derivatives are usually retained unless the carboxyl function is bound to a fully substituted carbon atom. The carbon skeleton of compounds to be dehydrogenated is frequently not altered unless the ring system to be aromatized

65. (a) E. B. Hershberg and L. F. Fieser, *Org. Syn.,* **Coll. Vol. 2,** 423 (1943). (b) See also R. Weiss, *Org. Syn.,* **Coll. Vol. 3,** 729 (1955).
66. H. A. Silverwood and M. Orchin, *J. Org. Chem.,* **27,** 3401 (1962).

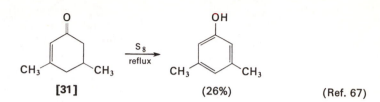

[31] (26%) (Ref. 67)

contains a quaternary carbon atom (e.g., **[32]**). In such cases, aromatization can occur only if one of the alkyl groups is lost or migrates to a nearby atom.[61,68]

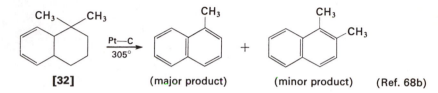

[32] (major product) + (minor product) (Ref. 68b)

Since the tendency toward rearrangement or loss of alkyl groups is normally enhanced when relatively high temperatures are used for dehydrogenation, it is often desirable in synthetic work to introduce multiple unsaturation prior to dehydrogenation. By this means, the final reaction leading to aromatization can be performed under relatively mild conditions which minimize the possibility of rearrangement. It should be noted that certain nitrogen heterocycles such as pyridine are coupled and dehydrogenated by treating them with a reactive grade of Raney nickel from which the adsorbed hydrogen has been removed.

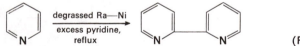

(Ref. 68e)

Dehydrogenation may also be achieved by treating partially unsaturated substrates with quinones containing electron-withdrawing substituents such as chloranil **[33]**,2,3-dichloro-5,6-dicyanobenzoquinone **[34]**, and others (e.g., **[35]** and **[36]**).[62] For the conversion of cyclohexenes or cyclohexadienes to benzene derivatives, these quinones are frequently used as solutions in boiling benzene (b.p. 80°), xylene (b.p. ca. 140°), chlorobenzene (b.p. 132°), or o-dichlorobenzene (b.p. 180°). Chloranil **[33]**, which is decidedly less reactive than the other quinones listed, is normally useful only for reaction with systems which are readily dehydrogenated. Among the more reactive quinones, **[34]**–**[36]**, the dichlorodicyanoquinone **[34]** has been most widely used.[62b] A solution of this quinone in

67. E. C. Horning, *J. Am. Chem. Soc.*, **67**, 1421 (1945). Also see Ref. 68c.
68. (a) R. P. Linstead, A. F. Millidge, S. L. S. Thomas, and A. L. Walpole, *J. Chem. Soc.*, 1146 (1937). (b) R. P. Linstead and S. L. S. Thomas, *J. Chem. Soc.*, 1127 (1940). (c) R. P. Linstead and K. O. A. Michaelis, *J. Chem. Soc.*, 1134 (1940). (d) R. P. Linstead, K. O. A. Michaelis, and S. L. S. Thomas, *J. Chem. Soc.*, 1139 (1940). (e) W. H. F. Sasse, *Org. Syn.*, **46**, 5 (1966).

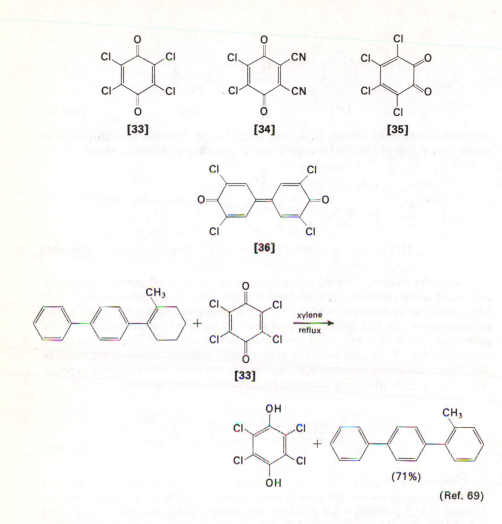

[33]   **[34]**   **[35]**

[36]

[33]

(71%)

(Ref. 69)

benzene is red because of the presence of a charge-transfer complex; as the dehydrogenation proceeds, the benzene-insoluble hydroquinone **[37]** separates as pale yellow solid. The following equation illustrates the probable course of this reaction involving either the indicated initial abstraction of a hydride ion[62,70] or a succession of hydrogen atom transfers. Similar reaction conditions are employed with the other reactive quinones (e.g., **[35]**) and the dehydrogenations are believed to occur by a comparable mechanism. The fact that these quinones oxidize

69. R. T. Arnold, C. Collins, and W. Zenk, *J. Am. Chem. Soc.,* **62**, 983 (1940).
70. (a) E. A. Braude, L. M. Jackman, R. P. Linstead, and J. S. Shannon, *J. Chem. Soc.,* 3116 (1960). (b) E. A. Braude, L. M. Jackman, R. P. Linstead, and G. Lowe, *ibid.,* 3123, 3133 (1960). (c) R. Mechoulam, B. Yagnitinsky, and Y. Gaoni, *J. Am. Chem. Soc.,* **90**, 2418 (1968).

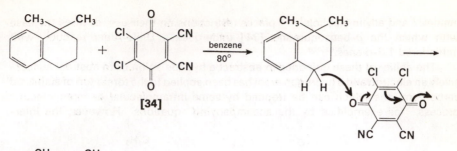

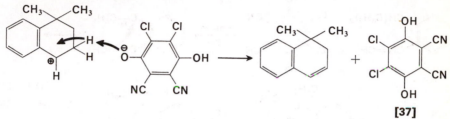

[37]

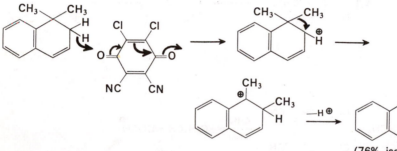

(76%, isolated as
the picrate)

(Ref. 62, 70b)

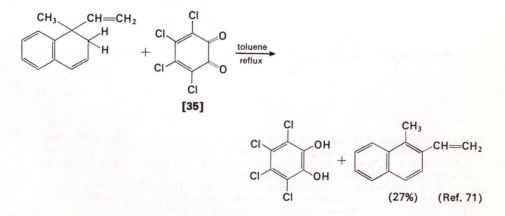

(27%)   (Ref. 71)

71. W. Herz and G. Caple, *J. Org. Chem.*, **29**, 1691 (1964).

amines[62] and allylic alcohols[72a-c] places restrictions on their use, as does the ease with which the *p*-benzoquinone **[34]** undergoes a Diels-Alder reaction with unhindered 1,3-dienes.[62,72d]

The ability of these quinones to abstract a hydride anion or to abstract successively an electron and a hydrogen atom has been applied to the formation of stabilized carbonium ions which can be trapped by some intramolecular or intermolecular process,[73] as exemplified by the accompanying equations. However, the inter-

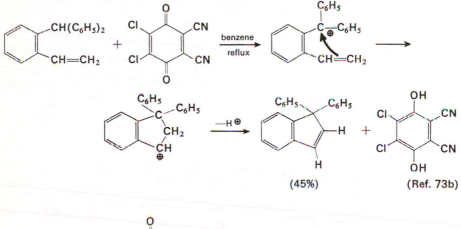

(45%)   (Ref. 73b)

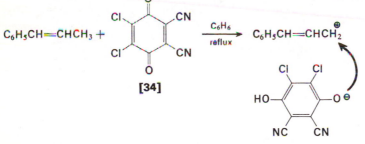

[34]

$$C_6H_5CH{=}CHCH_2OAr \xrightarrow{\text{[34]}} C_6H_5CH{=}CHCH(OAr)_2 \xrightarrow[\text{H}_2\text{O}]{\text{H}^{\oplus}} C_6H_5CH{=}CHCHO$$

(55%)

(Ref. 73e)

72. (a) E. A. Braude, R. P. Linstead, and K. R. Wooldridge, *J. Chem. Soc.,* 3070 (1956). (b) S. H. Burstein and H. J. Ringold, *J. Am. Chem. Soc.,* **86**, 4952 (1964). (c) D. Burn, V. Petrow, and G. O. Weston, *Tetrahedron Letters,* **No. 9,** 14 (1960). (d) A. E. Asato and E. F. Kiefer, *Chem. Commun.,* **No. 24,** 1685 (1968).
73. (a) A. M. Creighton and L. M. Jackman. *J. Chem. Soc.,* 3138 (1960). (b) R. F. Brown and L. M. Jackman, *ibid.,* 3144 (1960). (c) R. Foster and I. Horman, *ibid.,* B, 1049 (1966). (d) D. L. Coffen and P. E. Garrett, *Tetrahedron Letters,* **No. 25,** 2043 (1969). (e) I. H. Sadler and J. A. G. Stewart, *Chem. Commun.,* **No. 13,** 773 (1969). (f) E. S. Lewis, J. M. Perry, and R. H. Grinstein, *J. Am. Chem. Soc.,* **92**, 899 (1970).

and equilibration of the enols [42] and [43]. As a result, the bulk of the product is derived from the more stable enol [43]. It should be noted that in each case, the pseudoaxial hydrogen atom is the one transferred from the enol [42] or [43] to the

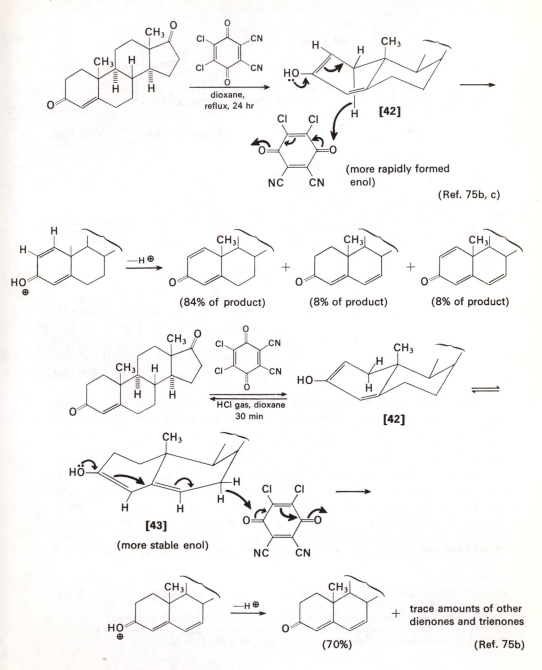

dioxane, reflux, 24 hr

[42]

(more rapidly formed enol)

(Ref. 75b, c)

(84% of product)

(8% of product)

(8% of product)

HCl gas, dioxane 30 min

[42]

[43]

(more stable enol)

(70%)

+ trace amounts of other dienones and trienones

(Ref. 75b)

quinone. As will be discussed subsequently (Chapters 8 and 9), loss of this hydrogen atom allows continuous overlap of the pi orbitals of the forming double bond
with the remainder of the conjugated system.

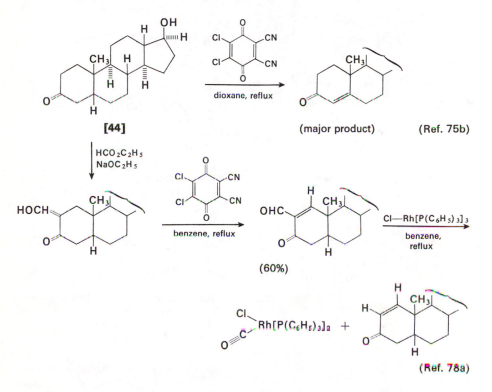

The dichlorodicyanoquinone **[34]** has also been used to convert saturated
cyclic ketones[75b] and saturated δ-lactones[76] to the corresponding α,β-unsaturated
derivatives as well as to dimerize phenols or enolizable ketones which have no
beta hydrogen atoms.[77] This latter reaction, which involves a transfer of one
electron from the substrate to the quinone **[34]**, raises the possibility that other
quinone oxidations formulated as hydride ion abstractions may involve free
radical intermediates.[77] The use of an α-formyl activating group (see Chapter 9)
has permitted the introduction of α,β-unsaturation into a cyclic ketone (e.g., **[44]**)
at a site different from the location of reaction in the absence of the activating
function. Of incidental interest in the accompanying reaction sequence is the use
of the previously discussed rhodium complex to remove the formyl group.[60]

76. B. Berkoz, L. Cuellar, R. Grezemkovsky, N. V. Avila, J. A. Edwards, and A. D. Cross,
Tetrahedron, **24**, 2851 (1968).
77. H.-D. Becker, *J. Org. Chem.*, **30**, 982, 989 (1965); **34**, 1198, 1211 (1969).
78. (a) Y. Shimizu, H. Mitsuhashi, and E. Caspi, *Tetrahedron Letters*, **No. 34**, 4113 (1966).
(b) J. A. Edwards and co-workers, *J. Org. Chem.*, **29**, 3481 (1964).

2

METAL HYDRIDE REDUCTIONS
AND RELATED REACTIONS

A serious shortcoming of catalytic hydrogenation, the inability to reduce selectively the carbonyl function of ketones, acids, esters, and amides in the presence of carbon-carbon double bonds, has led to the widespread use of certain complex metal hydrides[1] for reductions of carbonyl groups.

However, the metal hydrides encompass a wide variety of reducing agents which may be used for the selective reduction of groups in polyfunctional molecules. These reducing agents include not only the nucleophilic reagents which reduce polar multiple bonds such as carbonyl compounds, but also electrophilic reagents which react readily with nonpolar carbon-carbon multiple bonds. The first part of this chapter considers the reductions of polar multiple bonds by both nucleophilic and electrophilic reducing agents. The reactions of the electrophilic metal hydrides with relatively nonpolar carbon-carbon multiple bonds are discussed in the latter part of this chapter.

The most common commercially available[2] nucleophilic metal hydrides include, in order of decreasing activity, lithium aluminum hydride (also called lithium tetrahydroaluminate),[3] lithium borohydride, and sodium borohydride. Several of the reactions which have been used to produce these substances are indicated in the following equations.

$$4 \text{ LiH} + \text{AlCl}_3 \longrightarrow \text{LiAlH}_4 + 3 \text{ LiCl}$$

$$2 \text{ LiH} + 2 \text{ Al} + 3 \text{ H}_2 \longrightarrow 2 \text{ LiAlH}_4$$

$$\text{Na} + \text{Al} + 2 \text{ H}_2 \longrightarrow \text{NaAlH}_4$$

$$\text{NaAlH}_4 + \text{LiCl} \longrightarrow \text{LiAlH}_4 + \text{NaCl}$$

$$4 \text{ NaH} + (\text{CH}_3\text{O})_3\text{B} \longrightarrow \text{NaBH}_4 + 3 \text{ NaOCH}_3$$

1. (a) N. G. Gaylord, *Reduction with Complex Metal Hydrides,* Wiley-Interscience, New York, 1956. (b) M. N. Rerick in R. L. Augustine, ed., *Reduction,* Marcel Dekker, New York, 1968, pp. 1–94. (c) E. Schenker in W. Foerst, ed., *Newer Methods of Preparative Organic Chemistry,* Vol. 4, Academic Press, New York, 1968, pp. 197–335.
2. (a) Metal Chemicals Division, Ventron Corp., Congress St., Beverly, Mass., 01915. (b) Alfa Inorganics, Inc., Ventron Corp., P.O. Box 159, Beverly, Mass., 01915. (c) Research Organic-Inorganic Chemical Co., Sun Valley, Calif. (d) Texas Alkyls, Inc., P.O. Box 600, Deer Park, Texas.
3. (a) W. G. Brown, *Org. Reactions,* **6**, 469 (1951). (b) J. A. Dilts and E. C. Ashby, *Inorg. Chem.,* **9**, 855 (1970).

The complex anions in these salts may be considered to have been derived from an alkali metal hydride and either aluminum hydride (also called alane) or borane, as illustrated in the accompanying equations. These anions are nucleophilic reagents

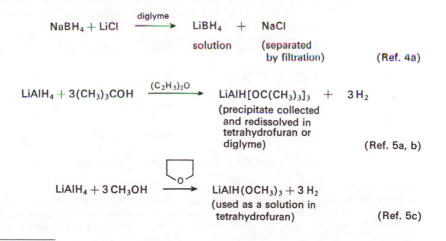

$$LiH + AlH_3 \longrightarrow Li^{\oplus} \quad H{-}\overset{\overset{\displaystyle H}{|}}{\underset{\underset{\displaystyle H}{|}}{Al}}{-}H \;\ominus$$

$$NaH + BH_3 \longrightarrow Na^{\oplus} \quad H{-}\overset{\overset{\displaystyle H}{|}}{\underset{\underset{\displaystyle H}{|}}{B}}{-}H \;\ominus$$

that normally attack polarized multiple bonds (e.g., $C{=}O$, $C{=}N$, $C{\equiv}N$, $N{=}O$) at the more positive atom but usually do not react with isolated carbon-carbon multiple bonds. As will be described subsequently, it is possible to convert these salts back to the trivalent aluminum and boron (e.g., BH_3) compounds by reaction with appropriate acids. Such trivalent materials, with only six electrons in their valence shell, have the requisite structure for Lewis acids and may behave as electrophilic reagents which may attack nonpolar carbon-carbon multiple bonds.

The various commercially available[2] metal hydrides which are particularly useful for reducing organic molecules are listed in Table 2-1. Certain of these reagents may be prepared easily from the commercially available materials, sodium borohydride and lithium aluminum hydride, as illustrated in the following equations.

$$NaBH_4 + LiCl \xrightarrow{\;\text{diglyme}\;} LiBH_4 \;+\; NaCl$$

solution   (separated
by filtration)   (Ref. 4a)

$$LiAlH_4 + 3(CH_3)_3COH \xrightarrow{\;(C_2H_5)_2O\;} LiAlH[OC(CH_3)_3]_3 \;+\; 3H_2$$

(precipitate collected
and redissolved in
tetrahydrofuran or
diglyme)   (Ref. 5a, b)

$$LiAlH_4 + 3CH_3OH \longrightarrow LiAlH(OCH_3)_3 + 3H_2$$

(used as a solution in
tetrahydrofuran)   (Ref. 5c)

4. (a) H. C. Brown, E. J. Mead, and B. C. Subba Rao, *J. Am. Chem. Soc.*, **77**, 6209 (1955).
(b) H. C. Brown and K. Ichikawa, *ibid.*, **83**, 4372 (1961).
5. (a) H. C. Brown and R. F. McFarlin, *J. Am. Chem. Soc.*, **80**, 5372 (1958). (b) M. H. A. Kader, *Tetrahedron Letters*, **No. 27**, 2301 (1969). (c) E. L. Eliel and Y. Senda, *Tetrahedron*, **26**, 2411 (1970).

Table 2–1 Commercially available metal hydrides which are useful reducing agents

Reagent	Useful reaction solvents[a]
NaBH$_4$ (also NaBD$_4$)	H$_2$O, CH$_3$OH,[b] EtOH,[b] i-PrOH, diglyme[c]
NaB(CN)H$_3$	H$_2$O(pH⩾3), CH$_3$OH, THF[c]
LiBH$_4$	THF, diglyme
(CH$_3$)$_4$NBH$_4$	EtOH, C$_6$H$_6$
(C$_2$H$_5$)$_4$NBH$_4$	EtOH, C$_6$H$_6$
BH$_3$ (1 *M* in THF)	Et$_2$O, THF, DME, diglyme
(C$_2$H$_5$N$\overset{\oplus}{}$—$\overset{\ominus}{}BH_3$	Et$_2$O, C$_6$H$_6$, hexane
[(CH$_3$)$_2$CHCH(CH$_3$)]$_2$BH	Et$_2$O, THF, DME, diglyme
(1 *M* in THF, called "disiamylborane")	
LiAlH$_4$ (also LiAlD$_4$)	Et$_2$O, THF, DME, diglyme
NaAlH$_4$	THF, DME, diglyme
LiAlH[OC(CH$_3$)$_3$]$_3$\{also	THF, diglyme
LiAlD[OC(CH$_3$)$_3$]$_3$\}	
LiAlH(OC$_2$H$_5$)$_3$	THF, diglyme
NaAlH$_2$(OCH$_2$CH$_2$OCH$_3$)$_2$ (available as a	C$_6$H$_6$, toluene, xylene
solution in benzene; called RED-AL)	
[(CH$_3$)$_2$CHCH$_2$]$_2$AlH (available pure or as a	Et$_2$O, hexane, benzene, toluene
solution in hexane, heptane, benzene or	
toluene; called DIBAL-H)	

[a]   The abbreviations used are: Et$_2$O = diethyl ether; EtOH = ethanol; i-PrOH = isopropyl alcohol; THF = tetrahydrofuran; DME = 1,2-dimethoxyethane; diglyme = CH$_3$OCH$_2$CH$_2$OCH$_2$-CH$_2$OCH$_3$.
[b]   Since the reaction of NaBH$_4$ with CH$_3$OH and EtOH is relatively rapid, especially at elevated temperatures, these solvents are only satisfactory for reductions which are complete in less than 30 min at temperatures below 25°.
[c]   For reasonably rapid reduction of a carbonyl group with NaBH$_4$, a protic solvent must be present in the reaction mixture.

$$3\,NaBH_4 + 4\,BF_3 \xrightarrow{\text{diglyme}} 3\,NaBF_4 + 2\,B_2H_6$$

(used as formed in the
diglyme solution or distilled
into tetrahydrofuran)

(Ref. 6)

Preparations are also described for the hydrocarbon-soluble tetraalkylammonium borohydrides[2,7] and the lithium[8] and sodium[2a,b,8] cyanoborohydrides which are stable in weakly acidic media.

6. (a) H. C. Brown and B. C. Subba Rao, *J. Org. Chem.*, **22**, 1135 (1957). (b) H. C. Brown, K. J. Murray, L. J. Murray, J. A. Snover, and G. Zweifel, *J. Am. Chem. Soc.*, **82**, 4233 (1960).
7. E. A. Sullivan and A. A. Hinckley, *J. Org. Chem.*, **27**, 3731 (1962).
8. R. F. Borch and H. D. Durst, *J. Am. Chem. Soc.*, **91**, 3996 (1969). R. F. Borch, M. D. Bernstein, and H. D. Durst, *ibid.*, **93**, 2897 (1971).

Just as borane (or its dimer, diborane) may be obtained by reaction of sodium borohydride with various Lewis acids such as boron trifluoride, aluminum hydride may be prepared by reaction of lithium aluminum hydride with various acids. Although aluminum hydride polymerizes in ether solution, stable solutions of this hydride can be obtained in tetrahydrofuran due to the formation of a covalent complex [1] from aluminum hydride and the Lewis base, tetrahydrofuran. An analogous monomeric complex is believed to form when diborane is dissolved in tetrahydrofuran. Preparative routes to these aluminum hydride solutions are indicated. Direct preparation of alane-amine complexes and aminoalanes have also been described.[9e] If molar ratios other than 3 moles of lithium aluminum

$$3 \text{ LiAlH}_4 + \text{AlCl}_3 \xrightarrow[-10 \text{ to } 25°]{\text{Et}_2\text{O}} 3 \text{ LiCl} + 4 \text{ AlH}_3$$

(precipitates as a
polymer on standing)   (Ref. 9)

$$2 \text{ LiAlH}_4 + \text{H}_2\text{SO}_4 \xrightarrow[25°]{} \text{Li}_2\text{SO}_4 \ + \ 2 \text{ H}_3\text{Al-O} \ + \ 2 \text{ H}_2$$

100%

(precipitate   **[1]**
separated by   (as a solution
filtration)   in tetrahydrofuran)

(Ref. 9a)

hydride to 1 mole of aluminum chloride are used, a series of intermediate chloro-aluminum hydrides (e.g., [2] and AlClH$_2$) are produced. The ability of these aluminum compounds to serve as electrophilic reagents (i.e., Lewis acids) appears to decrease in the order: AlCl$_2$H > AlClH$_2$ > AlH$_3$.[9b,d] Special properties which

$$\text{LiAlH}_4 + 3 \text{ AlCl}_3 \xrightarrow[-10 \text{ to } 25°]{(\text{C}_2\text{H}_5)_2\text{O}} 4 \text{ AlCl}_2\text{H} + \text{LiCl}$$

[2]   (believed to be
in solution as LiAlCl$_3$H)   (Refs. 9 and 10)

are useful in certain applications include the increased acid-stability of the cyano-borohydrides[8] and the hydrocarbon solubility of the tetraalkylammonium boro-hydrides,[7] the amine-borane complexes, and of sodium bis(2-methoxyethoxy)-aluminum hydride.

Lithium borohydride, borane, the aluminum hydrides, and especially lithium aluminum hydride react rapidly with hydroxylic compounds. Consequently, these metal hydride reagents must be used under anhydrous conditions with purified, nonhydroxylic solvents. For reductions with borane or lithium aluminum hydride,

9. (a) H. C. Brown and N. M. Yoon, *J. Am. Chem. Soc.*, **88**, 1464 (1966). (b) E. C. Ashby and J. Prather, *ibid.*, **88**, 729 (1966). (c) E. C. Ashby and W. E. Foster, *ibid.*, **88**, 3248 (1966). (d) E. C. Ashby and B. Cooke, *ibid.*, **90**, 1625 (1968). (e) E. C. Ashby, *ibid.*, **86**, 1882 (1964) : E. C. Ashby and R. Kovar, *J. Organometal. Chem.*, **22**, C34 (1970).
10. R. A. Daignault and E. L. Eliel, *Org. Syn.*, **47**, 37 (1967).

ether, tetrahydrofuran, 1,2-dimethoxyethane, and diglyme (the dimethyl ether of diethylene glycol) are the commonly employed solvents, whereas tetrahydrofuran and diglyme are useful solvents for lithium borohydride, lithium trimethoxyaluminum hydride, and lithium tri-*t*-butoxyaluminum hydride. Sodium borohydride reacts slowly enough with water, methanol, ethanol (provided the reaction mixture is kept cool—25° or less, and alkaline) to permit the use of these solvents for reductions that occur readily. For reductions requiring higher temperatures or long reaction periods, isopropyl alcohol is clearly preferable as a solvent.[4]

REDUCTION OF ALDEHYDES AND KETONES

The reaction of an aldehyde or ketone with lithium aluminum hydride involves the transfer of a hydride anion from the metal atom to the carbonyl carbon atom; this transfer is presumably assisted by the prior or concurrent association of the carbonyl oxygen atom with a lithium cation, as illustrated below. The resulting aluminum

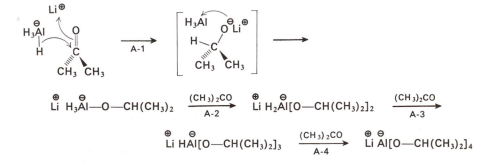

hydride-lithium alkoxide pair, which exits transiently if at all after transfer of the hydride anion, is converted to a lithium alkoxyaluminum hydride. This hydride transfer step (labelled A-1) may be followed by analogous steps (A-2, A-3, and A-4) such that as many as four molecules of ketone are reduced by one molecule of the metal hydride to form a tetralkoxyaluminum anion. There is considerable evidence[5,12] indicating each successive hydride ion transfer to be slower than the preceding step (rate A-1 > A-2 > A-3 > A-4). Removal of a hydride anion from the metal complex would be expected to become more difficult as the number of electronegative alkoxyl substituents in the complex is increased. Because of this reactivity difference it is possible to prepare reducing reagents such as lithium trimethoxyaluminum hydride and lithium tri-*t*-butoxyaluminum hydride, which are

11. (a) V. Bazant and co-workers, *Tetrahedron Letters,* **No. 29,** 3303 (1968). (b) M. Cerny and J. Malek, *ibid.,* **No. 22,** 1739 (1969).
12. (a) H. Haubenstock and E. L. Eliel, *J. Am. Chem. Soc.,* **84,** 2363, 2368 (1962). (b) H. C. Brown, P. M. Weissman, and N. M. Yoon, *ibid.,* **88,** 1458 (1966) and references therein. (c) H. C. Brown and C. J. Shoaf, *ibid.,* **86,** 1079 (1964). (d) P. T. Lansbury and R. E. MacLeay, *ibid.,* **87,** 831 (1965). (e) P. T. Lansbury, R. E. MacLeay, and J. O. Peterson, *Tetrahedron Letters,* **No. 6,** 311 (1964). (f) J. Klein, E. Dunkelblum, E. L. Eliel, and T. Senda, *ibid.,* **No. 58,** 6127 (1968).

less reactive (and more selective) than lithium aluminum hydride. The selective reduction of only the ketone function in the keto ester **[3]** with the tri-*t*-butoxyaluminum hydride illustrates the high degree of selectivity which may be obtained

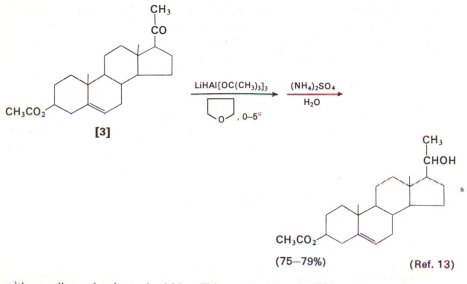

[3]

(75–79%)   (Ref. 13)

with an alkoxyaluminum hydride. This method for modifying the structural and stereochemical selectivity of lithium aluminum hydride as a reducing agent is complicated by the indicated disproportionation of the various alkoxyaluminum hydrides,[5b,c]

$$2\ H_3\overset{\ominus}{Al}\!-\!O\!-\!CH(CH_3) \;\rightleftharpoons\; \overset{\ominus}{AlH_4} + H_2\overset{\ominus}{Al}[O\!-\!CH(CH_3)_2]_2$$

a process which is thought to occur especially rapidly with secondary alcohol derivatives.[5c,12a,c] Such disproportionation can continually regenerate the reactive tetrahydridoaluminum anion and it accounts for many reports of comparable stereochemical results from ketone reductions irrespective of whether lithium aluminum hydride or a lithium alkoxyaluminum hydride is used as a reducing agent.[5c,12a,14] The stability of the alkoxyaluminum hydride complexes to exchange is substantially greater when they are derived from β-amino alcohols or 1,2-diols so that the amino alcohol or diol serves as a bidentate ligand in the metal complex.[14b,15] In such cases asymmetric reducing agents such as the complex **[4]** can be prepared and used to induce asymmetry in reduction products.

13. (a) K. Heusler, P. Weiland, and Ch. Meystre, *Org. Syn.,* **45**, 57 (1965). See also (b) S. G. Levine and N. H. Eudy, *J. Org. Chem.,* **35**, 549 (1970).
14. (a) D. C. Ayres and R. Sawdaye, *Chem. Commun.,* **No. 15**, 527 (1966). (b) S. R. Landor and J. P. Regan, *J. Chem. Soc.,* C, 1159 (1967).
15. (a) O. Cervinka and A. Fabryova', *Tetrahedron Letters,* **No. 13**, 1179 (1967). (b) S. R. Landor, B. J. Miller, and A. R. Tatchell, *J. Chem. Soc.,* C, 1822, 2280 (1966); *ibid.,* C, 197 (1967).

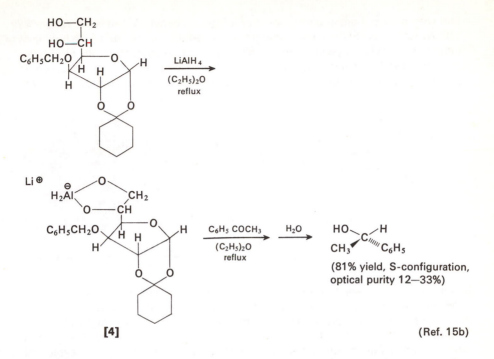

[4]   (Ref. 15b)

Reductions of aldehydes or ketones with sodium borohydride differ from lithium aluminum hydride reductions in requiring the presence of an electrophilic catalyst such as a protic solvent or a lithium or magnesium cation in order for reactions to occur.[4,12d,e] Most sodium borohydride reductions which have been reported to proceed at room temperature in anhydrous, alcohol free solvents such as pyridine or diglyme have probably actually occurred during the isolation procedure when the reaction mixture was mixed with water or an alcohol.[16] In solvents such as isopropyl alcohol, sodium borohydride reductions exhibit kinetics which is first order in ketone and first order in the borohydride anion; the reduction rate is retarded by changes in ketone structure such as increased steric hindrance at the carbonyl carbon or the presence of conjugating substituents (e.g., a phenyl group).[4,17] The reduction rate is also influenced by the changes in the borohydride anion such as the presence of an electron-withdrawing cyano group which retards the reaction) or the substitution of deuterium for hydrogen.[8,17f] These observations suggest that reduction occurs by transfer of a hydride ion to the carbonyl carbon

16. (a) C. D. Ritchie, *Tetrahedron Letters,* **No. 30,** 2145 (1963). (b) C. D. Ritchie and A. L. Pratt, *J. Am. Chem. Soc.,* **86,** 1571 (1964).
17. (a) H. C. Brown, O. H. Wheeler, and K. Ichikawa, *Tetrahedron,* **1,** 214 (1957). (b) H. C. Brown and K. Ichikawa, *ibid.,* **1,** 221 (1957). (c) K. Bowden and M. Hardy, *ibid.,* **22,** 1169 (1966). (d) H. C. Brown, E. J. Mead, and C. J. Shoaf, *J. Am. Chem. Soc.,* **78,** 3616 (1956). (e) H. C. Brown and J. Muzzio, *ibid.,* **88,** 2811 (1966). (f) D. G. Wigfield and D. J. Phelps, *Chem. Commun.,* **No. 18,** 1152 (1970). (g) B. Rickborn and M. T. Wuesthoff, *J. Am. Chem. Soc.,* **92,** 6894 (1970).

atom with prior or concurrent protonation of the carbonyl oxygen atom (structure [5]). As is the case of lithium aluminum hydride reductions, all four of the hydrogen atoms of sodium borohydride may be utilized in the reduction of ketone molecules. Consequently, borohydride reductions have been suggested[17d,g] to proceed in the series of steps illustrated (B-1, B-2, B-3, B-4). However, the reductions with

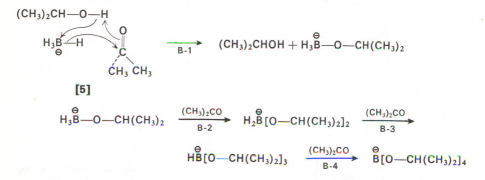

lithium aluminum hydride and sodium borohydride are not analogous, since the observed kinetics required the first hydride transfer step (B-1) to be rate-limiting. This fact has been interpreted to mean that reduction steps involving alkoxyboro-hydrides (B-2, B-3, B-4) are all faster than the step (B-1) involved in hydride transfer from sodium borohydride. The kinetic observations could also be explained, at least in part, by a rapid disproportionation of the initially formed alkoxyboro-hydride anion [6] so that the actual reducing agent is always the borohydride anion; this disproportionation appears most rapid with alkoxy groups from primary alcohols. A similar sequence leading to exchange between alkoxy groups in solution and bonded to alkoxyborohydride anions also appears likely. Although sodium triisopropoxyborohydride was found to be a more reactive reducing agent than sodium borohydride,[17d,g] the possibilities for disproportionation accompanied by the ready availability of other reactive metal hydrides (e.g., LiAlH$_4$ and LiBH$_4$) have resulted in little interest in the alkoxyborohydrides as useful reducing agents.

$$2\ H_3\overset{\ominus}{B}\!-\!OCH_3 \;\rightleftharpoons\; \overset{\ominus}{B}H_4 + H_2\overset{\ominus}{B}(OCH_3)_2$$

[6]

Although detailed studies of the reduction of ketones with lithium borohydride in tetrahydrofuran or diglyme are not available, it seems likely that these reductions are analogous to reductions with sodium borohydride (see structure [5]) except that lithium cation rather than a proton is the electrophile which is bonded to the carbonyl oxygen. Interestingly, even in isopropyl alcohol solution the addition of lithium salts (but not sodium salts) catalyzes the reduction of ketones with boro-hydride anion, suggesting that in this medium lithium cation is a better electrophilic catalyst than is the proton abstracted from isopropyl alcohol.[4b] This catalytic effect from lithium cation is no longer observed when the reduction is conducted in aqueous solution, where the lithium cation would be strongly solvated.

The following equations illustrate typical laboratory reductions with sodium borohydride and lithium aluminum hydride:

$$CH_3(CH_2)_5\text{—}CHO + LiAlH_4 \xrightarrow[\text{reflux}]{(C_2H_5)_2O} \xrightarrow{H_3O^\oplus} CH_3(CH_2)_5CH_2OH$$

(86%)

(Ref. 18a)

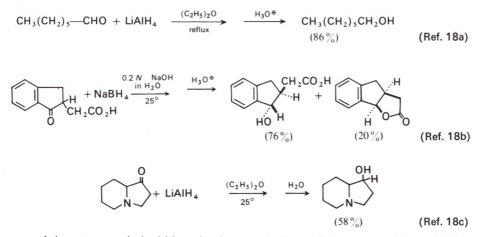

(76%)   (20%)

(Ref. 18b)

(58%)

(Ref. 18c)

It is customary in hydride reductions to employ a slight excess of the reducing reagent in the event that some of the metal hydride is unintentionally destroyed by reduction with hydroxyl-containing materials present in the reaction medium. Alternatively, a solution (filtered or decanted) of lithium aluminum hydride can be prepared and standardized prior to use, by titration with iodine,[19a] by an EDTA titration for aluminum, or by measuring the volume of hydrogen evolved when an aliquot of the solution is cautiously mixed with a solution of water in tetrahydrofuran. The isolation of organic products from sodium borohydride reductions is usually accomplished by diluting the reaction mixture with water or dilute aqueous base and then extracting the organic product from the aqueous solution containing boric acid salts. It should be noted that acidification of reaction solutions containing excess sodium borohydride can generate diborane which may cause further undesired reduction of groups in the initial reaction product.[19b] The reaction of the alkoxyboron intermediates with water presumably occurs as follows:

$$B^\ominus(OR)_4 \rightleftharpoons RO^\ominus + B(OR)_3 \xrightleftharpoons{H_2O} ROH + HO\text{—}B^\ominus(OR)_3$$

$$HO\text{—}B^\ominus(OR)_3 \rightleftharpoons ROH + (RO)_2B\text{—}O^\ominus \xrightarrow{H_2O} 2\,ROH + (HO)_2B\text{—}O^\ominus$$

Excess lithium aluminum hydride remaining in a reduction mixture is usually destroyed by the *dropwise* addition of ethanol or of ether saturated with water.

18. (a) R. F. Nystrom and W. G. Brown, *J. Am. Chem. Soc.,* **69**, 1197 (1947). (b) H. O. House, H. Babad, R. B. Toothill, and A. W. Noltes, *J. Org. Chem.,* **27**, 4141 (1962). (c) N. J. Leonard, S. Swann, Jr., and F. Figueras, Jr., *J. Am. Chem. Soc.,* **74**, 4620 (1952).
19. (a) H. Felkin, *Bull. Soc. Chim. France,* 347 (1951). (b) J. A. Marshall and W. S. Johnson, *J. Org. Chem.,* **28**, 595 (1963). (c) For a specific hydrolysis procedure, see L. F. Fieser and M. Fieser, *Reagents for Organic Synthesis,* Wiley, New York, 1967, p. 584.

Although the addition of ethyl acetate has also been employed, this procedure is not advisable, since both amine and alcohol products may react with ethyl acetate.[20] If the reduction product is nonbasic, reasonably insoluble in water, and reasonably stable to acid, the reaction mixture may then be poured into cold, dilute, aqueous acid and the organic product extracted with ether. When this acid isolation procedure is not applicable, it is customary to add to the mixture, dropwise and with stirring, the calculated amount of water or dilute aqueous sodium hydroxide[19c] required to convert the lithium and aluminum salts to lithium aluminate ($LiAlO_2$), a granular precipitate that can be filtered from the organic solution and extracted with boiling tetrahydrofuran to remove any occluded organic product. Some workers recommend the addition of solid sodium sulfate or magnesium sulfate to aid coagulation prior to the filtration of the lithium aluminate. In this isolation procedure it is important not to add excess water, since its presence will convert the reaction mixture to a gelatinous emulsion of the organic solvent, water, and aluminum hydroxide, which is very difficult either to filter or to extract. An alternative isolation procedure employs an aqueous solution of sodium potassium tartarate to hydrolyze and dissolve the aluminum salts present in the reaction mixture.

THE STEREOCHEMISTRY OF KETONE REDUCTIONS

The stereochemical course of the reduction of a carbonyl group by a metal hydride, like other nucleophilic additions to carbonyl functions, may be influenced by asymmetry in the molecule. The best-studied examples are those having an asymmetric carbon atom adjacent to the carbonyl group, as in the ketone [7]. Since the metal hydrides, particularly lithium aluminum hydride, may function as strong bases, one might expect an optically active ketone having an *alpha* hydrogen atom (e.g., [7]) to be converted to its enolate [8] and, consequently, racemized prior to reduction. Although small amounts of ketones, presumably formed by hydrolysis of an intermediate enolate anion such as [8], are often recovered from

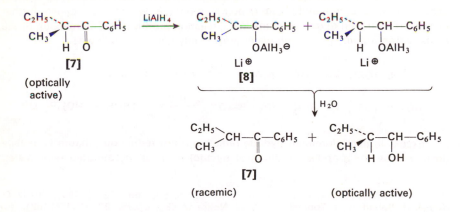

20. (a) W. B. Wright, *J. Org. Chem.*, **25**, 1033 (1960). (b) P. R. Stapp and N. Rabjohn, *ibid.*, **24**, 1798 (1959).

lithium aluminum hydride reactions, the reduced products are formed without racemization.[1]

This result is attributable to the fact that enolate anions react very slowly with lithium aluminum hydride, and the racemic ketone derived from [8] is usually produced by hydrolysis only during the isolation process when no excess hydride reducing agent remains. Since no protonic solvents are present during reductions with lithium aluminum hydride, the racemic ketone could be produced in the reaction mixture only by a proton transfer from the starting ketone [7] to the enolate [8] (see Chapter 9). The fact that little if any racemization is observed requires that the proton transfer must be slower than the reduction of the starting ketone [7] with lithium aluminum hydride. Since sodium borohydride reductions are typically carried out in hydroxylic solvents, a competing proton transfer to an enolate with racemization and subsequent reduction becomes very probable. However, the much lower base strength of the borohydride anion greatly diminishes the rate of formation of an enolate comparable to [8], and usually permits the reduction of a carbonyl function without racemization of an adjacent center of asymmetry. Exceptions are to be expected only with ketones which have relatively acidic alpha hydrogens (such as the α-aryl ketone shown in the following example) or for borohydride reductions performed under anhydrous conditions where sodium borohydride behaves as a strong base.[22f]

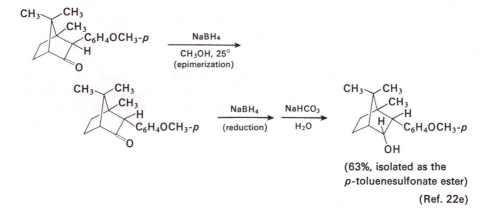

(63%, isolated as the
p-toluenesulfonate ester)

(Ref. 22e)

The following equation indicates the proportions of diastereoisomers [9] and [10] obtained by reduction of a ketone having three different *alpha* substituents (only one enantiomer of the racemic mixture employed is shown) with lithium aluminum hydride:

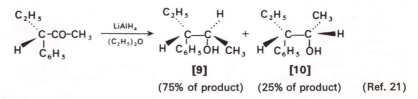

[9]   [10]

(75% of product)   (25% of product)   (Ref. 21)

The prediction of which diastereoisomer will predominate in such cases has been generalized as the Rule of Steric Control of Asymmetric Induction.[21,22] This rule is applied by considering that the starting ketone will react preferentially in the conformation illustrated in structure [11] (the conformation in which the carbonyl group is staggered between the medium and smallest substituents on the α-carbon) and that the carbonyl group in this conformation will be attacked by the metal hydride anion from the less hindered side as pictured. Similar examples of asymmetric induction have been observed in additions to ketones which have a chiral center at the carbon atom *beta* to the carbonyl group.[21b]

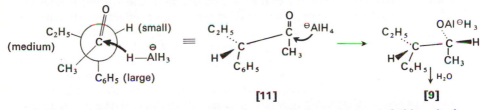

Although many examples support the empirical correctness of this rule for *acyclic* ketones with no polar α-substituents, the reason for the validity of the rule has received considerable discussion.[23] The most satisfactory explanation at the present time adopts the hypotheses that the transition state geometry for reactions of metal hydrides (and organometallic reagents) with carbonyl groups resembles the geometry of the starting ketone and that both steric interaction between nonbonded groups and torsional strain[24] are important factors in deciding the direction from which a carbonyl group will be attacked by a nucleophile. The application of these ideas to the formation of the epimeric alcohols [9] and [10] is illustrated with the transition states [12], [13], and [14]. These structures are considered to be of lower energy than other possible structures, because in no case are bonds eclipsed (dihedral angle = $0°$) and in all cases one of the large groups ($^{\ominus}AlH_4$ or CH_3)[25] bonded to the carbonyl carbon atom is approximately

21. (a) D. J. Cram and F. A. Abd Elhafez, *J. Am. Chem. Soc.*, **74**, 5828 (1952). (b) T. J. Leitereg and D. J. Cram, *ibid.*, **90**, 4011, 4019 (1968).
22. (a) D. J. Cram and K. R. Kopecky, *J. Am. Chem. Soc.*, **81**, 2748 (1959). (b) J. H. Stocker, P. Sidisunthorn, B. M. Benjamin, and C. J. Collins, *ibid.*, **82**, 3913 (1960). (c) D. J. Cram and D. R. Wilson, *ibid.*, **85**, 1245 (1963). (d) For recent reviews, see D. R. Boyd and M. A. McKervey, *Quart. Rev.*, **22**, 95 (1968); L. Velluz, J. Valls, and J. Mathieu, *Angew. Chem., Intern. Ed. Engl.*, **6**, 778 (1967). (e) W. F. Erman and T. J. Flautt, *J. Org. Chem.*, **27**, 1526 (1962). (f) V. Hach, E. C. Fryberg, and E. McDonald, *Tetrahedron Letters*, **No. 28**, 2629 (1971).
23. (a) M. Cherest, H. Felkin, and N. Prudent, *Tetrahedron Letters*, **No. 18**, 2199 (1968). (b) M. Cherest and H. Felkin, *ibid.*, **No. 18**, 2205 (1968). (c) G. J. Karabatsos, *J. Am. Chem. Soc.*, **89**, 1367 (1967).
24. Torsional strain, also called Pitzer strain, is the increase in the energy of a molecule observed when the dihedral angle defined by bonds to adjacent atoms approaches zero. For discussion, see E. L. Eliel, N. L. Allinger, S. J. Angyal, and G. A. Morrison, *Conformational Analysis*, Wiley-Interscience, New York, 1965, pp. 5–12.
25. In applying this hypothesis, the carbonyl oxygen atom is considered to have less steric bulk than either the attacking nucleophile (e.g., $^{\ominus}AlH_4$) or the group (e.g., CH_3) bound to carbonyl carbon atom.

trans (dihedral angle ~ 180°) to the largest α-substituent (C_6H_5). The unfavorable steric interactions between medium and large groups, represented by heavy arrows in structures [12]–[14], lead to the conclusion that structure [12] (with one interaction between a medium group and a large group) is of lower energy than [13] (two medium-large interactions) and [14] (one large-large interaction) and, consequently, that transition state [12] leading to alcohol [9] is the favored reaction path.

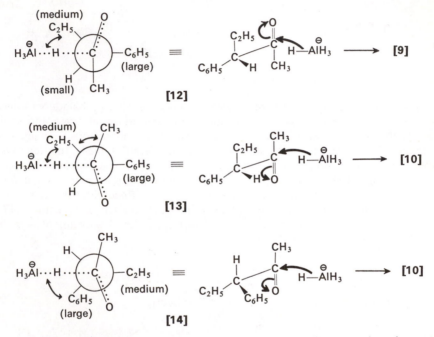

The rule of asymmetric induction usually requires modification when the asymmetric center bears polar substituents, presumably because the assumption that only torsional strain and steric interactions determine the favored transition state for reaction is no longer valid.[22,23] Reductions of α-hydroxy and α-amino ketones with lithium aluminum hydride often exhibit a relatively high degree of stereoselectivity, as illustrated in the accompanying example. These results have been attributed to the initial formation of a salt that serves as a bidentate ligand for the metal atom and reacts in conformation [15] with the metal hydride ion attacking om the less hindered side.[21b,22]

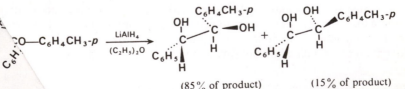

(85% of product)   (15% of product)

(Ref. 22b)

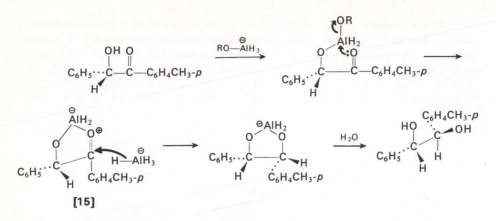

[15]

For acyclic ketones which contain polar α-substituents (e.g., halogens) which are unlikely to coordinate with metal atoms, not only torsional strain and steric interaction must be considered but also electrostatic interactions. Again, two interpretations of the experimental data have been offered. Initially, the data were explained by supposing that the reactive conformation of such ketones was a structure (e.g., **[16]**) in which the carbonyl function and the polar α-substituent were *trans* to minimize dipole-dipole repulsion.[22c,26] Recently the same data have been interpreted as supporting a transition state such as structure **[17]**, which allows maximum separation of the electronegative α-substituent and the negatively charged nucleophilic reagent.[23]

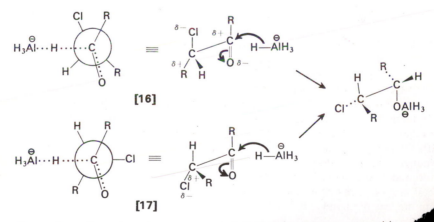

[16]

[17]

A number of workers have suggested that polar groups at positions more remote than *alpha* or *beta* to a carbonyl group may influence the stereochemistry of molecular reduction reactions. In certain cases involving hydroxy or amino substituents, the formation of an amine or alkoxymetal hydride complex followed

26. (a) J. W. Cornforth, R. H. Cornforth, and K. K. Mathew, (1959).
(b) R. Haller, *Tetrahedron Letters*, **No. 48**, 4347 (1965).

hydride transfer (e.g., [18]) has been suggested.[27] As the following example indicates, such an intramolecular process would be expected to be more important for reductions with lithium aluminum hydride in aprotic solvents than for reductions with sodium borohydride in protic solvents.

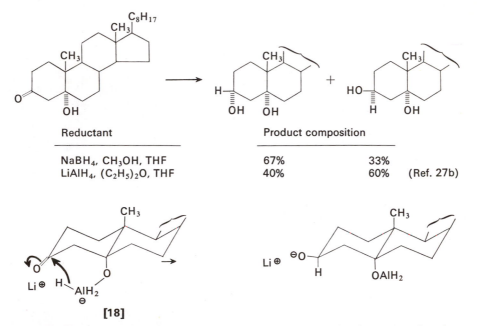

Reductant	Product composition		
NaBH$_4$, CH$_3$OH, THF	67%	33%	
LiAlH$_4$, (C$_2$H$_5$)$_2$O, THF	40%	60%	(Ref. 27b)

[18]

In other cases, the effect of polar substituents on reduction stereochemistry would appear better explained as a result of electrostatic effects.[28] At least for examples involving the reduction of flexible cyclohexanone derivatives, it would appear that not only the electrostatic effect but also the populations of various conformers of the starting ketone may be influencing the reduction reaction.[28c,d]

Although the reductions of a large number of cyclic ketones with metal hydrides have been studied, prediction of the stereochemical outcome of these reactions has been uncertain and has generated considerable discussion.[5c,17f,g,23,29]

27. For examples, see (a) P. T. Lansbury, J. F. Bieron, and M. Klein, *J. Am. Chem. Soc.,* **88,** 1477 (1966). (b) M. Akhtar and S. Marsh, *J. Chem. Soc.,* C, 937 (1966). (c) P. S. Portoghese and D. A. Williams, *Tetrahedron Letters,* No. 50, 6299 (1966). (d) S. Yamada and K. Koga, *ibid.,* No. 18, 1711 (1967). (e) E. C. Pesterfield and D. M. S. Wheeler, *J. Org. Chem.,* **30,** 1513 (1965). (f) See Ref. 37b for an application of the same idea to halohydrin reductions.
28. (a) H. B. Henbest and J. McEntee, *J. Chem. Soc.,* 4478 (1961); H. B. Henbest, *Proc. Chem. Soc.,* 159 (1963). (b) H. Kwart and T. Takeshita, *J. Am. Chem. Soc.,* **84,** 2833 (1962). (c) D. N. Kirk, *Tetrahedron Letters,* No. 22, 1727 (1969). (d) R. S. Monson, D. Przybycien, and A. Baraze, *J. Org. Chem.,* **35,** 1700 (1970).
29. (a) W. G. Dauben, G. J. Fonken, and D. S. Noyce, *J. Am. Chem. Soc.,* **78,** 2579 (1956). (b) O. R. Vail and D. M. S. Wheeler, *J. Org. Chem.,* **27,** 3803 (1962). (c) A. V. Kamernitzky and A. A. Akhrem, *Tetrahedron,* **18,** 705 (1962). (d) J. C. Richer, *J. Org. Chem.,* **30,** 324 (1965). (e) J. A. Marshall and R. D. Carroll, *ibid.,* **30,** 2748 (1965).

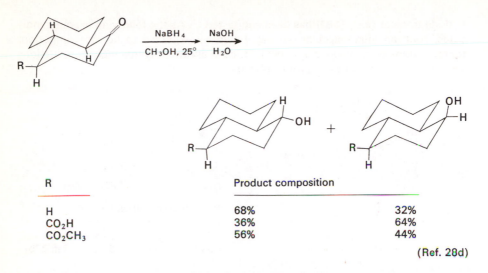

R	Product composition	
H	68%	32%
CO$_2$H	36%	64%
CO$_2$CH$_3$	56%	44%

(Ref. 28d)

In reductions where approach of the metal hydride to one side of the carbonyl function is clearly much more hindered than approach to the other side (e.g., [19]), the major product is formed by addition of a hydride ion from the less hindered side.

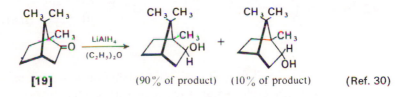

[19]   (90% of product)   (10% of product)   (Ref. 30)

When the steric environment is comparable on each side of the carbonyl group, as in ketone [20], the major reduction product is usually the more stable product; i.e., an equatorial alcohol such as [21] is usually the major product formed by reduction of a cyclohexanone.

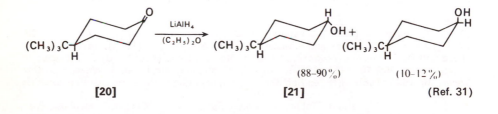

[20]   [21]   (Ref. 31)

(88–90%)   (10–12%)

30. (a) D. S. Noyce and D. B. Denney, *J. Am. Chem. Soc.*, **72**, 5743 (1950). (b) In reductions of norcamphor and its derivatives, where attack from the *exo* side is less hindered, the opposite stereochemical result is obtained; see C. H. DePuy and P. R. Story, *ibid.*, **82**, 627 (1960).
31. E. L. Eliel and M. N. Rerick, *J. Am. Chem. Soc.*, **82**, 1367 (1960).

These observations were initially interpreted by assuming that the transition state geometry for reduction of unhindered ketones resembled the geometry of the alcohol products. Consequently, the favored transition state was that leading to more stable alcohol, a circumstance described as "product development control."[29a] For reductions of hindered ketones (presumably a slower process), the transition state for reduction was suggested to resemble the starting ketone in geometry. In reductions of this type, the stereochemical outcome was attributed to attack of the metal hydride from the less hindered side of the carbonyl group (called "steric approach control"). Recent investigations[5c,12f,17f,23,29d,e] have supported the view that all metal hydride-ketone reductions have transition states that resemble the reactants in geometry, and the reduction stereochemistry is determined only by a combination of steric interference, torsional strain, and electrostatic effects in the transition state. This interpretation utilizes premises similar to those described for reductions of acyclic ketones (structures [12], [13], and [14]), but takes account of the fact that favorable transition states [12] and [13], with the attacking nucleophile *trans* to the largest α-substituent, are not feasible in a cyclohexane system. Reduction of a cyclohexanone to form an equatorial alcohol is suggested to involve the transition state [22], which is analogous to [14]. Although the structure [22] avoids torsional strain, substantial steric interference can result if an axial substituent larger than hydrogen is present at C-3 or C-5. This steric interference with an axial group R in [22] is illustrated with an arrow. The alternative transition state [23] leading to an axial alcohol

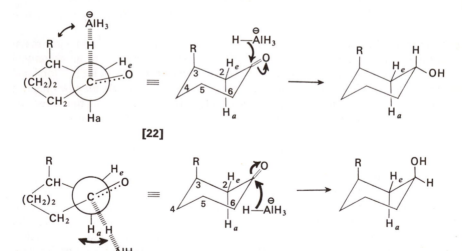

[22]

[23]

avoids steric interference with axial substituents at C-3 or C-5 but suffers from torsional strain (arrow in structure [23]) involving bonds to the entering nucleophile and the axial substituents (e.g., Ha in [23]) at C-2 and C-6. (For clarity only one of the two axial hydrogen atoms (at C-2) is shown in structures [22] and [23]). According to this interpretation, a cyclohexanone will normally prefer to

react by the path indicated in formula [22] in order to avoid the torsional strain present in the alternative structure [23]. However, if relatively large axial substituents are present at C-3 and/or C-5 (e.g., R in [22]) the transition state [23] will be favored. The fact that the favored transition state [22] for unhindered ketones usually leads to the more stable equatorial alcohol product is considered fortuitous.

The effect of an axial C-3 substituent is illustrated by comparison of the reduction of the unhindered ketone [20] with the reduction of dihydroisophorone [24], a ketone with an axial methyl group at C-3. In the latter case the more stable equatorial alcohol [25] is not the major product. As would also be expected from the transition

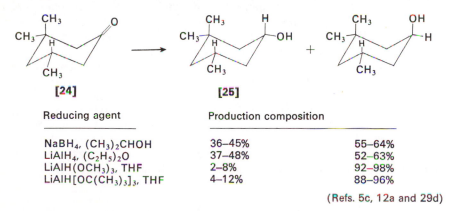

Reducing agent	Production composition	
NaBH₄, (CH₃)₂CHOH	36–45%	55–64%
LiAlH₄, (C₂H₅)₂O	37–48%	52–63%
LiAlH(OCH₃)₃, THF	2–8%	92–98%
LiAlH[OC(CH₃)₃]₃, THF	4–12%	88–96%

(Refs. 5c, 12a and 29d)

state models [22] and [23], an increase in the steric bulk of the attacking metal hydride anion would destabilize the transition state [22] leading to an equatorial alcohol more than the transition state [23] leading to an axial alcohol, since torsional strain (as in [23]) is influenced more by changes in dihedral angle than by changes in steric bulk. The following equations illustrate the realization of this effect by changing the reducing agent from lithium aluminum hydride to the more bulky lithium trialkoxyaluminum hydride.[12a,29d,32]

In each of these examples, reduction resulting from attack of the metal hydride from the less sterically hindered side (as in structure [23]) is enhanced as the reagent is changed from lithium aluminum hydride to the more bulky lithium trimethoxyaluminum hydride. The curious fact that lithium tri-*t*-butoxyaluminum hydride often behaves as a less hindered reducing agent than lithium trimethoxy-aluminum was initially suggested to indicate a change in reaction mechanism for the *t*-butoxy derivative, such as the involvement of di-*t*-butoxyalane as the actual reductant.[14a,32a,b] However, further study[32c,d] has indicated that this is not the case; instead the difference results from the tendency of lithium trimethoxyaluminum

32. (a) H. C. Brown and H. R. Deck, *J. Am. Chem. Soc.*, **87**, 5620 (1965). (b) A survey of the reducing properties of lithium trimethoxyaluminum hydride has been reported by H. C. Brown and P. M. Weissman, *ibid.*, **87**, 5614 (1965). (c) E. C. Ashby, J. P. Sevenair, and F. R. Dobbs, *J. Org. Chem.*, **36**, 197 (1971). (d) D. C. Ayres, D. N. Kirk, and R. Sawdaye, *J. Chem. Soc.*, B, 1133 (1970).

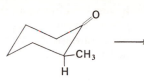 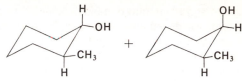

Reducing agent	Product composition	
LiAlH$_4$, THF	74–76%	24–26%
LiAlH(OCH$_3$)$_3$, THF[a]	31–72%	28–69%
LiAlH[OC(CH$_3$)$_3$]$_3$, THF	70%	30%
NaBH$_4$, (CH$_3$)$_2$CHOH	69%	31%

(Refs. 17e and 32)

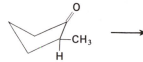 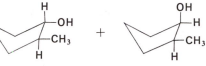

Reducing agent	Production composition	
LiAlH$_4$, THF	76–79%	21–24%
LiAlH(OCH$_3$)$_3$, THF	56%	44%
LiAlH[OC(CH$_3$)$_3$]$_3$, THF	72%	28%

(Ref. 32)

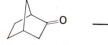 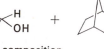

Reducing agent	Product composition	
LiAlH$_4$, THF	89%	11%
LiAlH(OCH$_3$)$_3$, THF	98%	2%
LiAlH[OC(CH$_3$)$_3$]$_3$, THF	93%	7%
NaBH$_4$, (CH$_3$)$_2$CHOH	86%	14%

(Refs. 17e and 32)

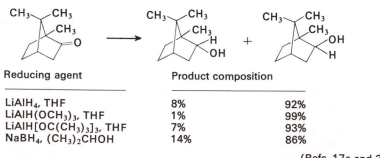

Reducing agent	Product composition	
LiAlH$_4$, THF	8%	92%
LiAlH(OCH$_3$)$_3$, THF	1%	99%
LiAlH[OC(CH$_3$)$_3$]$_3$, THF	7%	93%
NaBH$_4$, (CH$_3$)$_2$CHOH	14%	86%

(Refs. 17e and 32)

[a] The product composition is dependent on the concentration of the metal hydride.

hydride to aggregate into dimeric or trimeric species, whereas lithium tri-*t*-butoxy-aluminum hydride remains monomeric.[32c] As a result, the stereochemical outcome of ketone reductions with lithium trimethoxyaluminum hydride is dependent on the concentration of metal hydride used, with the hydride having greater effective steric requirements in relatively concentrated solutions.

The rate and stereospecificity of metal hydride-ketone reductions can also be influenced by the addition of Lewis acids to produce the previously described trivalent aluminum and boron hydride derivatives. Although studies of ketone reductions with diborane are limited,[33] the stereochemical outcome appears to be similar to the results obtained with the previously discussed metal hydride anions. This reduction, which may be catalyzed with the Lewis acid boron trifluoride, is

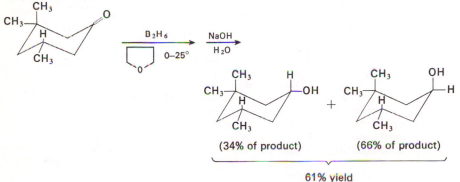

(34% of product)   (66% of product)

61% yield

(Ref. 33a)

believed to involve an initial electrophilic attack by borane (or boron trifluoride if this catalyst is present) to form a complex [26] which is then attacked by a molecule of diborane to form an alkoxyborane [27]. In subsequent reactions a dialkoxyborane [28] and finally a trialkylborate [29] may be formed. However, the final hydride transfer ([28] → [29]) is very slow. Since reactions with diborane and with the

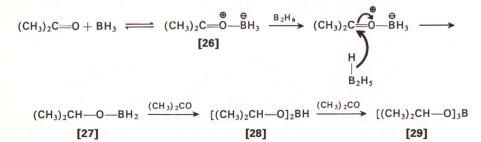

$(CH_3)_2C{=}O + BH_3 \rightleftharpoons (CH_3)_2C{=}\overset{\oplus}{O}{-}\overset{\ominus}{B}H_3 \xrightarrow{B_2H_6} (CH_3)_2C{=}\overset{\oplus}{O}{-}\overset{\ominus}{B}H_3 \longrightarrow$

[26]

$(CH_3)_2CH{-}O{-}BH_2 \xrightarrow{(CH_3)_2CO} [(CH_3)_2CH{-}O]_2BH \xrightarrow{(CH_3)_2CO} [(CH_3)_2CH{-}O]_3B$

[27]   [28]   [29]

33. (a) J. Klein and E. Dunkelblum, *Tetrahedron*, **23**, 205 (1967). (b) K. M. Biswas, L. E. Houghton, and A. H. Jackson, *ibid.*, **Suppl. No. 7**, 261 (1966). (c) H. C. Brown and B. C. Subba Rao, *J. Am. Chem. Soc.*, **82**, 681 (1960). (d) M. Stefanovic and S. Lajsic, *Tetrahedron Letters,* **No. 19**, 1777 (1967). (e) L. P. Kuhn and J. O. Doali, *J. Am. Chem. Soc.*, **92**, 5475 (1970). (f) S. S. White, Jr., and H. C. Kelly, *ibid.*, **92**, 4203 (1970). (g) D. J.

dialkoxyborane [28] have led to different stereochemical results, disproportionation of the intermediate alkoxyboranes is believed to occur much more slowly than
reduction.[33a,g] The reduction of ketones with amine-borane complexes has been
reported to occur by both an uncatalyzed process and an acid-catalyzed reaction;[33f]
the reaction of acetone with diborane in the gas phase is catalyzed by Lewis bases
such as tetrahydrofuran.[33e]

The use of the subsequently described boranes with alkyl substituents increases
the steric bulk of the reducing agent; attack by such bulky reducing agents tends
to occur from the less hindered side of relatively unhindered ketones such as [30].
Usually the less stable alcohol epimer is produced in such cases. These results

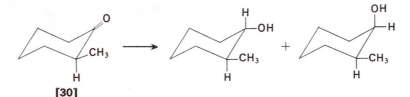

[30]

Reducing agent	Product composition	
LiAlH$_4$, THF	75%	25%
B$_2$H$_6$, THF	74%	26%
[(CH$_3$)$_2$CHCH(CH$_3$)]$_2$BH, THF	21%	79%
(diterpenylborane) BH, THF	6%	94%

(Ref. 34)

are again interpretable in terms of transition states like [22] and [23]; as the steric
bulk of the reductant is increased, the steric interactions which destabilize [22] are
enhanced, whereas the torsional strain which destabilizes [23] remains relatively
constant.

Although conjugation of a ketone with a double bond reduces the rate of
reduction of the carbonyl group by the nucleophilic reagent, sodium borohydride,[17a,b]
the opposite effect would be expected for the electrophilic borane reductions.
Thus in the reduction of the diketone, progesterone, formation of the presumed
intermediate complex (see [26]) would be more favorable at the conjugated
carbonyl group where more delocalization of the positive charge is possible. It is
presumably this factor (perhaps accompanied by steric hindrance at the saturated
ketone) which permits the indicated reduction of the conjugated ketone in preference to the saturated ketone.

Pasto, V. Balasubramaniyan, and P. J. Wojtkowski, *Inorg. Chem.,* **8**, 594 (1969). (h) For a
kinetic study of the reactions of ketones with bis(3-methyl-2-butyl) borane, see R. Fellous,
R. Luft, and A. Puill, *Tetrahedron Letters,* **No. 18,** 1509 (1970).
34. H. C. Brown and V. Varma, *J. Am. Chem. Soc.,* **88,** 2870 (1966).

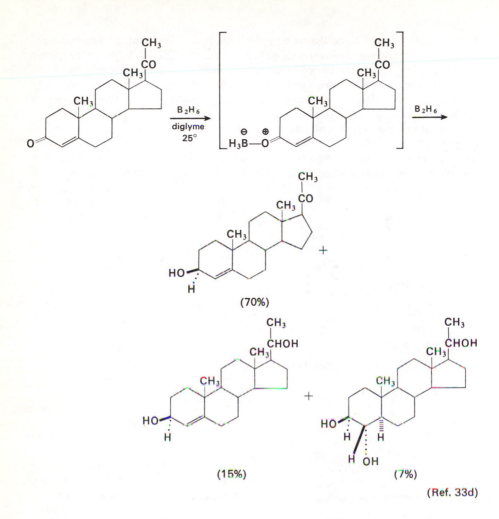

(70%)

(15%)   (7%)

(Ref. 33d)

The following equations illustrate typical reductions of ketones with various aluminum hydride derivatives.[32d,35,37b]

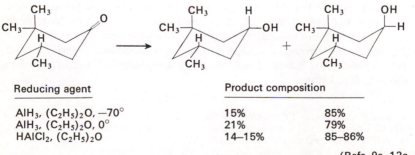

Reducing agent	Product composition	
AlH$_3$, (C$_2$H$_5$)$_2$O, −70°	15%	85%
AlH$_3$, (C$_2$H$_5$)$_2$O, 0°	21%	79%
HAlCl$_2$, (C$_2$H$_5$)$_2$O	14–15%	85–86%

(Refs. 9a, 12a, and 35)

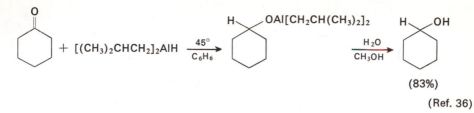

(83%)

(Ref. 36)

These reagents, like diborane, are believed to react initially as electrophiles which form a complex with the ketone analogous to structure [26]; it would appear that a second molecule of the aluminum hydride is required to donate a hydride ion to the carbonyl carbon atom. The stereochemical results obtained from reduction of cyclohexanones with aluminum hydride derivatives, under conditions where equilibration[31,36] of the initially formed alkoxyaluminum derivative was precluded, suggest that a relatively bulky reagent is involved in delivering the hydride ion; i.e., the transition state [23] to form an axial alcohol seems to be favored. This observation is better explained by supposing that a second molecule of the aluminum hydride reagent (presumably solvated or dimeric) is involved in the reduction step. In ketone reductions with triisobutylaluminum, a different reaction path is followed involving an intramolecular hydride transfer.

$$C_6H_5-CO-C_6H_5 + [(CH_3)_2CHCH_2]_3Al \xrightarrow[25°]{(C_2H_5)_2O}$$

$$(C_6H_5)_2C=O^\oplus$$

$$Al[CH_2CH(CH_3)_2]_2$$

$$H$$

$$(CH_3)_2C-CH_2 \longrightarrow$$

$$(C_6H_5)_2CH-OAl[CH_2CH(CH_3)_2] + (CH_3)_2C=CH_2 \xrightarrow{H_2O} (C_6H_5)_2CHOH$$

(97%)

(Ref. 36c)

Reductions with dichloroaluminum hydride[31,37] or trialkylaluminum reagents[36b,c] offer the special property of allowing the initially formed alkoxyaluminum intermediate [31] to be equilibrated if a small excess of ketone is added to the reaction mixture. This procedure appears to be one of the most satisfactory methods available for the formation of the more stable alcohol epimer from reduction of a ketone

35. (a) D. C. Ayres and R. Sawdaye, *J. Chem. Soc.*, B, 581 (1967). (b) D. C. Ayres, D. N. Kirk, and R. Sawdaye, *ibid.*, B, 505 (1970).
36. (a) A. E. G. Miller, J. B. Biss, and L. H. Schwartman, *J. Org. Chem.*, **24**, 627 (1959) and references therein. (b) For a discussion of the reduction of ketones with triisobutyl-aluminum, see H. Haubenstock and E. B. Davidson, *ibid.*, **28**, 2772 (1963). (c) E. C. Ashby and S. H. Yu, *ibid.*, **35**, 1034 (1970).
37. (a) E. L. Eliel, R. J. L. Martin, and D. Nasipuri, *Org. Syn.*, **47**, 16 (1967). (b) E. L. Eliel, *Rec. Chem. Progr.*, **22**, 129 (1961). (c) J. C. Richer and E. L. Eliel, *J. Org. Chem.*, **26**, 972 (1961). (d) E. L. Eliel and D. Nasipuri, *ibid.*, **30**, 3809 (1965).

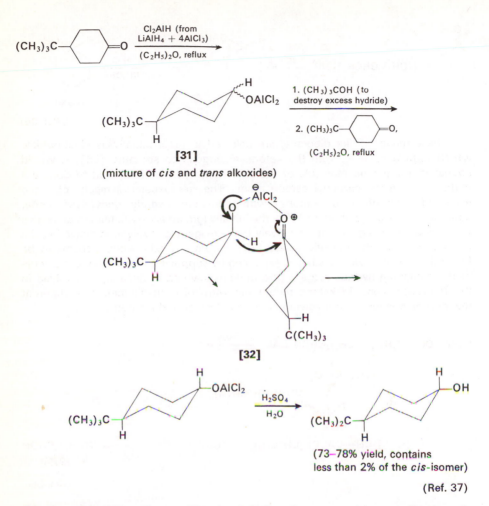

[31]

(mixture of *cis* and *trans* alkoxides)

[32]

(73–78% yield, contains
less than 2% of the *cis*-isomer)

(Ref. 37)

with a metal hydride. The equilibration is believed to proceed by coordination of the trivalent alkoxyaluminum intermediate with the excess ketone followed by hydride transfer as in structure **[32]**. This equilibration process is analogous to the Oppenauer oxidation of alcohols in the presence of a ketone and aluminum isopropoxide[38] and the related Meerwein-Ponndorf-Verley reduction of ketones with aluminum alkoxides.[39] It is apparent that this type of equilibrium (i.e., **[32]**) requires a trivalent aluminum species such as **[31]**, a Lewis acid, and would not be expected to intervene in lithium aluminum hydride reductions which involve tetravalent aluminum anions. The large proportion of equatorial alcohol formed

38. (a) J. F. Eastham and R. Teranishi, *Org. Syn.*, **Coll. Vol. 4**, 192 (1963). (b) C. Djerassi, *Org. Reactions*, **6**, 207 (1951).
39. (a) W. Chalmers, *Org. Syn.*, **Coll. Vol. 2**, 598 (1943). (b) A. L. Wilds, *Org. Reactions*, **2**, 178 (1944).

in this equilibration reflects the tendency of the bulky alcohol derivative, —OAlCl₂, to be in an equatorial position. This advantage is not offered in equilibrations effected with trialkoxyaluminum reagents. At the present time there is no indication that an analogous equilibration occurs at intermediate stages in reductions of ketones with diborane.

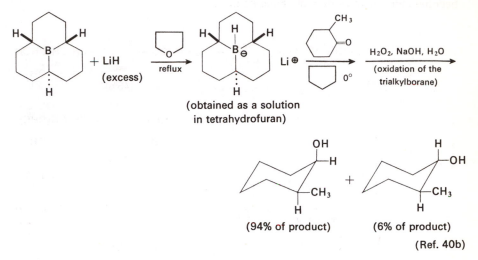

(94% of product)   (6% of product)

(Ref. 40b)

An exceedingly bulky borohydride reducing agent has recently been obtained by the reaction of lithium hydride with the subsequently described cyclic trialkylborane, *cis, cis, trans*-perhydro-9b-boraphenalene.[40b] The accompanying equations illustrate the preparation of this reagent and its use for the selective reduction of a cyclohexanone derivative from the less hindered side. After the reduction has been completed, the trialkylborane derivative is oxidized to facilitate separation of reaction product.

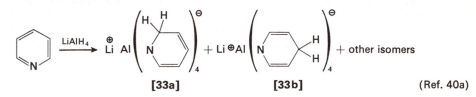

[33a]   [33b]   (Ref. 40a)

The reducing agent involved when lithium aluminum hydride is used in pyridine solution[12d,e,40a] is of special interest because of its ability to reduce aryl ketones more rapidly than aliphatic ketones. The reducing agent formed is a mixture of 1,2- and 1,4-dihydropyridine derivatives [33] which no longer contain

40. (a) P. T. Lansbury and J. O. Peterson, *J. Am. Chem. Soc.,* **85**, 2236 (1963). (b) H. C. Brown and W. C. Dickason, *ibid.,* **92**, 709 (1970) ; see also R. E. Ireland, D. R. Marshall, and J. W. Tilley, *ibid.,* **92**, 4754 (1970). (c) E. L. Eliel, T. W. Doyle, R. O. Hutchins, and E. C. Gilbert, *Org. Syn.,* **50**, 13 (1970). (d) H. B. Henbest and T. R. B. Mitchell, *J. Chem. Soc.,* C, 785 (1970).

metal hydrogen bonds. Reduction of the diketone **[34]** illustrates the selectivity of the reagent. One possible reason for the unusual selectivity of this reagent may be that this dihydropyridine derivative **[33]** is able to reduce ketones without assistance from an electrophilic catalyst such as lithium cation or a proton donor. As a result, the aryl ketones with relatively electron-deficient carbonyl groups become more reactive than aliphatic ketones.

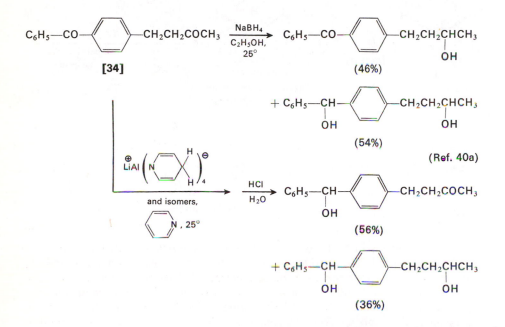

An alternative reduction method, which may involve an iridium hydride intermediate, consists of the reduction of ketones with phosphorous acid (or a trialkylphosphite) and a catalytic amount of an iridium salt in aqueous isopropyl alcohol.[44c,d] As the following example indicates, this reduction method shows a strong preference for the formation of axial alcohols from cyclohexanone derivatives.

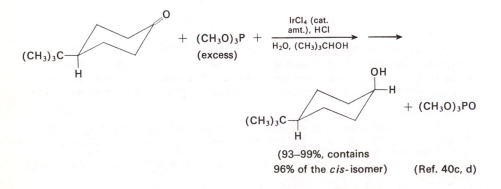

REDUCTION OF OTHER FUNCTIONAL GROUPS

Although the mild reducing agent sodium borohydride will reduce esters slowly[41] and lactones and acid chlorides rapidly, it is normally possible to reduce ketones and aldehydes selectively with this reagent in the presence of a variety of such functional groups such as esters, amides, cyano groups, nitro groups, and alkyl halides. The following examples are illustrative. Lithium borohydride will reduce esters more rapidly than sodium borohydride and has been used for this purpose.

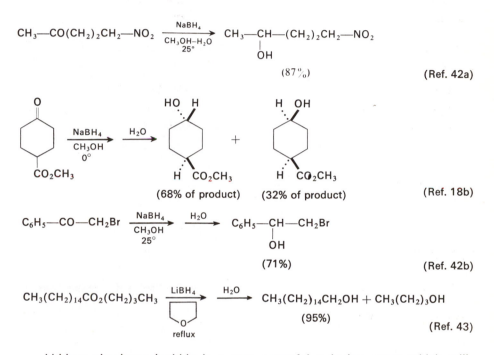

$$CH_3-CO(CH_2)_2CH_2-NO_2 \xrightarrow[\substack{CH_3OH-H_2O \\ 25°}]{NaBH_4} CH_3-\underset{\underset{OH}{|}}{CH}-(CH_2)_2CH_2-NO_2$$

(87%)   (Ref. 42a)

HO H   H OH

(68% of product)   (32% of product)   (Ref. 18b)

$$C_6H_5-CO-CH_2Br \xrightarrow[\substack{CH_3OH \\ 25°}]{NaBH_4} \xrightarrow{H_2O} C_6H_5-\underset{\underset{OH}{|}}{CH}-CH_2Br$$

(71%)   (Ref. 42b)

$$CH_3(CH_2)_{14}CO_2(CH_2)_3CH_3 \xrightarrow{LiBH_4} \xrightarrow{H_2O} CH_3(CH_2)_{14}CH_2OH + CH_3(CH_2)_3OH$$

reflux

(95%)   (Ref. 43)

Lithium aluminum hydride is a very powerful reducing agent which will reduce a variety of functional groups. Table 2-2 lists the common functional groups which may be reduced with ethereal solutions of lithium aluminum hydride. The groups are listed approximately in order of the ease with which they are reduced. In spite of reduction rate differences, attempts to effect selective reductions by

41. (a) M. S. Brown and H. Rapoport, *J. Org. Chem.*, **28**, 3261 (1963). (b) The ester function of α-cyano esters is reduced readily by sodium borohydride; see J. A. Meschino and C. H. Bond, *ibid.*, **28**, 3129 (1963). (c) The esters of phenols and other acidic hydroxy compounds are also reduced readily by sodium borohydride; S. Takahashi and L. A. Cohen, *ibid.*, **35**, 1505 (1970).

42. (a) H. Shechter, D. E. Ley, and L. Zeldin, *J. Am. Chem. Soc.*, **74**, 3664 (1952). (b) S. W. Chaikin and W. G. Brown, *ibid.*, **71**, 122 (1949). (c) Aromatic nitro compounds are reduced slowly to the corresponding anion radicals with sodium borohydride; M. G. Swanwick and W. A. Waters, *Chem. Commun.*, **No. 1**, 63, (1970).

43. R. F. Nystrom, S. W. Chaikin, and W. G. Brown, *J. Am. Chem. Soc.*, **71**, 3245 (1949). For a brief survey of the reactions of lithium borohydride, see Ref. 4a.

Table 2–2 Common functional groups reduced with lithium aluminum hydride
(The groups are listed approximately in order of the ease of reduction with lithium aluminum
hydride.)

Functional group	Product
—CHO	—CH$_2$OH
$>$C$=$O	$>$CH—OH
—COCl	—CH$_2$OH
—CH—C— (epoxide, O)	—CH$_2$—C— OH
—CO$_2$R	—CH$_2$OH + ROH
—CO$_2$H or —CO$_2$$^{\ominus}Li^{\oplus}$	—CH$_2$OH
—CO—NR$_2$	—CH$_2$—NR$_2$ or $\left[\begin{array}{c}\text{—CH—NR}_2 \\ | \\ \text{OH}\end{array}\right] \rightarrow$ —CHO + R$_2$NH
—CO—NH—R	—CH$_2$—NH—R
—C$\equiv$N	—CH$_2$—NH$_2$ or $\left[\text{—CH}=\text{NH}\right] \xrightarrow{\text{H}_2\text{O}}$ —CHO
$>$C$=$NOH	$>$CH—NH$_2$
—C—NO$_2$ (aliphatic)	—C—NH$_2$ and other products
—CH$_2$—O—SO$_2$—C$_6$H$_5$ or —CH$_2$Br	—CH$_3$
$>$CH—O—SO$_2$—C$_6$H$_5$ or $>$CH—Br	$>$CH$_2$

adding a limited amount of the metal hydride to solutions of a polyfunctional
reactant are often not satisfactory synthetic procedures unlike the selective catalytic
hydrogenations previously discussed (Chapter 1). Presumably, the difficulty with
selective lithium aluminum hydride reductions arises because the reduction rates
may be more rapid than the rate of mixing of the reactants. Consequently, the
local excess of metal hydride present during addition of the reagent is completely
consumed before uniform distribution of the reactants in solution can be attained.
For this reason, selective metal hydride reductions of polyfunctional compounds
are most often accomplished by use of various metal hydride reagents (e.g.,
lithium trialkoxyaluminum hydrides) which are less reactive than lithium aluminum

hydride. This procedure is illustrated by the previously discussed selective reduction of the keto ester **[3]** with lithium tri-*t*-butoxyaluminum hydride; a tetrahydrofuran solution of this reagent will reduce aldehydes, ketones, or acid chlorides

$$O_2N-\!\!\!\!\bigcirc\!\!\!\!-CO-Cl + 1 \text{ equiv. LiAlH[OC(CH}_3)_3]_3 \xrightarrow[-78°]{\text{diglyme}} O_2N-\!\!\!\!\bigcirc\!\!\!\!-CHO$$

(81%)

(Refs. 5a, 44)

without affecting most other functional groups. The accompanying example of
an acid chloride reduction illustrates a useful aldehyde synthesis. Although
lithium tri-*t*-butoxyaluminum hydride reacts very slowly with alkyl esters, the
more reactive aryl esters (e.g., **[35]**) are reduced by this reagent. Such reductions
of the aryl esters of *aliphatic* carboxylic acids are reported to offer an alternative
synthesis of aliphatic aldehydes.

$CH_3(CH_2)_4CO_2C_6H_5 + 1$ equiv. LiAlH[OC(CH$_3$)$_3$]$_3$
[35]

$$\xrightarrow[0°]{\text{O}} \left| CH_3(CH_2)_4CH-OC_6H_5 \atop O^\ominus Li^\oplus \right| \xrightarrow[H_2O]{H_2SO_4} CH_3(CH_2)_4CHO + C_6H_5OH$$

(62%) (Ref. 45)

Amines may be produced by the reduction of imines or iminium salts with
borohydride or aluminum hydride reagents. The accompanying examples illustrate
the need for neutral or acidic conditions to reduce imines with sodium borohydride;

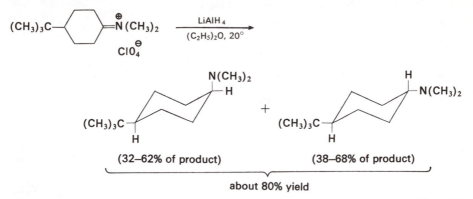

about 80% yield

(Ref. 46)

presumably an iminium salt such as **[36]** is the species attacked. Use of lithium or
sodium cyanoborohydride is advantageous in these reductions because of the
enhanced acidic stability of this reagent. Note should also be made of the reduction

44. H. C. Brown and B. C. Subba Rao, *J. Am. Chem. Soc.*, **80**, 5377 (1958).
45. P. M. Weissman and H. C. Brown, *J. Org. Chem.*, **31**, 283 (1966).

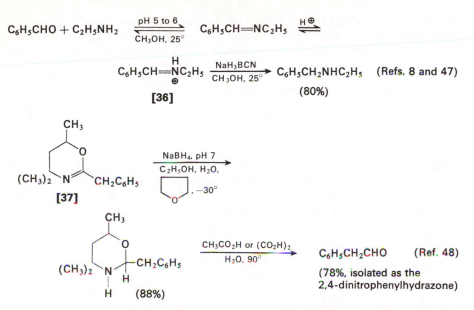

$$C_6H_5CHO + C_2H_5NH_2 \xrightleftharpoons[CH_3OH, 25°]{pH\ 5\ to\ 6} C_6H_5CH=NC_2H_5 \xrightleftharpoons{H^\oplus}$$

$$C_6H_5CH=\overset{H}{\underset{\oplus}{N}}C_2H_5 \xrightarrow[CH_3OH, 25°]{NaH_3BCN} C_6H_5CH_2NHC_2H_5 \quad \text{(Refs. 8 and 47)}$$

[36]   (80%)

[37]

NaBH₄, pH 7
C₂H₅OH, H₂O, , −30°

(88%)

CH₃CO₂H or (CO₂H)₂
H₂O, 90°

$C_6H_5CH_2CHO$ (Ref. 48)

(78%, isolated as the
2,4-dinitrophenylhydrazone)

and subsequent hydrolysis of dihydrooxazines such as [37] which offer a versatile synthetic route to aldehydes.[48]

Reactions related to these imine reductions include the reduction of N-alkyl-pyridinium salts with lithium aluminum hydride to form 1,2-dihydropyridines. The reductions of these same pyridinium salts with sodium borohydride in protonic solvents leads to 1,2,5,6-tetrahydropyridines or piperidines.[49] The probable course of this borohydride reduction is indicated below:

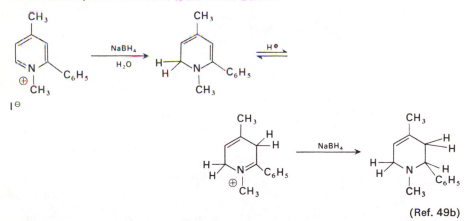

(Ref. 49b)

46. D. Cabaret, G. Chauviere, and Z. Welvart, *Tetrahedron Letters,* **No. 34,** 4109 (1966).
47. For examples of the reductions of anils with sodium borohydride in methanol, see J. R. Bull, D. G. Hey, G. D. Meakins, and E. E. Richards, *J. Chem. Soc.,* C, 2077 (1967).
48. (a) A. I. Meyers and A. Nabeya, *Chem. Commun.,* **No. 22,** 1163 (1967). (b) A. I. Meyers, A. Nabeya, H. W. Adickes, and I. R. Politzer, *J. Am. Chem. Soc.,* **91,** 763 (1969).

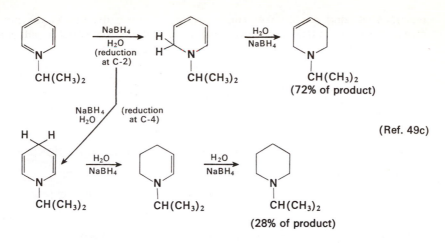

(Ref. 49c)

Although the reduction of *aliphatic* nitro compounds with lithium aluminum hydride has been used to prepare primary amines, the reduction of tertiary alkyl nitro compounds may be complicated by rearrangement of the intermediate hydroxylamine derivative. Such rearrangements, illustrated in the accompanying equation, lead to both primary and secondary amines.

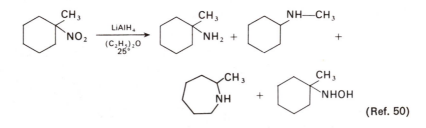

(Ref. 50)

A similar rearrangement of a hydroxylamine intermediate may complicate the reduction of oximes with lithium aluminum hydride to form primary amines. The accompanying equation illustrates this complication, which would appear to involve an electron-deficient nitrene [38] which may rearrange before a second hydride ion is transferred to the intermediate. In certain oxime reductions, particularly

(c) A. I. Meyers, A. Nabeya, H. W. Adickes, J. M. Fitzpatrick, G. R. Malone, and I. R. Politzer, *ibid.*, **91**, 764 (1969). (d) A. I. Meyers, H. W. Adickes, I. R. Politzer, and W. N. Beverung, *ibid.*, **91**, 765 (1969). (e) H. W. Adickes, I. R. Politzer, and A. I. Meyers, *ibid.*, **91**, 2155 (1969).
49. (a) R. E. Lyle, D. A. Nelson, and P. S. Anderson, *Tetrahedron Letters*, **No. 13**, 553 (1962). (b) P. S. Anderson and R. E. Lyle, *ibid.*, **No. 13**, 153 (1964). (c) P. S. Anderson, W. E. Krueger, and R. E. Lyle, *ibid.*, **No. 45**, 4011 (1965). (d) G. Palazzo and L. Baiocchi, *ibid.*, **No. 46**, 4739 (1968). (e) S. Yamada, M. Kuramoto, and Y. Kikugawa, *ibid.*, **No. 36**, 3101 (1969). (f) T. Miyadera and Y. Kishida, *Tetrahedron*, **25**, 209 (1969).
50. H. J. Barber and E. Lunt, *J. Chem. Soc.*, 1187 (1960).

reactions utilizing a chloroaluminum hydride reagent, the rearranged products may arise from an initial Beckmann rearrangement of the oxime prior to reduction.[51b]

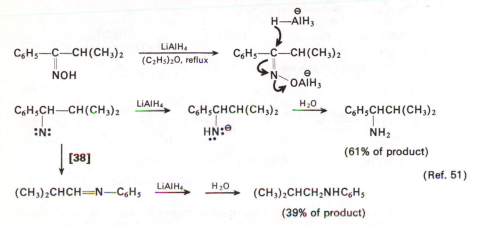

(61% of product)

(Ref. 51)

$(CH_3)_2CHCH=N-C_6H_5 \xrightarrow{LiAlH_4} \xrightarrow{H_2O} (CH_3)_2CHCH_2NHC_6H_5$

(39% of product)

Another seemingly related reaction which is observed particularly during the reduction of benzyl ketone oximes (and their O-alkyl or O-aryl derivatives as well as the related isoxazolines) with lithium aluminum hydride is the formation of aziridines.[52] The mechanism of this transformation appears to involve the indicated successive formation of an unsaturated nitrene and an azirine intermediate. The azirine is then reduced further to an aziridine.

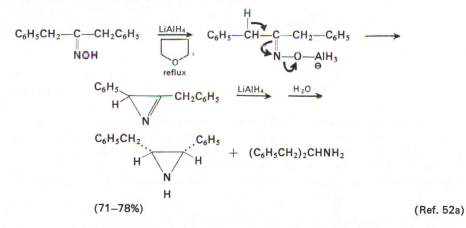

(71–78%)

(Ref. 52a)

51. (a) S. H. Graham and A. J. S. Williams, *Tetrahedron,* **21**, 3263 (1965) and references therein. (b) Rearrangement has also been noted to accompany the reduction of ketoximes with aluminum hydride or with one of the chloroaluminum hydrides; N. M. Yoon and H. C. Brown, *J. Am. Chem. Soc.,* **90**, 2927 (1968).
52. (a) K. Kotera and K. Kitahonoki, *Org. Syn.,* **48**, 20 (1968). (b) K. Kotera, S. Miyazaki, H. Takahashi, T. Okada, and K. Kitahonoki, *Tetrahedron,* **24**, 3681 (1968). (c) K. Kitahonoki, Y. Takano, and T. Takahashi, *ibid.,* **24**, 4605 (1968). (d) K. Kotera, T. Takano, A. Matsuura, and K. Kitahonoki, *ibid.,* **26**, 539 (1970); K. Kotera and K. Kitahonoki, *Organic Preparations and Procedures,* **1**, 305 (1969).

It would appear that most of the abnormal reactions encountered in reductions of oximes with lithium aluminum hydride can be avoided by the use of either aluminum hydride,[51b] or, preferably, diborane as the reductant. Reaction of either oximes or the salts of primary or secondary alkyl nitro compounds with diborane in tetrahydrofuran solution at 25° yields the corresponding hydroxylamines;[53] aryl nitroso compounds are reduced to amines under the same conditions.[53c] If the

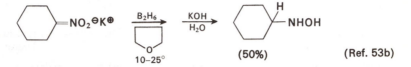

10–25°   (50%)   (Ref. 53b)

diborane-oxime reductions are performed at elevated temperatures in tetrahydrofuran-diglyme mixtures, complete reduction to the amines occurs. The same change results if the oximes are converted to their O-alkyl, O-acyl, or O-arenesulfonyl derivatives and then reduced with diborane at 25°.[54a,b] Diborane is also useful for the reduction of acyl hydrazides, hydrazones, and azines to form substituted hydrazines.[54c,b]

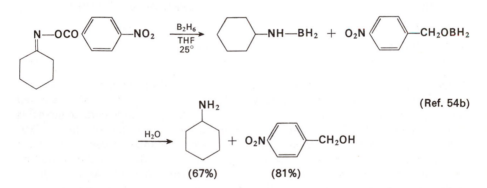

(Ref. 54b)

(67%)   (81%)

In addition to the subsequently discussed additions to olefins, diborane will reduce a variety of functional groups. The groups in Table 2-3 are listed approximately in the order of ease of reduction with diborane.[55] With the sterically

53. (a) H. Feuer, B. F. Vincent, Jr., and R. S. Bartlett, *J. Org. Chem.*, **30**, 2877 (1965). (b) H. Feuer, R. S. Bartlett, B. F. Vincent, Jr., and R. S. Anderson, *ibid.*, **30**, 2880 (1965). (c) H. Feuer and D. M. Braunstein, *ibid.*, **34**, 2024 (1969).
54. (a) A. Hassner and P. Catsoulacos, *Chem. Commun.*, **No. 12**, 590 (1967). (b) H. Feuer and D. M. Braunstein, *J. Org. Chem.*, **34**, 1817 (1969). (c) H. Feuer and F. Brown, Jr., *ibid.*, **35**, 1468 (1970). (d) J. A. Blair and R. J. Gardner, *J. Chem. Soc.*, C, 1714 (1970).
55. (a) H. C. Brown and W. Korytnyk, *J. Am. Chem. Soc.*, **82**, 3866 (1960); H. C. Brown, P. Heim, and N. M. Yoon, *ibid.*, **92**, 1637 (1970). (b) H. C. Brown and P. Heim, *ibid.*, **86**, 3566 (1964). (c) H. C. Brown, D. B. Bigley, S. K. Aurora, and N. M. Yoon, *ibid.*, **92**, 7161 (1970). (d) J. E. McMurry [*Chem. Commun.*, **No. 8**, 433 (1968)] has described the use of 2,4-dinitrophenylhydrazone derivatives to protect carbonyl groups during the reduction of C—C double bonds with diborane. (e) A. Pelter, M. G. Hutchings, T. E. Levitt, and K. Smith, *ibid.*, **No. 6**, 347 (1970).

Table 2–3 Common functional groups reduced with diborane or the borane-tetrahydrofuran complex
(The groups are listed approximately in order of the ease of reduction with diborane.)

Functional group	Product (after hydrolysis unless otherwise noted
—CO$_2$H	—CH$_2$OH
—CH=CH—	—CH$_2$—CH—BH$_2$ (product before hydrolysis)
—CHO, $>$C=O	—CH$_2$OH, $>$CHOH
—CO—NR$_2$, —CN	—CH$_2$NR$_2$, —CH$_2$NH$_2$
$>$C——C$<$ (epoxide)	$>$CH—C$<$ (very slow unless catalysts such as BF$_3$ are added) OH
—CO$_2$R	—CH$_2$OH + ROH
$>$C=NOR	$>$CHNHOH or $>$CHNH$_2$
—CO—Cl	—CH$_2$OH (very slow)
—CO$_2$$^{\ominus}M^{\oplus}$,—NO$_2$	inert

hindered di-(3-methyl-2-butyl)borane (also called disiamylborane), only the more reactive functional groups (aldehydes, ketones, olefins, and N,N-disubstituted amides) are reduced easily.[55e] The unusual reactivity of free carboxylic acids towards diborane has been attributed to the initial formation of a triacyl borate [39] (i.e., a mixed anhydride with boric acid) in which resonance interaction between the carbonyl group and the adjacent oxygen (structure [39b]) is diminished by interaction of the oxygen and boron atoms (structure [39c]). As a result the carbonyl function of the acyl derivative [39] is much more reactive toward nucleophiles than is the case with esters, free acids, and carboxylate anions. In some cases the triacylborate is found to disproportionate to form an acid anhydride which is reduced further.[55e]

$$3\ R—CO_2H + BH_3 \longrightarrow 3\ H_2 + R—\underset{\underset{O}{\|}}{C}—O—B(OCOR)_2 \xrightarrow{BH_3} \xrightarrow{H_2O} R—CH_2OH$$

[39a]

$$R—\underset{\underset{O}{\|}}{C}—\overset{\oplus}{O}{=}\overset{\ominus}{B}(OCOR)_2 \longleftrightarrow R—\underset{\underset{O^{\ominus}}{|}}{C}{=}\overset{\oplus}{O}—B(OCOR)_2$$

[39c]   **[39b]**

The reduction of substituted amides to amines, as exemplified in the following equations, is believed to proceed by an initial reduction to a geminal amino alcohol derivative (e.g., [40] or [41]), following by elimination and subsequent reduction of the resulting imine or iminium salt.

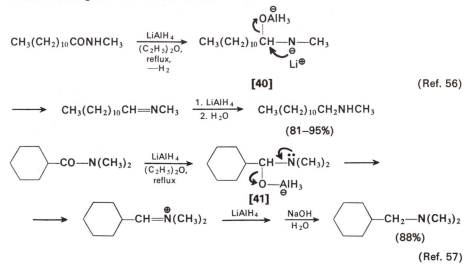

$$CH_3(CH_2)_{10}CONHCH_3 \xrightarrow[\substack{(C_2H_5)_2O,\\ reflux,\\ -H_2}]{LiAlH_4} CH_3(CH_2)_{10}CH-N-CH_3$$

[40]   (Ref. 56)

$$\longrightarrow CH_3(CH_2)_{10}CH{=}NCH_3 \xrightarrow[2.\ H_2O]{1.\ LiAlH_4} CH_3(CH_2)_{10}CH_2NHCH_3$$

(81–95%)

[41]

(88%)

(Ref. 57)

Although amides are normally not reduced by sodium borohydride, they can be alkylated to form easily reduced imino ester salts such as [42].

$$C_6H_5CON(C_2H_5)_2 \xrightarrow[CH_2Cl_2,\ 25°]{(C_2H_5)_3O^{\oplus}BF_4^{\ominus}} C_6H_5-C{=}N(C_2H_5)_2$$

[42]

$$\xrightarrow[C_2H_5OH,\ 0-25°]{NaBH_4} C_6H_5CH-N(C_2H_5)_2 \xrightarrow{NaBH_4} C_6H_5CH_2N(C_2H_5)_2$$

(75%)

(Refs. 8 and 58)

56. (a) C. V. Wilson and J. F. Stenberg, *Org. Syn.*, **Coll. Vol. 4**, 564 (1963). (b) M. S. Newman and T. Fukunaga [*J. Am. Chem. Soc.*, **82**, 693 (1960)] have presented evidence indicating that the reduction of unsubstituted amides with lithium aluminum hydride proceeds by the initial dehydration of the amide to form a nitrile. (c) Similarly, reaction of unsubstituted amides with sodium borohydride in refluxing diglyme (b.p. 161°) yields nitriles; S. E. Ellzey, Jr., C. H. Mack, and W. J. Connick, Jr., *J. Org. Chem.*, **32**, 846 (1967). (d) The reduction of amides, nitriles, or nitro compounds to amines with mixtures of sodium borohydride and various transition metal salts has been reported by T. Satoh, S. Suzuki, Y. Suzuki, Y. Miyaji, and Z. Imai, *Tetrahedron Letters*, **No. 52**, 4555 (1969). (e) The reduction of nitriles to amines with sodium borohydride in the presence of Raney nickel catalyst has also been described by R. A. Egli, *Helv. Chim. Acta*, **53**, 47 (1970).
57. A. C. Cope and E. Ciganek, *Org. Syn.*, **Coll. Vol. 4**, 339 (1963).
58. R. F. Borch, *Tetrahedron Letters*, **No. 1**, 61 (1968).

A similar sort of activation may be envisioned to explain the enhanced reactivity of N,N-disubstituted amides toward diborane[55a,b] and aluminum hydride.[9a,51b] Reaction of the electrophilic boron or aluminum reagent with the amide to form a reactive complex such as structure [43] is believed to account for the efficiency with which these compounds reduce amides and nitriles.[55a]

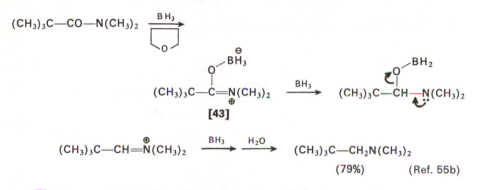

$(CH_3)_3C—CH=\overset{\oplus}{N}(CH_3)_2$ $\xrightarrow{BH_3}$ $\xrightarrow{H_2O}$ $(CH_3)_3C—CH_2N(CH_3)_2$

(79%)   (Ref. 55b)

The successful reduction of a urethane (e.g., [44]) to an N-methylamine with lithium aluminum hydride illustrates another useful application of an amide reduction.

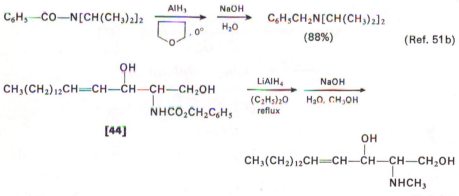

From consideration of the intermediates involved in the reduction of N,N-disubstituted amides to amines it is evident that if the reaction sequence could be stopped at the amino alcohol stage [41], hydrolysis of the reaction mixture would form an aldehyde. Although such a synthesis of aldehydes has been effected using a limited amount of lithium aluminum hydride and low reaction temperatures with various N,N-disubstituted amides,[60] better results have been obtained by reduction

59. (a) B. Weiss, *J. Org. Chem.*, **30**, 2483 (1965). (b) W. V. Curran and R. B. Angier [*ibid.*, **31**, 3867 (1966)] have reported the selective reduction of an amide function in the presence of a urethane by use of diborane in tetrahydrofuran.
60. E. Mosettig, *Org. Reactions*, **8**, 218 (1954).

of the amides derived from ethylenimine, carbazole, N-methylaniline, or imidazole.[61] With amides of these types the participation of the electron pair on nitrogen in the elimination process depicted in structure **[41]** is not favored.

$$\triangleright\text{-CO-N}\boxed{} \quad \xrightarrow[\substack{(C_2H_5)_2O \\ 0°}]{LiAlH_4} \quad \xrightarrow{H_3O^\oplus} \quad \triangleright\text{-CHO}$$

(60%)   (Ref. 61a)

In alternative procedures, the reduction of N,N-dimethylamides with limited amounts of reagents such as $LiAlH_2(OC_2H_5)_2$, $LiAlH(OC_2H_5)_3$, $NaAlH_4$, $NaAlH(OCH_3)_3$, or $NaAlH(OC_2H_5)_3$ has been found useful for the synthesis of aldehydes from amides. Of these reagents, the two whose use is illustrated in the following equations appear most convenient.

$$C_6H_5CON(CH_3)_2 \quad \xrightarrow{reduction} \quad \xrightarrow{H_2O} \quad C_6H_5CHO$$

Reducing agent	Yield
(0.25–0.40 mole per mole of amide)	(Ref. 62b)
NaAlH₄, THF, 0°	76–92%
LiAlH₄, THF, 0°	80%
LiAlH₄, (C₂H₅)₂O, 0°	50–60%

$$LiAlH_4 + 1 \text{ equiv. } CH_3CO_2C_2H_5 \xrightarrow{(C_2H_5)_2O} LiAlH_2(OC_2H_5)_2$$
$$\text{(or 2 equiv. } C_2H_5OH)$$

$$\bigcirc\text{-CO-N(CH}_3)_2 + LiAlH_2(OC_2H_5)_2 \xrightarrow[0°]{(C_2H_5)_2O} \xrightarrow{H_3O^\oplus} \bigcirc\text{-CHO}$$

(71%)   (Ref. 62a)

An aldehyde synthesis which utilizes the partial reduction of esters with diisobutylaluminum hydride or sodium aluminum hydride has also been reported.[63] This procedure is illustrated by the reduction of ethyl laurate.

$$CH_3(CH_2)_{10}CO_2C_2H_5 + [(CH_3)_2CHCH_2]_2AlH \xrightarrow[-70°]{hexane}$$

$$CH_3(CH_2)_{10}\text{-CH}\begin{smallmatrix} OC_2H_5 \\ OAl[CH_2CH(CH_3)_2]_2 \end{smallmatrix} \xrightarrow{H_2O} CH_3(CH_2)_{10}CHO$$

(88%)

(Ref. 63a)

61. (a) H. C. Brown and A. Tsukamoto, *J. Am. Chem. Soc.*, **83**, 4549 (1961). (b) G. Wittig and P. Hornberger, *Justus Liebigs Ann. Chem.*, **577**, 11 (1952). (c) F. Weygand and co-workers, *Angew Chem.*, **64**, 458 (1952); *ibid.*, **65**, 525 (1953); *ibid.*, **66**, 174 (1954). (d) H. A. Staab and H. Braunling, *Justus Liebigs Ann. Chem.*, **654**, 119 (1962).
62. (a) H. C. Brown and A. Tsukamoto, *J. Am. Chem. Soc.*, **81**, 502 (1959); *ibid.*, **86**, 1089 (1964). (b) L. I. Zakharkin, D. N. Maslin, and V. V. Gavrilenko, *Tetrahedron*, **25**, 5555 (1969).
63. (a) L. I. Zakharkin and I. M. Khorlina, *Tetrahedron Letters*, **No. 14**, 619 (1962). (b) L. I. Zakharkin, V. V. Gavrilenko, D. N. Maslin, and I. M. Khorlina, *ibid.*, **No. 29**, 2087 (1963).

A similar possibility of partial reduction exists in the reduction of nitriles to amines, since hydrolysis of the intermediate imine salt [45] would produce an aldehyde.[60] This possibility has been realized by the use of the reducing agent

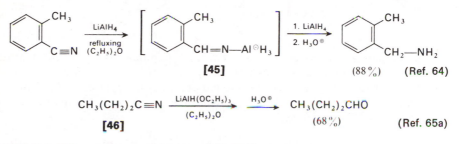

[45]   (88%)   (Ref. 64)

$$CH_3(CH_2)_2C{\equiv}N \xrightarrow[(C_2H_5)_2O]{LiAlH(OC_2H_5)_3} \xrightarrow{H_3O^\oplus} CH_3(CH_2)_2CHO$$

[46]   (68%)   (Ref. 65a)

$LiAlH(OC_2H_5)_3$, as illustrated for the nitrile [46]. A more satisfactory procedure for this partial reduction of a nitrile uses diisobutylaluminum hydride, as the following equation illustrates.

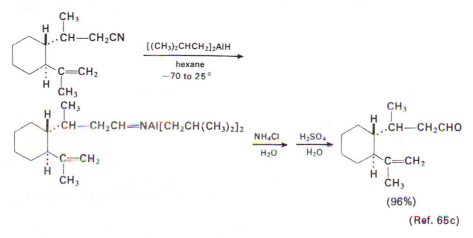

(96%)

(Ref. 65c)

Application of the same principle of partial reduction to acid chlorides with the reagent $LiAlH[OC(CH_3)_3]_3$ was described previously as a synthetic method for aldehydes.[44]

The reduction of enolizable 1,3-dicarbonyl compounds such as [47] is complicated by an elimination reaction similar to the one observed in the reduction of amides to amines. Thus, the elimination of aluminum alkoxides from the inter-

64. R. F. Nystrom and W. G. Brown, *J. Am. Chem. Soc.*, **70**, 3738 (1948).
65. (a) H. C. Brown, C. J. Shoaf, and C. P. Garg, *Tetrahedron Letters*, **No. 3**, 9 (1959); H. C. Brown and C. P. Garg, *J. Am. Chem. Soc.*, **86**, 1085 (1964). (b) For the use of sodium triethoxyaluminum hydride to reduce aromatic nitriles to aldehydes, see G. Hesse and R. Schrödel, *Justus Liebigs Ann. Chem.*, **607**, 24 (1957). (c) J. A. Marshall, N. H. Andersen, and J. W. Schlicher, *J. Org. Chem.*, **35**, 858 (1970); J. A. Marshall, N. H. Andersen, and P. C. Johnson, *ibid.*, **35**, 186 (1970).

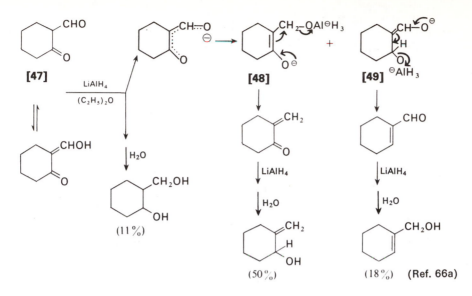

mediate products [48] and [49] produces allylic alcohols rather than 1,3-diols as the major reduction products.[66] This difficulty may be overcome with acyclic dicarbonyl compounds if sodium borohydride is used as the reducing agent in hydroxylic media. Under these reaction conditions protonation of intermediate enolates is possible and the tendency for enolate formation is reduced because of the low basicity of the borohydride anion. However, isolation of the diols is sometimes difficult because of the formation of relatively stable cyclic borate esters in the reaction mixture.

$$CH_3CO—CH_2—CO—CH_3 \xrightarrow[\substack{H_2O-CH_3OH \\ 15°}]{NaBH_4} CH_3—CH—CH_2—CH—CH_3$$

$$\underset{OH}{|} \qquad \underset{OH}{|}$$

(90% *meso* and 2% racemic) (Ref. 66d)

The formation of allylic alcohols can be also avoided by use of aluminum hydride as the reducing agent. Presumably this procedure is successful because elimination of a trivalent alkoxyaluminum species from the intermediate reduction products [50] is less favorable than the corresponding elimination (see structures [48] and [49]) in lithium aluminum hydride reductions. Consequently, hydrolysis of the reaction mixture yields a mixture of the diol and the partially reduced hydroxy carbonyl compounds [51]; the diol is obtained by further reduction of this mixture with sodium borohydride.

66. (a) A. S. Dreiding and J. A. Hartman, *J. Am. Chem. Soc.,* **75**, 939, 3723 (1953). (b) J. C. Richer and R. Clarke, *Tetrahedron Letters,* **No. 16**, 935 (1964). (c) E. Romann, A. J. Frey, P. A. Stadler, and A. Eschenmoser, *Helv. Chim. Acta,* **40**, 1900 (1957). (d) J. Dale, *J. Chem. Soc.,* 910 (1961).

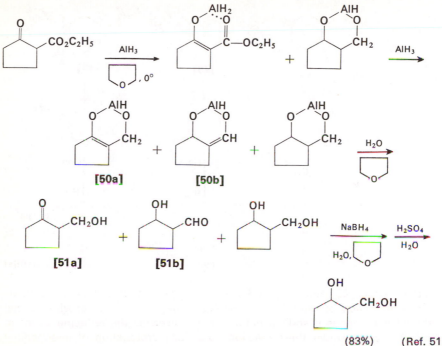

[50a]   [50b]

[51a]   [51b]

(83%)   (Ref. 51b)

As would be expected from the reaction pathway depicted for the lithium aluminum hydride reduction of the keto aldehyde [47], the proportion of the allylic alcohol in the product is controlled by the fraction of 1,3-dicarbonyl compound which is converted to its enolate prior to reduction. In cases where allylic alcohol formation is desired,[66c,67] the 1,3-dicarbonyl compound should be converted to its enolate anion (e.g., [52]) prior to reduction. The following examples illustrate this procedure for reduction of a malonic ester.

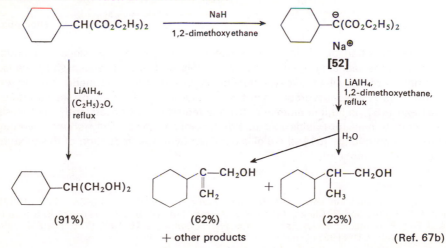

(91%)   (62%)   (23%)

+ other products   (Ref. 67b)

Reduction of the enol ethers of 1,3-dicarbonyl compounds provides a synthesis of α,β-unsaturated ketones, since the reduction product **[53]** undergoes an allylic rearrangement when treated with aqueous acid.

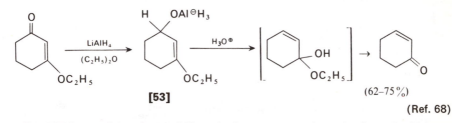

[53]

(62–75%)

(Ref. 68)

By use of one of the electrophilic reducing agents such as aluminum hydride or, particularly, dichloroaluminum hydride it is possible to effect the reductive cleavage of acetals, ketals, and telated compounds.[37b,69] This reductive cleavage, which is most rapid with the relatively strong Lewis acids chloroaluminum hydride and dichloroaluminum hydride,[69h] is believed to proceed by a mechanism resembling the reduction of N,N-disubstituted amides to amines (see structure **[41]**) in which an oxonium ion **[54]** is generated and then reduced by the metal hydride by either

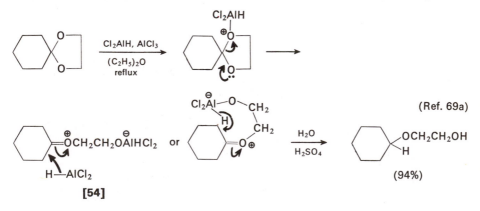

(Ref. 69a)

(94%)

[54]

67. (a) J. A. Marshall and N. Cohen, *J. Org. Chem.*, **30**, 3475 (1965). (b) J. A. Marshall, N. H. Andersen, and R. A. Hochstetler, *ibid.*, **32**, 113 (1967). In this study the hydrolysis procedure was found to be an important factor in determining the composition of the final mixture of products.

68. (a) W. F. Gannon and H. O. House, *Org. Syn.*, **40**, 14 (1960). (b) M. Stiles and A. L. Longroy, *J. Org. Chem.*, **32**, 1095 (1967).

69. (a) R. A. Daignault and E. L. Eliel, *Org. Syn.*, **47**, 37 (1967). (b) E. L. Eliel, V. G. Badding, and M. N. Rerick, *J. Am. Chem. Soc.*, **84**, 2371 (1962). (c) E. L. Eliel, L. A. Pilato, and V. G. Badding, *ibid.*, **84**, 2377 (1962). (d) E. L. Eliel, B. E. Nowak, R. A. Daignault, and V. G. Badding, *J. Org. Chem.*, **30**, 2441 (1965). (e) E. L. Eliel, B. E. Nowak, and R. A. Daignault, *ibid.*, **30**, 2448 (1965). (f) E. L. Eliel and R. A. Daignault, *ibid.*, **30**, 2450 (1965). (g) B. E. Leggetter and R. K. Brown, *Can. J. Chem.*, **41**, 2671 (1963). (h) U. E. Diner, H. A. Davies, and R. K. Brown, *ibid.*, **45**, 207 (1967). (i) W. W. Zajac, Jr., B. Rhee, and R. K. Brown, *ibid.*, **44**, 1547 (1966). (j) For the reductive cleavage of *ortho* esters, see E. L. Eliel and F. W. Nader, *J. Am. Chem. Soc.*, **92**, 3045 (1970). (k) W. W. Zajac, Jr. and K. Byrne, *J. Org. Chem.*, **35**, 3375 (1970).

an intermolecular or an intramolecular hydride transfer. The stereochemical outcome of this reduction at the ketal carbon atom is consistent with this view, as illustrated by the reduction of the ketal [55] to form the product expected from

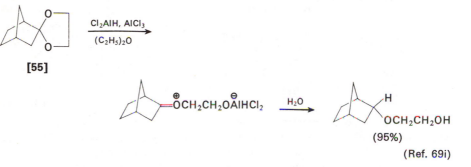

[55]

(95%)

(Ref. 69i)

addition of hydride ion to the oxonium ion from the less hindered side. The same dichloroaluminum hydride reagent may also be used to reductively remove the oxygen function from allylic alcohols and α-arylidene ketones such as [56].[70]

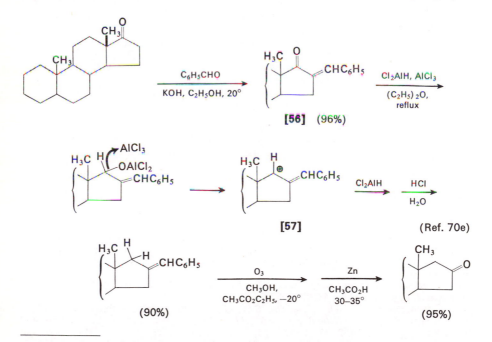

[56] (96%)

[57]

(Ref. 70e)

(90%)

(95%)

70. (a) J. H. Brewster and H. O. Bayer, *J. Org. Chem.*, **29**, 105, 116 (1964). (b) J. H. Brewster, H. O. Bayer, and S. F. Osman, *ibid.*, **29**, 110 (1964). (c) J. H. Brewster, S. F. Osman, H. O. Bayer, and H. B. Hopps, *ibid.*, **29**, 121 (1964). (d) J. H. Brewster and J. E. Privett, *J. Am. Chem. Soc.*, **88**, 1419 (1966). (e) J. E. Bridgeman, C. E. Butchers, E. R. H. Jones, A. Kasal, G. D. Meakins, and P. D. Woodgate, *J. Chem. Soc.*, C, 244 (1970). (f) M. Biswas and A. H. Jackson, *ibid.*, C, 1667 (1970).

This reduction procedure, believed to involve an allylic carbonium ion intermediate such as [57],[70d] has been applied to the unsaturated ketone [56] as part of a three-step sequence to transpose a carbonyl from position C-17 to the adjacent position C-16. A similar reductive cleavage of aromatic aldehydes and ketones has been accomplished with borane in tetrahydrofuran containing catalytic amounts of boron trifluoride.[70f]

The reductive cleavage of benzylic C—O bonds has been accomplished with nucleophilic metal hydride reagents by recourse to elevated reaction temperatures. The use of the thermally stable sodium bis(2-methoxyethoxy)-aluminumhydride[71] for this purpose is illustrated in the following equation.

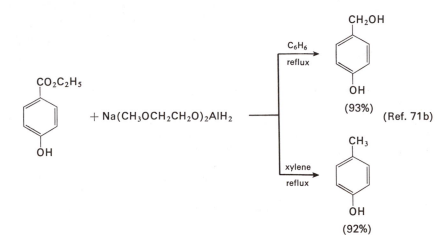

Several presumably related reductions have been reported[72] in which esters have been reduced to ethers with sodium borohydride and a large excess of a Lewis acid, boron trifluoride. This ester reduction, illustrated by the following examples, is presumably related mechanistically to the reduction of amides to amines with lithium aluminum hydride and the reduction of ketals to ethers with aluminum hydride-aluminum chloride mixtures. It is not clear in these cases whether the actual reducing agent is borane or a fluoroborane such as FBH_2 or F_2BH. In any event, it seems probable that the excess Lewis acid serves to form an oxonium ion from the intermediate reduction product. The same reaction conditions will also convert a thiol ester to a dialkylsulfide.[72e]

71. (a) V. Bazant and co-workers, *Tetrahedron Letters*, **No. 29**, 3303 (1968). (b) M. Cerny and J. Malek, *ibid.*, **No. 22**, 1739 (1969); *Coll. Czech. Chem. Commun.*, **35**, 1216 (1970).
72. (a) G. R. Pettit, U. R. Ghatak, B. Green, T. R. Kasturi, and D. M. Piatak, *J. Org. Chem.*, **26**, 1686 (1961). (b) G. R. Pettit and T. R. Kasturi, *ibid.*, **26**, 4557 (1961). (c) G. R. Pettit and D. M. Piatak, *ibid.*, **27**, 2127 (1962). (d) G. R. Pettit, B. Green, P. Hofer, D. C. Ayres, and P. J. S. Pauwels, *Proc. Chem. Soc.*, 357 (1962). (e) E. L. Eliel and R. A. Daignault, *J. Org. Chem.*, **29**, 1630 (1964). (f) The same types of reduction have been effected with diborane in the absence of boron trifluoride by the use of long reaction times or elevated reaction temperatures; G. R. Pettit and J. R. Dias, *Chem. Commun.*, **No. 15**, 901 (1970).

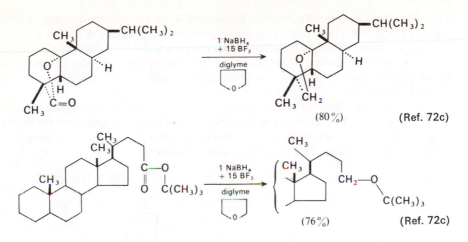

(80%)   (Ref. 72c)

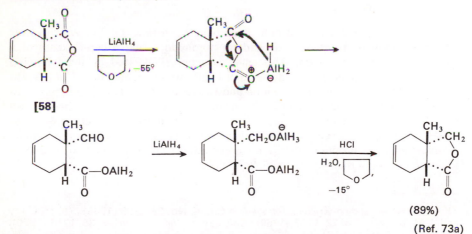

(76%)   (Ref. 72c)

Although the reductions of enol lactones and cyclic anhydrides (e.g., [58]) bear a superficial resemblance to the previous transformations, these reductions almost certainly involve the indicated acyclic intermediate.[73] A curious feature of these reactions, the selective reduction of the more hindered carbonyl function, has been explained to result from the initial complexing of the less hindered carbonyl function with the metal hydride followed by the transfer of a hydride anion (either intramolecular or intermolecular) to the more hindered carbonyl group.[73a] A similar selectivity is observed in reductions of cyclic anhydrides with lithium tri-*t*-butoxy-aluminum hydride; however, the product is a lactol resulting from closure of the intermediate carboxyaldehyde.[73f]

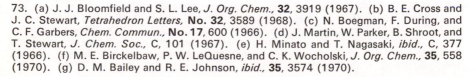

(89%)

(Ref. 73a)

73. (a) J. J. Bloomfield and S. L. Lee, *J. Org. Chem.,* **32**, 3919 (1967). (b) B. E. Cross and J. C. Stewart, *Tetrahedron Letters,* **No. 32**, 3589 (1968). (c) N. Boegman, F. During, and C. F. Garbers, *Chem. Commun.,* **No. 17**, 600 (1966). (d) J. Martin, W. Parker, B. Shroot, and T. Stewart, *J. Chem. Soc.,* C, 101 (1967). (e) H. Minato and T. Nagasaki, *ibid.,* C, 377 (1966). (f) M. E. Birckelbaw, P. W. LeQuesne, and C. K. Wocholski, *J. Org. Chem.,* **35**, 558 (1970). (g) D. M. Bailey and R. E. Johnson, *ibid.,* **35**, 3574 (1970).

Apart from its previously described use for reductions of amides, aluminum hydride is useful for other selective reductions. The common functional groups reduced by aluminum hydride are indicated in Table 2-4, the groups being listed in the order of ease of reduction.[9a,51b] The use of either preformed solutions of aluminum hydride in tetrahydrofuran or of the reagent prepared *in situ* from lithium aluminum hydride and aluminum chloride offers advantages in the reduction of α-β-unsaturated carbonyl compounds.

Table 2–4 Common function groups reduced with aluminum hydride
(The groups are listed approximately in order of the ease of reduction with aluminum hydride.)

Functional groups	*Product*
—CHO	—CH$_2$OH
$>$C$=$O	$>$CHOH
—CO—Cl	—CH$_2$OH
$>$C—C$<$ (epoxide)	$>$CH—C$<$ with OH
—CO$_2$H or —CO$_2$$^\ominus$M$^\oplus$	—CH$_2$OH
—CO—NR$_2$	—CH$_2$—NR$_2$
—C$\equiv$N	—CH$_2$NH$_2$
—CH$_2$—halogen	—CH$_3$ (slow)
—NO$_2$	reduced very slowly

In early studies[64] of metal hydride reductions, the observation was made that although carbon-carbon double bonds are generally not reduced with either sodium borohydride or lithium aluminum hydride, frequent exceptions to the rule were observed in the reductions of α-β-unsaturated carbonyl compounds, particularly β-aryl-α,β-unsaturated carbonyl compounds such as [59]. In these cases reduction of only the carbonyl function is possible if short reaction times (5 min or less)

C$_6$H$_5$—CH$=$CH—CHO $\xrightarrow[\text{(C}_2\text{H}_5)_2\text{O} \atop 25°]{\text{excess} \atop \text{LiAlH}_4}$ C$_6$H$_5$—CH$_2$CH$_2$CH$_2$OH (87%)

[59]

$\xrightarrow[\text{(C}_2\text{H}_5)_2\text{O} \atop -10°]{\text{LiAlH}_4 \text{ added} \atop \text{to aldehyde}}$ C$_6$H$_5$—CH$=$CH—CH$_2$OH (90%)

(Refs. 64, 74)

74. F. A. Hochstein and W. G. Brown, *J. Am. Chem. Soc.*, **70**, 3484 (1948).

and low reaction temperatures (25° or less) are employed. Sometimes addition of the hydride reagent to the carbonyl compound (inverse addition) has also proved beneficial. The further reduction of allylic alcohols to saturated alcohols with lithium aluminum hydride is believed to involve an intermediate organo-aluminum compound such as [60], since only one of the two hydrogen atoms added to the carbon-carbon double bond is derived from the metal hydride, and reaction of the complex (presumably [60]) with a deuterated solvent forms the indicated deuterated alcohol.

$$(C_6H_5\!-\!CH\!=\!CH\!-\!CH_2\!-\!O\!-\!)_2Al^{\ominus}H_2 \longrightarrow$$

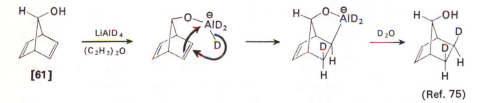

[60]   (Ref. 74)

Since cinnamaldehyde [59] is also reduced to the saturated alcohol by lithium trimethoxyaluminum hydride, one must conclude either that the tetraalkoxyaluminum anion disproportionates readily or that the indicated (see structure [60]) intramolecular addition of aluminum hydride to the carbon-carbon double bond is not the only reaction path available.[12b,32b] That the intramolecular addition of alkoxyaluminum hydride to carbon-carbon double bonds is not restricted to allylic alcohols was shown by the reaction of the bicyclic alcohol [61] with lithium

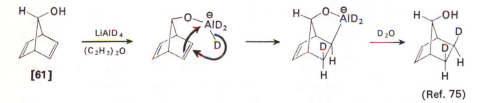

[61]   (Ref. 75)

aluminum deuteride. In at least some cases the addition to the carbon-carbon double bond has been shown to be intramolecular[75a,b] and to be accelerated by changing the solvent from ether to tetrahydrofuran or 1,2-dimethoxyethane[75d] or by adding other Lewis bases.[75f] It has further been shown that if the intermediate benzylaluminum intermediate [62] is heated in boiling tetrahydrofuran before hydrolysis, a cyclopropane is formed.[75c-e]

75. (a) B. Franzus and E. I. Snyder, *J. Am. Chem. Soc.*, **87**, 3423 (1965). (b) W. T. Borden, *ibid.*, **92**, 4898 (1970). (c) M. J. Jorgenson and A. W. Friend, *ibid.*, **87**, 1815 (1965). (d) M. J. Jorgenson and A. F. Thacher, *Chem. Commun.*, **No. 16**, 973 (1968); *ibid.*, **No. 21**, 1290 (1969). (e) R. T. Uyeda and D. J. Cram, *J. Org. Chem.*, **30**, 2083 (1965). (f) E. I. Snyder, *ibid.*, **32**, 3531 (1967).

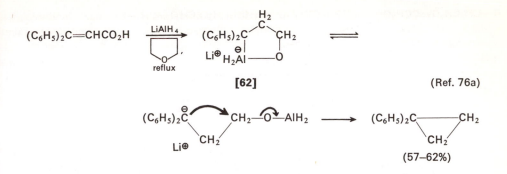

[62]   (Ref. 76a)

(57–62%)

The related intramolecular addition of alkoxy aluminum hydrides to carbon-carbon triple bonds also occurs. This reaction, which appears to be catalyzed by sodium methoxide,[75b,f,77b] yields a vinylaluminum intermediate **[63]**, presumably by either a concerted or a stepwise intramolecular *trans*-addition to the triple bond. (It will be noted than an intramolecular *cis*-addition of an Al-H grouping to an acetylene is geometrically very unfavorable.[75f]) Reaction of the organoaluminum intermediate with iodine served to form the vinyl iodide **[64]**. When aluminum chloride was added to the reaction mixture (presumably aluminum hydride is the reducing agent), the reaction took a different course (probably intermolecular) to form successively an isomeric vinylaluminum compound and an isomeric vinyl iodide **[65]**. The intermolecular *trans*-addition of lithium aluminum hydride to acetylenes (e.g., **[66]**) is known to occur at elevated reaction temperatures.

In cases where the reduction of an α,β-unsaturated ketone to an allylic alcohol without double bond reduction is desired, changes in reaction conditions with lithium aluminum hydride, such as the use of short reaction times and relatively low temperatures,[64,74] or the use of a slurry of this metal hydride in a nonbasic solvent such as benzene,[75f] may be beneficial. However, a far more satisfactory solution is the use of either aluminum hydride or diisobutylaluminum hydride as the reducing agent.[78] The results obtained from reduction of 2-cyclopentenone with a number of metal hydride reagents are indicated in the accompanying equation. In this case, and probably with other cycloalkenones, the previously described intramolecular

76. (a) M. J. Jorgenson and A. F. Thacher, *Org. Syn.,* **48**, 75 (1968). (b) In the reduction of α-chloro amides with aluminum hydride, an analogous ring closure occurs to produce aziridines; see Y. Langlois, H. P. Husson, and P. Potier, *Tetrahedron Letters,* **No. 25**, 2085 (1969).
77. (a) E. J. Corey, J. A. Katzenellenbogen, and G. H. Posner, *J. Am. Chem. Soc.,* **89**, 4245 (1967). (b) B. B. Molloy and K. L. Hauser, *Chem. Commun.,* **No. 17**, 1017 (1968). (c) For use of this reaction with polyunsaturated alcohols to form allenes, see R. J. D. Evans, S. R. Landor, and J. P. Regan, *ibid.,* **No. 17**, 397 (1965); see also S. R. Landor, E. S. Pepper, and J. P. Regan, *J. Chem. Soc.,* C, 189 (1967). (d) E. F. Magoon and L. H. Slaugh, *Tetrahedron,* **23**, 4509 (1967).
78. (a) M. J. Jorgenson, *Tetrahedron Letters,* **No. 13**, 559 (1962). (b) H. C. Brown and H. M. Hess, *J. Org. Chem.,* **34**, 2206 (1969). (c) W. L. Dilling and R. A. Plepys, *ibid.,* **35**, 2971 (1970). (d) K. E. Wilson, R. T. Seidner, and S. Masamune, *Chem. Commun.,* **No. 4**, 213 (1970).

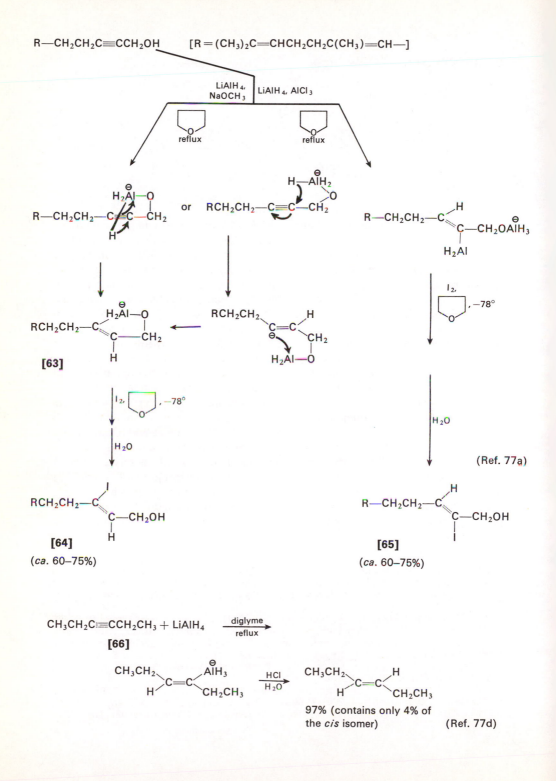

[63]

[64]
(*ca.* 60–75%)

[65]
(*ca.* 60–75%)

(Ref. 77a)

[66]

97% (contains only 4% of
the *cis* isomer)   (Ref. 77d)

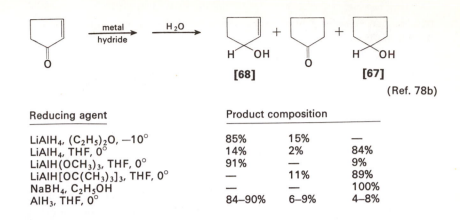

[68]   [67]

(Ref. 78b)

Reducing agent	Product composition		
LiAlH$_4$, (C$_2$H$_5$)$_2$O, $-10°$	85%	15%	—
LiAlH$_4$, THF, 0°	14%	2%	84%
LiAlH(OCH$_3$)$_3$, THF, 0°	91%	—	9%
LiAlH[OC(CH$_3$)$_3$]$_3$, THF, 0°	—	11%	89%
NaBH$_4$, C$_2$H$_5$OH	—	—	100%
AlH$_3$, THF, 0°	84–90%	6–9%	4–8%

hydride transfer (see structure [60]) is not the reaction path leading to the saturated alcohol [67] since the allylic alcohol [68] is not reduced by these reagents.[78b] It is apparent that a second reduction mechanism exists for both lithium aluminum hydride and sodium borohydride reductions in which the nucleophilic metal

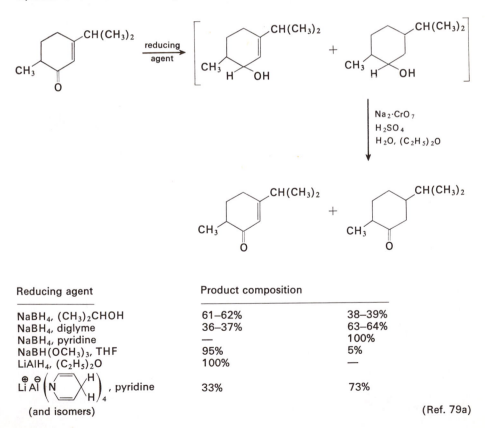

Reducing agent	Product composition	
NaBH$_4$, (CH$_3$)$_2$CHOH	61–62%	38–39%
NaBH$_4$, diglyme	36–37%	63–64%
NaBH$_4$, pyridine	—	100%
NaBH(OCH$_3$)$_3$, THF	95%	5%
LiAlH$_4$, (C$_2$H$_5$)$_2$O	100%	—
Li Al (N⊕ ⊖ H / H)$_4$, pyridine (and isomers)	33%	73%

(Ref. 79a)

hydride can transfer a hydride ion either to the carbonyl carbon atom or to the β-carbon atom of an α,β-unsaturated ketone (structure **[69]**.[78c,79]

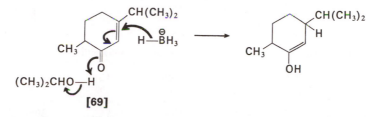

[69]

The accompanying equation compares the amounts of initial conjugate reduction (structure **[69]**) and initial reduction of the carbonyl group when carvenone was reduced with various metal hydrides. The tendency of the previously discussed nucleophilic dihydropyridine derivative (from initial reaction of lithium aluminum hydride with pyridine) to give conjugate reduction products is worthy of note, as is the tendency to favor initial reduction of the carbonyl group from reductions with sodium borohydride in methanol solution (or reduction with sodium trimethoxyborohydride[79a−c]) or lithium trimethoxyaluminum hydride.[78b] In some cases the rearrangement of initially formed allylic alcohols (e.g., **[70]**) in basic

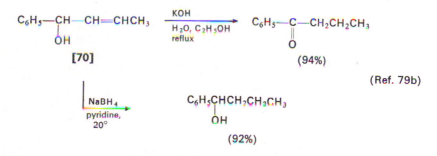

[70]

(94%)

(Ref. 79b)

(92%)

solution to form a saturated ketone which is then further reduced has been found to account for the saturated alcohol products.[79b,c] However, this reaction pathway appears to be important only when abstraction of a proton from the carbinol —CH bond is favored by activating substituents such as a phenyl group (as in **[70]**) or a carbonyl group. The effectiveness of diisobutylaluminum hydride for the selective reduction of the carbonyl group of α,β-unsaturated ketone is shown by the following example.

79. (a) W. R. Jackson and A. Zurqiyah, *J. Chem. Soc.*, 5280 (1965). (b) K. Iqbal and W. R. Jackson, *ibid.*, C, 616 (1968). (c) P. L. Southwick, N. Latif, B. M. Fitzgerald, and N. M. Zaczek, *J. Org. Chem.*, **31**, 1 (1966). (d) J. A. Marshall and R. D. Carroll, *ibid.*, **30**, 2748 (1965). (e) W. J. Bailey and M. E. Hermes, *ibid.*, **29**, 1254 (1964). (f) S. B. Kadin, *ibid.*, **31**, 620 (1966). (g) M. R. Johnson and B. Rickborn, *ibid.*, **35**, 1041 (1970). (h) For the reduction of α,β-unsaturated nitro compounds to saturated nitro compounds, see A. I. Meyers and J. C. Sircar, *ibid.*, **32**, 4134 (1967).

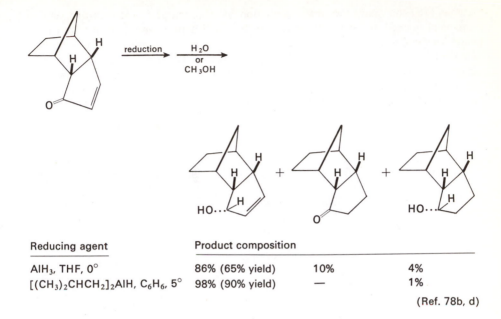

Reducing agent	Product composition		
AlH₃, THF, 0°	86% (65% yield)	10%	4%
[(CH₃)₂CHCH₂]₂AlH, C₆H₆, 5°	98% (90% yield)	—	1%

(Ref. 78b, d)

Reduction of a conjugated carbon-carbon double bond is observed in the reductions of alkylidenemalonic esters and alkylidenecyanoacetic esters with either sodium borohydride or lithium aluminum hydride.[79d–f]

$$CH_3CH{=}C(CO_2C_2H_5)_2 \xrightarrow[\substack{(C_2H_5)_2O, \\ 15°}]{LiAlH_4} \xrightarrow[H_2O]{HCl} CH_3CH_2CH(CO_2C_2H_5)_2$$

(46%) (Ref. 79e)

[71] $\xrightarrow[\substack{C_2H_5OH, 0° \\ 40\ min}]{NaBH_4}$ (85%)

$\xrightarrow[\substack{C_2H_5OH, 25°, \\ 18\ hr}]{NaBH_4}$ (54%)   (Ref. 79d)

As illustrated for the reduction of the cyanoacetate [71], prolonged reaction with sodium borohydride will reduce the ester as well as the carbon-carbon double bond.[41b] The stereochemistry of this carbon-carbon double bond reduction, illustrated with the ester [72], tends to favor the formation of more of the axial

isomer [73] than is found in the reduction of the analogous ketone; this stereo-chemical result has been rationalized in terms of a more reactant-like transition state for the C—C double bond reduction.[77d] However, a comparison of reduction stereochemistry for the cycloalkylidenecyanoacetate [72] and the analogous ketone is complicated by the possibility of conformation distortion in the cyano ester [72] to avoid a nonbonded steric interaction between the cyano and ester substituents and the adjacent equatorial hydrogen atoms.

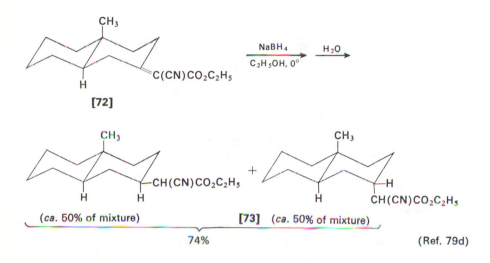

The previously described reductions of allylic alcohols involving the addition of an Al—H bond to a C—C double bond has an analogy in the reduction of enamines (e.g., [74]), enol ethers, and enol esters with aluminum hydride derivatives.[80]

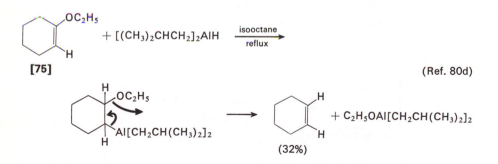

80. (a) W. G. Dauben and J. F. Eastham, *J. Am. Chem. Soc.,* **75,** 1718 (1953). (b) M. M. Rogic, *Tetrahedron,* **21,** 2823 (1965). (c) J. M. Coulter, J. W. Lewis, and P. P. Lynch, *ibid.,* **24,** 4489 (1968). (d) P. Pino and G. P. Lorenzi, *J. Org. Chem.,* **31,** 329 (1966).

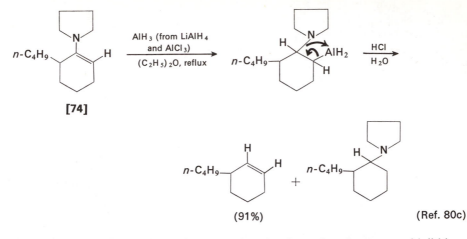

[74]

(91%)   (Ref. 80c)

Although the detailed reaction paths in reductions of enol acetates with lithium aluminum hydride are uncertain,[80b] appropriate deuterium-labelling studies[80a] have demonstrated that the saturated alcohol products are derived from an intermediate organoaluminum intermediate such as structure **[76]**.

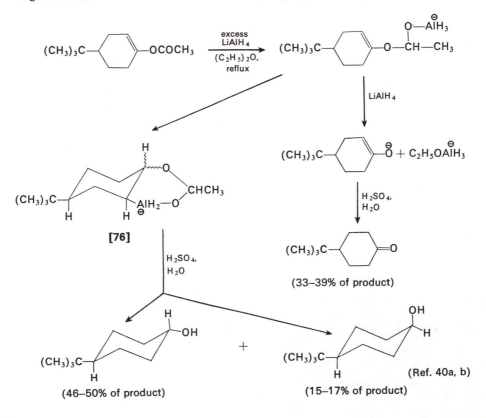

[76]

(46–50% of product)   (15–17% of product)

(33–39% of product)

(Ref. 40a, b)

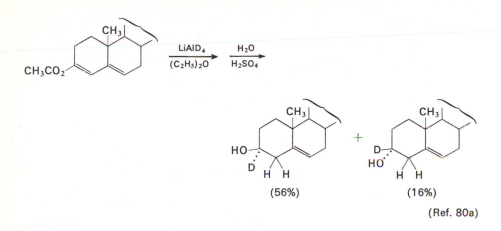

(56%)   (16%)

(Ref. 80a)

The use of aluminum hydride rather than lithium aluminum hydride as a reagent for reductions of halo ketones or halo esters is advantageous when loss of the halogen is to be minimized, since reductions of carbon-halogen bonds with aluminum hydride is a slow process. This advantage is illustrated by the reduction of 3-chloropropionic acid to the chloro alcohol.

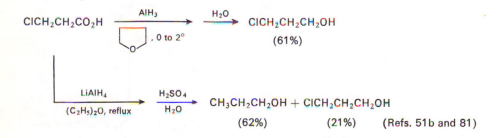

As is implied by these results, the reductions of primary and secondary halides or sulfonate esters by the nucleophilic metal hydride reagents to form a new C—H bond is a synthetically useful process. This reaction with lithium aluminum hydride in ethereal solvents is believed to be an S_N2 displacement in which the complex metal hydride anion is the nucleophile (see structure [77]). A formally similar

81. (a) E. L. Eliel and J. T. Traxler, *J. Am. Chem. Soc.,* **78**, 4049 (1956). (b) E. L. Eliel and T. J. Prosser, *ibid.,* **78**, 4045 (1956).
82. (a) E. J. Corey and D. Achiwa, *J. Org. Chem.,* **34**, 3667 (1969). (b) R. O. Hutchins, D. Hoke, J. Keogh, and D. Koharski, *Tetrahedron Letters,* **No. 40**, 3495 (1969). (c) S. Matsumura and N. Tokura, *ibid.,* **No. 5**, 363 (1969). (d) H. M. Bell and H. C. Brown, *J. Am. Chem. Soc.,* **88**, 1473 (1966). (e) L. J. Dolby and D. R. Rosencrantz, *J. Org. Chem.,* **28**, 1888 (1963). (f) H. M. Bell, C. W. Vanderslice, and A. Spehar, *ibid.,* **34**, 3923 (1969). (g) E. L. Eliel, *J. Am. Chem. Soc.,* **71**, 3970 (1949). (h) J. Jacobus, *Chem. Commun.,* **No. 6**, 338 (1970).

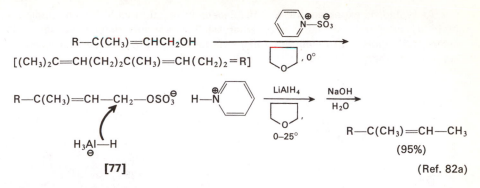

$$R—C(CH_3)=CHCH_2OH$$

$$[(CH_3)_2C=CH(CH_2)_2C(CH_3)=CH(CH_2)_2=R]$$

$$R—C(CH_3)=CH—CH_2—OSO_3^\ominus$$

$$H_3Al—H$$

[77]

R—C(CH_3)=CH—CH_3

(95%)

(Ref. 82a)

reduction can be effected by heating an alkyl halide with sodium borohydride in polar solvents such as dimethyl sulfoxide or sulfolane.[82b,f] It should be noted that a borane (or alane from LiAlH_4) is also generated in this reaction[82f] and may attack

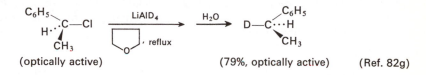

$$\begin{array}{c} C_6H_5 \\ \backslash \\ H \cdots \: C—Cl \\ \diagup \\ CH_3 \end{array} \xrightarrow[\text{, reflux}]{LiAlD_4} \xrightarrow{H_2O} \begin{array}{c} C_6H_5 \\ \diagup \\ D—C \cdots H \\ \backslash \\ CH_3 \end{array}$$

(optically active)

(79%, optically active)   (Ref. 82g)

$$CH_3(CH_2)_5CH—CH_3 + NaBH_4 \xrightarrow[45°]{(CH_3)_2SO} CH_3(CH_2)_6CH_3 + NaI + BH_3$$

(85%, isolated
after hydrolysis)   (Ref. 82f)

other functional groups present in the halide being reduced. A recent study of the sodium borohydride-dimethyl sulfoxide reaction has revealed that this reaction, at least with a tertiary alkyl halide, is not an S_N2 displacement but rather appears to involve the elimination-hydroboration sequence illustrated.[82h]

$$(CH_3)_2CH(CH_2)_3—\overset{\overset{\displaystyle CH_3}{|}}{\underset{\underset{\displaystyle Cl}{|}}{C}}—C_2H_5 \xrightarrow[\substack{CH_3SOCH_3, \\ 100°}]{NaBH_4}$$

(optically active)

$$(CH_3)_2CH(CH_2)_2CH=\overset{\overset{\displaystyle CH_3}{|}}{C}—C_2H_5 + BH_3 + H_2 + NaCl \xrightarrow{\text{hydroboration}}$$
(a mixture of olefin isomers)

$$(CH_3)_2CH(CH_2)_2CH—\overset{\overset{\displaystyle CH_3}{|}}{CH}—C_2H_5 \xrightarrow[\text{reflux}]{n\text{-}C_4H_9CO_2H} (CH_3)_2CH(CH_2)_3\overset{\overset{\displaystyle CH_3}{|}}{CH}—C_2H_5$$

$\underset{\displaystyle BH_2}{|}$

(a mixture of alkylboranes)

(racemic)

(Ref. 82h)

As would be expected, the S_N2 displacement reaction with lithium aluminum hydride is often not satisfactory with tertiary alkyl or neopentyl derivatives; sulfonate esters of alcohols with a neopentyl-like structure (e.g., [78]) may suffer attack at sulfur rather than carbon to regenerate the alcohol.

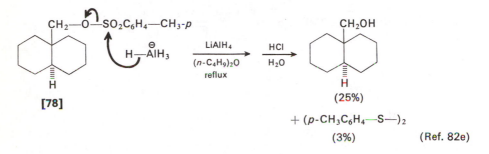

[78]   (25%)

$+ (p\text{-}CH_3C_6H_4\text{—}S\text{—})_2$

(3%)   (Ref. 82e)

Alternatively, secondary and tertiary alkyl halides which readily undergo S_N1 reactions can be reduced by trapping the intermediate carbonium ion with sodium borohydride[82d] or diborane.[82c]

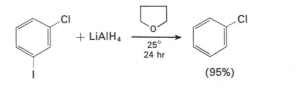

$(C_6H_5)_2CH\text{—}Cl \xrightarrow[\substack{H_2O\text{—}diglyme \\ 45^\circ}]{NaBH_4} (C_6H_5)_2CH^{\oplus} \longrightarrow (C_6H_5)_2CH_2$

(72%)

$\overset{\ominus}{H}\text{—}BH_4$ (Ref. 82d)

Although aryl halides are not expected to undergo S_N2 displacements, it is nonetheless possible to reduce these substances to hydrocarbons by reaction with lithium aluminum hydride in tetrahydrofuran or diglyme solution especially at elevated temperatures.[83] The order of reactivity of aryl halides, Ar-I > Ar-Br > Ar-Cl > Ar-F, permits selective reductions such as the partial reduction of *m*-chloroiodobenzene shown.

$+ \text{LiAlH}_4 \xrightarrow[\substack{25^\circ \\ 24 \text{ hr}}]{}$

(95%)   (Ref. 83b)

The use of trialkyltin hydrides for the reduction of carbon-halogen bonds to C—H bonds offers an alternative reduction procedure. Since this reaction is a

83. (a) G. J. Karabatsos and R. L. Shone, *J. Org. Chem.*, **33**, 619 (1968). (b) H. C. Brown and S. Krishnamurthy, *ibid.*, **34**, 3918 (1969). (c) P. Olavi, I. Virtanen, and P. Jaakkola, *Tetrahedron Letters*, **No. 16**, 1223 (1969). (d) For the use of mixtures obtained from sodium borohydride and transition metal salts to hydrogenolyze aryl halogen bonds, see R. A. Egli, *Helv. Chim. Acta*, **51**, 2090 (1968).

free-radical chain process, it is not restricted to halides that will undergo nucleophilic displacement and can be applied to both various alkyl (e.g., [79]) and aryl halides.[84]

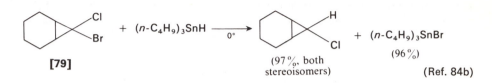

[79]   (97%, both
                                          stereoisomers)

+ $(n\text{-}C_4H_9)_3SnBr$

(96%)

(Ref. 84b)

Although the usual tin reagent, tri-*n*-butyltin hydride [80], may either be purchased or prepared by the reduction of tri-*n*-butyltin chloride or tri-*n*-butyltin oxide with lithium aluminum hydride,[84d,85] a simpler reaction procedure has been described in which the tin hydride is generated by reduction with the commercially

$$(n\text{-}C_4H_9)_3SnCl + LiAlH_4 \xrightarrow[\text{reflux}]{(C_2H_5)_2O} \xrightarrow{H_2O} (n\text{-}C_4H_9)_3SnH$$

[80]   (Ref. 85a)

available polymethylsiloxane and utilized *in situ* as illustrated in the accompanying equation. Since this free-radical chain process, which may be initiated thermally, photochemically, or with a radical initiator such as azobisisobutyronitrile[86b],

$$CH_3CO_2CH_2CH_2Br + \left(\begin{array}{c} CH_3 \\ | \\ -SiH-O- \end{array}\right)_n + [(n\text{-}C_4H_9)_3Sn]_2O \xrightarrow[\text{no solvent}]{0-25°}$$

$$(n\text{-}C_4H_9)_3Sn-H \xrightarrow{CH_3CO_2CH_2CH_2Br} CH_3CO_2CH_2CH_3 + (n\text{-}C_4H_9)_3SnBr$$

(63%)   (Ref. 86a)

84. (a) W. P. Neumann, *Angew. Chem., Intern. Ed. Engl.,* **2**, 165 (1963). (b) D. Seyferth, H. Yamazaki, and D. L. Alleston, *J. Org. Chem.,* **28**, 703 (1963). (c) H. G. Kuivila in F. G. A. Stone and R. West, eds., *Advances in Organometallic Chemistry*, Vol. 1, Academic Press, New York, 1964. (d) H. G. Kuivila, *Accts. Chem. Res.,* **1**, 299 (1968); *Synthesis*, **2**, 499 (1970).
85. (a) G. J. M. Van der Kerk, J. G. Noltes, and J. G. A. Luijten, *J. Appl. Chem.,* **7**, 366 (1957). (b) Tri-*n*-butyltin deuteride has been prepared by the reduction of hexa-*n*-butylditin with the sodium-naphthalene anion radical followed by reaction of the product (*n*-C$_4$H$_9$)$_3$SnNa with D$_2$O; K. Kühlein, W. P. Neumann, and H. Mohring, *Angew, Chem., Intern. Ed. Engl.,* **7**, 455 (1968).
86. (a) G. L. Grady and H. G. Kuivila, *J. Org. Chem.,* **34**, 2014 (1969). (b) F. D. Greene and N. N. Lowry, *ibid.,* **32**, 882 (1967). (c) D. J. Carlsson and K. U. Ingold, *J. Am. Chem. Soc.,* **90**, 7047 (1968). (d) L. Kaplan, *ibid.,* **88**, 4531 (1966). (e) T. Ando, F. Namigata, H. Yamanaka, and W. Funasaka, *ibid.,* **89**, 5719 (1967). (f) L. J. Altman and B. W. Nelson, *ibid.,* **91**, 5163 (1969). (g) R. J. Strunk, P. M. DiGiacomo, K. Aso, and H. G. Kuivila, *ibid.,* **92**, 2849 (1970). (h) J. W. Wilt, S. N. Massie, and R. B. Dabek, *J. Org. Chem.,* **35**, 2803 (1970).

involves intermediate alkyl or aryl radicals (e.g., **[81]**), the stereochemistry of the carbon-halogen bond and the final C—H bond need bear no particular relation to one another. However, because of the rapidity with which tin hydrides can donate a hydrogen atom to a carbon free radical,[86] in certain cases involving α-halo cyclopropyl radicals where inversion of a pyramidal radical might be slow, some stereospecificity has been reported.[86e,f]

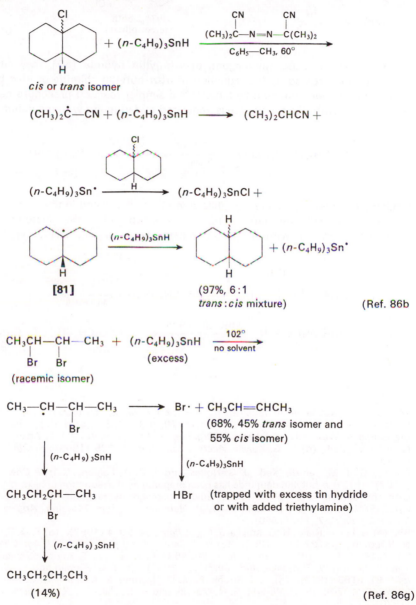

(97%, 6:1
trans:cis mixture)   (Ref. 86b

(trapped with excess tin hydride
or with added triethylamine)

(Ref. 86g)

The free radical intermediate involved in these reductions may sometimes undergo intramolecular addition reactions[86h] or elimination reactions[86g] at rates comparable to the intermolecular reaction with a molecule of tri-*n*-butyltin hydride. The competing elimination which may accompany the reduction of 1,2-dibromo-alkanes is illustrative.

$$(CH_3)_3C—CH—C(CH_3)_2 \xrightarrow[\text{(C}_2\text{H}_5)_2\text{O}]{\text{LiAlH}_4} (CH_3)_3C—CH_2—C(CH_3)_2$$

$$\underset{O}{\overset{}{}} \qquad\qquad \underset{OH}{\overset{}{}}$$

(21%; only alcohol formed)   (Ref. 87)

As would be anticipated for an S_N2 process, reaction of lithium aluminum hydride with epoxides occurs at the less highly substituted position to produce the more highly substituted alcohol. As illustrated by the reduction of the epoxide **[82]**,

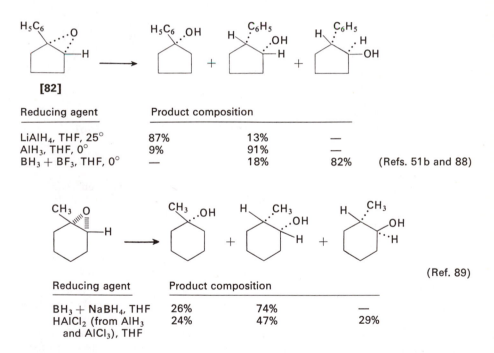

[82]

Reducing agent	Product composition		
LiAlH₄, THF, 25°	87%	13%	—
AlH₃, THF, 0°	9%	91%	—
BH₃ + BF₃, THF, 0°	—	18%	82% (Refs. 51b and 88)

(Ref. 89)

Reducing agent	Product composition		
BH₃ + NaBH₄, THF	26%	74%	—
HAlCl₂ (from AlH₃ and AlCl₃), THF	24%	47%	29%

87. (a) E. L. Eliel and M. N. Rerick, *J. Am. Chem. Soc.*, **82**, 1362 (1960). (b) For studies of this reduction with alkoxyaluminum hydrides and chloroaluminum hydrides, see E. C. Ashby and B. Cooke, *ibid.*, **90**, 1625 (1968); B. Cooke, E. C. Ashby, and J. Lott, *J. Org. Chem.*, **33**, 1132 (1968).
88. (a) H. C. Brown and N. M. Yoon, *Chem. Commun.*, **No. 23**, 1549 (1968). (b) P. T. Lansbury, D. J. Scharf, and V. A. Pattison, *J. Org. Chem.*, **32**, 1748 (1967). (c) R. E. Lyle and W. E. Krueger, *ibid.*, **32**, 2873 (1967).
89. H. C. Brown and N. M. Yoon, *J. Am. Chem. Soc.*, **90**, 2686 (1968).

$$(C_6H_5)_2C\!\!-\!\!CH\!\!-\!\!C_6H_5 \xrightarrow[\;(C_2H_5)_2O,\;\; 25°\;]{AlH_3\;(from\;AlCl_3\;+\;\;LiAlH_4)} (C_6H_5)_2CH\!\!-\!\!CH\!\!-\!\!C_6H_5$$

with the O bridging the two carbons on the left, and OH on the right product.

(90%)

(Ref. 90)

when the reducing agent is changed from the nucleophilic reagent, lithium aluminum hydride, to electrophilic reducing agents such as aluminum hydride or
diborane, very different mixtures of reaction products may result. At least three
different reaction paths appear to be involved. The nucleophilic displacement
(structure [83]) at the less highly substituted carbon atom usually seen with
lithium aluminum hydride is replaced by an assisted ring opening (structure [84])
where one of the electrophilic reagents such as aluminum hydride or borane is
bonded to the epoxide oxygen. Both intermolecular (e.g., [84a] and intramolecular
(e.g., [84b]) hydride ion transfers may be involved. In these cases, a ring opening

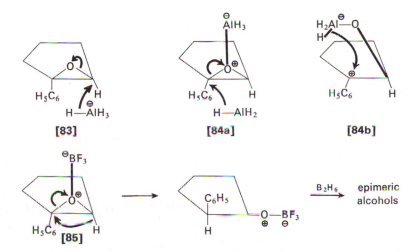

[83]   [84a]   [84b]

[85]

B₂H₆ → epimeric alcohols

appears to occur at the carbon atom that is better able to tolerate positive charge in
the transition state. Finally, if sufficiently strong Lewis acids (e.g., BF_3 or $AlCl_3$)
are present, rearrangement of the epoxide (structure [85]) to the alane or borane
complex of a ketone or aldehyde may precede reduction, or other changes may
occur.[88–90]

Like other openings of cyclohexene oxides by reactions with nucleophiles
(see Chapter 6), these derivatives are usually opened by metal hydrides to form
products in which the oxygen atom from the epoxide and the nucleophile (the
hydrogen atom from the metal hydride) are in axial positions. It will be noted in the
accompanying equation that this diaxial ring opening is the favored reaction path
even though the major product is formed by attack of the nucleophile at the more
highly substituted position.

90. M. N. Rerick and E. L. Eliel, *J. Am. Chem. Soc.*, **84**, 2356 (1962).

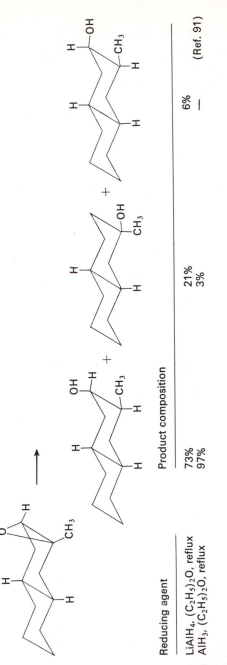

Reducing agent	Product composition			
LiAlH$_4$, (C$_2$H$_5$)$_2$O, reflux	73%	21%	6%	(Ref. 91)
AlH$_3$, (C$_2$H$_5$)$_2$O, reflux	97%	3%	—	

91. (a) D. K. Murphy, R. L. Alumbaugh, and B. Rickborn, *J. Am. Chem. Soc.*, **91**, 2649 (1969). (b) B. Rickborn and W. E. Lamke, II, *J. Org. Chem.*, **32**, 537 (1967). (c) For an exception in which an alkyl substituent with a 1,3-diaxial relationship to the epoxide prevents the usual diaxial opening of the epoxide, see J. A. Marshall, *Accts. Chem. Res.*, **2**, 33 (1969).

HYDROBORATION AND RELATED REACTIONS

Although diborane is capable of reducing a variety of functional groups (Table 2-3), the most useful synthetic applications of the reagent have involved addition to carbon-carbon multiple bonds, called hydroboration.[92] The rate of addition of diborane to carbon-carbon double bonds usually decreases as the number of alkyl substituents about the double bond increases.[93] With mono- and disubstituted

$$n\text{-}C_4H_9\text{---}CH\text{=}CH_2 \ + \ B_2H_6 \xrightarrow{} n\text{-}C_4H_9\text{---}CH_2\text{---}CH_2\text{---}BH_2 \xrightarrow{n\text{-}C_4H_9\text{---}CH\text{=}CH_2}$$
$$\text{C-1} \qquad\qquad\qquad\qquad\qquad \text{C-2}$$

$$(n\text{-}C_4H_9\text{---}CH_2\text{---}CH_2)_2BH \xrightarrow[\text{C-3}]{n\text{-}C_4H_9\text{---}CH\text{=}CH_2} (n\text{-}C_4H_9\text{---}CH_2\text{---}CH_2)_3B$$

olefins, all three hydrogen atoms of borane are used in the reduction.[94] The final addition step is slower than the first two reactions (i.e., C-1 and C-2 > C-3), a result that has been attributed to the increasing steric bulk of the partially alkylated boron reagent. In agreement with this idea, trisubstituted olefins normally react with borane to give dialkylboranes (e.g., [86]), and tetrasubstituted olefins form monoalkylboranes (e.g., [87]) with solutions of diborane at room temperature.[94] As with lithium aluminum hydride, this successive replacement of hydrogen atoms in borane permits the preparation of mono- and dialkylboranes that are less reactive and more selective than diborane itself.

The three alkylboranes whose preparations are illustrated in the accompanying equations have been used extensively for selective reductions; the first two of these derivatives have been given the trivial names disiamylborane [86] and thexylborane [87].

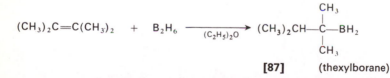

$$(CH_3)_2C\text{=}CH\text{---}CH_3 \ + \ B_2H_6 \xrightarrow{(C_2H_5)_2O} \left[(CH_3)_2CH\text{---}\overset{\displaystyle CH_3}{\underset{\displaystyle |}{CH}}\text{---}\right]_2 BH$$

[86]   (disiamylborane)

(Ref. 94a, b)

$$(CH_3)_2C\text{=}C(CH_3)_2 \ + \ B_2H_6 \xrightarrow{(C_2H_5)_2O} (CH_3)_2CH\text{---}\overset{\displaystyle CH_3}{\underset{\displaystyle \underset{\displaystyle CH_3}{|}}{\overset{\displaystyle |}{C}}}\text{---}BH_2$$

[87]   (thexylborane)

92. (a) H. C. Brown, *Tetrahedron*, **12**, 117 (1961). (b) G. Zweifel and H. C. Brown, *Org. Reactions*, **13**, 1 (1963). H. C. Brown, *Hydroboration*, Benjamin, New York, 1962.
93. (a) H. C. Brown and G. Zweifel, *J. Am. Chem. Soc.*, **82**, 3222, 3223, 4708 (1960). (b) H. C. Brown and R. L. Sharp, *ibid.*, **88**, 5851 (1966).
94. (a) H. C. Brown and B. C. Subba Rao, *J. Am. Chem. Soc.*, **81**, 6423, 6428 (1969). (b) H. C. Brown and A. W. Moerikofer, *ibid.*, **83**, 3417 (1961); *ibid.*, **84**, 1478 (1962); *ibid.*, **85**, 2063 (1963). (c) E. F. Knights and H. C. Brown, *ibid.*, **90**, 5280, 5281 (1968). (d) D. J. Sandman, K. Mislow, W. P. Giddings, J. Dirlam, and G. C. Hanson, *ibid.*, **90**, 4877 (1968). (e) R. Fellous and R. Luft, *Tetrahedron Letters*, **No. 18**, 1505 (1970).

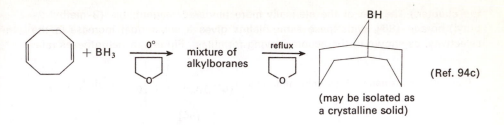

(may be isolated as
a crystalline solid)

(Ref. 94c)

The direction of addition of borane to unsymmetrical olefins is such that the predominant product usually has the boron atom bonded to the less highly substituted carbon atom. The following examples of product distributions[93] illustrate not only this generalization but also the fact that the direction of addition is influenced (though to a lesser extent) by electrical effects (e.g., [88]) and by the steric bulk of groups near the double bond (e.g., [90]). (The substantial effects observed when functional groups are bonded to the olefin are discussed later in

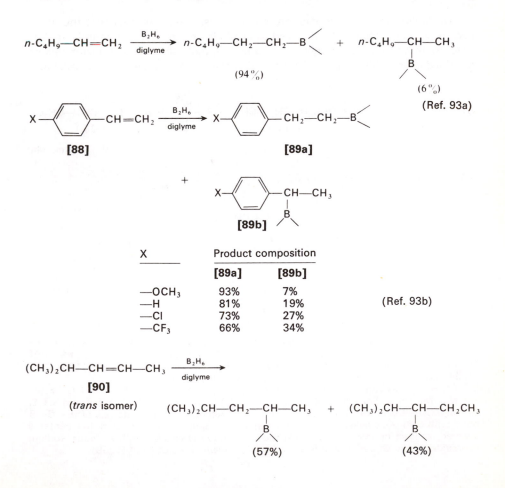

X	Product composition	
	[89a]	[89b]
—OCH$_3$	93%	7%
—H	81%	19%
—Cl	73%	27%
—CF$_3$	66%	34%

(Ref. 93b)

the chapter.) The use of the sterically more hindered reagent, bis (3-methyl-2-butyl) borane **[86]**, with these same olefins gives a substantial increase in the selectivity, as shown in the accompanying formulas. Figures in parentheses refer

$$\text{Product composition from hydroboration with} \quad \left[(CH_3)_2CH-\overset{\overset{\displaystyle CH_3}{|}}{CH}\right]_2 BH \quad \text{in diglyme}$$

[86]

$n\text{-}C_4H_9-CH{=}CH_2$ $C_6H_5-CH{=}CH_2$ $(CH_3)_2CH-CH{=}CH-CH_3$

$\quad\uparrow\quad\uparrow$ $\quad\uparrow\quad\uparrow$ $\qquad\quad\uparrow\qquad\uparrow$

$(1\%)\quad(99\%)$ $(2\%)\quad(98\%)$ $(5\%)\quad(95\%)$

to the fractions of product in which the boron atom is bonded to the carbon atom indicated.[95] The more rapid addition of bis (3-methyl-2-butyl) borane to less highly substituted carbon–carbon double bonds has also been used to separate trisubstituted olefins (which do react) from tetrasubstituted olefins (which do not react).[95e]

The utility of the hydroboration reaction arises from the fact that the intermediate alkylboranes can be oxidized to alcohols or carbonyl compounds, hydrolyzed to hydrocarbons, or utilized in other synthetically useful transformations described later in this chapter. The following examples illustrate these possibilities.

$$n\text{-}C_8H_{17}-CH{=}CH_2 \xrightarrow[\substack{\text{diglyme}\\25°}]{B_2H_6} (n\text{-}C_8H_{17}CH_2-CH_2-)_3B \xrightarrow[\substack{H_2O-\\\text{diglyme}}]{\substack{H_2O_2\\NaOH}} n\text{-}C_8H_{17}-CH_2-CH_2OH$$

$\qquad\qquad\qquad\qquad\qquad\qquad\qquad$ (not isolated) $\qquad\qquad\qquad\qquad\qquad$ (93%)

(Ref. 94a)

$$n\text{-}C_4H_9-CH{=}CH_2 \xrightarrow[\substack{\text{diglyme}\\25°}]{B_2H_6} (n\text{-}C_4H_9-CH_2-CH_2)_3B \xrightarrow[\text{reflux}]{CH_3CH_2CO_2H}$$

(not isolated)

[91]

$\qquad\qquad\qquad\qquad\longrightarrow\qquad n\text{-}C_4H_9-CH_2-CH_3$

$\qquad\qquad\qquad\qquad\qquad\qquad\qquad\qquad$ (91%)

(Ref. 96)

95. (a) H. C. Brown and G. Zweifel, *J. Am. Chem. Soc.*, **83**, 1241 (1961). (b) H. C. Brown and K. A. Keblys, *ibid.*, **86**, 1791 (1964). (c) H. C. Brown and O. J. Cope, *ibid.*, **86**, 1801 (1964). (d) M. Nussim, Y. Mazur, and F. Sondheimer, *J. Org. Chem.*, **29**, 1120, 1131 (1964). (e) E. M. Kaiser and R. A. Benkeser, *Org. Syn.*, **50**, 88 (1970).
96. (a) H. C. Brown and K. J. Murray, *J. Am. Chem. Soc.*, **81**, 4108 (1959). (b) H. C. Brown and K. J. Murray, *J. Org. Chem.*, **26**, 631 (1961). (c) L. H. Toporcer, R. E. Dessy, and S. I. E. Green, *J. Am. Chem. Soc.*, **87**, 1236 (1965). (d) Although the protonolysis of the C—B bond with carboxylic acids occurs with retention of configuration at carbon, this center is inverted, at least in some cases, when the C—B bond is cleaved with aqueous sodium hydroxide; see A. G. Davies and B. P. Roberts, *J. Chem. Soc.*, C, 1474 (1968).

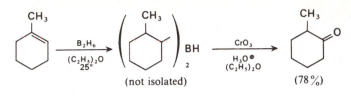

(not isolated)　　　　　　　　　　　　　　　　　(78%)

(Ref. 97)

It will be noted that hydroboration followed by oxidation of the alkylborane to an alcohol results in the overall addition of water to a double bond in a direction opposite to that obtained by the direct hydration of a double bond with aqueous acid, an addition which follows the Markownikoff rule. The formation of a hydrocarbon by hydroboration of an olefin and subsequent protonolysis normally offers an advantage over the simpler catalytic hydrogenation only in cases where the selective reduction of a double bond in the presence of some easily hydrogenated function (e.g., a nitro group) is desired or where a greater degree of stereospecificity can be obtained. However, this sequence for multiple-bond reduction is useful for the preparation of a hydrocarbon containing deuterium atoms in specific locations, since the deuterium-hydrogen scrambling often observed on catalytic hydrogenation appears not to be a problem in hydroboration reactions at room temperature. The protonolysis of alkylboranes is found to occur more rapidly with carboxylic acids than with water or aqueous mineral acids.[96] As illustrated in structure [91], this rapid protonolysis of carbon-boron bonds with carboxylic acids is believed attributable to the formation of an alkylborane-carboxylic acid complex which undergoes the indicated proton transfer.

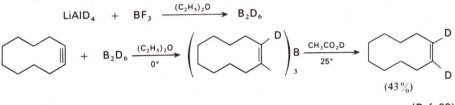

(43%)

(Ref. 98)

The addition of borane to a double bond has been found to occur in a *cis* manner from the less hindered side of the double bond, as indicated in the following

97. (a) H. C. Brown and C. P. Garg, *J. Am. Chem. Soc.*, **83**, 2951 (1961). (b) For example where this procedure for direct oxidation of an alkylborane to a ketone was accompanied by molecular rearrangement, see P. T. Lansbury and E. J. Nienhouse, *Chem. Commun.*, **No. 9**, 273 (1966).
98. A. C. Cope, G. A. Berchtold, P. E. Peterson, and S. H. Sharman, *J. Am. Chem. Soc.*, **82**, 6370 (1960).

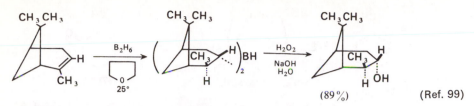

(89%)   (Ref. 99)

Product compositions from the hydroboration of substituted cyclohexenes in tetrahydrofuran solution. (The figures in parentheses refer to the fractions of the product in which the boron atom is bonded to each carbon from the direction indicated.)

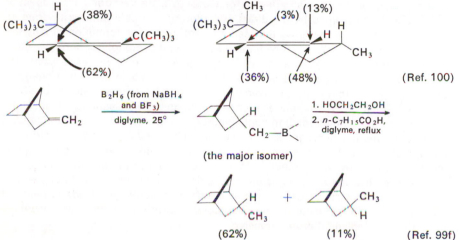

(Ref. 100)

(the major isomer)

(62%)   (11%)   (Ref. 99f)

examples.[99] Both the oxidation of the carbon-boron bond to form an alcohol[99,101] and the protonolysis (see structure [91]) of this bond to form a hydrocarbon[96] may be effected with retention of configuration.

99. (a) H. C. Brown and G. Zweifel, *J. Am. Chem. Soc.*, **83**, 2544 (1961); *ibid.*, **86**, 393 (1964). (b) H. C. Brown, N. R. Ayyangar and G. Zweifel, *ibid.*, **86**, 397, 1071 (1964); G. Zweifel, N. R. Ayyangar, T. Munekata and H. C. Brown, *ibid.*, **86**, 1076 (1964). (c) For the preparation of pure (−)-α-pinene to use in this procedure, see P. A. Spanninger and J. L. von Rosenberg, *J. Org. Chem.*, **34**, 3658 (1969). (d) For recent discussions of the transition state geometry in asymmetric hydroboration, see K. R. Varma and E. Caspi, *Tetrahedron*, **24**, 6365 (1968); D. R. Brown, S. F. A. Kettle, J. McKenna, and J. M. McKenna, *Chem. Commun.*, **No. 14**, 667 (1967). (e) D. R. Boyd, M. F. Grundon, and W. R. Jackson, [*Tetrahedron Letters*, **No. 22**, 2101 (1967)] have described the asymmetric reduction of imines with the complex formed from *n*-butyllithium and diisopinocampheylborane. (f) H. C. Brown and J. H. Kawakami, *J. Am. Chem. Soc.*, **92**, 1990 (1970). (g) J. Klein and D. Lichtenberg, *J. Org. Chem.*, **35**, 2654 (1970).
100. (a) D. J. Pasto and F. M. Klein, *J. Org. Chem.*, **33**, 1468 (1968). (b) For stereochemical studies of the hydroboration of various mono- and sesquiterpenes, see S. P. Acharya, H. C. Brown, A. Suzuki, S. Nozawa, and M. Itoh, *ibid.*, **34**, 3015 (1969) and references therein; S. P. Acharya and H. C. Brown, *ibid.*, **35**, 196 (1970).
101. (a) R. Köster and Y. Morita, *Angew. Chem., Intern. Ed. Engl.*, **5**, 580 (1966). (b) S. B. Mirviss, *J. Org. Chem.*, **32**, 1713 (1967). (c) A. G. Davies and B. P. Roberts, *Chem. Commun.*, **No. 10**, 298 (1966). (d) D. B. Bigley and D. W. Payling, *J. Chem. Soc.*, B, 1811 (1970). (e) A. G. Davies and R. Tudor, *ibid.*, B, 1815 (1970); H. C. Brown, M. M. Midland, and G. W. Kabalka, *J. Am. Chem. Soc.*, **93**, 1024 (1971).

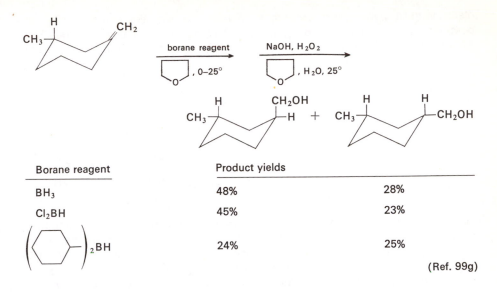

Borane reagent	Product yields	
BH_3	48%	28%
Cl_2BH	45%	23%
$\left(\bigcirc\right)_2 BH$	24%	25%

(Ref. 99g)

The oxidation of an alkylborane to the corresponding alcohol may be effected with amine oxides,[96d,101a] oxygen,[101b,c] peroxides, or most commonly with alkaline hydrogen peroxide.[99] The oxidations with amine oxides and *alkaline* hydrogen peroxide have been found to proceed with retention of configuration; a reasonable reaction path for these processes involves successive intramolecular rearrangements (i.e., [92]) of alkyl groups from boron to oxygen followed by hydrolysis. It should be noted that oxidation with oxygen or with *neutral* hydrogen peroxide follows a free radical path[101b−c] and is not stereospecific.

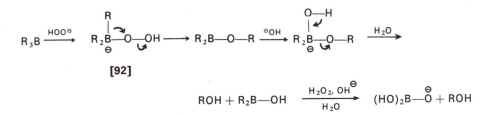

[92]

$$ROH + R_2B{-}OH \xrightarrow[\ H_2O\ ]{H_2O_2,\ OH^{\ominus}} (HO)_2B{-}\overset{\ominus}{O} + ROH$$

Studies of the reaction path followed in hydroboration reactions are complicated by the tendency of monoalkylboranes (e.g., [87]) and dialkylboranes (e.g., [86]) to disproportionate and exist as dimers in tetrahydrofuran solution;[33g,94d,102a] the reaction of monochloroborane with olefins is more satisfactory for kinetic study.[102b] All of the hydroboration reactions would appear to involve initial reaction of the

102. (a) H. C. Brown and G. J. Klender, *Inorg. Chem.,* **1**, 204 (1962). (b) D. J. Pasto and S. Z. Kang, *J. Am. Chem. Soc.,* **90**, 3797 (1968). (c) A. Streitwieser, Jr., L. Verbit, and R. Bittman, *J. Org. Chem.,* **32**, 1530 (1967). (d) J. Klein, E. Dunkelblum, and M. A. Wolff, *J. Organometal. Chem.,* **7**, 377 (1967).

olefin with a borane-tetrahydrofuran complex (or an alkylborane dimer),[94b] possibly via a pi-complex (e.g., [93]),[102c,d] leading to a four-centered transition state [94] by which the borane adds to the carbon-carbon double bond. Although

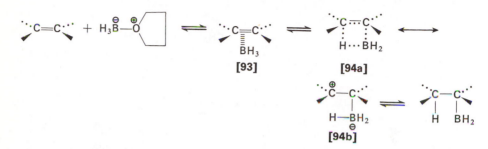

there is a slight tendency for the hydrogen to add to the carbon atom better able to tolerate positive charge (as though structure [94b] contributes to the transition state), the transition state (e.g., [94]) appears to be relatively nonpolar, so that the direction of addition to unsymmetrical olefins is determined principally by steric factors. Thus the rate of hydroboration of a carbon-carbon double bond is increased by increasing the electron density of the double bond, by increasing the strain present in the double bond, and, particularly, by decreasing the steric hindrance about the double bond.

Apart from the previously mentioned use of alkyl- or dialkylboranes to obtain increased structural selectivity in the hydroboration of olefins, stereospecificity is obtained in the hydroboration of *cis*-disubstituted olefins with the optically active

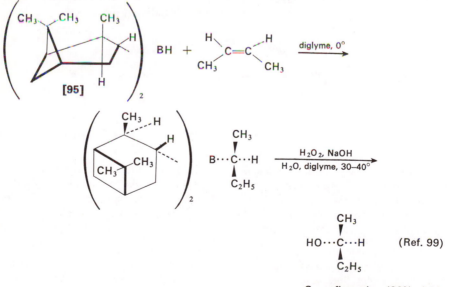

S-configuration (83% yield, 87% optically purity)

reagent diisopinocampheylborane [95]. As illustrated, a very high degree of asymmetry is induced into the 2-butylborane derivative formed in this way.[99]

Several possible transition states have been suggested[99b,d,102c] for this asymmetric hydroboration, which presumably involves reaction of the olefin with the dimer of the dialkylborane [95] (also called tetra-α-pinanyldiborane). It should be noted that both the degree and the direction of induced asymmetry obtained with a borane intermediate prepared from equimolar amounts of α-pinene and borane was found to vary with time.[94d] This result was interpreted to mean that the initially formed mixture ($R_2BH + BH_3$) disproportionated on standing (to form RBH_2).

The reversibility of the addition of borane to olefins has been demonstrated by heating initially formed alkylboranes to temperatures above 80° (usually in refluxing diglyme).[103] Since reversal of the hydroboration reaction can usually occur in two directions with the alkylboranes derived from nonterminal olefins, the net result of heating such an alkylborane is to isomerize it to the more stable terminal alkylborane, as illustrated below. A variety of applications of this isomerization process have been described.[92,103] It is apparent from the foregoing discussion

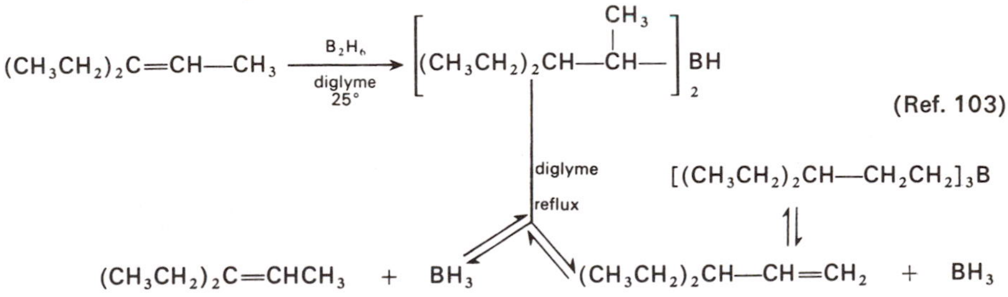

$$(CH_3CH_2)_2C{=}CH{-}CH_3 \xrightarrow[\substack{\text{diglyme} \\ 25°}]{B_2H_6} \left[(CH_3CH_2)_2CH{-}\overset{\overset{\displaystyle CH_3}{|}}{CH}{-} \right]_2 BH$$

(Ref. 103)

$$\downarrow \substack{\text{diglyme} \\ \text{reflux}} \qquad [(CH_3CH_2)_2CH{-}CH_2CH_2]_3B$$

$$(CH_3CH_2)_2C{=}CHCH_3 \;+\; BH_3 \qquad (CH_3CH_2)_2CH{-}CH{=}CH_2 \;+\; BH_3$$

that the isomerization of an alkylborane cannot occur past a carbon atom that is not bonded to at least one hydrogen atom. When the alkylborane [96], which cannot undergo the usual isomerization to a terminal alkylborane, was heated, the compound underwent an interesting insertion reaction to form the cyclic borane [97].

Cyclic dialkyl- and trialkylboranes may also be formed by reaction of borane or an alkylborane with dienes or trienes as illustrated in the accompanying equations. The initial products formed from dienes and diborane appear to be macrocyclic or polymeric dialkyl- or trialkylboranes which may be isomerized thermally to derivatives containing boron within a six-membered ring. Direct formation of cyclic trialkyl-

103. (a) H. C. Brown and G. Zweifel, *J. Am. Chem. Soc.*, **88**, 1433 (1966); **89**, 561 (1967). (b) H. C. Brown and M. V. Bhatt, *ibid.*, **82**, 2074 (1960). (c) H. C. Brown and B. C. Subba Rao, *ibid.*, **81**, 6434 (1959). (d) H. C. Brown, M. V. Bhatt, T. Munekata, and G. Zweifel, *ibid.*, **89**, 567 (1967). (e) J. E. Herz and L. A. Marquez, *J. Chem. Soc.*, C, 2243 (1969). (f) G. F. Hennion, P. A. McCusker, E. C. Ashby, and A. J. Rutkowski, *J. Am. Chem. Soc.*, **79**, 5190 (1957); E. C. Ashby, *ibid.*, **81**, 4791 (1959).

boranes appears to occur when the hindered monoalkylborane [87] is employed to hydroborate dienes.[105]

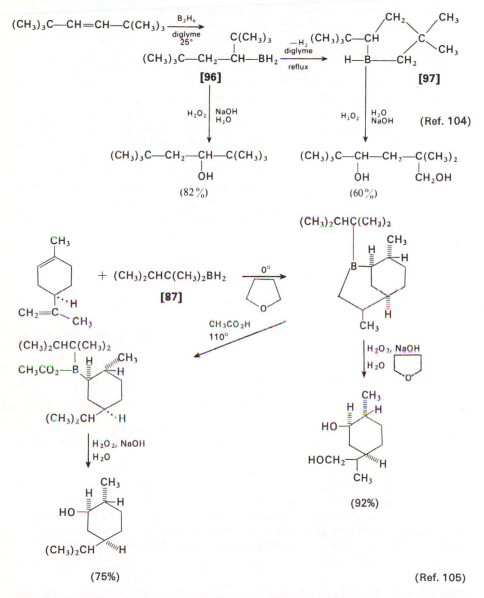

(82%)   (60%)

(Ref. 104)

(92%)

(75%)   (Ref. 105)

104. (a) T. J. Logan and T. J. Flautt, *J. Am. Chem. Soc.*, **82**, 3446 (1960). (b) For other examples of the formation of cyclic borane derivatives by heating alkylboranes, see R. Koster, *Angew. Chem., Intern. Ed. Engl.*, **3**, 174 (1964); H. C. Brown, K. J. Murray, H. Müller, and G. Zweifel, *J. Am. Chem. Soc.*, **88**, 1443 (1966).
105. H. C. Brown and C. D. Pfaffenberger, *J. Am. Chem. Soc.*, **89**, 5475 (1967).

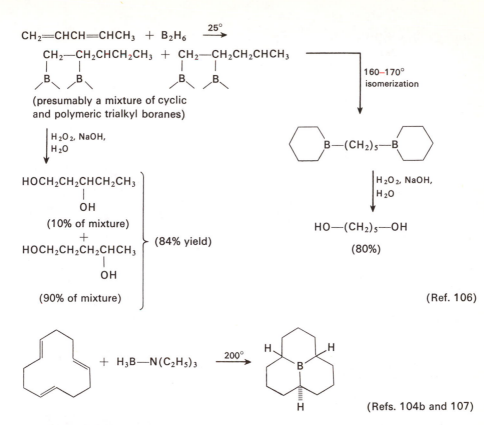

CH_2=$CHCH$=$CHCH_3$ + B_2H_6 $\xrightarrow{25°}$

CH$_2$—CH$_2$CHCH$_2$CH$_3$ + CH$_2$—CH$_2$CH$_2$CHCH$_3$
  | | | |
  B   B   B   B

(presumably a mixture of cyclic
and polymeric trialkyl boranes)

160–170°
isomerization

H$_2$O$_2$, NaOH,
H$_2$O

B—(CH$_2$)$_5$—B

HOCH$_2$CH$_2$CHCH$_2$CH$_3$
          |
          OH
(10% of mixture)
+
HOCH$_2$CH$_2$CH$_2$CHCH$_3$
          |
          OH
(90% of mixture)

(84% yield)

H$_2$O$_2$, NaOH,
H$_2$O

HO—(CH$_2$)$_5$—OH
(80%)

(Ref. 106)

+ H$_3$B—N(C$_2$H$_5$)$_3$ $\xrightarrow{200°}$

(Refs. 104b and 107)

The course of olefin hydroboration may be influenced by nearby functional groups such as halo, hydroxy, alkoxy, and acetoxy.[95c,108] This result, possibly attributable to the electron-attracting inductive effect of these substituents, is relatively small when such functional groups are separated from the C—C double bond by two (e.g., [98]) or more carbons, and the principal reaction course is determined by steric factors, especially if hindered alkylborane reagents are used.[108a] However, the directive effect is pronounced with the corresponding allylic derivatives, which yield mainly the vicinally disubstituted alkylboranes (e.g., [99]).

106. (a) K. A. Saegebarth, *J. Am. Chem. Soc.*, **82**, 2081 (1960). (b) For recent studies of the products obtained from 1,3-butadiene and diborane, see D. E. Young and S. G. Shore, *ibid.*, **91**, 3497 (1969); E. Breuer and H. C. Brown, *ibid.*, **91**, 4164 (1969); H. C. Brown, E. Negishi, and S. K. Gupta, *ibid.*, **92**, 2460, 6648 (1970) and references therein. (c) H. C. Brown, E. Negishi, and P. L. Burke, *ibid.*, **92**, 6649 (1970).
107. (a) G. W. Rotermund and R. Köster, *Justus Liebigs Ann. Chem.*, **686**, 153 (1965). (b) H. C. Brown and E. Negishi, *J. Am. Chem. Soc.*, **89**, 5478 (1967). (c) H. C. Brown and W. C. Dickason, *ibid.*, **91**, 1226 (1969); *ibid.*, **92**, 709 (1970).
108. (a) H. C. Brown and M. K. Unni, *J. Am. Chem. Soc.*, **90**, 2902 (1968). (b) H. C. Brown and R. M. Gallivan, Jr., *ibid.*, **90**, 2906 (1968). (c) H. C. Brown and R. L. Sharp, *ibid.*, **90**, 2915 (1968) and references therein. (d) K. H. Schulte-Elte and G. Ohloff, *Helv. Chim. Acta*, **50**, 153 (1967).

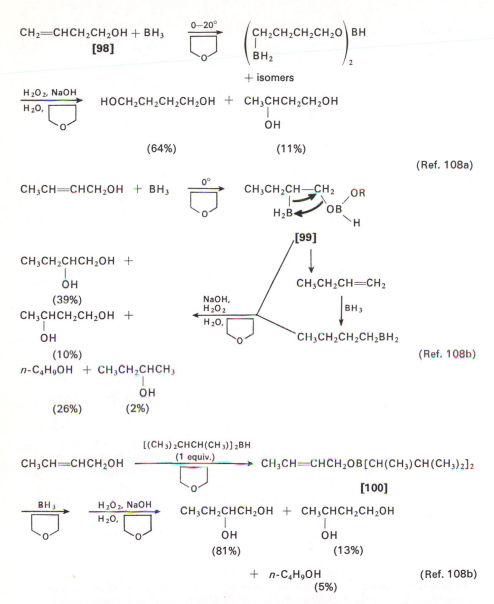

Although the reaction is complicated by the tendency of β-halo, β-hydroxy, or β-acetoxyalkylboranes to undergo elimination[95c,108,109] (see structure [99]), this tendency can be lessened by performing the hydroboration reaction at lower reaction temperatures in ether solution[109a] or by conversion of the hydroxy function

109. (a) D. J. Pasto and J. Hickman, *J. Am. Chem. Soc.*, **90**, 4445 (1968) and references therein. (b) H. C. Brown and E. F. Knights, *ibid.*, **90**, 4439 (1968).

to a poorer leaving group such as a dialkylborinate (e.g., **[100]**) or an ether.[108b] The combination of the directive influence of allylic substituents and the special ease of elimination of *cis*-β-alkoxyalkylboranes has resulted in a very useful synthesis of *trans*-1,2-diols or their derivatives.[109,110a,b] The following equations illustrate this procedure. Application of this procedure to appropriate cyclohexene

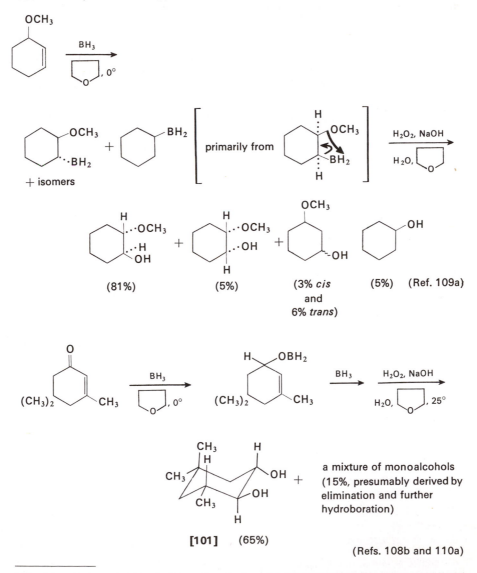

+ isomers

primarily from

(81%)   (5%)   (3% *cis* and 6% *trans*)   (5%)   (Ref. 109a)

[101]   (65%)

a mixture of monoalcohols (15%, presumably derived by elimination and further hydroboration)

(Refs. 108b and 110a)

110. (a) J. Klein and E. Dunkelblum, *Tetrahedron,* **24**, 5701 (1968). (b) J. P. Turnbull and J. H. Fried [*Tetrahedron Letters,* **No. 8**, 801 (1966)] have described the formation of a *cis*-1,3-diol from a rigid homoallyl ether. (c) R. E. Lyle and C. K. Spicer, *ibid.,* **No. 14**, 1133 (1970); R. E. Lyle, K. R. Carle, C. R. Ellefson, and C. K. Spicer, *J. Org. Chem.,* **35**, 802 (1970).

derivatives offers one of the few useful synthetic routes to *trans*-diequatorial-1,2-diols (e.g., **[101]**,[110a] derivatives which are not easily obtained in other synthesis of 1,2-diols (see Chapters 6 and 8). By modification of the reaction procedure to favor the elimination reaction (see structure **[99]**), the hydroboration-elimination sequence can be used as an olefin synthesis. This elimination reaction can occur as an intramolecular *cis*-elimination (see structure **[99]**) or by acid- or base-catalyzed processes in which *trans*-elimination is preferred.[111a]

Allylic amines such as the following example also exhibit some structural specificity in hydroboration reactions.[110c] In these reactions two equivalents of borane are required, since the first equivalent is consumed in forming a borane complex with the amino function.

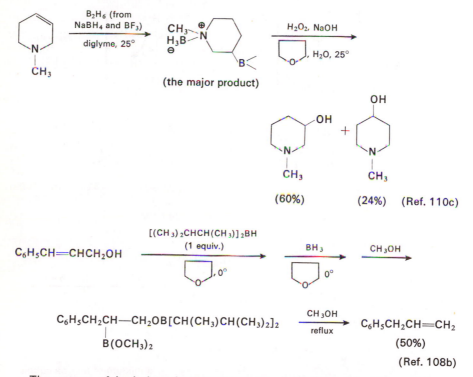

(the major product)

(60%)   (24%)   (Ref. 110c)

The course of hydroboration reactions may be influenced strikingly by the presence of substituents such as halogen, acyloxy, alkoxy, or amino groups directly attached to the carbon–carbon double bond.[108c] The hydroboration is retarded by

111. (a) D. J. Pasto and R. Snyder, *J. Org. Chem.*, **31**, 2773, 2777 (1966). (b) D. J. Pasto and J. Hickman, *J. Am. Chem. Soc.*, **89**, 5608 (1967). (c) A. Hassner, R. E. Barnett, P. Catsoulacos, and S. H. Wilen, *ibid.*, **91**, 2632 (1969). (d) D. B. Bigley and D. W. Payling, *J. Chem. Soc.*, 3974 (1965). (e) A. Suzuki, K. Ohmori, H. Takenaka, and M. Itoh, *Tetrahedron Letters*, **No. 47**, 4937 (1968). (f) A. Suzuki, K. Ohmori, and M. Itoh, *Tetrahedron*, **25**, 3707 (1969). (g) D. J. Pasto and C. C. Cumbo, *J. Am. Chem. Soc.*, **86**, 4343 (1964). (g) G. Zweifel and J. Plamondon, *J. Org. Chem.*, **35**, 898 (1970).

acetoxy and, especially, chloro substituents. Although the tendency of these two substituents to control the initial addition of borane to the carbon-carbon double bond is subject to some disagreement,[108c,111] it would appear that with symmetrically

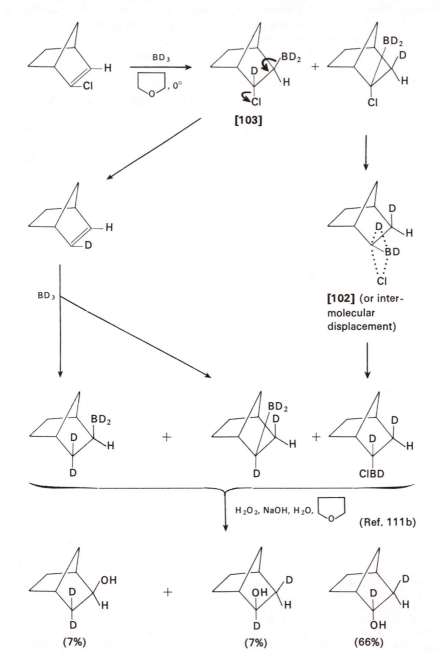

substituted olefins, mixtures of alkylboranes are usually obtained in which the boron has added both *alpha* and *beta* to the chloro or acetoxy substituent. In the case of chloro olefins, both initial hydroboration products can undergo further reaction. The α-chloroalkylborane can undergo either an intermolecular[108c] or an intramolecular[111a,b] displacement (see structure [102]) with inversion of configuration at carbon, to yield an alkylborane, and the β-chloroalkylborane can undergo an elimination reaction (see structure [103]) to form an olefin which reacts with more borane.

A similar initial reaction appears to occur with enol acetates. In some cases elimination (structure [104]) of the β-acetoxyalkylborane appears to occur; when the α-acetoxy derivative is allowed to react with excess diborane, the indicated [105] displacement (which may be either intermolecular or intramolecular) of the α-acetoxy function is observed.[111c,e,f]

Alkoxy and amino substituents exert a more powerful directive influence leading to predominant if not exclusive addition of boron to the carbon atom *beta* to the substituent even in cases (e.g., [106]) where this carbon is more hindered sterically.[111g,h,112] Whether this structural specificity should be attributed to an

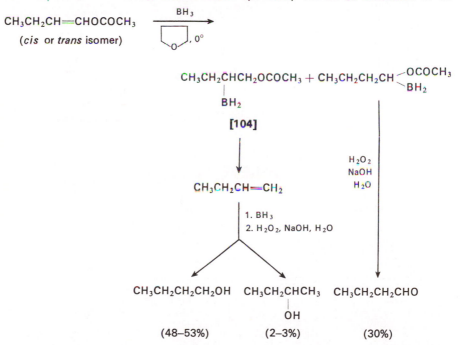

112. (a) I. J. Borowitz and G. J. Williams, *J. Org. Chem.*, **32**, 4157 (1967). (b) J. M. Coulter, J. W. Lewis, and P. P. Lynch, *Tetrahedron*, **24**, 4489 (1968). (c) J. W. Lewis and A. A. Pearce, *J. Chem. Soc.*, B, 863 (1969). (d) It is not clear whether the reduction of enamines to amines with diborane and acetic acid involves an alkylborane intermediate or is a reduction of an iminium salt; see J. A. Marshall and W. S. Johnson, *J. Org. Chem.*, **28**, 421 (1963). (e) J. Gore and J. J. Barieux, *Tetrahedron Letters*, **No. 32** 2849 (1970).

initial coordination of borane with the nitrogen or oxygen atom followed by an intramolecular addition of the boron atom to carbon is uncertain. The amount of elimination which occurs with the intermediate β-substituted alkylborane is determined in part by the reaction temperature, higher temperatures favoring elimination. It should be noted that although the reaction of enamines with diborane

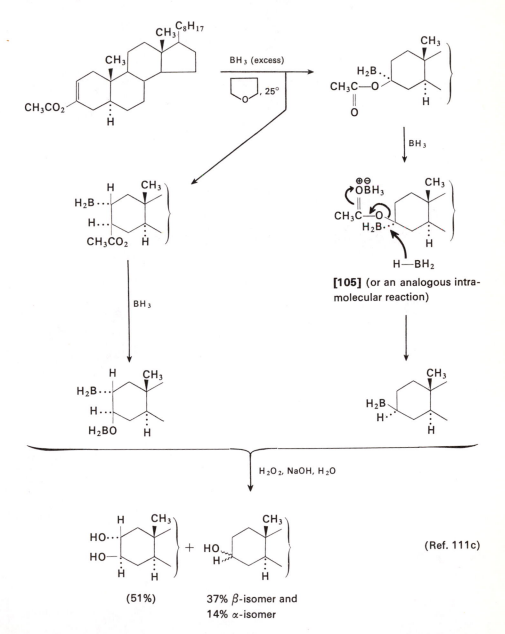

[105] (or an analogous intra-molecular reaction)

(Ref. 111c)

(51%)   37% β-isomer and
14% α-isomer

or aluminum hydride may be used to obtain β-amino alcohols or amines,[112b,d] the thermal or acid-catalyzed (e.g., [107]) elimination of the intermediate β-amino-alkylboranes or alanes offers a useful way to form olefins from ketones. Protonolysis of the intermediate β-amino alkylboranes to form saturated amines has been accomplished in boiling methanol or ethanol.[112e]

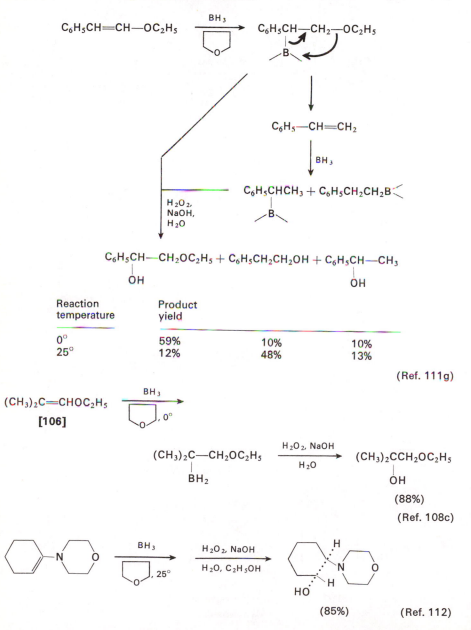

Reaction temperature	Product yield		
0°	59%	10%	10%
25°	12%	48%	13%

(Ref. 111g)

(Ref. 108c)

(Ref. 112)

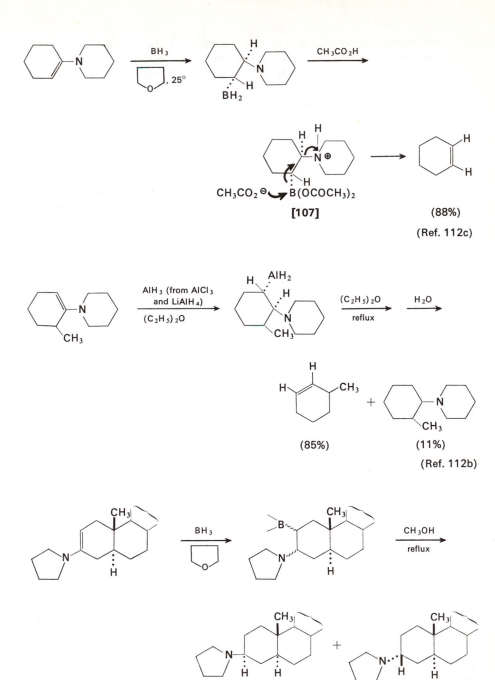

[107]   (88%)

(Ref. 112c)

(85%)   (11%)

(Ref. 112b)

(18% of the product)   (82% of the product)

(Ref. 112e)

The hydroboration of nonterminal acetylenes may be stopped at the mono-addition stage to form vinylborane derivatives (e.g., [108]).[113] Subsequent pro-tonolysis produces *cis*-olefins of purity as high as that obtained by the previously discussed catalytic hydrogenation of acetylenes. As would be expected, oxidation of the vinylborane derivatives produces carbonyl compounds. The hydroboration of terminal acetylenes with diborane has proved difficult to stop at the mono-addition stage. However, use of a more selective reagent, such as bis(3-methyl-2-butyl)borane [86], permits the reaction to be stopped after monoaddition. Sub-sequent hydrolysis yields a terminal olefin, and subsequent peroxide oxidation produces an aldehyde. The rapid cleavage of the geminal diboron derivative by reaction with bases such as hydroxide anion (structure [109]), alkoxide anion, or lithium reagents is worthy of note. The ease of this cleavage is attributed to the ability of boron to stabilize an adjacent negative charge as illustrated in structure [110].[113c] A related cleavage of the vinylborane intermediates obtained from pro-pargyl chlorides provides a synthesis for allenes.[113d]

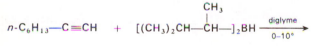

$$C_2H_5-C\equiv C-C_2H_5 \xrightarrow[\substack{diglyme \\ 0°}]{B_2H_6} \left(\begin{matrix} C_2H_5-\underset{\|}{C}-H \\ C_2H_5-C- \end{matrix} \right)_3 B \xrightarrow[25°]{CH_3CO_2H} \begin{matrix} C_2H_5-\underset{\|}{C}-H \\ C_2H_5-C-H \end{matrix}$$

[108]   (68%)   (Ref. 113a)

$(pH\ 8) \downarrow \substack{H_2O_2 \\ H_2O}$

$$C_2H_5-CH_2-CO-C_2H_5$$

(62%)

$$n\text{-}C_6H_{13}-C\equiv CH \quad + \quad [(CH_3)_2CH-\underset{\overset{|}{CH_3}}{CH}-]_2BH \xrightarrow[0-10°]{diglyme}$$

[86]

$$[(CH_3)_2CH-\underset{\overset{|}{CH_3}}{CH}-]_2B-CH=CH-C_6H_{13}\text{-}n$$

$\downarrow \substack{H_2O_2,\ NaOH \\ H_2O}$

$$n\text{-}C_6H_{13}-CH_2-CHO$$

(70%)   (Ref. 113a)

113. (a) H. C. Brown and G. Zweifel, *J. Am. Chem. Soc.*, **81**, 1512 (1959); *ibid.*, **83**, 3834 (1961); *ibid.*, **85**, 2066 (1963). (b) D. J. Pasto, *ibid.*, **86**, 3039 (1964). (c) G. Zweifel and H. Arzoumanian, *ibid.*, **89**, 291 (1967). (d) G. Zweifel, A. Horng, and J. T. Snow, *ibid.*, **92**, 1428 (1970). (e) For studies of the hydroboration of diacetylenes, see G. Zweifel and N. L. Polston, *ibid.*, **92**, 4068 (1970).

Acetylenes undergo a similar addition of dialkylaluminum hydrides.[114] This reaction, called hydroalumination, resembles hydroboration and appears to involve the *cis* addition of an Al—H bond to the carbon-carbon triple bond by way of a four-centered transition state; as with hydroboration reactions, a π-complex (e.g., [111]) precedes this four-center transition state.[114c] Although only the indicated 1:1 adducts of *cis*-addition are formed from a kinetically controlled reaction where the reactants are present in equimolar amounts, the vinylaluminum compounds

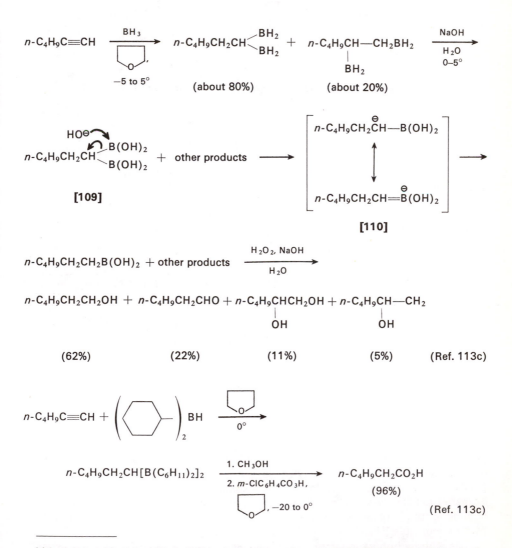

114. (a) J. J. Eisch and W. C. Kaska, *J. Am. Chem. Soc.*, **88**, 2213 (1966). (b) K. Ziegler, in H. Zeiss, ed., *Organometallic Chemistry*, Reinhold, New York, 1960, pp. 194–269. (c) J. J. Eisch, R. Amtmann, and M. W. Foxton, *J. Organometal. Chem.*, **16**, P55 (1969).

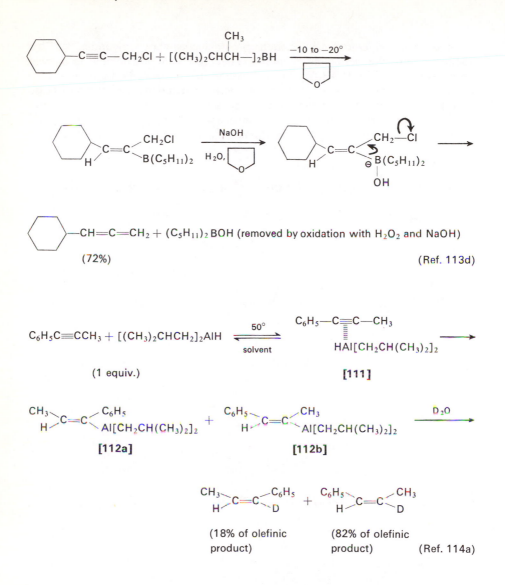

[112] are capable of adding to additional molecules of the acetylene to form dimeric adducts when excess acetylene is present. The initially formed *cis*-adducts [112] can also be isomerized to the *trans*-isomers upon prolonged reaction; this isomerization involves the addition of a second molecule of the dialkylaluminum hydride to form a dialuminoalkane followed by elimination of one dialuminoaluminum hydride molecule.[114]

    As with hydroboration, terminal acetylenes add either one or two molecules of dialkylaluminum hydrides; the additions occur predominantly to form 1-alumino-alkenes (e.g., [113]) or 1,1-dialuminoalkanes (e.g., [115]). The accompanying

equations illustrate the formation of typical organoaluminum derivatives [113] and [115] and their further stereospecific transformations to *trans* olefin derivatives. The 1,1-dialuminoalkanes (e.g., [115]) are cleaved with nucleophilic bases such as methyllithium or sodium methoxide in a reaction analogous to that described

$$n\text{-}C_4H_9C\!\equiv\!CH + [(CH_3)_2CHCH_2]_2AlH \xrightarrow[\text{heptane}]{40\text{–}50^\circ}$$

(1 equiv.)

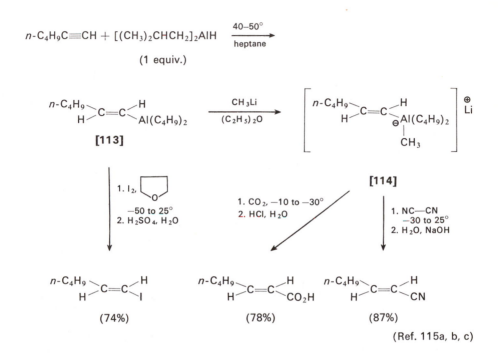

[113]

[114]

n-C₄H₉ ... (74%)   (78%)   (87%)

(Ref. 115a, b, c)

previously for 1,1-diboralkanes. Although the vinylaluminum reagents (e.g., [113]) normally do not react with carbon dioxide or cyanogen, formation of an "ate" complex [114] from the vinylalane and methyllithium increases the reactivity of vinylaluminum derivatives and offers a useful synthetic route to α,β-unsaturated acids and nitriles. Interestingly, if the dialkylaluminum hydride is converted to its "ate" complex [116], before reaction with the acetylene, then addition to the triple bond occurs in a *trans*-manner rather than the usual *cis*-addition observed with dialkylaluminum hydrides. The reaction path followed in the *trans* addition remains to be established.

The stereospecific reaction of a vinylalane [113] with iodine to form a *trans*-alkenyl iodide has prompted investigation of the corresponding reaction of

115. (a) G. Zweifel and C. C. Whitney, *J. Am. Chem. Soc.,* **89,** 2753 (1967). (b) G. Zweifel and R. B. Steele, *ibid.,* **89,** 2754, 5085 (1967). (c) G. Zweifel, J. T. Snow, and C. C. Whitney, *ibid.,* **90,** 7139 (1968). (d) G. Zweifel and R. B. Steele, *Tetrahedron Letters,* **No. 48,** 6021 (1966). (e) G. Cainelli, F. Bertini, P. Grasselli, and G. Zubiani, *ibid.,* **No. 17,** 1581 (1967).

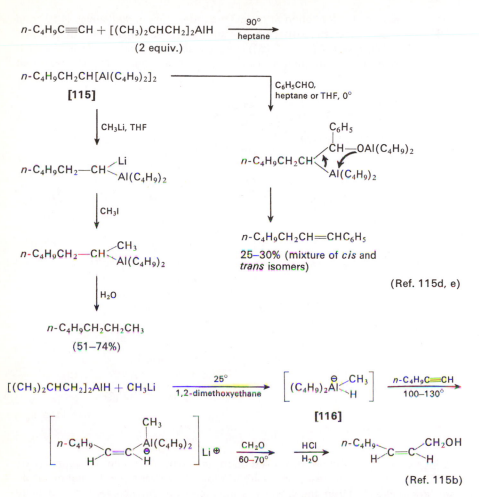

(Ref. 115d, e)

(Ref. 115b)

vinylboranes with halogens.[116] Reaction of 1-alkenylboranes with iodine leads to a rearrangement (see structure [117]), which appears comparable to the previously discussed reaction of α-haloalkylboranes. The divinylalkylboranes formed from 2,3-dimethyl-2-butylborane [87] and disubstituted acetylenes are coupled by treatment with iodine and base.[116d] The same conditions convert trialkylboranes to two molecules of alkyl halide and one molecule of the alkylboronic acid;[116e,f] reaction with anhydrous bromine forms an alkyl bromide and a bromodialkyl-

116. (a) G. Zweifel, H. Arzoumanian, and C. C. Whitney, *J. Am. Chem. Soc.,* **89,** 3652 (1967). (b) G. Zweifel and H. Arzoumanian, *ibid.,* **89,** 5086 (1967). (c) H. C. Brown, D. H. Bowman, S. Misumi, and M. K. Unni, *ibid.,* **89,** 4531 (1967). (d) G. Zweifel, N. L. Polston, and C. C. Whitney, *ibid.,* **90,** 6243 (1968). (e) H. C. Brown, M. W. Rathke, and M. M. Rogic, *ibid.,* **90,** 5038 (1968). (f) C. F. Lane and H. C. Brown, *ibid.,* **92,** 6660, 7212 (1970); **93,** 1025 (1971).

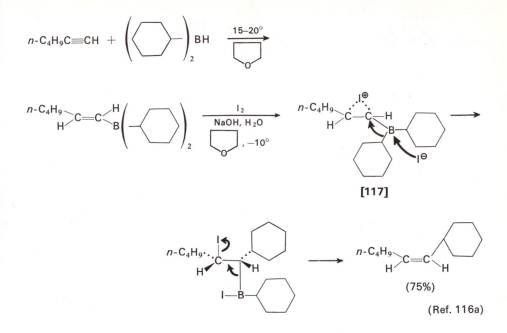

[117]

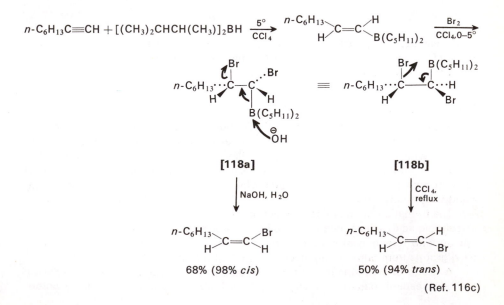

(75%)

(Ref. 116a)

borane.[116f] Reaction of vinylboranes with bromine affords an intermediate such as [118], whose decomposition to an olefin may be effected with either a *cis* or a *trans* elimination by a choice of reaction conditions. The reaction of a dialkylborane with haloacetylenes provides a route to *cis*-alkenyl halides; the intermediate

[118a]

[118b]

68% (98% *cis*)

50% (94% *trans*)

(Ref. 116c)

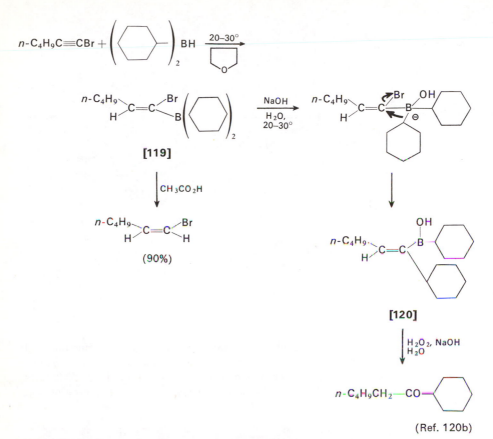

[119]

[120]

(Ref. 120b)

α-halovinylborane **[119]** undergoes rearrangement only if treated with a nucleophile. The rearranged vinylborane **[120]** may either be oxidized to yield a ketone or protonolyzed with acetic acid to form an olefin.[120b]

REACTIONS OF ALKYLBORANES

The previously discussed reactions of alkylboranes have included isomerization and elimination reactions, protonolysis to alkanes, oxidation to alcohols, and reaction with halogens. In addition, the alkylboranes can serve as intermediates for other synthetic processes, the most useful of which appear to be amination to form amines and various reactions which serve to replace the carbon-boron bond with a new carbon-carbon bond.

The formation of an amine by reaction of a trialkylborane with hydroxylamine-O-sulfonic acid is illustrated in the following equation. As indicated, this reaction occurs with retention of configuration and presumably involves the migration of alkyl groups from boron to an electron-deficient nitrogen atom.

A group of reactions leading to new carbon-carbon bond formation follows a similar reaction pattern involving attack by a nucleophilic reagent at the boron

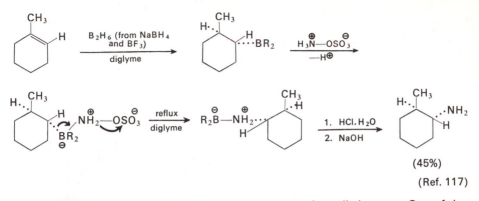

(45%)

(Ref. 117)

atom of a trialkylborane followed by rearrangement of an alkyl group. One of the most versatile members of this group is the reaction of carbon monoxide with trialkylboranes.[118] With unhindered trialkylboranes the reaction proceeds at a convenient rate at 125–150° with carbon monoxide at atmospheric pressure to form a trialkylcarbinol. As the accompanying example illustrates, the transformation involves a series of three successive intramolecular rearrangements of alkyl groups. The reaction is effected in the presence of ethylene glycol to trap the final boronic

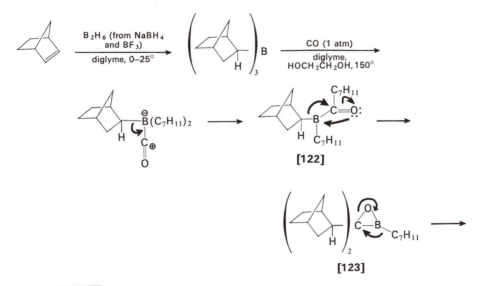

[122]

[123]

117. (a) H. C. Brown, W. R. Heydkamp, E. Breuer, and W. S. Murphy, *J. Am. Chem. Soc.,* **86**, 3565 (1964). (b) M. W. Rathke, N. Inone, K. R. Varma, and H. C. Brown, *ibid.,* **88**, 2870 (1966). (c) L. Verbit and P. J. Heffron, *J. Org. Chem.,* **32**, 3199 (1967). (d) For an example of an abnormal rearrangement during an attempt to prepare a benzylamine derivative by this procedure, see L. A. Levy and L. Fishbein, *Tetrahedron Letters,* **No. 43**, 3773 (1969).
118. (a) H. C. Brown, *Accts. Chem. Res.,* **2**, 65 (1969). (b) H. C. Brown and M. W. Rathke, *J. Am. Chem. Soc.,* **89**, 2737, 2738, 4528 (1967). (c) H. C. Brown, G. W. Kabalka, and M. W. Rathke, *ibid.,* **89**, 4530 (1967). (d) M. W. Rathke and H. C. Brown, *ibid.,* **89**, 2740 (1967); *ibid.,* **88**, 2606 (1966). (e) H. C. Brown and E. Negishi, *ibid.,* **89**, 5285 (1967).

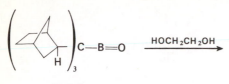

[121]

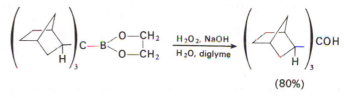

(80%)

(Ref. 118b)

acid derivative **[121]** as a boronic ester rather than allowing it to trimerize or polymerize. This reaction can also be applied to trialkylboranes containing three different alkyl groups without obtaining the mixture of different trialkylcarbinols which might be expected from the previously discussed reversal of the hydroboration reaction at 125–150°. Evidently the initial reaction with carbon monoxide to form **[122]** is much faster than reversal of the hydroboration. The subsequent rearrangements (i.e., **[122]**→**[123]**→**[121]**) must proceed at successively slower rates, since each of the intermediates **[122]** and **[123]** can be intercepted if appropriate reagents are added to the reaction mixture. The boraepoxide intermediate **[123]** can be intercepted if water is present in the mixture during the addition of carbon monoxide. This modification in reaction procedure is believed to form a diol such as **[124]**, which yields a ketone upon oxidation. The use of this

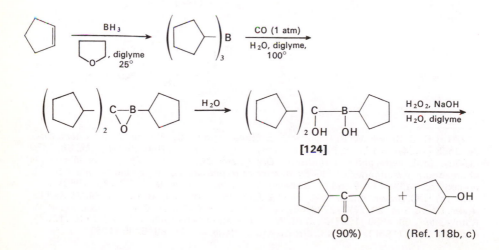

[124]

(90%)   (Ref. 118b, c)

synthetic route for the preparation of unsymmetrical ketones can obviously be complicated by formation of two different ketones (e.g., [125]), with the proportions being determined by the relative ease of migration of different alkyl groups from boron to carbon. Since the ease of these alkyl group migrations appears to be in the following order: primary > secondary > tertiary, good yields of unsymmetrical

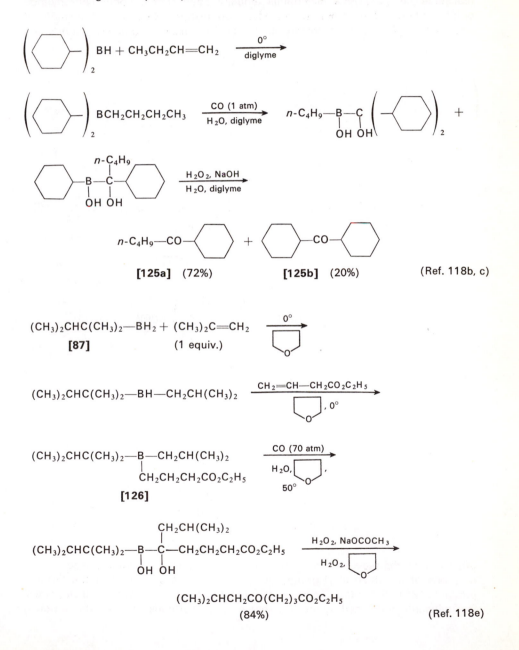

[125a] (72%)   [125b] (20%)   (Ref. 118b, c)

$(CH_3)_2CHC(CH_3)_2\!-\!BH_2 + (CH_3)_2C\!\!=\!\!CH_2$
   [87]   (1 equiv.)

$(CH_3)_2CHC(CH_3)_2\!-\!BH\!-\!CH_2CH(CH_3)_2$

$(CH_3)_2CHC(CH_3)_2\!-\!B\!-\!CH_2CH(CH_3)_2$
         |
         $CH_2CH_2CH_2CO_2C_2H_5$
      [126]

$(CH_3)_2CHCH_2CO(CH_2)_3CO_2C_2H_5$
         (84%)   (Ref. 118e)

ketones can be obtained if a tertiary alkylborane such as [87] is used as the hydroborating agent. In such cases the intermediate trialkylborane [126] is sufficiently
hindered that it is necessary to use carbon monoxide under pressure (70 atm or
1000 psi) to obtain reaction in a reasonable period of time.

The previously described rearrangement sequence can also be stopped at the
first stage (i.e., [122]) if a nucleophilic reducing agent such as sodium borohydride
or lithium borohydride is present in the reaction mixture. Hydrolysis of the reduced
intermediate [127] yields a primary alcohol [128]. This procedure is obviously less

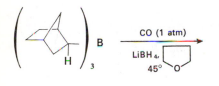

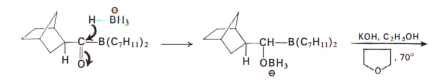

[122]   [127]

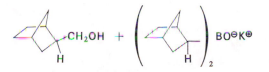

[128]   (85%)   (Ref. 118d)

efficient than the previously described ketone and trialkylcarbinol syntheses, since
only one of the three alkyl groups bound to boron is converted to the desired
product [128]. Since the rate of reaction of carbon monoxide with trialkylboranes
is significantly enhanced by the presence of metal hydrides such as lithium boro-

hydride[118d] and lithium trialkoxyaluminum hydride,[119] it is not yet clear in these cases whether the carbon monoxide is reacting with the trialkylborane or with a trialkylborohydride anion.

If the intermediate product [127] formed in these combined reduction-carbonylation processes is oxidized in a buffered medium instead of being hydrolyzed by base, aldehydes are formed.[119] With the further modification of using the bicyclic dialkylborane [129] as the hydroborating agent, the aforementioned problem of "wasting" two of the three alkyl groups bound to boron in the trialkylborane is also avoided.[119b]

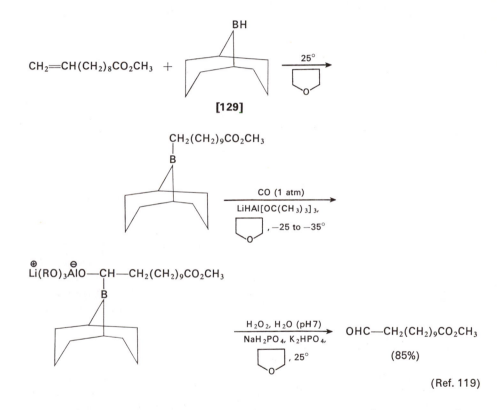

$$CH_2=CH(CH_2)_8CO_2CH_3 +$$ [129]

$$OHC-CH_2(CH_2)_9CO_2CH_3$$

(85%)

(Ref. 119)

The carbonylation of trialkylboranes in the presence of water to produce ketones has also been applied to cyclic boranes to form cyclic ketones;[107c] in the absence of water trialkylcarbinols may be formed. The following examples are illustrative.

119. (a) H. C. Brown, R. A. Coleman, and M. W. Rathke, *J. Am. Chem. Soc.,* **90,** 499 (1968). (b) H. C. Brown and R. A. Coleman, *ibid.,* **91,** 4606 (1969).
120. (a) H. C. Brown and E. Negishi, *J. Am. Chem. Soc.,* **89,** 5477, 5478 (1967); **91,** 1224 (1969). (b) E. F. Knights and H. C. Brown, *ibid.,* **90,** 5283 (1968). (c) H. C. Brown and E. Negishi, *Chem. Commun.,* **No. 11,** 594 (1968).

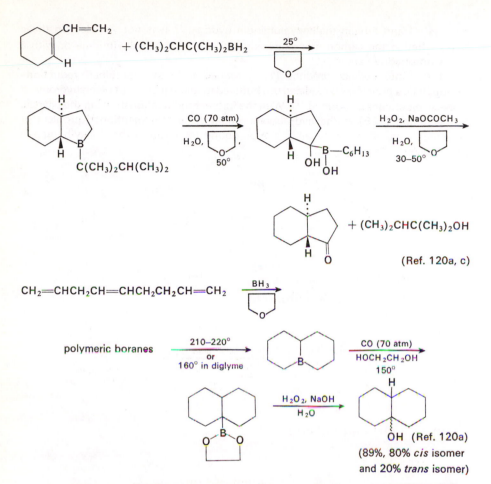

(Ref. 120a, c)

(89%, 80% *cis* isomer
and 20% *trans* isomer)

Another process leading to carbon-carbon bond formation which appears to follow a similar pathway is the reaction of the enolate anions of α-halocarbonyl compounds or α-halonitriles with trialkylboranes.[121] Although this reaction can be performed in a tetrahydrofuran-*t*-buryl alcohol mixture with α-halo esters and α,α-dihalo esters, the more reactive α-halo ketones and α-halo nitriles are used in tetrahydrofuran solution. This preparative sequence also suffers from the disadvantage that only one of the three alkyl groups bound to boron is used to form the product. The second of the two preparations illustrated in the accompanying equations indicates again the use of the bicyclic dialkylborane [129] as a precursor

121. (a) H. C. Brown, M. M. Rogic, M. W. Rathke, and G. W. Kabalka, *J. Am. Chem. Soc.,* **90**, 818, 1911 (1968) ; **91**, 2150 (1969). (b) H. C. Brown, M. M. Rogic, and M. W. Rathke, *ibid.,* **90**, 6218 (1968). (c) H. C. Brown and M. M. Rogic, *ibid.,* **91**, 2146, 4304 (1969). (d) H. C. Brown, M. M. Rogic, H. Nambu, and M. W. Rathke, *ibid.,* **91**, 2147 (1969). (e) H. C. Brown, H. Nambu, and M. M. Rogic, *ibid.,* **91**, 6852, 6854, 6855 (1969). (f) H. C. Brown and H. Nambu, *ibid.,* **92**, 1761, 5790 (1970).

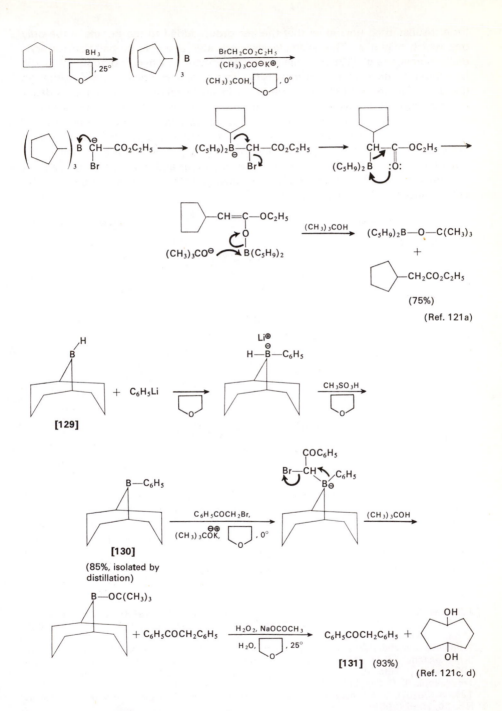

(Ref. 121a)

(Ref. 121c, d)

for a trisubstituted borane so that the last group added to the borane is the only one which migrates. This example also indicates a simple preparation for aryl-dialkylboranes (e.g., [130]) which cannot be obtained by hydroboration.[121c] It will be noted that the reaction mixture was oxidized prior to product isolation; although the desired product [131] was produced in the reaction mixture, this oxidative isolation procedure has been found advantageous for destroying various borane derivatives present which may be difficult to separate from the desired product.[121a]

Recently, a further modification in the reaction procedure has been recommended[121e,f] in which the hindered phenoxide [132] is used as a base rather than potassium t-butoxide. By use of this base, the order and time of mixing reagents becomes a less critical factor, and very reactive α-halo ketones and nitriles can be utilized successfully in the reaction.

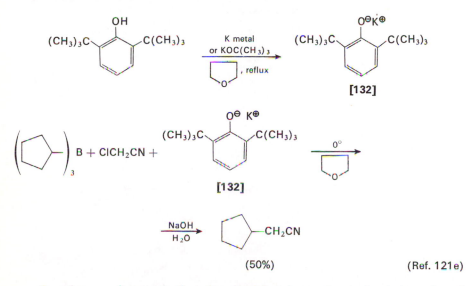

(50%)   (Ref. 121e)

Reactions analogous to that described for the enolates of α-halo carbonyl compounds are observed when trialkylboranes are treated with α-diazo carbonyl compounds (e.g., [133]),[122] ylids (e.g., [134]),[123] and α-halo organometallic compounds.[124] In all of these cases, it appears that the isolable intermediate present prior to hydrolysis is an enol ether which has been formed by the indicated rearrangement of the first formed α-keto alkylborane.[122c]

122. (a) J. Hooz and S. Linke, *J. Am. Chem. Soc.*, **90**, 5936 (1968). (b) J. Hooz and D. M. Gunn, *ibid.*, **91**, 6195 (1969); *Tetrahedron Letters*, **No. 40**, 3455 (1969). Reaction of trialkyl derivatives formed from the bicyclic borane [129] with diazo carbonyl compounds yields cyclooctyl derivatives from migration of one of the alkyl groups which is part of the cyclic system. (c) D. J. Pasto and P. W. Wojtkowski, *ibid.*, **No. 3**, 215 (1970).
123. (a) J. J. Tufariello and L. T. C. Lee, *J. Am. Chem. Soc.*, **88**, 4757 (1966). (b) J. J. Tufariello, L. T. C. Lee, and P. Wojtkowski, *ibid.*, **89**, 6804 (1967).
124. A. Suzuki, S. Nozawa, N. Miyaura, M. Itoh, and H. C. Brown, *Tetrahedron Letters*, **No. 34**, 2955 (1969).

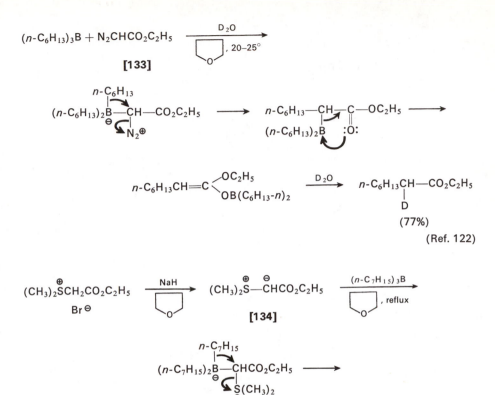

$$(n\text{-}C_6H_{13})_3B + N_2CHCO_2C_2H_5 \xrightarrow[\text{, 20–25°}]{D_2O}$$

[133]

In additon to these alkyl group rearrangements observed with α-haloalkylborane derivatives and related compounds and the previously discussed elimination reactions of β-haloalkylborane derivatives, reactions involving displacement of a leaving group are also observed with γ-haloalkylboranes and related substances. Such displacements, promoted by attack of a nucleophilic reagent at the boron atom of the trialkylborane (see structure **[135]**), provide a synthesis for cyclo-propanes[125] and cyclobutane derivatives.[125b]

In more rigid cyclic systems (e.g., **[136]**) fragmentation rather than cyclo-butane formation appears to be the preferred reaction path.[126] The conversion of

125. (a) M. F. Hawthorne, *J. Am. Chem. Soc.*, **82**, 1886 (1960). (b) H. C. Brown and S. P. Rhodes, *ibid.*, **91**, 2149, 4306 (1967).
126. (a) J. A. Marshall and G. L. Bundy, *J. Am. Chem. Soc.*, **88**, 4291 (1966); *Chem. Commun.*, **No. 17**, 854 (1967). (b) J. A. Marshall and J. H. Babler, *ibid.*, **No. 16**, 993 (1968). (c) J. A. Marshall, *Rec. Chem. Progr.*, **30**, 3 (1969).

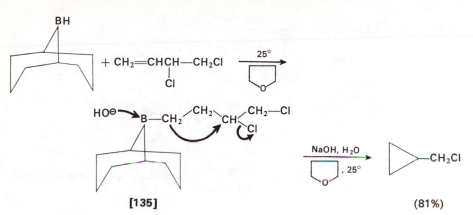

[135]

(81%)

(Ref. 125b)

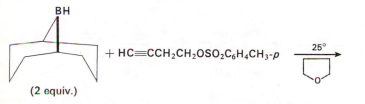

(2 equiv.)

$$(C_8H_{14}B)_2HCCH_2CH_2CH_2OSO_2C_7H_7 \xrightarrow[\text{, 25°}]{CH_3Li}$$

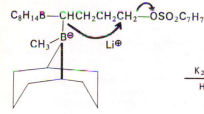

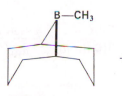

(60%, isolated by distillation)

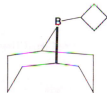

(65%, isolated by distillation)

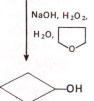

(65%, formed by oxidation without isolation of the cyclobutylborane)

(Ref. 125b)

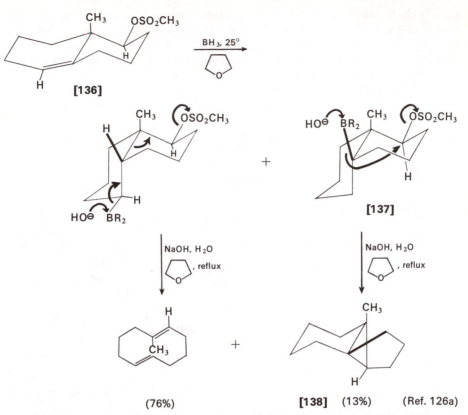

(76%)   **[138]**   (13%)   (Ref. 126a)

the minor hydroboration product **[137]** to form the cyclopropane **[138]** is of interest because the stereochemical course of the change **[137]** → **[138]** requires inversion of the configuration at both carbon atoms. This result is analogous to the *trans* elimination observed in the previously discussed elimination of α-substituted alkylboranes in the presence of nucleophiles.

A different type of reaction which has been used to form new carbon-carbon bonds from trialkylboranes arises from the ability of alkoxy free radicals (e.g., **[139]**) to react with trialkylboranes to liberate free alkyl radicals. This process is

$$(CH_3)_3C—O—O—C(CH_3)_3 + (n\text{-}C_4H_9)_3B \xrightarrow[\text{isooctane}]{\text{light, 25}^\circ}$$

$$(CH_3)_3C—O\cdot \longrightarrow B(C_4H_9\text{-}n)_3 \longrightarrow (CH_3)_3C—\overset{\oplus}{\underset{}{O}}—\overset{\ominus}{\underset{}{B}} (C_4H_9\text{-}n)_3 \longrightarrow$$

[139]

$$(CH_3)_3C—O—B(C_4H_9\text{-}n)_2 + n\text{-}C_4H_9\cdot$$

(identified by its
e.p.r. spectrum)   (Ref. 127)

---

127. (a) A. G. Davies and B. P. Roberts, *Chem. Commun.,* **No. 13,** 699 (1969). (b) P. J. Krusic and J. K. Kochi, *J. Am. Chem. Soc.,* **91,** 3942 (1969).

sufficiently favorable to support a radical chain reaction in which one of the alkyl groups of a trialkylborane can be added to an α,β-unsaturated carbonyl compound or an analogous acetylenic ketone.[128] This reaction was originally described as an addition of trialkylboranes to α,β-unsaturated carbonyl compounds containing no β-substitutents and with no intentionally added catalyst. Presumably these reactions owe their success to contact of the reaction mixture with oxygen, since subsequent studies have shown that the reaction is inhibited by galvinoxyl [140], a good

$$(C_2H_5)_3B + CH_2\!\!=\!\!CH\!\!-\!\!CHO \xrightarrow[\substack{H_2O,}]{25^\circ} (C_2H_5)_3\dot{B}\!\!-\!\!OR \longrightarrow$$

RO· (from reaction with
oxygen of the air)

$$C_2H_5\!\cdot\; + \;CH_2\!\!=\!\!CH\!\!-\!\!CHO \longrightarrow C_2H_5\!\!-\!\!CH_2\!\!-\!\!CH\!\!=\!\!CH\!\!-\!\!O\cdot \xrightarrow{(C_2H_5)_3B}$$

(Ref. 128d)

$$C_2H_5\!\!-\!\!CH_2\!\!-\!\!CH\!\!=\!\!CH\!\!-\!\!O\!\!-\!\!B(C_2H_5)_2 \xrightarrow{H_2O} C_2H_5CH_2CH_2CHO$$

$+\; C_2H_5\!\cdot$ (used to continue
the chain reaction)

(94%, gas chromato-
graphic analysis)

reaction inhibited by adding

[140] (galvinoxyl, a stable
free radical)

radical scavenger, and the reaction is strongly catalyzed by acyl peroxides, light, and oxygen.[128c,d,f] With these added catalysts, the addition reaction is successful with α,β-unsaturated carbonyl compounds that have β-substituents. Of the various reaction procedures described, catalysis of the reaction by passing a *slow* stream of air through the reaction solution appears to be the most convenient; with this procedure some of the trialkylborane is lost in an oxidative side reaction so that an excess of the borane derivative is required.[128d] The α,β-unsaturared carbonyl component can either be added to the reaction mixture or it can be generated in the reaction mixture from the quaternary salt derivative of a Mannich base (see Chapter 10). Presumably the related alkylation of *p*-benzoquinones[129] with

128. (a) A. Suzuki, A. Arase, H. Matsumoto, M. Itoh, H. C. Brown, M. M. Rogic, and M. W. Rathke, *J. Am. Chem. Soc.*, **89**, 5708 (1967). (b) H. C. Brown, M. M. Rogic, M. W. Rathke, and G. W. Kabalka, *ibid.*, **89**, 5709 (1967); **90**, 4165, 4166 (1968). (c) G. W. Kabalka, H. C. Brown, A. Suzuki, S. Honma, A. Arase, and M. Itoh, *ibid.*, **92**, 710 (1970). (d) H. C. Brown and G. W. Kabalka, *ibid.*, **92**, 712, 714 (1970). (e) A. Suzuki, S. Nozawa, M. Itoh, H. C. Brown, E. Negishi, and S. K. Gupta, *Chem. Commun.*, **No. 17**, 1009 (1967). (f) A. Suzuki, S. Nozawa, M. Itoh, H. C. Brown, G. W. Kabalka, and G. W. Holland, *J. Am. Chem. Soc.*, **92**, 3503 (1970).
129. M. F. Hawthorne and M. Reintjes, *J. Am. Chem. Soc.*, **87**, 4585 (1965).

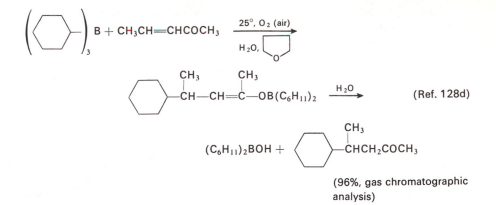

(Ref. 128d)

$$(C_6H_{11})_2BOH +$$

(96%, gas chromatographic analysis)

trialkylboranes follows a reaction path comparable to that described for α,β-unsaturated carbonyl compounds. Since all of these appear to involve a chain reaction mechanism in which a free alkyl radical is liberated from the trialkylborane, it can be anticipated that this reaction of alkylboranes will differ from reactions previously discussed in that the configuration at the α-carbon of the alkyl group will not be retained. It would further be expected that the ease of removal of alkyl radicals from the boron intermediate would be: tertiary > secondary > primary.

The intervention of a free radical mechanism has also been suggested to account for the coupling of two alkyl groups which is observed when trialkylboranes are treated with silver or gold salts in alkaline solution.[130] As indicated in the accompanying equation, this reaction is thought to proceed by the formation of an alkyl-silver intermediate [141] which appears to decompose, forming a free alkyl

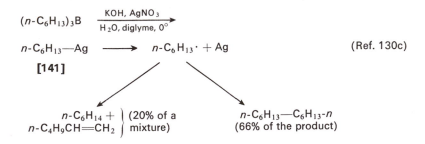

radical. Since the thermal decomposition of alkenylsilver complexes to dimers occurs stereospecifically (and presumably does not involve a free vinyl radical),[131] the application of this silver-catalyzed coupling reaction to alkenylboranes would appear profitable to explore.

130. (a) H. C. Brown, C. Verbrugge, and C. H. Snyder, *J. Am. Chem. Soc.,* **83**, 1001 (1961). (b) H. C. Brown, N. C. Hébert, and C. H. Snyder, *ibid.,* **83**, 1001 (1961). (c) H. C. Brown and C. H. Snyder, *ibid.,* **83**, 1002 (1961).
131. G. M. Whitesides and C. P. Casey, *J. Am. Chem. Soc.,* **88**, 4541 (1966).

Primary alkylboranes also react readily with either mercury(II) acetate or mercury(II) oxide to form alkylmercury derivatives.[132] The dialkylmercury compounds formed by this procedure are readily converted to alkyl bromides, as shown in the following example. The selective transfer of only the primary alkyl group from boron to mercury should also be noted.

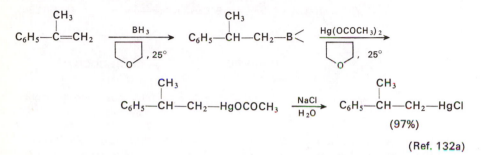

(Ref. 132a)

$$n\text{-}C_6H_{13}\ B \left(\underset{2}{\bigcirc} \right) + HgO \xrightarrow{OH^{\ominus}} (n\text{-}C_6H_{13}\text{---})_2 Hg \xrightarrow{Br_2} n\text{-}C_6H_{13}\text{---}Br$$

(75%)

(Ref. 132b)

132. (a) R. C. Larock and H. C. Brown, *J. Am. Chem. Soc.*, **92**, 2467 (1970). (b) J. J. Tufariello and M. M. Hovey, *Chem. Commun.*, **No. 6**, 372 (1970).

3

DISSOLVING METAL REDUCTIONS AND RELATED REACTIONS

A variety of organic molecules are reduced by reaction with a metal, either in the presence of a proton donor or followed by treatment with a proton donor. Although this method of reduction, which was one of the first used for organic compounds, has been replaced by catalytic hydrogenation and metal hydride reduction for some classes of compounds, there remains a substantial group of dissolving metal reductions that are currently used synthetically because of advantages offered in selectivity of reduction or in stereoselectivity. The metals commonly involved include the alkali metals—lithium, sodium, and potassium—as well as calcium, zinc, magnesium, tin, and iron. The alkali metals and calcium have been used as *solutions* in liquid ammonia, b.p. $-33°$ (the Birch reduction[1]); in low-molecular-weight aliphatic amines;[2] in hexamethylphosphoramide;[3] as very dilute solutions in ethers such as 1,2-dimethoxyethane;[2a,4a,b] or as solutions in ether or tetrahydrofuran of certain alkali metal (potassium and cesium) complexes with macrocyclic polyethers[4c] (sometimes called crown ethers[4d]). Reactions with the metal solutions in liquid ammonia often use a cosolvent, such as ether, tetrahydrofuran, or 1,2-dimethoxyethane, to increase the solubility of the organic substrate in the reaction mixture. These same metals as well as zinc and magnesium have been used as suspensions in invert solvents such as ether, toluene, or xylene. For both

1. (a) A. J. Birch, *Quart. Rev.*, **4**, 69 (1950); (b) A. J. Birch and H. Smith, *ibid.*, **12**, 17 (1958). (c) G. W. Watt, *Chem. Rev.*, **46**, 317 (1950). (d) C. Djerassi, ed., *Steroid Reactions*, Holden-Day, Inc., San Francisco, 1963, pp. 267–288, 299–325. (e) H. Smith, *Organic Reactions in Liquid Ammonia, Chemistry in Non-aqueous Ionizing Solvents*, Vol. 1, part 2, Wiley, New York, 1963. (f) M. Smith, in R. L. Augustine, ed., *Reduction*, Marcel Dekker, New York, 1968, pp. 95–170. W. Reusch, *ibid.*, pp. 186–194.
2. For recent reviews of the properties of solutions of alkali metals in ammonia and low-molecular-weight amines see (a) M. C. R. Symons, *Quart. Rev.*, **13**, 99 (1959). (b) U. Schindewolf, *Angew. Chem., Intern. Ed. Engl.*, **7**, 190 (1968). (c) J. L. Dye, *Accts. Chem. Res.*, **1**, 306 (1968).
3. (a) H. Normant, *Angew. Chem., Intern. Ed. Engl.*, **6**, 1046 (1967). (b) H. Normant, *Bull. Soc. Chim. France*, 791 (1968). (c) K. W. Bowers, R. W. Giese, J. Grimshaw, H. O. House, N. H. Kolodny, K. Kronberger, and D. K. Roe, *J. Am. Chem. Soc.*, **92**, 2783 (1970). (d) M. Larcheveque, *Ann. Chim.* (Paris), **5 (14)**, 129 (1970). (e) For the reduction of olefins, see G. M. Whitesides and W. J. Ehmann, *J. Org. Chem.*, **35**, 3565 (1970).
4. (a) J. L. Down, J. Lewis, B. Moore, and G. Wilkinson, *J. Chem. Soc.*, 3767 (1959). (b) For a review of the use of 1,2-dimethoxyethane as a solvent, see C. Agami, *Bull. Soc. Chim. France*, 1205 (1968). (c) J. L. Dye, M. G. DeBacker, and V. A. Nicely, *J. Am. Chem. Soc.*, **92**, 5226 (1970). (d) C. J. Pedersen, *ibid.*, **89**, 7017 (1967); **92**, 386, 391 (1970).

procedures a proton source (frequently ethanol, isopropyl alcohol, *t*-butyl alcohol, or water) is present in the reaction medium, is added concurrently with the compound to be reduced, or is added during the isolation. Finally, sodium amalgam, aluminum amalgam, zinc, zinc amalgam, tin and iron have been added directly to solutions of the compound being reduced in hydroxylic solvents such as ethanol, isopropyl alcohol, *n*-butyl alcohol, isoamyl alcohol, acetic acid, water, or an aqueous mineral acid.

   The early hypothesis that dissolving metal reductions were effected by reaction of the "nascent" hydrogen (liberated from the metal and the hydroxylic solvent) with the molecule being reduced appears to be untenable.[1a,5] In fact, the formation of hydrogen gas during these reductions is normally an undesirable side reaction.

   With hydroxylic solvents, and especially with relatively acidic solvents, metal amalgams are often used rather than free metals to minimize the amount of hydrogen gas produced as a side reaction from reduction of protons. The dissolving metal reductions are better considered as "internal" electrolytic reductions (cf. Ref. 5), in which an electron is transferred from the metal surface (or from the metal in solution) to the organic molecule being reduced. Alternatively, the electron may be transferred from one of the lower valence states of certain metal ions; the most notable examples involve reductions with solutions of chromium(II) salts.[6] The

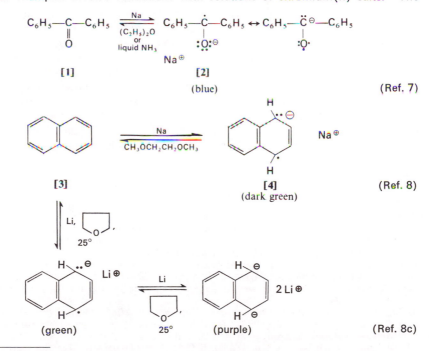

5. (a) J. H. Brewster, *J. Am. Chem. Soc.,* **76**, 6361 (1954). (b) F. D. Popp and H. P. Schultz, *Chem. Rev.,* **62**, 19 (1962). (c) M. J. Allen, *Organic Electrode Processes,* Reinhold, New York, 1958. (d) J. H. P. Utley, *Ann. Repts. Chem. Soc.,* **65B**, 231 (1968); **66B**, 217 (1969). (e) C. L. Perrin, *Progs. Phys. Org. Chem.,* **3**, 165 (1965).
6. J. R. Hanson and E. Premuzic, *Angew. Chem., Intern. Ed. Engl.,* **7**, 247 (1968).

anion radicals[9] (usually colored) produced by this electron transfer are illustrated by the reaction of benzophenone [1] to form the ketyl [2] and by the reaction of naphthalene [3] to form the naphthalide ion radical [4]. These two examples represent cases where relatively high concentrations of stable anion radicals are obtained as solutions in nonprotonic solvents. Even relatively stable ion radicals such as these are bases of sufficient strength to readily abstract a proton from materials with relatively acidic hydrogen atoms such as alcohols and aliphatic ketones.[10]

The presence in these anion radicals of an unpaired electron, which interacts with the atoms in the conjugated system, has been established by measurement of the e.p.r. spectra of various anion radical solutions.[9,11] The more reactive anion radicals formed from systems that offer less possibility for resonance stabilization normally either react with the solvent to abstract a proton or dimerize, polymerize, or disproportionate in the absence of a protonic solvent. However, it has been possible to intercept a reaction intermediate with negative charge at the terminal carbon of the conjugated system such as [5] prior to its reaction with other molecules in the reaction medium by use of intramolecular displacement or elimination

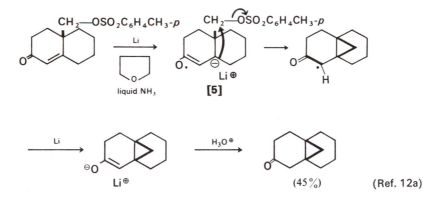

7. (a) W. E. Bachmann, *J. Am. Chem. Soc.,* **55**, 1179 (1933). (b) C. B. Wooster, *ibid.,* **59**, 377 (1937). (c) N. Hirota, *ibid.,* **89**, 32 (1967) and references therein. (d) D. H. Eargle and E. N. Cox, *Spec. Pub. No. 22,* Chemical Society, London, 1967, pp. 116–124. (e) J. F. Garst, D. Walmsley, C. Hewitt, W. R. Richards, and E. R. Zabolotny, *J. Am. Chem. Soc.,* **86**, 412 (1964) ; J. F. Garst, R. A. Klein, D. Walmsley, and E. R. Zabolotny, *ibid.,* **87**, 4080 (1965).

8. (a) N. D. Scott, J. F. Walker, and V. L. Hansley, *J. Am. Chem. Soc.,* **58**, 2442 (1936). (b) J. F. Walker and N. D. Scott, *ibid.,* **60**, 951 (1938). (c) J. Smid, *ibid.,* **87**, 655 (1965). (d) G. Henrici-Olive and S. Olive, *Z. Physik. Chem.,* **43**, 327, 334, 340 (1964).

9. For reviews and discussion of the formation and properties of anion radicals, see (a) E. T. Kaiser and L. Kevan, ed., *Radical Ions,* Wiley-Interscience, New York, 1968. (b) B. J. McClelland, *Chem. Rev.,* **64**, 301 (1964). (c) M. Szwarc, *Progr. Phys. Org. Chem.,* **6**, 323 (1968). (d) M. Szwarc, *Accts. Chem. Res.,* **2**, 87 (1969).

10. (a) H. Normant and B. Angelo, *Bull. Soc. Chim. France,* 354 (1960) ; 810 (1962). (b) J. J. Eisch and W. C. Kaska, *J. Org. Chem.,* **27**, 3745 (1962).

11. (a) A. Carrington, *Quart. Rev.,* **17**, 67 (1963). (b) K. W. Bowers, *Adv. Magnetic Resonance,* **1**, 317 (1965). (c) F. Schneider, K. Möbius, and M. Plato, *Angew. Chem., Intern. Ed. Engl.,* **4**, 856 (1965).

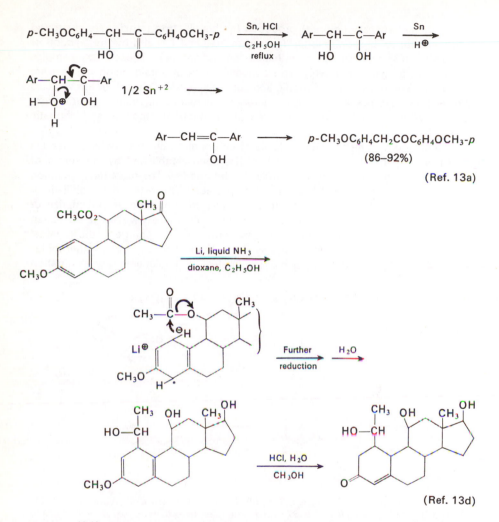

(Ref. 13a)

(Ref. 13d)

reactions.[12,13] The intermolecular transfer of an electron from one anion radical to another conjugated system in the same solution, as illustrated in the equations below, has also been demonstrated.

12. (a) G. Stork, P. Rosen, N. Goldman, R. V. Coombs, and J. Tsuji, *J. Am. Chem. Soc.*, **87**, 275 (1965). (b) S. Rakhit and M. Gut, *ibid.*, **86**, 1432 (1964). (c) P. S. Venkataramani and W. Reusch, *Tetrahedron Letters*, **No. 51**, 5283 (1968). (d) R. G. Carlson and R. G. Blecke, *Chem. Commun.*, **No. 3**, 93 (1969). (e) P. Weiland and G. Anner, *Helv. Chim. Acta*, **51**, 1932 (1968); **53**, 116 (1970). (f) T. A. Spencer, K. K. Schmiegel, and W. W. Schmiegel, *J. Org. Chem.*, **30**, 1626 (1965).
13. (a) P. H. Carter, J. C. Craig, R. E. Lack, and M. Moyle, *Org. Syn.*, **40**, 16 (1960). (b) For an example of the loss of methyllithium from an ion radical, see H. L. Dryden, Jr., G. M. Webber, and J. J. Wieczorek, *J. Am. Chem. Soc.*, **86**, 742 (1964). (c) Although the e.p.r. hyperfine coupling constants of α-keto radicals suggest that the unpaired electron is

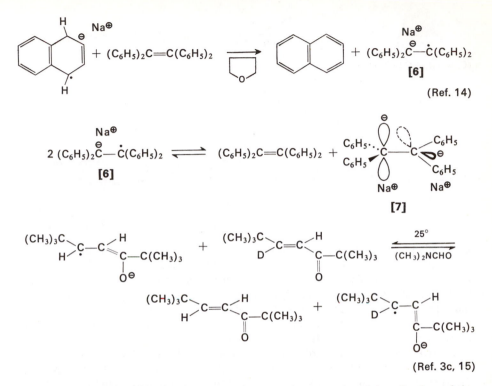

(Ref. 14)

[6]

[7]

(Ref. 3c, 15)

If an anion radical has a structure that permits extensive delocalization of the extra electron present, it is possible under certain conditions to add a second electron to the system to form a dianion. This possibility is illustrated by the previously illustrated reduction of the lithium naphthalene radical anion to the dilithium-napthalene dianion and by the formation of dianions [8] and [9]. In general dianion formation is favored by the use of solvents of low dielectric constant and

located primarily on carbon [G. A. Russell and J. Lokensgard, *ibid.,* **89**, 5059 (1967)], the temperature dependence of these spectra suggests some delocalization of the unpaired electron, resulting in an appreciable energy barrier (*ca.* 9 kcal/mole) to rotation about the carbon–carbon bond [G. Golde, K. Möbius, and W. Kaminski, *Z. Naturforsch.,* **24A**, 1214 (1969). However, more recent kinetic investigations suggest the absence of resonance, stabilization in the α-keto radical derived from acetone: K. D. King, D. M. Golden, and S. W. Benson, *J. Am. Chem. Soc.,* **92**, 5541 (1970). (d) B. J. Magerlein and J. A. Hogg, *ibid.,* **80**, 2220 (1958); also see D. J. Cram and C. K. Dalton, *ibid.,* **85**, 1268 (1963); W. A. Remers and M. J. Weiss, *Tetrahedron Letters,* **No. 1**, 81 (1968); M. Tanabe, J. W. Chamberlain, and P. Y. Nishiura, *ibid.,* **No. 17**, 601 (1961).
14. (a) J. E. Bennett, A. G. Evans, J. C. Evans, E. D. Owen, and B. J. Tabner, *J. Chem. Soc.,* 3954 (1963). (b) A. G. Evans and J. C. Evans, *ibid.,* 6036 (1963). (c) A. G. Evans and B. J. Tabner, *ibid.,* 4613, 5560 (1963). (d) J. F. Garst, J. G. Pacifici, and E. R. Zabolotny, *J. Am. Chem. Soc.,* **88**, 3872 (1966) and references therein.
15. The bimolecular rate constant for exchange of an electron between benzene and its radical anion in a mixture of 1,2-dimethoxyethane and tetrahydrofuran is approximately 10^6 M^{-1} sec^{-1}. G. L. Malinoski and W. H. Bruning, *J. Am. Chem. Soc.,* **89**, 5063 (1967).

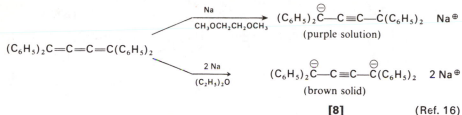

[8]   (Ref. 16)

(Refs. 16b and 17)

poor solvating ability (e.g. diethyl ether) and by the presence of lithium rather than sodium or potassium as a counter ion of the anion.[5e,9c,d,14,16,17] Under these circumstances the dianion derivative may exist as ion aggregates rather than free ions.[17c] Although the addition of one electron to these conjugated systems normally produces the radical anion which may be in equilibrium with its dimer,[9c,d,17] in certain cases (e.g., **[6]**), the radical anion is less stable than the dianion **[7]** and disproportionation occurs. It is believed that this rather unusual order of stability arises when the dianion is not closely associated with its metal cations and steric factors in the molecule favor a twisted geometry. In such a twisted arrangement **[7]** electrostatic repulsion between regions of high electron density in these dianions are minimized.[14d] The same possibility of twisted structures obviously exists for the dianions **[8]** and **[9]**.

REDUCTION OF CARBONYL FUNCTIONS

Esters may be successfully reduced with lithium or sodium (either the free metal or its solution in liquid ammonia) and an alcohol. Unconjugated carbon-carbon double bonds are normally inert to these reducing conditions. Although this method (the Bouveault-Blanc reduction) for the reduction of esters offers occasional advantages, it has largely been replaced in the laboratory by the more convenient reductions of esters with lithium aluminum hydride or by catalytic hydrogenation. One of the advantages offered by the metal reductions is the ability to selectively reduce only the ester formation in a diacid monoester such as **[10]**. [Recall (see Chapter 2) that diborane may be used for the selective reduction of a carboxylic acid in the presence of an ester.] Although the salts of carboxylic acids are normally not reduced by solutions of alkali metals in liquid ammonia, solutions of lithium in the higher boiling solvents, methylamine or ethylamine, will

16. (a) A. Zweig and A. K. Hoffmann, *J. Am. Chem. Soc.*, **84**, 3278 (1962). (b) R. Nahon and A. R. Day, *J. Org. Chem.*, **30**, 1973 (1965).
17. (a) D. A. Dadley and A. G. Evans, *J. Chem. Soc.*, B, 418 (1967); *ibid.*, B, 107 (1968). (b) G. Levin, J. Jagur-Grodzinski, and M. Szwarc, *J. Am. Chem. Soc.*, **92**, 2268 (1970); *J. Org. Chem.*, **35**, 1702 (1970).

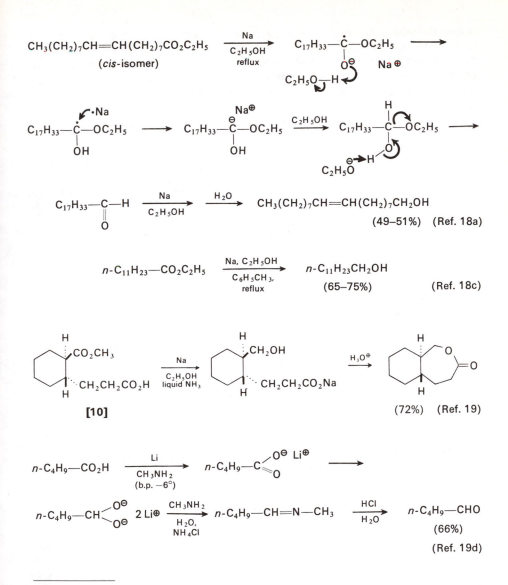

18. H. Adkins and R. H. Gillespie, *Org. Syn.*, **Coll. Vol. 3**, 671 (1955). (b) E. E. Reid, R. O. Cockerille, J. D. Meyer, W. M. Cox, Jr., and J. R. Ruhoff, *ibid.*, **Coll. Vol. 2**, 468 (1943). (c) S. G. Ford and C. S. Marvel, *ibid.*, **Coll. Vol. 2**, 372 (1943). (d) The reaction of hindered esters with a solution of lithium in an ammonia-tetrahydrofuran mixture in the absence of a proton donor has been found to yield the salt of the corresponding carboxylic acid; E. Wenkert and B. G. Jackson, *J. Am. Chem. Soc.*, **80**, 217 (1958).

19. (a) L. A. Paquette and N. A. Nelson, *J. Org. Chem.*, **27**, 2272 (1962). (b) A. W. Burgstahler, L. R. Worden, and T. B. Lewis, *ibid.*, **28**, 2918 (1963). (c) A. W. Burgstahler and L. R. Worden, *J. Am. Chem. Soc.*, **86**, 96 (1964). (d) A. O. Bedenbaugh, J. H. Bedenbaugh, W. A. Bergin, and J. D. Adkins, *ibid.*, **92**, 5774 (1970). (e) R. A. Benkeser, H. Watanabe, S. J. Mels, and M. A. Sabol, *J. Org. Chem.*, **35**, 1210 (1970).

reduce the salts of carboxylic acids to aldehyde derivatives.[19b,c] A possible reaction path for this reaction is indicated in the accompanying equation. A related aldehyde synthesis utilizes the reduction of amides with a solution of lithium in methylamine containing ethanol as a proton donor; the lithium solution is generated in the reaction vessel by the electrolysis of lithium chloride.[19e]

The carbonyl group of aldehydes and ketones is reduced to an alcohol by reaction with a variety of metals in protic solvents, as the following equations illustrate.

$$(C_6H_5)_2CO \xrightarrow[\substack{C_2H_3OH \\ 70°}]{\text{Zn, NaOH}} (C_6H_5)_2CH{-}OH$$

$$(96\text{–}97\%) \qquad \text{(Ref. 20)}$$

$$n\text{-}C_6H_{13}CHO \xrightarrow[\substack{H_2O, 100°}]{\text{Fe, CH}_3CO_2H} n\text{-}C_6H_{13}CH_2OH$$

$$(75\text{–}81\%) \qquad \text{(Ref. 21)}$$

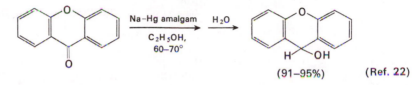

$$(91\text{–}95\%) \qquad \text{(Ref. 22)}$$

$$n\text{-}C_5H_{11}COCH_3 \xrightarrow[\substack{H_2O}]{\text{Na, C}_2H_5OH} n\text{-}C_5H_{11}{-}\underset{\underset{OH}{|}}{CH}{-}CH_3$$

$$(62\text{–}65\%) \qquad \text{(Ref. 23)}$$

One of the more convenient procedures for the reduction of ketones (e.g., **[11]**) uses sodium and isopropyl alcohol in refluxing toluene. The reduction of ketones

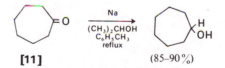

[11] $(85\text{–}90\%)$ (Ref. 24)

with metals offers the stereochemical advantage that the more stable alcohol is frequently (but not always) the predominant product. The following equations illustrate the stereochemical results which have been obtained in reductions of various ketones with metals or electrolytically; generally unhindered ketones yield

20. F. Y. Wiselogle and H. Sonneborn, III, *Org. Syn.*, **Coll. Vol. 1**, 90 (1944).
21. H. T. Clarke and E. E. Dreger, *Org. Syn.*, **Coll. Vol. 1**, 304 (1944).
22. A. F. Holleman, *Org. Syn.*, **Coll. Vol. 1**, 554 (1944).
23. F. C. Whitmore and T. Otterbacher, *Org. Syn.*, **Coll. Vol. 2**, 317 (1943).
24. S. Dev, *J. Indian Chem. Soc.*, **33**, 769 (1956).

the more stable isomeric alcohol, whereas strained or sterically hindered ketones may be reduced to mixtures of alcohols in which the less-stable isomer may predominate.[25,26]

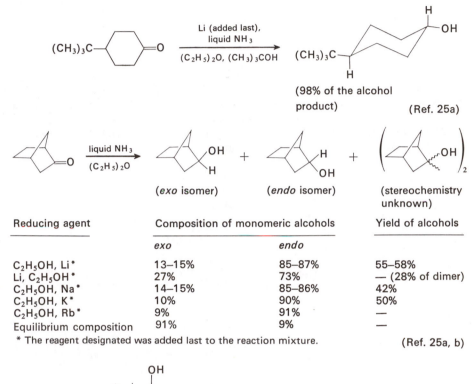

(98% of the alcohol product)   (Ref. 25a)

Reducing agent	Composition of monomeric alcohols		Yield of alcohols
	exo	*endo*	
C$_2$H$_5$OH, Li*	13–15%	85–87%	55–58%
Li, C$_2$H$_5$OH*	27%	73%	— (28% of dimer)
C$_2$H$_5$OH, Na*	14–15%	85–86%	42%
C$_2$H$_5$OH, K*	10%	90%	50%
C$_2$H$_5$OH, Rb*	9%	91%	—
Equilibrium composition	91%	9%	—

* The reagent designated was added last to the reaction mixture.   (Ref. 25a, b)

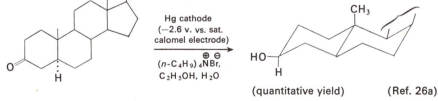

(quantitative yield)   (Ref. 26a)

25. (a) J. W. Huffman and J. T. Charles, *J. Am. Chem. Soc.,* **90,** 6486 (1968). (b) A. Coulombeau and A. Rassat, *Chem. Commun.,* **No. 24,** 1587 (1968). (c) D. A. H. Taylor, *ibid.,* **No. 9,** 476 (1969). (d) D. N. Kirk and A. Mudd, *J. Chem. Soc.,* C, 968 (1969); D. C. Ayres, D. N. Kirk, and R. Sawdaye, *ibid.,* B, 505 (1970). (e) J. W. Huffman, D. M. Alabran, T. W. Bethea, and A. C. Ruggles, *J. Org. Chem.,* **29,** 2963 (1964). (f) D. M. S. Wheeler, M. M. Wheeler, M. Fetizon, and W. H. Castine, *Tetrahedron,* **23,** 3909 (1967).
26. (a) P. Kabasakalian, J. McGlotten, A. Basch. and M. D. Yudis, *J. Org. Chem.,* **26,** 1738. (1961). (b) L. Horner and D. Degner, *Tetrahedron Letters,* **No. 56,** 5889 (1968). (c) L. Mandell, R. M. Powers, and R. A. Day, Jr., *J. Am. Chem. Soc.,* **80,** 5284 (1958). (d) For recent electrochemical studies of ketone reductions, see R. F. Michielli and P. J. Elving, *ibid.,* **90,** 1989 (1968); B. E. Conway, E. J. Judd, and L. G. M. Gordon, *Dis. Faraday Soc.,* **No. 45,** 87 (1968).

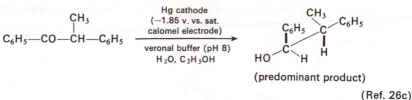

(predominant product)

(Ref. 26c)

In ketone reductions which yield predominantly the less stable alcohol (e.g., [12]), it is sometimes possible to epimerize the initially formed mixture so that the more stable alcohol (e.g., [13]) becomes the major product. One useful method for accomplishing this epimerization involves adding certain ketones such as benzophenone or fluorenone[27b,c] to the alkoxide solution *after the sodium has been consumed* and then refluxing the solution.

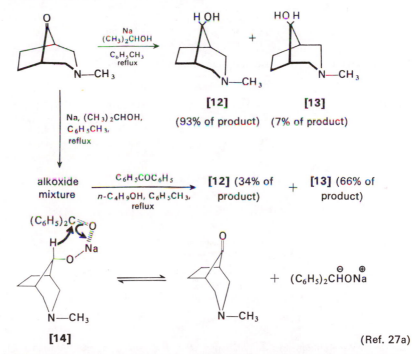

(Ref. 27a)

As indicated in structure [14], this epimerization is believed to occur by a process analogous to the Meerwein-Ponndorf-Verley reduction,[28] in which the hydrogen atom bound to the carbinol carbon is transferred from the alkoxide to the free ketone. Although this equilibration process involving a ketone and a metal

27. (a) H. O. House, H. C. Müller, C. G. Pitt, and P. P. Wickham, *J. Org. Chem.*, **28**, 2407 (1963). (b) W. E. Doering, G. Cortes, and L. H. Knox, *J. Am. Chem. Soc.*, **69**, 1700 (1947). (c) W. von E. Doering and T. C. Aschner, *ibid.*, **71**, 838 (1949); **75**, 393 (1953).
28. A. L. Wilds, *Org. Reactions*, **2**, 178 (1944).

alkoxide or the related reduction of a ketone with a metal isopropoxide may occur to a limited extent in some dissolving-metal reductions of ketones, in most ketone reductions the formation of the more stable alcohol product is the result of a kinetic-ally controlled process and does not involve equilibration of an initially formed less-stable isomer.[25a,b]

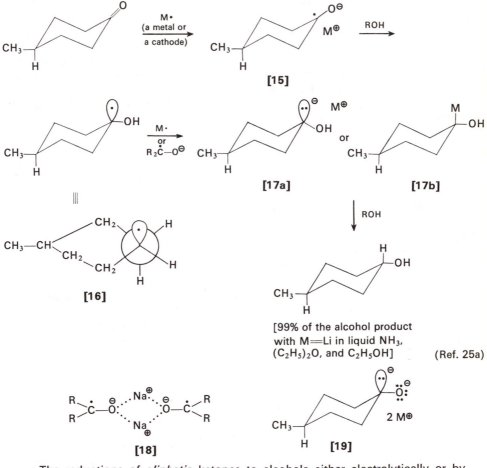

[99% of the alcohol product with M=Li in liquid NH_3, $(C_2H_5)_2O$, and C_2H_5OH]   (Ref. 25a)

The reductions of *aliphatic* ketones to alcohols either electrolytically or by reaction with metals are believed to follow the reaction path indicated in the above equation. The initially formed anion radical or ketyl is best represented by a resonance structure such as **[15]**, since about 70% of the unpaired electron density is located on carbon.[29a,b] If the counter ion for the radical ion (e.g. $M^{\oplus}$ in

29. (a) N. Steinberger and G. K. Fraenkel, *J. Chem. Phys.*, **40**, 723 (1964). (b) B. Mile, *Angew. Chem., Intern. Ed. Engl.*, **7**, 507 (1968). (c) N. Hirota, *J. Am. Chem. Soc.*, **89**, 32 (1967) and references therein. (d) The special ability of lithium cation in aprotic media to associate with anion radicals has been demonstrated in several studies. See Ref. 3c and B. R. Eggins, *Chem. Commun.*, **No. 21**, 1267 (1969).

structure [15]) is a relatively small alkali metal cation such as lithium or sodium, the initial product will have a strong tendency to exist as a dimeric ion pair (e.g., [18]) especially in relatively nonpolar media.[29c,b] Protonation of the radical ion [15] forms a free radical intermediate such as [16]. Evidence derived from e.p.r. studies suggests that such α-hydroxy or α-alkoxy carbon radicals are nonplanar;[30] these radicals would be expected to adopt the more stable configuration as indicated in structure [16], with the hydroxyl group in an equatorial position. In a medium containing excess reducing species (either a metal or the ketone anion), rapid reduction of the neutral radical [16] to form an anion [17] or an organometallic intermediate is to be expected. This species is again expected to adopt (or retain) the indicated more stable geometry [17] which will normally place the hydroxyl substituent in an equatorial conformation. The final protonation of the anionic (or organometallic) intermediate with retention of configuration at carbon leads to the observed more stable product.

As this reaction scheme would suggest, it is important that the protonation and reduction steps ([15]→[16]→[17]) which follow radical anion formation be relatively rapid, a circumstance that is favored by conducting the reduction *in the presence of* a proton donor and excess reducing agent. Failure to observe these precautions can result in the formation of substantial amounts of pinacols from the subsequently discussed bimolecular reduction. These dimeric products can arise from competing dimerization of the ion-pair dimers [18] formed during reductions with metals if no proton donor is present during the reduction to form the radical [16].[25a,b,31] The pinacol products may also arise from dimerization of the uncharged radicals [16]; this latter path becomes particularly likely in the electrolytic or dissolving metal reduction of aryl or diaryl ketones, since the intermediate benzylic radicals (analogous to [16]) have sufficient stability to be present in relatively high concentrations in the region near the surface of the electrode or the dissolving metal. Consequently, the bimolecular combination of two radicals to form a pinacol becomes competitive with the further reduction of the radical, which requires diffusion back to the metal surface or an encounter with a radical anion.

A number of authors have discussed reduction pathways for ketones which involve the formation of dianionic intermediates such as structure [19].[25] The

30. (a) A. J. Dobbs, B. C. Gilbert, and R. O. C. Norman, *Chem. Commun.,* **No. 23**, 1353 (1969). (b) A. Hudson and K. D. J. Root, *Tetrahedron,* **25**, 5311 (1969). (c) D. R. G. Brimage, J. D. P. Cassell, J. H. Sharp, and M. C. R. Symons, *J. Chem. Soc.,* A, 2619 (1969).
31. J. Fried and N. A. Abraham, *Tetrahedron Letters,* **No. 28**, 1879 (1964).
32. A number of workers have described the preparation and subsequent alkylation or acylation of diaryl ketone dianions. For examples, see (a) P. J. Hamrick, Jr., and C. R. Hauser, *J. Am. Chem. Soc.,* **81**, 493 (1959). (b) D. V. Ioffe, *J. Gen. Chem. U.S.S.R.,* **34**, 3960 (1964). (c) B. Z. Askinazi, D. V. Ioffe, and S. G. Kuznetsov, *J. Org. Chem. U.S.S.R.,* **4**, 1934 (1968). (d) E. L. Anderson and J. E. Casey, Jr., *J. Org. Chem.,* **30**, 3959 (1965). (e) H. E. Zaugg and R. J. Michaels, *ibid.,* **33**, 2167 (1968). (f) S. Selman and J. F. Eastham, *ibid.,* **30**, 3804 (1965). (g) G. O. Schenck and G. Matthias, *Tetrahedron Letters,* **No. 8**, 699 (1967). (h) W. S. Marphy and D. J. Buckley, *ibid.,* **No. 35**, 2975 (1969). (i) R. Turle and J. G. Smith [*ibid.,* **No. 27**, 2227 (1969)] have described an analogous reaction of disodiobenzophenone anil with alkyl halides. Also see M. Winn, D. A. Dunnigan, and H. E. Zaugg, *J. Org. Chem.,* **33**, 2388 (1968).

formation of such dianion intermediates (probably as ion pairs or ion aggregates) seems possible from reductions of certain aryl ketones[26d,32c] and appears well established in reductions of diaryl ketones such as benzophenone, particularly when the reductions are performed with sodium or lithium in a relatively nonpolar solvent.[7,32] Dianion formation is feasible in such cases, since the energetically unfavorable electrostatic interaction expected in structure [19] can be lowered both by formation of tight ion-pairs or covalent bonds with the small alkali metal cations and by extensive delocalization of the negative charge in one or both aromatic rings. This delocation of negative charge is indicated by the isolation of ring-substituted products from the alkylation or acylation of these dianions.[32c,g]

However, the formation of dianionic intermediates such as [19] from aliphatic ketones, or even from α,β-unsaturated ketones, seems unlikely. The reduction potentials for forming even the singly charged radical anions (e.g., [15]) from such compounds are in the range -2.1 to -2.6 v. (vs. a saturated calomel electrode) and the potential value expected for addition of a second electron to such systems would be significantly more negative than -3 v.[3c,33] The reduction potentials available in preparative electrolyses or with solutions of alkali metals in liquid ammonia or hexamethylphosphoramide are in the range -2.9 v. or less (vs. a saturated calomel electrode).[1,2,3c,5] Consequently, these commonly used reducing systems do not have a sufficient reduction potential to add a second electron to an aliphatic anion radical (e.g., [15]) unless the negative charge is first neutralized by protonation (to form [16]) or possibly by tight association (or covalent bonding) with a metal cation such as lithium cation[29d] in a nonpolar solvent to form the equivalent of structure [16] with an O—Li bond replacing the O—H bond.

In the foregoing reaction scheme, the stereochemical course of the reaction is determined by some combination of two factors. The intermediate radical [16] presumably has a sufficient lifetime to adopt the more stable configuration before it is reduced (to [17]) and protonated. The lifetime of the anionic or organometallic intermediate [17] is almost certainly very short; it is presently not clear whether or not equilibration of this anion is faster or slower than protonation. If equilibration of the anion stereoisomer is possible, then the relative rates of protonation of the two anion stereoisomers will be an important factor in determining the reduction stereochemistry.[25c] However, the alternative view that protonation of the anionic or organometallic intermediate [17] is faster than conformational inversion presently appears more probable. If this assumption is correct, then the stereochemistry of these reductions is established by the intermediate radical [16] adopting the more favorable configuration indicated prior to reduction and protonation and by the relative rates of reduction of the isomeric radicals to the corresponding anions. Cases in which the metal reductions do not lead to the more stable alcohol appear explicable in the same terms if the steric interference and torsional strain (see Chapter 2) present in the various conformations of the intermediate radical or anion are considered.[25b] Thus, in the reduction of norbornanone,

---

33. A relatively recent compilation of reduction potentials for organic compounds is available in L. Meites, *Polarographic Techniques,* 2d ed., Wiley-Interscience, New York, 1965, pp. 671–711. It should be noted that a change in reduction potential of 1.0 v. corresponds to a free energy change of 23 kcal/mole.

where the angle of bonds to carbon atoms 1 and 4 are distorted, the conformations of the two intermediates are illustrated in structures [20] and [21]. The torsional strain (arrow in structure [20]) present in the intermediate [20] leading to the more

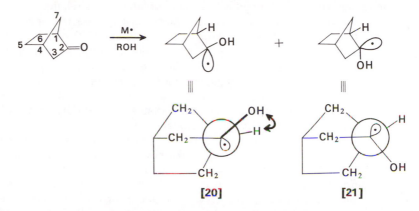

[20]   [21]

stable *exo*-alcohol is diminished in the intermediate [21] leading to the *endo*-alcohol, the major product of this reduction. Similar considerations of conformational stabilities appear sufficient to explain other abnormal reductions, such as the formation of mainly an axial alcohol product in the reduction of certain 12-keto steroids which possess a C-17 side chain capable of interfering sterically with a C-12 equatorial substituent.[25e] It will be noted that changes in the nature of the cation present prior to protonation (i.e. M in [17]) could influence the stereo-specificity of these reductions, either by influencing the relative rates of reduction of the stereoisomeric radicals or by retarding the rate of the final protonation step. These factors should be influenced by the size, solvation, and extent of bonding of the metal. Although changes in the degree of stereospecificity of ketone reductions have been noted with different metals,[25a,b,d] the magnitude and nature of the effect are presently uncertain.

As noted earlier, the reduction of an α-substituted ketone (e.g., [22]) to its anion [23] may be followed by the loss of the alpha-substituent if the latter is a reasonably good leaving group. This elimination to form an enol [24] is followed by tautomerization to form a ketone or by acetylation (if the reaction solvent is acetic anhydride) to form an enol acetate. Typical reactions of this type include the reductions of α-halo, α-amino, α-acyloxy, and α-hydroxy ketones, as illustrated below. These reductions are most often effected with either zinc or zinc amalgam in acetic anhydride, acetic acid, or aqueous mineral acid or with calcium in liquid ammonia.[34b]

In order for this type of reduction to proceed, it is probably necessary that the alpha substituent be able to occupy a conformation in which it is perpendicular to the plane of the original carbonyl group, as illustrated in formula [25]. This

34. (a) R. S. Rosenfeld and T. F. Gallagher, *J. Am. Chem. Soc.*, **77**, 4367 (1955). (b) J. H. Chapman, J. Elks, G. H. Phillipps, and L. J. Wyman [*J. Chem. Soc.*, 4344 (1956)] describe the use of calcium in liquid ammonia to achieve the same type of reduction.

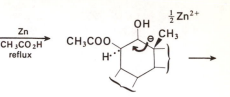

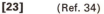

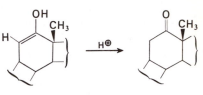

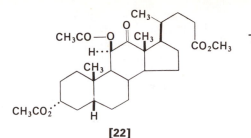

[22]   [23]   (Ref. 34)

[24]

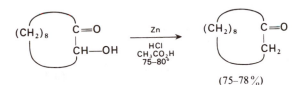

(75–78%)

(Ref. 35)

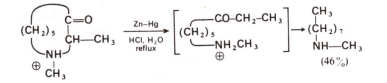

(46%)

(Ref. 36)

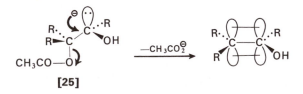

[25]

35. A. C. Cope, J. W. Barthel, and R. D. Smith, *Org. Syn.,* **Coll. Vol. 4,** 218 (1963).
36. (a) N. J. Leonard and R. C. Sentz, *J. Am. Chem. Soc.,* **74,** 1704 (1952). (b) Phenacyl ethers and esters or phenacylsulfonamides can serve as blocking groups for phenols, acids and esters. These blocking groups are removed by reductive cleavage with zinc and acetic acid. J. B. Hendrickson and C. Kandall, *Tetrahedron Letters,* **No. 5.** 343 (1970); J. B. Hendrickson and R. Bergeron, *ibid.,* **No. 5,** 345 (1970).

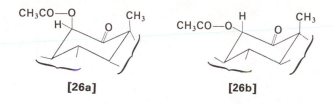

[26a]   [26b]

conformation allows a minimum expenditure of energy for the elimination, since breaking the carbon-substituent bond may be accompanied by a continuous overlap of the developing p orbital at the alpha carbon atom with the pi-orbital system of the anion. A consequence of this stereoelectronic requirement is that cyclohexanones with axial alpha substituents (e.g., [26a]) are reductively cleaved more readily than analogous compounds (e.g., [26b]) with equatorial alpha substituents.[34]

Although the reductive cleavage of α-halo ketones (e.g., [27]) would appear to be analogous to the reduction of other α-substituted ketones, the fact that the reduction potential for carbon-halogen bonds is usually less negative than the

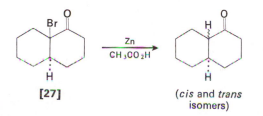

[27]   (*cis* and *trans* isomers)   (Ref. 37a)

potential required for reduction of aliphatic ketones[33] suggests that the reduction of α-halo ketones may involve an initial interaction with the carbon-halogen bond.[38] This idea is supported by the observation that α-halo carbonyl compounds are normally reduced at potentials significantly less negative than the potential required for either the corresponding ketone or the halide.[33,37c] The reaction scheme illustrated in the following equation in which a halide anion is lost from an anion radical intermediate[38] such as [28] to form an α-keto radical [29] appears to account for the experimental observations. This halide-ion elimination (e.g., [28]) is expected to have stereoelectronic requirements similar to the previously discussed

37. (a) H. E. Zimmerman and A. Mais, *J. Am. Chem. Soc.,* **81**, 3644 (1959). (b) A. K. Bose and M. S. Tibbetts, *Tetrahedron,* **23**, 3887 (1967). (c) J. H. Stocker and R. M. Jenevein, *Chem. Commun.,* **No. 16**, 934 (1968). (d) A. M. Wilson and N. L. Allinger, *J. Am. Chem. Soc.,* **83**, 1999 (1961).
38. Since the intramolecular exchange of an electron is probably a very rapid process, it is not clear whether it is meaningful to discuss the following radical anions as different intermediates.

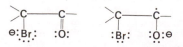

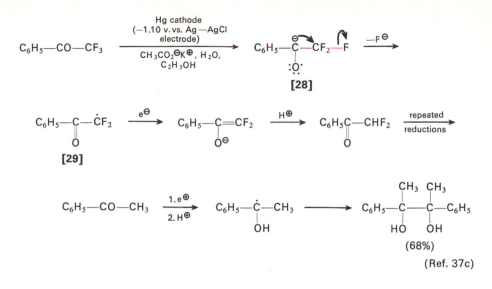

[28]

[29]

(68%)

(Ref. 37c)

acetoxy ketone reductions (i.e., [25]). In agreement with this idea, the potentials for the reduction of axial α-halocyclohexanone derivatives are found to be less negative (i.e., more favorable for reduction) than the values for the corresponding equatorial isomers.[37d]

The same type of reductive cleavage of α-substituted ketones has been achieved with chromium(II) salts in acetone or acetic acid;[6,39] the mechanism of this reductive cleavage is thought to be analogous to that of the dissolving metal reductions.

Studies with the acyloxy ketone [30] suggested that reductions with chromium(II) salts were more satisfactory than reductions with zinc or calcium for cyclohexanone derivatives with equatorial α-substituents.[39c]

Bond formation involving carbon-halogen and carbonyl groups has also been observed when chromium(II)[39d] or metal reductions[40] are applied to halo ketones in which the halogen substituent is not alpha to the carbonyl group. The examples in the accompanying equations illustrate such reactions as involving initial reduction by the metal at the more easily reduced[33] carbon-halogen bond. However, intermediates involving electron exchange between the carbon-halogen bond and

39. (a) G. Rosenkranz, O. Mancera, J. Gatica, and C. Djerassi *J. Am. Chem. Soc.,* **72** 4077 (1950). (b) W. Cole and P. L. Julian, *J. Org. Chem.,* **19**, 131 (1954). (c) H. O. House and R. G. Carlson, *ibid.,* **29**, 74 (1964). (d) C. H. Robinson, O. Gnoj, E. P. Oliveto, and D. H. R. Barton, *ibid.,* **31**, 2749 (1966). (e) D. H. R. Barton and J. T. Pinhey, *Proc. Chem. Soc.,* 279 (1960). (f) L. H. Slaugh and J. H. Raley, *Tetrahedron,* **20**, 1005 (1964); T. Shirafuji, Y. Yamamoto, and H. Nozaki, *Tetrahedron Letters,* **No. 47**, 4097 (1969).
40. (a) D. P. G. Hamon and R. W. Sinclair, *Chem. Commun.,* **No. 15**, 890 (1968). (b) K. M. Baker and B. R. Davis, *Tetrahedron,* **24**, 1655 (1968). (c) Y. Leroux, *Bull. Soc. Chim. France,* 359 (1968). (d) H. O. House, J. J. Riehl, and C. G. Pitt, *J. Org. Chem.,* **30**, 650 (1965). (e) S. Danishefsky and R. Dumas, *Chem. Commun.,* **No. 21**, 1287 (1968). (f) For the use of organonickel or organocopper intermediates to effect intramolecular cyclization, see E. J. Corey and I. Kuwajima, *J. Am. Chem. Soc.,* **92**, 395 (1970).

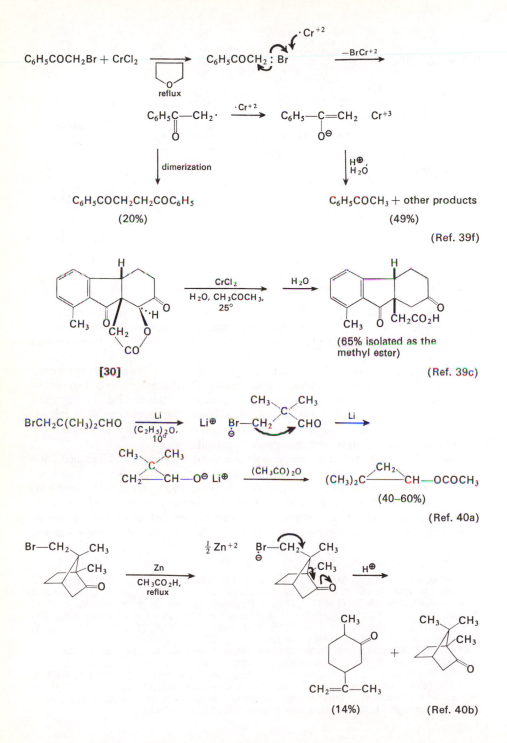

$C_6H_5COCH_2Br + CrCl_2 \xrightarrow[\text{reflux}]{} C_6H_5COCH_2 \cdot Br \xrightarrow{-BrCr^{+2}}$

$\cdot Cr^{+2}$

$C_6H_5C{-}CH_2\cdot \xrightarrow{\cdot Cr^{+2}} C_6H_5{-}C{=}CH_2 \quad Cr^{+3}$

dimerization $H^{\oplus}, H_2O$

$C_6H_5COCH_2CH_2COC_6H_5$

(20%)

$C_6H_5COCH_3 +$ other products

(49%)

(Ref. 39f)

$\xrightarrow[\text{H}_2\text{O, CH}_3\text{COCH}_3,]{CrCl_2}$ $\xrightarrow{H_2O}$

[30]

(65% isolated as the methyl ester)

(Ref. 39c)

$BrCH_2C(CH_3)_2CHO \xrightarrow[(C_2H_5)_2O,]{Li} Li^{\oplus} \; Br{-}CH_2 \;\underset{\ominus}{C}(CH_3)_2 \; CHO \xrightarrow{Li}$

10°

$CH_2{-}CH{-}O^{\ominus} Li^{\oplus} \xrightarrow{(CH_3CO)_2O} (CH_3)_2C{-}CH{-}OCOCH_3$

(40–60%)

(Ref. 40a)

$\tfrac{1}{2} Zn^{+2}$ $\underset{\ominus}{Br}{-}CH_2$ $H^{\oplus}$

$\xrightarrow[\text{reflux}]{Zn, CH_3CO_2H}$

$CH_2{=}C{-}CH_3$ +

(14%)

(Ref. 40b)

the carbonyl group may be involved.[38] Related examples are found in the carbon-carbon bond cleavage or bond formation observed during reductions of 1,4-dicarbonyl compounds (e.g., [31]),[41] and the bond formation observed, or inferred, from product studies, when 1,3-dicarbonyl compounds are reduced.[42,43]

The reduction of ketones with zinc amalgam in strong aqueous acids, the Clemmensen reduction[1f,43,44] has been used to convert ketones to methylene

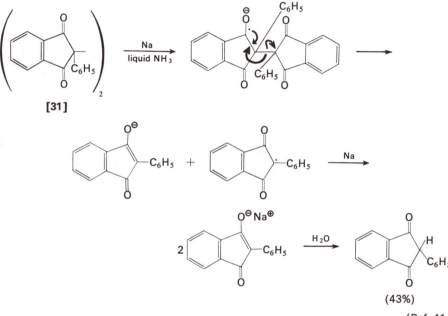

[31]

(43%)

(Ref. 41a)

41. (a) F. M. Beringer, S. A. Galton, and S. J. Huang, *Tetrahedron,* **19**, 809 (1963); (b) J. G. St. C. Buchanan and B. R. Davis, *J. Chem. Soc.,* C, 1340 (1967). (c) A. Eschenmoser, *19th Intern. Congress Pure and Appl. Chem.,* London., 1963, Congress Lectures, p. 297. (d) E. Wenkert and J. E. Yoder, *J. Org. Chem.,* **35**, 2986 (1970). (e) D. R. Crump and B. R. Davis, *Chem. Commun.,* **No. 12**, 768 (1970).
42. (a) W. Reusch and D. B. Priddy, *J. Am. Chem. Soc.,* **91**, 3677 (1969). (b) T. J. Curphey, C. W. Amelotti, T. P. Layloff, R. L. McCartney, and J. H. Williams, *ibid.,* **91**, 2817 (1969). (c) T. J. Curphey and R. L. McCartney, *J. Org. Chem.,* **34**, 1964 (1969). (d) M. L. Kaplan, *ibid.,* **32**, 2346 (1967). (e) D. E. Evans and E. C. Woodbury, *ibid.,* **32**, 2158 (1967). (f) N. J. Cusack and B. R. Davis, *ibid.,* **30**, 2062 (1965). (g) E. Wenkert and E. Kariv, *Chem. Commun.,* **No. 22**, 570 (1965). (h) V. T. C. Chuang and R. B. Scott, Jr., *ibid.,* **No. 13**, 758 (1969). (i) R. LeGoaller, M. Rougier, C. Zimero, and P. Arnaud, *Tetrahedron Letters,* **No. 48**, 4193 (1969). (j) D. B. Priddy and W. Reusch, *ibid.,* **No. 60**, 5291 (1969). (k) T. J. Curphey and R. L. McCartney, *ibid.,* **No. 60**, 5295 (1969).
43. For a review of the Clemmensen reduction of difunctional ketones, see J. G. St. C. Buchanan and P. D. Woodgate, *Quart. Rev.,* **23**, 522 (1969).
44. (a) E. L. Martin, *Org. Reactions,* **1**, 155 (1942). (b) D. Staschewski, *Angew. Chem.,* **71**, 726 (1959). (c) T. Nakabayashi, *J. Am. Chem. Soc.,* **82**, 3900, 3906, 3909 (1960). (d) J. H. Brewster, *ibid.,* **76**, 6364 (1954). (e) J. H. Brewster, J. Patterson, and D. A. Fidler, *ibid.,* **76**, 6368 (1954).

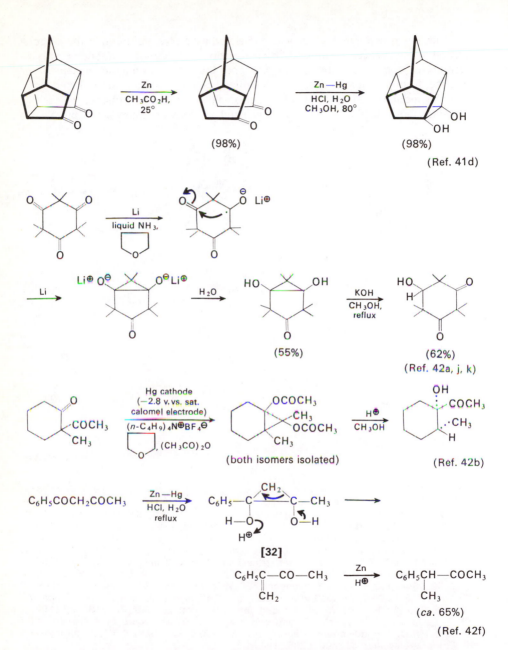

groups. As illustrated in the following equation, the reaction has commonly been run as a three-phase mixture of toluene, aqueous hydrochloric acid, and zinc amalgam. Under these conditions, a low concentration of the ketone and its conjugate acid are present in the aqueous phase, which is in contact with the zinc, and bimolecular reduction (to be discussed subsequently) is minimized.

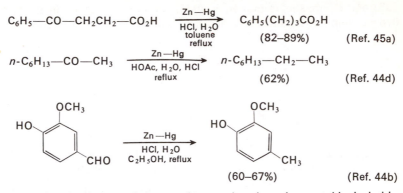

$$C_6H_5-CO-CH_2CH_2-CO_2H \xrightarrow[\substack{HCl, H_2O \\ toluene \\ reflux}]{Zn-Hg} C_6H_5(CH_2)_3CO_2H$$

(82–89%)   (Ref. 45a)

$$n\text{-}C_6H_{13}-CO-CH_3 \xrightarrow[\substack{HOAc, H_2O, HCl \\ reflux}]{Zn-Hg} n\text{-}C_6H_{13}-CH_2-CH_3$$

(62%)   (Ref. 44d)

(60–67%)   (Ref. 44b)

As noted previously, the use of zinc amalgam rather than zinc metal is desirable to minimize the side reaction leading to formation of hydrogen from reaction of zinc with the aqueous acid. Recently, another experimental procedure for the Clemmensen reduction has been described in which a solution of the ketone in acetic anhydride or ether is treated with powdered zinc and hydrogen chloride gas.[46]

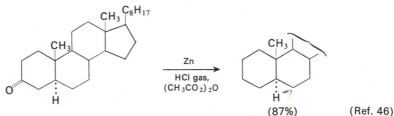

(87%)   (Ref. 46)

A transformation analogous to the Clemmensen reduction may also be accomplished electrolytically at a lead cathode, as the following equation illustrates. In both this electrolytic reduction and the related Clemmensen reduction, the use of O-deuterated solvents permits the synthesis of dideuterated products such as

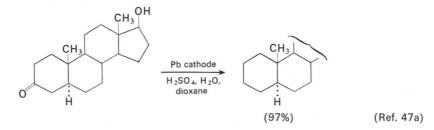

(97%)   (Ref. 47a)

45. (a) E. L. Martin, *Org. Syn., Coll. Vol. 2*, 499 (1943). (b) R. Schwarz and H. Hering, *ibid.,* **Coll. Vol. 4**, 203 (1963). (c) For other examples, see R. R. Read and J. Wood, Jr., *ibid.,* **Coll. Vol. 3**, 444 (1955); K. H. Meyer, *ibid.,* **Coll. Vol. 1**, 60 (1944).
46. (a) S. Yamamura and Y. Hirata, *J. Chem. Soc., C*, 2887 (1968). (b) M. Toda, Y. Hirata, and S. Yamamura, *Chem. Commun.,* **No. 16**, 919 (1969).
47. (a) L. Throop and L. Tökes, *J. Am. Chem. Soc.,* **89**, 4789 (1967). (b) C. R. Enzell, *Tetrahedron Letters,* **No. 12**, 1285 (1966). (c) I. Elphimoff-Felkin and P. Sarda, *Chem. Commun.,* **No. 18**, 1065 (1969).

[33] and demonstrates that both hydrogen atoms involved in the reduction are derived from proton donors, rather than from hydrogen-atom donors.

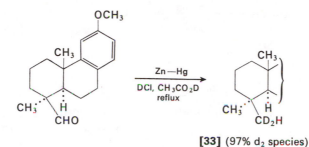

[33] (97% d_2 species)   (Ref. 47b)

The reduction of ketones or aldehydes to methylene or methyl groups requires both the use of a strongly acidic medium and the use of a metal, such as zinc or lead, as a cathode or source of electrons. For example, reduction of a ketone under comparable conditions at a mercury cathode yields a pinacol rather than the product of Clemmensen reduction.[44c] Since a number of experiments suggest that an alcohol or an olefin is not an intermediate in the Clemmensen reduction, one of the intermediates **[15]**, **[16]**, or **[17]** in the previously discussed path for the reduction of a ketone to an alcohol must be intercepted to form different products. The potential required for reduction of ketones in strongly acidic media is substantially less negative (*ca.* −1.0 v. vs. a saturated calomel electrode)[44] than the ketone reduction potential (−2.0 to −2.5 v.) in neutral media, suggesting that the sequence of proton and electron transfers also differs from the previous scheme. Although certain aspects of the reaction mechanism remain uncertain, the pathway indicated in the following equations appears compatible with the data presently available. Formation of the protonated ketone **[34]** prior to the first electron transfer accounts for the less negative reduction potential required. Presumably the resulting hydroxy radical **[35]** is reduced further, to a species such as **[36]**, more rapidly than it diffuses from the metal surface and dimerizes. Protonation of the intermediate **[36]** to cleave the carbon-metal bond would correspond to the previously discussed pathway leading to an alcohol. However, in the strongly acid medium the indicated (structure **[36]**) protonation of the hydroxyl group with loss of water and successive transfers of two more electrons leads to intermediates such as **[37]** and **[38]**. Reaction of intermediate **[38]** with two protons produces the final reduction product. It has been suggested that a series of reduction intermediates such as **[35]** through **[38]** are all bonded to zinc atoms at the metal surface and leave this surface only after at least one proton has been added to carbon.[44d,e] Although experimental evidence bearing on this idea is difficult to obtain, a recent report[47c] describes the isolation of a cyclopropane product **[39]** which would appear to have been formed from a carbene-like intermediate. This observation suggests that the intermediate **[38]** may diffuse from the metal surface before protonation. The intermediate **[38]**, which could be considered to be the product of reaction of a carbene with metallic zinc, is perhaps stabilized, as are

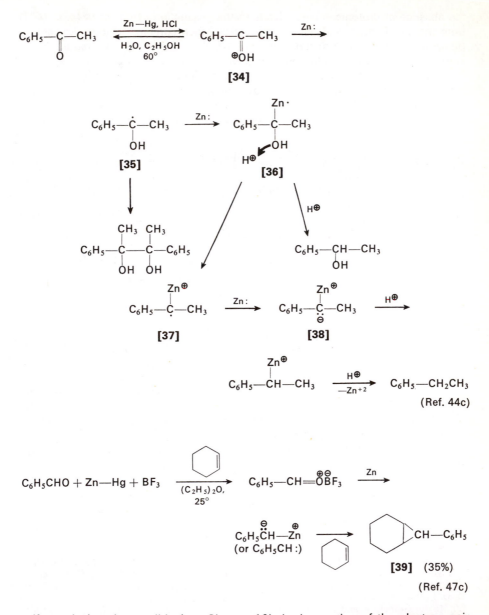

sulfur and phosphorus ylids (see Chapter 10), by interaction of the electron pair at carbon with the zinc atom.

The reaction of ketones with metals, particularly magnesium (or magnesium plus iodine),[48] magnesium amalgam, zinc, zinc amalgam, or aluminum amalgam, in

48. M. D. Rausch, W. E. McEwen, and J. Kleinberg, *Chem. Rev.*, **57**, 417 (1957).

the absence of protonic solvents leads to the production of ion pairs (e.g., [40]) from the metal cation and the anion radicals which combine to form the salts of pinacols. Although this dimerization may be observed even in the reductions of ketones with alkali metals in nonpolar solvents,[31] the reaction becomes especially favorable with metals such as zinc, magnesium, and aluminum. In nonpolar media these oxygen-metal bonds are either covalent or are tight ion pairs, and the electrostatic repulsion which would impede dimerization of free anion radicals is diminished. This process, illustrated in the following equations and called *bimolecular re-*

$$CH_3{-}CO{-}CH_3 \xrightarrow[\substack{C_6H_6 \\ reflux}]{Mg{-}Hg} \left(\begin{matrix}CH_3 \\ CH_3\end{matrix}\!\!\dot{C}{-}\overset{\ominus}{O}\right)_2 Mg^{+2} \longrightarrow$$

$$[40]$$

$$\begin{matrix}(CH_3)_2C{-}\overset{\ominus}{O}{\cdot}{\cdot} \\ | \qquad\qquad {:}Mg^{+2} \\ (CH_3)_2C{-}\underset{\ominus}{O}{\cdot}{\cdot}\end{matrix} \xrightarrow{H_2O} \begin{matrix}(CH_3)_2C{-}C(CH_3)_2 \\ | \qquad\quad | \\ OH \ \ OH\end{matrix}$$

$$(43{-}50\%)$$

(Ref. 49a)

$$C_6H_5{-}CO{-}CH(CH_3)_2 \xrightarrow[\substack{C_6H_6,\ C_2H_5OH, \\ 50°}]{Al{-}Hg} \xrightarrow[H_2O]{HCl} \begin{matrix} \qquad\quad C_6H_5\ \ C_6H_5 \\ \qquad\qquad | \qquad | \\ (CH_3)_2CH{-}C{-}{-}C{-}CH(CH_3)_2 \\ \qquad\quad | \qquad | \\ \qquad\quad OH \quad OH\end{matrix}$$

(45% yield, 60%
meso isomer and
40% *dl* isomer)   (Ref. 49b)

duction or *hydrodimerization*, is often a competing reaction in other dissolving metal reductions, such as the Clemmensen reduction, and in the electrolytic reduction of aryl or diaryl ketones.[50] The electrolytic reductions in protic media illustrate the fact that pinacols may also arise from dimerization of the carbinol radical [41].[50d] The predominant formation of the less stable racemic isomer [42b], especially in alkaline media, has been attributed to dimerization of a hydrogen-bonded species such as [43] in a conformation that minimizes steric interference between the two large phenyl groups.[50c]

49. (a) R. Adams and E. W. Adams, *Org. Syn.,* **Coll. Vol. 1,** 459 (1944). (b) A. Vigevani, R. Pasqualucci, G. G. Gallo, and G. Pifferi, *Tetrahedron,* **25,** 573 (1969). (c) A comparable dimerization is observed when anils are reduced with metals in aprotic media; J. J. Eisch, D. D. Kaska, and C. J. Peterson, *J. Org. Chem.,* **31,** 453 (1966).
50. (a) J. Grimshaw and J. S. Ramsey, *J. Chem. Soc.,* C, 653 (1966). (b) R. N. Gourley and J. Grimshaw, *ibid.,* C, 2388 (1968). J. H. Stocker and R. M. Jenevein, *J. Org. Chem.,* **33,** 294, 2145 (1968), *ibid.,* **34,** 2807 (1969). (d) The bimolecular rate constant (*ca.* $10^7\ M^{-1}\ sec^{-1}$) for the dimerization or disproportionation of hydroxy radicals similar to [41] is only slightly slower than a diffusion-controlled reaction. S. A. Weiner, E. J. Hamilton, Jr., and B. M. Munroe, *J. Am. Chem. Soc.,* **91,** 6350 (1969). (e) For the hydrodimerization of aromatic aldehydes with chromium(II) salts in acidic aqueous alcohol solutions, see D. D. Davis and W. B. Bigelow, *ibid.,* **92,** 5127 (1970).

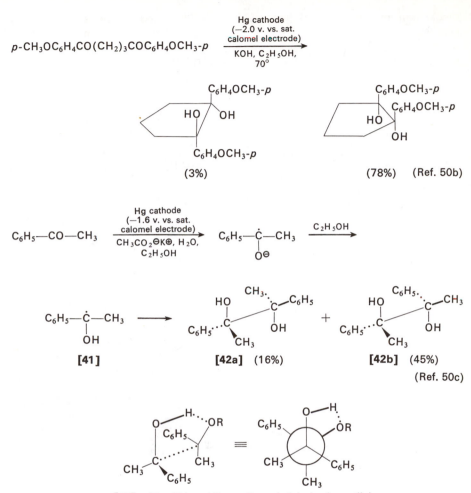

$$p\text{-}CH_3OC_6H_4CO(CH_2)_3COC_6H_4OCH_3\text{-}p$$

Hg cathode
(−2.0 v. vs. sat.
calomel electrode)
────────────
KOH, C₂H₅OH,
70°

C₆H₄OCH₃-*p*

HO   OH

C₆H₄OCH₃-*p*

(3%)

C₆H₄OCH₃-*p*

C₆H₄OCH₃-*p*

HO

OH

(78%)   (Ref. 50b)

$$C_6H_5\text{—}CO\text{—}CH_3$$

Hg cathode
(−1.6 v. vs. sat.
calomel electrode)
────────────
CH₃CO₂⊖K⊕, H₂O,
C₂H₅OH

$$C_6H_5\text{—}\overset{\cdot}{C}\text{—}CH_3$$
$$\overset{|}{O}\ominus$$

C₂H₅OH

$$C_6H_5\text{—}\overset{\cdot}{C}\text{—}CH_3$$
$$\overset{|}{OH}$$

[41]

CH₃

HO   C—C₆H₅

C₆H₅ C   OH

CH₃

[42a] (16%)

+

C₆H₅

HO   C—CH₃

C₆H₅ C   OH

CH₃

[42b] (45%)

(Ref. 50c)

O─H···OR

C₆H₅

CH₃ C

C₆H₅ CH₃

≡

O─H

C₆H₅   ÖR

CH₃   C₆H₅

CH₃

[43] (R═H in acidic media and ⊖ in basic media)

An analogous bimolecular reduction is found in the reduction of esters. The previous discussion of reduction of esters to alcohols noted the successive protonating and further reduction of the initially formed anion radical (e.g., **[44]**) if a monomeric product was desired. In the absence of a proton donor, the metal cation-radical anion pairs dimerize (see structure **[45]**) and undergo the further changes indicated to form an α-hydroxy ketone after hydrolysis. This bimolecular reduction of esters, called the acyloin reaction,[51] has proved of special value for the

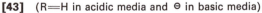

51. (a) S. M. McElvain, *Org. Reactions,* **4**, 256 (1948). (b) K. T. Finley, *Chem. Rev.,* **64**, 573 (1964). (c) J. M. Snell and S. M. McElvain, *Org. Syn.,* **Coll. Vol. 2**, 114 (1943). (d) D. E. Ames, G. Hall, and B. T. Warren, *J. Chem. Soc.,* C, 2617 (1968).

preparation of medium and large rings,[52] as illustrated in the accompanying equation.

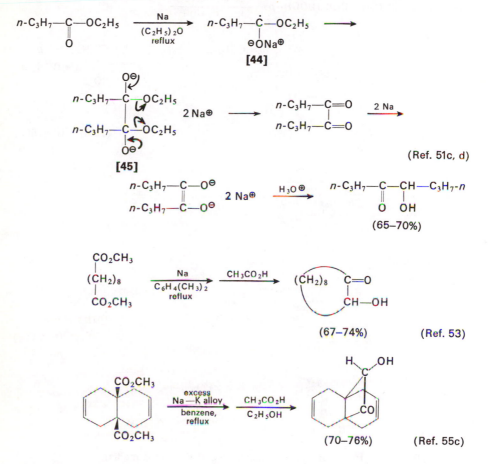

(Ref. 51c, d)

(65–70%)

(67–74%)   (Ref. 53)

(70–76%)   (Ref. 55c)

52. (a) V. Prelog, *J. Chem. Soc.*, 420 (1960). (b) K. Ziegler in E. Muller, ed., *Methoden der organischen Chemie (Houben-Weyl)*, Vol. 4, Part 2, Georg Thieme Verlag, Stuttgart, Germany, 1955, pp. 729–822.
53. (a) N. L. Allinger, *Org. Syn.*, **Coll. Vol. 4**, 840 (1963). (b) K. T. Finley and N. A. Sasaki, *J. Am. Chem. Soc.*, **88**, 4267 (1966). (c) A. T. Blomquist, R. E. Burge, Jr., and A. C. Sucsy, *ibid.*, **74**, 3636 (1952).
54. (a) J. C. Sheehan, R. C. O'Neill, and M. A. White, *J. Am. Chem. Soc.*, **72**, 3376 (1950). (b) J. C. Sheehan and R. C. Coderre, *ibid.*, **75**, 3997 (1953). (c) J. C. Sheehan and W. F. Erman, *ibid.*, **79**, 6050 (1957).
55. (a) A. C. Cope and E. C. Herrick, *J. Am. Chem. Soc.*, **72**, 983 (1950). (b) A. de Groot, D. Oudman, and H. Wynberg, *Tetrahedron Letters*, **No. 20**, 1529 (1969). (c) J. J. Bloomfield and J. R. S. Irelan, *J. Org. Chem.*, **31**, 2017 (1966). (d) For an example of a reductive cleavage of a carbon–carbon bond, analogous to the previously described reductive cleavage of other 1,4-dicarbonyl compounds with metals, rather than ring closure, see J. J. Bloomfield, R. G. Todd, and L. T. Takahashi, *ibid.*, **28**, 1474 (1963).

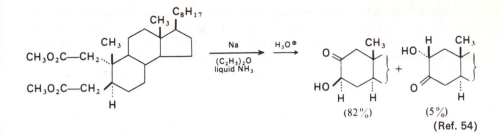

(Ref. 54)

Although four-, five-, six-, and seven-membered rings have been obtained by reaction of diesters with solutions of sodium in liquid ammonia[54] or with dispersions of sodium or sodium-potassium alloy in benzene, toluene, or xylene,[55] these reactions are often complicated by a competing Dieckmann ring closure[56] (see Chapter 11) or by instability of the endiolate dianion present before hydrolysis. These problems have been alleviated by conducting the acyloin reaction in the presence of chlorotrimethylsilane to form a relatively stable *bis* silyl enol ether [46], which can be isolated and subsequently hydrolyzed to the hydroxy ketone with aqueous acid.[57] This chlorosilane also reacts rapidly with the metal alkoxides formed (see structure [45]) in the acyloin reaction to remove these bases which are catalysts for the competing Dieckmann ring closure.[56]

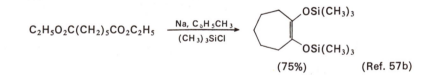

(75%) (Ref. 57b)

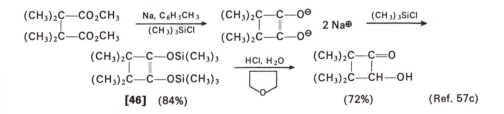

Other reductive cyclization reactions which involve a carbonyl group and a second functional group include the previously discussed alkylations or ring closures found in the reduction of certain halo ketones, and reductions of certain

56. J. P. Schaefer and J. J. Bloomfield, *Org. Reactions,* **15**, 1 (1967).
57. (a) U. Schräpler and K. Rühlmann, *Chem. Ber.,* **97**, 1383 (1964). (b) J. J. Bloomfield, *Tetrahedron Letters,* No. **5**, 587, 591 (1968). (c) G. E. Gream and S. Worthley, *ibid.,* No. **29**, 3319 (1968). (d) C. Ainsworth and F. Chen, *J. Org. Chem.,* **35**, 1272 (1970).

keto esters (e.g. [47]) and lactone esters.[58] In special cases keto olefins such as [48] have been cyclized, presumably because the geometry of the starting material is especially favorable; however, this reaction does not ordinarily occur. The analogous ring closure of a keto acetylene illustrated in the following equation would appear to be more general. In this latter case it has been suggested that the initial reduction site is the acetylene rather than the carbonyl group.

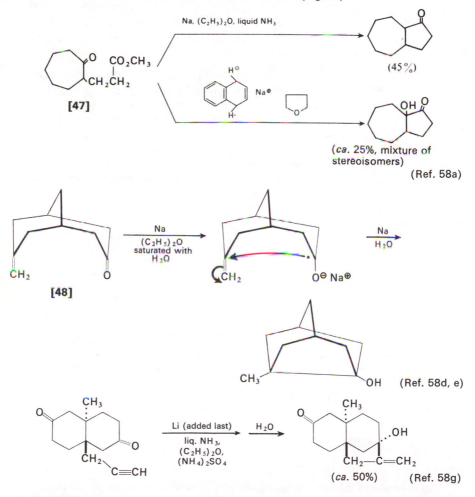

(45%)

[47]

(*ca.* 25%, mixture of stereoisomers)

(Ref. 58a)

[48]

(Ref. 58d, e)

(*ca.* 50%)

(Ref. 58g)

58. (a) C. D. Gutsche, I. Y. C. Tao, and J. Kozma, *J. Org. Chem.*, **32**, 1782 (1967). (b) R. G. Carlson and R. G. Blecke, *ibid.*, **32**, 3538 (1967). (c) E. E. Van Tamelen, T. A. Spencer, Jr., D. S. Allen, Jr., and R. L. Orvis, *Tetrahedron*, **14**, 8 (1961). (d) M. Eakin, J. Martin, and W. Parker, *Chem. Commun.*, **No. 11**, 206 (1965). (e) J. M. Greenwood, I. H. Qureshi, and J. K. Sutherland, *J. Chem. Soc.*, 3154 (1965). (f) For studies of intermolecular coupling between ketones and conjugated olefins, see J. K. Kochi, *J. Org. Chem.*, **28**, 1960, 1969 (1963); O. R. Brown and K. Lister, *Dis. Faraday Soc.*, **No. 45**, 106 (1968). (g) G. Stork, S. Malhotra, H. Thompson, and M. Uchibayashi, *J. Am. Chem. Soc.*, **87**, 1148 (1965).

REDUCTION OF CONJUGATED SYSTEMS

The most common reactions in this category are the dissolving metal reductions of carbon-carbon double bonds conjugated with carbonyl groups, aromatic systems, or other carbon-carbon multiple bonds; and the dissolving metal reductions of aromatic systems to form dihydro or tetrahydro aromatic systems. Although α,β-unsaturated carbonyl compounds are initially reduced to saturated carbonyl compounds with zinc in acetic acid or with zinc amalgam in aqueous acid (Clemmensen reduction conditions), reductions with zinc or zinc-amalgam in acidic media may be complicated by olefin formation or by rearrangement,[43,59] as the following equations illustrate. The first of the two examples illustrates the general tendency of $\Delta^{2,3}$-cyclohexenones with a 3-alkyl substituent to form $\Delta^{1,2}$-olefins under the conditions of the Clemmensen reduction.[43,59b] The mechanism involved in this transformation is not clear. The carbon skeleton rearrangement, illustrated by the second example, is a general reaction observed when either cyclic or acyclic α,β-unsaturated ketones are subjected to the Clemmensen reduction.[43,59b] This reaction appears to involve cyclization of some intermediate such as the organozinc compound **[48]** to form a cyclopropanol intermediate. This cyclopropanol intermediate may undergo an acid-catalyzed ring opening[59d] in either of two directions

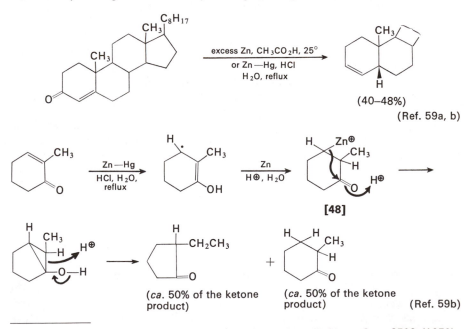

(40–48%)

(Ref. 59a, b)

[48]

(*ca.* 50% of the ketone product)   (*ca.* 50% of the ketone product)   (Ref. 59b)

59. (a) J. McKenna, J. K. Norymberski, and R. D. Stubbs, *J. Chem. Soc.*, 2502 (1959). (b) B. R. Davis and P. D. Woodgate, *ibid.*, 5943 (1965); *ibid.*, C, 2006 (1966). (c) I. Elphimoff-Felkin and P. Sarda, *Tetrahedron Letters*, **No. 35**, 3045 (1969). (d) For a discussion of the opening of the cyclopropanol intermediates, see C. H. DePuy, *Accts. Chem. Res.*, **1**, 33 (1968). (e) The reduction of enediones with zinc and acetic acid or acetic anhydride to form saturated diketone derivatives is apparently not complicated by such rearrangements. S. K. Pradhan, G. Subrahmanyam, and H. J. Ringold, *J. Org. Chem.*, **32**, 3004 (1967); J. Scotney and E. V. Truter, *J. Chem. Soc.*, C, 1079 (1969).

to form the structurally isomeric products indicated. This carbon-skeleton re-arrangement is not observed[60] under more typical reduction conditions[1] with an alkali metal (usually lithium or sodium) and a proton donor in liquid ammonia.

This metal-ammonia reduction is characterized by its ability to selectively reduce conjugated carbon-carbon double bonds to form unconjugated ketones such as [49]. However, the accompanying example shows that use of prolonged

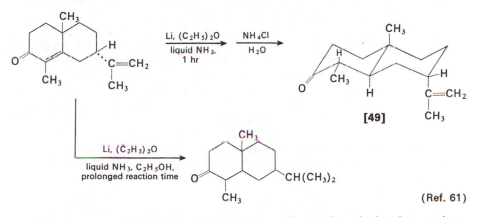

[49]

(Ref. 61)

reaction times may also lead to reduction of nonconjugated terminal carbon-carbon double bonds. The reduction is frequently performed by adding an ether or tetra-hydrofuran solution of the unsaturated ketone (e.g., [50]) and one equivalent (or more) or a proton donor such as ethanol or t-butyl alcohol to a solution containing more than two equivalents of sodium or lithium in liquid ammonia or a low-molecular-weight amine. If only one equivalent of a proton donor is employed, the product of this process is an enolate anion such as [51], which may be protonated, carbonated, or alkylated (for further discussion, see Chapter 9), as shown.[12a,62] As discussed previously, the further reduction of an aliphatic radical anion such as [52] is unlikely until it has been protonated to form an allylic radical (e.g., [53]). This circumstance is indicated by the titration of a standardized solution of sodium in hexanethylphosphoramide with an unsaturated ketone. In the absence of a proton donor only one equivalent of sodium is consumed per mole of enone added, whereas two equivalents of sodium are consumed when a proton donor (triethyl-carbinol) is present.[3c] One possible exception to this requirement for a proton donor involves reductions with lithium metal in nonpolar solvents. It may be possible in such circumstances for the initially formed anion radical (e.g., [52]) to form a covalent bond with the relatively small lithium cation. Such an intermediate (e.g., [53] with OH replaced by OLi) may be able to accept a second electron even

60. H. O. House, R. W. Giese, K. Kronberger, J. P. Kaplan, and J. F. Simeone, *J. Am. Chem. Soc.*, **92**, 2800 (1970).
61. T. G. Halsall, D. W. Theobald, and K. B. Walshaw, *J. Chem. Soc.*, 1029 (1964).
62. (a) H. A. Smith, B. J. L. Huff, W. J. Powers, III, and D. Caine, *J. Org. Chem.*, **32**, 2851 (1967). (b) L. E. Hightower, L. R. Glasgow, K. M. Stone, D. A. Albertson, and H. A. Smith, *ibid.*, **35**, 1881 (1970). (c) M. J. Weiss and co-workers, *Tetrahedron*, **20**, 357 (1964).

in the absence of a proton donor.[60] In many studies no proton donor (such as *t*-butyl alcohol) was added during the initial phase of the reduction.[63] In these cases, the initial protonation of the radical anion **[52]**→**[53]** was presumably

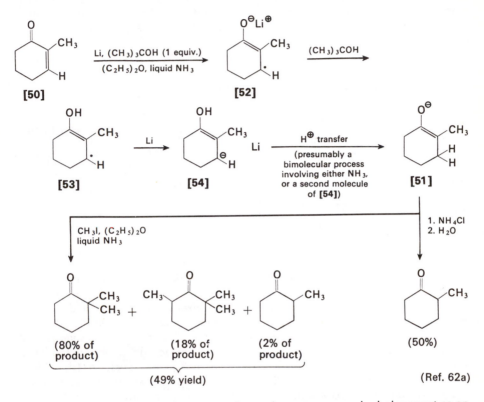

[50]

Li, (CH₃)₃COH (1 equiv.)
(C₂H₅)₂O, liquid NH₃

[52]

(CH₃)₃COH

[53]

Li

[54]

Li

H⊕ transfer

(presumably a bimolecular process involving either NH₃, or a second molecule of **[54]**)

[51]

1. NH₄Cl
2. H₂O

CH₃I, (C₂H₅)₂O
liquid NH₃

(80% of product)

(18% of product)

(2% of product)

(50%)

(49% yield)

(Ref. 62a)

effected by small amounts of an ammonium salt, water, or an alcohol present as an impurity in one of the reactants or solvents or by a second molecule of the starting ketone. [It should be noted that ammonia and primary amines (pK_a 35–36) are much weaker acids than ammonium salts (pK_a 10), alcohols (pK_a 16–19), and enolizable ketones (pK_a 19–23, see Chapter 9) ; consequently, ammonia is unlikely to serve as a proton donor when any of these other materials are present.] This practice of not adding an equivalent amount of a proton donor frequently leads to lower yields of monomeric reduction products for one of two reasons. Any starting ketone which serves as a proton donor for the radical anion will be converted to the corresponding enolate anion. This anion is resistant to further reduction and

63. (a) G. Stork and S. D. Darling, *J. Am. Chem. Soc.*, **86**, 1761 (1964). (b) D. H. R. Barton and C. H. Robinson, *J. Chem. Soc.*, 3045 (1954). (c) M. Fetizon and J. Gore, *Tetrahedron Letters*, **No. 5**, 471 (1966). (d) P. Angibeaud, M. Larcheveque, H. Normant, and B. Tchoubar [*Bull. Soc. Chim. France*, 595 (1968)] suggest the abstraction of a hydrogen atom from hexamethylphosphoramide when no proton donor is present.

will either reform the starting material when the reaction mixture is hydrolyzed or may undergo condensation reactions with a second molecule of the starting ketone. In circumstances in which the initial anion radical (e.g., [52]) exists as a tight ion pair with the metal cation, this species tends to yield dimeric products unless a proton donor is present so that protonation and further reduction of this intermediate will be rapid. The accompanying equation indicates the variation in yields of monomeric and dimeric reduction products which are observed with an unsaturated ketone [55] which lacks acidic hydrogen atoms and cannot function as

$$(CH_3)_3CCH{=}CHCOC(CH_3)_3 \xrightarrow{\text{reduction}} \xrightarrow{H_2O}$$

[55]

$$(CH_3)_3CCH_2CH_2COC(CH_3)_3 \quad + \quad \begin{array}{c} (CH_3)_3C{-}CHCH_2COC(CH_3)_3 \\ | \\ (CH_3)_3C{-}CHCH_2COC(CH_3)_3 \end{array}$$

(Ref. 3c)

Reduction conditions	Yields	
Na, THF, 0°	2%	98%
Na, THF, $(CH_3)_3COH$, 0°	73%	27%
Na, $[(CH_3)_2N]_3PO$, THF, −33°	18–25%	70%
Na, $[(CH_3)_2N]_3PO$, THF, $(CH_3)_3COH$, −33°	68%	28%

a proton donor. For these reasons, a number of workers recommend the addition of one equivalent or more of a proton donor during reductions of unsaturated ketones.[60,62a,b,64] Often in these reductions an excess of the proton donor and metal are used, with the result that the initially formed enolate anion is protonated to form a saturated ketone in the reaction mixture. This saturated ketone is then further reduced to the corresponding alcohol. For this reason, the crude reduction product is often oxidized with chromic acid in aqueous acetone (the Jones oxidation procedure, see Chapter 5) before purification is attempted, so that any saturated alcohol formed will be reoxidized to a saturated ketone.[60,62a,64] The use of triphenylcarbinol as a proton donor in these reductions has been reported to diminish greatly the further reduction of the saturated ketone to an alcohol.[62b] However, the difficulty of separating the desired reaction product from triphenylcarbinol would appear to make this procedure of questionable value for preparative work.

The metal reductions of α,β-unsaturated ketones, like the ketone reductions described earlier, usually yield a saturated ketone which has the more stable configuration at the β-carbon atom. The configuration at the α-carbon atom, which is controlled by the nature of the protonation of the intermediate enol or enolate (see Chapter 9), is usually of less concern, since the stereochemistry at

64. (a) M. J. T. Robinson, *Tetrahedron*, **21**, 2475 (1965). (b) S. K. Malhotra, D. F. Moakley, and F. Johnson, *Tetrahedron Letters*, **No. 12**, 1089 (1967). (c) S. D. Darling, O. N. Devgan, and R. E. Cosgrove, *J. Am. Chem. Soc.*, **92**, 696 (1970). For a recent study of the preparation and properties of the anion radical from tri-1-naphthylborane, see H. J. Shine, L. D. Hughes, and P. Gesting, *J. Organometal. Chem.*, **24**, 53 (1970).

this center is subject to change if the initial product is subjected to acid- or base-catalyzed enolization. As indicated in the foregoing discussion, the reduction stereochemistry at the β-position is established by protonation of an allylic anion (or organometallic intermediate) such as [54]. Although the geometry of such allylic carbanions is uncertain and various viewpoints have been expressed,[60,63a,64] the existing evidence[65] suggests that these intermediates are probably pyramidal [56] rather than planar, especially when they are associated with a metal cation. These anions are stabilized by overlap of the carbanion sp^3 orbital with the adjacent pi-orbital of the double bond, so that the conformations [56] are favored energetically with respect to other conformations obtained by rotation about the $C\alpha$—$C\beta$ bond.[63a,65a] It seems likely that these anions [56] can equilibrate (by rotation, inversion, or both) more rapidly than they are protonated and that at least the protonations of those anions which have a metal counter ion with simple alcohols in relatively nonpolar media occur with retention of configuration, as implied in structure [56b].

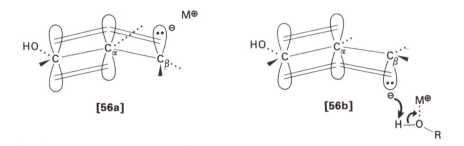

[56a]   [56b]

The foregoing considerations suggest that the stereochemical outcome at the β-carbon in these reductions will be determined by the allylic anion (or perhaps its precursor, the allylic radical) adopting the conformation of lowest energy prior to protonation. The usual stereochemical outcome from metal reductions of $\Delta^{1,9}$-octal-2-one derivatives is illustrated by the formation of the trans-decalones [49] and [57].

Of the three probable conformations [58] for a pyramidal allylic anion intermediate, conformer [58c] has less serious 1,3-diaxial interactions than do the other conformers, but in this conformation the carbanion sp^3 orbital is approximately at right angles to the double-bond pi-orbital, so that stabilization by orbital overlap is seriously impaired. Of the two conformations where overlap is effective, conformer [58a] has fewer 1,3-diaxial interactions (see heavy arrows in [58a] and [58b]) than conformer [58b]. Hence conformer [58a] would be expected to be most stable and serve as the precursor for the observed product [57].

65. (a) D. J. Cram, *Fundamentals of Carbanion Chemistry,* Academic Press, New York, 1965, pp. 47–84. (b) P. E. Verkade, K. S. DeVries, and B. M. Wepster, *Rec. Trav. Chem.,* **84**, 1295 (1965). (c) Appropriate labeling studies have demonstrated that allyl carbanions become symmetrically solvated faster than they react with liquid ammonia. E. Grovenstein, Jr., S. Chandra, C. E. Collum, and W. E. Davis, Jr., *J. Am. Chem Soc.,* **88**, 1275 (1966).

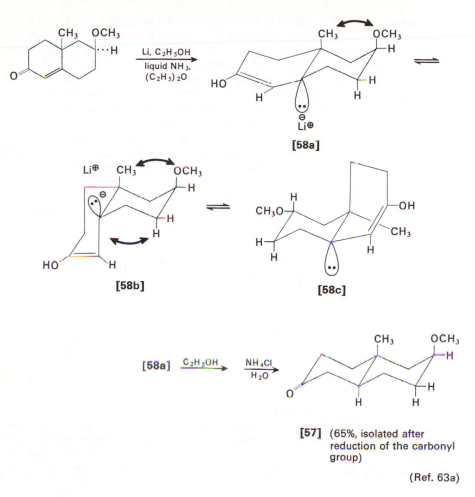

[57] (65%, isolated after reduction of the carbonyl group)

(Ref. 63a)

Recently reductions of enones effected with solutions of sodium in hexamethyl-phosphoramide[60] or with ethereal solutions of the radical ion derived from metals and trimesitylboron[64c,d] have demonstrated that the stereochemistry of these enone reductions may be modified by changing the metal atom,[64c,66] the polarity of the reaction solvent, and the reaction temperature. The outcome of untried modifications in reaction conditions is difficult to predict in the absence of firm knowledge about the geometry, extent of ion-pairing or covalent bonding, and state of aggregation of the intermediate allylic anions (or organometallic derivatives), such as **[54]** and **[59]**. However, it would appear that changes in reaction conditions which will favor tight ion-pairs or covalent bonding between the metal cation and allylic anion (e.g., use of lithium cation, use of nonpolar ethereal solvents, use of

66. G. E. Arth and co-workers, *J. Am. Chem. Soc.,* **76,** 1715 (1954).

low reaction temperatures) all tend to yield more of the less stable reduction product than is obtained in a typical metal-ammonia reduction. This result is what might be expected for intermediates such as [59] where the steric bulk of the metal atom M and its associated solvent will tend to increase the relative stability of the conformer [59b] in which the large solvated metal substituent, M, is equatorial to one of the two rings. This isomer [59b] would be expected, upon protonation, to yield an atypical *cis*-fused decalone product.

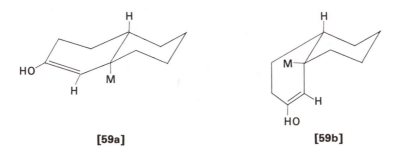

[59a]   [59b]

Just as was true for the metal reductions of non-conjugated ketones, metal reductions of certain enones (e.g., [60] and [61]) give predominantly either the less stable epimer of the final reduction product or the atypical reduction product (e.g. a *cis*-fused decalone derivative).[67] Such cases are characterized by the fact

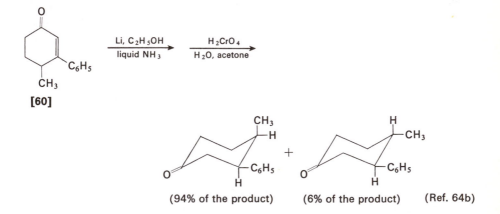

(94% of the product)   (6% of the product)   (Ref. 64b)

67. (a) H. O. House and H. W. Thompson, *J. Org. Chem.*, **28**, 360 (1963). (b) W. Herz and J. J. Schmid, *ibid.*, **34**, 3473 (1969); P. A. Mayor and G. D. Meakins, *J. Chem. Soc.*, 2800 (1960). For other discussion of the effect of conformational factors on reduction stereochemistry, see (c) F. Johnson, *Chem. Rev.*, **68**, 375 (1968). (d) L. Velluz, J. Valls, and G. Nominé, *Angew. Chem., Intern. Ed. Engl.*, **4**, 181 (1965).

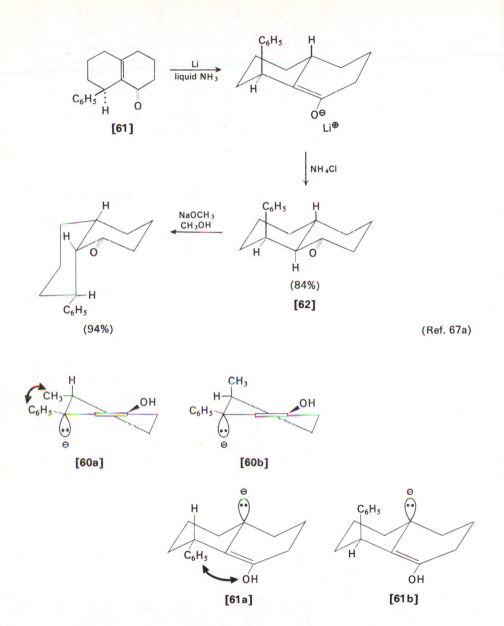

(Ref. 67a)

that conformations, such as **[60a]** and **[61a]**, which would lead to the more stable product with all substituents equatorial (or the more typical product), suffer from a serious nonbonded repulsion in the intermediate allylic anion (arrows in structures **[60a]** and **[61a]**) which may be avoided in the alternative conformations which lead to the observed products. With specially constituted conjugated systems, it appears that the reduction stereochemistry can also be influenced by

electrostatic destabilization of one conformer of the allylic anion intermediate[68c] or by providing an alcohol function which can serve as an intramolecular proton donor during the reduction.[68d]

The final protonation at the alpha carbon atom of the enolate anion to form the product is usually a kinetically controlled process. The kinetic protonation of an enolate anion at carbon must, for stereoelectronic reasons, occur from a direction perpendicular to the plane of the enolate anion. Although this protonation frequently occurs from the less hindered side of the enolate anion, no such generalization always is valid since other factors, such as the steric requirements of the proton donor, may influence the direction from which the proton is added.[68a,b] In a number of cases (e.g., [62]) kinetic protonation leads to a product that does not have the more stable configuration at the alpha carbon atom. However, the initial product may be subsequently equilibrated in the presence of acid or base to produce the diastereoisomer which does have the more stable configuration at the alpha carbon.

The enol ethers of 1,3-dicarbonyl compounds may also be reduced with solutions of metals in ammonia.[68e] The following example illustrates the use of this method to convert a β-keto ester to a saturated ester. This reaction utilizes the ability of the initially formed enolate anion to undergo an elimination in the reaction mixture; further reduction of the α,β-unsaturated ester then occurs.

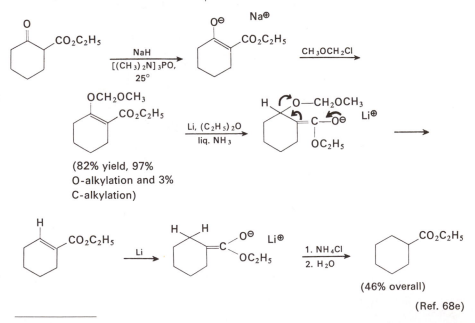

(82% yield, 97% O-alkylation and 3% C-alkylation)

(46% overall)

(Ref. 68e)

68. (a) H. E. Zimmerman In P. deMayo, ed., *Molecular Rearrangements*, Vol. 1, Wiley-Interscience, New York, 1963, p. 345. (b) E. J. Corey and R. A. Sneen, *J. Am. Chem. Soc.,* **78**, 6269 (1956). (c) U. R. Ghatak, N. R. Chatterjee, A. K. Banerjee, J. Chakravarty, and R. E. Moore, *J. Org. Chem.,* **34**, 3739 (1969). (d) L. H. Knox and co-workers, *ibid.,* **30**, 2198 (1965). (e) R. M. Coates and J. E. Shaw, *ibid.,* **35**, 2597, 2601 (1970). Other examples are given in Ref. 75c.

The carbon-carbon double bonds of some α,β-unsaturated carbonyl compounds can be reduced with chromium(II) salts in aqueous dimethylformamide solution.[69] Although the applicability of this reduction procedure is rather limited, it should be noted that the reducing power of the chromium(II) species toward unsaturated ketones can be enhanced significantly by coordination of the metal ion with ammonia or ethylenediamine.[69f,g] α,β-Unsaturated carbonyl compounds can also be reduced electrochemically, as the accompanying equations indicate.

$$Cr_2(SO_4)_3 + Zn\!-\!Hg \xrightarrow[\text{H}_2\text{O, 80}^\circ]{\text{N}_2 \text{ atmosphere}} CrSO_4 + ZnSO_4$$

(blue aqueous solution which
is stable if protected from
oxygen)

$$C_2H_5O_2C\!-\!CH\!=\!CH\!-\!CO_2C_2H_5 \xrightarrow[\substack{(\text{CH}_3)_2\text{NCHO,} \\ \text{N}_2 \text{ atmosphere}}]{\text{CrSO}_4, \text{ H}_2\text{O}} \xrightarrow[\text{H}_2\text{O}]{(\text{NH}_4)_2\text{SO}_4}$$

$$C_2H_5O_2C\!-\!CH_2CH_2CO_2C_2H_5$$

(88–94%)   (Ref. 69a)

$$(CH_3)_2C\!=\!CH\!-\!CO\!-\!CH_3 + Cr(NH_3)_4{}^{+2} \xrightarrow[\text{H}_2\text{O, 25}^\circ]{\text{NH}_3} (CH_3)_2CH\!-\!CH_2COCH_3$$

(33%, isolated as the 2,4-
dinitrophenylhydrazone)

(Ref. 69g)

$$C_6H_5CH\!=\!CHCO{}^\ominus{}_2Na{}^\oplus \xrightarrow[\text{H}_2\text{O, Na}_2\text{SO}_4]{\text{Hg cathode}} \xrightarrow{\text{H}_2\text{SO}_4} C_6H_5CH_2CH_2CO_2H$$

(80–90%)   (Ref. 69d)

The electrochemical reductions of α,β-unsaturated ketones and related compounds in aprotic media and in the absence of metal cations can in some instances lead to the formation of relatively stable anion radicals such as [63].[3c,70] However, when proton donors are present in the reaction medium, the anion radicals are protonated to form allylic radicals such as [64]. Although these intermediates appear to have relatively long lifetimes[71] when compared with many

69. (a) A. Zurqiyah and C. E. Castro, *Org. Syn.*, **49**, 98 (1969). (b) C. E. Castro, R. D. Stephens, and S. Moje, *J. Am. Chem. Soc.*, **88**, 4964 (1966). (c) For the reduction of enediones, see J. R. Hanson and E. Premuzic, *J. Chem. Soc.*, C, 1201 (1969). (d) A. W. Ingersoll, *Org. Syn.*, **Coll. Vol. I,** 311 (1944). (e) S. Ono and T. Hayashi, *Bull. Chem. Soc. Japan*, **26**, 11 (1953). (f) H. O. House and N. P. Peet, to be published. (g) K. D. Kopple, *J. Am. Chem. Soc.*, **84**, 1586 (1962).
70. A. Berndt, *Angew. Chem., Intern. Ed. Engl.*, **6**, 251 (1967); *Tetrahedron Letters*, **No. 2,** 177, (1970).
71. (a) G. Rämme, R. L. Strong, and H. H. Richtol, *J. Am. Chem. Soc.*, **91**, 5711 (1969). (b) T. J. Bennett, R. C. Smith, and T. H. Wilmshurst, *Chem. Commun.*, **No. 11**, 513 (1967).

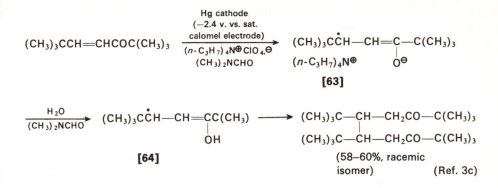

[63]

[64]

(58–60%, racemic
isomer)   (Ref. 3c)

free radical intermediates, these radicals **[64]** still normally dimerize or disproportionate more rapidly than they can diffuse back to an electrode or metal surface
to undergo further reduction. Although these allyl radical intermediates (e.g., **[64]**)
have usually been observed to dimerize by coupling at the β-position as indicated,
when two β-substituents are present and the beta position is relatively hindered
as with the enone **[65]**, coupling at the carbonyl carbon to form a pinacol may be
observed. Occasionally, examples of "mixed" dimers such as **[66]** are found. As

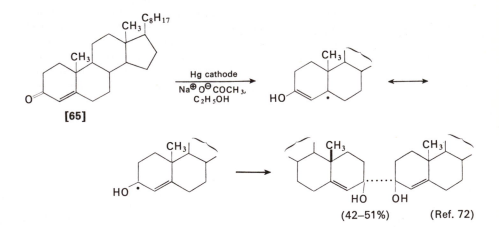

[65]

(42–51%)   (Ref. 72)

noted previously, these same types of reductive dimerizations have been observed
when unsaturated carbonyl compounds are allowed to react with various metals
such as lithium, potassium, sodium, sodium amalgam, aluminum amalgam, zinc,

---

72. (a) P. Bladon, J. W. Cornforth, and R. H. Jaeger, *J. Chem. Soc.*, 863 (1958). (b) H.
Lund, *Acta Chim. Scand.*, **11**, 283 (1957). (c) This same dimer has been isolated as a by-
product from the reduction of the enone **[65]** with lithium in liquid ammonia; private
communication from Professor T. A. Spencer, Dartmouth College.

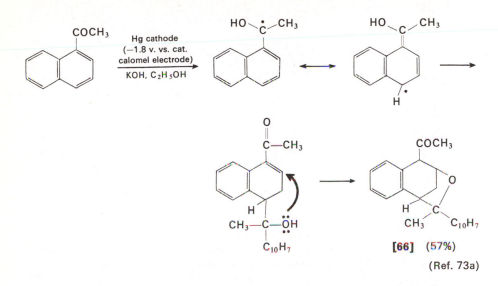

[66] (57%)

(Ref. 73a)

and magnesium.[73d,74] The formation of monomeric reduction products in these reductions over a metal surface is impeded by the necessity for the intermediate allylic radical (e.g., [64]) to diffuse back to the surface of the electrode or metal for further reduction. With metal-ammonia or chromium(II)-salt reductions in which the metal is in solution, this difficulty is not encountered and monomeric reduction products almost always predominate. A possible solution to this problem involves the concurrent electrolytic generation of a soluble reducing agent which can intercept the radical intermediates before dimerization. This idea is exemplified by the subsequently discussed reductions of aromatic systems by the electrolytic reduction of lithium salts in amine solvents, by the electrolytic generation of solutions of magnesium in liquid ammonia,[75c] and by the reduction of unsaturated carbonyl compounds such as [67] in the presence of amine salts.[75a,b] This latter

73. (a) J. Grimshaw and E. J. F. Rea, *J. Chem. Soc.,* C, 2628 (1967). (b) For other examples of the more typical dimerization at the β-carbon, see J. F. Archer and J. Grimshaw, *ibid.,* B, 266 (1969); S. Wawzonek and A. Gundersen, *J. Electrochem. Soc.,* **111,** 324 (1964). (c) J. Wiemann and M. L. Bouguerra, *Compt. rend.,* **265,** 751 (1967). (d) For a review of reductive dimerizations involving coupling at aromatic nuclei, see R. C. Fuson, *Rec. Chem. Progr.,* **12,** 1 (1951).
74. (a) C. G. Overberger and A. M. Schiller, *J. Org. Chem.,* **26,** 4230 (1961). (b) E. L. Totton, N. C. Camp, III, G. M. Cooper, B. D. Haywood, and D. P. Lewis, *ibid.,* **32,** 2033 (1967). (c) H. Rosen, Y. Arad, M. Levy, and D. Vofsi, *J. Am. Chem. Soc.,* **91,** 1425 (1969). (d) P. Matsuda, *Tetrahedron Letters,* **No. 49,** 6193 (1966). (e) A. Zysman, G. Dana, and J. Wiemann, *Bull. Soc. Chim. France,* 1019 (1967); J. Wiemann, M. R. Monot, G. Dana, and J. Chuche, *ibid.,* 3293 (1967); E. Touboul, F. Weisbuch, and J. Wiemann, *ibid.,* 4291 (1967). (f) C. Glacet, *Compt. rend.,* **227,** 480 (1948); J. Wiemann and R. Nahum, *ibid.,* **238,** 2091 (1954).
75. (a) R. N. Gourley, J. Grimshaw, and P. G. Miller, *J. Chem. Soc.,* C, 2318 (1970). (b) For a comparable example of asymmetric induction in other electrochemical reductions, see L. Horner and D. H. Skaletz, *Tetrahedron Letters,* **No. 42,** 3679 (1970), and references therein. (c) A. Spassky-Pasteur, *Bull. Soc. Chim. France,* 2900 (1969).

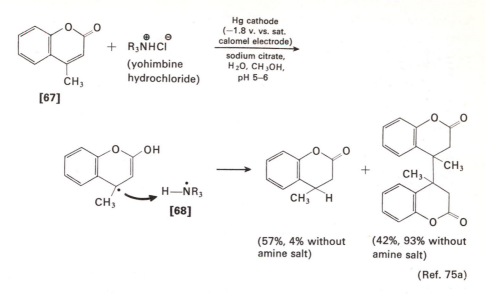

(57%, 4% without
amine salt)

(42%, 93% without
amine salt)

(Ref. 75a)

process is thought to involve the concurrent reduction of the tertiary amine salt to
a radical **[68]** which transfers a hydrogen atom to the intermediate allylic radical.
In agreement with this hypothesis, the use of an optically active ammonium salt
induces some asymmetry in the reduced product.[75a,b]

The previously described reductive dimerizations of α,β-unsaturated carbonyl
compounds to form a new carbon-carbon bond at the β-carbon atom has been
shown to have considerable synthetic utility.[76] This coupling, which has been
called hydrodimerization or electrohydrodimerization, is illustrated by the accompanying equations.

The first example, involving a mixed coupling of two conjugated olefins **[69]**
and **[70]** that are reduced at different potentials, demonstrates the occurrence of
some coupling mechanism in addition to the previously discussed coupling of
allylic radicals (e.g., **[71]**→**[72]**). Since a mixed coupling product **[73]** was
formed, although the reduction potential used was insufficient to reduce acrylonitrile **[70]**, it is apparent that some intermediate formed in the reduction of the
diester **[69]** is intercepted by the acrylonitrile **[70]** present at high concentration
in the reaction mixture. Although several reaction paths have been suggested,[76] it

76. (a) M. M. Baizer, *J. Org. Chem.,* **29**, 1670 (1964); *ibid.,* **31**, 3847 (1966). (b) M. M.
Baizer and J. D. Anderson, *ibid.,* **30**, 1348, 1351, 1357, 3138 (1965). (c) J. D. Anderson,
M. M. Baizer, and E. J. Prill, *ibid.,* **30**, 1645 (1965). (d) J. D. Anderson, M. M. Baizer,
and J. P. Petrovich, *ibid.,* **31**, 3890, 3897 (1966). (e) J. H. Wagenknecht and M. M. Baizer,
ibid., **31**, 3885 (1966). (f) M. R. Ort and M. M. Baizer, *ibid.,* **31**, 1646 (1966). (g) M. M.
Baizer and J. D. Anderson, *J. Electrochem. Soc.,* **111**, 223, 226 (1964); M. M. Baizer, *ibid.,*
111, 215 (1964). (h) For reviews, see M. M. Baizer, J. D. Anderson, J. H. Wagenknecht,
M. R. Ort, and J. P. Petrovich, *Progr. Electrochem. Acta.,* **12**, 1377 (1967); J. D. Anderson,
J. P. Petrovich, and M. M. Baizer, *Adv. Org. Chem.,* **6**, 257 (1969); M. M. Baizer and J. P.
Petrovich, *Progr. Phys. Org. Chem.,* **7**, 189 (1970).

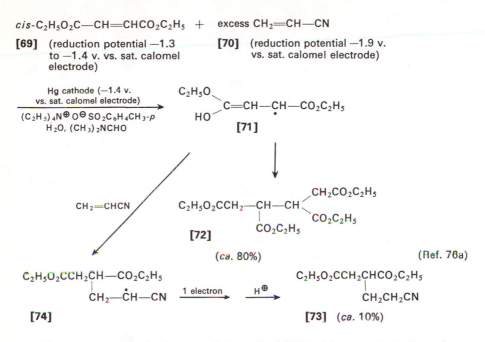

$cis\text{-}C_2H_5O_2C\!-\!CH\!=\!CHCO_2C_2H_5$ + excess $CH_2\!=\!CH\!-\!CN$

[69] (reduction potential −1.3
to −1.4 v. vs. sat. calomel
electrode)

[70] (reduction potential −1.9 v.
vs. sat. calomel electrode)

Hg cathode (−1.4 v.
vs. sat. calomel electrode)
$(C_2H_5)_4N^{\oplus}O^{\ominus}SO_2C_6H_4CH_3\text{-}p$
H_2O, $(CH_3)_2NCHO$

$\begin{array}{c}C_2H_5O\\ HO\end{array}\Big\rangle C\!=\!CH\!-\!\overset{\cdot}{C}H\!-\!CO_2C_2H_5$

[71]

$CH_2\!=\!CHCN$

$C_2H_5O_2CCH_2\!-\!CH\!-\!CH\Big\langle\begin{array}{c}CH_2CO_2C_2H_5\\ CO_2C_2H_5\end{array}$

$\underset{CO_2C_2H_5}{|}$

[72]

(*ca.* 80%)   (Ref. 76a)

$\underset{\underset{CH_2-\overset{\cdot}{C}H-CN}{|}}{C_2H_5O_2CCH_2CH\!-\!CO_2C_2H_5}$

[74]

$\xrightarrow{\text{1 electron}}$ $H^{\oplus}$ →

$\underset{\underset{CH_2CH_2CN}{|}}{C_2H_5O_2CCH_2CHCO_2C_2H_5}$

[73] (*ca.* 10%)

would appear most likely that the allylic radical **[71]** adds to acrylonitrile to form
an α-cyano radical **[74]** which is rapidly reduced and protonated to yield the mixed
coupling product **[73]**. The formation of minor amounts of unsymmetrical dimers
(e.g., **[75]**) is consistent with the occurrence of this radical addition process.

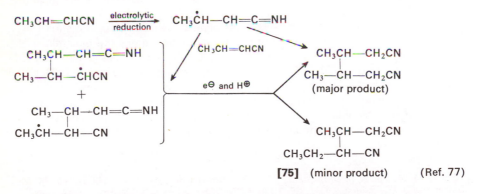

$CH_3CH\!=\!CHCN$ $\xrightarrow[\text{reduction}]{\text{electrolytic}}$ $CH_3\overset{\cdot}{C}H\!-\!CH\!=\!C\!=\!NH$

$\underset{\underset{CH_3-\overset{\cdot}{C}H-\overset{\cdot}{C}HCN}{|}}{CH_3CH\!-\!CH\!=\!C\!=\!NH}$

$+$

$\underset{\underset{CH_3\overset{\cdot}{C}H-\overset{\cdot}{C}H-CN}{|}}{CH_3\!-\!CH\!-\!CH\!=\!C\!=\!NH}$

$CH_3CH\!=\!CHCN$

$e^{\ominus}$ and $H^{\oplus}$

$\underset{\underset{CH_3-\overset{\cdot}{C}H-CH_2CN}{|}}{CH_3CH\!-\!CH_2CN}$
(major product)

$\underset{\underset{CH_3CH_2-\overset{\cdot}{C}H-CN}{|}}{CH_3CH\!-\!CH_2CN}$

[75] (minor product)   (Ref. 77)

   A similar reaction is the intramolecular hydrodimerization to form cyclic
products such as **[76]**. It will be noted that the formation of cyclic products **[76]**
(mixtures of *cis*- and *trans*-isomers were obtained) is only useful for the formation
of three- to six-membered rings where cyclization is kinetically favorable. Although

77. G. C. Jones and T. H. Ledford, *Tetrahedron Letters,* **No. 7**, 615 (1967).

these intramolecular reductive dimerizations may also involve the previously discussed allylic radical addition indicated in structure [78], the cyclization of the anion radical [79] before protonation is a reasonable alternative. The fact that diesters which cyclize (e.g., [80], $n = 3$) are reduced at potentials ca. 0.2–0.3 v. less negative than esters (e.g., [80], $n = 6$) which give noncyclic products such as [77] is best explained by an irreversible chemical step following the reversible reduction

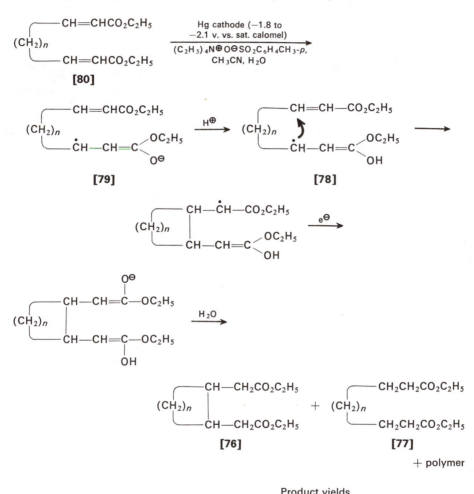

Product yields

Value of n	[76]	[77]
2	41%	48%
3	ca. 100%	—
4	90%	—
6	—	43%
8	—	50%

(Ref. 76d)

of the esters [80] to their radical anions [79]. Thus, if cyclization involves the neutral radical [78], the preceding protonation step must be rapid and reversible.

Carbon-carbon double bonds conjugated with aromatic systems or with other multiple bonds may also be reduced with metals. The nature of the reduction product is dependent on the availability of a proton donor in the reaction medium, a fact illustrated in the accompanying examples. In the absence of excess proton donor, dimerization of the initially formed anion radical (e.g., [81]) is observed.

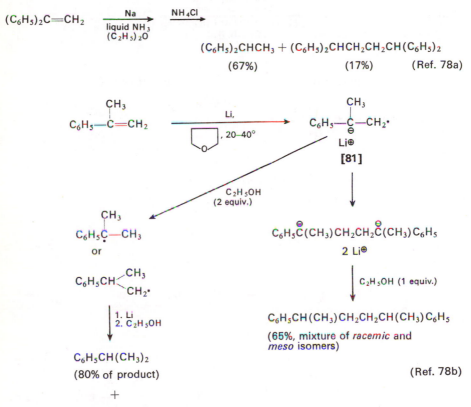

78. (a) H. Gilman and J. C. Bailie, *J. Am. Chem. Soc.*, **65**, 267 (1943). (b) D. R. Weyenberg, *J. Org. Chem.*, **30**, 3236 (1965). (c) As implied in structure [81], the highest unpaired electron density in styrene anion radicals is found at the β-carbon atom; A. R. Buick, T. J. Kemp, G. T. Neal, and T. J. Stone, *Chem. Commun.*, No. 5, 282 (1970). (d) For examples in which similar anion radical intermediates have been trapped with aziridine radical, acetone, trimethylsilyl chloride, or carbon dioxide, see R. K. Razdan, *ibid.*, No. 13, 770 (1969); R. Dietz and M. E. Peover, *Dis. Faraday Soc.*, No. 45, 154 (1968); J. K. Kochi, *J. Org. Chem.*, **28**, 1960, 1969 (1963); D. R. Weyenberg, L. H. Toporcer, and A. E. Bey, *ibid.*, **30**, 4096 (1965).

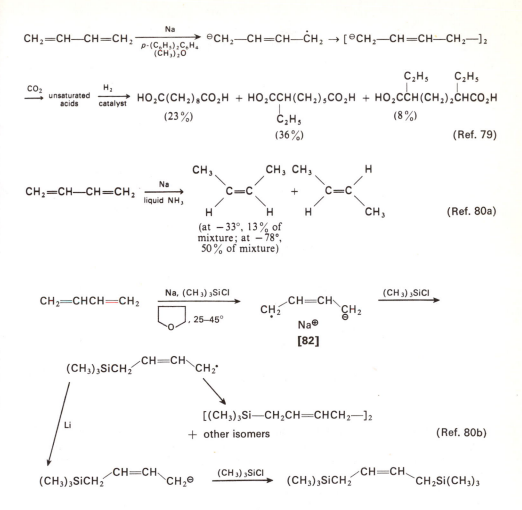

Both the reduction of 1,3-dienes and trapping experiments with trimethylsilyl chloride have suggested that the initial anion radical formed from these acyclic dienes has the indicated *cis*-configuration **[82]** when it is formed at low temperatures or in nonpolar media. This configuration for the tight-ion-pair intermediate is presumably favored for electrostatic reasons. These reductions of conjugated double bonds can also be effected electrolytically. The following equation illustrates both the reduction of a conjugated olefin and partial trapping of the intermediate anion radical to yield a mixed hydrodimerization product.

79. National Distillers and Chemical Corp., Brit. Pat. No. 756,385 dated Sept. 5, 1956; *CA*, **51**, 15557 (1957).

80. (a) N. L. Bauld, *J. Am. Chem. Soc.*, **84**, 4345, 4347 (1962). (b) D. R. Weyenberg, L. H. Toporcer, and L. E. Nelson, *J. Org. Chem.*, **33**, 1975 (1968).

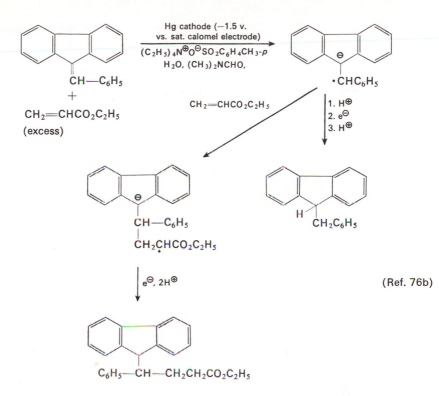

Hg cathode (−1.5 v.
vs. sat. calomel electrode)

$(C_2H_5)_4N^{\oplus}O^{\ominus}SO_2C_6H_4CH_3$-$p$

H_2O, $(CH_3)_2NCHO$,

$CH_2=CHCO_2C_2H_5$

1. $H^{\oplus}$
2. $e^{\ominus}$
3. $H^{\oplus}$

$e^{\ominus}$, $2H^{\oplus}$

(Ref. 76b)

The ability of alkali metals to partially reduce aromatic systems has proved especially useful in organic synthesis.[1] Illustrative is the reduction of benzene to 1,4-dihydrobenzene **[86]** with lithium and ethanol in liquid ammonia. In reductions of benzene derivatives, the anion radical **[83]** is formed reversibly in low concentration and then reacts further with the protonic solvent to form the radical **[84]** and, subsequently, the anion **[85]** and the dihydro derivatives.[81] Thus one important function of the alcohol in the reduction of benzene derivatives is to provide a proton source that is more acidic than ammonia. A number of polycyclic aromatic systems, which have less negative reduction potentials than benzene, are reduced

81. (a) A. P. Krapcho and A. A. Bothner-By, *J. Am. Chem. Soc.*, **81**, 3658 (1959); **82**, 751 (1960). (b) J. F. Eastham and D. R. Larkin, *ibid.*, **81**, 3652 (1959). (c) A. P. Krapcho and M. E. Nadel, *ibid.*, **86**, 1096 (1964); Although internal olefins are normally not reduced with metals in ammonia, certain non-conjugated olefins with low-lying antibonding orbitals are exceptions. B. R. Ortiz de Montellano, B. A. Loving, T. C. Shields, and P. D. Gardner, *ibid.*, **89**, 3365 (1967). (d) O. J. Jacobus and J. F. Eastham, *ibid.*, **87**, 5799 (1965). (e) A *bis*-trimethylsilyl derivative of dihydrobenzene is formed when the reduction is effected in the presence of trimethylsilyl chloride. D. R. Weyenberg and L. H. Toporcer, *ibid.*, **84**, 2844 (1962). (f) The e.p.r. spectra of the radical anions have been described by T. R. Tuttle, Jr., and S. I. Weissman, *ibid.*, **80**, 5342 (1958).

with an alkali metal in liquid ammonia in the absence of an alcohol, with suspensions of sodium in an alcohol, or with less reactive metals such as calcium or magnesium.[82]

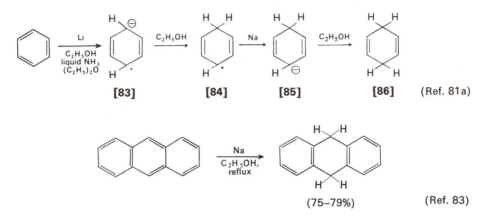

[83]   [84]   [85]   [86]   (Ref. 81a)

(75–79%)   (Ref. 83)

With attention to the detailed reaction conditions employed, it has been possible to effect selective reductions of various polycyclic aromatic systems with metal-ammonia solutions to form dihydro, tetrahydro, or hexahydro derivatives.[84] The presence of iron compounds, common impurities in *undistilled* liquid ammonia, lowers the overall reducing ability of lithium-ammonia solutions presumably by catalyzing the reaction of lithium with ammonia to form hydrogen and lithium amide. Consequently, in the presence of iron salts (which are reduced to metallic iron) and no alcohol cosolvent, anthracene was converted to its dianion [87][85] which was protonated to give the dihydro product [88]. When no iron was present, the excess lithium was not destroyed and further reaction occurred when a proton donor was added. The analogous reduction of 9,10-dialkylanthracenes yields primarily *trans*-dihydro derivatives, whereas reduction of 9,10-dihydro-phenanthrenes produces mainly *cis*-dihydro compounds.[84c-f] Both these stereo-chemical results are explicable in terms of previously discussed pyramidal carbanion intermediates such as [89] which is protonated in its favored conformation (e.g., [89b]) to avoid eclipsing adjacent substituents (arrows in [89a]).

82. (a) D. Bryce-Smith and A. C. Skinner, *J. Chem. Soc.*, C, 154 (1966). (b) A. R. Utke and R. T. Sanderson, *J. Org. Chem.*, **29**, 1261 (1964). (c) P. Markov and C. Ivanov, *Compt. rend.*, **264C**, 1605 (1967).
83. (a) K. C. Bass, *Org. Syn.*, **42**, 48 (1962). (b) E. B. H. Waser and H. Möllering, *ibid.*, **Coll. Vol. 1**, 499 (1944).
84. (a) R. G. Harvey, *J. Org. Chem.*, **32**, 238 (1967). (b) R. G. Harvey and K. Urberg, *ibid.*, **33**, 2206, 2570 (1968). (c) P. W. Rabideau and R. G. Harvey, *ibid.*, **35**, 25 (1970). (d) R. G. Harvey and C. C. Davis, *ibid.*, **34**, 3607 (1969). (e) R. G. Harvey and L. Arzadon, *Tetrahedron*, **25**, 4887 (1969). (f) R. G. Harvey, L. Arzadon, J. Grant, and K. Urberg, *J. Am. Chem. Soc.*, **91**, 4535 (1969). (g) R. H. Harvey and P. W. Rabideau, *Tetrahedron Letters*, **No. 42**, 3695 (1970).
85. It should be noted that the reduction potential for anthracene is significantly less negative than that for naphthalene, phenanthrene, or benzene so that formation of the dianion [87] is a favorable process.

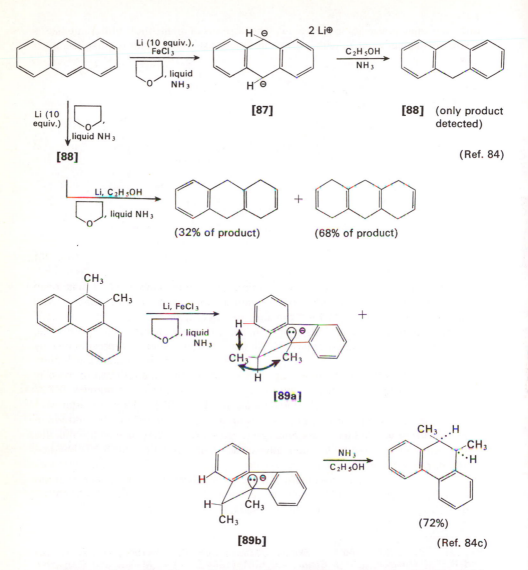

[87]

[88]   (only product detected)

[88]

(Ref. 84)

(32% of product)   (68% of product)

[89a]

[89b]   (72%)

(Ref. 84c)

Among the reductions of polycyclic aromatic compounds, studies with naphthalene have been most extensive. The reversible formation of the naphthalene anion radical [90] and its reaction with proton donors have been supported by e.p.r. data, tracer studies, and electrochemical studies.[8,86] A reaction which would appear analogous to the protonation illustrated in the accompanying equation has

86. (a) F. J. Smentowski and G. R. Stevenson, *J. Am. Chem. Soc.*, **90**, 4661 (1968) and references therein. (b) S. Bank and W. D. Closson, *Tetrahedron Letters*, **No. 19**, 1349 (1965). (c) P. H. Given and M. E. Peover, *J. Chem. Soc.*, 385 (1960).

been reported when the anion radical [90] is treated with trimethylsilyl chloride.[87] However, recent studies of the reaction of this anion radical [90] with alkyl halides have indicated that the alkylated products are derived from radical coupling reactions.[88]

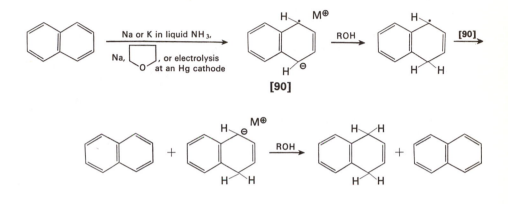

Since the reduction of benzene derivatives occurs only very slowly in the absence of an alcohol (or some other relatively acidic proton donor) in the reaction medium, it is possible to effect the selective reduction of an α,β-unsaturated carbonyl system (e.g., [91]) in the presence of a benzene ring if no alcohol is added. It should be noted that catalytic reduction of the unsaturated ketone [91] leads to a diastereoisomer [92] of the product [93] produced by a dissolving metal reduction.

The kinetically controlled protonation or alkylation of a pentadienyl anion such as [85] to form the nonconjugated diene [86] rather than the conjugated

87. D. R. Weyenberg and L. H. Toporcer, *J. Org. Chem.*, **30**, 943 (1965).
88. (a) W. Hückel and J. Wolfering, *Justus Liebigs Ann. Chem.*, **686**, 34 (1965). (b) W. Hückel and M. Wartini, *ibid.*, **686**, 40 (1965). (c) G. D. Sargent, J. N. Cron, and S. Bank, *J. Am. Chem. Soc.*, **88**, 5363 (1966). (d) J. F. Garst, P. W. Ayers, and R. C. Lamb, *ibid.*, **88**, 4260 (1966). (e) S. J. Cristol and R. V. Barbour, *ibid.*, **88**, 4262 (1966). (f) J. F. Garst, J. T. Barbas, and F. E. Barton, *ibid.*, **90**, 7159 (1968). (g) G. D. Sargent and G. A. Lux, *ibid.*, **90**, 7160 (1968). (h) J. F. Garst and J. T. Barbas, *ibid.*, **91**, 3385 (1969); *Tetrahedron Letters*, **No. 36**, 3125 (1969). (i) J. F. Garst and F. E. Barton, *ibid.*, **No. 7**, 587 (1969). (j) S. Bank and J. F. Bank, *ibid.*, **No. 52**, 4533 (1969). (k) J. F. Garst, R. H. Cox, J. T. Barbas, R. D. Roberts, J. I. Morris, and R. C. Morrison, *J. Am. Chem. Soc.*, **92**, 5761 (1960). (l) For an analogous reaction of an anion radical with α-dialkylamino nitriles, see C. Fabre, H. Ali, and Z. Welvart, *Chem. Commun.*, **No. 18**, 1149 (1970). (m) J. F. Garst and R. H. Cox, *J. Am. Chem. Soc.*, **92**, 6389 (1970). (n) P. W. Rabideau and R. G. Harvey, *Tetrahedron Letters*, **No. 48**, 4139 (1970).

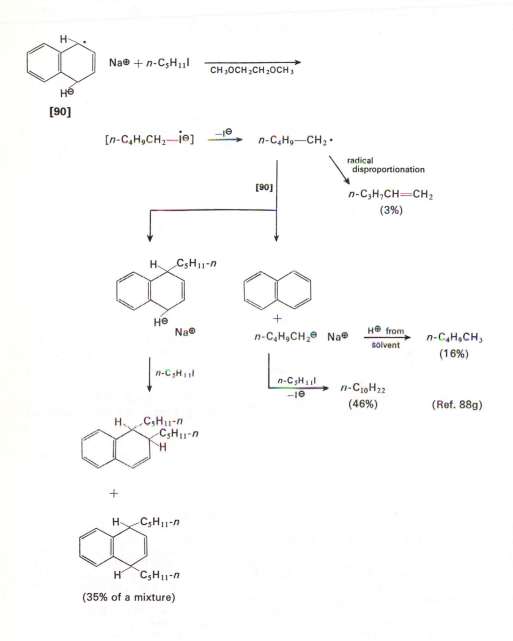

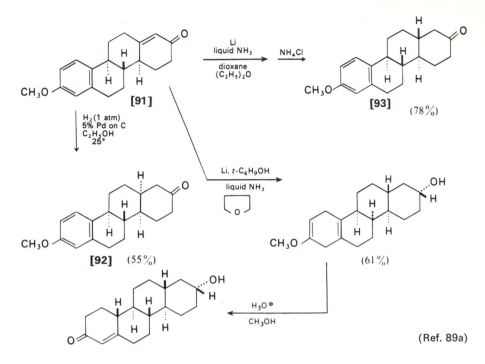

(Ref. 89a)

isomer is representative of a general phenomenon observed in the kinetic protonation of conjugated anions of the type [94].[68a,90] The reaction course indicates that the energy barrier for proton addition (or proton removal) at the center of the system [94] to form the nonconjugated isomer [95] is lower than the energy barrier for the addition (or removal) of a proton to the terminal carbon atom to form the conjugated isomer [96], even though the latter is frequently more stable. It is generally true that the more stable isomer in a set of tautomers is also the less acidic species.

For the reduction of an aromatic system to be stopped at the dihydro stage it is necessary that the nonconjugated diene (e.g., [86]) initially formed not be

89. (a) W. F. Johns, *J. Org. Chem.*, **28**, 1856 (1963). (b) The addition of *t*-butyl alcohol allowed the successful reduction of the aromatic ring in benzofurans without concurrent base-catalyzed cleavage of the heterocyclic ring. S. D. Darling and K. D. Wills, *J. Org. Chem.*, **32**, 2794 (1967).

90. (a) H. J. Ringold and S. K. Malhotra, *Tetrahedron Letters*, **No. 15**, 669 (1962). (b) H. J. Ringold and S. K. Malhotra, *J. Am. Chem. Soc.*, **84**, 3402 (1962). (c) H. J. Ringold and S. K. Malhotra, *ibid.*, **85**, 1538 (1963). (d) R. B. Bates, R. H. Carnighan, and C. E. Staples, *ibid.*, **85**, 3030–3032 (1963). (e) R. B. Bates, D. W. Gosselink, and J. A. Kaczynski, *Tetrahedron Letters*, **No. 3**, 199, 205 (1967). (f) R. B. Bates, E. S. Caldwell, and H. P. Klein, *J. Org. Chem.*, **34**, 2615 (1969). (g) This phenomenon has been discussed in a more general way: G. S. Hammond, *J. Am. Chem. Soc.*, **77**, 334 (1955); J. Hine, *J. Org. Chem.*, **31**, 1236 (1966). The favored geometries of these pentadienyl anions are discussed by R. Hoffmann and R. A. Olofson, *J. Am. Chem. Soc.*, **88**, 943 (1966). (h) J. Hine and D. B. Knight, *J. Org. Chem.*, **35**, 3946 (1970).

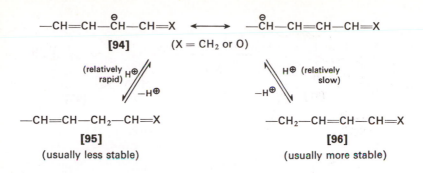

$$-CH=CH-\overset{\ominus}{C}H-CH=X \quad\longleftrightarrow\quad -\overset{\ominus}{C}H-CH=CH-CH=X$$

[94]   (X = CH$_2$ or O)

(relatively rapid)   (relatively slow)

$$-CH=CH-CH_2-CH=X \qquad\qquad -CH_2-CH=CH-CH=X$$

[95]   **[96]**

(usually less stable)   (usually more stable)

reconverted to the anion (e.g., **[85]**). Such reconversion would allow the non-conjugated diene to be in equilibrium with the conjugated diene, which would be reduced further to the tetrahydro stage. Because bases such as NH$_2^{\ominus}$ and R—NH$^{\ominus}$ are sufficiently strong to effect this equilibration, a second function of the alcohol present in metal-ammonia reduction mixtures is to ensure the absence of an appreciable concentration of amide ion (NH$_2^{\ominus}$), a substantially stronger base than alkoxide ion (R—O$^{\ominus}$). The presence of alcohol has the same effect on reductions of aromatic systems by solutions of sodium or lithium in low-molecular-weight amines (e.g., methylamine, b.p. −6°, or ethylamine, b.p. 17°),[91] as illustrated in the accompanying equations.

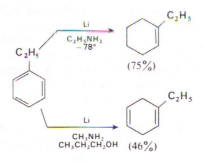

(Ref. 91 a–c)

91. (a) R. A. Benkeser, R. E. Robinson, D. M. Sauve, and O. H. Thomas, *J. Am. Chem. Soc.,* **77**, 3230 (1955). (b) R. A. Benkeser, M. L. Burrous, J. J. Hazdra, and E. M. Kaiser, *J. Org. Chem.,* **28**, 1094 (1963); R. A. Benkeser and co-workers, *ibid.,* **29**, 1313 (1964). (c) Similar reductions may be effected in an autoclave at 60° or above with alkali metals and ammonia with or without a cosolvent. L. H. Slaugh and J. H. Raley, *ibid.,* **32**, 369, 2861 (1967). (d) For examples of reductions with lithium in ethylenediamine, see L. Reggel, R. A. Friedel, and I. Wender, *ibid.,* **22**, 891 (1957); A. W. Burgstahler and L. R. Worden, *J. Am. Chem. Soc.,* **86**, 96 (1964). (e) E. M. Kaiser and R. A. Benkeser, *Org. Syn.,* **50**, 88 (1970). (f) In the reductions of polycyclic aromatic compounds with sodium in an amine solvent, the dihydroaromatic product has been found to add the amide ion formed from the amine solvent to form a secondary or tertiary amine; E. J. Eisenbraun, R. C. Bansal, D. V. Hertzler, W. P. Duncan, P. W. K. Flanagan, and M. C. Hamming, *J. Org. Chem.,* **35**, 1265 (1970).

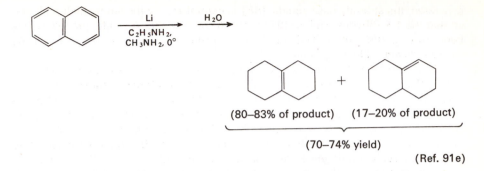

(80–83% of product)   (17–20% of product)

(70–74% yield)

(Ref. 91e)

A comparable phenomenon is seen in the reductions of aromatic systems with the lithium-amine solutions generated by the electrolysis of solutions of lithium chloride.[92] In this procedure, electrolysis with a barrier to prevent mixing of products formed at the anode (a triazine derivative and methylamine hydrochloride) and at the cathode (initially lithium metal) results in the formation of a 1,4-dihydrobenzene derivative [97] and lithium methylamide [98], a strong base. Base-catalyzed isomerization to a conjugated diene (e.g., [99]) results in reduction to a cyclohexene derivative. When the anode and cathode components are not

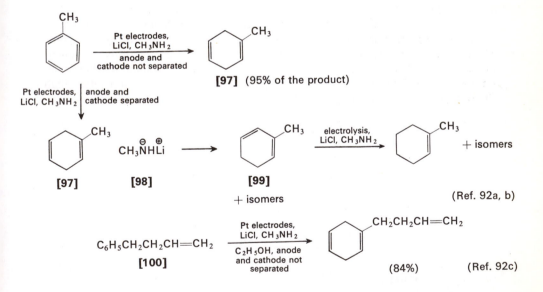

92. (a) R. A. Benkeser and E. M. Kaiser, *J. Am. Chem. Soc.*, **85**, 2858 (1963). (b) R. A. Benkeser, E. M. Kaiser, and R. F. Lambert, *ibid.*, **86**, 5272 (1964). (c) R. A. Benkeser and S. J. Mels, *J. Org. Chem.*, **34**, 3970 (1969). (d) Similar reductions have been effected by the electrolysis of solutions of lithium chloride in hexamethylphosphoramide containing ethanol; H. W. Sternberg, R. E. Markby, I. Wender, and D. M. Mohilner, *J. Am. Chem. Soc.*, **89**, 186 (1967).

separated, the strongly basic amide [98] is neutralized by the amine hydrochloride so that the 1,4-dihydro product [97] is the major product. A similar procedure has been used to effect the selective reduction of the phenyl-substituted olefin [100] to the indicated dihydro product.

As would be anticipated from the previous discussion, substituents which can aid in charge delocation (e.g. phenyl) or electron withdrawing substituents (e.g. —COONa) on an aromatic system will favor the acceptance of an electron to form an anion radical and accelerate reduction, whereas electron-donating substituents (e.g. —CH$_3$, —NH$_2$, —O$^{\ominus}$) will retard reduction.[1,81a,93c] Also, the presence of bulky substituents on an aromatic ring retards reduction, presumably because of steric interference with solvation of the radical anion.[81a,93c] Thus, the order of stability for the anion radicals from alkyl substituted benzenes is:[93c]

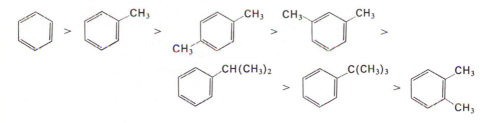

The selective reduction of the non-substituted ring of 1-naphthol, indicated in the accompanying equation, illustrates the deactivating influence of an electron-donating alkoxide substituent. However, it will be noted that the salts of some phenols (e.g., [101]) can be reduced by the use of ammonia solutions containing higher (4 to 5 M) than normal (1.5 M) concentrations of lithium.[93b] These very concentrated solutions of lithium in ammonia take on a bronze color rather than the deep blue color characteristic of dilute solutions.

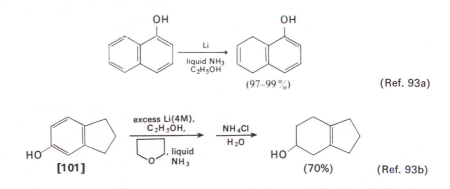

93. (a) C. D. Gutsche and H. H. Peter, *Org. Syn., Coll. Vol. 4*, 887 (1963). (b) J. Fried, N. A. Abraham, and T. S. Santhanakrishnan, *J. Am. Chem. Soc.*, **89**, 1044 (1967). (c) The relative stabilities of alkylbenzene anion radicals in tetrahydrofuran-1,2-dimethoxyethane mixtures have been measured by R. G. Lawler and C. T. Tabit, *ibid.*, **91**, 5671 (1969).

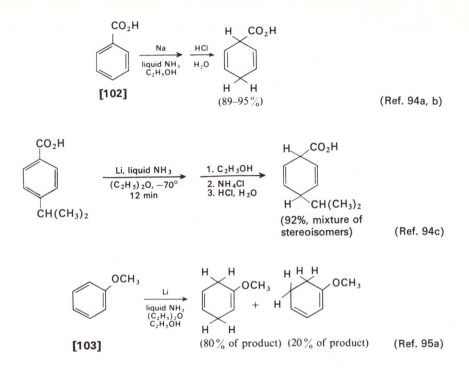

[102]

(89–95%) (Ref. 94a, b)

(92%, mixture of stereoisomers) (Ref. 94c)

[103] (80% of product) (20% of product) (Ref. 95a)

The dihydrobenzenes produced from benzene derivatives having an electron-withdrawing substituent (e.g., [102]) have been found to be 1-substituted-1,4-dihydrobenzenes, whereas the products from benzene derivatives having electron-donating substituents (e.g., [103]) and ([104]) are usually 1-substituted-2,5-dihydrobenzenes.[1] A notable exception is the reduction of N,N-dimethylaniline derivatives where the major product contains the indicated conjugated 1-dimethyl-amino-1,3-diene system.[95d] Presumably, the ready acid-catalyzed interconversion of enamines (see Chapter 9) allows equilibration to occur during the isolation of the product if not earlier. Among the electron-donating groups, the directive effects of —O—CH$_3$ and —N(CH$_3$)$_2$ substituents outweigh the directive effect of an alkyl group.[1] When both an electron-withdrawing and an electron-donating substituent are present, the activating effect of the electron-withdrawing substituent usually controls the course of the reaction.[1]

94. (a) M. E. Kuehne and B. F. Lambert, *Org. Syn.*, **43**, 22 (1963). (b) M. E. Kuehne and B. F. Lambert, *J. Am. Chem. Soc.*, **81**, 4278 (1959). (c) F. Camps, J. Coll, and J. Pascual, *J. Org. Chem.*, **32**, 2563 (1967). (d) The reduction of aromatic rings is also facilitated by trimethylsilyl and phenyl substituents; P. J. Grisdale, T. H. Regan, J. C. Doty, J. Figueras, and J. L. R. Williams, *ibid.*, **33**, 1116 (1968); H. Alt, E. R. Franke, and H. Bock, *Angew. Chem., Intern. Ed. Engl.*, **8**, 525 (1969).
95. (a) A. L. Wilds and N. A. Nelson, *J. Am. Chem. Soc.*, **75**, 5360, 5366 (1953). (b) L. A. Paquette and J. H. Barrett, *Org. Syn.*, **49**, 62 (1969). (c) R. N. McDonald and C. E. Reineke, *ibid.*, **50**, 50 (1970). (d) A. J. Birch, E. G. Hutchinson, and G. Subba Rao, *Chem. Commun.*, **No. 11**, 657 (1970) and references therein.

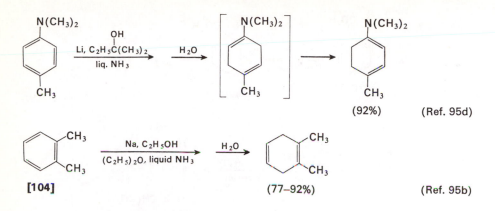

(92%)   (Ref. 95d)

(77–92%)   (Ref. 95b)

These directive effects have been rationalized[1] by suggesting that in an anion radical the electron density is greatest at the atom bearing the substituent carbon and at the position para to the substituent (structures [104]) for electron-with-drawing groups and at the ortho and meta positions (structures [105]) for electron-donating groups. These suggestions are supported by the e.p.r. spectra

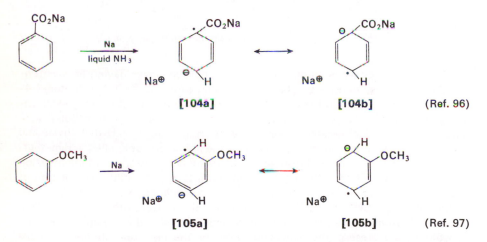

of various anion radicals obtained from benzene derivatives.[96,97] Such radical anions are suggested to be protonated initially at the sites of highest electron density, leading, after further electron transfer and protonation, to the observed products. Molecular orbital calculations of the sites of highest electron density

96. (a) A. R. Buick, T. J. Kemp, G. T. Neal, and T. J. Stone, *Chem. Commun.*, **No. 21**, 1331 (1968). (b) Comparable studies of the benzonitrile radical anion are described by P. H. Rieger, I. Bernal, W. H. Reinmuth, and G. K. Fraenkel, *J. Am. Chem. Soc.*, **85**, 683 (1963).
97. (a) J. K. Brown, D. R. Burnham, and N. A. J. Rogers, *J. Chem. Soc.*, B, 1149 (1969). (b) D. R. Burnham, *Tetrahedron*, **25**, 897 (1969). (c) The e.p.r. spectrum of toluene anion radical is described in Ref. 81f.

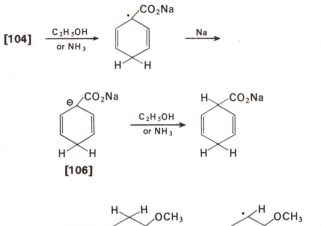

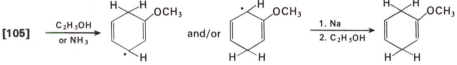

in anion radicals have successfully explained the orientation effects observed in dissolving metal reductions of aromatic systems.[98] However, the site of the first protonation in some cases (e.g., **[105]**) is uncertain.[97b] A very common side reaction in the reduction of methoxy substituted aromatic systems is reductive cleavage of the methoxyl function.[1,98b] One possible reaction path for this cleavage, indicated in the following equation, would suggest that at least part of the initially formed radical anion is protonated at the carbon atom bonded to the methoxyl function (or rearranges to this intermediate). It should be noted that, for the case cited, the amount of reductive cleavage is diminished substantially by per- forming the reduction in the presence of ethanol as a proton donor.

The protonation of the sodium benzoate radical anion (and presumably other radical anions with electron withdrawing groups) appears to follow the pathway indicated to form the intermediate anion **[106]** since in analogous systems (e.g., **[107]**) this anion can be alkylated prior to protonation. This procedure is feasible because the activation afforded by the carboxyl substituent permits the reduction to occur even in the absence of a relatively acidic proton donor such as an alcohol. The ease of reduction of an aromatic ring with carboxyl substituents is further illustrated by the following reduction effected with the relatively weak reducing agent, sodium amalgam.

Two frequently encountered problems in the usual procedure for Birch reductions (addition of sodium to a solution of the aromatic compound and an alcohol in liquid ammonia[1]) have been the insolubility of the aromatic compounds in the reaction medium and the failure of this reaction system to reduce aromatic compounds having several electron-donating or bulky substituents. Solubility

98. (a) A. Streitwieser, Jr., *Molecular Orbital Theory for Organic Chemists,* Wiley, New York, 1962, pp. 425–431. (b) H. E. Zimmerman, *Tetrahedron,* **16**, 169 (1961).

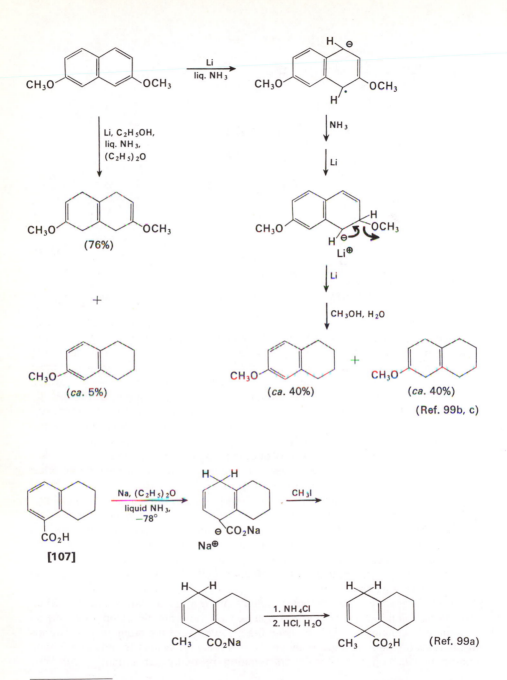

99. (a) M. D. Bachi, J. W. Epstein, Y. Herzberg-Minzly, and H. J. E. Lowenthal, *J. Org. Chem.*, **34**, 126 (1969). (b) B. Weinstein and A. H. Fenselau, *ibid.*, **29**, 2102 (1964). (c) J. A. Marshall and N. H. Andersen, *ibid.*, **30**, 1292 (1965).

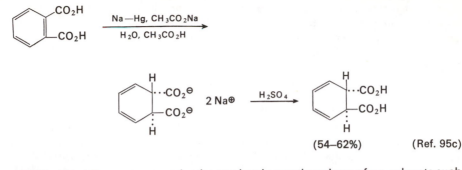

(54–62%)   (Ref. 95c)

problems have been overcome by the previously mentioned use of co-solvents such as ether, tetrahydrofuran, ethanol, or *t*-butyl alcohol; by the use of low-molecular-weight amines; or by conversion of the aromatic compounds to the more soluble β-hydroxyethyl ethers or glyceryl ethers.[1]

A modification[93b,95a] of this reduction procedure, which consists of adding the aromatic compound to a solution of lithium and a nonprotonic solvent (e.g., ether, tetrahydrofuran, dioxane, or 1,2-dimethoxyethane) in liquid ammonia and then adding an alcohol (ethanol, isopropyl alcohol, *t*-butyl alcohol), was found to facilitate the reduction of difficultly reduced aromatic systems. The use of a larger excess of alcohol, accompanied by the continuous addition of lithium to provide a high metal concentration in the reaction medium,[100] has permitted the successful reaction of systems (e.g., [108]) that are essentially inert to the usual conditions

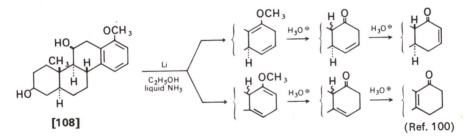

[108]   (Ref. 100)

of the Birch reduction.[101] The enhanced reducing power of such reaction systems employing lithium rather than sodium was originally attributed[95a] to the greater solubility of lithium in liquid ammonia and to a higher reduction potential for lithium than for sodium. Subsequent study showed that at least part of the advantage of lithium over sodium or potassium has been found to arise from the fact that traces of iron (present in undistilled liquid ammonia) catalyze the

100. (a) W. S. Johnson, B. Bannister, and R. Pappo, *J. Am. Chem. Soc.,* **78**, 6331 (1956). (b) W. S. Johnson, R. Pappo, and W. F. Johns, *ibid.,* **78**, 6339 (1956). (c) W. S. Johnson, J. A. Marshall, J. F. W. Keana, R. W. Franck, D. G. Martin, and V. J. Bauer, *Tetrahedron,* **Suppl. 8, Pt. 2**, 541 (1966).
101. It will be noted that in systems such as [108], the anion radical produced (see structure [105]) necessarily must have an energetically unfavorable structure with high electron density at one of the carbon atoms bonded to an electron donating substituent.

unwanted reaction of sodium or potassium with an alcohol (to form hydrogen) to a much greater extent than the corresponding reaction with lithium.[102] In the absence of iron, lithium, sodium, and potassium are all effective reducing agents. In fact, even very difficultly reduced substances such as the ether [108] can be successfully reduced with potassium in ammonia when care is taken to exclude iron compounds from the reaction mixture.

Although the relative rates of reduction of benzene with the alkali metals and ethanol in liquid ammonia follow the order lithium > sodium > potassium,[81a] all three are sufficiently rapid to be synthetically useful. In certain difficult cases, the selection of an alcohol of the correct acidity to serve as a proton donor may be important for successful reduction.[19c,103] Utilizing these facts, a procedure has been developed[102] employing a solution of the aromatic compound (e.g., [109]) in a mixture of distilled liquid ammonia (2 parts), tetrahydrofuran (1 part), and *t*-butyl alcohol (1 part), to which either sodium or lithium metal is added.

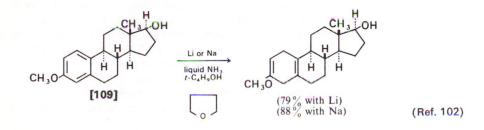

(79 % with Li)
(88 % with Na)   (Ref. 102)

Pyridine and its derivatives readily accept an electron from a metal to form an anion radical, and consequently these materials are easily reduced with metals in

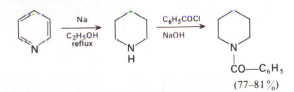

(77–81 %)   (Ref. 103)

protonic solvents. However, if the pyridine anion radical is formed in the absence of a protonic solvent, it very rapidly dimerizes,[104] as the following equations illustrate.

---

102. H. L. Dryden, Jr., G. M. Webber, R. R. Burtner, and J. A. Cella, *J. Org. Chem.*, **26**, 3237 (1961).
103. C. S. Marvel and W. A. Lazier, *Org. Syn.*, **Coll. Vol. 1**, 99 (1944).
104. (a) R. L. Ward, *J. Am. Chem. Soc.*, **83**, 3623 (1961). (b) J. W. Dodd, F. J. Hopton, and N. S. Hush, *Proc. Chem. Soc.*, 61 (1962). (c) C. D. Schmulbach, C. C. Hinckley, and D. Wasmund, *J. Am. Chem. Soc.*, **90**, 6600 (1968).

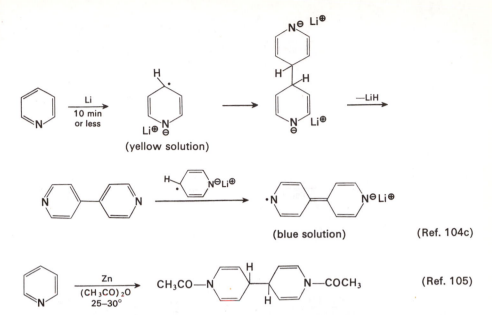

(yellow solution)

(blue solution)   (Ref. 104c)

(Ref. 105)

The reduction of certain pyridinium salts leads to the production of stable free radicals which can be isolated.[106]

The presence or absence of methanol as a proton donor has been found to have an interesting effect on the reduction of quinoline and indole derivatives (e.g., [110]).[107a,c] The authors suggest that the reduction of the heterocyclic ring, observed in the absence of methanol, may be the result of dianion formation. Finally, it is to be noted that dihydropyridines formed by Birch reduction can serve as precursors for 1,5-dicarbonyl compounds or their condensation products.[107b] The use of this intermediate in an annelation procedure is discussed in Chapter 10.

REDUCTION OF OTHER FUNCTIONAL GROUPS

Although nonconjugated, nonterminal[81c,92c] olefins are normally stable to solutions of alkali metals in liquid ammonia or low-molecular-weight amines, disubstituted acetylenes (e.g., [111]) are readily reduced to form *trans*-olefins. This reduction, which serves as a useful synthesis for *trans*-olefins, is in contrast to the catalytic hydrogenation or hydroboration of acetylenes, which yields *cis*-olefins. The reaction presumably proceeds by addition of one electron to form a linear anion

105. (a) R. L. Frank and P. V. Smith, *Org. Syn.,* **Coll. Vol. 3,** 410 (1955). (b) A. T. Nielsen, D. W. Moore, G. M. Muha and K. H. Berry, *J. Org. Chem.,* **29,** 2175 (1964).
106. (a) E. M. Kosower and E. J. Poziomek, *J. Am. Chem. Soc.,* **86,** 5515 (1964). (b) E. M. Kosower and J. L. Cotter, *ibid.,* **86,** 5524 (1964). (c) E. M. Kosower and I. Schwager, *ibid.,* **86,** 5528 (1964).
107. (a) W. A. Remers, G. J. Gibs, C. Pidacks, and M. J. Weiss, *J. Org. Chem.,* **36,** 279 (1971). (b) S. Danishefsky and R. Cavanaugh, *J. Am. Chem. Soc.,* **90,** 520 (1968). (c) The 2,3-double bond in 2-carboethoxyindole has been reduced selectively with tin and hydrogen chloride in ethanol solution; E. J. Corey, R. J. McCaully, and H. S. Sachdev, *ibid.,* **92,** 2476 (1970).

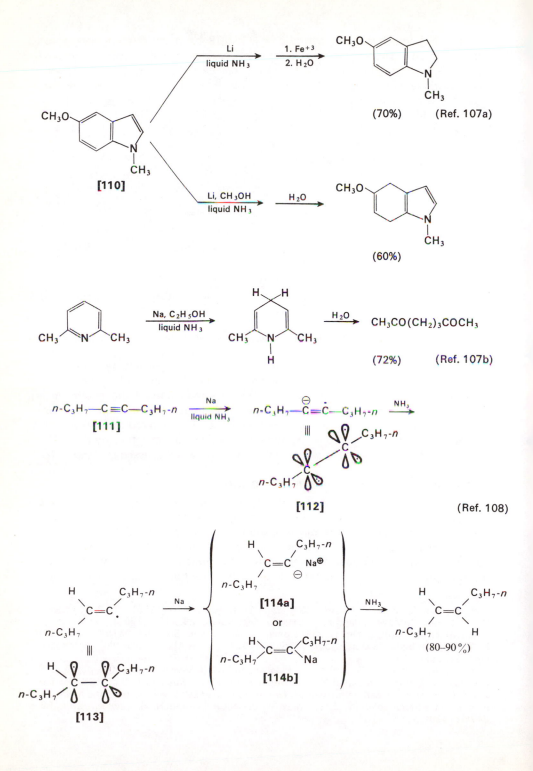

(70%)   (Ref. 107a)

(60%)

(72%)   (Ref. 107b)

(Ref. 108)

(80–90%)

radical [112], which is protonated to give [113]. Subsequent electron transfer to form the anion [114], followed by protonation, leads to the olefin. Both e.p.r. and electrochemical studies support a reaction scheme of this type.[17c,109] Although diarylacetylenes are exceptional cases in which reduction to dianions is possible prior to protonation in relatively non-polar solvents at low temperatures,[17o] even with these substances the formation of radical anions followed by protonation and further reduction (analogous to the scheme [111]→[112]→[113]→[114]) appears more likely for reductions performed under the usual conditions for preparative reactions.[109b] As the following equations indicate, the geometry of these diarylacetylene dianions is determined by the nature of the metal cation. The lithium derivative of diphenylacetylene dianion (which may be dimeric or polymeric) has the *cis*-geometry, whereas the corresponding sodium derivative is *trans*.

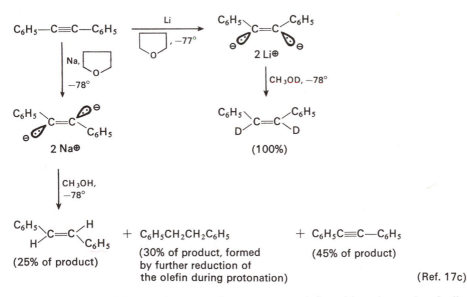

It is also possible to reduce acetylenes to trans-olefins either electrochemically or with a solution of lithium generated electrochemically as the following equations illustrate. For the reductions with lithium solutions (generated electrolytically

108. (a) K. N. Campbell and L. T. Eby, *J. Am. Chem. Soc.*, **63**, 216, 2683 (1941). (b) A. L. Henne and K. W. Greenlee, *ibid.*, **65**, 2020 (1943). (c) R. A. Benkeser, G. Schroll, and D. M. Sauve, *ibid.*, **77**, 3378 (1955). (d) For reductions of acetylenes to *trans*-olefins with chromium(II) salts see C. E. Castro and R. D. Stephens, *ibid.*, **86**, 4358 (1964).
109. (a) J. G. Broadhurst and E. Warhurst, *J. Chem. Soc.*, A, 351 (1966). (b) R. E. Sioda, D. O. Cowan, and W. S. Koski, *J. Am. Chem. Soc.*, **89**, 230 (1967). (c) R. A. Benkeser and C. A. Tincher, *J. Org. Chem.*, **33**, 2727 (1968). (d) L. Horner and H. Röder, *Justus Liebigs Ann. Chem.*, **723**, 11 (1969). (e) For a review of the electrochemical reduction of acetylenes, see A. P. Tomilov, *Russ. Chem. Rev.*, **31**, 569 (1962). (f) H. E. Zimmerman and J. R. Dodd, *J. Am. Chem. Soc.*, **92**, 6507 (1970). (g) For a review of the base-catalyzed isomerization of acetylenes, see R. J. Bushby, *Quart. Rev.*, **24**, 585 (1970).

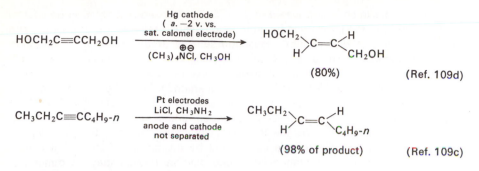

HOCH$_2$C≡CCH$_2$OH $\xrightarrow[\substack{\oplus\ominus\\ (CH_3)_4NCl,\ CH_3OH}]{\substack{Hg\ cathode\\(\ a.\ -2\ v.\ vs.\\ sat.\ calomel\ electrode)}}$

$\underset{H}{\overset{HOCH_2}{>}}C=C\underset{CH_2OH}{\overset{H}{<}}$

(80%)   (Ref. 109d)

CH$_3$CH$_2$C≡CC$_4$H$_9$-n $\xrightarrow[\substack{anode\ and\ cathode\\not\ separated}]{\substack{Pt\ electrodes\\LiCl,\ CH_3NH_2}}$

$\underset{H}{\overset{CH_3CH_2}{>}}C=C\underset{C_4H_9\text{-}n}{\overset{H}{<}}$

(98% of product)   (Ref. 109c)

from lithium chloride), it is important to use the previously discussed undivided cell so that the strongly basic lithium methylamide, formed by protonation of the radical anion, is neutralized by the anodic products before this base can catalyze isomerization of the starting acetylene to isomers which would yield a mixture of olefinic products.[109c,g]

The reason for the high degree of stereospecificity in these various reductions of acetylenes to *trans* olefins is still unclear. The protonation of the presumably linear[109a] anion radical [112] leads to an alkyl-substituted vinyl radical. Such alkyl-substituted vinyl radicals possess a non-linear structure implied in structure [113] with a relatively low energy barrier to inversion: aryl-substituted vinyl radicals are either linear or have a very low barrier to inversion.[110] Presumably, vinyl anions such as [114a] (or the vinyl organometallic compound [114b] in reductions done with metals) without aryl substituents are capable of retaining their stereochemistry for the periods of time required for protonation. It appears possible that the reduction of an alkyl-substituted vinyl radical [113] by electron transfer may be significantly faster than the rate of inversion of the vinyl radical.[110c] If the foregoing assumptions are correct, the *trans* stereochemistry obtained in reductions of alkylacetylenes appears to be attributable to the protonation of the linear anion radical [112] to give the trans-vinyl radical [113] as the initial product; this result could be considered analogous to the *trans* addition of electrophilic agents to acetylenes. Presumably the stereospecificity associated with reductions of arylacetylenes is attributable to rapid inversion (or planarity) of the vinyl radical, the vinyl anion, or both so that the sterically less hindered vinyl anion is the major material present at the time of protonation. The reductive cyclization of γ-keto acetylenes[58g] described previously in this chapter presumably also involves the formation of a vinyl anion which adds intramolecularly to the carbonyl function more rapidly than it is protonated.

Although terminal acetylenes may be reduced to terminal olefins with sodium in liquid ammonia,[108b] the acetylide anions produced from terminal acetylenes are inert to these reducing conditions because of the difficulty of adding an electron

---

110. (a) L. A. Singer and N. P. Kong, *J. Am. Chem. Soc.*, **88**, 5213 (1966); *ibid.*, **89**, 5251 (1967). (b) R. M. Fantazier and J. A. Kampmeier, *ibid.*, **88**, 5219 (1966). (c) G. D. Sargent and M. W. Browne, *ibid.*, **89**, 2788 (1967). (d) A. J. Fry and M. A. Mitnick, *ibid.*, **91**, 6207 (1969). (e) L. A. Singer and J. Chen. *Tetrahedron Letters*, **No. 55**, 4849 (1969).

to the multiple bond, which already carried a full negative charge. This fact permits the selective reduction of a disubstituted acetylene in the presence of a terminal acetylene, as illustrated in the accompanying equation. Allenes (e.g., **[115]**) are also reduced by sodium in liquid ammonia and have been shown to be intermediates

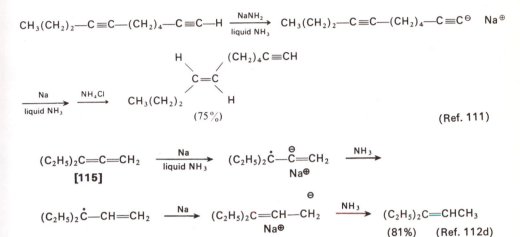

(Ref. 111)

in the reduction of certain cyclic acetylenes with sodium and liquid ammonia to form *cis*-olefins rather than the expected *trans*-olefins.[112b] The ability of the strong base sodium amide, produced in these reductions, to interconvert allenes and acetylenes by base-catalyzed isomerization[109g] has been suggested to cause the isomerization of many allenes to acetylenes prior to reduction. As a result, *trans*-olefins are often the major products from the reductions of acyclic allenes with metals in ammonia.[109c,112d] Studies of the electrochemical and metal reductions of tetraphenylallene also indicate a rapid reaction of the initially formed anion radical (or possibly a dianion in the metal reduction) to abstract a proton from the solvent followed by further reduction.[113]

Certain other functional groups—nitro groups, nitroso groups, imino groups, and oximino groups among them—may be reduced electrochemically or by reaction

111. N. A. Dobson and R. A. Raphael, *J. Chem. Soc.*, 3558 (1955).
112. (a) D. Devaprabhakara and P. D. Gardner, *J. Am. Chem. Soc.*, **85**, 648 (1963). (b) M. Svoboda, J. Sicher, and J. Zavada, *Tetrahedron Letters*, **No. 1**, 15 (1964); M. Svoboda, J. Zavada, and J. Sicher, *Coll. Czech. Chem. Commun.*, **30**, 413, 421 (1965). (c) J. M. Brown, *Chem. Ind. (London)*, **No. 42**, 1689 (1963). (d) G. Nagendrappa, R. K. Srivastava, and D. Devaprabhakara, *J. Org. Chem.*, **35**, 347 (1970). (e) For the isomerization of an allene to a terminal acetylene, see H. W. Thompson, *ibid.*, **32**, 3712 (1967).
113. (a) R. Dietz, M. E. Peover, and R. Wilson, *J. Chem. Soc.*, B, 75 (1968). (b) P. Dowd, *Chem. Commun.*, **No. 22**, 568 (1965).
114. S. A. Mahood and P. V. L. Schaffner, *Org. Syn.*, **Coll. Vol. 2**, 160 (1943).
115. O. Kamm, *Org. Syn.*, **Coll. Vol. 1**, 445 (1944).
116. H. E. Bigelow and D. B. Robinson, *Org. Syn.*, **Coll. Vol. 3**, 103 (1955).

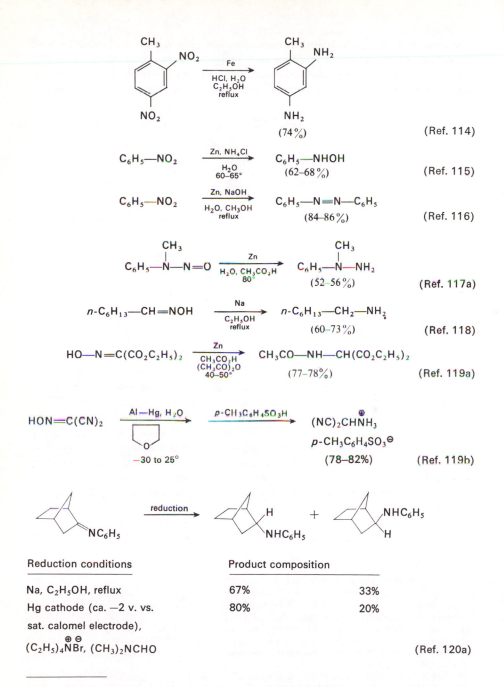

Reduction conditions	Product composition	
Na, C_2H_5OH, reflux	67%	33%
Hg cathode (ca. -2 v. vs.	80%	20%
sat. calomel electrode),		
$(C_2H_5)_4\overset{\oplus}{N}\overset{\ominus}{Br}$, $(CH_3)_2NCHO$		(Ref. 120a)

117. (a) W. W. Hartman and L. J. Roll, *Org. Syn.,* **Coll. Vol. 2**, 418 (1943). (b) See also B. T. Hayes and T. S. Stevens, *J. Chem. Soc.,* C, 1088 (1970).
118. W. H. Lycan, S. V. Puntambeker, and C. S. Marvel, *Org. Syn.,* **Coll. Vol. 2**, 318 (1943).

with metals in protonic solvents; examples are provided in the accompanying equations. Both from e.p.r. studies with aromatic nitro compounds[11] and from other observations[121] there is reason to believe that these compounds form relatively stable anion radicals such as [116]. The further protonation and reduction of such

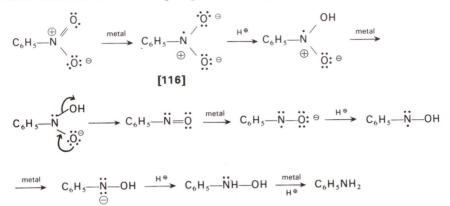

[116]

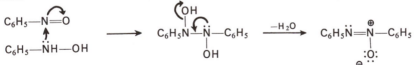

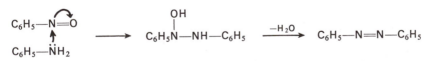

intermediates are presumably analogous to the dissolving metal reductions previously discussed. A possible reaction path for the reductions of aromatic nitro compounds is indicated above. The final step, involving cleavage of the N—O bond is analogous to the subsequently discussed cleavage of benzylic C—O bonds in dissolving metal reductions. The complete reduction of these functional groups to amino groups is generally more easily accomplished by catalytic hydrogenation. However the partial reduction of aryl nitro compounds to azo compounds or hydroxyalmine derivatives is best accomplished with metals such as zinc in neutral or basic media where the intermediate reduction products may undergo the reactions indicated below.

C_6H_5—N=O

C_6H_5—NH—OH $\longrightarrow$ C_6H_5N—N—C_6H_5 OH OH $\xrightarrow{-H_2O}$ $C_6H_5\overset{..}{N}$=N—C_6H_5 :O:$^\ominus$

C_6H_5—N=O

C_6H_5—NH$_2$ $\longrightarrow$ C_6H_5N—NH—C_6H_5 OH $\xrightarrow{-H_2O}$ C_6H_5—N=N—C_6H_5

119. (a) A. J. Zambito and E. E. Howe, *Org. Syn.,* **40**, 21 (1960). (b) J. P. Ferris, R. A. Sanchez, and R. W. Mancuso, *ibid.,* **48**, 1 (1968).
120. (a) A. J. Fry and R. G. Reed, *J. Am. Chem. Soc.,* **91**, 6448 (1969). Also see Ref. 75b. (b) A. J. Fry and J. H. Newberg, *ibid,* **89**, 6374 (1967). (c) Ketones have been converted to N-methylamine derivatives by reduction with solutions of lithium (generated electrolytically) in methylamine in a reaction which apparently involves reduction of an imine intermediate. R. A. Benkeser and S. J. Mels, *J. Org. Chem.,* **35**, 261 (1970). (d) For studies of the electrolytic reduction of azo compounds, see J. L. Sadler and A. J. Bard, *J. Am. Chem. Soc.,* **90**, 1979 (1968). P. E. Iversen and H. Lund, *Tetrahedron Letters,* **No. 40**, 3523 (1969). (e) For the reduction of alkyl azides to amines with chromium(II) salts, see D. N. Kirk and M. A. Wilson, *Chem. Commun.,* **No. 1**, 64 (1970).

Aliphatic nitro groups are also readily reduced to radical anions (e.g., **[117]**) electrolytically or with metals.[121c,d,122] However, unless these anion radicals are intercepted by protonation and further reduction, they decompose to alkyl radicals **[118]** which couple with the nitroalkane anion radicals to form insoluble inter-

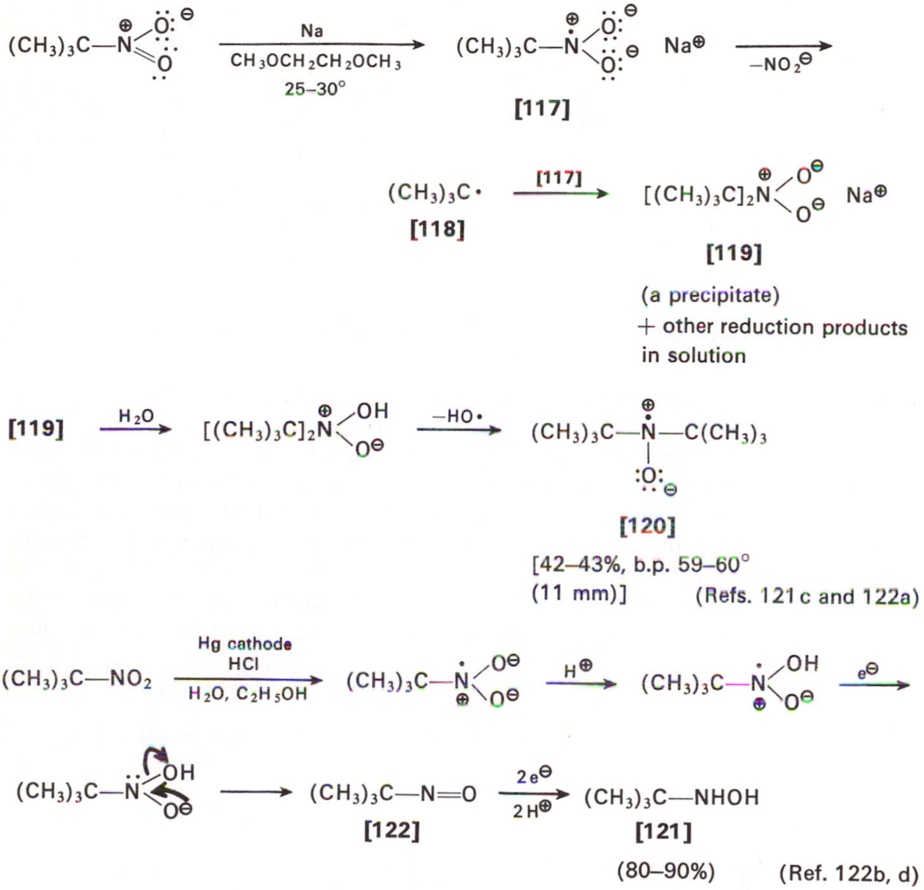
Na
$CH_3OCH_2CH_2OCH_3$
25–30°
Na⊕
$-NO_2^{\ominus}$
[117]
[118]
[119]
(a precipitate)
+ other reduction products
in solution
H_2O
$-HO\cdot$
[120]
[42–43%, b.p. 59–60°
(11 mm)]
(Refs. 121c and 122a)
Hg cathode
HCl
H_2O, C_2H_5OH
[122]
[121]
(80–90%)
(Ref. 122b, d)

121. (a) G. A. Russell and E. G. Janzen, *J. Am. Chem. Soc.,* **84**, 4153 (1962). (b) G. A. Russell, E. G. Janzen, and E. T. Strom, *ibid.,* **84**, 4155 (1962); *ibid.,* **86**, 1807 (1964). (c) A. K. Hoffmann and co-workers, *ibid.,* **86**, 631, 639, 646 (1964). (d) R. H. Gibson and J. C. Crosthwaite, *ibid.,* **90**, 7373 (1968). (e) P. L. Kolker and W. A. Waters, *J. Chem. Soc.,* 1136 (1964). (f) V. Kalyanaraman, C. N. R. Rao, and M. V. George, *Tetrahedron Letters,* **No. 55**, 4889 (1969).

122. (a) A. K. Hoffmann, A. M. Feldman, E. Gelblum, and A. Henderson, *Org. Syn.,* **48**, 62 (1968). (b) P. E. Iversen and H. Lund, *Tetrahedron Letters,* **No. 41**, 4027 (1967). (c) H. Lund, *ibid.,* **No. 33**, 3651 (1968). (d) H. Sayo, Y. Tsukitani, and M. Masui, *Tetrahedron,* **24**, 1717 (1968). (e) For the reduction of nitro compounds to oximes with chromium(II) salts, see J. R. Hanson and E. Premuzic, *ibid.,* **23**, 4105 (1967); J. R. Hanson and T. D. Organ, *J. Chem. Soc.,* C, 1182 (1970). (f) E. J. Corey and J. E. Richman, *J. Am. Chem. Soc.,* **92**, 5276 (1970).

mediates **[119]**. These intermediates, when separated and hydrolyzed, yield nitroxides such as **[120]**; these stable free radical species have found extensive use as spin labels for e.p.r. studies to characterize free-radical intermediates.[123]

The electrochemical reductions of nitroalkanes in the presence of a proton source may yield either N-alkylhydroxylamines (e.g., **[121]**) or amines depending on the fate of the intermediate nitroso compound (e.g., **[122]**). If a primary or secondary nitrosoalkane is formed, isomerization to an oxime is possible and the use of acid media and elevated temperatures can promote the indicated elimination (structure **[123]**) to an imine which is reduced to an amine. In alkaline media and at lower temperatures, reduction to the hydroxylamine is favored. It should be noted that, unlike the previously discussed N—O cleavage observed in the reduction of arylhydroxylamines, alkylhydroxylamines are usually stable to further reduction.[122c] However, it has been possible to cleave the N—O bond of both aliphatic and aromatic oxime acetates by reduction with chromium(II) acetate.[122f]

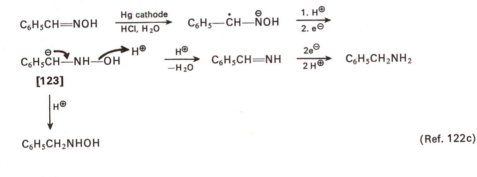

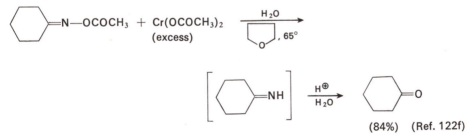

Various single bonds may be cleaved by dissolving metal reductions. The cleavage of benzyl or allyl alcohol derivatives (e.g., **[124]** and **[125]**), quaternary ammonium salts (e.g., **[126]**), and thioethers occurs with particular ease, making this reductive cleavage valuable for the removal of benzyl protecting groups.

123. (a) O. H. Griffith and A. S. Waggoner, *Accts. Chem. Res.*, **2**, 17 (1969). (b) E. G. Janzen, *ibid.*, **2**, 279 (1969); **4**, 31 (1971). (c) M. J. Perkins, P. Ward, and A. Horsfield, *J. Chem. Soc.*, B, 395 (1970); G. R. Chalfont, M. J. Perkins, and A. Horsfield, *ibid.*, B, 401 (1970).

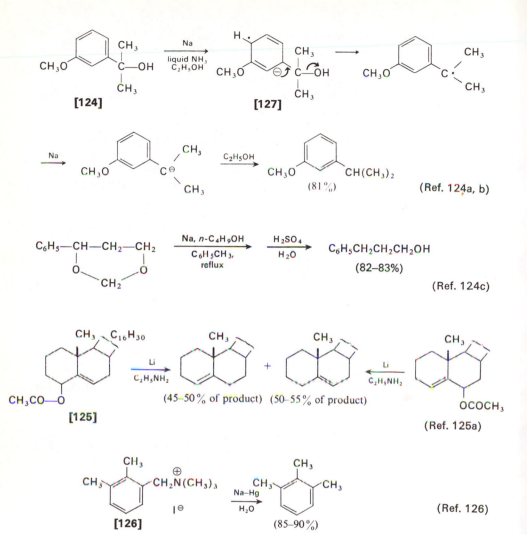

(Ref. 124a, b)

(Ref. 124c)

(Ref. 125a)

(Ref. 126)

Such reactions are believed to proceed by formation of a radical anion (e.g., [127]) followed by elimination of a stable anion or some good leaving group. Further reaction of the resulting radical with the metal and subsequent protonation give the product. Similar reaction conditions have been used to cleave dialkyl

124. (a) A. J. Birch, *J. Chem. Soc.*, 809 (1945). (b) A. J. Birch and S. M. Mukherji, *ibid.*, 2531 (1949). (c) R. L. Shriner and P. R. Ruby, *Org. Syn.*, **Coll. Vol. 4**, 798 (1963).
125. (a) A. S. Hallsworth, H. B. Henbest, and T. I. Wrigley, *J. Chem. Soc.*, 1969 (1957). (b) R. E. Ireland, T. I. Wrigley, and W. G. Young, *J. Am. Chem. Soc.*, **80**, 4604 (1958).
126. W. R. Brasen and C. R. Hauser, *Org. Syn.*, **Coll. Vol. 4**, 508 (1963).

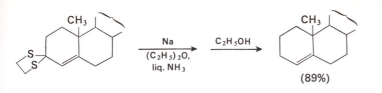

Na → C_2H_5OH → (89%)   (Ref. 125b)

$(C_2H_5)_2O,$ liq. NH_3

sulfides,[127] sulfones,[127a] sulfonic acid derivatives,[128] quaternary ammonium salts,[129] sulfonium and phosphonium salts,[76a,d,130] and nitriles.[131] The ease of these reductive cleavage reactions has allowed certain of the foregoing groups to serve as protective groups which can be removed by reduction at a later stage in a synthesis. The accompanying equations illustrate these cleavage reactions effected electro-lytically or by reductions with metals. It will be noted that unsymmetrical compounds are normally cleaved by fragmentation of the anion radical (e.g., **[128]**) in the direction which would yield the more stable carbon radical.

Aryl ethers (e.g., **[129]** and **[130]**) and aryl thioethers (e.g., **[131]**) are also cleaved by reaction with metals although with more difficulty than benzyl and allyl systems. In these cases it is not certain whether the anion radical undergoes cleavage or whether a second electron is added and the resulting dianion then cleaves.[99b,c,132] Alternatively, the dimer derived from the anion radical may be

127. (a) W. E. Truce and F. J. Frank, *J. Org. Chem.,* **32,** 1918 (1967) ; W. E. Truce, D. P. Tate, and D. N. Burdge, *J. Am. Chem. Soc.,* **82,** 2872 (1960). (b) E. L. Eliel, T. W. Doyle, R. A. Daignault, and B. C. Newman, *ibid.,* **88,** 1828 (1966) ; E. L. Eliel and T. W. Doyle, *J. Org. Chem.,* **35,** 2716 (1970). (c) E. D. Brown, S. M. Iqbal, and L. N. Owen, *J. Chem. Soc.,* C, 415 (1966). (d) B. C. Newman and E. L. Eliel, *J. Org. Chem.,* **35,** 3641 (1970).
128. (a) W. D. Closson, P. Wriede, and S. Bank, *J. Am. Chem. Soc.,* **88,** 1581 (1966). (b) For similar reductive cleavages of sulfonamides see S. Ji, L. B. Gortler, A. Waring, A. Battisti, S. Bank, W. D. Closson, and P. Wriede, *ibid.,* **89,** 5311 (1967) ; W. D. Closson, S. Ji, and S. Schulenberg, *ibid.,* **92,** 650 (1970) ; J. Kovacs and U. R. Ghatak, *J. Org. Chem.,* **31,** 119 (1966). (c) For studies of the electrochemical cleavage of sulfonates and sulfona-mides, see P. Yousefzadeh and C. K. Mann, *ibid.,* **33,** 2716 (1968) ; L. Horner and R. J. Singer, *Tetrahedron Letters,* **No. 20,** 1545 (1969). (d) R. Adams and C. S. Marvel, *Org. Syn.,* **Coll. Vol. 1,** 504 (1944) ; F. C. Whitmore and F. H. Hamilton, *ibid.,* **Coll. Vol. 1,** 492 (1944) ; P. D. Caesar, *ibid.,* **Coll. Vol. 4,** 695 (1963).
129. (a) E. Grovenstein, Jr., and L. C. Rogers, *J. Am. Chem. Soc.,* **86,** 854 (1964) ; E. Grovenstein, Jr., S. Chandra, C. E. Collum, and W. E. Davis, Jr., *ibid.,* **88,** 1275 (1966). (b) R. R. Dewald and K. W. Browall, *Chem. Commun.,* **No. 23,** 1511 (1968). (c) For the electrochemical conversion of quaternary ammonium cations, $R_4N^{\oplus}$, to amalgams of the neutral species, R_4N, see J. D. Littlehailes and B. J. Woodhall, *Dis. Faraday Soc.,* **No. 45,** 187 (1968). (d) S. D. Ross, M. Finkelstein, and R. C. Petersen, *J. Am. Chem. Soc.,* **92,** 6003 (1970).
130. (a) T. Shono and M. Mitani, *J. Am Chem. Soc.,* **90,** 2728 (1968) ; *Tetrahedron Letters,* **No. 9,** 687 (1969). (b) A. D. Britt and E. T. Kaiser, *J. Phys. Chem.,* **69,** 2775 (1965). (c) J. Grimshaw and J. S. Ramsey, *J. Chem. Soc.,* B, 63 (1968).
131. (a) P. G. Arapakos, M. K. Scott, and F. E. Huber, Jr., *J. Am. Chem. Soc.,* **91,** 2059 (1969). (b) C. Fabre and Z. Welvart, *Tetrahedron Letters,* **No. 39,** 3801 (1967).
132. (a) D. H. Eargle, Jr., *J. Org. Chem.,* **28,** 1703 (1963) ; D. H. Eargle, Jr., and W. B. Moniz, *ibid.,* **32,** 2227 (1967). (b) H. O. House and V. Kramar, *ibid.,* **27,** 4146 (1962). (c) For the reductive cleavage of epoxides with alkali metals in ammonia, see E. M. Kaiser, C. G. Edmonds, S. D. Grubb, J. W. Smith, and D. Tramp, *ibid.,* **36,** 330 (1971).

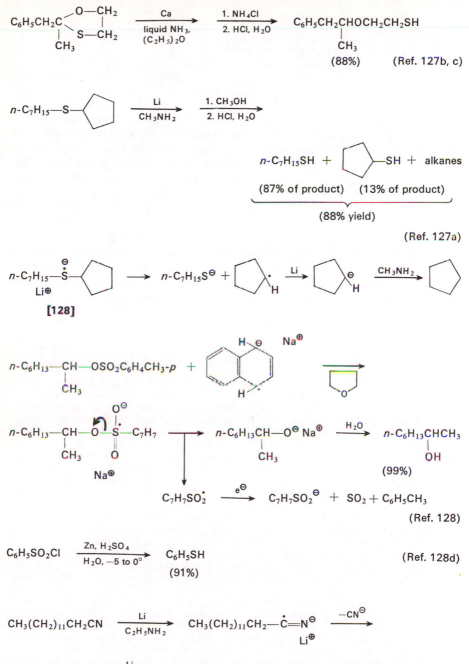

$$C_6H_5SO_2Cl \xrightarrow[H_2O, -5 \text{ to } 0°]{Zn, H_2SO_4} C_6H_5SH \qquad\qquad (Ref. 128d)$$
(91%)

$$CH_3(CH_2)_{11}CH_2CN \xrightarrow[C_2H_5NH_2]{Li} CH_3(CH_2)_{11}CH_2-\overset{\displaystyle .}{C}=N^{\ominus} \xrightarrow{-CN^{\ominus}}$$
$$\underset{Li^{\oplus}}{}$$

$$CH_3(CH_2)_{11}CH_2^{\boldsymbol{\cdot}} \xrightarrow[C_2H_5NH_2]{Li} CH_3(CH_2)_{11}CH_3 + CH_3(CH_2)_{11}CH_2CH_2NH_2$$
                                        (35%)   (65%)   (Ref. 131a)

cleaved in certain cases.[132b] The cleavage of vinyl or aryl diethyl phosphates and aryl sulfonates with sodium or lithium in liquid ammonia or an amine to form hydrocarbons[133] is also worthy of note, particularly the use of this method to convert unsaturated ketones such as **[132]** to olefins without ambiguity concerning the location of the double bond.

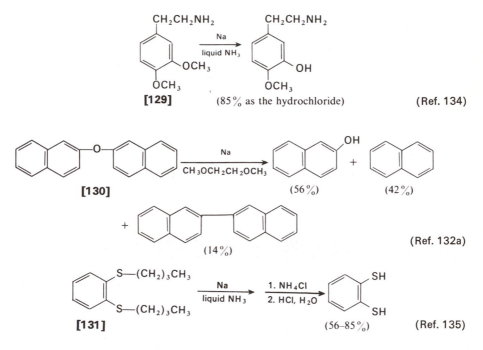

$[129]$ (85% as the hydrochloride)   (Ref. 134)

[130]   (56%)   (42%)

(14%)   (Ref. 132a)

[131]   (56–85%)   (Ref. 135)

Similarly, carbon-halogen bonds may be reductively cleaved electrolytically, by treatment with chromium(II) salts, or by reaction with metals in protonic solvents, as shown in the accompanying examples. Such reactions are probably related to the formation of organometallic derivatives (or dimeric products) when alkyl or aryl halides are allowed to react with metals. The possibility that free radicals or organometallic derivatives may be short-lived intermediates in these reductive cleavage reactions is suggested, since elimination rather than cleavage is usually

133. (a) G. W. Kenner and N. R. Williams, *J. Chem. Soc.*, 522 (1955). (b) R. E. Ireland and G. Pfister, *Tetrahedron Letters*, **No. 26**, 2145 (1969). (c) M. Fetizon, M. Jurion, and N. T. Anh, *Chem. Commun.*, **No. 3**, 112 (1969).
134. K. E. Hamlin and F. E. Fischer, *J. Am. Chem. Soc.*, **75**, 5119 (1953).
135. (a) A. Ferretti, *Org. Syn.*, **42**, 54 (1962). (b) For studies of the electrochemical cleavage of aryl sulfides, see R. Gerdil and E. A. C. Lucken, *J. Chem. Soc.*, 3916 (1964); R. Gerdil, *ibid.*, B, 1071 (1966).
136. (a) P. A. Levene, *Org. Syn.*, **Coll. Vol. 2**, 320 (1943). (b) S. Gronowitz and T. Raznikiewicz, *ibid.*, **44**, 9 (1964). (c) P. G. Gassman and J. L. Marshall, *ibid.*, **48**, 68 (1968). (d) D. Bryce-Smith and B. J. Wakefield, *ibid.*, **47**, 103 (1967). (e) A. J. Fry, M. A. Mitnick, and R. G. Reed, *J. Org. Chem.*, **35**, 1232 (1970).

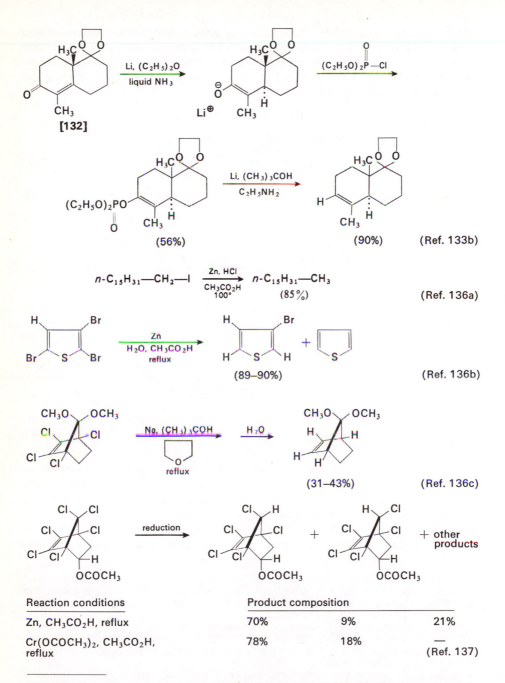

$$n\text{-}C_{15}H_{31}\text{---}CH_2\text{---}I \xrightarrow[\substack{CH_3CO_2H \\ 100°}]{Zn,\ HCl} n\text{-}C_{15}H_{31}\text{---}CH_3$$

(85%)   (Ref. 136a)

(89–90%)   (Ref. 136b)

(31–43%)   (Ref. 136c)

Reaction conditions	Product composition		
Zn, CH₃CO₂H, reflux	70%	9%	21%
Cr(OCOCH₃)₂, CH₃CO₂H, reflux	78%	18%	—
			(Ref. 137)

137. (a) C. F. Wilcox, Jr., and F. G. Zajacek, *J. Org. Chem.*, **29**, 2209 (1964). (b) K. L. Williamson, Y. F. L. Hsu, and E. I. Young, *Tetrahedron*, **24**, 6007 (1968). (c) C. E. Moppett and J. K. Sutherland, *J. Chem. Soc.*, C, 3040 (1968).

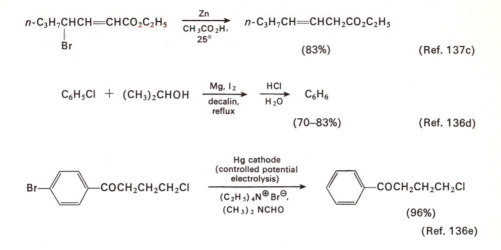

$n\text{-}C_3H_7CHCH{=}CHCO_2C_2H_5 \xrightarrow[\substack{25°}]{\substack{Zn \\ CH_3CO_2H,}} n\text{-}C_3H_7CH{=}CHCH_2CO_2C_2H_5$

|
Br

(83%)   (Ref. 137c)

$C_6H_5Cl + (CH_3)_2CHOH \xrightarrow[\substack{decalin, \\ reflux}]{\substack{Mg, I_2}} \xrightarrow[\substack{H_2O}]{HCl} C_6H_6$

(70–83%)   (Ref. 136d)

Br——⟨ ⟩——$COCH_2CH_2CH_2Cl$ $\xrightarrow[\substack{(C_2H_5)_4N^{\oplus}Br^{\ominus}, \\ (CH_3)_2NCHO}]{\substack{Hg\ cathode \\ (controlled\ potential \\ electrolysis)}}$ ⟨ ⟩——$COCH_2CH_2CH_2Cl$

(96%)

(Ref. 136e)

observed if a substituent (halogen, —OH, —OR, —OCO—R) that can be lost as a stable anion is present at an adjacent carbon atom (e.g., [132]).[138,139] One of the following examples illustrates a comparable elimination effected by reaction of a β-halo ether with n-butyllithium.[138g]

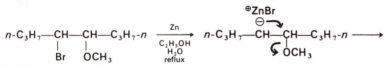

$n\text{-}C_3H_7{-}CH{-}CH{-}C_3H_7\text{-}n \xrightarrow[\substack{C_2H_5OH \\ H_2O \\ reflux}]{\substack{Zn}} n\text{-}C_3H_7{-}CH{-}CH{-}C_3H_7\text{-}n \longrightarrow$

| |
Br OCH_3

[132]

(either *erythro* or *threo*)

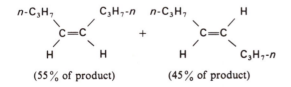

(55% of product)   (45% of product)   (Ref. 138a, b)

138. (a) H. O. House and R. S. Ro, *J. Am. Chem. Soc.,* **80**, 182 (1958). (b) S. J. Cristol and L. E. Rademacher, *ibid.,* **81**, 1600 (1959). (c) W. M. Schubert, B. S. Rabinovitch, N. R. Larson, and F. A. Sims, *ibid.,* **74**, 4590 (1952). (c) J. Sicher, M. Havel, and M. Svoboda, *Tetrahedron Letters,* **No. 40**, 4269 (1968). (e) C. G. Scouten, F. E. Barton, Jr., J. R. Burgess, P. R. Story, and J. F. Garst, *Chem. Commun.,* **No. 2**, 78 (1969). (f) I. M. Mathai, K. Schug, and S. I. Miller, *J. Org. Chem.,* **35**, 1733 (1970). (g) U. Schöllkopf, J. Paust, and M. R. Patsch, *Org. Syn.,* **49**, 86 (1969).
139. (a) D. M. Singleton and J. K. Kochi, *J. Am. Chem. Soc.,* **89**, 6547 (1967); **90**, 1582 (1968). (b) W. C. Kray, Jr., and C. E. Castro, *ibid.,* **86**, 4603 (1964). (c) D. H. R. Barton, N. K. Basu, R. H. Hesse, F. S. Morehouse, and M. M. Pechet, *ibid.,* **88**, 3016 (1966). (d) M. C. Cabaleiro and M. D. Johnson, *J. Chem. Soc.,* B, 565 (1967). (e) J. K. Kochi, D. M. Singleton, and L. J. Andrews, *Tetrahedron,* **24**, 3503 (1968).

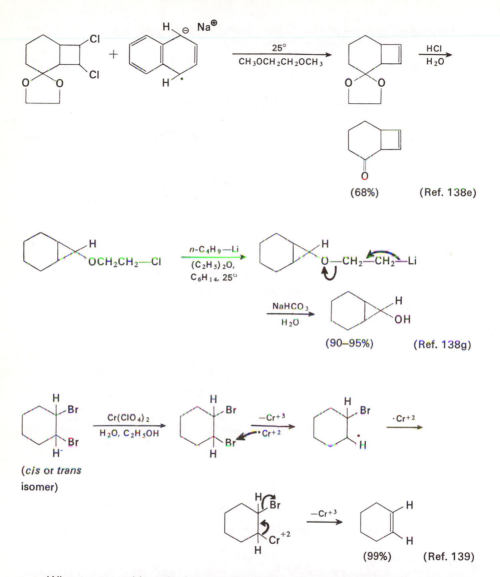

(68%)   (Ref. 138e)

(90–95%)   (Ref. 138g)

(cis or trans
isomer)

(99%)   (Ref. 139)

When comparable reducing conditions are applied to dihalides in which the halogens are separated by three or four carbon atoms, ring closure may be observed;[140] alternatively, 1,4-dihalides may undergo a fragmentation[141] reaction as

140. (a) T. F. Corbin, R. C. Hahn, and H. Shechter, *Org. Syn.,* **44**, 30 (1964). (b) J. K. Kochi and D. M. Singleton, *J. Org. Chem.,* **33**, 1027 (1968). (c) J. T. Gragson, K. W. Greenlee, J. M. Derfer, and C. E. Boord, *ibid.,* **20**, 275 (1955). (d) M. R. Rifi, *J. Am. Chem. Soc.,* **89**, 4442 (1967); *Tetrahedron Letters,* **No. 13**, 1043 (1969).
141. (a) F. N. Stepanov and V. D. Sukhoverkhov, *Angew. Chem., Intern. Ed. Engl.,* **6**, 864 (1967). (b) C. A. Grob and W. Baumann, *Helv. Chim. Acta,* **38**, 594 (1955).

illustrated below. Usually the cyclization reactions are best suited for the preparation of three- and four-membered rings and give poor yields of larger sized rings.

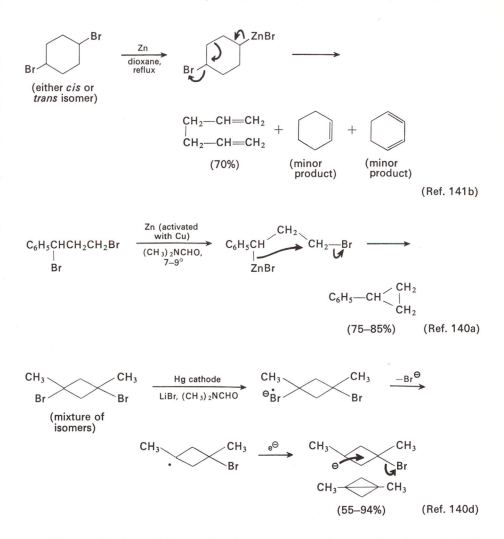

(70%)   (minor product)   (minor product)

(Ref. 141b)

(75–85%)   (Ref. 140a)

(55–94%)   (Ref. 140d)

The mechanism of the carbon-halogen bond cleavage has been studied extensively for reductions effected with solutions of alkali metals,[142] with chromium-

---

142. (a) P. E. Verkade, K. S. deVries, and B. M. Wepster, *Rec. Trav. Chim.*, **83**, 367, 1149 (1964); **84**, 1295 (1965). (b) J. Jacobus and J. F. Eastham, *Chem. Commun.*, **No. 4**, 138 (1969). (c) J. Jacobus and D. Pensak, *ibid.*, **No. 8**, 400 (1969). (d) H. M. Walborsky, F. P. Johnson, and J. B. Pierce, *J. Am. Chem. Soc.*, **90**, 5222 (1968). (e) S. J. Cristol and R. W. Gleason, *J. Org. Chem.*, **34**, 1762 (1969). (f) D. B. Ledlie and S. MacLean, *ibid.*, **34**, 1123 (1969).

(II) salts,[139,143] and by electrolysis.[144] Although there exists some disagreement about interpretation of the data, the reaction paths indicated in the following equations (where M is a metal, a chromium(II) species, or an electrode surface) appear to be consistent with most observations. The formation of an intermediate free radical (e.g., [133]) accounts for the lack of stereospecificity indicated in the

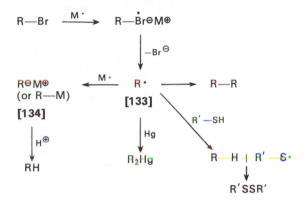

various elimination reactions described previously and the partial or complete racemization observed in various carbon-halogen bond reductions. In cases (e.g., [135]) where radical inversion may be retarded and further reduction to form [134] is rapid, partial retention of configuration may be observed in the reduction process. In two cases involving reduction with a chromium(II) salt or the more reactive[143b,c] *bis*-ethylenediamine complex [136], it has been possible to intercept the intermediate radical (e.g., [137]) with added *n*-butyl mercaptan, an efficient hydrogen atom donor.[139c,143d]

In several electrolytic reductions at mercury cathodes these radical intermediates have been intercepted by reaction with mercury to form dialkylmercury derivatives such as [139].[144a,b]

It should be noted that these reduction conditions have also been effective for the partial reduction of geminal dihalogen compounds such as [140].[144e,f,145]

143. (a) C. E. Castro and W. C. Kray, Jr., *J. Am. Chem. Soc.,* **85**, 2768 (1963) ; **88**, 4447 (1966). (b) J. K. Kochi and P. E. Mocadlo, *ibid.,* **88**, 4094 (1966). (c) J. K. Kochi and J. W. Powers, *ibid.,* **92**, 137 (1970). (d) R. E. Erickson and R. K. Holmquist, *Tetrahedron Letters,* **No. 48**, 4209 (1969).
144. (a) J. L. Webb, C. K. Mann, and H. M. Walborsky, *J. Am. Chem. Soc.,* **92**, 2042 (1970). (b) J. Grimshaw and J. S. Ramsey, *J. Chem. Soc.,* B, 60 (1968) ; J. A. Azoo, F. G. Coll, and J. Grimshaw, *ibid.,* C, 2521 (1969). (c) F. H. Covitz, *J. Am. Chem. Soc.,* **89**, 5403 (1967). (d) J. G. Lawless, D. E. Bartak, and M. D. Hawley, *ibid.,* **91**, 7121 (1969). (e) R. Annino, R. E. Erickson, J. Michalovic, and B. McKay, *ibid.,* **88**, 4424 (1966) ; R. E. Erickson, R. Annino, M. D. Scanlon, and G. Zon, *ibid.,* **91**, 1767 (1969). (f) A. J. Fry and R. H. Moore, *J. Org. Chem.,* **33**, 1283 (1968). (g) L. Eberson, *Acta Chem. Scand.,* **22**, 3045 (1968).
145. (a) M. Nagao, N. Sato, T. Akashi, and T. Yoshida, *J. Am. Chem. Soc.,* **88**, 3447 (1966). (b) H. Nozaki, T. Aratani, and R. Noyori, *Tetrahedron,* **23**, 3645 (1967). (c) For a survey of the related reaction of 1,1-dihalocyclopropanes with alkyllithium reagents to form allenes, see W. R. Moore and H. R. Ward, *J. Org. Chem.,* **27**, 4179 (1962), and L. Skattebøl and S. Solomon, *Org. Syn.,* **49**, 35 (1969).

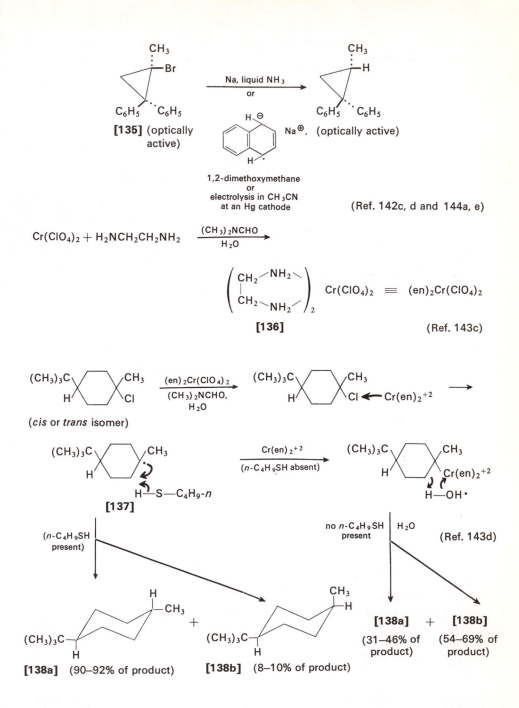

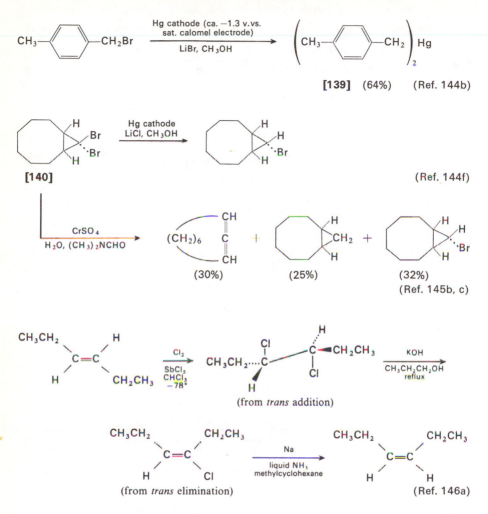

[139]   (64%)   (Ref. 144b)

[140]   (Ref. 144f)

(30%)   (25%)   (32%)
                                 (Ref. 145b, c)

(from *trans* addition)

(from *trans* elimination)   (Ref. 146a)

Although the reductive cleavage of alkyl halides may or may not occur with retention of configuration, the cleavage of vinyl chlorides with sodium in liquid ammonia has been found to yield olefins having the same configuration as the starting halide.[146a] Although vinyl carbanions are conformationally more stable than alkyl carbanions, vinyl radicals invert rapidly.[110] Thus, this stereospecific bond cleavage implies that reaction of the intermediate vinyl radical with solutions of sodium in liquid ammonia must be very rapid as was noted in the previously discussed acetylene reduction. In studies of the reduction of vinyl halides such as [141] by other methods, only partial stereospecificity was observed implying that

146. (a) M. C. Hoff, K. W. Greenlee, and C. E. Boord, *J. Am. Chem. Soc.,* **73**, 3329 (1951). (b) For studies of the electrolytic reduction of vinyl halides, see ref. 110d and L. L. Miller and E. Riekena, *J. Org. Chem.,* **34**, 3359 (1969).

the intermediate radical **[142]** was inverting at a rate comparable to the rate of further reduction.

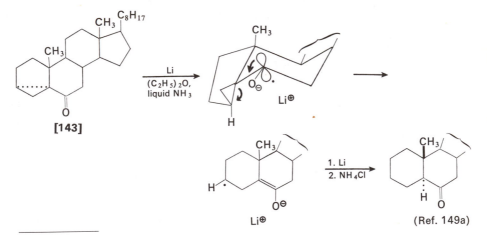

Reduction conditions	Product composition	
X = Cl, Na, naphthalene, $CH_3OCH_2CH_2OCH_3$, 0°	44%	56%
X = I, Hg cathode, $(CH_3)_2NCHO$, $(C_2H_5)_4N^{\oplus}Br^{\ominus}$	30%	70%
		(Ref. 110c, d)

Various metal reductions have also been used to cleave carbon-carbon bonds in strained ring systems such as bicyclobutane derivatives[147] or cyclopropane derivatives,[148] especially cyclopropyl ketones.[149] As illustrated in the following equations which carbon-carbon bond is broken is determined at least in part by stereoelectronic factors; in each of the isomeric ketones **[143]** and **[144]** the bond

(Ref. 149a)

147. W. R. Moore, S. S. Hall, and C. Largman, *Tetrahedron Letters,* **No. 50,** 4353 (1969).
148. (a) H. M. Walborsky and J. B. Pierce, *J. Org. Chem.,* **33,** 4102 (1968). (b) S. W. Staley and J. J. Rocchio, *J. Am. Chem. Soc.,* **91,** 1565 (1969). (c) L. L. Miller and L. J. Jacoby, *ibid.,* **91,** 1130 (1969).
149. (a) W. G. Dauben and E. J. Deviny, *J. Org. Chem.,* **31,** 3794 (1966). (b) W. G. Dauben and R. E. Wolf, *ibid.,* **35,** 374, 2361 (1970). (c) H. E. Zimmerman, K. G. Hancock, and G. C. Licke, *J. Am. Chem. Soc.,* **90,** 4892 (1968); H. E. Zimmerman, R. D. Riéke, and J. R. Scheffer, *ibid.,* **89,** 2033 (1967); H. E. Zimmerman and R. L. Morse, *ibid.,* **90,** 954 (1968). (d) R. Fraisse-Jullien, C. Fiejaville, and V. Toure, *Bull. Soc. Chim. France,* 3725 (1966). (e) A. J. Bellamy and G. H. Whitham, *Tetrahedron,* **24,** 247 (1968). (f) S. B. Laing and P. J. Sykes, *J. Chem. Soc.,* C, 937 (1968). (g) S. A. Monti, D. J. Bucheck, and J. C. Shepard, *J. Org. Chem.,* **34,** 3080 (1969). (h) R. K. Hill and J. W. Morgan, *ibid.,* **33,** 927 (1968).

broken is the one most nearly parallel to the pi orbital of the carbonyl anion radical.[149a,b,g,h] With more flexible ketones such as [145] there is a preference for cleavage of the more highly substituted C—C bond, a result which would appear to be attributable to a combination of a stereoelectronic factor and the increased stability of the initial anion radical cleavage product [146a].[149b,d]

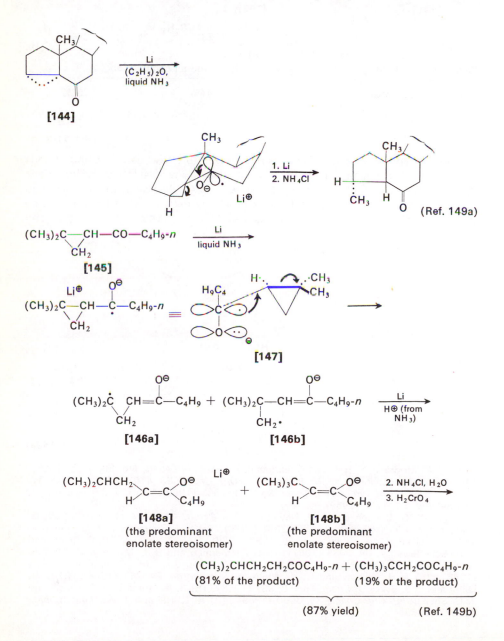

[144]

(Ref. 149a)

$(CH_3)_2C$—CH—CO—C_4H_9-n $\xrightarrow[\text{liquid NH}_3]{\text{Li}}$
 $\ |$
 CH_2

[145]

$(CH_3)_2C$—CH—C—C_4H_9-n ≡
 $\ |$
 CH_2

[147]

$(CH_3)_2\overset{\cdot}{C}$ CH=C—C_4H_9 + $(CH_3)_2C$—CH=C—C_4H_9-n $\xrightarrow[\text{NH}_3)]{\text{Li}}$
     $\ |$ $\ |$
     CH_2 $CH_2\cdot$

[146a]   [146b]

$(CH_3)_2CHCH_2$ 〉C=C〈 $O^\ominus$ + $(CH_3)_3C$ 〉C=C〈 $O^\ominus$ $\xrightarrow[\text{3. H}_2\text{CrO}_4]{\text{2. NH}_4\text{Cl, H}_2\text{O}}$
     H C_4H_9 H C_4H_9

[148a]   [148b]
(the predominant   (the predominant
enolate stereoisomer)   enolate stereoisomer)

$(CH_3)_2CHCH_2CH_2COC_4H_9$-$n$ + $(CH_3)_3CCH_2COC_4H_9$-n
(81% of the product)   (19% or the product)

(87% yield)   (Ref. 149b)

With flexible cyclopropyl ketones, the cleavage of the anion radical is believed to involve the *cisoid* conformers such as **[147]** which allow continuous overlap of the anion radical pi orbital with the new *p* orbital formed as the cyclopropane carbon-carbon bond breaks. The enolate anions **[148]** formed have the geometry expected from cleavage of the cisoid conformers such as **[147]**. A similar reductive cleavage of cyclopropyl ketones has been effected by irradiating a methanolic solution of a cyclopropyl ketone and tri-*n*-butyltin hydride (see Chapter 2) with ultraviolet light.[150]

One carbon-oxygen bond of epoxides may also be cleaved by reduction with solutions of lithium in ammonia,[132c] ethylamine, or, preferably, ethylenediamine.[151] These reductions usually occur with predominant cleavage at the less hindered epoxide carbon atom as one of the following examples indicates. Reduction of

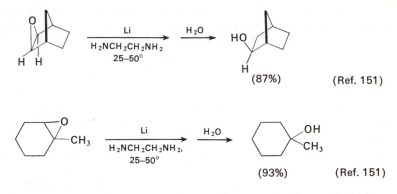

epoxides by this procedure is reported to be superior to reduction with lithium aluminum hydride (see Chapter 2).

---

150. M. Pereyre and J. Y. Godet, *Tetrahedron Letters,* **No. 42,** 3653 (1970).
151. H. C. Brown, S. Ikegami, and J. H. Kawakami, *J. Org. Chem.,* **35,** 3243 (1970).

4

REDUCTIONS WITH HYDRAZINE AND
ITS DERIVATIVES

The selective reduction of certain unsaturated functions can be accomplished with hydrazine or one of its derivatives. By proper choice of reactants and conditions, compounds containing either carbon-carbon multiple bonds or carbonyl groups may be reduced to hydrocarbons. Hydrazine[1] is most commonly used as the hydrate ($N_2H_4 \cdot H_2O$ containing 85% hydrazine, b.p. 119°). However, in certain circumstances use of the anhydrous liquid (b.p. 114°)[1] is preferred; a product containing more than 95% hydrazine is commercially available.

THE WOLFF-KISHNER REDUCTION AND RELATED REACTIONS

Reaction of the hydrazone of an aldehyde or a ketone with a relatively strong base; usually an alkali metal alkoxide at 180–200°, results in reduction of the carbonyl function to a methylene group.[2] Although the same change may be accomplished by the Clemmensen reduction (see Chapter 3), the two methods compliment one another since the Clemmensen procedure utilizes a strongly acidic medium rather than the strongly basic conditions of the Wolff-Kishner method. The original procedures for the alkaline decomposition of hydrazones involved addition of the hydrazone to hot potassium hydroxide in the absence of a solvent or heating a solution of either the preformed hydrazone or the carbonyl compound and hydrazine in ethanolic sodium ethoxide to 160–200° in a sealed tube or an autoclave. In the example illustrating this latter procedure, reduction of the ketone group in the starting material [1] is accompanied by loss of the carboethoxy function in the strongly basic medium (see Chapter 11).

To avoid the use of pressure equipment and, also, to permit the removal of water at the start of the reaction, these procedures have been replaced by the Huang-Minlon modification in which a mixture of the carbonyl compound, excess hydrazine hydrate, sodium hydroxide or potassium hydroxide, and diethylene glycol (b.p. 245°) is refluxed for approximately one hour and then heated to 190–200° while the water and excess hydrazine are allowed to distil from the reaction flask. The

---

1. For the preparation of anhydrous hydrazine see (a) L. I. Smith and K. L. Howard, *Org. Syn.*, **24**, 53 (1944). (b) C. C. Clark, *Hydrazine,* Mathieson Chemical Corp., Baltimore, Maryland, 1953. (c) L. F. Audrieth and B. A. Ogg, *The Chemistry of Hydrazine,* Wiley, New York, 1951.
2. (a) D. Todd, *Org. Reactions,* **4**, 378 (1948). (b) H. H. Szmant, *Angew. Chem., Intern. Ed. Engl.,* **7**, 120 (1968). (c) W. Reusch in R. L. Augustine, ed., *Reduction,* Marcel Dekker, New York, 1968, pp. 171–185.

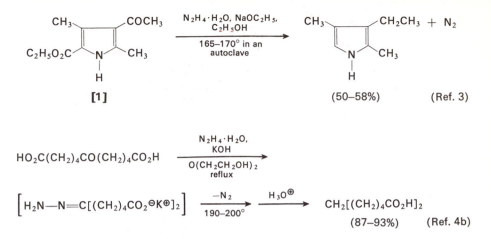

$$HO_2C(CH_2)_4CO(CH_2)_4CO_2H \xrightarrow[\substack{O(CH_2CH_2OH)_2 \\ reflux}]{\substack{N_2H_4 \cdot H_2O, \\ KOH}}$$

$$\left[H_2N-N=C[(CH_2)_4CO_2^{\ominus}K^{\oplus}]_2\right] \xrightarrow[190-200°]{-N_2} \xrightarrow{H_3O^{\oplus}} CH_2[(CH_2)_4CO_2H]_2$$

(87–93%)   (Ref. 4b)

reaction mixture is then heated to 190–200° until the evolution of nitrogen is complete.[4] If the ketone to be reduced is very hindered (e.g., [2]), this procedure may be modified by the use of anhydrous hydrazine and the sodium alkoxide obtained from reaction of sodium with diethylene glycol so that the intital reaction mixture is completely free of water.[5a,b,e] An alternative procedure, which is especially valuable for hindered ketones, consists of forming the hydrazone under conditions of acid catalysis with a mixture of hydrazine and hydrazine dihydro-chloride in triethylene glycol at 130°; the mixture is then made basic with potassium hydroxide and heated to 210–220°.[2c,5f]

These forcing conditions have led to successful reductions in cases where the normal Huang-Minlon procedure failed. The concurrent saponification of the ester function during the reduction of the diketo ester [2] is a typical side reaction which is to be expected whenever molecules containing base-sensitive functional groups are subjected to Wolff-Kishner reduction. Base-catalyzed epimerization of asymmetric centers[5c–e] and cleavage of methyl ethers may also be observed under the strongly basic conditions of the Wolff-Kishner reduction. As is illustrated by the accompanying example, the occurrence of base-catalyzed epimerization during the Wolff-Kishner reduction of a ketone need not lead to the formation of the most stable stereoisomer of the hydrocarbon product. Instead, the stereochemistry of the product appears to be determined by the relative rates of formation and decomposition of the hydrazone intermediates.[5c–e]

3. H. Fischer, *Org. Syn.*, **Coll. Vol. 3**, 513 (1955).
4. (a) Huang-Minlon, *J. Am. Chem. Soc.*, **68**, 2487 (1946); **71**, 3301 (1949). (b) L. J. Durham, D. J. McLeod, and J. Cason, *Org. Syn.*, **Coll. Vol. 4**, 510 (1963). (c) S. Hünig, E. Lücke, and W. Brenninger, *Org. Syn.*, **43**, 34 (1963).
5. (a) D. H. R. Barton, D. A. J. Ives, and B. R. Thomas, *J. Chem. Soc.*, 2056 (1955). (b) see also R. A. Bell, R. E. Ireland, and R. A. Partyka, *J. Org. Chem.*, **31**, 2530 (1966). (c) J. W. Huffman, T. Kamiya, L. H. Wright, J. J. Schmid, and W. Herz, *ibid.*, **31**, 4128 (1966). (d) C. Djerassi, T. T. Grossnickle and L. B. High, *J. Am. Chem. Soc.*, **78**, 3166 (1956). (e) R. L. Clarke, *ibid.*, **83**, 965 (1961). (f) W. Nagata and H. Itazaki, *Chem. Ind.* (London), 1194 (1964).

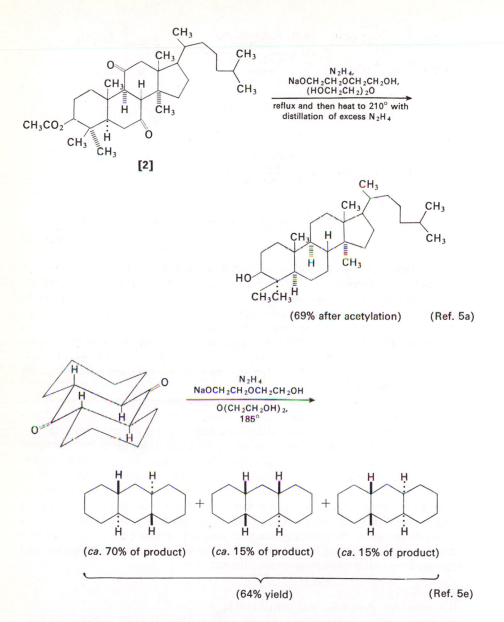

(69% after acetylation)   (Ref. 5a)

(*ca.* 70% of product)   (*ca.* 15% of product)   (*ca.* 15% of product)

(64% yield)   (Ref. 5e)

This reduction method is also applicable to aldehydes, a particularly useful example being the conversion of a hindered carbomethoxyl group (i.e., [3]) to a methyl group by the sequence of reductions and oxidation indicated below. This reaction scheme is distinctly superior to a possible alternative route in which the intermediate alcohol [4] would be converted to an arylsulfonate ester and then reduced with lithium aluminum hydride (see Chapter 2) because the hydride

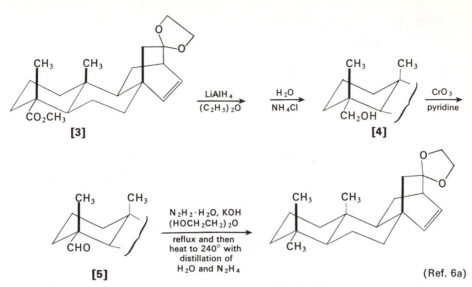

[5]

(Ref. 6a)

reduction would entail a bimolecular nucleophilic displacement at a very un-favorable neopentyl positon. Also worthy of note is the survival of the acid sensitive ketal function during the Wolff-Kishner reduction of the aldehyde [5].

The carbonyl compound being reduced may be introduced into the reaction mixture in the form of its hydrazone, semicarbazone, or azine derivative. Alternatively, semicarbazide rather than hydrazine may be added as the reducing agent as illustrated in the following example. In all cases, the actual intermediate preced-

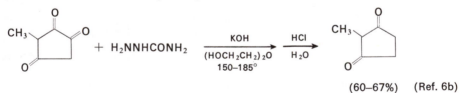

(60–67%)   (Ref. 6b)

ing reduction is relieved to be the hydrazone which is formed from the carbonyl compound, semicarbazone, or azine (e.g., [6]) precursor. Since azines are usually not decomposed by treatment with the bases used in the Wolff-Kishner reduction, an excess of hydrazine is used in the initial stage of the reaction to favor the formation of the hydrazone (e.g., [7]) rather than the azine (e.g., [6]). Subsequent steps are believed[2b,7,8] to involve formation of an alkyl diimide (e.g., [8]) whose

6. (a) L. H. Zalkow and N. N. Girotra, *J. Org. Chem.,* **29**, 1299 (1964). (b) J. P. John, S. Swaminathan, and P. S. Venkataramani, *Org. Syn.,* **47**, 83 (1967).
7. (a) H. H. Szmant and C. M. Harmuth, *J. Am. Chem. Soc.,* **86**, 2909 (1964). (b) H. H. Szmant and M. N. Roman, *ibid.,* **88**, 4034 (1966). (c) E. M. Kaiser, F. E. Henoch, and C. R. Hauser, *ibid.,* **90**, 7287 (1968).
8. (a) D. J. Cram, M. R. V. Sahyun, and G. R. Knox, *J. Am. Chem. Soc.,* **84**, 1734 (1962). (b) D. J. Cram, *Fundamentals of Carbanion Chemistry,* Academic Press, New York, 1965, pp. 159–164, 212–215. (c) D. J. Cram and J. S. Bradshaw, *J. Am. Chem. Soc.,* **85**, 1108 (1963). (d) M. F. Grundon, H. B. Henbest, and M. D. Scott, *J. Chem. Soc.,* 1855 (1963). (e) A. J. Parker, *Adv. Org. Chem.,* **5**, 1 (1965).

$$(C_6H_5)_2C\!\!=\!\!N\!\!-\!\!N\!\!=\!\!C(C_6H_5)_2 + H_2NNH_2 \;\rightleftharpoons\; 2\,(C_6H_5)_2C\!\!=\!\!N\!\!-\!\!NH_2$$

$$\mathbf{[6]} \hspace{7cm} \mathbf{[7]}$$

$$(C_6H_5)_2C\!\!=\!\!NNH_2 \;\xrightarrow[\text{base}]{-H^{\oplus}}\; (C_6H_5)_2C\!\!=\!\!N\!\!-\!\!\overset{\ominus}{N}H \;\xrightarrow{\text{ROH}}$$

$$(C_6H_5)_2CH\!\!-\!\!N\!\!=\!\!N\!\!-\!\!H \;\xrightarrow[\text{base}]{-H^{\oplus}}\; (C_6H_5)_2CH\!\!-\!\!N\!\!=\!\!N^{\ominus} \;\xrightarrow{-N_2}$$

$$\mathbf{[8]} \hspace{7cm} \mathbf{[9]}$$

$$(C_6H_5)_2CH^{\ominus} \;\xrightarrow{\text{ROH}}\; (C_6H_5)_2CH_2$$

conjugate base [9] looses nitrogen to form a carbanion which is protonated by the solvent. There is evidence indicating that several of these proton transfer steps involving intermediates [8] and [9] may be concerted so that a free alkyldiimide intermediate [8] is not necessarily involved.[7,8] The use of dimethyl sulfoxide as the predominant solvent has been found[7b,8a] to lower substantially the temperature required for base-catalyzed decomposition of hydrazones. The reaction is also facilitated by the use of anhydrous conditions such as treatment of the hydrazone with potassium *t*-butoxide in boiling toluene.[8d] Since the formation of the anion [9] is believed to be the rate-limiting step in the previously outlined sequence, the effect of dimethylsulfoxide is attributed to the ability of this solvent to solvate the

$$(C_6H_5)_2C\!\!=\!\!N\!\!-\!\!NH_2 \;\xrightarrow[\substack{(CH_3)_2SO \\ 25°}]{t\text{-}C_4H_9O^{\ominus}K^{\oplus}}\; (C_6H_5)_2CH_2 + (C_6H_5)_2C\!\!=\!\!N\!\!-\!\!N\!\!=\!\!C(C_6H_5)_2$$

$$\text{(90\%)} \hspace{6cm} \text{(Ref. 8a)}$$

$$(C_6H_5)_2C\!\!=\!\!NNH_2 \;\xrightarrow[C_6H_5CH_3,\ \text{reflux}]{t\text{-}C_4H_9O^{\ominus}K^{\oplus}}\; (C_6H_5)_2CH_2$$

$$\text{(85\%)} \hspace{6cm} \text{(Ref. 8d)}$$

metal cation and to aid proton removal.[2b] The base-catalyzed decomposition of semicarbazones (e.g., [10]) is believed to follow a similar sequence in which the initially formed hydrazone (e.g., [11]) undergoes further reaction with base to form the hydrocarbon product. The initial conversion of the semicarbazone to the hydrazone may involve the elimination of isocyanic acid rather than saponification of the carboxamide moiety.[9]

Two procedures have been developed for the reductive deamination of amines[10,11] which bear a mechanistic relationship to the Wolff-Kishner reduction in that alkyl diimide intermediates (or the related anions) are apparently involved.

---

9. For a study of a related process, see W. M. Jones and D. L. Muck, *J. Am. Chem. Soc.,* **88,** 3798 (1966).

10. A. Nickon and A. S. Hill, *J. Am. Chem. Soc.,* **86,** 1152 (1964).

11. (a) C. L. Bumgardner, K. J. Martin, and J. P. Freeman, *J. Am. Chem. Soc.,* **85,** 97 (1963). (b) C. L. Bumgardner and J. P. Freeman, *ibid.,* **86,** 2233 (1964). (c) C. L. Bumgardner and J. P. Freeman, *Tetrahedron Letters,* **No. 13,** 737 (1964).

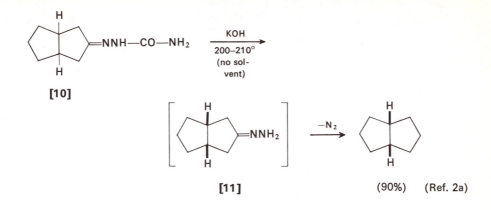

[10]

[11]   (90%)   (Ref. 2a)

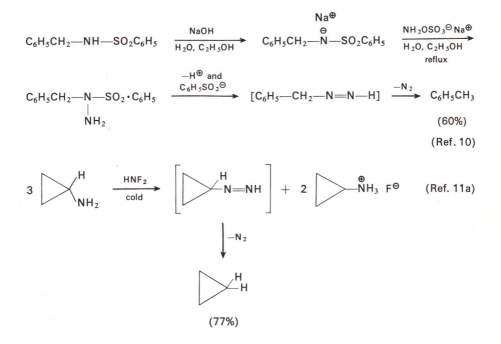

$$C_6H_5CH_2-NH-SO_2C_6H_5 \xrightarrow[H_2O, C_2H_5OH]{NaOH} C_6H_5CH_2-\overset{\ominus}{\underset{Na^\oplus}{N}}-SO_2C_6H_5 \xrightarrow[\text{reflux}]{\overset{NH_2OSO_3^\ominus Na^\oplus}{H_2O, C_2H_5OH}}$$

$$C_6H_5CH_2-\underset{\underset{NH_2}{|}}{N}-SO_2 \cdot C_6H_5 \xrightarrow[\substack{C_6H_5SO_2^\ominus}]{-H^\oplus \text{ and}} [C_6H_5-CH_2-N=N-H] \xrightarrow{-N_2} C_6H_5CH_3$$

(60%)

(Ref. 10)

(77%)

However, differences have been observed when the reactions have been applied to reactants which could yield intermediate cyclopropylcarbinyl diimides since the two deamination procedures favor ring-opened products in cases where no ring opening is observed in the Wolff-Kishner reduction.[8b,11] The following examples illustrate the various possibilities in the Wolff-Kishner procedure. The absence of ring-opened products in some cases suggests that if a free carbanion intermediate is involved, it must have a very short lifetime.

The reduction of α,β-unsaturated ketones or aldehydes by the Wolff-Kishner method may be complicated in one of two ways. If cyclization of the intermediate

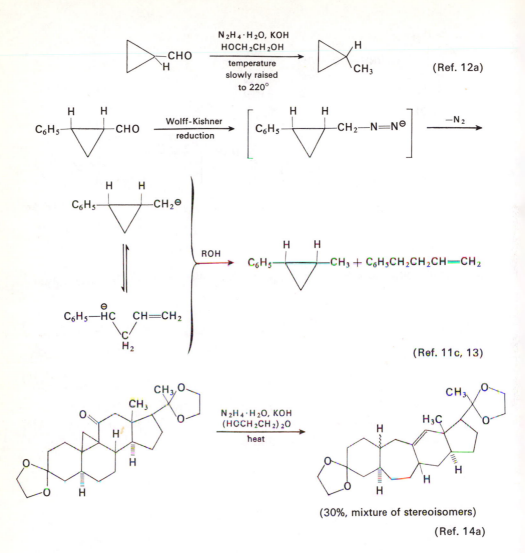

(30%, mixture of stereoisomers)

(Ref. 14a)

12. (a) E. Renk, P. R. Shafer, W. H. Graham, R. H. Mazur, and J. D. Roberts, *J. Am. Chem. Soc.*, **83**, 1987 (1961). (b) U. Biethan, H. Klusacek, and H. Musso, *Angew. Chem., Intern. Ed. Engl.*, **6**, 176 (1967).
13. For a general discussion of the opening of the cyclopropylcarbinyl carbanion with leading references, see (a) M. E. H. Howden, A. Maercker, J. Burdon, and J. D. Roberts, *J. Am. Chem. Soc.*, **88**, 1732 (1966). (b) A. Maercker and J. D. Roberts, *J. Am. Chem. Soc.*, **88**, 1742 (1966).
14. (a) S. M. Kupchan, E. Abushanab, K. T. Shamasundar, and A. W. By, *J. Am. Chem. Soc.*, **89**, 6327 (1967). (b) R. Ramamoorthy and G. S. Krishna Rao, *Tetrahedron Letters*, **No. 51**, 5145 (1967). (c) H. G. Heller and R. A. N. Morris, *J. Chem. Soc.*, C, 1004 (1966). (d) D. E. Evans, G. S. Lewis, P. J. Palmer, and D. J. Weyell, *ibid.*, C, 1197 (1968). (e) R. J. Petersen and P. S. Skell, *Org. Syn.*, **47**, 98 (1967).

hydrazone to a pyrazoline (e.g., [12]) is sterically favorable, then formation of this pyrazoline will frequently precede base-catalyzed decomposition. Tautomerization and subsequent loss of nitrogen offers a synthetic route to cyclopropane derivatives (e.g., [13]. In cases (e.g., [14]) where closure of the intermediate hydrazone to

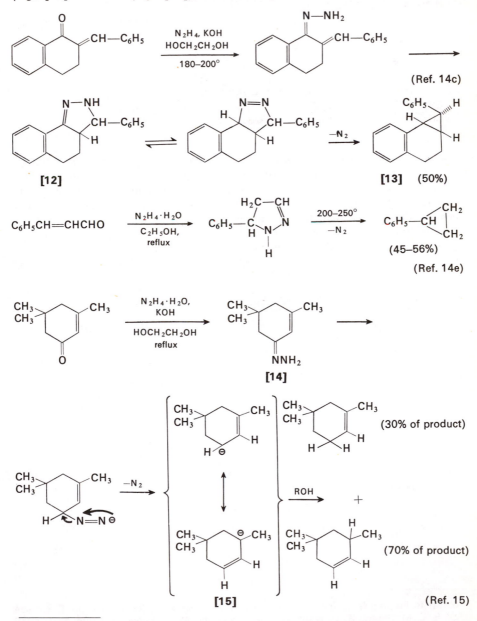

15. R. A. Sneen and N. P. Matheny, *J. Am. Chem. Soc.*, **86**, 5503 (1964).

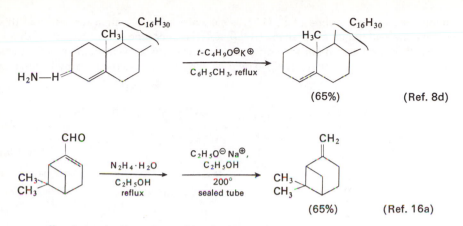

(65%)   (Ref. 8d)

(65%)   (Ref. 16a)

a pyrazoline is sterically unfavorable, the base-catalyzed decomposition is thought to follow the previously discussed pathway. However, it will be noted that an allylic carbanionic intermediate is formed (e.g., [15]) which may be protonated at one of two different carbon atoms to yield a mixture of isomeric olefins. In other cases Wolff-Kishner reduction has apparently yielded a single olefin formed with or without migration of the carbon-carbon double bond. For example, the reaction of α,β-unsaturated ketone hydrazones with potassium t-butoxide in toluene has been reported to yield reduction products in which the double bond has not moved.[8d]

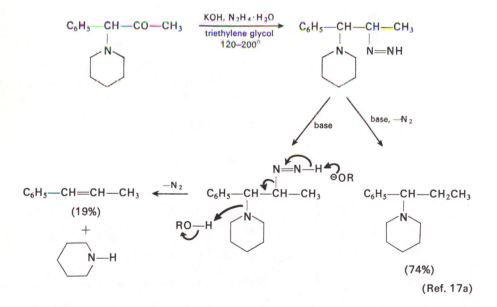

(74%)

(Ref. 17a)

16. (a) R. Fischer, G. Lardelli, and O. Jeger, *Helv. Chim. Acta,* **34,** 1577 (1951). (b) see also E. L. McGinnis, G. D. Meakins, J. E. Price, and M. C. Styles, *J. Chim. Soc.,* 4379 (1965).

$(CH_3)_3C—CO—CH_2—O—C_6H_5$ $\xrightarrow[\substack{\text{triethylene} \\ \text{glycol} \\ 120–130°}]{N_2H_4 \cdot H_2O}$

$\underset{\underset{\displaystyle (CH_3)_3C—\overset{\displaystyle \|}{C}—CH_2—OC_6H_5}{}}{\overset{N—NH_2}{}}$ $\xrightarrow[\substack{\text{triethylene} \\ \text{glycol,} \\ 200–210°}]{KOH, \ N_2H_4}$ $(CH_3)_3C—CH{=}CH_2$

(80%)   (Ref. 17a)

The presence of a reasonably reactive leaving group at the carbon *alpha* to an aldehyde or ketone carbonyl function also leads to complications if a Wolff-Kishner reduction is attempted because olefinic products often accompany or replace the expected reduction products.[2b,17] If the alpha substituent is a relatively poor leaving group such as an α-amino function, the normal Wolff-Kishner reduction product is usually formed accompanied by lesser amounts of olefinic products. The probable mechanism of olefin formation, illustrated in the accompanying example, involves elimination of the alpha substituent during the base-catalyzed decomposition of the alkyl diimide. The extent of this elimination side reaction may be diminished by conducting the reaction at the lower temperatures possible when the hydrazone is treated with potassium *t*-butoxide in toluene under an-hydrous conditions.[8d] As the ability of the alpha substituent to serve as a leaving group increases, the unsaturated compound becomes the major or only product. With a good leaving group (e.g. a halogen or an epoxide oxygen) at the *alpha* position the nature of the reaction changes such that the initial reaction of the α-substituted ketone with hydrazine leads directly to elimination even in the absence of a stronger base. The illustrated reaction of α,β-epoxy ketones with hydrazine constitutes a very useful synthesis of allylic alcohols.[18]

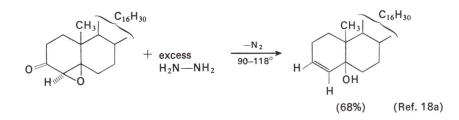

(68%)   (Ref. 18a)

17. (a) N. J. Leonard and S. Gelfand, *J. Am. Chem. Soc.*, **77**, 3269, 3272 (1955). (b) For an example where rearrangement was observed when a Wolff-Kishner reduction of a β-amino ketone was attempted, see C. G. Overberger, J. Reichenthal, and J. P. Anselme, *J. Org. Chem.*, **35**, 138 (1970). (c) D. F. Morrow, M. E. Butler, W. A. Neuklis, and R. M. Hofer, *ibid.*, **32**, 86 (1967).
18. (a) P. S. Wharton and D. H. Bohlen, *J. Org. Chem.*, **26**, 3615 (1961). (b) W. R. Benn and R. M. Dobson, *ibid.*, **29**, 1142 (1964); G. Ohloff and G. Uhde, *Helv. Chim. Acta*, **53**, 531 (1970).

The reaction of α-halo ketones (e.g., **[16]**) with hydrazine to form olefins also occurs, but this reaction would appear to be a less satisfactory synthetic sequence because of a number of possible side reactions between the α-halo ketone and hydrazine which can compete with olefin formation.[19] The probable course of this reaction is illustrated in the following example. Support for this scheme is

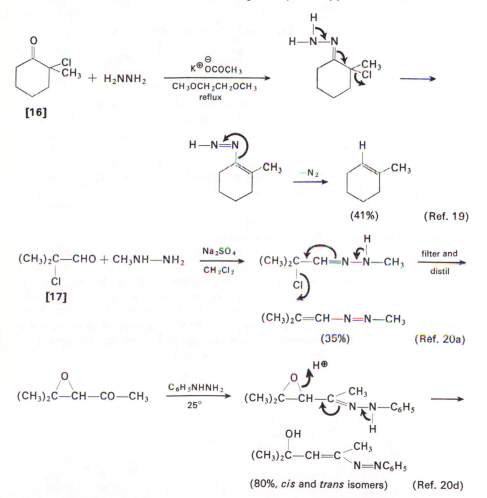

(41%)   (Ref. 19)

(35%)   (Ref. 20a)

(80%, *cis* and *trans* isomers)   (Ref. 20d)

19. P. S. Wharton, S. Dunny, and L. S. Krebs, *J. Org. Chem.,* **29**, 958 (1964).
20. (a) B. T. Gillis and J. D. Hagarty, *J. Am. Chem. Soc.,* **87**, 4576 (1965). (b) L. Caglioti, G. Rosini, and F. Rossi, *ibid.,* **88**, 3865 (1966). (c) J. Buckingham and R. D. Guthrie, *J. Chem. Soc.,* C, 1700 (1967) ; *ibid.,* C, 3079 (1968) ; J. Buckingham, G. J. F. Chittenden, and R. D. Guthrie, *ibid.,* C, 1703 (1967). (d) A. Dondoni, G. Rossini, G. Mossa, and L. Caglioti, *ibid.,* B, 1404 (1968). (e) E. Gründemann, *Angew. Chem., Intern. Ed., Engl.,* **8**, 459 (1969). (f) A. Hassner and P. Catsoulacos, *Tetrahedron Letters,* **No. 6**, 489 (1967) ; *Chem. Commun.,* **No. 3**, 121 (1967). (g) For a study of the base-catalyzed rearrangement of N,N,N-trialkylhydrazonium salts, see G. R. Newkome, *ibid.,* **No. 20**, 1227 (1969).

found in the reaction of α-substituted carbonyl compounds (e.g., **17**), with substituted hydrazines to form α,β-unsaturated azo compounds.[20] Perhaps the best known applications of this reaction are the formation of osazones (bis-1,2-phenylhydrazones)[20] and the dehydrohalogenation of α-halo ketones (e.g., [**18**])

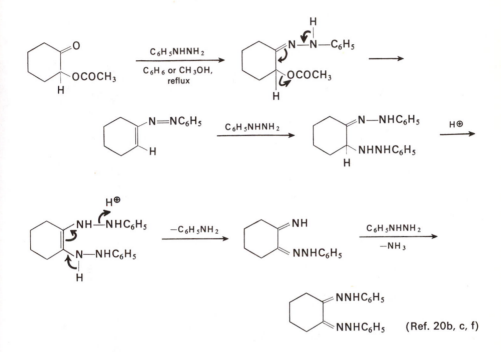

by reaction with 2,4-dinitrophenylhydrazine or semicarbazide to form the derivative of the corresponding α,β-unsaturated ketone. This dehydrohalogenation method (the Mattox-Kendall procedure)[21] is a useful alternative to the other methods which have been employed (cf. Chapter 8) with α-halo ketones. An example of a more complex fragmentation-elimination sequence is provided by the attempted reduction of the bromo ketone [**19**] by the Wolff-Kishner procedure.

REDUCTIONS WITH ARENESULFONYL DERIVATIVES OF HYDRAZINE

The Wolff-Kishner procedures are normally not applicable to the reduction of base-sensitive compounds such as 1,3-dicarbonyl compounds (β-diketones or β-keto esters) or β-keto sulfones[17] because of side reactions resulting in base-catalyzed cleavage (see Chapter 11) or in the formation of pyrazoles or pyrazolones.

21. (a) C. Djerassi, *J. Am. Chem. Soc.*, **71**, 1003 (1949). (b) V. R. Mattox and E. C. Kendall, *ibid.*, **72**, 2290 (1950). (c) W. F. McGuckin and E. C. Kendall, *ibid.*, **74**, 5811 (1952).
22. D. H. Gustafson and W. F. Erman, *J. Org. Chem.*, **30**, 1665 (1965).

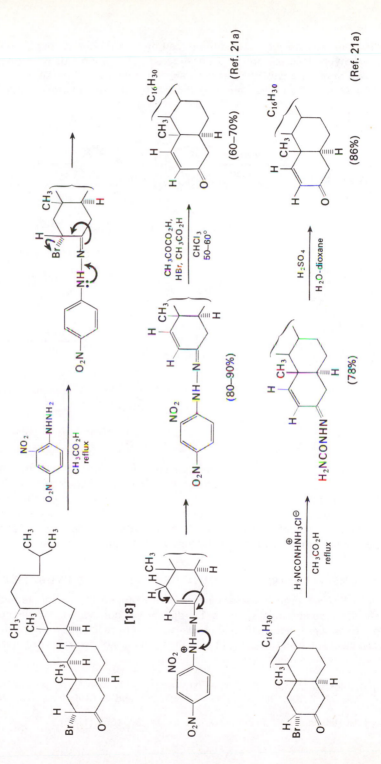

[18]

CH_3CO_2H reflux

CH_3COCO_2H, HBr, CH_3CO_2H
$CHCl_3$
50–60°

(80–90%)

(60–70%)   (Ref. 21a)

$H_2NCONHNH_3Cl^{\oplus}$
CH_3CO_2H
reflux

H_2SO_4
H_2O–dioxane

(78%)

(86%)   (Ref. 21a)

$C_{16}H_{30}$

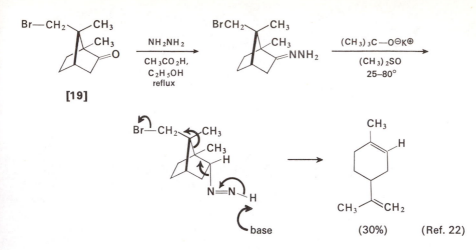

[19]

(30%)   (Ref. 22)

A useful variant of the Wolff-Kishner method which may overcome these difficulties in some cases is based on the fact that arylsulfonylhydrazine derivatives
(e.g., [20]) are converted to alkyl or aryl diimides[23] and subsequently to alkanes
by treatment with bases. By utilizing the ability of metal hydrides such as lithium

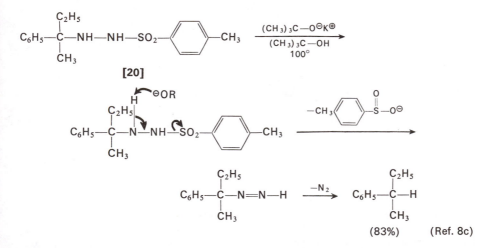

[20]

(83%)   (Ref. 8c)

23. For preparation and study of alkyl and aryl diimides (R—N=NH) see (a) P. C. Huang
and E. M. Kosower, *J. Am. Chem. Soc.,* **90**, 2354, 2362, 2367 (1968); E. M. Kosower,
P. C. Huang, and T. Tsuji, *ibid.,* **91**, 2325 (1969); T. Tsuji and E. M. Kosower, *ibid.,* **91**,
3375 (1969); **92**, 1429 (1970). (b) phenyldiazene has been isolated as its copper(I)
chloride complex by D. Petredis, A. Burke, and A. L. Balch, *ibid.,* **92**, 428 (1970). (c) S.
Hünig and G. Büttner, *Angew. Chem., Intern. Ed. Engl.,* **8**, 451 (1969). (d) R. W. Hoffmann
and K. R. Eicken, *Chem. Ber.,* **100**, 1465 (1967); R. W. Hoffmann and G. Guhn, *ibid.,* **100**,
1474 (1967). (e) R. Fuchs, *Tetrahedron Letters,* **No. 25**, 2419 (1967). (f) G. Fraenkel and
E. Pecchold, *ibid.,* **No. 54**, 4821 (1969). (g) A. Heesing and B. U. Kaiser, *ibid.,* **No. 32**,
2845 (1970). (h) R. O. Hutchins, B. E. Maryanoff, and C. A. Milewski, *J. Am. Chem. Soc.,*
93, 1793 (1971).

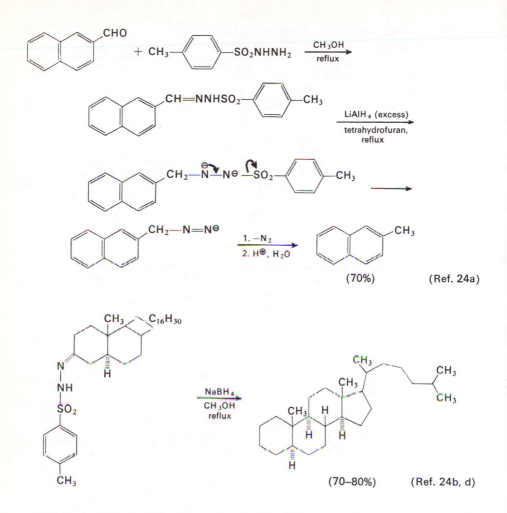

(70%)   (Ref. 24a)

(70–80%)   (Ref. 24b, d)

aluminum hydride or, preferably, sodium borohydride or sodium cyanoborohydride[23h] (see Chapter 2) to selectively reduce the C=N bond of tosylhydrazones (p-toluenesulfonylhydrazones) or their anions, the combined sequence of reactions serves to reduce the carbonyl group of an aldehyde or a ketone to a methylene group.[24] The fact that only one of the hydrogen atoms of the methylene group in the product has been derived from the metal hydride can be utilized to prepare

24. (a) L. Caglioti and M. Magi, *Tetrahedron,* **19**, 1127 (1963). (b) L. Caglioti, *ibid.,* **22**, 487 (1966). (c) M. Fischer, Z. Pelah, D. H. Williams, and C. Djerassi, *Chem. Ber.,* **98**, 3236 (1965). (d) L. Caglioti and P. Grasselli, *Chem. Ind.* (London), 153 (1964). (e) I. Elphimoff-Felkin and M. Verrier, *Tetrahedron Letters,* **No. 12**, 1515 (1968). (f) For an example where cleavage was observed when the reaction was applied to a β-keto sulfone see H. O. House and J. K. Larson, *J. Org. Chem.,* **33**, 61 (1968). (g) J. E. Herz and E. Gonzalez, *Chem. Commun.,* **No. 23**, 1395 (1969).

monodeuterated alkanes.[24c] Application of this reduction procedure to α,β-unsaturated carbonyl compounds has yielded products in which carbon-carbon double bond migration is observed as the following equation indicates. The success of this reduction method is dependent on the reduction of the C=N bond before

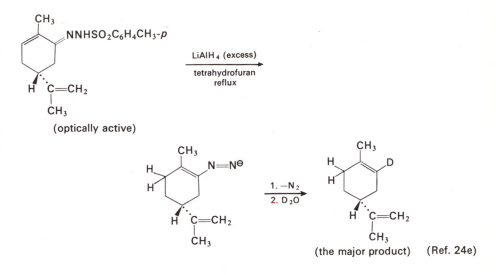

(the major product)   (Ref. 24e)

decomposition of the arylsulfonylhydrazone; with hindered ketones, where this reduction may be slow, the subsequently described Bamford-Stevens reaction to yield an olefin may occur even when an excess of the metal hydride reagent is employed.[25m] This problem may be avoided by reducing the arenesulfonylhydrazone with sodium cyanoborohydride in a slightly acidic medium[23h] as illustrated below.

$$n\text{-}C_9H_{19}\text{---}CO\text{---}CH_3 + p\text{-}CH_3C_6H_4SO_2NHNH_2 \xrightarrow[\substack{(CH_3)_2NCHO,}]{p\text{-}CH_3C_6H_4SO_3H \text{ (cat. amt.)}}$$

$$n\text{-}C_9H_{19}\overset{\overset{\displaystyle CH_3}{|}}{C}=NNHSO_2C_6H_4CH_3\text{-}p \xrightarrow[\substack{p\text{-}CH_3C_6H_4SO_3H \text{ (cat. amt.)}\\(CH_3)_2NCHO,\\100\text{--}105°}]{NaH_3BCN} n\text{-}C_9H_{19}CH_2CH_3$$

(86%)

(Ref. 23h)

Although a similar reaction, resulting in the addition of a butyl group has been realized when certain tosylhydrazones have been treated with excess n-butyllithium,[24g] a more common process involves reaction of a tosylhydrazone with a limited amount of a base such as a metal alkoxide, an alkyllithium reagent, or even lithium aluminum hydride[25m] to form a diazo compound or an olefin as the reaction

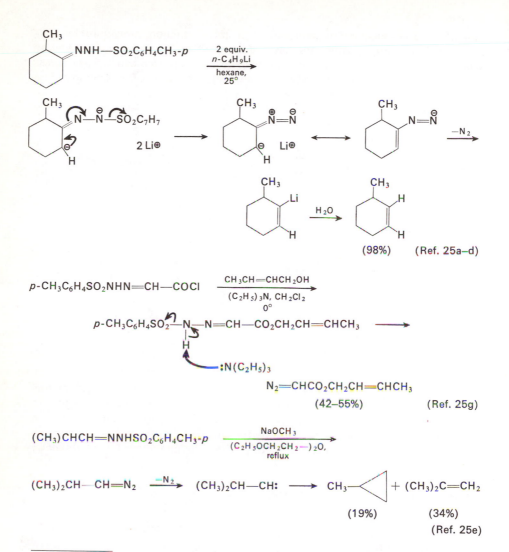

(98%)   (Ref. 25a–d)

$p\text{-CH}_3\text{C}_6\text{H}_4\text{SO}_2\text{NHN}{=}\text{CH}{-}\text{COCl}$ $\xrightarrow[\substack{(\text{C}_2\text{H}_5)_3\text{N, CH}_2\text{Cl}_2 \\ 0°}]{\text{CH}_3\text{CH}{=}\text{CHCH}_2\text{OH}}$

$p\text{-CH}_3\text{C}_6\text{H}_4\text{SO}_2{-}\overset{\text{H}}{\underset{}{\text{N}}}{-}\text{N}{=}\text{CH}{-}\text{CO}_2\text{CH}_2\text{CH}{=}\text{CHCH}_3 \longrightarrow$

:$\text{N}(\text{C}_2\text{H}_5)_3$

$\text{N}_2{=}\text{CHCO}_2\text{CH}_2\text{CH}{=}\text{CHCH}_3$

(42–55%)   (Ref. 25g)

$(\text{CH}_3)\text{CHCH}{=}\text{NNHSO}_2\text{C}_6\text{H}_4\text{CH}_3\text{-}p$ $\xrightarrow[\substack{(\text{C}_2\text{H}_5\text{OCH}_2\text{CH}_2-)_2\text{O,} \\ \text{reflux}}]{\text{NaOCH}_3}$

$(\text{CH}_3)_2\text{CH}{-}\text{CH}{=}\text{N}_2$ $\xrightarrow{-\text{N}_2}$ $(\text{CH}_3)_2\text{CH}{-}\text{CH:} \longrightarrow$ $\text{CH}_3{-}\triangleleft$ $+ (\text{CH}_3)_2\text{C}{=}\text{CH}_2$

(19%)   (34%)

(Ref. 25e)

25. (a) R. H. Shapiro, *Tetrahedron Letters,* **No. 29**, 3401 (1966); **No. 3**, 345 (1968). (b) R. H. Shapiro and M. J. Heath, *J. Am. Chem. Soc.,* **89**, 5734 (1967). (c) G. Kaufman, F. Cook, H. Shechter, J. Bayless, and L. Friedman, *ibid.,* **89**, 5736 (1967). (d) W. G. Dauben, M. E. Lorber, N. D. Vietmeyer, R. H. Shapiro, J. H. Duncan, and K. Tomer, *ibid.,* **90**, 4762 (1968). (e) L. Friedman and H. Shechter, *ibid.,* **81**, 5512 (1959); **83**, 3159 (1961); J. H. Bayless, L. Friedman, F. B. Cook, and H. Shechter, *ibid.,* **90**, 531 (1968); R. H. Shapiro, J. H. Duncan, and J. C. Clopton, *ibid.,* **89**, 1442 (1967). (f) G. L. Closs, L. E. Closs, and W. A. Böll, *ibid.,* **85**, 3796 (1963). (g) A. C. Cope and S. S. Hecht, *ibid.,* **89**, 6920 (1967). (h) J. W. Powell and M. C. Whiting, *Tetrahedron,* **7**, 305 (1959). (i) J. M. Coxon, M. P. Hartshorn, D. N. Kirk, and M. A. Wilson. *ibid.,* **25**, 3107 (1969). (j) C. J. Blankley, F. J. Sauter, and H. O. House, *Org. Syn.,* **49**, 22 (1969). (k) J. A. Landgrebe and A. G. Kirk, *J. Org. Chem.,* **32**, 3499 (1967). (l) P. K. Freeman and R. C. Johnson, *ibid.,* **34**, 1746, 1751 (1969). (m) F. Y. Edamura and A. Nickon, *ibid.,* **35**, 1509 (1970).

product.[25] The formation and decomposition of a tosylhydrazone salt in this way, a process called the Bamford-Stevens reaction, is most often used to convert carbonyl compounds to olefins or diazo compounds as the accompanying examples illustrate. Since the final olefinic C—H bond is formed by reaction of an anion with a proton donor (usually water), the isolation procedure may be modified to provide a useful synthesis for monodeuterated olefins as shown in the next equation.

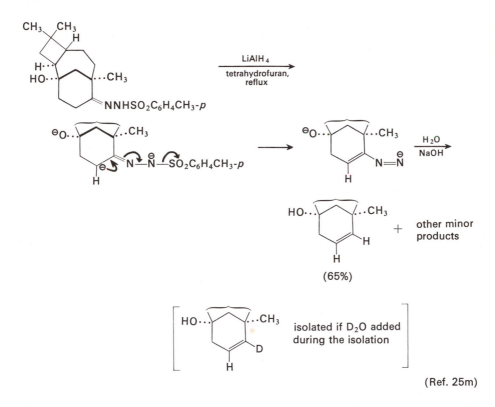

(65%)

isolated if D_2O added during the isolation

(Ref. 25m)

A procedure very closely related to the foregoing reduction and/or decomposition of tosylhydrazones is the base-catalyzed decomposition of 1-acyl-2-arylsulfonylhydrazines which serves as one step in the McFadyen-Stevens method for the reduction of carboxylic acids to aldehydes. The probable course of the reaction is illustrated in the following example. As noted previously, an analogous base-catalyzed reaction serves to convert both alkylhydrazines and arylhydrazines via their arylsulfonyl derivatives to the corresponding hydrocarbons.[8,23,26] Although the McFadyen-Stevens procedure was initially regarded as being applicable to the preparation of aromatic aldehydes or aldehydes which possessed no *alpha* C—H bonds, it has recently been found that aliphatic aldehydes may be prepared in

26. (a) E. Mosettig, *Org. Reactions*, **8**, 232 (1954). (b) T. A. Geissman, *ibid.*, **2**, 94 (1944).

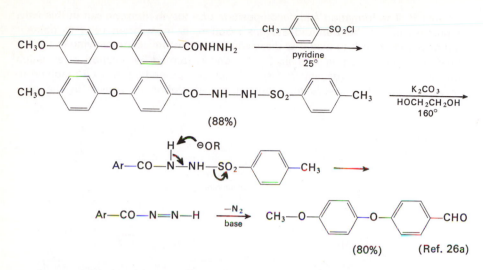

(Ref. 26a)

moderate yields by this general method if the aldehyde formed can be distilled from the reaction mixture sufficiently rapidly to prevent further transformation from aldol condensation (see Chapter 10) or a Cannizzaro reaction.[27]

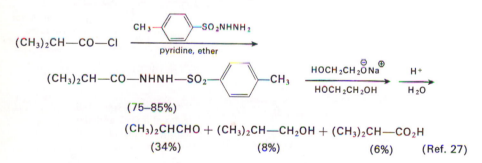

(CH₃)₂CHCHO + (CH₃)₂CH—CH₂OH + (CH₃)₂CH—CO₂H

$$(CH_3)_2CHCHO + (CH_3)_2CH{-}CH_2OH + (CH_3)_2CH{-}CO_2H$$

(34%)   (8%)   (6%)   (Ref. 27)

Carbonyl compounds with heteroatom substituents at the *alpha* carbon form tosylhydrazones which may undergo elimination of the α-substituent during base-catalyzed decomposition.[28a-c] One of the most interesting applications of this reaction has been to α,β-epoxy ketones.[28d-f] In these cases the intermediate azo

27. H. Babad, W. Herbert, and A. W. Stiles, *Tetrahedron Letters*, **No. 25**, 2927 (1966).
28. (a) R. K. Bartlett and T. S. Stevens, *J. Chem. Soc.*, C, 1964 (1967). (b) L. Caglioti, P. Grasselli, F. Morlacchi, and G. Rosini, *Chem. Ind.* (London), 25 (1968). (c) T. Iwadare, I. Adachi, M. Hayashi, A. Matsunaga, and T. Kitai, *Tetrahedron Letters*, **No. 51**, 4447 (1969). (d) M. Tanabe, D. F. Crowe, R. L. Dehn, and G. Detre, *ibid.*, **No. 38**, 3739 (1967); M. Tanabe, D. F. Crowe, and R. L. Dehn, *ibid.*, **No. 40**, 3943 (1967). (e) A. Eschenmoser, D. Felix, and G. Ohloff, *Helv. Chim. Acta*, **50**, 708 (1967); J. Schrieber and co-workers, *ibid.*, **50**, 2101 (1967). (f) For other modifications of this acetylene synthesis, see P. Wieland, H. Kaufmann, and A. Eschenmoser, *ibid.*, **50**, 2108 (1967); D. Felix, J. Schreiber, K. Piers, U. Horn, and A. Eschenmoser, *ibid.*, **51**, 1461 (1968); P. Weiland, *ibid.*, **53**, 171 (1970). R. K. Müller, D. Felix, J. Schreiber, and A. Eschenmoser, *ibid.*, **53**, 1479 (1970).

olefin undergoes further fragmentation to form an acetylene and a carbonyl compound as illustrated in the following examples. This reaction may also be effected by reaction of the epoxy ketone with other nitrogen-containing reactants[28f] such as 1-amino-2-phenylaziridine.

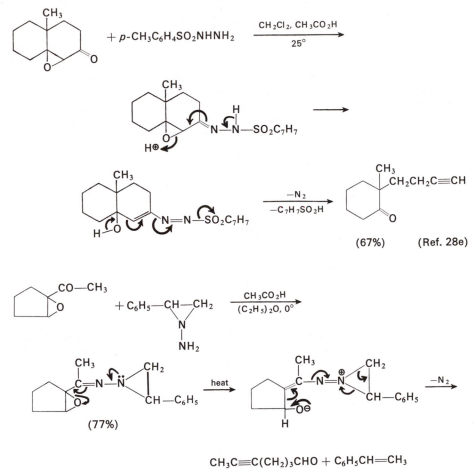

(67%) (Ref. 28e)

(77%)

$$CH_3C \equiv C(CH_2)_3CHO + C_6H_5CH = CH_3$$
(94%) (Ref. 28f)

REDUCTIONS EMPLOYING HYDRAZINE AND A HYDROGENATION CATALYST

A variety of functional groups (e.g. nitro groups, nitriles, carbon-carbon multiple bonds, carbon-halogen bonds) have been reduced by treatment with hydrazine at elevated temperatures.[29] The most widely used application of this type of reduction has been the reduction of aromatic nitro compounds by reaction with

29. A. Furst, R. C. Berlo, and S. Hooton, *Chem. Rev.*, **65**, 51 (1965).

hydrazine in the presence of hydrogenation catalysts such as Raney nickel or a palladium-on-carbon catalyst (see Chapter 1).[29-30] The general procedure is illustrated by the accompanying example. Since the heterogeneous metal catalysts

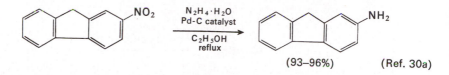

(93–96%)   (Ref. 30a)

used promote the decomposition of hydrazine to nitrogen (or ammonia) and hydrogen,[29] it would appear that this reduction procedure is equivalent to a catalytic hydrogenation in which the decomposing hydrazine serves as a source of hydrogen. In a number of instances where the amount of hydrazine has been limited, it has been possible to isolate reduction products intermediate between the nitro compound and the final reduction product, the amine.[29] The further reduction of these intermediate products by reaction with more hydrazine and a catalyst is frequently possible. This possibility is illustrated by the reduction of ketoximes (e.g., [22]) to amines with hydrazine and Raney nickel.[31]

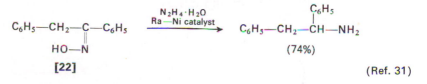

[22]   (74%)

(Ref. 31)

REDUCTIONS WITH DIIMIDE

Although many early studies included reports of the reduction of carbon-carbon multiple bonds by reaction with a mixture of hydrazine and some oxidizing agent, only more recently have these procedures been recognized as reductions involving diimide [23].[32] As indicated in the accompanying equations, diimide is also

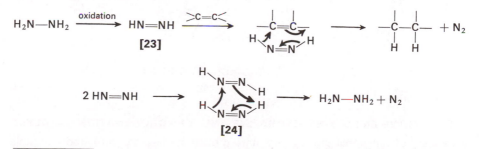

[23]

[24]

30. (a) P. M. G. Bavin, *Org. Syn.*, **40**, 5 (1960). (b) Also see P. M. G. Bavin, *Can. J. Chem.*, **36**, 238 (1958).
31. D. Lloyd, R. H. McDougall, and F. I. Wasson, *J. Chem. Soc.*, 822 (1965).
32. For reviews of the chemistry of diimide, see (a) S. Hünig, H. R. Müller, and W. Thier, *Angew. Chem., Intern. Ed. Engl.*, **4**, 271 (1965). (b) C. E. Miller, *J. Chem. Ed.*, **42**, 254 (1965).

capable of reacting with itself to form the disproportionation products, hydrazine and nitrogen. As a result of this disproportionation, the common procedures for reduction with diimide involve generation of diimide in the presence of the molecule to be reduced rather than attempting to preform this unstable reducing agent. It is presently believed that the intermediate diimide is a mixture of *cis* ([23a]) and *trans* ([23b]) isomers which equilibrate rapidly with one another. The fact that reduction of carbon-carbon double bonds with diimide involves the stereospecific *cis* addition of hydrogen and appears to involve a nonpolar transition state suggests that the reduction proceeds by the cyclic process indicated in structure [25]. Such a process requires the *cis* form [23a] of diimide.

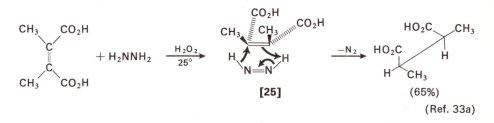

[25]   (65%)

(Ref. 33a)

If the transition state for disporportionation of diimide to nitrogen and hydrazine involves a similar transition state as seems likely, then both *cis* [23a] and *trans* [23b] isomers could be involved in the disproportionation process as indicated in structure [24]. In practice, an excess of the diimide precursors is often used for the reduction of carbon-carbon multiple bonds to allow for loss of part of the diimide formed by concurrent disproportionation. Consequently, it is difficult to determine what fraction of the diimide generated is potentially useful as a reductant for multiple bonds. It is possible that the common preparative procedures for diimide lead to mixtures of *cis* [23a] and *trans* [23b] diimide with only the *cis* isomer [23a] functioning as a reducing agent. Alternatively, the *cis* and *trans* forms may be rapidly equilibrated by the reversible protonation indicated in the following equation[33d] or by thermal isomerization.[33e]

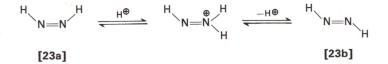

[23a]   [23b]

A variety of methods have been used for the generation of diimide in reaction mixtures containing substrates to be reduced.[32,34] The most useful methods for

33. (a) E. J. Corey, D. J. Pasto, and W. L. Mock, *J. Am. Chem., Soc.* **83**, 2957 (1961). See also (b) E. J. Corey, W. L. Mock, and D. J. Pasto, *Tetrahedron Letters,* No. 11, 347 (1961). (c) S. Hünig, H.-R. Müller, and W. Thier, *ibid.,* **No. 11**, 353 (1961). (d) F. Aylward and M. H. Sawistowska, *J. Chem. Soc.,* 1435 (1964). (e) For a discussion of the thermal isomerization of diimide, see J. Alster and L. A. Burnell, *J. Am. Chem. Soc.,* **89**, 1261 (1967).
34. J. Wolinsky and T. Schultz, *J. Org. Chem.,* **30**, 3980 (1965).

synthetic work appear to involve one of the following: (1) oxidation of hydrazine with one of several oxidants, (2) the base-catalyzed (or thermal) elimination of a proton and the substituent from an acyl or sulfonyl hydrazide, (3) the decomposition of azodicarboxylic acid, or (4) the thermal decomposition of the anthracene-diimide adduct [27]. These procedures are illustrated in the following equations.

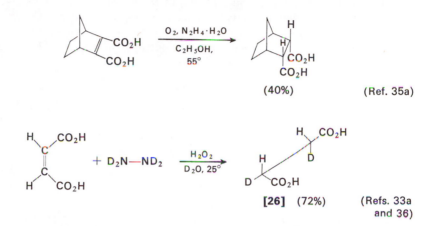

(40%)   (Ref. 35a)

[26]   (72%)   (Refs. 33a and 36)

The formation of the dideuterated product [26] illustrates the utility of this reduction method for adding deuterium to a carbon-carbon double bond without the complication of exchange which is often observed under the conditions of catalytic hydrogenation (see Chapter 1).

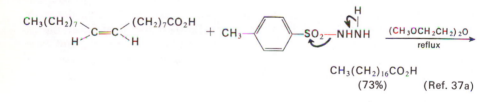

$$CH_3(CH_2)_{16}CO_2H$$
(73%)   (Ref. 37a)

---

35. (a) E. E. van Tamelen and R. J. Timmons, *J. Am. Chem. Soc.*, **84**, 1067 (1962). The addition of catalytic amounts of copper(II) salts to catalyze air oxidation of hydrazine has been recommended. See Ref. 33b, 33d, and (b) M. Ohno and M. Okamoto, *Tetrahedron Letters*, No. 35, 2423 (1964); *Org. Syn.*, **49**, 30 (1969).

36. (a) In addition to hydrogen peroxide, potassium ferricyanide has also served as an oxidant for hydrazine. See Ref. 33c. (b) For other examples of the use of perdeuterio-hydrazine, see U. Eppenberger, M. E. Warren, and H. Rapoport, *Helv. Chim. Acta*, **51**, 381 (1968).

37. (a) R. S. Dewey and E. E. van Tamelen, *J. Am. Chem. Soc.*, **83**, 3729 (1961). (b) Benzenesulfonylhydrazide has been converted to diimide by treatment with base at elevated temperatures; see Ref. 33c. (c) E. W. Garbisch, Jr., S. M. Schildcrout, D. B. Patterson, and C. M. Sprecher, *J. Am. Chem. Soc.*, **87**, 2932 (1965). (d) Reaction of α-chloroacetyl-hydrazide with base to form diimide and ketene was found to occur at 0°. R. Buyle, A. Van Overstraeten, and F. Eloy, *Chem. Ind.* (London), 839 (1964). (e) J. A. Deyrup and S. C. Clough, *J. Am. Chem. Soc.*, **90**, 3592 (1968).

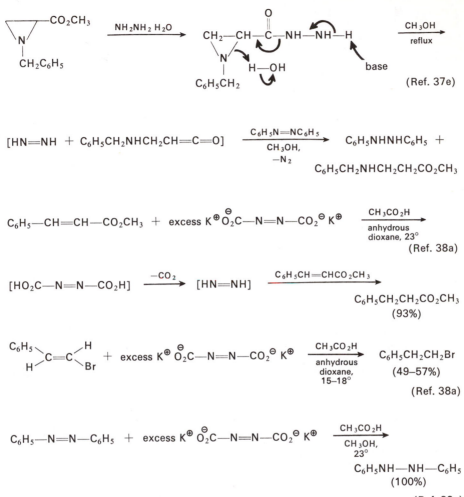

(Ref. 37e)

$$[HN{=}NH \ + \ C_6H_5CH_2NHCH_2CH{=}C{=}O] \xrightarrow[\substack{CH_3OH, \\ -N_2}]{C_6H_5N{=}NC_6H_5} C_6H_5NHNHC_6H_5 \ +$$

$$C_6H_5CH_2NHCH_2CH_2CO_2CH_3$$

$$C_6H_5{-}CH{=}CH{-}CO_2CH_3 \ + \ \text{excess } K^\oplus\ ^\ominus O_2C{-}N{=}N{-}CO_2\ ^\ominus K^\oplus \xrightarrow[\substack{\text{anhydrous} \\ \text{dioxane, } 23^\circ}]{CH_3CO_2H}$$

(Ref. 38a)

$$[HO_2C{-}N{=}N{-}CO_2H] \xrightarrow{-CO_2} [HN{=}NH] \xrightarrow{C_6H_5CH{=}CHCO_2CH_3} $$

$$C_6H_5CH_2CH_2CO_2CH_3$$
$$(93\%)$$

$$\overset{C_6H_5}{\underset{H}{>}}C{=}C\overset{H}{\underset{Br}{<}} \ + \ \text{excess } K^\oplus\ ^\ominus O_2C{-}N{=}N{-}CO_2\ ^\ominus K^\oplus \xrightarrow[\substack{\text{anhydrous} \\ \text{dioxane,} \\ 15{-}18^\circ}]{CH_3CO_2H} \ \begin{array}{c} C_6H_5CH_2CH_2Br \\ (49{-}57\%) \end{array}$$

(Ref. 38a)

$$C_6H_5{-}N{=}N{-}C_6H_5 \ + \ \text{excess } K^\oplus\ ^\ominus O_2C{-}N{=}N{-}CO_2\ ^\ominus K^\oplus \xrightarrow[\substack{CH_3OH, \\ 23^\circ}]{CH_3CO_2H}$$

$$C_6H_5NH{-}NH{-}C_6H_5$$
$$(100\%)$$

(Ref. 38a)

Of these procedures, the use of potassium azodicarboxylate for the generation of diimide would appear to be the most useful. A study of conditions for reductions with this reagent indicated that the use of pyridine or dioxane as a solvent under anhydrous conditions led to the highest yields of reduction products.[38a] The ease with which diimide reacts with olefins is influenced by the electrical nature of substituents on the olefin. As a general rule, multiple bonds that are symmetrical with regard to electronic distribution are more reactive than multiple bonds in

38. (a) J. W. Hamersma and E. I. Snyder, *J. Org. Chem.*, **30**, 3985 (1965); also see Ref. 33.
(b) E. E. van Tamelen, R. S. Dewey, and R. J. Timmons, *J. Am. Chem. Soc.*, **83**, 3725 (1961).
(c) S. G. Cohen, R. Zand, and C. Steel, *ibid.*, **83**, 2895 (1961).

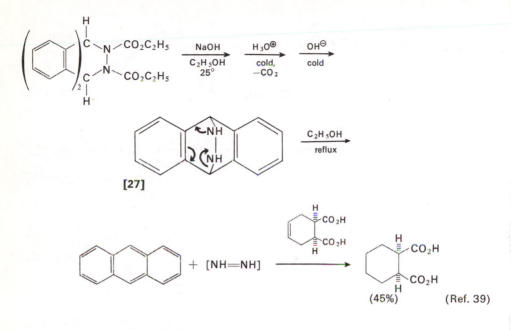

[27]

(45%)   (Ref. 39)

which the electrons of the double bond are polarized.[38a,40] For example, sym-metrical multiple bonds such as C=C, C≡C, and N=N are reduced much more readily by diimide than are unsymmetrical bonds such as C=N, C≡N, C=O, S=O, and C—S. Consequently, it is practical to reduce carbon-carbon multiple bonds in the presence of carbonyl functions, nitro groups, nitrile groups, sulfoxide groups, or sulfide functions.[40] Also, the reactivity of olefinic double bonds is diminished by the presence of substituents such as carbonyl functions, halogens, ethers, and amines.[38a,40b] These and related observations are the basis for the previous statement that the preferred transition state for diimide reductions (e.g., [28]) is nonpolar with a synchronous transfer of the two hydrogen atoms.[40,41] Although the proper choice of reaction conditions allows the successful reduction of the carbon-carbon double bond of α,β-unsaturated acids and esters and of vinyl halides,[38a] the reduction of the carbon-carbon double bond of α,β-unsaturated

[28]

39. E. J. Corey and W. L. Mock, *J. Am. Chem. Soc.*, **84**, 685 (1962).
40. (a) E. E. van Tamelen, R. S. Dewey, M. F. Lease, and W. H. Pirkle, *J. Am. Chem. Soc.*, **83**, 4302 (1961). (b) E. E. van Tamelen, M. Davis, and M. F. Deem, *Chem. Commun.*, **No. 4**, 71 (1965); D. C. Curry, B. C. Uff, and N. D. Ward, *J. Chem. Soc.*, C, 1120 (1967); J. J. Looker, *J. Org. Chem.*, **32**, 472 (1967).
41. S. Hünig and H. R. Müller, *Angew. Chem., Intern. Ed. Engl.*, **1**, 213 (1962).

ketones with diimide does not appear to be a satisfactory procedure.[40b] Interestingly, diimide has been found to reduce the C=O bond of aromatic aldehydes and ketones but to reduce the carbonyl function of aliphatic aldehydes and ketones slowly if at all.[40b] Although this observation is suggested[40b] to reflect a diminished

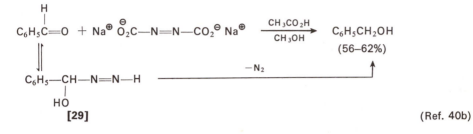

(Ref. 40b)

polarity of the carbonyl group (i.e. $>C=O \leftrightarrow >C^{\oplus}—O^{\ominus}$) when it is conjugated with an aromatic ring, it seems more likely that this reduction is occurring by a different mechanism in which diimide serves as a nucleophile which adds reversibly to the carbonyl group to form the hydroxy diazene [29]. This diazene intermediate [29] could decompose with loss of nitrogen to form the alcohol product by a process analogous to the previously discussed decomposition of other alkyl-diazenes.[23] The nucleophilic addition of diimide to form [29] is analogous to the known addition of alkyldiazenes to aldehydes to form hydroxyalkyl azo compounds.[23c]

The reactivity of diimide with olefins is also influenced by the number and nature of alkyl substituents present. *Trans* olefins are usually reduced slightly more rapidly than *cis* olefins and the ease of olefin reduction diminishes as more alkyl substituents are added.[32a,37c,41] These reactivity differences have been attributed to differences in strain (torsional, bond angle, and eclipsing) in the various olefins with the more strained olefins reacting more rapidly.[37c] The relative rates of reaction of a number of olefins with diimide are summarized in Table 4–1. The selective reduction of strained carbon-carbon double bonds in the presence of less strained olefinic bonds is illustrated by the reduction of the *trans* double bonds in the cyclic triene [30].

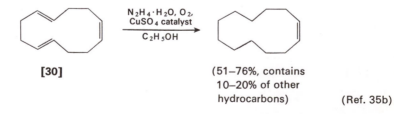

[30]   (51–76%, contains 10–20% of other hydrocarbons)   (Ref. 35b)

As in the case of catalytic hydrogenation (see Chapter 1), addition of hydrogen from diimide usually occurs from the less hindered side of an unsymmetrical olefin (e.g., [31]). The following equations provide a comparison of the proportions of stereoisomers produced when several olefins were reduced with diimide

Table 4–1 Relative rates of reaction of diimide with olefins (Ref. 37c)

Olefin	Relative rate	Olefin	Relative rate
(norbornadiene structure)	4.5×10^2	(methylenecyclohexane, CH_2)	3.3
(bicyclic diene structure)	29	(1-methylcyclohexene, CH_3)	0.11
(cyclopentene)	15.5	$CH_3(CH_2)_2CH{=}CH_2$	20.2
(cyclohexene)	1.00	$\underset{H}{\overset{CH_3}{\diagdown}}C{=}C\underset{H}{\overset{C_2H_5}{\diagup}}$	2.65
(cycloheptene)	12.1	$\underset{H}{\overset{CH_3}{\diagdown}}C{=}C\underset{C_2H_5}{\overset{H}{\diagup}}$	2.59
(cyclooctene)	17.0	$CH_3(CH_2)_2\overset{\overset{\textstyle CH_3}{|}}{C}{=}CH_2$	2.04
		$(CH_3)_2C{=}CHCH_3$	0.28

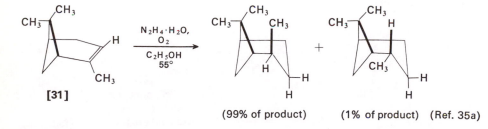

[31] $\xrightarrow[\substack{C_2H_5OH \\ 55°}]{\substack{N_2H_4 \cdot H_2O, \\ O_2}}$ (99% of product)   +   (1% of product)   (Ref. 35a)

and by catalytic hydrogenation.[35a] Although individual differences exist, the degree of stereoselectivity obtained by the two reduction procedures are frequently similar. An interesting exception is found in the partial reduction of norbornadiene

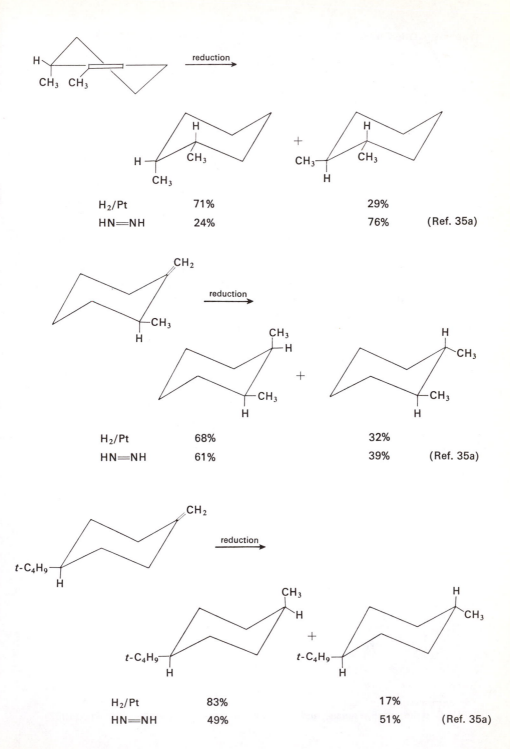

H₂/Pt   71%   29%
HN=NH   24%   76%   (Ref. 35a)

H₂/Pt   68%   32%
HN=NH   61%   39%   (Ref. 35a)

H₂/Pt   83%   17%
HN=NH   49%   51%   (Ref. 35a)

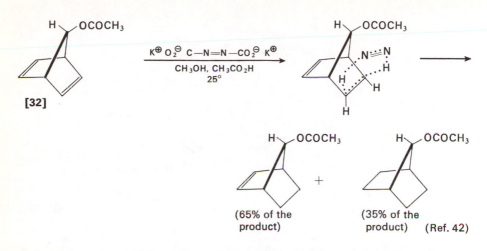

(65% of the product) + (35% of the product)   (Ref. 42)

derivatives such as [32] with an electronegative substituent at C-7.[42] In these cases, it was clear that reduction was occurring from an *exo* direction at the sterically more hindered carbon-carbon double bond. This result may indicate that the transition state for transfer of hydrogen atoms from diimide to the double bond can be stabilized by electrostatic interaction with a suitably located substituent.

42. W. C. Baird, Jr., B. Franzus, and J. H. Surridge, *J. Am. Chem. Soc.*, **89**, 410 (1967).

5

OXIDATIONS WITH CHROMIUM AND MANGANESE COMPOUNDS

Among the variety of agents available for the oxidation of organic compounds,[1] the most commonly used are derivatives of hexavalent chromium (Cr^{VI}) or heptavalent manganese (Mn^{VII}). Chromium trioxide (CrO_3) and sodium dichromate ($Na_2Cr_2O_7$) are converted to the chromium(III) ion (Cr^{3+}) in the course of such oxidations, for a net transfer of three electrons to each chromium atom. The oxidation of organic compounds with potassium permanganate ($KMnO_4$) in acidic media produces the manganese(II) ion (Mn^{2+}) for a net transfer of five electrons to each manganese atom; in neutral or basic media, manganese dioxide (MnO_2) is formed, with a corresponding net transfer of three electrons. The over-all stoichiometry for the oxidation of a secondary alcohol to a ketone (transfer of two electrons) with these reagents is illustrated in the accompanying equations. It will be noted that in each case acid is consumed (or hydroxide ion is produced) during the reaction.

$$2\,CrO_3 \;+\; 3\,R_2CHOH \;+\; 6\,H^\oplus \;\longrightarrow\; 3\,R_2C{=}O \;+\; 2\,Cr^{\oplus} \;+\; 6\,H_2O$$

$$2\,MnO_4{}^\ominus \;+\; 5\,R_2CHOH \;+\; 6\,H^\oplus \;\longrightarrow\; 5\,R_2C{=}O \;+\; 2\,Mn^{\ominus} \;+\; 8\,H_2O$$

$$2\,MnO_4{}^\ominus \;+\; 3\,R_2CHOH \;\longrightarrow\; 3\,R_2C{=}O \;+\; 2\,MnO_2 \;+\; 2\,H_2O \;+\; 2\,OH^\ominus$$
$$\text{(ppt.)}$$

Potassium permanagnate is normally employed as a neutral or alkaline aqueous solution. If the organic compound to be oxidized is not soluble in water, it may be oxidized as a suspension in the aqueous permanganate solution, or a co-solvent such as t-butyl alcohol or acetic acid may be employed; ethanol or acetone are less desirable co-solvents, since they are rapidly attacked by permanganate in aqueous alkaline solution. Solutions of potassium permanganate in dry acetone, acetic anhydride, or in pyridine have also been used. Chromium trioxide, a polymer, is insoluble in glacial acetic acid but dissolves (with reaction leading to depolymerization) in water, acetic anhydride, pyridine, and t-butyl alcohol.[1e,2] Although

1. (a) W. A. Waters in H. Gilman, ed., *Organic Chemistry*, Vol. 4, Wiley, New York, 1953, pp. 1120–1245. (b) R. Stewart, *Oxidation Mechanisms*, Benjamin, New York, 1964. (c) K. B. Wiberg in A. F. Scott, ed., *Survey of Progress in Chemistry*, Vol. I, Academic Press, New York, 1963, pp. 211–248. (d) R. Stewart in K. B. Wiberg, ed., *Oxidation in Organic Chemistry*, Part A, Academic Press, New York, 1965, pp. 1–68. (e) K. W. Wiberg, *ibid.*, pp. 69–184. (f) D. G. Lee in R. L. Augustine, ed., *Oxidation*, Vol. 1, Marcel Dekker, New York, 1969, pp. 1–118.
2. (a) J. C. Collins, W. W. Hess, and F. J. Frank, *Tetrahedron Letters*, No. 30, 3363 (1968). (b) W. G. Dauben, M. Lorber, and D. S. Fullerton, *J. Org. Chem.*, **34**, 3587 (1969). (c) R. Ratcliffe and R. Rodehurst, *ibid.*, **35**, 4000 (1970).

a suspension of this reagent in glacial acetic acid has been used for oxidations, a solution in aqueous sulfuric acid is more commonly employed. A comparable solution may be prepared by adding sodium or potassium dichromate to aqueous sulfuric acid. Co-solvents most frequently used are acetic acid or acetone. Oxidations have also been run in a two-phase system consisting of the aqueous chromic acid layer and a second layer which is either the pure organic reactant or a solution of the organic reactant in benzene, methylene chloride, or ether. Alternatively, hexavalent chromium oxidations have been run with a solution of sodium dichromate dihydrate in glacial acetic acid,[3] with a solution of chromium trioxide in acetic anhydride, with a solution of the chromium trioxide dipyridine complex in excess pyridine or in methylene chloride,[2] and with a solution of chromyl chloride (CrO_2Cl_2) in carbon tetrachloride or carbon disulfide.[4] The anhydrous chromium trioxide-dipyridine complex, whose preparation is illustrated in the following equation, would appear to be especially useful. The complex may be isolated as a *hygroscopic* red crystalline solid which is soluble both in pyridine and in various chlorinated hydrocarbons such as methylene chloride. Upon exposure to water this anhydrous complex rapidly forms a yellow crystalline hydrate ($C_{10}H_{12}Cr_2N_2O_7$) which is no longer soluble in chlorinated hydrocarbons.

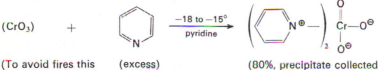

(CrO_3)   + $\xrightarrow[\text{pyridine}]{-18 \text{ to } -15°}$

(To avoid fires this   (excess)   (80%, precipitate collected
  reagent is *added to*   by filtration)   (Ref. 2b)
  the pyridine.)

Solutions of hexavalent chromium compounds in aqueous mineral acids or in mixtures of acetic acid and aqueous mineral acids contain an equilibrating mixture of the acid chromate ion and the dichromate ion, as indicated in the equation below.[1e,5] In aqueous acetic acid solutions the acetylchromic anion ($CH_3CO_2CrO_3^{\ominus}$

$$2\ HCrO_4{}^{\ominus} \;\rightleftharpoons\; H_2O\ +\ Cr_2O_7{}^{\ominus}$$

may also be present.[5c] The predominant species in solutions of chromium trioxide in acetic acid is chromyl acetate [$CrO_2(OCOCH_3)_2$, the analog of CrO_2Cl_2.][1e,4,6] The same material may be formed by solution of sodium dichromate in acetic acid.

3. L. F. Fieser. *J. Am. Chem. Soc.,* **75**, 4377 (1953).
4. W. H. Hartford and M. Darrin, *Chem. Rev.,* **58**, 1 (1958).
5. F. H. Westheimer, *Chem. Rev.,* **45**, 419 (1949). (b) K. B. Wiberg and T. Mill, *J. Am. Chem. Soc.,* **80**, 3022 (1958). (c) G. T. E. Graham and F. H. Westheimer, *ibid.,* **80**, 3030 (1958). (d) K. B. Wiberg and P. A. Lepse, *ibid.,* **86**, 2612 (1964). (e) D. G. Lee and R. Stewart, *ibid.,* **86**, 3051 (1964) ; *J. Org. Chem.,* **32**, 2868 (1967).
6. H. L. Krauss, *Angew. Chem.,* **70**, 502 (1958).

Studies of the mechanisms of oxidations with chromium and manganese compounds[1e,5,7,8] have been complicated by the fact that each stage in the oxidation of most organic compounds is accompanied by the net transfer of two electrons although the oxidizing agents normally accept a total of three or five electrons. It is therefore evident that intermediate valence states of chromium and manganese are important in the overall process. The problem is well illustrated by the oxidation of a secondary alcohol with a hexavalent chromium compound, one of the many possible reaction schemes for which is presented in the accompanying equations.

$$Cr^{VI} + R_2CHOH \longrightarrow R_2C{=}O + 2H^{\oplus} + Cr^{IV}$$

$$Cr^{IV} + Cr^{VI} \longrightarrow 2 Cr^{V}$$

$$Cr^{V} + R_2CHOH \longrightarrow R_2C{=}O + 2H^{\oplus} + Cr^{III}$$

Both the Cr^{V} and the Cr^{IV} species appear to be more powerful oxidizing agents than Cr^{VI}.[5,8] It is evident from the equations that these intermediate valence states may be responsible for as much as two-thirds of the total oxidation and in some cases may lead to unwanted side reactions such as carbon-carbon bond cleavage.[5,8] Although the presence of intermediate valence states of chromium in a reaction mixture can be minimized if manganese(II) or cerium(II) ion is added to the solution,[5,8] such a procedure has not yet been widely used in synthetic work.

OXIDATION OF ALCOHOLS

The oxidation of a secondary alcohol to a ketone has usually been accomplished with a solution of the alcohol and aqueous acidic chromic acid in acetic acid or acetone, with a solution of sodium dichromate in acetic acid, or by reaction of the alcohol (e.g., [1]) with aqueous acidic chromic acid as a heterogeneous system.

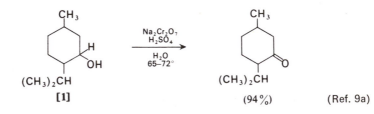

[1]   (94%)   (Ref. 9a)

7. (a) J. W. Ladbury and C. F. Cullis, *Chem. Rev.,* **58**, 403 (1958). (b) W. A. Waters, *Quart. Rev.,* **12**, 277 (1958).

8. (a) K. B. Wiberg and R. J. Evans, *Tetrahedron,* **8**, 313 (1960). (b) K. B. Wiberg and H. Schäfer, *J. Am. Chem. Soc.,* **91**, 927, 933 (1969). (c) J. Roček and A. E. Radkowsky, *ibid.,* **90**, 2986 (1968). (d) P. M. Nave and W. S. Trahanovsky, *ibid.,* **92**, 1120 (1970). (e) F. B. Beckwith and W. A. Waters, *J. Chem. Soc.,* B, 929 (1969).

9. (a) A. S. Hussey and R. H. Baker, *J. Org. Chem.,* **25**, 1434 (1960). (b) L. T. Sandborn, *Org. Syn.,* **Coll. Vol. 1**, 340 (1944). (c) L. F. Fieser, *ibid.,* **Coll. Vol. 4**, 189 ,195 (1963).

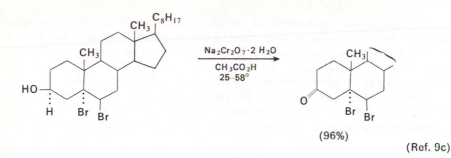

(96%)

(Ref. 9c)

Alternatively, the oxidation may be effected by stirring a solution of the compound to be oxidized (e.g., **[2]**, **[3]**, or **[4]**) in benzene, methylene chloride, or ether with a second liquid phase containing an acidic aqueous solution of chromic acid. When the oxidation is relatively slow, the use of ether as a solvent is undesirable because the latter is slowly oxidized, consuming the chromic acid. The course of

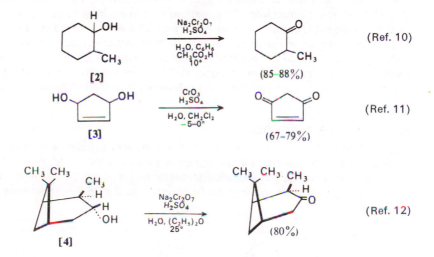

these oxidations may be followed spectrophotometrically as the yellow-orange absorption at 350 mμ of the Cr^{VI} reactants is converted to the green of the hydrated chromium(III) ion.[13] The presence of unchanged Cr^{VI} species may be detected by adding a few drops of the reaction mixture to an aqueous solution of sodium

10. (a) E. W. Warnhoff, D. G. Martin, and W. S. Johnson, *Org. Syn.*, **Coll. Vol. 4**, 162 (1963). (b) for similar procedures where benzene is used as a co-sovent, see W. F. Bruce, *ibid.*, **Coll. Vol. 2**, 139 (1943); W. S. Johnson, C. D. Gutsche, and D. K. Banerjee, *J. Am. Chem. Soc.*, **73**, 5464 (1951).
11. G. H. Rasmusson, H. O. House, E. F. Zaweski, and C. H. DePuy, *Org. Syn.*, **42**, 36 (1962).
12. (a) H. C. Brown and C. P. Garg, *J. Am. Chem. Soc.*, **83**, 2952 (1961). (b) H. C. Brown, C. P. Garg, and K. T. Liu, *J. Org. Chem.*, **35**, 387 (1971).
13. (a) F. H. Westheimer and N. Nicolaides, *J. Am. Chem. Soc.*, **71**, 25 (1949). (b) J. Schreiber and A. Eschenmoser, *Helv. Chim. Acta.*, **38**, 1529 (1955).

diphenylaminesulfonate; a blue-violet color develops. Excess oxidant remaining in an oxidation reaction mixture may be destroyed by the dropwise addition of methanol or isopropyl alcohol. For homogeneous oxidations in aqueous acetic acid solution, the rate of reaction may be increased by decreasing the concentration or water and by increasing the concentration of acid.

The probable mechanism of oxidation of alcohols by Cr^{VI} species is outlined in the accompanying equations.[1e,5,7,8] Currently available evidence[1,5,8] suggests that the Cr^{IV} species formed in the first step reacts with a Cr^{VI} species to form two equivalents of a Cr^V derivative which also oxidizes alcohols to ketones. The oxidation of alcohols with Cr^V intermediates appears to be analogous to the Cr^{VI} oxidation.[8b] With unhindered alcohols the initial equilibrium to form the chromate ester [5] is fast and the subsequent decomposition of the chromate ester is the rate-limiting step.

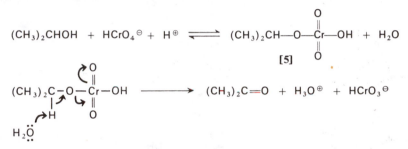

However, if the formation of the chromate ester results in a serious steric interaction (e.g., [6]), then its decomposition rate is accelerated because steric strain is relieved in going from the reactant to the product.[13b,14] In extreme cases such as [7], the

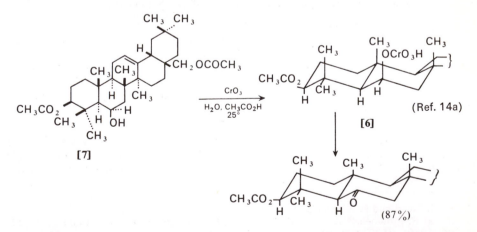

14. (a) J. Roček, F. H. Westheimer, A. Eschenmoser, L. Moldovanyi, and J. Schreiber, *Helv. Chim. Acta,* **45**, 2554 (1962). (b) H. Kwart and P. S. Francis, *J. Am. Chem. Soc.,* **81**, 2116 (1959). (c) A. K. Awasthy, J. Roček, and R. M. Moriarty, *ibid.,* **89**, 5400 (1967). (d) C. F. Wilcox, Jr., M. Sexton, and M. F. Wilcox, *J. Org. Chem.,* **28**, 1079 (1963). (e) L. F. Fieser and M. Fieser, *Steroids,* Reinhold, New York, 1959, pp. 202–225. (f) R. Baker and T. J. Mason, *Tetrahedron Letters,* **No. 57**, 5013 (1969).

initial step esterification becomes the rate-limiting step in the oxidation. Measurement of the relative rates of chromic acid oxidations of epimeric alcohols to their corresponding ketones has proved useful for making stereochemical assignments since, *in the absence of competing side reactions*, the alcohol with the more hindered hydroxyl group is oxidized more rapidly. Examples of relative oxidation rates for epimeric pairs of alcohols are provided in the accompanying formulas. The rate of oxidation of benzylic secondary alcohols (e.g., **[8]**) has been found to be accelerated by electron-donating substituents.[15]

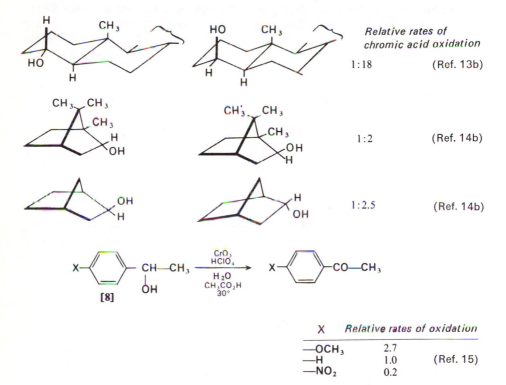

	Relative rates of chromic acid oxidation	
	1:18	(Ref. 13b)
	1:2	(Ref. 14b)
	1:2.5	(Ref. 14b)

X	*Relative rates of oxidation*	
—OCH₃	2.7	
—H	1.0	(Ref. 15)
—NO₂	0.2	

The previously discussed reaction conditions with chromic acid are sufficiently vigorous to effect the slow oxidation of ethers,[16] amines,[17] carbon-carbon multiple bonds, enolizable ketones, and benzylic and allylic C—H bonds.[9c] Hydroxyl

15. H. Kwart and P. S. Francis, *J. Am. Chem. Soc.*, **77**, 4907 (1955).

16. (a) R. Brownell, A. Leo, Y. W. Chang, and F. H. Westheimer, *J. Am. Chem. Soc.*, **82**, 406 (1960). (b) The oxidative cleavage of methyl ethers with chromium trioxide has been described by I. T. Harrison and S. Harrison, *Chem. Commun.*, No. **20**, 752 (1966). (c) For the oxidation of thio ethers to sulfoxides with chromium trioxide, see R. G. Hiskey and M. A. Harpold, *J. Org. Chem.*, **32**, 3191 (1967).

17. (a) A. T. Bottini and R. E. Olsen, *J. Org. Chem.*, **27**, 452 (1962). (b) For the oxidation of N-methylamines to formamide derivatives with chromium trioxide in pyridine, see A. Cavé, C. Kan-Fan, P. Potier, J. LeMen, and M. M. Janot, *Tetrahedron*, **23**, 4691 (1967).

groups may be protected from oxidation by acetylation (e.g., [7]), and acylation or salt formation may be used to protect primary and secondary amines.[18] However, a milder method of oxidation, which at least in part avoids these complications, consists of adding, dropwise, an aqueous chromic acid solution (the Jones reagent) to an acetone solution of the compound to be oxidized. A stoichiometric amount of the aqueous chromic acid solution (which is 8 N in chromic acid and contains the calculated amount of sulfuric acid) is normally utilized. The progress of the oxidation is indicated by a color change from the orange Cr^{VI} oxidant to the green color of the hydrated chromium(III) ion product. The reaction is run at or below room

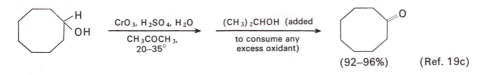

(92–96%)   (Ref. 19c)

temperature[19] and has permitted the oxidation of alcohols (e.g., [9]) to ketones without appreciable oxidation or rearrangement of double bonds. It has also permitted the oxidation of β,γ-unsaturated alcohols such as [10] to β,γ-unsaturated ketones without subsequent isomerization to the more stable α,β-unsaturated ketones.[20a] The same procedure has been used for the oxidation of alcohols to enolizable ketones without epimerization of an asymmetric center *alpha* to the ketone function.[20b] Alternatively, epimerization has been avoided by oxidizing an alcohol to a ketone in a two-phase reaction mixture consisting of benzene and aqueous chromic acid.[10b]

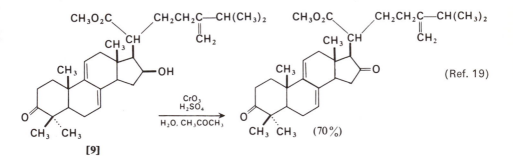

(Ref. 19)

[9]

18. (a) J. F. W. McOmie, *Adv. Org. Chem.,* **3**, 191 (1963). (b) the rate of oxidation of a secondary alcohol with chromic acid appears to be retarded by the presence of a tertiary amine function in the same molecule: see G. I. Poos and M. M. Lehman, *J. Org. Chem.,* **26**, 2576 (1961).
19. (a) A. Bowers, T. G. Halsall, E. R. H. Jones, and A. J. Lemin, *J. Chem. Soc.,* 2548 (1953). (b) D. C. Kleinfelter and P. von R. Schleyer, *Org. Syn.,* **42**, 79 (1962). (c) E. J. Eisenbraun, *ibid.,* **45**, 28 (1965). (d) J. Meinwald, J. Crandall, and W. E. Hymans, *ibid.,* **45**, 77 (1965).
20. (a) C. Djerassi, R. R. Engle, and A. Bowers, *J. Org. Chem.,* **21**, 1547 (1956). (b) C. Djerassi, P. A. Hart, and E. J. Warawa, *J. Am. Chem. Soc.,* **86**, 78 (1964).

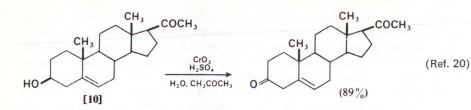

(Ref. 20)

For compounds that contain acid-sensitive functions (e.g., acetals or ketals), oxidation has been effected with the previously described chromium trioxide–pyridine complex ($CrO_3 \cdot 2C_5H_5N$).[2,21] This reagent is prepared by *adding chromium trioxide to pyridine* (adding pyridine to chromium trioxide frequently starts a fire) to form a pyridine solution (or partial solution–partial suspension) of the complex. (The hydrated complex has also been prepared by the addition of a concentrated aqueous solution of chromic acid to pyridine.[21c]) In one procedure,[21a] the alcohol to be oxidized is introduced and the mixture is allowed to stand at room temperature. The crude product may be isolated by diluting the reaction mixture with water and extracting with an organic solvent; but because this extraction is made difficult by the presence of insoluble basic chromium salts, two alternative isolation procedures have been developed: dilution of the pyridine solution with ether followed by filtration of the insoluble chromium-pyridine complex prior to the addition of water, and dilution of the reaction mixture with ethyl acetate followed by filtration through an alumina-celite column.[22b]

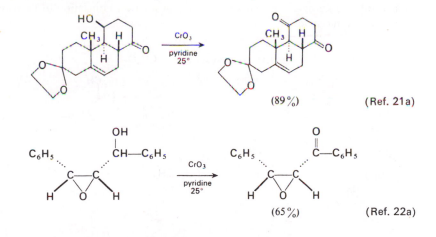

(Ref. 21a)

(Ref. 22a)

21. (a) G. I. Poos, G. E. Arth, R. E. Beyler, and L. H. Sarett, *J. Am. Chem. Soc.*, **75**, 422 (1953). (b) J. R. Holum, *J. Org. Chem.*, **26**, 4814 (1961). (c) R. H. Cornforth, J. W. Cornforth, and G. Popjak, *Tetrahedron*, **18**, 1351 (1962). (d) The Jones oxidation procedure (aqueous chromic acid in acetone) has also been used successfully for the oxidation of alcohols to ketones in the presence of ketal functions: see R. F. Church, R. E. Ireland, and D. R. Shridhar, *J. Org. Chem.*, **27**, 707 (1962).
22. (a) H. H. Wasserman and N. E. Aubrey, *J. Am. Chem. Soc.*, **77**, 590 (1955). (b) R. K. Hill, J. A. Joule, and L. J. Loeffler, *ibid.*, **84**, 4951 (1962).

The difficulty of product isolation is avoided with a recently developed procedure[2] which involves the separation of the anhydrous crystalline chromium trioxide-dipyridine complex. A solution containing a six-fold excess of this anhydrous complex in methylene chloride is then mixed with the alcohol to be oxidized. The alcohol oxidations are usually complete in 5–15 min.; after the brownish-black insoluble reduced chromium species have been separated by filtration, the

$$n\text{-}C_6H_{13}\text{---}\underset{\underset{\text{OH}}{|}}{\text{CH}}\text{---}CH_3 \xrightarrow[\text{CH}_2\text{Cl}_2,\ 25°]{(C_5H_5N)_2CrO_3} n\text{-}C_6H_{13}\text{---}\underset{\underset{\text{O}}{\|}}{C}\text{---}CH_3$$

$$(97\%) \hspace{4cm} (\text{Ref. 2a})$$

oxidation product is recovered from the methylene chloride solution. This oxidation procedure appears to be as mild and as selective as the previously described Jones procedure (chromic acid in aqueous acetone) and, at least in cases where nonacidic reaction conditions are desirable, it appears to be the optimum procedure for the oxidation of relatively small amounts of alcohols to the corresponding carbonyl compounds.

Many allylic alcohols may be oxidized to α,β-unsaturated ketones with chromic acid (especially with a two phase ether-aqueous chromic acid system or in acetone solution at low temperature).[23] The oxidation of these allylic alcohols is believed to follow the previously discussed reaction path in which decomposition of a chromate ester intermediate is rate-limiting.[23b,c] The especially rapid oxidation observed for equatorial allylic alcohols (e.g., [11])[23b,c] has been attributed at least in part to a stereoelectronic factor, namely that breaking of an axial C—H bond will be facilitated by overlap of the developing carbon-orbital with the adjacent pi-orbital of the carbon-carbon double bond.[23c] The corresponding oxidation of axial allylic alcohols is slower and in some cases (e.g., [12]) is sufficiently slow that the intermediate chromate ester attacks the carbon-carbon double bond in a subsequently discussed reaction to form an epoxide. As a result the final oxidation product is the indicated α,β-epoxy ketone rather than an α,β-unsaturated ketone.[23b]

A milder, more selective method for the oxidation of allylic and benzylic alcohols, however, consists of stirring a pentane solution of the alcohol with a large excess of specially prepared manganese dioxide.[24] Whether the actual oxidizing agent is manganese dioxide or some other manganese compound adsorbed on the surface

23. (a) A. E. Vanstone and J. S. Whitehurst, *J. Chem. Soc.*, C, 1972 (1966). (b) E. Glotter, S. Greenfield, and D. Lavie, *ibid.*, C, 1646 (1968). (c) S. H. Burstein and H. J. Ringold, *J. Am. Chem. Soc.*, **89**, 4722 (1967).

24. (a) R. J. Gritter and T. J. Wallace, *J. Org. Chem.*, **24**, 1051 (1959). (b) R. M. Evans, *Quart. Rev.*, **13**, 61 (1959). (c) S. P. Korshunov and L. I. Vereshchagin, *Russ. Chem. Rev.*, **35**, 942 (1966). (d) R. Giovanoli, K. Bernhard, and W. Feitknecht, *Helv. Chim. Acta*, **51**, 355 (1968); R. Giovanoli, E. Stähl, and W. Feitknecht, *ibid.*, **53**, 453 (1970). (e) D. Dollimore and K. H. Tonge, *J. Chem. Soc.*, B, 1380 (1967). (f) E. P. Papadopoulos, A. Jarrar and C. H. Issidorides, *J. Org. Chem.*, **31**, 615 (1966). (g) I. M. Goldman, *ibid.*, **34**, 3289 (1969). (h) L. A. Carpino, *ibid.*, **35**, 3971 (1970).

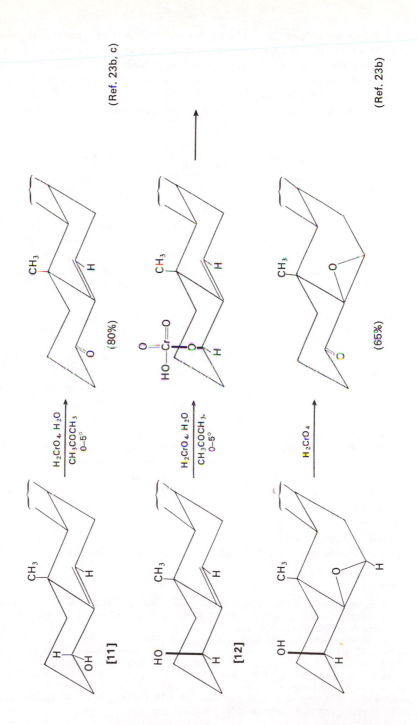

of the manganese dioxide is not clear. In any event the procedure offers an unusually mild method for the preparation of unsaturated aldehydes and ketones, as is illustrated by the following equations. The major disadavntages of the method

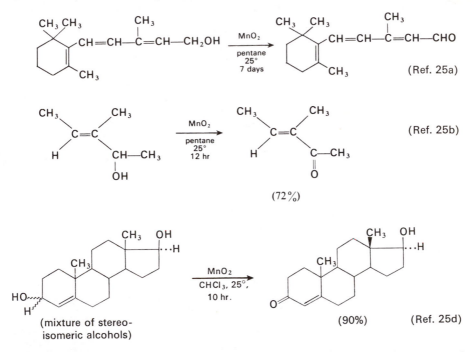

(Ref. 25a)

(Ref. 25b)

(72%)

(mixture of stereo-isomeric alcohols)

(90%)   (Ref. 25d)

are that it is general only for allylic and benzylic alcohols[25c] and that large excesses of oxidant and relatively long reaction times are often required. Manganese dioxide has also been found to oxidize N-alkylamines to form various products, depending on the nature of the N-alkyl group.[26]

$$(CH_3)_3CCH_2C(CH_3)_2N(CH_3)_2 \xrightarrow[\text{cyclohexane, } 20°]{MnO_2}$$

$$(CH_3)_3CCH_2C(CH_3)_2N\begin{subarray}{l}CH_3\\CH_2OH\end{subarray} \xrightarrow{-CH_2O} (CH_3)_3CCH_2C(CH_3)_2NH-CH_3$$

(58%)   (Ref. 26b)

25. (a) N. L. Wendler, H. L. Slates, N. R. Trenner, and M. Tishler, *J. Am. Chem., Soc.* **73**, 719 (1951). (b) H. O. House and R. S. Ro, *ibid.,* **80**, 2428 (1958); E. J. Corey, N. W. Gilman, and B. E. Ganem, *ibid.,* **90**, 5616 (1968). (c) For examples of the oxidation of saturated alcohols with manganese dioxide, see M. Z. Barakat, M. F. Abdel-Wahab, and M. M. El-Sadr, *J. Chem. Soc.,* 4685 (1956); L. Crombie and J. Crossley, *ibid.,* 4983 (1963); I. T. Harrison, *Proc. Chem. Soc.,* 110 (1964). (d) F. Sondheimer, C. Amendolla, and G. Rosenkranz, *J. Am. Chem. Soc.,* **75**, 5930, 5932 (1953).
26. (a) H. B. Henbest, and A. Thomas, *J. Chem. Soc.,* 3032 (1957). (b) H. B. Henbest and M. J. W. Stratford, *ibid.,* C, 995 (1966). (c) I. Bhatnagar and M. V. George, *Tetrahedron,* **24**, 1293 (1968); *J. Org. Chem.,* **33**, 2407 (1968).

The following equations illustrate the fact that aqueous solutions of potassium permanganate may also be used to oxidize secondary alcohols to ketones and to oxidize amines. Although *t*-alkylamines may be oxidized to nitro compounds, alkylamines possessing C—H bonds *alpha* to the nitrogen atom are normally

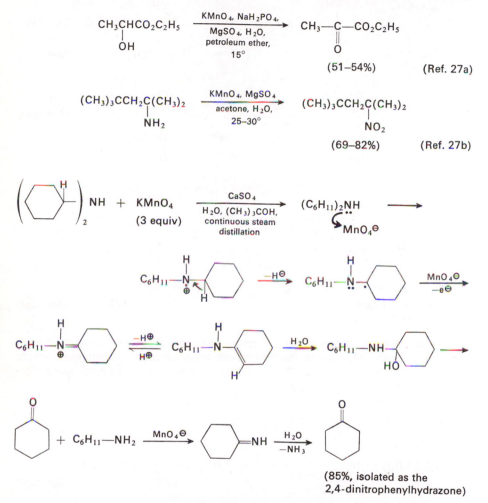

(51–54%)   (Ref. 27a)

(69–82%)   (Ref. 27b)

(85%, isolated as the
2,4-dinitrophenylhydrazone)

(Ref. 27e)

---

27. (a) J. W. Cornforth, *Org. Syn.*, **Coll. Vol. 4**, 467 (1963). (b) N. Kornblum and W. J. Jones, *ibid.*, **43**, 87 (1963). (c) M. M. Wei and R. Stewart, *J. Am. Chem. Soc.*, **88**, 1974 (1966). (d) H. Shechter, S. S. Rawalay, and M. Tubis, *ibid.*, **86**, 1701 (1964). (e) H. Shechter and S. S. Rawalay, *ibid.*, **86**, 1706 (1964); *J. Org. Chem.*, **32**, 3129 (1967). (f) D. H. Rosenblatt, G. T. Davis, L. A. Hull, and G. D. Forberg, *ibid.*, **33**, 1649 (1968). (g) For a discussion of the mechanism of the permanganate oxidation of alcohols and ethers, see R. M. Barter and J. S. Littler, *J. Chem. Soc.*, B, 205 (1967); F. Freeman and M. A. H. Scott, *J. Org. Chem.*, **35**, 2989 (1970).

oxidized to imines or to products derived from imines.[27b-f] The indicated oxidation of amines to carbonyl compounds with potassium permanganate in neutral solution constitutes a useful degradative method.[27e] In this procedure, calcium sulfate is added to maintain the neutrality of the reaction mixture by forming calcium hydroxide, which is relatively insoluble in the reaction medium. The aldehyde or ketone products are steam distilled from the reaction mixture as they form to avoid further oxidation. The mechanism suggested for the amine oxidation, involving successive electron transfer and deprotonation steps, is analogous to the scheme believed operative in the oxidation of amines with mercuric acetate (see Chapter 7).

Although the oxidation of primary alcohols to aldehydes may be accomplished with chromic acid, the reaction may be complicated by the fact that the aldehyde formed can be further oxidized to the corresponding carboxylic acid, though at a rate slower than that of the oxidation of the alcohol.[28] A more serious complication is the reaction of the aldehyde and the alcohol in the reaction mixture to form a hemiacetal (e.g., [13]), which is oxidized rapidly (cf. the effect of electron-donating

$$n\text{-}C_3H_7\text{---}CH_2OH \xrightarrow[\substack{H_2O \\ 35°}]{\substack{Na_2Cr_2O_7 \\ H_2SO_4}} n\text{-}C_3H_7\text{---}CHO \xrightleftharpoons[]{n\text{-}C_3H_7CH_2OH}$$

$$n\text{-}C_3H_7\text{---}\underset{\underset{OH}{|}}{CH}\text{---}OCH_2\text{---}C_3H_7\text{-}n \xrightarrow{H_2CrO_4} n\text{-}C_3H_7\text{---}\underset{\underset{O}{\|}}{C}\text{---}O\text{---}CH_2\text{---}C_3H_7\text{-}n$$

[13]   (41–47%)   (Ref. 29)

$$C_2H_5\text{---}CH_2OH \xrightarrow[\substack{H_2O}]{\substack{Na_2Cr_2O_7 \\ H_2SO_4}} C_2H_5\text{---}CHO \qquad\qquad (\text{Ref. 31a})$$

[14]   (45–49%, distilled from mixture as formed)

$$HC\equiv C\text{---}CH_2OH \xrightarrow[\substack{H_2O}]{\substack{CrO_3 \\ H_2SO_4}} HC\equiv C\text{---}CHO \qquad\qquad (\text{Ref. 31b})$$

           (35–41%, distilled from
[15]   mixture as formed)

groups in Ref. 15) to an ester. However, moderate yields of aldehydes have been obtained from oxidations of primary alcohols with chromic acid, especially in such cases as [14] and [15] where the volatile aldehyde can be distilled from the reaction mixture as it is formed. This oxidation procedure is also satisfactory for the preparation of sterically hindered aldehydes such as those obtained from alcohols of the neopentyl type.[30]

28. W. A. Mosher and D. M. Preiss, *J. Am. Chem. Soc.*, **75**, 5605 (1953).
29. G. R. Robertson, *Org. Syn.*, **Coll. Vol. 1**, 138 (1944).
30. For example, see R. E. Ireland and P. W. Schiess, *J. Org. Chem.*, **28**, 6 (1963).
31. (a) C. D. Hurd and R. N. Meinert, *Org. Syn.*, **Coll. Vol. 2**, 541 (1943). (b) J. C. Sauer, *ibid.*, **Coll. Vol. 4**, 813 (1963). (c) The oxidation of primary alcohols to aldehydes by use of *t*-butyl chromate has been reported by T. Suga and T. Matsuura, *Bull. Chem. Soc. Japan*, **38**, 1503 (1965). (d) R. E. Ireland, M. I. Dawson, J. Bordner, and R. E. Dickerson, *J. Am. Chem. Soc.*, **92**, 2568 (1970).

Reasonable yields of aldehydes have been obtained by the oxidation of primary alcohols with chromium trioxide in pyridine, especially in cases where the alcohol is allylic or benzylic. The use of a methylene chloride solution of the chromium trioxide-dipyridine complex appears to be the method of choice for the oxidation of primary alcohols to aldehydes.[2a] The success of these procedures may be attributable in part to the fact that the oxidation is done in the absence of an appreciable concentration of water, a necessary reactant for at least one of the mechanistic pathways for the conversion of an aldehyde to an acid.

$$C_6H_5-CH=CH-CH_2OH \xrightarrow[\substack{pyridine \\ 25°}]{CrO_3} C_6H_5-CH=CH-CHO \qquad (Ref.\ 21b)$$
$$(81\%)$$

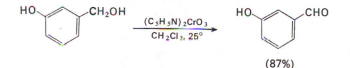

$$(87\%) \qquad\qquad (Ref.\ 2a)$$

$$\underset{\substack{| \\ CH_3}}{H_2C=CHCH_2CH_2C} \overset{CH_3}{=} C(CH_2)_3CH_2OH \xrightarrow[CH_2Cl_2,\ 25°]{(C_5H_5N)_2CrO_3}$$

$$H_2C=CHCH_2CH_2C\underset{\substack{| \\ CH_3}}{\overset{\substack{CH_3 \\ |}}{=}}C(CH_2)_3CHO$$

$$(45\%) \qquad (Ref.\ 31d)$$

Tertiary alcohols are usually relatively inert to oxidation by chromic acid;[32] failure of an alcohol to be oxidized by chromic acid is often cited as evidence that it is tertiary. However, tertiary 1,2-diols are very rapidly cleaved by chromic acid, provided they are sterically capable of forming cyclic chromate esters such as [16].[33] The rate-limiting step in this oxidative cleavage appears to be the initial formation of the monochromate ester.[33b,c] Whether formation of the cyclic ester [16] is an essential step in the cleavage is not certain.[33c] Although this cleavage reaction appears similar to that of 1,2-diols with lead tetraacetate and periodic acid (see Chapter 7), the cleavage with chromic acid is not completely analogous because 1,2-diols, having at least one *alpha* hydrogen atom (e.g., [17]), are normally oxidized by chromic acid to form α-hydroxyketones which may be accompanied by cleavage products.[34]

32. For an exception see J. Roček and A. Radkowsky, *Tetrahedron Letters,* **No. 24**, 2835 (1968).
33. (a) J. Roček and F. H. Westheimer, *J. Am. Chem. Soc.,* **84**, 2241 (1962). (b) H. Kwart and D. Bretzger, *Tetrahedron Letters,* **No. 45**, 3985 (1965). (c) N. D. Heindel, E. S. Hanrahan, and R. J. Sinkovitz, *J. Org. Chem.,* **31**, 2019 (1966).
34. B. H. Walker, *J. Org. Chem.,* **32**, 1098 (1967).

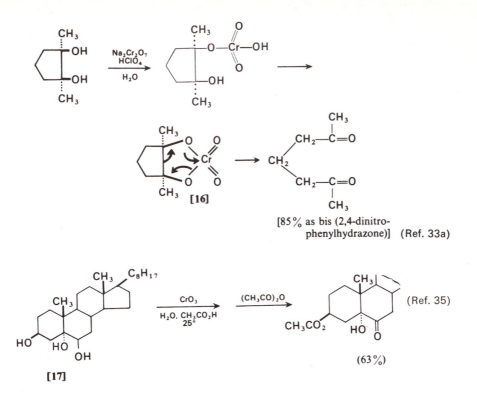

[85% as bis (2,4-dinitro-phenylhydrazone)]   (Ref. 33a)

(Ref. 35)

(63%)

[17]

As the following equation illustrates, the amount of cleavage observed with such 1,2-diols may be significantly lowered if a manganese(II) salt is added to the reaction mixture to reduce the Cr^{IV} species produced to Cr^{III} salts. This observation suggests that the bulk of the oxidative cleavage observed with such 1,2-diols is not analogous to the cleavage of ditertiary 1,2-diols involving a cyclic chromium(VI) ester such as [16].

Instead, the cleavage of 1,2-diols having at least one alpha hydrogen atom is apparently analogous to the carbon-carbon bond cleavage observed during the chromic acid oxidation of certain secondary alcohols such as [18], which contain an *alpha* substituent (*t*-butyl group in [18]) that can be eliminated as a relatively stable free radical. This cleavage reaction is of particular interest because it provides an example of an alcohol that is oxidized normally with Cr^{VI} species to form the ketone, whereas reaction of the same alcohol with Cr^{IV} species results in cleavage.[8,36] As would be anticipated from the previous discussion, the yield of cleavage products can be substantially lowered by performing such oxidations in the presence of manganese(II) or cerium(II) ion.[8,36]

35. B. Ellis and V. Petrow, *J. Chem. Soc.*, 1078 (1939).
36. (a) J. Hampton, A. Leo, and F. H. Westheimer, *J. Am. Chem. Soc.*, **78**, 306 (1956). (b) J. J. Cawley and F. H. Westheimer, *ibid.*, **85**, 1771 (1963). (c) W. A. Mosher and G. L. Driscoll, *ibid.*, **90**, 4189 (1968).

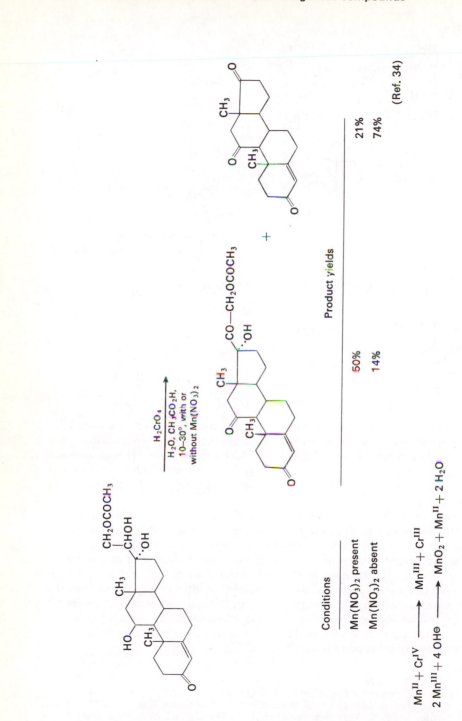

Conditions	Product yields	
$Mn(NO_3)_2$ present	50%	21%
$Mn(NO_3)_2$ absent	14%	74%

$$Mn^{II} + Cr^{IV} \longrightarrow Mn^{III} + Cr^{III}$$

$$2\,Mn^{III} + 4\,OH^{\ominus} \longrightarrow MnO_2 + Mn^{II} + 2\,H_2O$$

(Ref. 34)

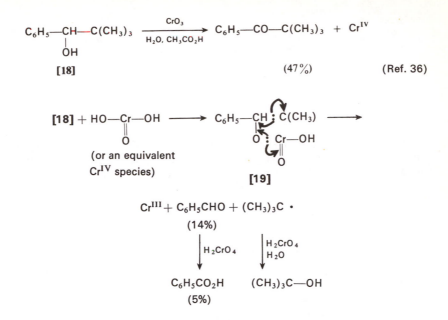

As the equations summarizing the oxidation of alcohol **[18]** indicate, it is currently believed[8c,d] that an intermediate Cr^{IV} ester such as **[19]** undergoes the indicated transfer of one electron to eliminate a carbon free radical whereas the analogous Cr^{VI} and Cr^{V} esters both undergo transfer of two electrons and a proton (or a hydride anion) to give carbonyl compounds without carbon-carbon bond cleavage. It is apparent that the process **[19]** involving a Cr^{IV} ester may also be involved in the aforementioned cleavage of ditertiary 1,2-diols.[8c]

OXIDATION OF ALDEHYDES

Although both chromium and manganese compounds can be used for the oxidation of aldehydes to carboxylic acids,[1,5,7,8,37] an aqueous solution of potassium permanganate under either acidic or basic conditions is the more commonly employed reagent.

$$n\text{-}C_6H_{13}\text{---CHO} \xrightarrow[\substack{H_2O \\ 20°}]{\substack{KMnO_4 \\ H_2SO_4}} n\text{-}C_6H_{13}\text{---CO}_2H \qquad \text{(Ref. 38)}$$
$$(76\text{--}78\%)$$

37. (a) K. B. Wiberg and R. Stewart, *J. Am. Chem. Soc.*, **77**, 1786 (1955). (b) F. Freeman and co-workers, *J. Org. Chem.*, **35**, 982 (1970). (c) For an example utilizing the chromium trioxide-pyridine complex with added water for the oxidation of an aldehyde to a carboxylic acid, see W. S. Johnson and co-workers. *J. Am. Chem. Soc.*, **85**, 1409 (1963). (d) The oxidation of aldehydes to acids with diacetyl chromate has been studied by K. B. Wiberg and P. A. Lepse, *ibid.*, **86**, 2612 (1964). (e) D. G. Lee and U. A. Spitzer, *J. Org. Chem.*, **35**, 3589 (1970).
38. J. R. Ruhoff, *Org. Syn.*, **Coll. Vol. 2**, 315 (1943).

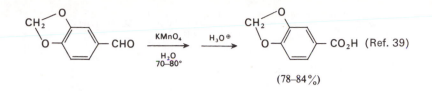

(78–84%)

Permanganate oxidations in neutral or acidic media are catalyzed by acid, and the reaction is accelerated slightly by the presence of electron-donating groups.[37] The oxidation mechanism is believed to involve the formation and subsequent decomposition of the permanganate ester [20]. Oxidation in basic media is

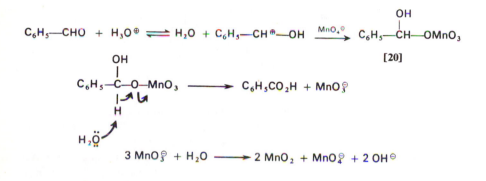

apparently more complex. Under these conditions the reaction, which is accelerated by electron-withdrawing groups, has been suggested to be a free-radical chain process,[37a] or to involve a hydride transfer.[37b] The following equations illustrate the later possibility.

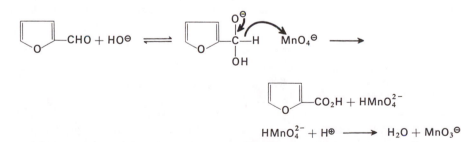

The oxidation of aldehydes with chromic acid, a reaction catalyzed by acid and accelerated by electron-withdrawing groups, is believed to involve the for-

39. R. L. Shriner and E. C. Kleiderer, *Org. Syn.,* **Coll. Vol. 2,** 538 (1943).

mation and decomposition of the chromate ester of the aldehyde hydrate, as illustrated below.[5b−d]

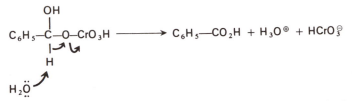

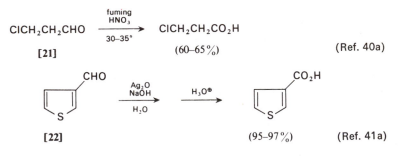

Other agents that have been used for the oxidation of aldehydes include fuming nitric acid (e.g., [21]) and a suspension of silver oxide in aqueous alkali (e.g., [22]). The use of the latter provides a very mild and selective method for this type of oxidation.

$$CICH_2CH_2CHO \xrightarrow[\text{30–35°}]{\substack{\text{fuming} \\ \text{HNO}_3}} CICH_2CH_2CO_2H$$

[21]   (60–65%)   (Ref. 40a)

[22]   (95–97%)   (Ref. 41a)

OXIDATION OF CARBON-CARBON DOUBLE BONDS

An aqueous solution of potassium permanganate reacts with olefins to add two hydroxyl functions to the double bond in a *cis* manner, provided the reaction mixture is alkaline. If the reaction mixture is kept neutral, either by the continuous addition of acid or by adding magnesium sulfate (hydroxide ion is removed by the precipitation of magnesium hydroxide), the permanganate oxidation results in cleavage or in the formation of α-hydroxyketones.[1,5,7,42] These reactions, illustrated by the

40. (a) C. Moureu and R. Chaux, *Org. Syn.*, **Coll. Vol. 1**, 166 (1944). (b) J. English, Jr., and J. E. Dayan, *ibid.*, **Coll. Vol. 4**, 499 (1963).
41. (a) E. Campaigne and W. M. LeSuer, *Org. Syn.*, **Coll. Vol. 4**, 919 (1963). (b) see also K. J. Clark, G. I. Fray, R. N. Jaeger, and R. Robinson, *Tetrahedron*, **6**, 217 (1959). (c) I. A. Pearl, *Org. Syn.*, **Coll. Vol. 4**, 972 (1963). (d) For the use of oxygen with a silver oxide-copper oxide catalyst, see R. J. Harrisson and M. Moyle, *ibid.*, **Coll. Vol. 4**, 493 (1963).
42. (a) J. E. Coleman, C. Ricciuti, and D. Swern, *J. Am. Chem. Soc.*, **78**, 5342 (1956). (b) K. B. Wiberg and K. A. Saegebarth, *ibid.*, **79**, 2822 (1957). (c) K. B. Wiberg and R. D. Geer, *ibid.*, **88**, 5827 (1966). (d) H. B. Henbest, W. R. Jackson, and B. C. G. Robb, *J. Chem. Soc.*, B, 803 (1966). (e) For the use of solutions of potassium permanganate in acetic anhydride to oxidize olefins to α-diketones, see K. B. Sharpless, R. F. Lauer, O. Repic, A. Y. Teranishi, and D. R. Williams, *J. Am. Chem. Soc.*, **93**, 3303 (1971).

following equations, are believed to proceed by a very rapid initial formation of a
cyclic manganese ester **[23]**, which undergoes the further changes indicated.[42b]
Although oxidation with potassium permanganate is almost always inferior to

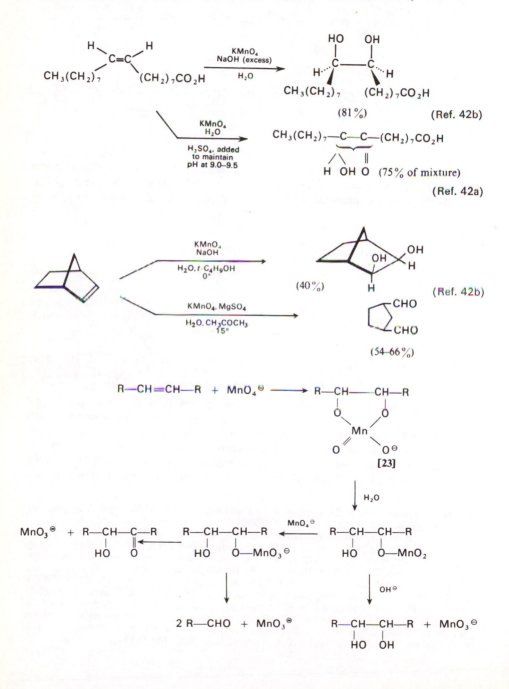

(81%)   (Ref. 42b)

(75% of mixture)

(Ref. 42a)

(40%)   (Ref. 42b)

(54–66%)

[23]

oxidation with osmium tetroxide[42d,43] for selective conversion of olefins (e.g., [24]) to *cis*-1,2-diols in high yield, the permanganate oxidation is less hazardous (osmium tetroxide is very toxic) and decidedly less expensive for reactions that are to be run on a relatively large scale.

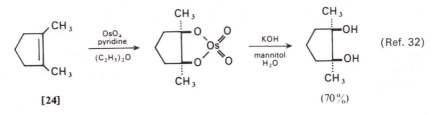

[24]   (70%)

The permanganate and osmium tetroxide oxidations of olefins differ in that electron withdrawing substituents in the olefin appear to accelerate the permanganate oxidations whereas the same substituents retard the reaction of olefins with osmium tetroxide.[42d] Strained and unhindered olefins usually react with osmium tetroxide more rapidly than unstrained[43d] or sterically hindered olefins. Both the osmium and the manganese oxidants can normally be expected to attack a double bond from its less hindered side as the accompanying equations indicate.

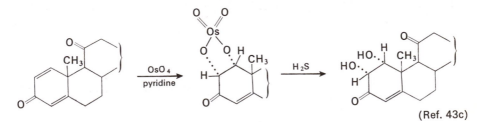

(Ref. 43c)

In some cases it has been possible to use osmium tetroxide in catalytic amounts with a less expensive oxidant, such as aqueous sodium chlorate[43a] or oxygen in an alkaline medium,[43e] being used to continuously reoxidize the Os[VI] species to Os[VIII].

An interesting modification of the permanganate hydroxylation procedure utilizes a mixture of potassium permanganate and potassium periodate in aqueous

43. (a) F. D. Gunstone, *Adv. Org. Chem.*, **1**, 103 (1960) ; K. Wiesner, K. K. Chan, and C. Demerson, *Tetrahedron Letters*, **No. 33**, 2893 (1965) ; K. Wiesner and J. Santroch, *ibid.*, **No. 47**, 5939 (1966). (b) Alternative methods for the isolation of the diol from the intermediate cyclic osmium ester include reduction with hydrogen sulfide [D. H. R. Barton and D. Elad, *J. Chem. Soc.*, 2085 (1956)] and reduction with lithium aluminum hydride in tetrahydrofuran [J. Castells, G. D. Meakins, and R. Swindells, *ibid.*, 2917 (1962)]. (c) T. Kubota and F. Hayashi, *Tetrahedron*, **23**, 995 (1967). (d) R. E. Erickson and R. L. Clark, *Tetrahedron Letters*, **No. 45**, 3997 (1969). (e) J. F. Cairns and H. L. Roberts, *J. Chem. Soc.*, C, 640 (1968). (f) For examples of the formation of epoxides from the reaction of olefins with osmium tetroxide and sodium chlorate, see W. Kruse, *Chem. Commun.*, **No. 24**, 1610 (1968). (g) For a general discussion of the addition of oxidants to carbon-carbon double bonds, see J. S. Littler, *Tetrahedron*, **27**, 81 (1971).

solution at pH 7.7 for the oxidation of olefins.[44] Under these conditions, the
α-hydroxyketone and 1,2-diol resulting from reaction of the olefin (e.g., [25])
with permanganate are cleaved by the periodate to form carbonyl compounds;
any aldehydes produced by this cleavage are oxidized to carboxylic acids. Only
a catalytic amount of the permanganate is required because the lower-valent
derivatives of manganese are reoxidized to permanganate ion by the periodate in
the reaction mixture. A related reaction, the use of a catalytic amount of osmium
tetroxide in the presence of sodium periodate, to effect the oxidative cleavage of
olefinic double bonds is discussed in Chapter 7. This process differs from the
permanganate-periodate oxidation in permitting the isolation of any aldehydes
formed.

$$CH_3(CH_2)_7CH\!=\!CH(CH_2)_7CO_2H \xrightarrow[\substack{K_2CO_3 \\ H_2O}]{\substack{NaIO_4 \\ KMnO_4}} \xrightarrow{H_3O^\oplus} CH_3(CH_2)_7CO_2H$$

[25] $+ \; HO_2C(CH_2)_7CO_2H$ (Ref. 44)

The oxidation of carbon-carbon double bonds with chromic acid is often
complicated by a competing oxidation that occurs at allylic C—H bonds, leading to
mixtures of products, particularly when cyclic olefins are involved. In the oxidation
of olefins such as [26] allylic oxidation becomes especially favorable when the
chromium trioxide-dipyridine complex is used. As will be discussed subsequently,

$$\xrightarrow[\substack{H_2O, \; CH_3CO_2H \\ 25-35°}]{CrO_3}$$

(33–38%)   (25%)   (Ref. 45a)

this reaction is believed to proceed by the indicated formation of allylic radical
intermediates; it should be noted that the major product is derived by removal of a
tertiary allylic hydrogen atom. When an increasing number of phenyl groups are
attached to the double bond (e.g., [27] and [28]), oxidation at the double bond
rather than attack at an allylic position becomes the predominant reaction. The
oxidative degradation of the ester [29] via the olefin [28], known as the Barbier-
Wieland degradation, illustrates the utility of this oxidation when two phenyl
substituents are present. Even in cases such as [30] where no allylic C—H bonds

44. (a) R. U. Lemieux and E. von Rudloff, *Can. J. Chem.,* **33**, 1701, 1710 (1955); E. von
Rudloff, *ibid.,* **33**, 1714 (1955). (b) S. W. Pelletier, K. N. Iyer, and C. W. J. Chang, *J. Org.
Chem.,* **35**, 3535 (1970).
45. (a) F. C. Whitmore and G. W. Pedlow, Jr., *J. Am. Chem. Soc.,* **63**, 758 (1941). (b)
D. T. Cropp, J. S. E. Holker, and W. R. Jones, *J. Chem. Soc.,* C, 1443 (1966). (c) For
examples of oxidation of cyclic polyolefins with chromium trioxide, see R. C. Cambie,
V. F. Carlisle, C. J. LeQuesne, and T. D. R. Manning, *ibid.,* C, 1234 (1969). (d) For similar
studies with *t*-butyl chromate, see T. Matsuura and T. Suga, *J. Org. Chem.,* **30**, 518 (1965).

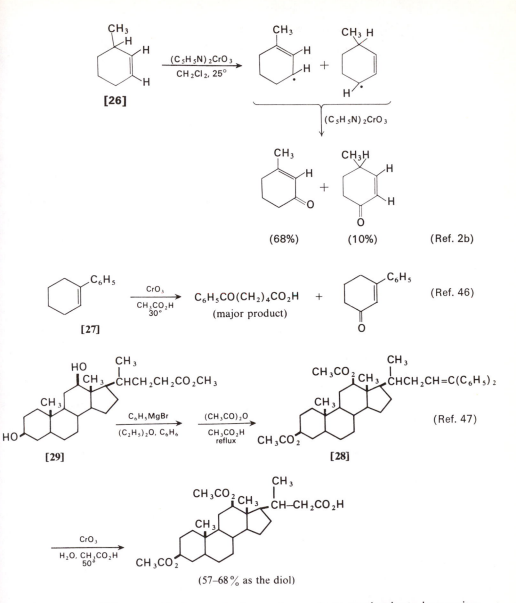

(68%)   (10%)   (Ref. 2b)

(major product)

(Ref. 46)

(Ref. 47)

(57–68 % as the diol)

are present, different oxidation products can be made to predominate by varying the reaction conditions. A thorough study of the oxidation of cholesterol (partial structure [31])[9c,14e,45b] has revealed that, although the major course of the reaction is the one indicated in the accompanying equations, a variety of minor by-products are also formed.

46. L. F. Fieser and J. Szmuskovicz, *J. Am. Chem. Soc.*, **70**, 3352 (1948).
47. B. Riegel, R. B. Moffett, and A. V. McIntosh, *Org. Syn.*, **Coll. Vol. 3**, 234, 237 (1955).

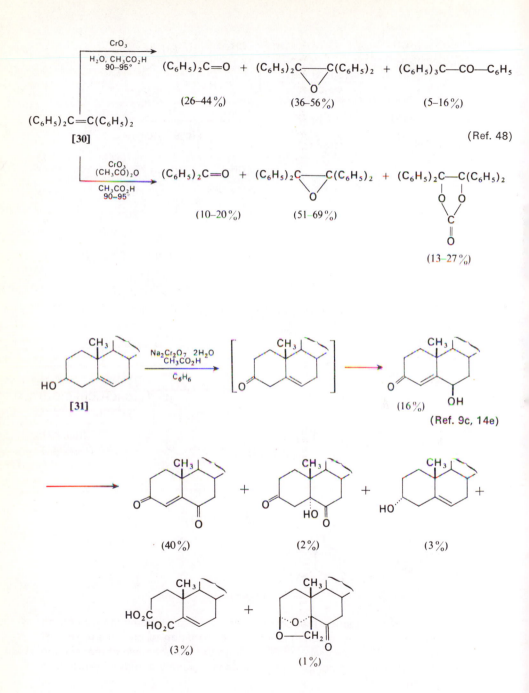

$(C_6H_5)_2C$=O + $(C_6H_5)_2C$——$C(C_6H_5)_2$ + $(C_6H_5)_3C$—CO—C_6H_5

(26–44 %)   (36–56 %)   (5–16 %)

(Ref. 48)

$(C_6H_5)_2C$=O + $(C_6H_5)_2C$——$C(C_6H_5)_2$ + $(C_6H_5)_2C$—$C(C_6H_5)_2$

(10–20 %)   (51–69 %)

(13–27 %)

(16 %)

(Ref. 9c, 14e)

(40 %)   (2 %)   (3 %)

(3 %)   (1 %)

48. W. A. Mosher, F. W. Steffgen, and P. T. Lansbury, *J. Org. Chem.*, **26**, 670 (1961).

In general, the use of chromic acid in a partially aqueous medium favors oxidative cleavage of carbon-carbon double bonds, whereas anhydrous conditions (chromium trioxide in glacial acetic acid or acetic anhydride,[49a] sodium dichromate dihydrate in glacial acetic acid, t-butyl chromate in carbon tetrachloride–acetic acid,[45d,49b] or chromyl chloride in methylene chloride or carbon tetrachloride[50]) favor either attack at allylic positions or partial oxidation of the double bond to form epoxides, diol derivatives, or ketols; rearrangements of the intermediate oxidation products may also be observed.[50,51] Since many studies of olefin oxidations with chromic acid[7,51] have dealt primarily with the composition of the product mixtures, a good deal of uncertainty exists about the mechanism of these reactions.

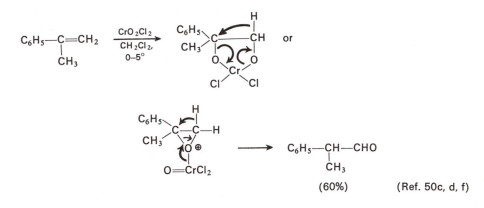

(60%)   (Ref. 50c, d, f)

The formations of epoxides and diol carbonates are known not to be stereospecific processes (e.g., the oxidation of [32]), indicating that a cyclic chromate ester, analogous to intermediate [23] in permanganate oxidations, is not formed directly from the olefin and the Cr^{VI} reactant (cf. Ref. 51c). The reaction has been suggested to proceed by an electrophilic attack of the Cr^{VI} species (represented as chromic acid in the accompanying equations) on the olefin to form a transient carbonium ion such as [33].[7,51a,51b] (An initial one-electron transfer to form the analogous radical has not been excluded.) The intermediate [33] may then either

49. (a) H. Schildknecht and W. Föttinger, *Justus Liebigs Ann. Chem.,* **659**, 20 (1962). (b) D. Ginsburg and R. Pappo, *J. Chem. Soc.,* 516 (1951). (c) For the preparation and use of bistriphenylsilyl chromate to cleave olefins see L. M. Baker and W. L. Carrick, *J. Org. Chem.,* **35**, 774 (1970).
50. (a) S. J. Cristol and K. R. Eilar, *J. Am. Chem. Soc.,* **72**, 4353 (1950). (b) F. Freeman, P. J. Cameron, and R. H. DuBois, *J. Org. Chem.,* **33**, 3970 (1968). (c) F. Freeman, R. H. DuBois, and N. J. Yamachika, *Tetrahedron,* **25**, 3441 (1969). (d) F. Freeman and N. J. Yamachika, *Tetrahedron Letters,* **No. 41,** 3615 (1969). (e) A. K. Awasthy and J. Roček, *J. Am. Chem. Soc.,* **91**, 991 (1969). (f) F. Freeman and N. J. Yamachika, *ibid.,* **92**, 3730 (1970); F. Freeman, P. D. McCart, and N. J. Yamachika, *ibid.,* **92**, 4621 (1970). (g) J. Roček and J. C. Drozd, *ibid.,* **92**, 6668 (1970).
51. (a) W. J. Hickinbottom, D. Peters, and D. G. M. Wood, *J. Chem. Soc.,* 1360 (1955); (b) M. A. Davis and W. J. Hickinbottom, *ibid.,* 2205 (1958). (c) M. S. Carson, W. Cocker, D. H. Grayson, and P. V. R. Shannon, *ibid.,* C, 2220 (1969). (d) H. H. Zeiss and F. R. Zwanzig, *J. Am. Chem. Soc.,* **79**, 1733 (1957).

lose a proton (as in the conversion of cholesterol [31] to 6-hydroxy-4-cholesten-3-one) or react with a nucleophile (e.g., water, acetate ion, or acetic acid) in the reaction mixture to form intermediates such as [34] and [35]. It is not clear whether

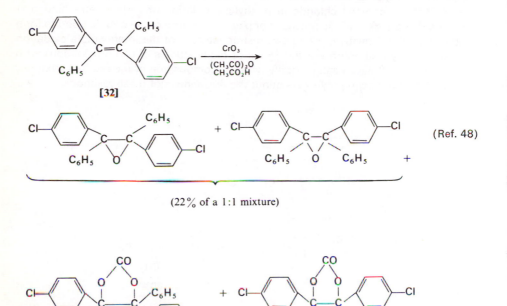

(Ref. 48)

(22% of a 1:1 mixture)

(53% of a 1:1 mixture)

the various products with rearranged carbon skeletons that have been isolated from chromic acid oxidations[48,51] result from rearrangement of initially formed oxidation products or from rearrangement of the ion [33]. In support of the above mechanistic scheme, olefins (e.g., [39]) have reacted with chromyl chloride[50] to form chromium esters of chlorohydrins that could be reduced and hydrolyzed with aqueous sodium bisulfite to chlorohydrins. In the presence of water and excess chromic acid, the intermediates [34] and [35] could be converted to the cyclic chromate ester [36] (analogous to the intermediate [16] postulated for the cleavage of glycols) and then cleaved or converted to an α-hydroxyketone. Alternatively, the intramolecular displacements represented in structures [35] and [37] could lead to an epoxide or to a precursor [38] for a diol carbonate. This general scheme serves to account for the ready oxidative cleavage of 1,1-diphenylethylene derivatives such as [28] with aqueous chromic acid since, initially, attack at the carbon-carbon double bond would be favored by the formation of a carbonium ion (i.e., [33]) that could be stabilized by the two adjacent phenyl rings.

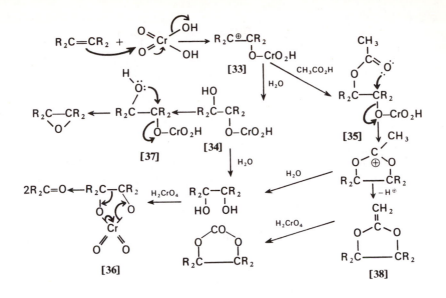

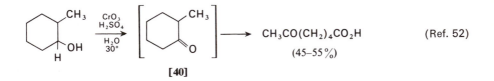

Such a reaction scheme also accounts for the oxidative cleavage of ketones (e.g., [40]), which is believed to proceed by electrophilic attack on the enol[53] as illustrated in the accompanying equations. The successful oxidation to diketones of

certain diterpene derivatives having rings A and B *cis*-fused (e.g., [41]), under conditions where corresponding *trans*-fused compounds (e.g., [42]) form only monoketones, has been attributed to steric hindrance to attack of the enol in the latter case.

52. J. R. Schaeffer and A. O. Snoddy, *Org. Syn.*, **Coll. Vol. 4**, 19 (1963).
53. (a) P. A. Best, J. S. Littler, and W. A. Waters, *J. Chem. Soc.*, 822 (1962). (b) E. Wenkert and B. G. Jackson, *J. Am. Chem. Soc.*, **80**, 211 (1958). (c) J. Roček and A. Riehl, *ibid.*, **89**, 6691 (1967); *J. Org. Chem.*, **32**, 3569 (1967).

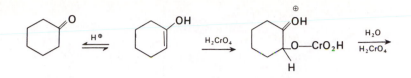

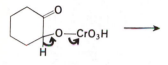

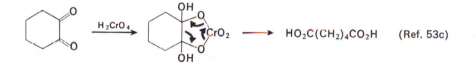

HO₂C(CH₂)₄CO₂H (Ref. 53c)

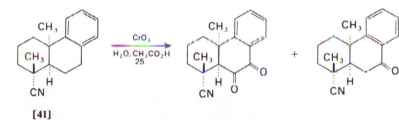

(Ref. 53b)

[41]

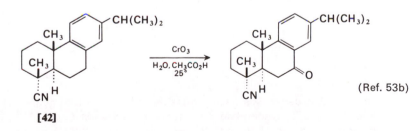

(Ref. 53b)

[42]

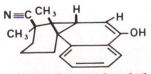

intermediate enol from [41]

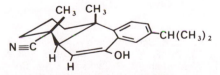

intermediate enol from [42]

Alkaline potassium permanganate also effects the rapid oxidative cleavage of enolizable ketones as well as primary and secondary nitroalkanes and alkane-sulfonic acid derivatives.[54a-d] As the accompanying equations indicate, these reactions are believed to proceed by an initial reaction of the substrate with base to form an anion which is attacked by permanganate anion.[54a-d]

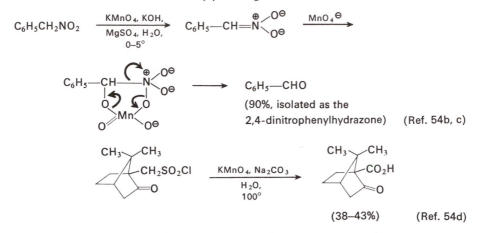

OXIDATION OF CARBON-HYDROGEN BONDS IN HYDROCARBONS

Perhaps the most common application of this process is the oxidation of the side chain of benzene derivatives with aqueous potassium permanganate (e.g., [43]), with chromic acid (e.g., [44]) in aqueous solution or in acetic acid, or with aqueous

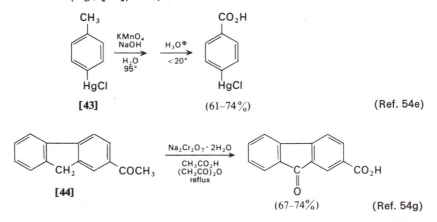

54. (a) K. B. Wiberg and R. D. Geer, *J. Am. Chem. Soc.,* **87**, 5202 (1965). (b) H. Shechter and F. T. Williams, Jr., *J. Org. Chem.,* **27**, 3699 (1962). (c) F. Freeman and A. Yeramyan, *Tetrahedron Letters,* **No. 46**, 4783 (1968) ; *J. Org. Chem.,* **35**, 2061 (1970). (d) P. D. Bartlett and L. H. Knox, *Org. Syn.,* **45**, 55 (1965). (e) F. C. Whitmore and G. E. Woodward, *ibid.,* **Coll. Vol. 1**, 159 (1944). (f) A. W. Singer and S. M. McElvain, *ibid.,* **Coll. Vol. 3**, 740 (1955). (g) G. Rieveschl, Jr. and F. E. Ray, *ibid.,* **Coll. Vol. 3**, 420 (1955). (h) R. A. Pacaud and C. F. H. Allen, *ibid.,* **Coll. Vol. 2**, 336 (1943). (i) C. F. H. Allen and J. A. Van Allan, *ibid.,* **Coll. Vol. 3**, 1 (1955).

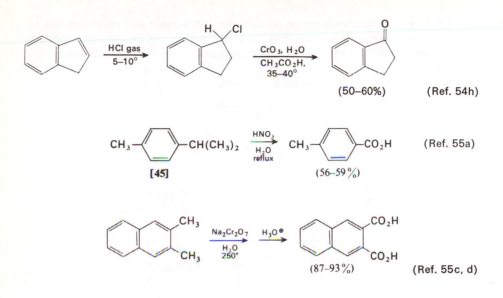

(50–60%)   (Ref. 54h)

(Ref. 55a)

[45]   (56–59%)

(87–93%)   (Ref. 55c, d)

sodium dichromate at elevated temperatures. This type of oxidation has also been effected with nitric acid (e.g., **[45]**). The previously discussed chromic acid oxidation of olefins (e.g., **[46]**) at the allylic position, although probably mechanistically related to the side-chain oxidation, is complicated by the concurrent oxidation that occurs at the double bond and by the fact that the intermediate allylic radical (e.g., **[47]**) or the corresponding cation) can often lead to structurally isomeric

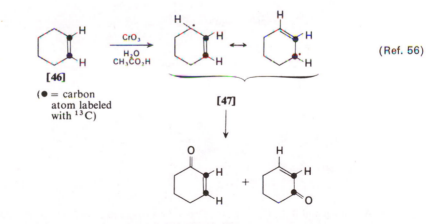

[46]

(● = carbon
atom labeled
with ¹³C)

[47]

(Ref. 56)

55. (a) W. F. Tuley and C. S. Marvel, *Org. Syn.*, **Coll. Vol. 3**, 822 (1955). (b) S. M. McElvain, *ibid.*, **Coll. Vol. 1**, 385 (1944). (c) L. Friedman, *ibid.*, **43**, 80 (1963). (d) L. Friedman, D. L. Fishel, and H. Shechter, *J. Org. Chem.*, **30**, 1453 (1965). (e) D. G. Lee and U. A. Spitzer, *ibid.*, **34**, 1493 (1969).
56. K. B. Wiberg and S. D. Nielsen, *J. Org. Chem.*, **29**, 3353 (1964).

products. For the successful oxidation of side chains attached to benzene rings, it is necessary that the ring contain no hydroxyl or amino substituent since aromatic compounds of these types (e.g., **[48]**) are readily oxidized to quinones and, with

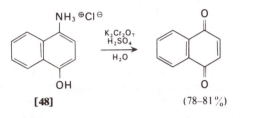

[48]   (78–81 %)   (Ref. 57)

excess oxidant, to carbon dioxide and water. Phenolic hydroxyl groups may be protected as their methyl ethers; aniline derivatives are usually converted to the corresponding acetanilide or benzanilide derivatives, which are then oxidized with aqueous potassium permanganate.[18] For the oxidation of acetanilides, magnesium sulfate is sometimes added to the reaction mixture to prevent it from becoming strongly basic and, consequently, to avoid hydrolysis of the amide. The selective oxidation of the side chains on polycyclic aromatic compounds with chromic acid is usually an unsatisfactory synthetic procedure because of competing oxidation of the aromatic system.[2,58] However as indicated in an earlier equation, successful side-chain oxidations of such systems have been accomplished using aqueous sodium dichromate at 250°.[55c−e] The ready oxidation of polycyclic aromatic systems to quinones[57] is illustrated by the following example.

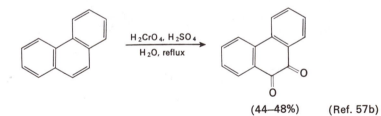

(44–48%)   (Ref. 57b)

Since the side chain of *t*-butylbenzene is not oxidized by chromic acid, oxidation of aromatic side chains is believed to be initiated by attack at a benzylic C—H bond. However, occasionally products in which the oxygen has been introduced at the carbon atom beta to the aromatic ring are formed; such products

57. (a) L. F. Fieser, *Org. Syn.,* **Coll. Vol. 1**, 383 (1944). (b) R. Wendland and J. LaLonde, *ibid.,* **Coll. Vol. 4**, 757 (1963). (c) E. A. Braude and J. S. Fawcett, *ibid.,* **Coll. Vol. 4**, 698 (1963). (d) For the isolation of dimeric coupled products from the oxidation of β-naphthylamines with potassium permanganate, see R. F. Bridger, D. A. Law, D. F. Bowman, B. S. Middleton, and K. U. Ingold, *J. Org. Chem.,* **33**, 4329 (1968). (e) For the use of chromyl chloride to oxidize phenols to quinones, see J. A. Strickson and M. Leigh, *Tetrahedron,* **24**, 5145 (1968).
58. For example, see L. F. Fieser, W. P. Campbell, E. M. Fry, and M. D. Gates, Jr., *J. Am. Chem. Soc.,* **61**, 3216 (1939).

are presumably the result of initial attack at the benzylic position followed by rearrangement.[55e,59] The chromic acid oxidation of aromatic side chains is accelerated by electron-donating substituents, and the rate of reaction in aqueous acetic acid is increased by increasing the concentration of mineral acid or by decreasing the concentration of water.[60] The accompanying equations illustrate the reaction path that is believed to be operative. Whether the initial hydrogen transfer involves a hydride ion (two-electron transfer) or a hydrogen atom (one-electron transfer, as illustrated) has been suggested to depend upon the substituents present.[60] One solution to this ambiguity was offered in the proposal[60b] that the transition state for hydrogen transfer is a resonance hybrid (e.g., [49]), in which the carbon atom has both radical and carbonium ion character. The oxidations of aromatic nucleii would appear to involve initial electron transfer from the aromatic nucleus to the oxidant. Further reaction of the resulting aryl cation radical (or dication) with nucleophilic compounds in the reaction mixture then leads to the observed products.[61]

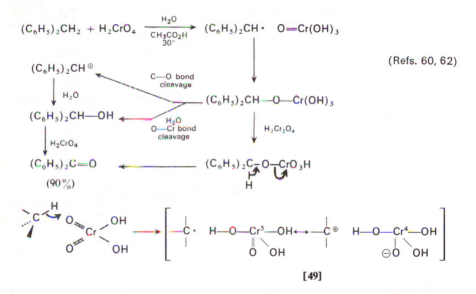

[49]

From studies of the oxidation of the tertiary C—H bond of certain alkanes (e.g., [50] and [51]) by either chromic acid or potassium permanganate, the initial conversion of the C—H bond to a C—O bond is known to proceed with at least partial retention of configuration. The tendency of chromic acid to attack secondary and, especially, tertiary C—H bonds serves as the basis for the Kuhn-Roth

59. K. B. Wiberg, B. Marshall, and G. Foster, *Tetrahedron Letters,* **No. 8,** 345 (1962).
60. (a) K. B. Wiberg and R. J. Evans, *Tetrahedron,* **8,** 313 (1960). (b) J. Roček, *Tetrahedron Letters,* **No. 4,** 135 (1962).
61. For recent reviews of this process, particularly in related electrochemical oxidations, see (a) N. L. Weinberg and H. R. Weinberg, *Chem. Rev.,* **68,** 449 (1969). (b) O. C. Musgrave, *ibid.,* **69,** 499 (1969). (c) R. N. Adams, *Accts. Chem. Res.,* **2,** 175 (1967).

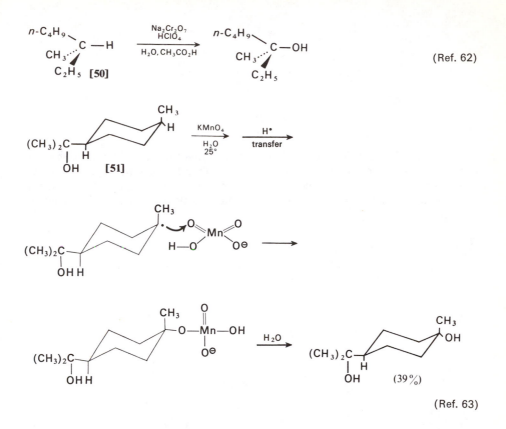

(Ref. 62)

(Ref. 63)

determination of the number of C-methyl groups in a molecule. In this determination, a sample is oxidized with hot aqueous chromic acid and acetic acid and its higher homologs are distilled from the reaction mixture. The yields of these acids provide a measure of the number of C-methyl groups present. Several modifications of this classical analytical procedure have been recommended to increase the information obtained from such determinations.[62b,c]

The conversion of methyl groups bonded to aromatic rings (e.g., [52]) to the corresponding aldehydes has been accomplished in moderate yield by oxidation with chromium trioxide in acetic anhydride. The success of this procedure has been attributed[7] to the formation of the diacetate [53], which is relatively stable to the reaction conditions. The same type of conversion has been achieved by treating toluene derivatives (e.g., [54]) with chromyl chloride in either carbon tetrachloride

62. (a) K. B. Wiberg and G. Foster, *J. Am. Chem. Soc.*, **83**, 423 (1961). (b) A. K. Awasthy, R. Belcher, and A. M. G. Macdonald, *J. Chem. Soc.*, C, 799 (1967). (c) V. S. Pansare and S. Dev, *Tetrahedron*, **24**, 3767 (1968).
63. (a) R. H. Eastman and R. A. Quinn, *J. Am. Chem. Soc.*, **82**, 4249 (1960). (b) K. B. Wiberg and A. S. Fox, *ibid.*, **85**, 3487 (1963). (c) J. I. Brauman and A. J. Pandell, *ibid.*, **92**, 329 (1970).

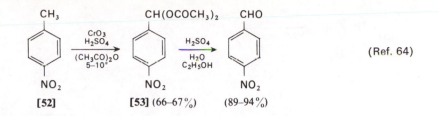

(Ref. 64)

[52]   [53] (66–67%)   (89–94%)

or carbon disulfide. This reaction, known as the Etard reaction,[2,4,65] produces an initial precipitate (containing two atoms of chromium for each hydrocarbon molecule), which yields an aldehyde or ketone upon treatment with water. Alkanes containing tertiary C—H bonds may be converted to alcohols by the same reaction procedure.[65]

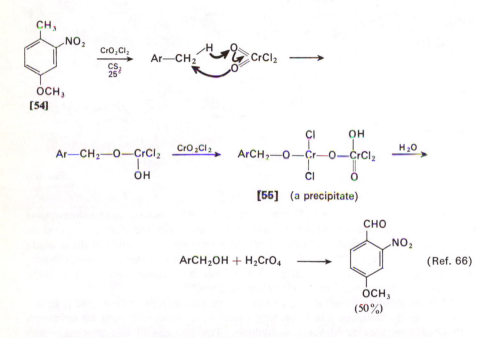

[54]

[55]   (a precipitate)

$ArCH_2OH + H_2CrO_4 \longrightarrow$ (Ref. 66)

(50%)

64. (a) T. Nishimura, *Org. Syn.,* **Coll. Vol. 4**, 713 (1963). (b) S. V. Lieberman and R. Connor, *ibid.,* **Coll. Vol. 2**, 441 (1943).
65. I. Necsoiu, A. T. Balaban, I. Pascaru, E. Sliam, M. Elian, and C. D. Nenitzescu, *Tetrahedron,* **19**, 1133 (1963). (b) O. H. Wheeler, *Can. J. Chem.,* **42**, 706 (1964). (c) R. A. Stairs, *ibid.,* **42**, 550 (1964). (d) K. B. Wiberg and R. Eisenthal, *Tetrahedron,* **20**, 1151 (1964). (e) I. Necsoiu, V. Przemetchi, A. Ghenciulescu, C. N. Rentea, and C. D. Nenitzescu, *ibid.,* **22**, 3037 (1966). (f) C. N. Rentea, I. Necsiou, M. Rentea, A. Ghenciulescu, and C. D. Nenitzescu, *ibid.,* **22**, 3501 (1966). (g) C. N. Rentea, M. Rentea, I. Necsoiu, and C. D. Nenitzescu, *ibid.,* **24**, 4667 (1968). (h) H. C. Duffin and R. B. Tucker, *ibid.,* **23**, 2803 (1967); *ibid.,* **24**, 389, 6999 (1968).
66. W. R. Boon, *J. Chem. Soc.,* S230 (1940).

Although the mechanism of these transformations has been the subject of some disagreement,[65] the pathway indicated in the accompanying equation appears to be in best agreement with the data currently available. The presence of an intermediate benzyl (or alkyl) ester of a chromium acid serves to explain a number of side reactions observed such as molecular rearrangements,[65g] the formation of alcohols rather than carbonyl compounds when the insoluble intermediates are hydrolyzed in the presence of a reducing agent such as sulfur dioxide,[65e,f] and the formation in certain cases of alkyl chlorides rather than alcohols or carbonyl compounds.

6

OXIDATIONS WITH PERACIDS AND OTHER PEROXIDES

Peracids have been used most extensively for the selective oxidation of carbon-carbon double bonds; they also slowly convert ketones to esters. Other peroxide oxidations to be considered here include the copper-catalyzed reactions of peresters that allow selective oxidation of allylic and other activated C—H bonds as well as certain reactions of peracids, hydrogen peroxide, alkyl hydroperoxides, and oxygen with other functional groups. In general, the peroxide reagents accomplish the oxidation of functions that are frequently difficult to oxidize selectively with chromium and manganese compounds.

Below are listed formulas for the peracids commonly used in the organic laboratory.[1] The most useful reagents in this group are peracetic acid [2], peroxytrifluoroacetic acid [3], and the perbenzoic acids [4]. Performic acid [1] is

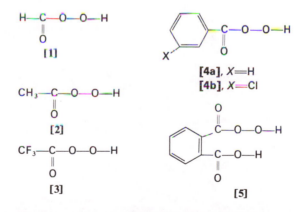

prepared by the addition of hydrogen peroxide to excess formic acid; the resulting solution, which loses oxygen on standing must be used immediately.

Equilibria comparable to that illustrated for performic acid occur with all the peracids. However, with most higher-molecular-weight acids it is necessary to

---

1. (a) D. Swern, *Chem. Rev.,* **45**, 1 (1949). (b) D. Swern, *Org. Reactions,* **7**, 378 (1953). (c) R. Criegee in E. Müller, ed., *Methoden der organischen Chemie (Houben-Weyl),* Vol. 8, Georg Thieme Verlag, Stuttgart, Germany, 1952, pp. 1–74. (d) S. N. Lewis in R. L. Augustine, ed., *Oxidation,* Vol. 1, Marcel Dekker, New York, 1969, pp. 213–258. (e) J. O. Edwards, ed., *Peroxide Reaction Mechanisms,* Wiley-Interscience, New York, 1962.

$$H{-}CO_2H + H_2O_2 \ \rightleftharpoons \ H{-}CO_3H + H_2O \qquad \text{(Ref. 2)}$$

(30% aqueous
solution)

add a mineral acid such as sulfuric acid as a catalyst to establish equilibrium rapidly.[1,3] Peracetic acid [2] is commercially available as a 40 per cent solution in acetic acid containing a small amount of sulfuric acid.[4] Similar solutions may be prepared by reaction of hydrogen peroxide with acetic acid in the presence of sulfuric acid or by reaction of hydrogen peroxide with acetic anhydride.[1] The concentrations of peracetic acid and hydrogen peroxide in such solutions may be determined by a double titration procedure.[5] The hydrogen peroxide content is measured by titrating a weighted sample of the peracid in cold aqueous sulfuric acid with a standard solution of ceric sulfate until the salmon color of Ferroin indicator is just discharged. The resulting solution is then treated with excess potassium iodide, and the iodine liberated by reaction of the peracid with iodide ion is titrated with standard sodium thiosulfate. The relative proportions of peracetic acid and acetic acid present in commercially available solutions may be determined from the n.m.r. spectrum of these solutions; the methyl peaks for peracetic acid and acetic acid are found at $\delta 2.22$ and 2.16, respectively. Since the solutions of peracetic acid (and other peracids) slowly lose oxygen on standing, standardization prior to use is always desirable. As will be discussed later in the chapter, the salts of peracids formed in alkaline solution decompose relatively rapidly.

Solutions of peracetic acid that do not contain acetic acid and a mineral acid have been obtained in relatively inert solvents such as ethyl acetate by air oxidation of acetaldehyde and subsequent thermal decomposition of the intermediate oxidation product [6].[6] Since this procedure is not readily adapted to laboratory

$$CH_3CHO \xrightarrow[\substack{CH_3CO_2C_2H_5 \\ 0°}]{O_2} CH_3{-}CH\overset{\text{OH}}{\underset{O{-}O{-}\overset{O}{\overset{\|}{C}}{-}CH_3}{\diagup}} \xrightarrow[\text{distil}]{100°} CH_3CO_3H \qquad \text{(Ref. 6)}$$

[6]   (as a solution
in $CH_3CO_2C_2H_5$)

preparations, however, an alternative method has been described in which water is distilled as its azeotrope with ethyl acetate to displace to the right the equilibrium illustrated in the accompanying equations. The residual liquid is then distilled under reduced pressure to separate a solution of peracetic acid in ethyl acetate.

2. (a) A. Roebuck and H. Adkins, *Org. Syn.,* **Coll. Vol. 3,** 217 (1955). (b) J. E. Horan and R. W. Schiessler, *ibid.,* **41,** 53 (1961).
3. (a) L. S. Silbert, E. Siegel, and D. Swern, *J. Org. Chem.,* **27,** 1336 (1962). (b) L. S. Silbert, E. Siegel, and D. Swern, *Org. Syn.,* **43,** 93 (1963). (c) E. Koubek and J. O. Edwards, *J. Org. Chem.,* **28,** 2157 (1963). (d) Y. Ogata and Y. Sawaki, *Tetrahedron,* **21,** 3381 (1965); **23,** 3327 (1967).
4. The FMC Corp., Inorganic Chemical Div., New York, N.Y.
5. F. P. Greenspan and D. G. MacKellar, *Anal. Chem.,* **20,** 1061 (1948).
6. (a) B. Phillips, F. C. Frostick, Jr., and P. S. Starcher, *J. Am. Chem. Soc.,* **79,** 5982 (1957). (b) P. S. Starcher, B. Phillips, and F. C. Frostick, Jr., *J. Org. Chem.,* **26,** 3568 (1961).

$$CH_3CO_2H + H_2O_2 \underset{\underset{50°}{CH_3CO_2C_2H_5}}{\overset{H_2SO_4}{\rightleftharpoons}} H_2O + CH_3CO_3H \qquad \text{(Ref. 7)}$$

$$\xrightarrow{\text{distil}} CH_3CO_3H$$
$$(87\% \text{ as a solution} \atop \text{in } CH_3CO_2C_2H_5)$$

Care must be exercised in this preparation to avoid very concentrated solutions of peracetic acid, which will detonate on impact. An alternative procedure involves the illustrated reaction of acetic anhydride with 90% hydrogen peroxide to give a solution of peracetic acid in 1,2-dichloroethane.

$$(CH_3CO)_2O + H_2O_2 \xrightarrow[\underset{0°}{ClCH_2CH_2Cl,}]{CF_3CO_2H \text{ (cat. amt.)}} CH_3CO_3H$$
$$(90\% \qquad\qquad\qquad\qquad (\text{a solution in}$$
$$\text{solution}) \qquad\qquad\qquad\qquad ClCH_2CH_2Cl) \qquad \text{(Ref. 8b)}$$

Peroxytrifluoroacetic acid [3], the most powerful oxidizing agent of the peracids previously listed, is prepared by an analogous reaction of trifluoroacetic anhydride with hydrogen peroxide in methylene chloride. Perbenzoic acid [4a]

$$(CF_3CO)_2O + H_2O_2 \xrightarrow{CH_2Cl_2} CF_3CO_2H + CF_3CO_3H \qquad \text{(Ref. 8)}$$
$$(90\% \text{ aqueous} \atop \text{solution})$$

$$(C_6H_5-CO-O-)_2 + NaOCH_3 \xrightarrow[\underset{0°}{CH_2Cl_2,}]{CH_3OH} C_6H_5CO_2CH_3 + C_6H_5CO_3Na$$
$$\text{[7]}$$

$$\Big\downarrow H_3O^\oplus$$

$$C_6H_5CO_3H$$
$$(83–86\% \text{ as a} \atop CH_2Cl_2 \text{ solution})$$
$$\text{(Ref. 9, 11a)}$$

has been prepared by the reaction of benzoyl peroxide [7] with methanolic sodium methoxide.[1,9] Although the early procedures specified the use of chloroform as a co-solvent, the rapid rate at which chloroform reacts with alkoxide ions[10] makes preferable a modification employing methylene chloride rather than chloroform.[11a] A more satisfactory preparative route involves the reaction of benzoyl peroxide[3d] or benzoyl chloride[3d,11b] with sodium hydroperoxide followed by acidification and

7. B. Phillips, P. S. Starcher, and B. D. Ash, *J. Org. Chem.*, **23**, 1823 (1958).
8. (a) W. D. Emmons, *J. Am. Chem., Soc.* **76**, 3468, 3470 (1954). (b) M. F. Hawthorne, W. D. Emmons, and K. S. McCullum, *ibid.*, **80**, 6393 (1958).
9. G. Braun, *Org. Syn.*, **Coll. Vol. 1**, 431 (1944).
10. J. Hine, *Physical Organic Chemistry*, 2d ed., McGraw-Hill, New York, 1962, p. 484.
11. (a) R. F. Kleinschmidt and A. C. Cope, *J. Am. Chem. Soc.*, **66**, 1929 (1944). (b) J. R. Moyer and N. C. Manley, *J. Org. Chem.*, **29**, 2099 (1964). (c) Aldrich Chemical Company Inc. (d) N. N. Schwartz and J. H. Blumbergs, *J. Org. Chem.*, **29**, 1976 (1964). (e) R. N. McDonald, R. N. Steppel, and J. E. Dorsey, *Org. Syn.*, **50**, 15 (1970).

$(C_6H_5\!-\!CO\!-\!O)_2 + NaOOH \xrightarrow[20°]{H_2O,\ CH_3OH}$

          (from NaOH
          and H_2O_2)

$C_6H_5CO_3H + C_6H_5CO_3^{\ominus}Na^{\oplus} \xrightarrow[H_2O]{H_2SO_4} C_6H_5CO_3H$

                                  (90% as a $CHCl_3$ solution)   (Ref. 3d)

extraction with chloroform. Perhaps the most convenient method of preparation
for perbenzoic acid (and other higher-molecular-weight peracids) is the reaction
of benzoic acid with hydrogen peroxide in methanesulfonic acid; the sulfonic acid

$C_6H_5\!-\!CO_2H + H_2O_2 \underset{25-30°}{\overset{CH_3SO_3H}{\rightleftharpoons}} H_2O + C_6H_5\!-\!CO_3H$ (Ref. 3b)

        (70%   (85–90%, as a
        aqueous   solution in C_6H_6)
        solution)

serves as both the solvent and the acid catalyst. The reaction mixture is diluted
with ice and water and the peracid is isolated by extraction (or filtration if the
peracid is a solid). One of the most convenient peracids to use is *m*-chloroper-
benzoic acid [4b] which is commercially available[11c] as a stable crystalline solid
containing 85% of the peracid. Washing this solid with an aqueous phosphate
buffer (pH 7.5) removes the *m*-chlorobenzoic acid contaminant to leave the pure
peracid which can be used in inert solvents such as 1,2-dichloroethane.[11d] A
convenient preparation of *m*-chloroperbenzoic acid from the corresponding acid
chloride is indicated below.

m-ClC_6H_4COCl + NaOOH $\xrightarrow[15-25°]{MgSO_4 \atop H_2O,}$

          (from NaOH
          and H_2O_2)

$m\text{-}Cl\!-\!C_6H_4\!-\!CO_3^{\ominus}Na^{\oplus} \xrightarrow[H_2O]{H_2SO_4} m\text{-}ClC_6H_4CO_3H$

                                 (isolated as a solution
                                 in CH_2Cl_2)   (Ref. 11e)

Monoperphthalic acid [5] is most conveniently prepared by the reaction of
phthalic anhydride with hydrogen peroxide in ether solution. A comparable

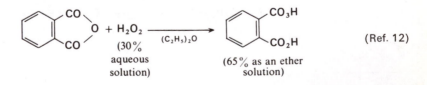

                    (30% $(C_2H_5)_2O$ (Ref. 12)
                    aqueous   (65% as an ether
                    solution)   solution)

12. E. E. Royals and L. L. Harrell, Jr., *J. Am. Chem. Soc.,* **77**, 3405 (1955): this procedure
is superior to the reaction of phthalic anhydride with sodium hydroperoxide described by
H. Böhme, *Org. Syn., Coll. Vol. 1,* 619 (1955); see also G. B. Payne, *ibid.,* **42**, 77 (1962).

method has been used to prepare monopermaleic acid [8]. This peracid appears to be a more powerful oxidant than most common peracids—with the exception of peroxytrifluoroacetic acid—and is useful for the oxidation of ketones and amines.[13]

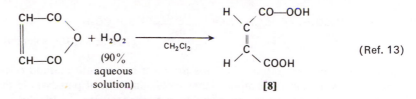

(Ref. 13)

That the peracids are substantially weaker than the corresponding carboxylic acids is illustrated by the pK_a values for formic acid (3.6), performic acid (7.1), acetic acid (4.8), and peracetic acid (8.2).[14] This fact permits the use of various inorganic buffers that do not react with the weakly acidic peracids but do react with the corresponding carboxylic acids, a technique illustrated both in the subsequent discussion and by the previously described purification of *m*-chloroperbenzoic acid [4b].[11d] Although carboxylic acids exist in solution as hydrogen-bonded dimers (e.g., [9]), the corresponding peracids appear to be hydrogen-bonded intramolecularly (e.g., [10]) and exist as monomers,[1,3,15] As a result, the peracids are more volatile than the corresponding carboxylic acids. In solvents containing relatively basic oxygen atoms (e.g. ethers, esters, and amides), the peracids exist as monomers hydrogen bonded to the solvent rather than as the intramolecularly hydrogen-bonded structures [10].[15b]

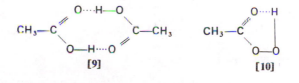

OXIDATION OF CARBON-CARBON DOUBLE BONDS

The reaction of olefins (e.g., [11]) with peracids to form epoxides (called epoxidation)[1] provides a convenient and selective method for the oxidation of carbon-carbon double bonds in the presence of hydroxyl (e.g., [12]) and carbonyl functions.

$$C_6H_5-CH=CH_2 + C_6H_5-CO_3H \xrightarrow[0°]{CHCl_3} C_6H_5-CH-CH_2$$

[11]

(Ref. 16)

(69–75%)

13. R. H. White and W. D. Emmons, *Tetrahedron,* **17**, 31 (1962).
14. A. J. Everett and G. S. Minkoff, *Trans. Faraday Soc.,* **49**, 410 (1953).
15. (a) B. M. Lynch and K. H. Pausacker, *J. Chem. Soc.,* 1525 (1955). (b) R. Kavcic and B. Plesnicar, *J. Org. Chem.,* **35**, 2033 (1970).
16. H. Hibbert and P. Burt, *Org. Syn.,* **Coll. Vol. 1,** 494 (1944).

Although amines are readily attacked by peracids, unsaturated amides (e.g., [13]) may be epoxidized without difficulty. Even the acid-sensitive oxaziranes may be obtained by the illustrated reaction of imines with peracids under carefully controlled conditions. The differences between this peracid reaction and the epoxidation of olefins will be discussed later in this chapter.

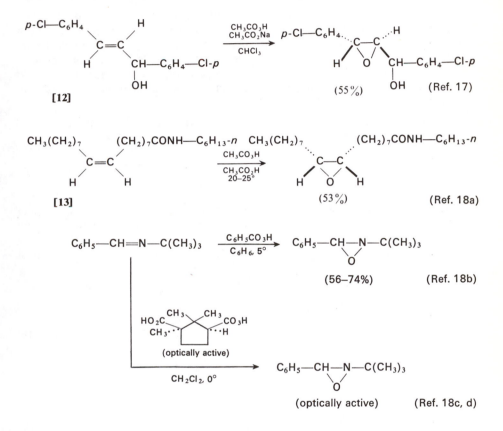

The epoxides are susceptible to attack by carboxylic acids to form the mono-esters of 1,2-diols, a reaction catalyzed by mineral acids.[1,19] Consequently, epoxidation reactions (e.g. with olefin [14]) run with a peracid in the presence of

17. H. O. House, *J. Am. Chem. Soc.*, **78**, 2298 (1956).
18. (a) E. T. Roe, J. T. Scanlan, and D. Swern, *J. Am. Chem. Soc.*, **71**, 2219 (1949). (b) W. D. Emmons and A. S. Pagano, *Org. Syn.*, **49**, 13 (1969). (c) D. R. Boyd and R. Graham, *J. Chem. Soc.*, C, 2648 (1969). (d) For examples in which optically active epoxides have been produced by the epoxidation of olefins with percamphoric acid, see F. Montanari, I. Moretti, and G. Torre, *Chem. Commun.*, **No. 3**, 135 (1969).
19. (a) R. E. Parker and N. S. Isaac, *Chem. Rev.*, **59**, 737 (1959). (b) S. Winstein and R. B. Henderson in R. C. Elderfield, ed., *Heterocyclic Compounds*, Vol. 1, Wiley, New York, 1950, pp. 1–60.

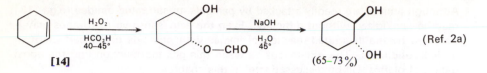

[14]   (65–73%)   (Ref. 2a)

an excess of the corresponding carboxylic acid frequently yield the hydroxy esters derived from the initially formed epoxide. This is especially true for reactions run in formic acid and for reaction mixtures that contain a strongly acidic mineral acid. Reactions run either with monopermaleic acid in methylene chloride[13] or with mixtures of peroxytrifluoroacetic acid and the strongly acidic trifluoroacetic acid (pK_a 0.3) in methylene chloride[20] (e.g., [15]) also usually produce 1,2-diol

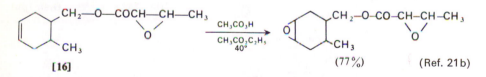

[15]   (mixture)   (Ref. 20a)

deratives. However, the use of perbenzoic acid or m-chloroperbenzoic acid (e.g. with [16]) in chloroform or methylene chloride, of monoperphthalic acid in ether, or of peracetic acid in ethyl acetate[21] permits the isolation of the initially formed epoxide. Reaction of the olefin (e.g., [17]) with a mixture of the commercial

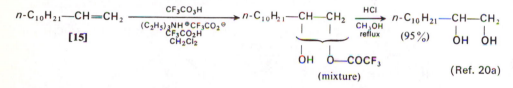

[16]   (77%)   (Ref. 21b)

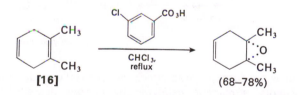

[16]   (68–78%)   (Ref. 22c)

20. (a) W. D. Emmons, A. S. Pagano, and J. P. Freeman, *J. Am. Chem. Soc.*, **76**, 3472 (1954). (b) W. D. Emmons and A. S. Pagano, *ibid.*, **77**, 89 (1955).
21. (a) D. L. MacPeek, P. S. Starcher, and B. Phillips, *J. Am. Chem. Soc.*, **81**, 680 (1959). (b) P. S. Starcher, F. C. Frostick, Jr., and B. Phillips, *J. Org. Chem.*, **25**, 1420 (1960).
22. (a) D. J. Reif and H. O. House, *Org. Syn.*, **Coll. Vol. 4**, 860 (1963). (b) M. Korach, D. R. Nielsen, and W. H. Rideout, *ibid.*, **42**, 50 (1962). (c) L. A. Paquette and J. H. Barrett, *ibid.*, **49**, 62 (1969). (d) also see T. H. Kinstle and P. J. Ihrig, *J. Org. Chem.*, **35**, 257 (1970). (e) R. L. Camp and F. D. Greene, *J. Am. Chem. Soc.*, **90**, 7349 (1968). (f) J. K. Crandall and W. H. Machleder, *ibid.*, **90**, 7292, 7347 (1968); J. K. Crandall, W. H. Machleder, and M. J. Thomas, *ibid.*, **90**, 7346 (1968). (g) J. Grimaldi and M. Bertrand, *Tetrahedron Letters*, **No. 38**, 3269 (1969).

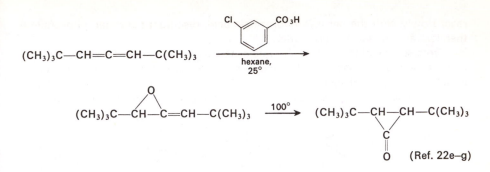

$(CH_3)_3C—CH—C=CH—C(CH_3)_3$ $\xrightarrow{100°}$ $(CH_3)_3C—CH—CH—C(CH_3)_3$ (Ref. 22e–g)

40 per cent peracetic acid–acetic acid reagent plus sodium acetate (to neutralize the sulfuric acid present) in methylene chloride solution also permits isolation of epoxides. Although the latter procedure is less satisfactory than that utilizing

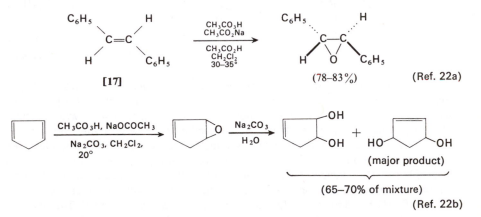

[17]   (78–83%)   (Ref. 22a)

(major product)

(65–70% of mixture)

(Ref. 22b)

a pure peracid in an inert solvent, it is nonetheless frequently used because of the ready commercial availability of the peracetic acid–acetic acid mixture. In certain cases peracetic acid has decided advantages over perbenzoic acid or mono-perphthalic acid because both the peracid and the corresponding carboxylic acid are volatile and can readily be removed.

By the addition of anhydrous sodium carbonate to a reaction mixture containing peroxytrifluoroacetic acid and trifluoroacetic acid in methylene chloride, the concentration of the trifluoroacetic acid (which reacts immediately with the sodium carbonate) may be lowered, permitting the successful conversion of olefins (e.g., [18]) to epoxides.[20b] However, the peroxytrifluoroacetic acid also reacts slowly with the sodium carbonate to form a salt (which decomposes); with olefins that

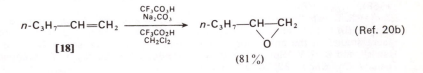

(Ref. 20b)

[18]

(81%)

react slowly with the perviously cited peroxytrifluoroacetic acid, disodium hydrogen phosphate is therefore a more satisfactory buffer.

From several of the perviously cited examples it will be noted that the epoxidation reaction is stereospecific, leading to a *cis* addition of the oxygen atom to the double bond. The subsequent opening of the epoxide (e.g., **[19]**) by reaction with some nucleophilic reagent usually proceeds with inversion of configuration at the

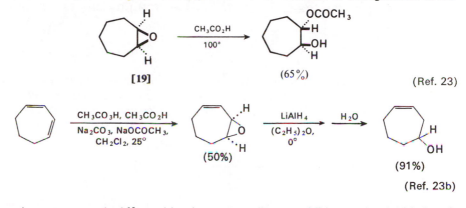

[19]   (65%)   (Ref. 23)

(50%)   (91%)

(Ref. 23b)

carbon atom attacked,[19] resulting in an over-all *trans* addition to the double bond. Thus, the preparation of *trans*-1,2-diols by this sequence complements the previously discussed oxidations with alkaline potassium permanganate and with osmium tetroxide, both of which result in *cis* hydroxylation. The epoxide ring may be opened by a variety of nucleophilic reagents,[19] among them metal hydrides (see Chapter 2), hydrogen halides, enolate anions (e.g., **[20]**), and aqueous acids (e.g., **[21]**). The presence, in the molecule being epoxidized, of a hydroxyl, ketone, or carboxyl function may lead to the intramolecular displacement of the epoxide C—O bond to form a hydroxy ether or a hydroxy lactone.[24] The opening

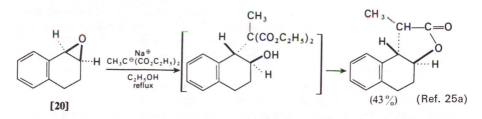

[20]   (43%)   (Ref. 25a)

23. (a) L. N. Owen and G. S. Saharia, *J. Chem. Soc.,* 2582 (1953). (b) J. K. Crandall, D. B. Banks, R. A. Colyer, R. J. Watkins, and J. P. Arrington, *J. Org. Chem.,* **33,** 423 (1968).
24. For examples, see (a) G. Berti, *J. Org. Chem.,* **24,** 934 (1959). (b) Y. Gaoni, *J. Chem. Soc.,* C, 2925, 2934 (1968). (c) J. A. Marshall and M. T. Pike, *J. Org. Chem.,* **33,** 435 (1968).
25. (a) E. E. van Tamelen, G. Van Zyl, and G. D. Zuidema, *J. Am. Chem. Soc.,* **72,** 488 (1950). (b) H. B. Henbest, M. Smith, and A. Thomas, *J. Chem. Soc.,* 3293 (1958). (c) For recent studies of the analogous reaction with styrene oxide, see P. M. G. Bavin, D. P. Hansell, and R. G. W. Spickett, *ibid.,* 4535 (1964) ; C. H. DePuy, F. W. Breitbeil, and K. L. Eilers, *J. Org. Chem.,* **29,** 2810 (1964).

of the three-membered ring in compound **[21]** is of particular interest since it illustrates the general rule[26] that a cyclohexene epoxide ring is opened predominantly in such a way as to produce two new substituents, which are *trans* and diaxial (as in **[22]**) rather than *trans* and diequatorial (as in **[23]**). The same phenomenon is seen in the reaction of the epoxide **[24]** with hydrogen bromide.

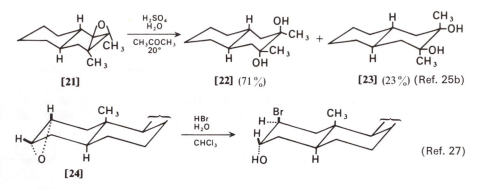

[21]   [22] (71%)   [23] (23%) (Ref. 25b)

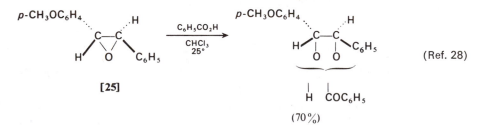

[24]   (Ref. 27)

In certain cases (e.g., **[25]**) where at least one aryl substituent is bonded to the epoxide ring, opening of the heterocyclic ring with retention of configuration has

been observed. In other similar systems (e.g., **[26]**), ring opening with either retention or inversion of configuration may be obtained by the appropriate choice of reaction conditions.[29] The direction of ring opening when an unsymmetrical

26. (a) D. H. R. Barton and R. C. Cookson, *Quart. Rev.,* **10**, 44 (1956). (b) E. L. Eliel in M. S. Newman, ed., *Steric Effects in Organic Chemistry,* Wiley, New York, 1956, pp. 130–134. (c) E. L. Eliel, *Stereochemistry of Carbon Compounds,* McGraw-Hill, 1962, pp. 229–231.
27. G. H. Alt and D. H. R. Barton, *J. Chem. Soc.,* 4284 (1954).
28. (a) D. Y. Curtin, A. Bradley, and Y. G. Hendrickson, *J. Am. Chem. Soc.,* **78**, 4064 (1956). (b) J. H. Brewster, *ibid.,* **78**, 4061 (1956). (c) G. Berti, F. Bottari, and B. Macchia, *Tetrahedron,* **20**, 545 (1964). (d) W. E. Rosen, L. Dorfman, and M. P. Linfield [*J. Org. Chem.,* **29**, 1723 (1964)] have reported that indene reacts with performic acid to form *cis*-2-formyloxy-1-hydroxyindane and suggest a concerted addition of the elements of performic acid to the olefin.
29. (a) H. O. House, *J. Org. Chem.,* **21**, 1306 (1956). (b) C. C. Tung and A. J. Speziale, *ibid.,* **28**, 2009 (1963). (c) H. H. Wasserman and N. E. Aubrey, *J. Am. Chem. Soc.,* **78**, 1726 (1956). (d) G. Berti, B. Macchia and F. Macchia, *Tetrahedron,* **24**, 1755 (1968); G. Bertie, F. Bottari, G. Lippi, and B. Macchia, *ibid.,* **24**, 1959 (1968); and references cited in these papers. (e) J. R. Doherty and co-workers, *ibid.,* **26**, 2545 (1970).

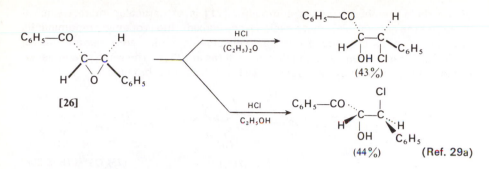

[26]

(43%)

(44%)   (Ref. 29a)

epoxide is allowed to react with a nucleophile is also determined, at least in part, by the reaction conditions. In general, nucleophiles attack at the less highly substituted carbon atom of the epoxide ring in neutral or basic media, as would be anticipated for a normal S_N2 process; in acidic media, the proportion of attack at the more highly substituted carbon is increased[19] and may become the predominant reaction. In cases where equal numbers of substituents are present at each carbon of the epoxide (e.g., [20]), attack of the nucleophile at an allylic or benzylic position is favored.

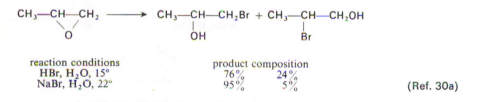

reaction conditions	product composition	
HBr, H_2O, 15°	76%	24%
NaBr, H_2O, 22°	95%	5%

(Ref. 30a)

The epoxidation of olefins is believed to proceed by an electrophilic attack, as indicated in the accompanying equations.[15,30b] In accord with this mechanism, the peracid usually attacks the olefin (e.g., [27]) from its less hindered side to produce

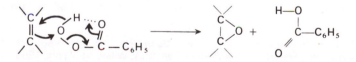

the less hindered epoxide as the major product. The stereospecificity of the epoxidation may be influenced by changes in the reaction solvent, the effective steric bulk of the peracid in the transition state being somewhat greater in relatively

---

30. (a) C. A. Stewart and C. A. Vanderwerf, *J. Am. Chem. Soc.*, **76**, 1259 (1954). (b) Although an alternative mechanism involving a 1,3-dipolar addition has been discussed, the presently available data are in better agreement with the reaction path indicated in the accompanying equation. H. Kwart, P. S. Starcher, and S. W. Tinsley, *Chem. Commun.*, No. **7**, 335 (1967); A. Azman, B. Borstnik, and B. Plesnicar, *J. Org. Chem.*, **34**, 971 (1969).

polar solvents such as methylene chloride or chloroform.[31e,f] However, the epoxidations of allylic alcohols and olefins such as [29] illustrate the fact that the direction of attack by the peracid may be influenced by nearly polar substituents.[32]

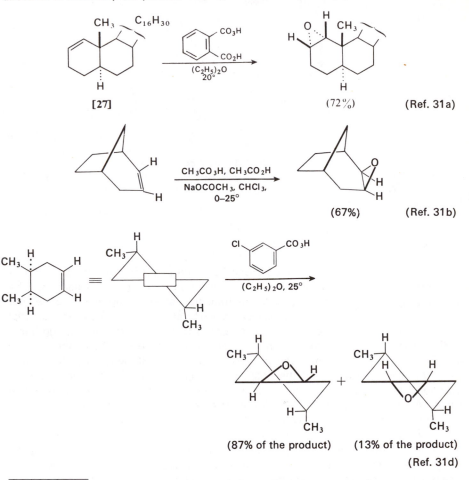

[27]   (72%)   (Ref. 31a)

(67%)   (Ref. 31b)

(87% of the product)   (13% of the product)

(Ref. 31d)

31. (a) H. B. Henbest and R. A. L. Wilson, *J. Chem. Soc.*, 3289 (1956). (b) R. R. Sauers, H. M. How, and H. Feilich, *Tetrahedron*, **21**, 983 (1965). (c) For a study of the effect of steric hindrance from alkyl substituents on the rate of olefin epoxidation, see M. S. Newman, N. Gill, and D. W. Thomson, *J. Am. Chem. Soc.*, **89**, 2059 (1967). (d) B. Rickborn and S. Y. Lwo, *J. Org. Chem.*, **30**, 2212 (1965). (e) R. C. Ewins, H. B. Henbest, and M. A. McKervey, *Chem. Commun.*, **No. 21**, 1085 (1967). (f) R. G. Carlson and N. S. Behn, *J. Org. Chem.*, **32**, 1363 (1967).
32. (a) H. B. Henbest and J. J. McCullough, *Proc. Chem. Soc.*, 74 (1962). (b) H. B. Henbest, *ibid.*, 159 (1963). (c) N. S. Crossley, A. C. Darby, H. B. Henbest, J. J. McCullough, B. Nicholls, and M. F. Stewart, *Tetrahedron Letters*, **No. 12**, 398 (1961). (d) H. B. Henbest and R. A. L. Wilson, *J. Chem. Soc.*, 1958 (1957). (e) G. I. Fray, R. J. Hilton, and J. M. Teice, *ibid.*, C, 592 (1966). (f) A. P. Gray and D. E. Heitmeier, *J. Org. Chem.*, **30**, 1226 (1965). (g) P. Chamberlain, M. L. Roberts, and G. H. Whitham, *J. Chem. Soc.*, B, 1374 (1970) and references therein.

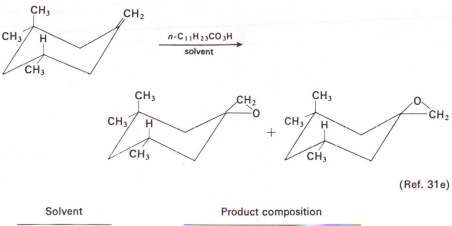

(Ref. 31e)

Solvent	Product composition	
$(C_2H_5)_2O$	65%	35%
CCl_4	75%	25%
C_6H_6	80%	20%
CH_2Cl_2 or $CHCl_3$	83%	17%

The directive effect of the hydroxyl group has been suggested to arise because of hydrogen bonding between the hydroxyl group and the attacking peracid. The optimum geometry for this epoxidation transition state is suggested to be that illustrated in structure [28] obtained from an allylic alcohol with a pseudoequatorial hydroxyl group.[32g] With olefins such as [29] in which a hydrogen bonding inter- action is not important, the electrostatic effect of the polar substituent (i.e. the cyano group in [29]) is believed to favor a transition state leading to the isomer with the epoxide group and the polar electronegative substituent *trans*. Such a reaction path minimizes an unfavorable electrostatic interaction in the transition state. As the accompanying equation illustrates, this electrostatic factor becomes less important in determining product stereochemistry as solvents of relatively high dielectric constant (e.g. acetonitrile) are used.

The epoxidation reaction is not catalyzed by acids[15,16b] and, as noted previously, the presence of strong acids in the reaction mixture is usually undesirable. Also, the use of solvents capable of hydrogen bonding to the peracid retards the epoxidation of olefins.[11d,e] However, the rate of epoxidation is enhanced by the presence either of electron-withdrawing groups in the peracid (e.g. CF_3CO_3H is more

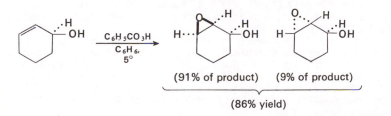

(91% of product)   (9% of product)

(86% yield)

(Ref. 32b, g)

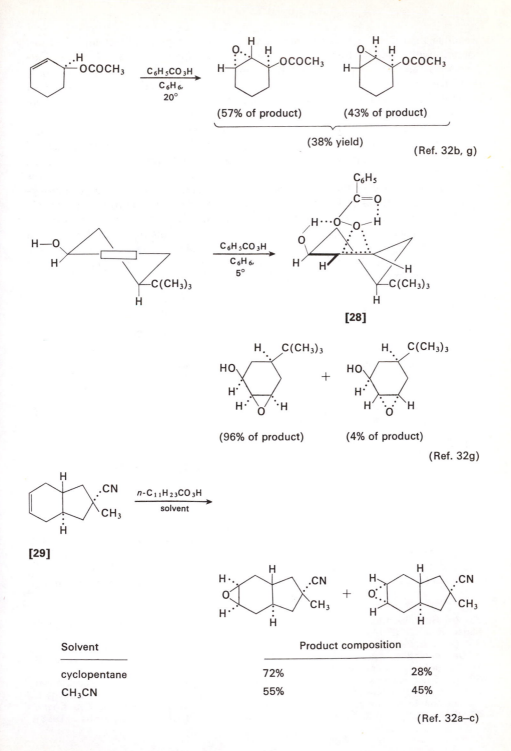

(57% of product)   (43% of product)

(38% yield)

(Ref. 32b, g)

[28]

(96% of product)   (4% of product)

(Ref. 32g)

[29]

Solvent	Product composition	
cyclopentane	72%	28%
CH$_3$CN	55%	45%

(Ref. 32a–c)

reactive than CH_3CO_3H) or of electron-donating groups in the olefin.[1,15] Thus, olefins with three or four alkyl substituents are rapidly epoxidized by peracids whereas terminal, monosubstituted olefins react very slowly[1] unless a highly reactive peracid such as peroxytrifluoroacetic acid is used.[22,31c] This effect is well illustrated by the previously described selective epoxidation of only the more highly substituted double bond in the diene **[16]**. Conjugation of the olefin with aromatic rings or with other multiple bonds also reduces the rate of epoxidation, since the delocalization of pi electrons possible in the conjugated systems reduces the electron density at the double bond undergoing electrophilic attack.[15] Although the rate of reaction of α,β-unsaturated esters with peracids is relatively slow, it is possible to epoxidize the double bond in these compounds either with peroxy-trifluoroacetic acid (e.g., **[30]**), with peracetic acid, or with m-chloroperbenzoic acid (e.g., **[31]**). On the other hand, the reaction of α,β-unsaturated ketones with peracids usually does not lead to epoxidation of the double bond; as illustrated in the accompanying equations, reaction with the double bond is retarded sufficiently that the subsequently discussed reaction of the peracid with the ketone usually becomes the predominant process.[33,34]

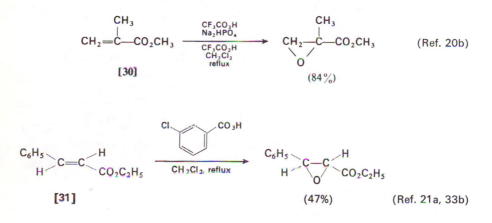

However, α,β-unsaturated ketones (e.g., **[32]**) can be epoxidized, using nucleophilic reagents such as the sodium salt of hydrogen peroxide (NaOOH)[35a] or the sodium salt of t-butyl hydroperoxide[35b] rather than a peracid. The reaction is believed to proceed by nucleophilic addition of the hydroperoxide anion at the

33. (a) H. M. Walton, *J. Org. Chem.*, **22**, 1161 (1957). (b) V. R. Valente and J. L. Wolf-hagen, *ibid.*, **31**, 2509 (1966).
34. (a) G. B. Payne and P. H. Williams, *J. Org. Chem.*, **24**, 284 (1959). (b) M. Gorodetsky, N. Danieli, and Y. Mazur, *ibid.*, **32**, 760 (1967). (c) C. R. Zanesco, *Helv. Chim. Acta*, **49**, 1002 (1966). (d) For an exception to this generality, see Ref. 40i.
35. (a) Since such salts of hydrogen peroxide are thermally unstable, it is necessary to control the temperature of reactions in which they are used. For a study of the stability of alkaline solutions of hydrogen peroxide, see W. D. Nicoll and A. F. Smith, *Ind. Eng. Chem.*, **47**, 2548 (1955). (b) N. C. Yang and R. A. Finnegan, *J. Am. Chem. Soc.*, **80**, 5845 (1958).

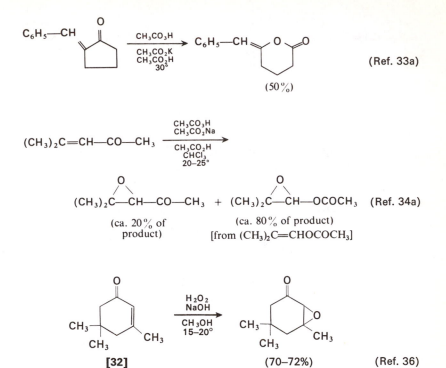

(Ref. 33a)

(50%)

(Ref. 34a)

[32]   (70–72%)   (Ref. 36)

beta carbon of the unsaturated ketone followed by intramolecular displacement of hydroxide ion, as illustrated.[37]

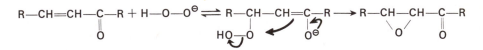

Similar reaction conditions have been used for the epoxidation of alkylidene-malonic esters[38] and α,β-unsaturated aldehydes (e.g., [33]) as well as α,β-unsaturated nitro compounds, α,β-unsaturated sulfones, and vinyl phosphonates.[39]

Epoxidation with alkaline hydrogen peroxide differs from peracid epoxidation in that the former is not stereospecific (i.e. the stereochemistry of the reactant and

36. R. L. Wasson and H. O. House, *Org. Syn.*, **Coll. Vol. 4**, 552 (1963).
37. C. A. Bunton and G. J. Minkoff, *J. Chem. Soc.*, 665 (1949).
38. G. B. Payne, *J. Org. Chem.*, **24**, 2048 (1959).
39. (a) G. B. Payne, *J. Org. Chem.*, **25**, 275 (1960). (b) G. B. Payne, *J. Am. Chem. Soc.*, **81**, 4901 (1959). (c) C. E. Griffin and S. K. Kundu, *J. Org. Chem.*, **34**, 1532 (1969). (d) H. Newman and R. B. Angier, *Tetrahedron*, **26**, 825 (1970). (e) B. Zwanenburg and J. ter Wiel, *Tetrahedron Letters*, **No. 12**, 935 (1970). (f) T. Durst and K. C. Tin, *ibid.*, **No. 27**, 2369 (1970); also see D. F. Tavares, R. E. Estep, and M. Blezard, *ibid.*, **No. 27**, 2373 (1970).

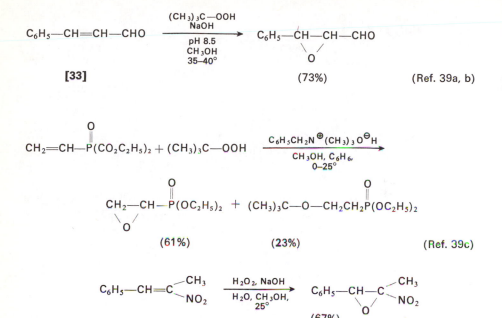

the product do not bear a definite relationship to one another) but is stereoselective (i.e., a single stereoisomer of the product is formed, which bears no definite relationship to the stereochemistry of the reactant).[40a,b,i] Thus, epoxidation of either isomer of the unsaturated ketone [34] yields a single epoxyketone in which

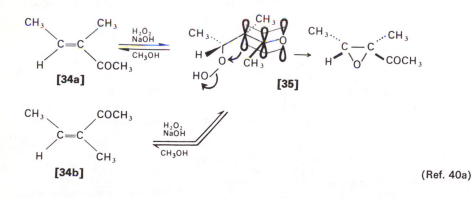

(Ref. 40a)

40. (a) H. O. House and R. S. Ro, *J. Am. Chem. Soc.*, **80**, 2428 (1958). (b) H. E. Zimmerman, L. Singer, and B. S. Thyagarajan, *ibid.*, **81**, 108 (1959). (c) H. B. Henbest and W. R. Jackson, *J. Chem. Soc.*, C, 2459 (1967). (d) H. B. Henbest, W. R. Jackson, and I. Malunowicz, *ibid.*, C, 2469 (1967). (e) H. O. House and R. L. Wasson, *J. Org. Chem.*, **22**, 1157

the ketone function and the larger group on the beta carbon atom are *trans* to one another. This stereoselectivity has been justified on the basis that the intermediate enolate anion [35] will be more stable if it is not eclipsed with a large *beta* substituent in the transition state, leading to formation of the three-membered ring.[40b] It should also be noted that, at least for the example cited, the *cis* starting material [34b] is isomerized to the more stable *trans* isomer [34a] in the reaction mixture at approximately the same rate as it is epoxidized; consequently, stereospecific epoxidation would not be expected in any case. Just as was observed in the previously described epoxidations of olefins with peracids, the stereochemistry of epoxidation of enones (e.g., [36]) with sodium hydroperoxide is influenced by the presence of remote polar substituents. This effect is presumably attributable to an electrostatic interaction between the polar substituent and the intermediate enolate anion (e.g., [35]). Since the formation of these stereoisomeric intermediate enolate anions from [36] is reversible, such an electrostatic interaction could influence the relative rates at which the intermediates either give epoxy ketone products or return to starting materials.

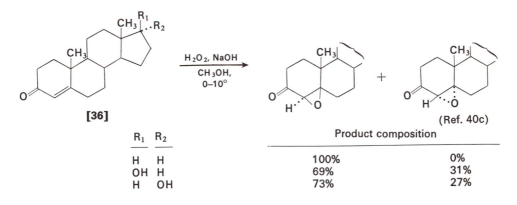

[36]

H_2O_2, NaOH
CH_3OH,
0–10°

(Ref. 40c)

Product composition

R_1	R_2		
H	H	100%	0%
OH	H	69%	31%
H	OH	73%	27%

In cases where the carbon-carbon double bond of the enone is highly substituted and the carbonyl group is relatively reactive, or where reaction with sodium hydroperoxide is continued for relatively long periods of time, the usual epoxide formation may be accompanied by or superseded by oxidative cleavage. This cleavage, illustrated by the following examples, appears to involve the subsequently discussed attack of the hydroperoxide anion at the carbonyl function.[44e–h] The epoxy ketone products from a normal epoxidation may also be cleaved by further reaction with alkaline hydrogen peroxide.[40k]

(1957); D. L. Coffen and D. G. Korzan, *ibid.,* **36**, 390 (1971). (f) S. M. Marmor and M. M. Thomas, *ibid.,* **32**, 252 (1967). (g) W. S. Johnson, B. Bannister, R. Pappo, and J. E. Pike, *J. Am. Chem. Soc.,* **78**, 6354 (1956). (h) S. D. Levine, *J. Org. Chem.,* **31**, 3189 (1966). (i) D. D. Keane, W. I. O'Sullivan, E. M. Philbin, R. M. Simons, and P. C. Teague, *Tetrahedron,* **26**, 2533 (1970). (j) For oxidative degradations of enones with a mixture of hydrogen peroxide and selenium dioxide, see E. Caspi and Y. Shimizu, *J. Org. Chem.,* **30**, 223 (1965). (k) R. D. Temple, *ibid.,* **35**, 1275 (1970).

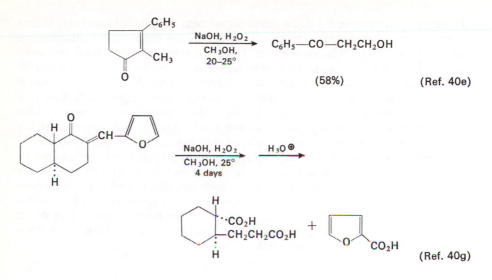

(Ref. 40g)

The reaction of α,β-unsaturated nitriles (e.g., **[37]**) with alkaline hydrogen peroxide usually yields epoxyamides. The preparation of tetracyanoethylene epoxide indicated in the accompanying equations represents an exceptional case in which the highly electron-deficient double bond is attacked by neutral hydrogen peroxide.[41f] Although the epoxy amide products would appear to result from the

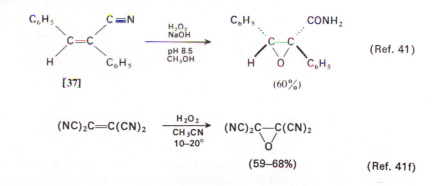

41. (a) G. B. Payne and P. H. Williams, *J. Org. Chem.*, **26**, 651 (1961). (b) G. B. Payne, P. H. Deming, and P. H. Williams, *ibid.*, **26**, 659 (1961). (c) G. B. Payne, *ibid.*, **26**, 663, 668 (1961). (d) G. B. Payne, *Tetrahedron*, **18**, 763 (1962). (e) Y. Ogata and Y. Sawaki, *ibid.*, **20**, 2065 (1964). (f) W. J. Linn, *Org. Syn.*, **49**, 103 (1969). (g) The sodium tungstate-catalyzed epoxidation of α,β-unsaturated nitriles has been studied by M. Igarashi and H. Midorikawa, *J. Org. Chem.*, **32**, 3399 (1967). (h) Olefins have also been epoxidized by reaction with peroxides in the presence of vanadium salts [E. S. Gould, R. R. Hiatt, and K. C. Irwin, *J. Am. Chem. Soc.*, **90**, 4573 (1968)], tungstic acid [H. C. Stevens and A. J. Kaman, *ibid.*, **87**, 734 (1965)], and borate esters [P. F. Wolf and R. K. Barnes, *J. Org. Chem.*, **34**, 3441 (1969)]. (i) The product from phenylisocyanate and hydrogen peroxide has been used to epoxidize olefins; N. Matsumura, N. Sonoda, and S. Tsutsumi, *Tetrahedron Letters*, **No. 23**, 2029 (1970).

previously discussed base-catalyzed epoxidation and a subsequent known[42] (e.g., [38]) rapid conversion of the epoxynitrile to the epoxyamide by further reaction with the salt of hydrogen peroxide, detailed examination revealed a very

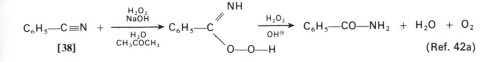

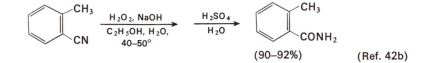

different reaction course.[41] The probable path, illustrated in the accompanying equations, involves a peroxyimidic acid intermediate [39], which acts as an electrophilic reagent. This intermediate either may be intercepted by a more reactive olefin such as cyclohexene to produce cyclohexene epoxide and the unsaturated amide, or may react with more of the hydroperoxide anion to form the unsaturated amide plus oxygen. Although the origin of the oxygen formed in the

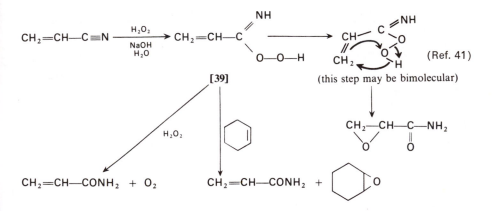

process is not certain, it could arise from reaction of the peroxy imidic acid with hydrogen peroxide to form the amide, water, and the subsequently discussed singlet oxygen. The intermediate peroxyimidic acid derived from benzonitrile has been used to epoxidize olefins (e.g., [40]) under neutral conditions. This procedure would appear to be useful for the preparation of epoxides that are very unstable to acid, in spite of the difficulty of separating the product from benzamide. An accompanying equation also illustrates the increased tendency of peroxybenzimidic

42. (a) K. B. Wiberg, *J. Am. Chem. Soc.*, **75**, 3961 (1953); **77**, 2519 (1955); (b) C. R. Noller, *Org. Syn.*, **Coll. Vol. 2**, 586 (1943). (c) J. S. Buck and W. S. Ide, *ibid.*, **Coll. Vol. 2**, 44 (1943).

acid to attack an exocyclic olefin to form an epoxide with an equatorial oxygen atom.[31f] These results suggest that the peroxybenzimidic acid has a greater steric bulk than a peracid and consequently attack from an axial direction is less favorable (see Chapter 2, structures [22] and [23]).

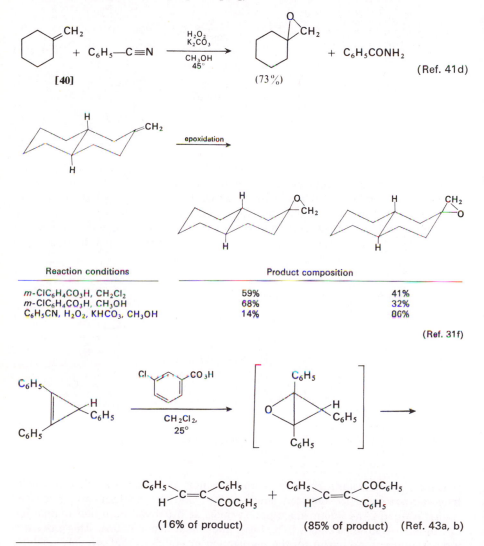

Reaction conditions	Product composition	
m-ClC$_6$H$_4$CO$_3$H, CH$_2$Cl$_2$	59%	41%
m-ClC$_6$H$_4$CO$_3$H, CH$_3$OH	68%	32%
C$_6$H$_5$CN, H$_2$O$_2$, KHCO$_3$, CH$_3$OH	14%	86%

(Ref. 31f)

(16% of product)   (85% of product)   (Ref. 43a, b)

43. (a) L. E. Friedrick and R. A. Cormier, *J. Org. Chem., 35*, 450 (1970). (b) J. Ciabattoni and P. J. Kocienski, *J. Am. Chem. Soc., 91*, 6534 (1969). (c) R. N. McDonald and P. A. Schwab, *ibid., 85*, 820, 4004 (1963). (d) R. N. McDonald and T. E. Tabor, *ibid., 89*, 6573 (1967) ; *J. Org. Chem., 33*, 2934 (1968). (e) A. Hassner and P. Catsoulacos, *ibid., 32*, 549 (1967). (f) R. N. McDonald and R. N. Steppel, *ibid., 35*, 1250 (1970) ; *J. Am. Chem. Soc., 92*, 5664 (1970). (g) J. Ciabattoni, R. A. Campbell, C. A. Renner, and P. W. Concannon, *ibid., 92*, 3826 (1970).

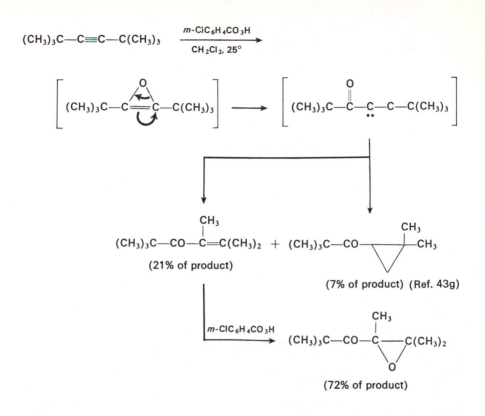

(21% of product)

(7% of product) (Ref. 43g)

(72% of product)

Acetylenes, highly strained olefins, and olefins with chlorine substituents, alkoxy substituents (i.e., enol ethers), or acyloxy substituents (i.e. enol esters) are readily epoxidized with peracids. However, the products of these reactions are unstable, especially to acid and heat, and frequently undergo further change in the reaction mixture. The following examples are illustrative. The reaction of a peracid with acetylenes affords primary products which appear to be derived from the indicated intermediate keto carbene.[43g] Whether an acetylene epoxide is a true intermediate preceding this keto carbene is not certain. Reaction of the chloro-olefin [41] with peracid produced the chloroketone [42] by initial formation of a

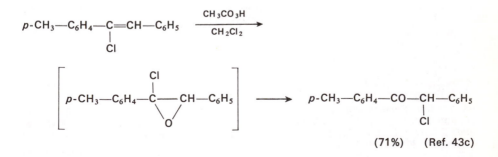

(71%)   (Ref. 43c)

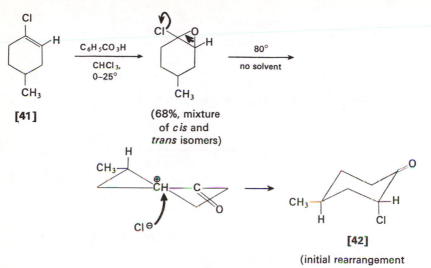

[41]

(68%, mixture
of *cis* and
trans isomers)

[42]

(initial rearrangement
product)   (Ref. 43d)

chloroepoxide and subsequent rearrangement. As indicated in the accompanyng equation, various lines of evidence indicate that this rearrangement with chlorine migration proceeds by the formation and subsequent collapse of an ion pair composed of a chloride anion and an α-keto carbonium ion.[43c,d,f] Although certain enol ethers (e.g., [43]) have been successfully converted to epoxy ethers by the use of low temperatures and short reaction times, other enol ethers (e.g., [44]) give intermediates which appear to react with the carboxylic acid in the reaction mixture too rapidly to permit isolation. As indicated in the following equation this type of reaction may be used with a bicyclic enol ether to form a macrocyclic lactone.

$$C_6H_5-C=C(CH_3)_2 \xrightarrow[\substack{(C_2H_5)_2O \\ 0°, 30\ sec}]{C_6H_5CO_3H} C_6H_5-C-C(CH_3)_2$$

with OC_2H_5 below

[43]   (70%)   (Ref. 44a, b)

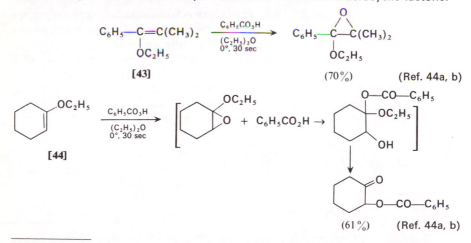

[44]

(61%)   (Ref. 44a, b)

44. (a) C. L. Stevens and J. Tazuma, *J. Am. Chem. Soc.*, **76**, 715 (1954). (b) For a discussion of the epoxy ether–carboxylic acid reaction, see C. L. Stevens and S. J. Dykstra, *ibid.*, **75**, 5975 (1953). (c) I. J. Borowitz, G. J. Williams, L. Gross, and R. Rapp, *J. Org. Chem.*, **33**, 2013 (1968).

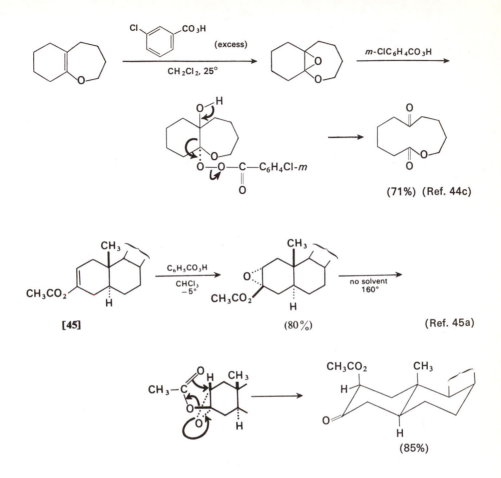

(71%) (Ref. 44c)

[45]   (80%)   (Ref. 45a)

(85%)

The epoxidation of aliphatic enol acetates (e.g., [45]) leads to the formation of epoxyacetates, which undergo either an intramolecular thermal isomerization or an acid-catalyzed isomerization to form α-acetoxyketones. Enol acetates with aryl substituents (e.g., [46]) appear to undergo rearrangement more readily since acetoxyketones have been isolated directly from the epoxidation mixture. By an appropriate choice of reaction conditions, it is possible to convert an unsymmetrical ketone (e.g., [47]) predominantly to either the more highly substituted

45. (a) K. L. Williamson and W. S. Johnson, *J. Org. Chem.*, **26**, 4563 (1961). (b) P. D. Gardner, *J. Am. Chem. Soc.*, **78**, 3421 (1956). (c) A. H. Soloway, W. J. Considine, D. K. Fukushima, and T. F. Gallagher, *ibid.*, **76**, 2941 (1954). (d) N. S. Leeds, D. K. Fukushima, and T. F. Gallagher, *ibid.*, **76**, 2943 (1954). (e) H. J. Shine and G. E. Hunt, *ibid.*, **80**, 2434 (1958). (f) A. L. Draper, W. J. Heilman, W. E. Schaefer, H. J. Shine, and J. N. Shoolery, *J. Org. Chem.*, **27**, 2727 (1962). (g) K. L. Williamson, J. I. Coburn, and M. F. Herr, *ibid.*, **32**, 3934 (1967). (h) D. N. Kirk, and J. M. Wiles, *Chem. Commun.*, **No. 9**, 518 (1970); **No. 16**, 1015 (1970).

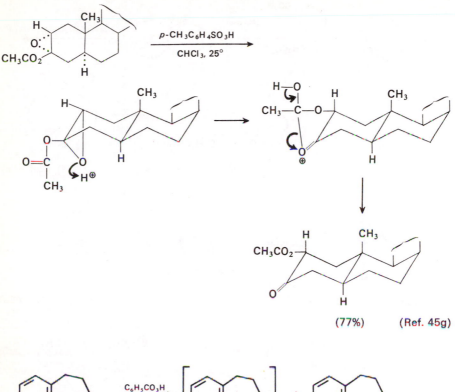

(77%)   (Ref. 45g)

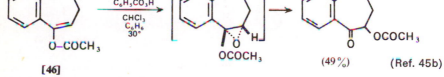

[46]

(49%)   (Ref. 45b)

or the less highly substituted enol acetate.[46] Consequently, by preparation of the appropriate enol acetate and subsequent epoxidation and rearrangement it is possible to introduce an α-acetoxyl substituent on either side of the carbonyl function in an unsymmetrical ketone.

The dienol ethers or dienol acetates derived from α,β-unsaturated ketones are usually attacked by electrophilic agents at the double bond not attached to the oxygen function (see Chapter 9). Although this generalization appears applicable to the

46. (a) H. O. House and V. Kramar, *J. Org. Chem.*, **28**, 3362 (1963); H. O. House and B. M. Trost, *ibid.*, **30**, 1341, 2502 (1965); H. O. House, L. J. Czuba, M. Gall, and H. D. Olmstead, *ibid.*, **34**, 2324 (1969). (b) W. B. Smith and T. K. Chen, *ibid.*, **30**, 3095 (1965). (c) B. E. Edwards and P. N. Rao, *ibid.*, **31**, 324 (1966). (d) A. J. Liston, *ibid.*, **31**, 2105 (1966). (e) A. J. Liston and P. Toft, *ibid.*, **34**, 2288 (1969); *Chem. Commun.*, **No. 2**, 111 (1970).

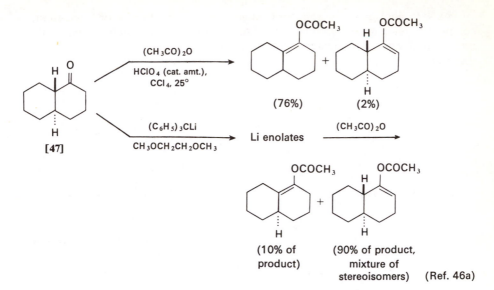

reactions of dienol acetates with peracids, the position of attack on dienol ethers has been found to vary with the nature of the reaction conditions as the following examples indicate.[45h]

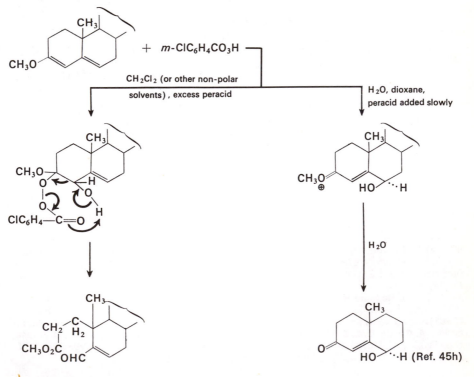

As with enol acetates and enol ethers, the enol tautomers of β-dicarbonyl compounds react rapidly with peracids. As the following equation illustrates, the presumed intermediate hydroxy epoxides rapidly rearrange to 2-hydroxy-1,3-dicarbonyl compounds which are cleaved by reaction with excess peracid.

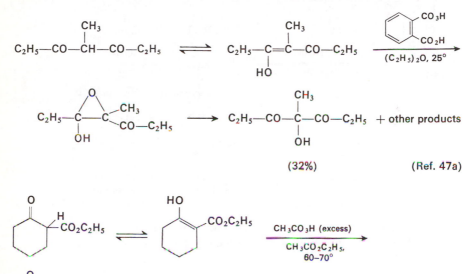

(32%)   (Ref. 47a)

(54%) (Ref. 47b)

Among the synthetic uses of epoxidation products, the acid-catalyzed rearrangements of epoxides[19] (e.g., [48]) or of the corresponding 1,2-diol monoesters (e.g., [49]) are of special interest since the over-all reaction scheme provides a

$$C_6H_5-CH-CH-C_6H_5 \xrightarrow[\substack{C_6H_6 \\ 25°}]{BF_3-(C_2H_5)_2O} (C_6H_5)_2CHCHO$$

[48]

(73–83%)

(Ref. 48a)

47. (a) H. O. House and W. F. Gannon, *J. Org. Chem.,* **23,** 879 (1958). (b) A. J. Hubert and P. S. Starcher, *J. Chem. Soc.,* C, 2500 (1968). (c) For cleavages effected with hydrogen peroxide, see S. I. Zavialov, L. P. Vinogradova, and G. V. Kondratieva, *Tetrahedron,* **20,** 2745 (1964).
48. (a) D. J. Reif and H. O. House, *Org. Syn.,* **Coll. Vol. 4,** 375 (1963). (b) H. O. House, *J. Am. Chem. Soc.,* **77,** 3070, 5083 (1955); H. O. House and G. D. Ryerson, *ibid.,* **83,** 979 (1961). (c) S. M. Naqvi, J. P. Horwitz, and R. Filler, *ibid.,* **79,** 6283 (1957). (d) B. Rickborn and R. M. Gerkin, *ibid.,* **93,** 1693 (1971). (e) G. D. Ryerson, R. L. Wasson, and H. O. House, *Org. Syn.,* **Coll. Vol. 4,** 957 (1963). (f) R. N. McDonald and D. G. Hill, *J. Org. Chem.,* **35,** 2942 (1970). (g) A. J. Sisti, *ibid.,* **35,** 2670 (1970).

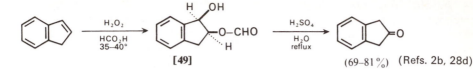

$$(69–81\%) \quad \text{(Refs. 2b, 28d)}$$

method for the conversion of an olefin to a carbonyl compound. The acid catalysts commonly employed are aqueous mineral acid, boron trifluoride etherate in benzene, anhydrous magnesium bromide in benzene or ether, or various lithium salts in benzene.[19,39f,48,49] Either the mineral acid or the boron trifluoride catalyst gives satisfactory results with 1,1-dialkylepoxides, with trisubstituted epoxides, and with epoxides that have at least one aryl or vinyl substituent.[48,49] The procedure has proved useful for the conversion of substituted cyclohexene derivatives (e.g. [50]) to the corresponding cyclohexanones and for the conversion of exocyclic

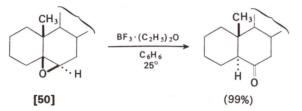

[50]   (99%)

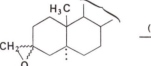

(either stereoisomer)

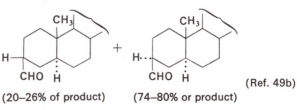

(20–26% of product)   (74–80% or product)   (Ref. 49b)

49. (a) H. B. Henbest and T. I. Wrigley, *J. Chem. Soc.*, 4596, 4765 (1957). (b) B. N. Blackett, J. M. Coxon, M. P. Hartshorn, B. L. J. Jackson, and C. N. Muir, *Tetrahedron*, **25**, 1479 (1969); B. N. Blackett, J. M. Coxon, M. P. Hartshorn, and K. E. Richards, *ibid.*, **25**, 4999 (1969); J. M. Coxon, M. P. Hartshorn, and W. J. Rae, *ibid.*, **26**, 1091 (1970). (c) H. Hart, R. M. Lange, and P. M. Collins, *Org. Syn.*, **48**, 87 (1968). (d) H. Hart and L. R. Lerner, *J. Org. Chem.*, **32**, 2669 (1967). (e) H. Hart and R. K. Murray, Jr., *ibid.*, **32**, 2448 (1967); H. Hart and D. C. Lankin, *ibid.*, **33**, 4398 (1968). (f) J. K. Crandall and L. H. C. Lin, *ibid.*, **33**, 2375 (1968). (g) B. Rickborn and R. P. Thummel, *ibid.*, **34**, 3583 (1969). (h) A. C. Cope and J. K. Heeren, *J. Am. Chem. Soc.*, **87**, 3125 (1965). (i) For the use of aluminum isopropoxide to convert epoxides to allylic alcohols, see E. H. Eschinasi, *J. Org. Chem.*, **35**, 1598 (1970). (j) J. A. Marshall and R. A. Ruden, *Tetrahedron Letters*, **No. 15**, 1239 (1970).

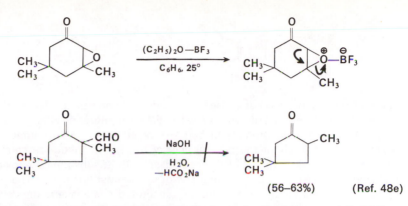

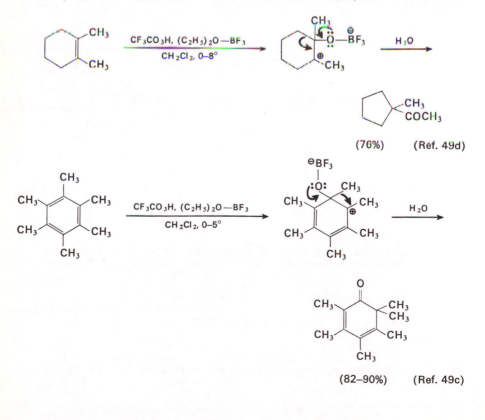

olefin derivatives to cycloalkylcarboxaldehydes or cycloalkyl ketones. 1,2-Dialkyl-epoxides usually give satisfactory yields of rearrangement products with magnesium bromide, a reagent that converts the epoxides to bromohydrin derivatives prior to rearrangement.[48]

This conversion of an olefin to a carbonyl compound may be effected in one reaction vessel by the use of a mixture of peroxytrifluoracetic acid and boron trifluoride etherate.[49c−e] This oxidizing system is sufficiently powerful so that even polyalkyl benzene derivatives react to form substituted cyclohexadienones.

Ethylene oxide derivatives also serve as useful intermediates for the formation of allylic alcohols. This transformation is best accomplished by the base-catalyzed elimination reaction illustrated in the following equations employing an ethereal solution of lithium diethylamide, a strong base acid and a relatively poor nucleophile.[49f-1] The indicated preference for the formation of an axial allylic alcohol from a cyclohexene oxide derivative is believed to be the result of a concerted *trans* elimination involving abstraction of an equatorial proton by the strong base.[49g]

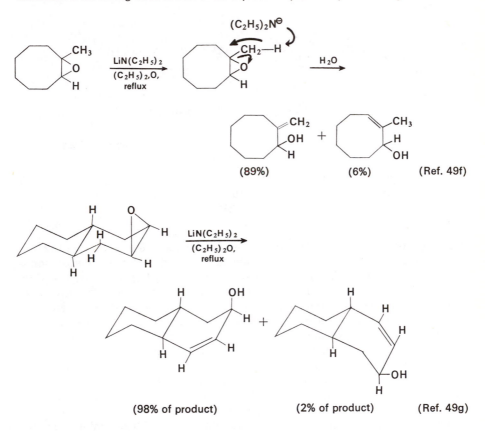

(89%)   (6%)   (Ref. 49f)

(98% of product)   (2% of product)   (Ref. 49g)

OXIDATION OF CARBONYL COMPOUNDS

Although the reaction of peracids with ketones is normally much slower than the epoxidation of olefins, relatively long reaction times, strong acids as catalysts, or very reactive peracids permit the conversion of carbonyl compounds to esters in good yield. This conversion, known as the Baeyer-Villiger reaction,[1,50] is usually accomplished either by the use of a solution of peracetic acid in acetic acid containing sulfuric acid (e.g., [51]) or *p*-toluenesulfonic acid as a catalyst, or by the

50. (a) C. H. Hassall, *Org. Reactions,* **9**, 73 (1957). (b) P. A. S. Smith in P. deMayo, ed., *Molecular Rearrangements,* Vol. 1, Wiley-Interscience, New York, 1963, pp. 568–591.

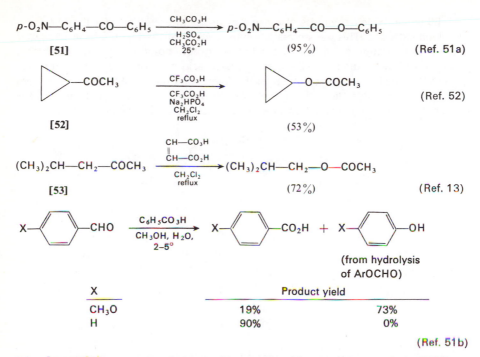

$p\text{-}O_2N\text{—}C_6H_4\text{—}CO\text{—}C_6H_5 \xrightarrow[\substack{H_2SO_4 \\ CH_3CO_2H \\ 25°}]{CH_3CO_3H} p\text{-}O_2N\text{—}C_6H_4\text{—}CO\text{—}O\text{—}C_6H_5$

$[51]$ (95%) (Ref. 51a)

$[52]$ (53%) (Ref. 52)

$(CH_3)_2CH\text{—}CH_2\text{—}COCH_3 \xrightarrow[\substack{CH_2Cl_2 \\ reflux}]{\substack{CH\text{—}CO_3H \\ \| \\ CH\text{—}CO_2H}} (CH_3)_2CH\text{—}CH_2\text{—}O\text{—}COCH_3$

$[53]$ (72%) (Ref. 13)

$X\text{—}\langle\text{—}\rangle\text{—}CHO \xrightarrow[\substack{CH_3OH,\ H_2O, \\ 2\text{-}5°}]{C_6H_5CO_3H} X\text{—}\langle\text{—}\rangle\text{—}CO_2H + X\text{—}\langle\text{—}\rangle\text{—}OH$

(from hydrolysis of ArOCHO)

X	Product yield	
CH$_3$O	19%	73%
H	90%	0%

(Ref. 51b)

use of one of the more reactive peracids—peroxytrifluoroacetic acid (e.g., [52]) or monopermaleic acid (e.g., [53])—in methylene chloride solution. Peroxymonosulfuric acid (Caro's acid, $H_2S_2O_8$)[53c] or a mixture of hydrogen peroxide and an acid or a base have also been utilized, as have such other peracids as perbenzoic and monoperphthalic and the hydrogen peroxide-hexafluoroacetone adduct.[53d]

Peroxytrifluoroacetic acid is generally the reagent of choice because of the speed with which the oxidation occurs. However, use of this peracid is complicated by transesterification, which occurs between the initial ester product and trifluoroacetic acid as illustrated in the accompanying equation.[8b,52,53a] Although this transesterification is of no concern if the crude ester product is to be hydrolyzed prior to the isolation of products, a buffer such as solid disodium hydrogen phosphate must be added to the reaction mixture if isolation of the ester is desired; the buffer reacts with the trifluoroacetic acid to form a salt and hence minimize transesterification.

$$CH_3CO_2C_2H_5 + CF_3CO_2H \rightleftharpoons CH_3CO_2H + CF_3CO_2C_2H_5$$

51. (a) W. von E. Doering and L. Speers, *J. Am. Chem. Soc.*, **72**, 5515 (1950). (b) Y. Ogata and Y. Sawaki, *J. Org. Chem.*, **34**, 3985 (1969).
52. W. D. Emmons and G. B. Lucas, *J. Am. Chem. Soc.*, **77**, 2287 (1965).
53. (a) M. F. Hawthorne and W. D. Emmons, *J. Am. Chem. Soc.*, **80**, 6398 (1958). (b) E. E. Smissman, J. P. Li, and Z. H. Israili, *J. Org. Chem.*, **33**, 4231 (1968). (c) The conversion of aldehydes to esters has been effected by oxidation of aldehydes with persulfuric acid in alcohol solution. A. Nishihara and I. Kubota, *ibid.*, **33**, 2525 (1968). (d) R. D. Chambers and M. Clark, *Tetrahedron Letters*, **No. 32**, 2741 (1970).

The oxidation of cyclic ketones (e.g., **[54]** and **[55]**), with peracids serves as a useful preparative route to lactones. Since the six-membered- and larger-ring lactones are equilibrated with the corresponding dimers, trimers, and linear

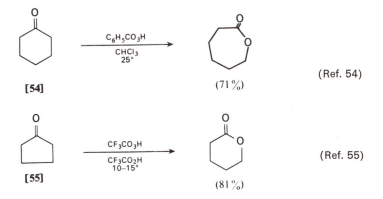

(Ref. 54)

(Ref. 55)

polymeric esters in the presence of strong acids, the use of distilled solutions of peracetic acid in ethyl acetate has been recommended for this type of oxidation (e.g., **[56]**). Under these conditions, however, elevated temperatures were required for reasonably rapid reactions with the less reactive[54] seven- and eight-membered-ring ketones, and a substantial amount of further oxidation to form dicarboxylic acids was observed.

A variety of studies[51,53,54,57] of the Baeyer-Villiger reaction indicate that the mechanism is that shown in the accompanying equations. The reaction is catalyzed by acid and the rate of oxidation is accelerated by electron-donating groups in the ketone and by electron-withdrawing groups in the peracid. Furthermore, the

54. (a) S. L. Friess, *J. Am. Chem. Soc.,* **71**, 2571 (1949); see also S. L. Friess and P. E. Frankenburg, *ibid.,* **74**, 2679 (1952). (b) see also J. L. Mateos and H. Menchaca, *J. Org. Chem.,* **29**, 2026 (1964).
55. W. F. Sager and A. Duckworth, *J. Am. Chem. Soc.,* **77**, 188 (1955).
56. P. S. Starcher and B. Phillips, *J. Am. Chem. Soc.,* **80**, 4079 (1958).
57. (a) W. von E. Doering and E. Dorfman, *J. Am. Chem. Soc.,* **75**, 5595 (1953). (b) S. L. Friess and N. Franham, *ibid.,* **72**, 5518 (1950). (c) S. L. Friess and A. H. Soloway, *ibid.,* **73**, 3968 (1951). (d) S. L. Friess and R. Pinson, Jr., *ibid.,* **74**, 1302 (1952). (e) B. W. Palmer and A. Fry, *ibid.,* **92**, 2580 (1970). (f) R. R. Sauers and R. W. Ubersax, *J. Org. Chem.,* **30**, 3939 (1965). (g) J. C. Robertson and A. Swelim, *Tetrahedron Letters,* **No. 30**, 2871 (1967).

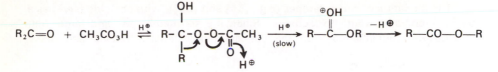

reaction has been demonstrated to occur with retention of configuration,[58] as is illustrated by the oxidation of ketone **[57]**. It will be noted from these examples

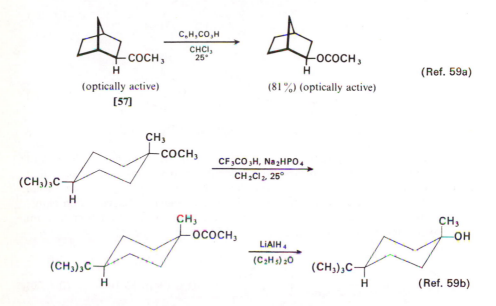

that the oxidation not only provides a useful method for the conversion of a methyl ketone to an alcohol but also offers a method for relating the stereochemistry of a ketone and an alcohol. The following equations illustrate comparable processes that have been used to interrelate the stereochemistry of alcohols and carboxylic acids.

It is apparent that the oxidation of an unsymmetrical ketone (e.g., **[58]**) can lead to two isomeric esters. From a study of a series of alkyl aryl ketones[53] as well as from other studies,[51a,57f,g] the relative ease of migration (i.e., the migratory

58. (a) R. B. Turner, *J. Am. Chem. Soc.,* **72**, 878 (1950). (b) T. F. Gallagher and T. H. Kritchevsky, *ibid.,* **72**, 882 (1950). (c) K. Mislow and J. Brenner, *ibid.,* **75**, 2318 (1953). 59. (a) J. A. Berson and S. Suzuki, *J. Am. Chem. Soc.,* **81**, 4088 (1959). (b) H. O. House and T. M. Bare, *J. Org. Chem.,* **33**, 943 (1968). (c) W. G. Dauben, R. C. Tweit, and C. Mannerskantz, *J. Am. Chem. Soc.,* **76**, 4420 (1954); T. M. Bare and H. O. House, *Org. Syn.,* **49**, 81 (1969). (d) D. B. Denney and N. Sherman, *J. Org. Chem.,* **30**, 3760 (1965). (e) The thermal decomposition of peracids in boiling hexane or heptane also yields the corresponding alcohols; however, since this process involves a free alkyl radical intermediate, it is not stereospecific; M. Gruselle, J. Fossey, and D. Lefort, *Tetrahedron Letters,* **No. 24**, 2069 (1970).

aptitude) of various groups in the Baeyer-Villiger reaction has been found to be *t*-alkyl > cyclohexyl ~ *sec*-alkyl ~ benzyl ~ phenyl > primary alkyl > cyclopropyl > methyl. Interestingly, even a bridgehead *t*-alkyl group (e.g., **[59]**) migrates readily, providing a useful synthetic route to bridgehead alcohols. The migration of a phenyl ring is facilitated by the presence of electron-donating substituents and retarded by the presence of electron-withdrawing substituents. In certain cases involving *ortho* substituents, this order may not be followed because of the operation of an alternative reaction mechanism.[53b] The relative ease of migration of the various

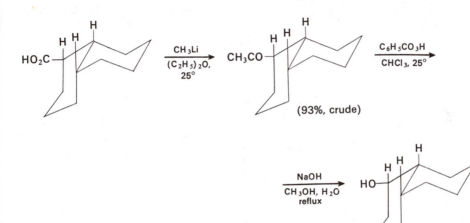

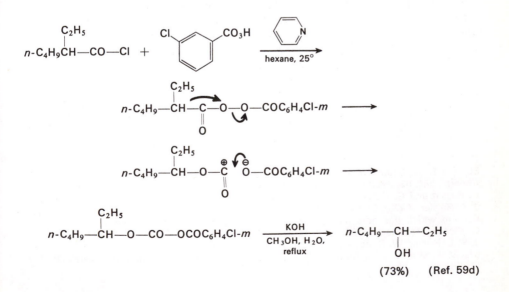

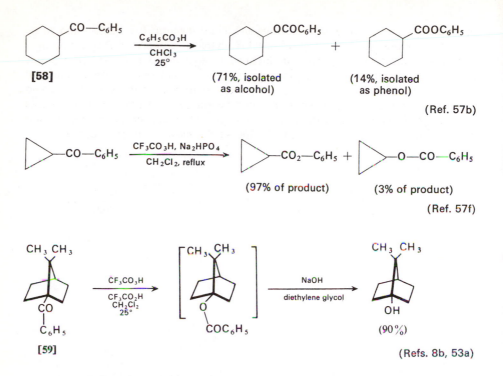

[58]

(71%, isolated as alcohol)

(14%, isolated as phenol)

(Ref. 57b)

(97% of product)

(3% of product)

(Ref. 57f)

[59]

(90%)

(Refs. 8b, 53a)

groups in the Baeyer-Villiger reaction has been suggested to reflect the ability of the migrating group to accept a partial positive charge in the transition state. However, this property, arising from the electronic distribution in the migrating group, cannot be the only factor determining migratory aptitude. In rearrangements involving the migration of groups to electron-deficient carbon atoms, different reactions and different reaction conditions result in different orders of migratory aptitude. For example, a recent study of the oxidation of substituted benzophenones has suggested that although the migrating aryl group bears a partial positive charge when peroxytrifluoroacetic acid is the oxidant, a transition state having free radical behavior appears to be operative when the oxidation is effected with peracetic acid.[57g]

The reactions of peracids with several cyclic ketones ([60]–[62]) have been studied[60] to examine the suggestion[61] that a chairlike transition state (e.g., [63]) should be favored over a boatlike transition state (e.g., [64]) in the Baeyer-Villiger

60. (a) R. R. Sauers, *J. Am. Chem. Soc.,* **81**, 925 (1959). (b) R. R. Sauers and G. P. Ahearn, *ibid.,* **83**, 2759 (1961). (c) J. Meinwald and E. Frauenglass, *ibid.,* **82**, 5235 (1960); A. Rassat and G. Ourisson, *Bull. Soc. Chim. France,* 1133 (1959). (d) R. R. Sauers and J. A. Beisler, *J. Org. Chem.,* **29**, 210 (1964). (e) For other examples of cyclic ketone oxidations in which the more highly substituted alkyl group migrates, see P. A. Tardella and G. DiMaio, *Tetrahedron,* **23**, 2285 (1967); D. Rosenthal, *J. Org. Chem.,* **32**, 4084 (1967); J. S. E. Holker, W. R. Jones, and R. J. Ramm, *J. Chem. Soc.,* C, 357 (1969).
61. M. F. Murray, B. A. Johnson, R. L. Pederson, and A. C. Ott, *J. Am. Chem. Soc.,* **78**, 981 (1956).

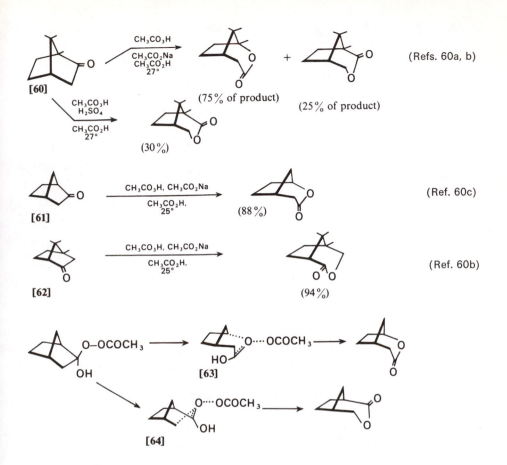

migration step. Although the mechanistic interpretation of the results obtained is not clear, it is apparent that factors other than the electronic distribution in the migrating group may influence the relative ease of migration. Furthermore, the different results obtained by oxidizing camphor [60] on the one hand in neutral and on the other in strongly acidic media offer further illustration of the fact that reaction conditions may influence the reaction course.

As illustrated previously, aldehydes react with peracids to form either carboxylic acids or formate esters; 1,2-dicarbonyl compounds are oxidized to acid anhydrides (or acids after hydrolysis) with peracids.[50] These two classes of compounds are also frequently oxidized with hydrogen peroxide or peracids in alkaline solution,[1d,50,51b,64] as illustrated by the oxidations of [65] and [66]. The reaction of a carbonyl group with alkaline hydrogen peroxide is believed to proceed as indicated in the accompanying equations; in the oxidation of α-keto acids, the intermediate [67] presumably undergoes protonation followed by loss of carbon dioxide and hydroxide ion rather than rearrangement. Monoketones are also cleaved by

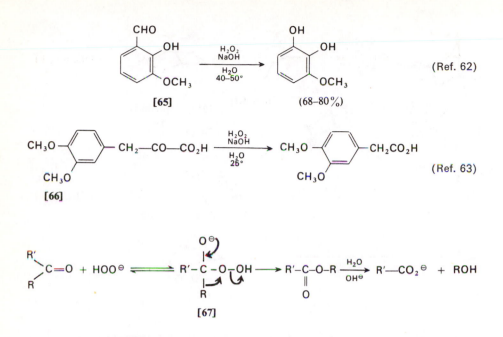

(Ref. 62)

[65]   (68–80%)

(Ref. 63)

[66]

[67]

reaction with hydrogen peroxide or an alkyl hydroperoxide in alkaline solution.[40e,64] Although this oxidation is of interest because the direction of cleavage sometimes differs from that observed in a Baeyer-Villiger reaction (e.g., [68]), it is currently of little preparative value because of the poor yields obtained.

$$C_6H_5—CO—CH_2CH_2CH_3 \xrightarrow[\substack{CH_3OH \\ reflux}]{\substack{H_2O_2 \\ NaOH}} C_6H_5CO_2H + CH_3CH_2CH_2OH + CH_3CH_2CO_2H$$

[68]   (11%)   + starting ketone
(81% recovered)

(Ref. 64a)

Unhindered ketones also react with hydrogen peroxide under various conditions. The following equations illustrate reactions of this type which appear to have preparative utility.

62. A. R. Surrey, *Org. Syn.,* **Coll. Vol. 3**, 759 (1955).
63. H. R. Snyder, J. S. Buck, and W. S. Ide, *Org. Syn.,* **Coll. Vol. 2**, 333 (1943).
64. (a) K. Maruyama, *Bull. Chem. Soc. Japan,* **33**, 1516 (1960). (b) J. E. Leffler, *J. Org. Chem.,* **16**, 1785 (1951). (c) M. S. Kharasch and G. Sosnovsky, *ibid.,* **23**, 1322 (1958). (d) P. R. Story, D. D. Denson, C. E. Bishop, B. C. Clark, Jr., and J. C. Farine, *J. Am. Chem. Soc.,* **90**, 817 (1968); P. R. Story, B. Lee, C. E. Bishop, D. D. Denson, and P. Busch, *J. Org. Chem.,* **35**, 3059 (1970). (e) G. B. Payne and C. W. Smith, *ibid.,* **22**, 1680 (1957). (f) H. M. Hellman and R. A. Jerussi, *Tetrahedron,* **20**, 741 (1964). (g) E. Caspi, S. K. Malhotra, Y. Shimizu, K. Maheshwari, and M. J. Gasic, *ibid.,* **22**, 595 (1966).

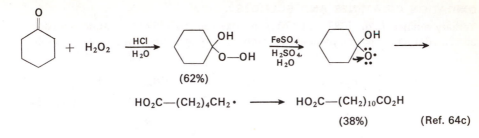

(62%)

$$HO_2C\!-\!(CH_2)_4CH_2\cdot \longrightarrow HO_2C\!-\!(CH_2)_{10}CO_2H$$

(38%)   (Ref. 64c)

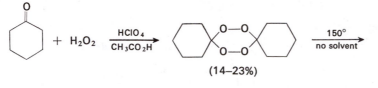

(14–23%)

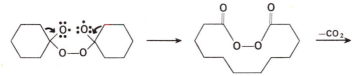

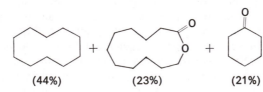

(44%)   +   (23%)   +   (21%)   (Ref. 64c, d)

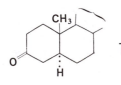 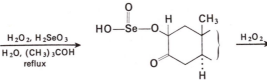

(35%, isolated as the methyl
ester)

+   other products

(Ref. 64e–g)

OXIDATION OF AMINES AND SULFIDES

Tertiary amines (e.g., **[69]** and **[70]**) are oxidized by either hydrogen peroxide, alkyl hydroperoxides, or peracids to produce amine oxides that serve as useful

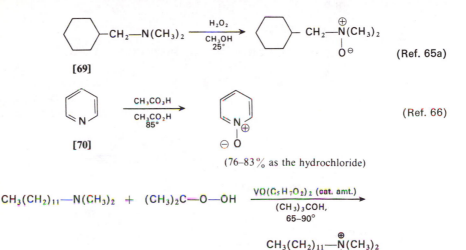

[69]

(Ref. 65a)

[70]

(Ref. 66)

(76–83 % as the hydrochloride)

$$CH_3(CH_2)_{11}-N(CH_3)_2 \; + \; (CH_3)_2C-O-OH \xrightarrow[\substack{(CH_3)_3COH, \\ 65-90°}]{VO(C_5H_7O_2)_2 \text{ (cat. amt.)}}$$

$$CH_3(CH_2)_{11}-\overset{\oplus}{N}(CH_3)_2$$
$$\underset{O^{\ominus}}{|}$$

(76–83%)   (Ref. 65b)

synthetic intermediates.[67,68] The accompanying example illustrating the formation and decomposition of an aziridine N-oxide to form an olefin and a nitroso compound[67b−d] is of interest both as a synthetic method and because of the apparent relationship of this reaction to the subsequently described olefin-singlet oxygen reaction. The mechanism of the oxidation of tertiary amines is presumably analogous to the epoxidation of olefins, involving an electrophilic attack by the peracid to transfer an oxygen atom to the amine. Although the same initial step may be involved in the reaction of peracids with primary and secondary amines, the initial product, a hydroxylamine derivative after transfer of a proton, is susceptible to further oxidation. Nitrones are formed by the oxidation of secondary aliphatic amines with peracids, and nitroso compounds result from the oxidation of primary amines (e.g., **[71]**); the nitroso compounds (or their oxime tautomers) may be

65. (a) A. C. Cope and E. Ciganek, *Org. Syn.,* **Coll. Vol. 4**, 612 (1963). (b) M. N. Sheng and J. G. Zajacek, *ibid.,* **50**, 56 (1970) ; *J. Org. Chem.,* **33**, 588 (1968).
66. (a) H. S. Mosher, L. Turner, and A. Carlsmith, *Org. Syn.,* **Coll. Vol. 4**, 828 (1963). (b) E. C. Taylor, Jr., and A. J. Crovetti, *ibid.,* **Coll. Vol. 4**, 704 (1963). (c) J. C. Craig and K. K. Purushothaman, *J. Org. Chem.,* **35**, 1721 (1970).
67. (a) A. R. Katritzky, *Quart. Rev.,* **10**, 395 (1956). (b) A. Padwa and L. Hamilton, *J. Org. Chem.,* **31**, 1995 (1966). (c) H. W. Heine, J. D. Myers, and E. T. Peltzer, *Angew. Chem., Intern. Ed. Engl.,* **9**, 374 (1970). (d) J. E. Baldwin, A. K. Bhatnagar, S. C. Choi, and T. J. Shortridge, *J. Am. Chem. Soc.,* **93**, 4082 (1971) ; suitably substituted aziridine N-oxides may also rearrange to form N-allyl hydroxylamine derivatives.
68. A. C. Cope and E. R. Trumbull, *Org. Reactions,* **11**, 317 (1960).

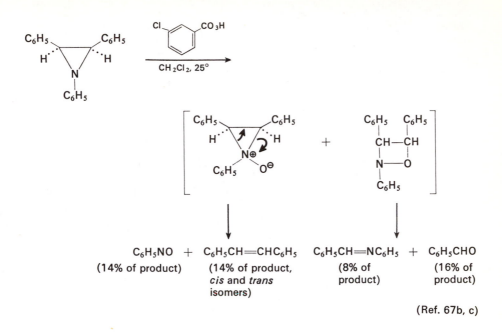

$$C_6H_5NO \ + \ C_6H_5CH{=}CHC_6H_5 \quad C_6H_5CH{=}NC_6H_5 \ + \ C_6H_5CHO$$

| (14% of product) | (14% of product, *cis* and *trans* isomers) | (8% of product) | (16% of product) |

(Ref. 67b, c)

oxidized further to nitro compounds.[69] Peroxytrifluoroacetic acid has been found to be a useful reagent for the direct oxidation of primary aromatic amines (e.g., [72]) to nitro compounds[70] as well as for the oxidation of other nitrogen-containing compounds.[70c-h] Although peroxytrifluoroacetic acid has proved unsuitable for

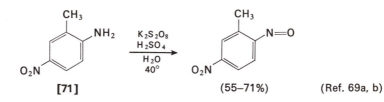

[71] (55–71%) (Ref. 69a, b)

69. (a) W. D. Langley, *Org. Syn., Coll. Vol. 3*, 334 (1955). (b) J. W. Gorrod, *Tetrahedron Letters,* No. 59, 6155 (1968). (c) R. J. Sundberg and P. A. Bukowick, *J. Org. Chem., 33*, 4098 (1968) ; see also Ref. 2h.
70. (a) A. S. Pagano and W. D. Emmons, *Org. Syn., 49*, 47 (1969) ; W. D. Emmons, *J. Am. Chem. Soc., 76*, 3470 (1954) ; *79*, 5528 (1957). (b) C. H. Robinson, L. Milewich, and P. Hofer, *J. Org. Chem., 31*, 524 (1966). (c) W. D. Emmons, *ibid., 77*, 4557 (1955). (d) W. D. Emmons and A. S. Pagano, *ibid., 77*, 4557 (1955). (e) For a discussion of the oxidation of primary amines to nitro compounds with peracids and with potassium permanganate, see N. Kornblum, *Org. Reactions, 12*, 101 (1962). (f) V. Madan and L. B. Clapp, *J. Am. Chem. Soc., 91*, 6078 (1969) ; *92*, 4902 (1970). (g) A. Padwa, *ibid., 87*, 4365 (1965). (h) For a study of the reaction of ketone and aldehyde hydrazone derivatives with peracetic acid, see B. T. Giliis and K. F. Schimmel, *J. Org. Chem., 32*, 2865 (1967). (i) D. R. Boyd, W. B. Jennings, R. Spratt, and D. M. Jerina, *Chem. Commun.,* No. 12, 745 (1970).

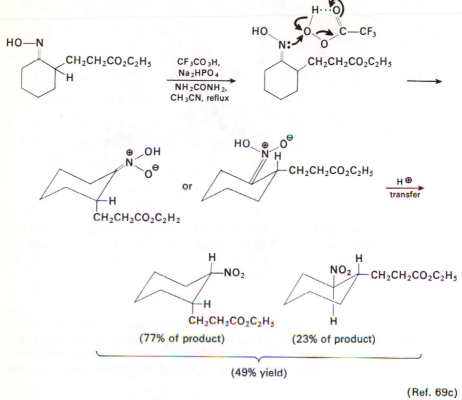

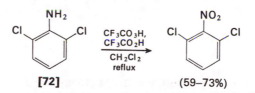

(77% of product)   (23% of product)

(49% yield)

(Ref. 69c)

[72]   (59–73%)   (Ref. 70a)

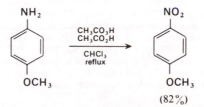

(82%)   (Ref. 70a)

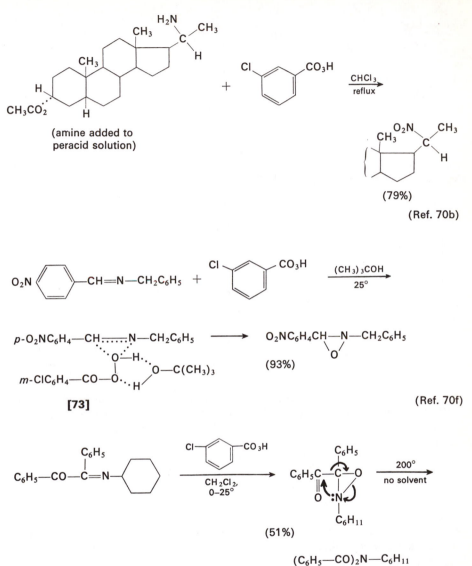

the oxidation of primary aromatic amines that contain electron-donating groups, amines of this type have been oxidized to nitro compounds with anhydrous peracetic acid. m-Chloroperbenzoic acid has been used to oxidize primary aliphatic amines to the corresponding nitro compounds.

A study of the previously illustrated reaction[18c,d] of imines with peracids to form oxaziranes[70f] has indicated the mechanism of this reaction to resemble the

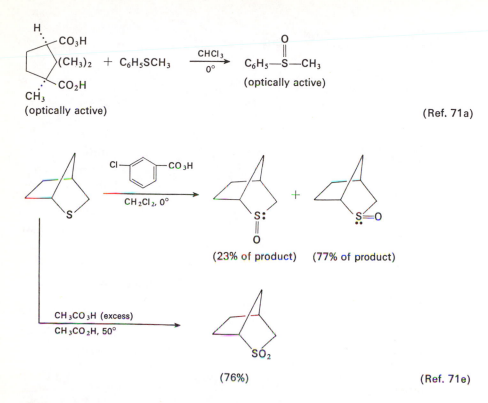

(Ref. 71a)

(23% of product)   (77% of product)

(76%)   (Ref. 71e)

mechanism followed in the epoxidation of olefins. However, the transition state [73] suggested for this reaction differs because a molecule of a proton-donating solvent or reactant is involved.[70f] A recent report[70i] indicating that the peracid oxidation of aldimines to oxaziranes is not stereospecific also indicates a further difference between this reaction and olefin epoxidation and raises a question about the correctness of the suggested transition state [73].

Peracids, alkyl hydroperoxides, and hydrogen peroxide are all capable of oxidizing dialkyl or diaryl sulfides to sulfoxides. The sulfoxides are more slowly oxidized by peracids to form sulfones. If optically active peracids are used for these oxidations, the sulfoxide products formed are found to be optically active although the optical purity is low.[71a,b,d] Studies of the mechanism of this oxidation[71f]

71. (a) U. Folli, D. Iarossi, F. Montanari, and G. Torre, *J. Chem. Soc., C,* 1371 (1968); U. Folli, D. Iarossi, and F. Montanari, *ibid., C,* 1372 (1968). (b) C. J. M. Stirling, *ibid.,* 5741 (1963). (c) D. N. Jones, M. J. Green, and R. D. Whitehouse, *ibid., C,* 1166 (1969). (d) K. Mislow, M. M. Green, and M. Raban, *J. Am. Chem. Soc.,* **87,** 2761 (1965) and references therein. (e) C. R. Johnson, H. Diefenbach, J. E. Keiser, and J. C. Sharp, *Tetrahedron,* **25,** 5649 (1969). (f) R. Curci, R. A. DiPrete, J. O. Edwards, and G. Modena, *J. Org. Chem.,* **35,** 740 (1970) and references therein. (g) For peracid oxidations of compounds with carbon-sulfur double bonds, see G. Opitz, *Angew. Chem., Intern. Ed. Engl.,* **6,** 107 (1967) ; B. Zwanenburg, L. Thys, and J. Strating, *Tetrahedron Letters,* **No. 36,** 3453 (1967).

suggest that the oxygen is transferred from the peracid to sulfur by processes which resemble the previously discussed oxidations of olefins to epoxides and amines to amine oxides.

OXIDATION AT ALLYLIC CARBON-HYDROGEN BONDS

Although allylic C—H bonds are susceptible to attack by a wide variety of free radicals to form allylic radicals and, subsequently, allyl derivatives, perhaps the most convenient route to allyl alcohol derivatives involves the reaction of the hydrocarbon (e.g., [74]) with a perester in the presence of copper(I) ion. Other oxidations at allylic positions with derivatives of selenium, lead, and mercury are discussed in Chapter 7. The peresters commonly employed for allylic oxidation are the commercially available[73] t-butyl esters of perbenzoic and peracetic acid.

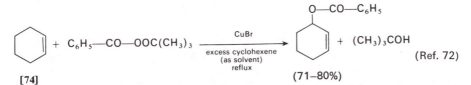

[74] (71–80%)

Mechanism studies point to the operation of a chain reaction following the path indicated in the accompanying equations.[74] In the absence of copper(I) salts, the reactions with both toluene derivatives[74i] and olefins with allylic C—H bonds[74j] yield primarily dimeric hydrocarbon products. As would be anticipated in the case of an intermediate allylic free radical, reaction with an unsymmetrical olefin (e.g.,

$$CH_3CO—O—OC(CH_3)_3 + Cu^{\oplus} \longrightarrow CH_3CO_2^{\ominus}Cu^{2\oplus} + (CH_3)_3C—O \cdot$$

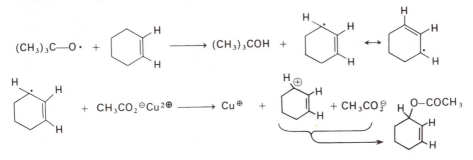

---

72. (a) M. S. Kharasch, G. Sosnovsky, and N. C. Yang, *J. Am. Chem. Soc.*, **81**, 5819 (1959). (b) G. Sosnovsky and N. C. Yang, *J. Org. Chem.*, **25**, 899 (1960). (c) K. Pedersen, P. Jakobsen, and S. O. Lawesson, *Org. Syn.*, **48**, 18 (1968).
73. Lucidol Div., Wallace and Tiernan Corp., Buffalo, N.Y.
74. (a) J. K. Kochi, *Tetrahedron*, **18**, 483 (1962). (b) J. K. Kochi, *J. Am. Chem. Soc.*, **84**, 774, 2785, 3271 (1962); *ibid.*, **85**, 1958 (1963). (c) J. K. Kochi and H. E. Mains, *J. Org. Chem.*, **30**, 1862 (1965). (d) J. K. Kochi and A. Bemis, *Tetrahedron*, **24**, 5099 (1968). (e) C. Walling and A. A. Zavitsas, *J. Am. Chem. Soc.*, **85**, 2084 (1963). (f) D. B. Denney, R. Napier, and A. Cammarata, *J. Org. Chem.*, **30**, 3151 (1965). (g) M. E. Kurz and P. Kovacic, *ibid.*, **33**, 1950 (1968). (h) C. G. Reid and P. Kovacic, *ibid.*, **34**, 3308 (1969). (i) H. H. Huang and P. K. K. Lim, *J. Chem. Soc.*, C, 2432 (1967). (j) J. R. Shelton and C. W. Uzelmeier, *J. Org. Chem.*, **35**, 1576 (1970).

[75]) leads to a mixture of isomeric products. It is of interest that the major product formed from a terminal olefin such as **[75]** is the allylic ester with a terminal double bond. This product distribution is suggested to arise from the indicated collapse

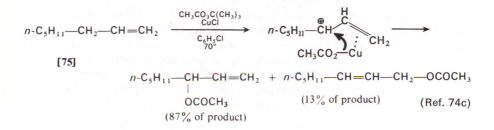

[75]

$$n\text{-}C_5H_{11}\text{—CH—CH}=CH_2 \quad + \quad n\text{-}C_5H_{11}\text{—CH}=CH\text{—CH}_2\text{—OCOCH}_3$$

$$\underset{\text{OCOCH}_3}{|}$$

$$(87\% \text{ of product}) \qquad\qquad (13\% \text{ of product}) \qquad (\text{Ref. 74c})$$

of a complex formed from the intermediate allylic cation and copper (I) acetate.[74b,c] The predominant copper (I)-olefin pi complex is expected to be that shown in which the metal is bonded to the least highly substituted double bond. If the copper salt of an optically active acid is employed, the allylic alcohol, formed by saponification of the initially formed allylic ester, is found to have some optical activity.[74f]

From the oxidation of the cyclohexyl system **[76]** the major product isolated was the ester of an axial alcohol. Both the initial abstraction of a hydrogen atom and the final attack of the benzoate anion on the allylic carbonium ion **[77]** would be expected to occur from an axial direction in order that continuous pi-orbital overlap be maintained (i.e., stereoelectronic control).

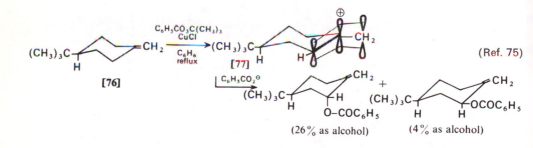

(26% as alcohol)   (4% as alcohol)

Reaction of the diene **[78]** with *t*-butyl perbenzoate was found to yield an ether that appeared to arise from attack at a nonallylic position. However, subsequent investigation demonstrated that the reaction occurred by initial addition of the *t*-butoxy radical to the olefin followed by oxidation of the carbon radical with copper(II) ion and rearrangement, as illustrated. The copper-catalyzed reaction of peresters has also proved useful for the oxidation of benzylic C—H bonds

75. (a) B. Cross and G. H. Whitham, *J. Chem. Soc.*, 1650 (1961). (b) See also H. L. Goering and U. Mayer, *J. Am. Chem. Soc.*, **86**, 3753 (1964).

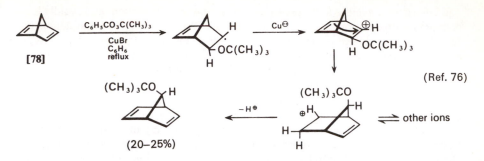

(Ref. 76)

(20–25%)   ⇌ other ions

(e.g., **[79]**),[74e,g–i] of alpha C—H bonds in ethers[77] and thioethers,[78] as well as of other activated C—H bonds.[72b]

$$C_6H_5-CH(CH_3)_2 \xrightarrow[\substack{CH_3CO_2H \\ C_6H_6 \\ reflux}]{\substack{Cu(OCOCH_3)_2 \\ CH_3CO_3C(CH_3)_3}}$$

[79]

$$C_6H_5C(CH_3)_2 \; + \; C_6H_5C{=}CH_2 \; + \; C_6H_5C(CH_3)_2{-}C(CH_3)_2C_6H_5$$
$$\underset{OCOCH_3}{|} \qquad\qquad \underset{CH_3}{|}$$

(28%)   (10%)   (Ref. 74e)

OXIDATION WITH MOLECULAR OXYGEN

An alternative method for effecting oxidation at an allylic carbon-hydrogen bond involves reaction of the olefin with singlet oxygen, one of the excited states **[80]** of molecular oxygen which normally exists as a triplet (i.e. a diradical **[81]**) in its ground state.[79]

$$(\downarrow) \; \cdot\ddot{O}{-}\ddot{O}\cdot \; (\uparrow) \qquad \ddot{O}{=}\ddot{O} \qquad (\uparrow) \; \cdot\ddot{O}{-}\ddot{O}\cdot \; (\uparrow)$$

[80a]   **[80b]**   **[81]**

[second excited   [first excited   [ground state ($^3\Sigma g$)]
state ($^1\Sigma g$)]   state ($^1\Delta g$)]

---

76. (a) P. R. Story, *J. Org. Chem.,* **26**, 287 (1961). (b) P. R. Story, *Tetrahedron Letters,* **No. 9**, 401 (1962). (c) P. R. Story and S. R. Fahrenholtz, *Org. Syn.,* **44**, 12 (1964). (d) See also M. E. Brennan and M. A. Battiste, *J. Org. Chem.,* **33**, 324 (1968).

77. (a) S. O. Lawesson and C. Berglund, *Tetrahedron Letters,* **No. 2**, 4 (1960). (b) G. Sosnovsky, *Tetrahedron,* **13**, 241 (1961); *J. Org. Chem.,* **28**, 2934 (1963).

78. (a) G. Sosnovsky, *Tetrahedron,* **18**, 15 (1962); G. Sosnovsky, and H. J. O'Neill, *J. Org. Chem.,* **27**, 3469 (1962). (c) S. O. Lawesson and C. Berglund, *Acta Chem. Scand.,* **15**, 36 (1961).

79. (a) C. S. Foote, *Accts. Chem. Res.,* **1**, 104 (1968). (b) K. Gollnick and G. O. Schenck in J. Hamer, ed., *1,4-Cycloaddition Reactions,* Academic Press, New York, 1967, pp. 255–344. (c) Yu. A. Arbuzov, *Russ. Chem. Rev.,* **34**, 558 (1965). (d) K. Gollnick, *Adv. Photochem.,* **6**, 1 (1968). (e) D. R. Kearns, *Chem. Rev.,* **71**, 395 (1971).

The second excited state **[80a]** of oxygen has a short lifetime and rapidly loses energy to form the more stable first excited state **[80b]**. This form of singlet oxygen **[80b]** has a half-life in excess of 10^{-6} sec. and is the usual form of singlet oxygen involved in reactions although, by use of high energy photosensitizers and high reactant concentrations, the second excited state **[80a]** of singlet oxygen can be involved in certain photooxidations.[80]

For the remainder of this discussion, the singlet oxygen reactant will be considered to be the more stable first excited state **[80b]** and will be designated in equations as $\overset{..}{\underset{..}{O}}=\overset{..}{\underset{..}{O}}$. Even this more stable form **[80b]** of singlet oxygen is a transient intermediate and rapidly looses its energy by light emission (called chemiluminescence if **[80b]** is formed by a chemical reaction)[81] unless it is trapped by a *reactive substrate in the same reaction vessel*. Although certain experiments have been performed in which singlet oxygen has been transferred from one region of a reaction vessel to another region, such procedures usually have little value for preparative work because most of the singlet oxygen **[80b]** is converted to ground state oxygen during the transfer.[79,82] Even when singlet oxygen is generated in the presence of a reactive substrate, the efficiency of this oxidative process is not likely to be high and an amount of singlet oxygen in excess of the stoichiometric amount will be required.[79,81,83]

Among the methods which have been used to generate singlet oxygen,[79] those procedures which offer the most preparative utility appear to be the passage of ground state oxygen through a microwave discharge,[82b,f] the oxidation of hydrogen peroxide with oxidants which abstract two electrons such as sodium hypochlorite[81a,83] bromine in alkaline solution,[81a] or peracids in alkaline solution,[81] the thermal decomposition of the ozone-triphenylphosphite adduct **[82]**[82c,84a,c−f] or 9,10-diphenylanthracene peroxide **[83]**,[84b] and, especially, the photoexcitation of ground state oxygen in a solution containing a photosensitizer such as naphthalene[82e] or one of the number of dyes including rose bengal **[84]**, eosin **[85]**,

80. (a) D. R. Kearns, R. A. Hollins, A. U. Khan, R. W. Chambers, and P. Radlick, *J. Am. Chem. Soc.*, **89**, 5455 (1967). (b) D. R. Kearns, R. A. Hollins, A. U. Khan, and P. Radlick, *ibid.*, **89**, 5456 (1967).

81. (a) E. McKeown and W. A. Waters, *J. Chem. Soc.*, B, 1040 (1966). (b) K. Akiba and O. Simamura, *Tetrahedron, 26*, 2519, 2527 (1970).

82. (a) C. S. Foote and S. Wexler, *J. Am. Chem. Soc.*, **86**, 3879, 3880 (1964). (b) E. J. Corey and W. C. Taylor, *ibid.*, **86**, 3881 (1964). (c) R. W. Murray and M. L. Kaplan, *ibid.* **91**, 5358 (1969). (d) D. R. Kearns, A. U. Khan, C. K. Duncan, and A. H. Maki, *ibid.*, **91**, 1039 (1969). (e) E. Wasserman, V. J. Kuck, W. M. Delavan, and W. A. Yager, *ibid.*, **91**; 1040 (1967). (f) J. R. Scheffer and M. D. Ouchi, *Tetrahedron Letters,* No. 3, 223 (1970).

83. (a) C. S. Foote, S. Wexler, W. Ando, and R. Higgins, *J. Am. Chem. Soc.*, **90**, 975 (1968). (b) C. S. Foote, S. Wexler, and W. Ando, *Tetrahedron Letters,* No. 46, 4111 (1965).

84. (a) P. D. Bartlett and G. D. Mendenhall, *J. Am. Chem. Soc.*, **92**, 210 (1970). (b) H. H. Wasserman and J. R. Scheffer, *ibid.*, **89**, 3073 (1967). (c) E. Koch, *Tetrahedron, 26*, 3503 (1970). (d) P. D. Bartlett and A. P. Schaap, *J. Am. Chem. Soc.*, **92**, 3223, 6055 (1970). (e) L. J. Bollyky, *ibid.*, **92**, 3230 (1970). (f) For the use of the product (possibly a hydrotrioxide) from diisopropyl ether and ozone as a source of singlet oxygen, see R. W. Murray, W. C. Lumma, Jr., and J. W. P. Lin, *ibid.*, **92**, 2305 (1970). (g) S. Mazur and C. S. Foote, *ibid.*, **92**, 3225 (1970). (h) A. G. Shultz and R. H. Schlessinger, *Tetrahedron Letters,* No. 32, 2731 (1970).

methylene blue **[86]**, hematoporphyrin, or other porphyrin derivatives. Of these procedures, the photosensitized oxidations have been the most extensively studied[79] and a number of more recent studies have been devoted to verifying the

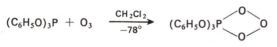

$$\ddot{O}=\ddot{O} \;+\; C_6H_5-\overset{\displaystyle O}{\underset{}{C}}-OH \;+\; Na^{\oplus}OH^{\ominus}$$

$$C_6H_5CO_2{}^{\ominus} \; Na^{\oplus} \;+\; H_2O$$

(Ref. 81b)

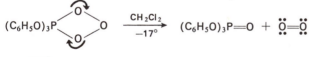

[82]   (a solution in CH_2Cl_2 from which the excess O_3 is removed by purging with N_2)

[82]

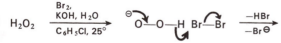

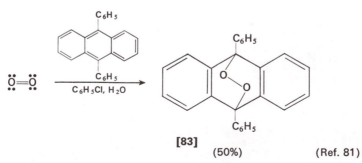

[83]

(50%)   (Ref. 81)

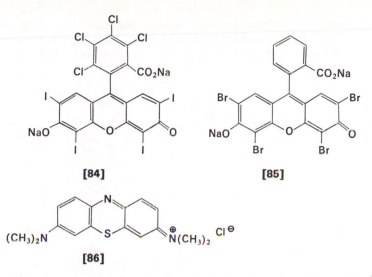

idea that the same reactive intermediate is involved in most, if not all, of these oxidation methods.[79,82,83,84b,85]

Examples of these various methods for generating and using singlet oxygen are provided in the following equations. In the reaction of various substrates with

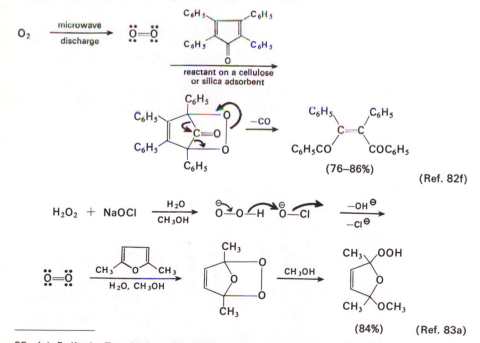

(Ref. 82f)

(84%)   (Ref. 83a)

85. (a) E. Koch, *Tetrahedron*, **24**, 6295 (1968). (b) T. Wilson, *J. Am. Chem. Soc.*, **88**, 2898 (1966). (c) For the inhibition of singlet oxygen reactions with 1,4-diazabicyclo-[2,2,2]octane, see C. Ouannes and T. Wilson, *ibid.*, **90**, 6527 (1968).

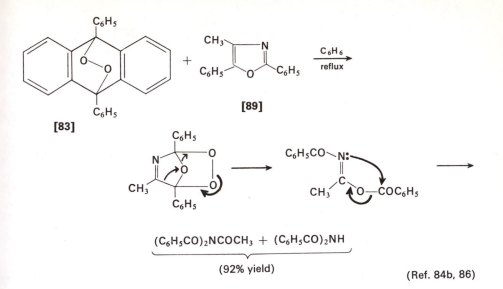

[83] + [89]

$(C_6H_5CO)_2NCOCH_3 + (C_6H_5CO)_2NH$

(92% yield)

(Ref. 84b, 86)

the phosphite-ozone complex [82], there appear to be two reaction paths, one of which involves free singlet oxygen while the other involves the ozone complex.[84a,c,d]

The accompanying equations illustrate the two reactions of singlet oxygen which have found most preparative use, namely the addition to 1,3-dienes such as [87] to form cyclic peroxide adducts and the reaction with olefins possessing three or four alkyl substituents (e.g., [88]) to form allylic hydroperoxides. The cycloaddition reaction is applicable not only to 1,3-dienes but also to reactive aromatic nuclei such as anthracenes (e.g. the formation of [83]), oxazines (e.g.,

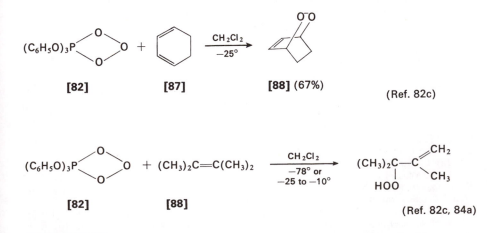

$(C_6H_5O)_3P$ [82] + [87] $\xrightarrow[-25°]{CH_2Cl_2}$ [88] (67%)

(Ref. 82c)

$(C_6H_5O)_3P$ [82] + $(CH_3)_2C\!=\!C(CH_3)_2$ [88] $\xrightarrow[\substack{-78° \text{ or} \\ -25 \text{ to } -10°}]{CH_2Cl_2}$ $(CH_3)_2C\!-\!C\substack{=CH_2 \\ \diagdown CH_3}$ $\underset{HOO}{|}$

(Ref. 82c, 84a)

86. For the analogous cleavage of bicyclic oxazole derivatives to form ω-cyano acids, see H. H. Wasserman and E. Druckrey, *J. Am. Chem. Soc.*, **90**, 2240 (1968).

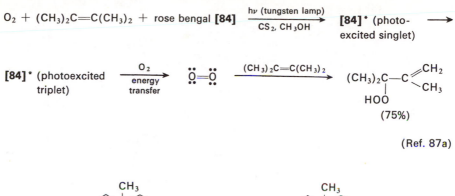

(Ref. 87a)

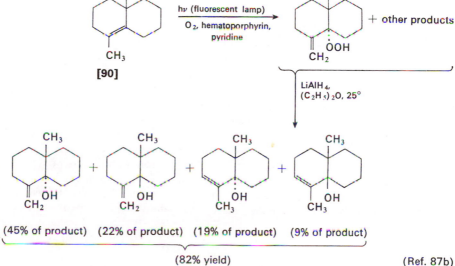

(Ref. 87b)

[89]), furans,[79,82c,83a] substituted thiophenes,[88a,b] purines,[88e] imidazoles,[88f] and oxygenated aromatic systems.[88c]

The conversion of substituted olefins to allylic hydroperoxides and, after reduction, to allylic alcohols is of special interest because, as the previous examples

87. (a) E. J. Forbes and J. Griffiths, *Chem. Commun.*, No. 9, 427 (1967). (b) J. A. Marshall and A. R. Hochstetler, *J. Org. Chem.*, 31, 1020 (1966).
88. (a) C. N. Skold and R. H. Schlessinger, *Tetrahedron Letters*, No. 10, 791 (1970). (b) H. H. Wasserman and W. Strehlow, *ibid.*, No. 10, 795 (1970). (c) J. E. Baldwin, H. H. Basson, and J. Krauss, Jr., *Chem. Commun.*, No. 16, 984 (1968). (d) For the photochemical oxidation of certain amines, see M. H. Fisch, J. C. Gramain, and J. A. Oleson, *ibid.*, No. 1, 13 (1970). (e) T. Matsuura and I. Saito, *Tetrahedron Letters*, No. 29, 3273 (1968). (f) H. H. Wasserman, K. Stiller, and M. B. Floyd, *ibid.*, No. 29, 3277 (1968). (g) R. F. Bartholomew and R. S. Davidson, *Chem. Commun.*, No. 18, 1174 (1970). (h) For a study of the quenching of singlet oxygen with amines, see E. A. Ogryzlo and C. W. Tang, *J. Am. Chem. Soc.*, 92, 5034 (1970).

illustrate, the oxygen is bonded to one of the carbon atoms of the original double bond.[79,83,89]

Consequently, this method differs from the previously discussed allylic oxidations with peroxide derivatives in that a resonance stabilized allylic radical is not involved; thus the singlet-oxygen oxidation of (+)-limonene [91] affords optically active *trans*-carveol [92] among other products indicating that the symmetrical allyl radical [93] is not an intermediate.[90]

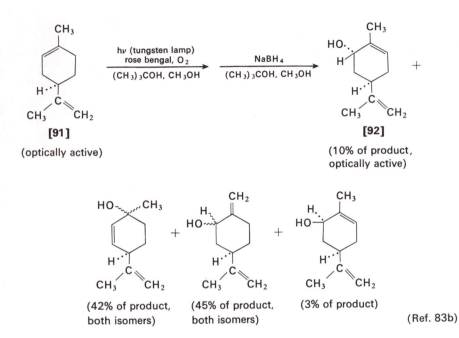

[91]

(optically active)

[92]

(10% of product, optically active)

(42% of product, both isomers)

(45% of product, both isomers)

(3% of product)

(Ref. 83b)

The selective oxidation of only one double bond in limonene [91] also illustrates the tendency of singlet oxygen to attack only electron-rich double bonds possessing three or four alkyl substituents.

89. (a) C. S. Foote and M. Brenner, *Tetrahedron Letters,* **No. 57,** 6041 (1968). (b) M. Mousseron-Canet, J. P. Dalle, and J. C. Mani, *ibid.,* **No. 57,** 6037 (1968). (c) P. Bladon and T. Sleigh, *J. Chem. Soc.,* 6991 (1965). (d) E. Klein and W. Rojahn, *Tetrahedron,* **21,** 2173 (1965). (e) K. H. Schulte-Elte and G. Ohloff, *Helv. Chim. Acta,* **51,** 494 (1968); G. Ohloff, H. Strickler, B. Willhalm, C. Borer, and M. Hinder, *ibid.,* **53,** 623 (1970). (f) A Nickon, N. Schwartz, J. B. DiGiorgio, and D. A. Widdowson, *J. Org. Chem.,* **30,** 1711 (1965) and references therein.
90. For example, the oxidation of the same olefin [91] with oxygen and a tris-(triphenyl-phosphine)rhodium chloride catalyst yields racemic products indicating the intervention of a symmetrical intermediate such as [93]. J. E. Baldwin and J. C. Swallow, *Angew Chem., Intern. Ed. Engl.,* **8,** 601 (1969). For a review of oxidations effected with oxygen and transition metal complexes, see S. Fallab, *ibid.,* **6,** 496 (1967).

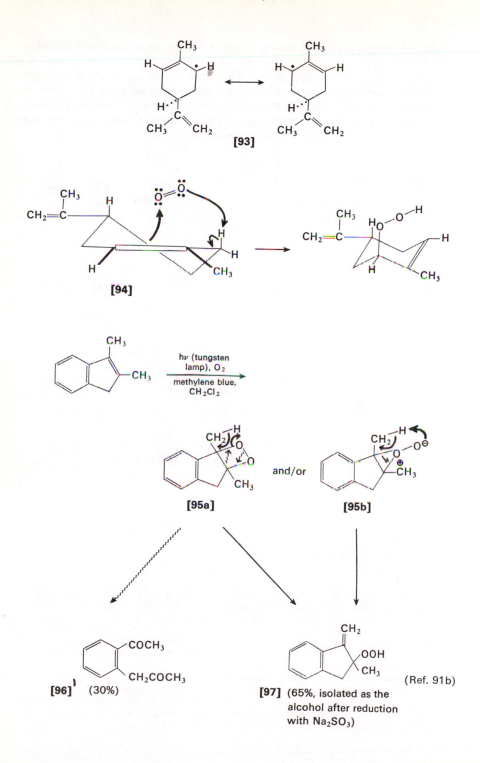

[93]

[94]

[95a]   and/or   [95b]

[96]   (30%)

[97] (65%, isolated as the
alcohol after reduction
with Na₂SO₃)

(Ref. 91b)

These results were originally interpreted to mean that the singlet oxygen reaction with the olefin by the concerted process illustrated in structure **[94]**.[79,83,89] More recently evidence has been offered indicating a different reaction path[91] in which an initial cycloaddition of singlet oxygen to an electron-rich olefin forms either a dioxetane intermediate **[95a]** or a three-membered ring dipolar intermediate **[95b]** analogous to the previously discussed azirane N-oxides. Decomposition of this intermediate by hydrogen transfer (solid arrows in **[95a]** or **[95b]**) affords an allylic hydroperoxide **[97]** whereas a dicarbonyl compound **[96]** may be formed by an alternative decomposition of the dioxetane. When photosensitized oxidations were conducted in the presence of sodium azide, the intermediate was intercepted as indicated in the following equations.

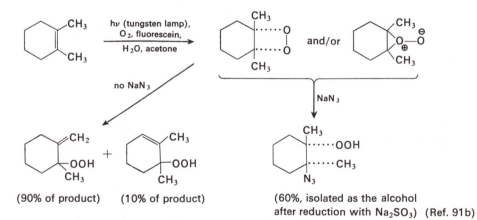

(90% of product)   (10% of product)

(60%, isolated as the alcohol after reduction with Na_2SO_3) (Ref. 91b)

Either of these reaction paths is in agreement with many experiments indicating that the *alpha* hydrogen atom to be removed is the one which is approximately perpendicular (i.e. pseudoaxial) to the original olefinic double bond; also the hydroperoxide group is added to the molecule on the same side from which the hydrogen atom is removed.[79,89] The following equation illustrates this stereospecificity.

The previously noted cleavage of olefinic double bonds to form carbonyl compounds (e.g., **[96]**) becomes the major reaction path for singlet oxygen and electron rich olefins such as enol ethers[84d,g] and enamines (e.g., **[99]**);[92] this mode of reaction also becomes important for phenyl substituted olefins.[84e,h,91b] One possible course for these reactions is the initial formation of a dipolar oxyepoxide

91. (a) D. R. Kearns, *J. Am. Chem. Soc.,* **91**, 6554 (1969). (b) W. Fenical, D. R. Kearns, and P. Radlick, *ibid.,* **91**, 3396, 7771 (1969).
92. (a) C. S. Foote and J. W. P. Lin, *Tetrahedron Letters,* **No. 29**, 3267 (1968). (b) J. E. Huber, *ibid.,* **No. 29**, 3271 (1968). (c) For other reactions of enamines with molecular oxygen (some of which presumably involve oxygen in its ground state) see S. K. Malhotra, J. J. Hostynek, and A. F. Lundin, *J. Am. Chem. Soc.,* **90**, 6565 (1968); R. A. Jerussi, *J. Org. Chem.,* **34**, 3648 (1969); W. Carpenter and E. M. Bens, *Tetrahedron,* **26**, 59 (1970); V. Van Rheenen, *Chem. Commun.,* **No. 6**, 314 (1969).

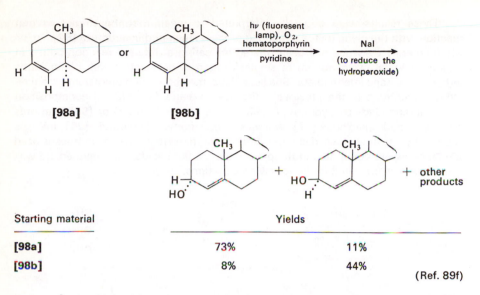

Starting material	Yields	
[98a]	73%	11%
[98b]	8%	44%

(Ref. 89f)

intermediate which either abstracts an adjacent hydrogen (see structure **[95b]**) or, in favorable cases, undergoes the rearrangement indicated in the following equation to form a dioxetane intermediate. If such a path is followed, the lifetime

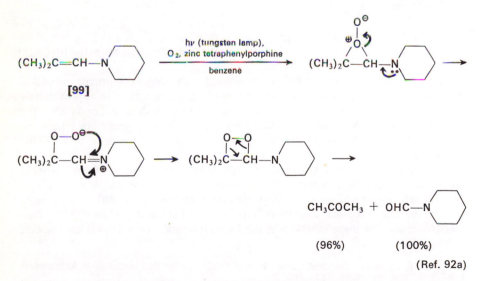

of the open chain intermediate must be short since the formation of the oxetane from singlet oxygen is a stereospecific process.[84d] The accompanying equations illustrate this stereospecificity as well as the non-stereospecific reaction with the triphenylphosphite-ozone complex. In at least one case, the formation of a dipolar oxyepoxide intermediate is favored in a polar solvent whereas the dioxetane

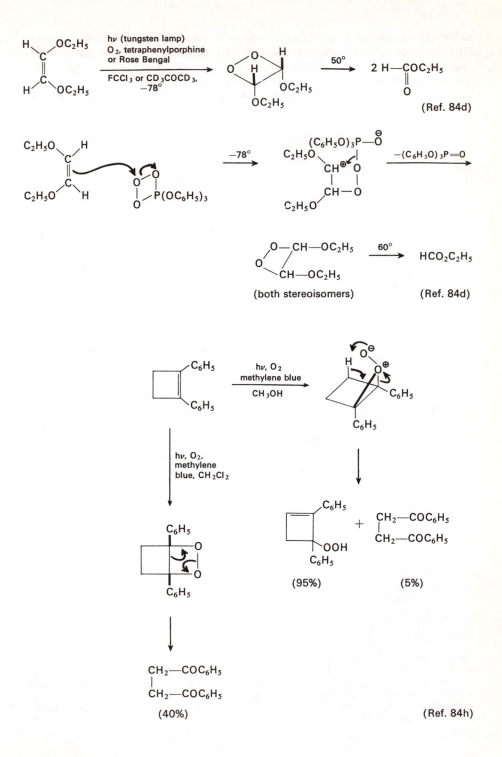

(Ref. 84d)

(both stereoisomers)   (Ref. 84d)

(95%)   (5%)

(40%)   (Ref. 84h)

intermediate leading to a diketone product is favored in a non-polar solvent.[84h]
The oxidation of allylic alcohols (e.g., **[100]**) with singlet oxygen also follows a
modified reaction course leading to α,β-epoxy ketones as the principal products.[93]
In such reactions evidence has been presented indicating that this reaction involves
the first excited state of oxygen **[80b]** and the enone **[101]** is formed from reaction

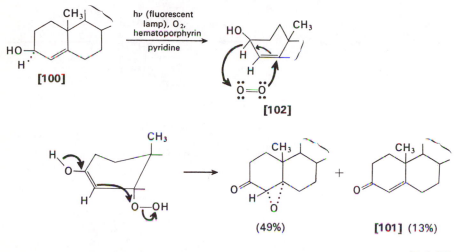

(Ref. 93)

with the second excited state **[80a]**.[80]   Although the initial step in the oxidation is
formulated (see **[102]**) as a concerted reaction of the allylic system with singlet
oxygen, the involvement of a dioxetane intermediate has not been excluded.

The oxidation of C—H bonds *alpha* to carbonyl groups with peroxides or with
molecular oxygen bears at least a formal resemblance to the previously discussed

$$C_2H_5CH(CO_2C_2H_5)_2 \xrightarrow[C_6H_6]{NaH} C_2H_5{-}\overset{\ominus}{C}(CO_2C_2H_5)_2\overset{\oplus}{Na} \xrightarrow[C_6H_6]{(C_6H_5CO{-}O{-})_2}$$

[103]

$$C_2H_5{-}\overset{\ominus}{C}(CO_2C_2H_5)$$
$$C_6H_5{-}CO{-}O{-}O{-}CO{-}C_6H_5 \longrightarrow$$

$$C_2H_5C(CO_2C_2H_5)_2$$
$$|$$
$$O{-}COC_6H_5$$
$$(75{-}78\%)$$

(Ref. 94)

93. A. Nickon and W. L. Mendelson, *J. Am. Chem. Soc.*, **87**, 3921 (1965).
94. (a) E. H. Larsen and S. O. Lawesson, *Org. Syn.*, **45**, 37 (1965). (b) S. O. Lawesson,
C. Frisell, D. Z. Denney, and D. B. Denney, *Tetrahedron*, **19**, 1229 (1963). (c) For examples
of mechanistically similar reactions in which Grignard reagents are treated with *t*-butyl
perbenzoate to form *t*-butyl ethers, see C. Frisell and S. O. Lawesson, *Org. Syn.*, **41**, 91
(1961) ; **43**, 55 (1963).

oxidations of allylic C—H bonds. Most of these oxidations are base catalyzed and are believed to require the initial formation of an enolate anion (e.g. [103], see Chapter 9).

The reaction of enolate anions with molecular oxygen to yield cleavage products has been known for some time.[95] The following equations provide examples of these oxidative cleavage reactions. The probable mechanism involves oxidation of an enolate anion [104] to an α-keto radical [105] which reacts with the excess oxygen present in the indicated chain process to form the anion [106] of an α-keto hydroperoxide.[96] Stabilized carbanions also appear to react very rapidly with singlet oxygen produced photochemically.[96e,f]

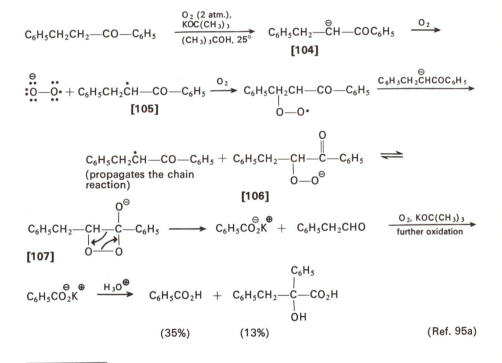

(35%)   (13%)   (Ref. 95a)

---

95. (a) W. von E. Doering and R. M. Haines, *J. Am. Chem. Soc.,* **76**, 482 (1954). (b) T. J. Wallace, H. Pobiner, and A. Schriesheim, *J. Org. Chem.,* **30**, 3768 (1965). (c) R. C. P. Cubbon and C. Hewlett, *J. Chem. Soc.,* C, 2978, 2983, 2986 (1968). (d) J. E. Baldwin, D. H. R. Barton, D. J. Faulkner, and J. F. Templeton, *ibid.,* 4743 (1962). (e) For the oxidative cleavage of aldehyde in the presence of a copper(II) catalyst, see V. Van Rheenen, *Tetrahedron Letters,* **No. 12**, 985 (1969).

96. (a) E. J. Bailey, D. H. R. Barton, J. Elks, and J. F. Templeton, *J. Chem. Soc.,* 1578 (1962). (b) G. A. Russell and G. Kaupp, *J. Am. Chem. Soc.,* **91**, 3851 (1969). (c) G. A. Russell, E. T. Strom, E. R. Talaty, K. Y. Chang, R. D. Stephens, and M. C. Young, *Rec. Chem. Progress,* **27**, 3 (1966); G. A. Russell and co-workers in *Selective Oxidation Processes, Advances in Chemistry Series,* No. 51, American Chemical Society, Washington, D. C., 1965, pp. 112–171. (d) O. H. Mattsson and C. A. Wachtmeister, *Tetrahedron Letters,* **No. 20**, 1855 (1967). (e) R. H. Young and H. Hart, *Chem. Commun.,* **No. 16**, 827 (1967). (f) D. Bethell and R. G. Wilkinson, *ibid.,* **No. 18**, 1178 (1970).

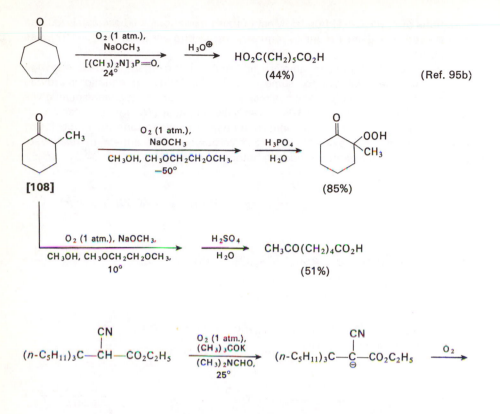

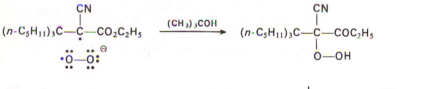

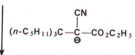

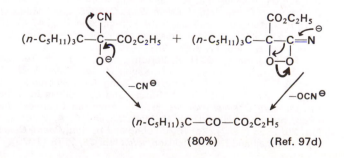

(80%)   (Ref. 97d)

Under certain conditions the α-keto radical intermediates formed by oxidation of the enolates may couple;[96b,d] also, the α-keto hydroperoxide products formed from α-methylene ketones can yield α-diketones which, in the absence of excess oxygen, react with the starting enolate anion to form semidione anion radicals.[96c] In the presence of base at room temperature the α-keto hydroperoxides may either yield 1,2-dicarbonyl compounds or they undergo the cleavage indicated in structure [107];[95,96,97] the products of this cleavage may be oxidized further if excess oxygen is present. Alternatively, the α-keto hydroperoxides may rearrange in a manner similar to the previously discussed Baeyer-Villiger reaction if they are treated with acids[97c] or they may decompose thermally as the following equation illustrates.

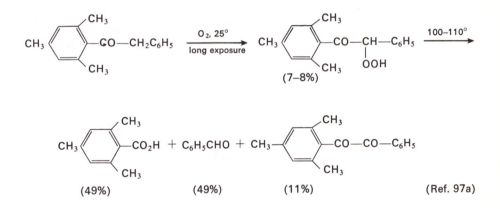

The most efficient procedure for the base-catalyzed oxidation of secondary alkyl ketones (e.g., [108] and [109]) to α-keto hydroperoxides is the use of low reaction temperatures (-25 to $-50°$) with dipolar aprotic solvents. Although the α-keto hydroperoxides may be isolated[98] and subsequently reduced to the α-hydroxy ketone with zinc dust,[96a] an especially efficient procedure involves oxidation of the ketone (e.g., [109]) to the hydroperoxide in the presence of triethyl phosphite.

---

97. (a) A. G. Pinkus, W. C. Servoss, and K. K. Lum, *J. Org. Chem.*, **32**, 2649 (1967); A. G. Pinkus, M. Z. Haq, and J. G. Lindberg, *ibid.*, **35**, 2555 (1970). (b) P. Aeberli and W. J. Houlihan, *ibid.*, **33**, 1640 (1968). (c) J. N. Gardner, F. E. Carlon, and O. Gnoj, *ibid.*, **33**, 1566, 3294 (1968). (d) N. Rabjohn and C. A. Harbert, *ibid.*, **35**, 3240 (1970).
98. W. H. Richardson and R. F. Steed [*J. Org. Chem.*, **32**, 771 (1967)] have studied infrared and nmr spectra of α-keto hydroperoxides and conclude that the materials exist in the forms indicated in the following equations.

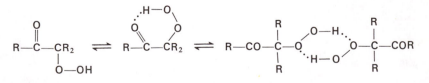

By this procedure, the α-keto hydroperoxide is reduced as it is produced in the reaction mixture (see structure **[110]**) and subsequent rearrangement or cleavage of the hydroperoxide is minimized.

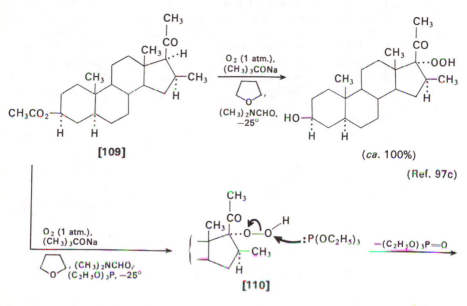

[109]

(ca. 100%)

(Ref. 97c)

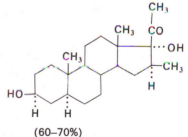

[110]

(60–70%)

7

OTHER METHODS OF OXIDATION

Apart from the previously discussed (Chapters 5 and 6) common methods of oxidation with peroxidic compounds and with compounds of chromium and manganese, a variety of other reagents and procedures have been utilized for the oxidation of organic compounds.[1] In this chapter consideration is given to certain of these oxidation procedures which have in common the property of effecting selective oxidations which would be difficult if not impossible to accomplish in good yield with the usual oxidizing reagents discussed in earlier chapters.

OXIDATION WITH PERIODIC ACID

Aqueous solutions of paraperiodic acid, H_5IO_6, potassium metaperiodate, or sodium metaperiodate, $NaIO_4$, serve as remarkably selective reagents for the cleavage of 1,2-diols and related compounds.[2,3] Although periodic acid will oxidize polycyclic aromatic hydrocarbons,[4c] hydrazines,[4c] certain active methylene compounds, and phenols and will convert sulfides to sulfoxides,[4a,b] the primary use of this oxidant is for the cleavage of 1,2-diols, 1-amino-2-hydroxy compounds, α-hydroxy ketones, and 1,2-diketones. The necessity of employing water as a solvent or as a cosolvent mixed with methanol, ethanol, t-butyl alcohol, dioxane, or acetic acid places some restrictions on the use of periodate oxidations. For water-insoluble compounds the subsequently discussed oxidative cleavage with lead tetraacetate is often more suitable. However, the periodate procedures are

1. (a) W. A. Waters in H. Gilman, ed., *Organic Chemistry,* Vol. 4, Wiley, New York, 1953, pp. 1120–1245. (b) R. Stewart, *Oxidation Mechanisms,* Benjamin, New York, 1964. (c) K. B. Wiberg, ed., *Oxidation in Organic Chemistry*, Part A, Academic Press, New York, 1965. (d) K. B. Wiberg in A. F. Scott, ed., *Survey of Progress in Chemistry,* Vol. I, Academic Press, New York, 1963, pp. 211–248. (e) R. L. Augustine, ed., *Oxidation,* Marcel Dekker, New York, 1969.
2. (a) E. L. Jackson, *Org. Reactions,* **2,** 341 (1944). (b) C. A. Bunton in K. B. Wiberg, ed., *Oxidation in Organic Chemistry,* Part A, Academic Press, New York, 1965, p. 367. (c) A. S. Perlin in R. L. Augustine, ed., *Oxidation,* Marcel Dekker, New York, 1969, pp. 189–212.
3. R. Criegee in *Newer Methods of Preparative Organic Chemistry,* Vol. 1, Wiley-Interscience, New York, 1948, p. 1.
4. (a) N. J. Leonard and C. R. Johnson, *J. Org. Chem.,* **27,** 282 (1962). (b) C. R. Johnson and J. E. Keiser, *Org. Syn.,* **46,** 78 (1966). (c) A. J. Fatiadi, *Chem. Commun.,* **No. 21,** 1087 (1967); *J. Org. Chem.,* **35,** 831 (1970).

ideal for water-soluble polyfunctional compounds such as the carbohydrates and certain amino acids.

$$C_6H_5—S—CH_3 \xrightarrow[\text{H}_2\text{O, 0°}]{\text{NaIO}_4} \underset{\overset{\|}{O}}{C_6H_5—S—CH_3}$$

(91%)   (Ref. 4a, b)

The following equation illustrates a typical procedure and the probable mechanism[2b] involved in this cleavage reaction. The cleavage of 1,2-diols is usually most rapid in the acidity range pH 1–6 where the monoanion of paraperiodic acid appears to be the principal reactant. This oxidative cleavage may also be used as

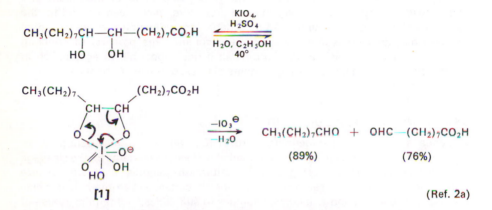

$$CH_3(CH_2)_7CH—CH—(CH_2)_7CO_2H \xrightleftharpoons[\substack{\text{H}_2\text{O, C}_2\text{H}_5\text{OH} \\ 40°}]{\substack{\text{KIO}_4, \\ \text{H}_2\text{SO}_4}}$$
$$\underset{\overset{|}{\text{HO}} \quad \overset{|}{\text{OH}}}{}$$

$$\xrightarrow[-\text{H}_2\text{O}]{-\text{IO}_3^\ominus} CH_3(CH_2)_7CHO \ + \ OHC—(CH_2)_7CO_2H$$

(89%)   (76%)

[1]   (Ref. 2a)

the basis for an analytical procedure since each 1,2-diol function converts one molar equivalent of periodate ion to iodate anion. In neutral solution, as in the presence of excess sodium bicarbonate as a buffer, the consumption of periodate ion may be followed by use of the fact that iodide ion is oxidized to free iodine by

$$IO_4^\ominus + 2\ I^\ominus + 2\ H^\oplus \xrightarrow[\text{H}_2\text{O}]{\text{pH} \sim 7} IO_3^\ominus + I_2 + H_2O$$

$$HAsO_3^\ominus + I_2 + H_2O \longrightarrow HAsO_4^\ominus + 2I^\ominus + 2H^\oplus$$

periodate but not by iodate. The liberated iodine is then titrated with a standard solution of sodium arsenite.[1e,5a] Alternatively, the iodate concentration may be determined polarographically[5b] or spectrophotometrically.[1e]

The apparent necessity to form a cyclic ester, e.g. [1], prior to the oxidative cleavage results in significant differences in the rates of cleavage of diastereoiso-

5. (a) D. M. Lemal, P. D. Pacht, and R. B. Woodward, *Tetrahedron,* **18**, 1275 (1962).
(b) W. G. Breck, R. D. Corlett, and G. H. Hay, *Chem. Commun.,* **No. 12**, 604 (1967).

meric 1,2-diols. For example, the oxidation of racemic 2,3-butanediol is more rapid than oxidation of the *meso*-isomer and the oxidation of *cis*-1,2-cyclohexane-diol is more rapid than oxidation of the *trans*-isomer.[2b,6] The formation of the complex [2] derived from the racemic diol requires less non-bonded interaction of

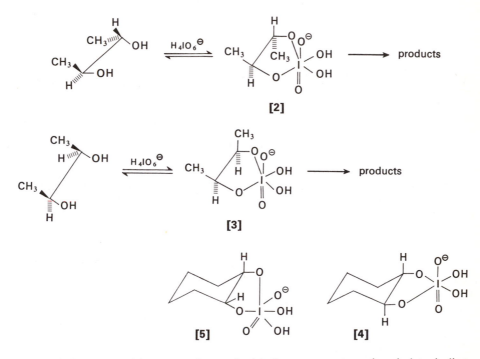

the methyl groups with one another and with the oxygen atoms bonded to iodine than is the case with the complex [3] from the *meso*-diol. The formation of a complex [4] from the *trans*-1,2-cyclohexanediol is apparently less favorable than the formation of complex [5] because of the increased puckering of the six-membered ring required to form [4].[2b,7]

The oxidative cleavage of 1,2-diketones is believed to involve the formation of a similar cyclic intermediate, e.g. [6], prior to cleavage.[8] The following equations

---

6. (a) G. J. Buist, C. A. Bunton, and J. H. Miles, *J. Chem. Soc.,* 743 (1959). (b) G. J. Buist, C. A. Bunton, and J. Lomas, *ibid.,* B, 1094, 1099 (1966). (c) G. J. Buist and J. D. Lewis, *ibid.,* B, 90 (1968). (d) A discussion of the mechanistic differences between the cleavage of glycols with chromic acid (see Chapter 5) and periodic acid has been given by M. C. R. Symons, *ibid.,* 4331 (1963).

7. For further discussion, see E. L. Eliel and C. Pillar, *J. Am. Chem. Soc.,* **77**, 3600 (1955).

8. (a) V. J. Shiner, Jr., and C. R. Wasmuth *J. Am. Chem. Soc.,* **81**, 37 (1959). (b) C. A. Bunton and V. J. Shiner, Jr., *J. Chem. Soc.,* 1593 (1960). (c) G. Dahlgren and K. L. Reed, *J. Am. Chem. Soc.,* **89**, 1380 (1967). (d) Certain 2-methoxy-1,3-dicarbonyl compounds are also oxidized by periodic acid. J. P. Girma, M. T. Rokicka, and P. Szabo, *J. Chem. Soc.,* C, 909 (1969). (e) G. W. Perold and K. G. R. Pachler, *ibid.,* C, 1918 (1966). (f) Y. Yanuka, R. Katz, and S. Sarel, *Tetrahedron Letters,* **No. 14**, 1725 (1968).

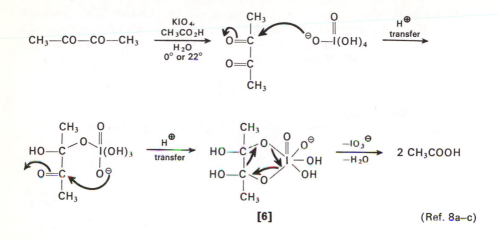

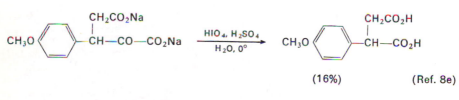

[6]   (Ref. 8a–c)

illustrate the oxidative cleavage of α-keto acids and α-hydroxy acids. The latter process is relatively slow and requires extended reaction periods at elevated temperatures.

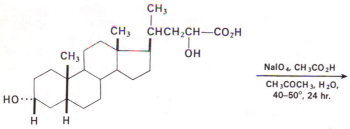

(16%)   (Ref. 8e)

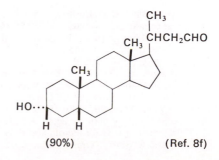

(90%)   (Ref. 8f)

Presumably intermediates comparable to those formed with diols are involved in the cleavage of α-hydroxy ketones, 1-amino-2-hydroxy compounds, and α-amino ketones

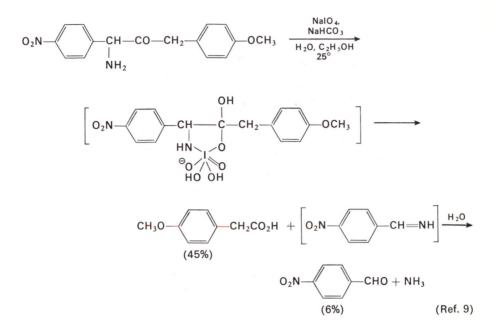

A useful modification of these periodate cleavage reactions involves the reaction of olefins, e.g. [7], with sodium periodate and a catlaytic amount of osmium tetroxide. This process, which is reminiscent of the previously discussed (see Chapter 5) cleavage of olefins with sodium periodate and a catalytic amount of potassium permanganate, relies upon the continuous reoxidation of the reduced osmium intermediates with sodium periodate to reform osmium tetroxide which can react with more olefin. This procedure offers an advantage over the per-manganate-periodate method in permitting the isolation of aldehydes which

9. (a) H. O. House and W. F. Berkowitz, *J. Org. Chem.*, **28**, 2271 (1963). (b) For use of the rates of oxidation of amino alcohols with sodium periodate to determine configurations, see J. Kovar, J. Jary, and K. Blaha, *Coll. Czech. Chem. Commun.*, **28**, 2199 (1963).
10. (a) R. Pappo, D. S. Allen, Jr., R. U. Lemieux, and W. S. Johnson, *J. Org. Chem.*, **21**, 478 (1956). (b) S. H. Graham and A. J. S. Williams, *J. Chem. Soc.*, C, 655 (1966). (c) For reviews, see P. S. Bailey, *Chem. Rev.*, **58**, 925 (1958); J. S. Belew in R. L. Augustine, ed., *Oxidation*, Marcel Dekker, New York, 1969, pp. 259–335; R. W. Murray, *Accts. Chem. Res.*, **1**, 313 (1968); R. Criegee, *Chimia*, **22**, 392 (1968); R. Criegee in J. O. Edwards, ed., *Peroxide Reaction Mechanisms*, Wiley-Interscience, New York, 1962, pp. 29–39. (d) P. S. Bailey and R. E. Erickson, *Org. Syn.*, **41**, 41, 46 (1961). (e) R. E. Dessy and M. S. Newman, *ibid.*, **Coll. Vol. 4**, 484 (1963). (f) L. I. Smith, F. L. Greenwood, and O. Hudrlik, *ibid.*, **Coll. Vol. 3**, 673 (1955). (g) S. Fliszar and M. Granger, *J. Am. Chem. Soc.*, **92**, 3361 (1970). (h) S. W. Pelletier, K. N. Iyer, and C. W. J. Chang, *J. Org. Chem.*, **35**, 3535 (1970).

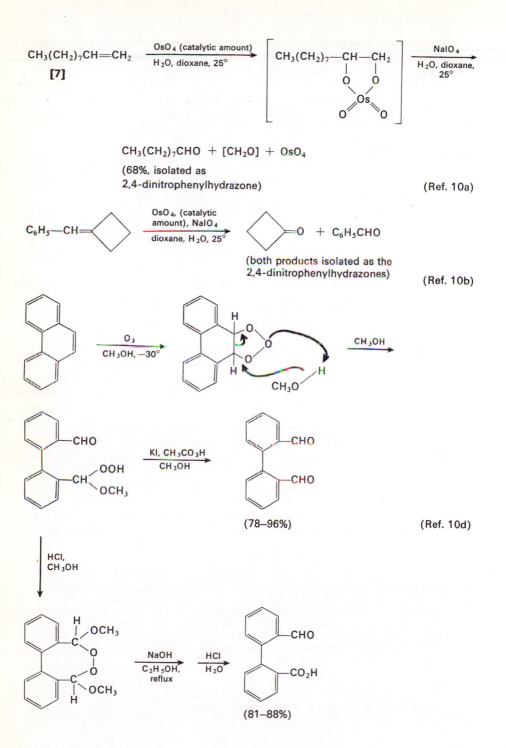

$$CH_3(CH_2)_7CHO \ + \ [CH_2O] \ + \ OsO_4$$

(68%, isolated as
2,4-dinitrophenylhydrazone)   (Ref. 10a)

(both products isolated as the
2,4-dinitrophenylhydrazones)   (Ref. 10b)

(78–96%)   (Ref. 10d)

(81–88%)

would be further oxidized by potassium permanganate. Procedures such as these and the subsequently discussed ruthenium tetroxide oxidation for cleaving carbon-carbon double bonds have frequently replaced the older direct cleavage of carbon-carbon double bonds by ozonolysis.[10c-h] However, as the following examples illustrate, the use of the commercially available ozone generators makes this oxidation procedure at least as convenient as other oxidative cleavage methods discussed in this chapter and in Chapter 5.

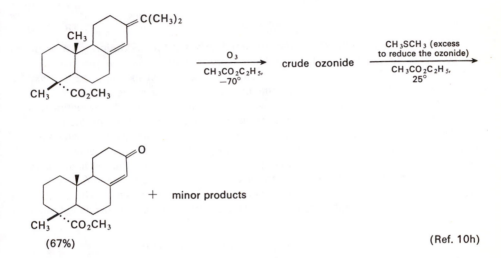

crude ozonide

+ minor products

(67%)

(Ref. 10h)

OXIDATION WITH LEAD TETRAACETATE

Lead tetraacetate, a commercially available crystalline solid which decomposes at approximately 140°,[11] is commonly used as a solution in acetic acid or benzene for the oxidation of organic compounds.[2b,c,3,12]

The solubility and conductance properties[11b] of lead tetraacetate indicate it to be a covalent compound; the infrared absorption of the acetate (or other

11. (a) For preparations of this reagent, see J. C. Bailar, Jr., *Inorg. Syn.*, **1**, 47 (1939); L. F. Fieser, *Experiments in Organic Chemistry*, 3rd ed., D. C. Heath, Boston, 1955, p. 325. (b) R. Partch and J. Monthony, *Tetrahedron Letters*, **No. 45**, 4427 (1967). (c) K. Heusler and H. Loeliger, *Helv. Chim. Acta*, **52**, 1495 (1969). (d) H. Loeliger, *ibid.*, **52**, 1516 (1969); (e) K. Heusler, *ibid.*, **52**, 1520 (1969). (e) L. F. Fieser and M. Fieser, *Reagents for Organic Synthesis*, Wiley, New York, 1967, pp. 537–563.
12. (a) R. Criegee in W. Foerst, ed., *Newer Methods of Preparative Organic Chemistry*, Vol. 2, Academic Press, New York, 1963, p. 367. (b) R. Criegee in K. Wiberg, ed., *Oxidation in Organic Chemistry*, Part A, Academic Press, New York, 1965, p. 277. (c) E. Mosettig, *Org. Reactions*, **8**, 225 (1954).

acyloxy) groups[11b,c] suggests that both oxygen atoms of each acyloxy group are coordinated with the lead atom. The following equation indicates a probable structure for the lead tetraacetate and also the structure of the crystalline complex formed when the tetraacetate is treated with a good donor ligand such as pyridine.[11b]

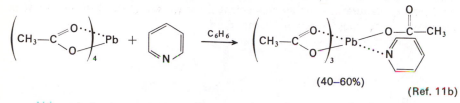

(40–60%)

(Ref. 11b)

Although lead tetraacetate may be employed for the oxidation of a wide variety of organic compounds,[3,12a,b] its primary use has been for the oxidative cleavage of 1,2-diols, α-hydroxy ketones, 1,2-diketones, and α-hydroxy acids.[2b,c,3,12c] In this respect, lead tetraacetate serves as a reagent used in non-aqueous media to accomplish the same types of reactions effected by periodates with water-soluble compounds. Being a more active oxidizing agent, lead tetraacetate will cleave certain classes of compounds (e.g. α-hydroxy acids) which are oxidized only very slowly by periodates. Although small amounts of water appear not to be deleterious to oxidations with lead tetraacetate, especially in acetic acid solution, the rapid reaction of this reagent with any molecule containing hydroxyl functions prevents the use of water or alcohols as reaction solvents. The acetoxy residues of lead tetraacetate also equilibrate rapidly with the acyloxy groups of other carboxylic acids.

$$Pb(OCOCH_3)_4 + 2 H_2O \longrightarrow PbO_2 + 4 CH_3CO_2H$$

$$Pb(OCOCH_3)_4 + CH_3OH \longrightarrow (CH_3COO)_3Pb-OCH_3 + CH_3CO_2H$$

$$Pb(OCOCH_3)_4 + C_6H_5CO_2H \rightleftharpoons CH_3CO_2H + (CH_3COO)_3Pb-OCOC_6H_5 \rightleftharpoons$$

further equilibration

As in the case of periodate oxidations, the preferred pathway for cleavage of 1,2-diols seems to involve formation of a cyclic intermediate which subsequently decomposes as indicated in structure [8]. Since lead tetraacetate may react

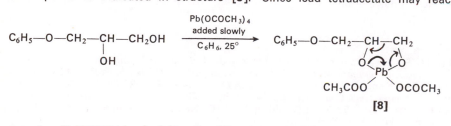

[8]

$$\longrightarrow Pb(OCOCH_3)_2 + C_6H_5-O-CH_2-CHO + [CH_2O]$$

(Ref. 13)

13. R. J. Speer and H. R. Mahler, *J. Am. Chem. Soc.,* **71**, 1133 (1949).

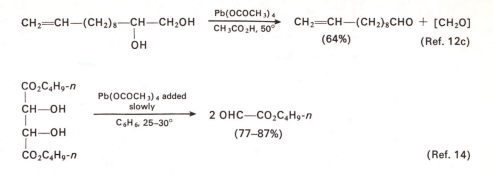

$$CH_2{=}CH{-}(CH_2)_8{-}\underset{\underset{OH}{|}}{CH}{-}CH_2OH \xrightarrow[\text{CH}_3\text{CO}_2\text{H, 50°}]{Pb(OCOCH_3)_4} \quad CH_2{=}CH{-}(CH_2)_8CHO + [CH_2O]$$

(64%) (Ref. 12c)

$$\underset{\underset{CO_2C_4H_9\text{-}n}{|}}{\overset{\overset{CO_2C_4H_9\text{-}n}{|}}{\underset{|}{CH{-}OH}}} \xrightarrow[\text{C}_6\text{H}_6,\ 25\text{–}30°]{\overset{Pb(OCOCH_3)_4 \text{ added}}{\text{slowly}}} \quad 2\ OHC{-}CO_2C_4H_9\text{-}n$$

(77–87%)

(Ref. 14)

further with the initial cleavage product, it is normally best to add the oxidant slowly so that the initially formed oxidation products are not in contact with excess lead tetraacetate throughout the course of the reaction. The preference for the oxidative cleavage to proceed via a cyclic intermediate can lead to significant differences in the rates of oxidation of diastereoisomeric 1,2-diols. As with periodate oxidations, *threo*-1,2-diols are normally oxidized more rapidly than the corresponding *erythro*-isomers and *cis* 1,2-diols in the cyclohexane and cyclopentane systems are oxidized more rapidly than the corresponding *trans* isomers. This latter difference is especially striking if the ring system is relatively rigid so that the hydroxy functions of the *trans*-1,2-diol could not be incorporated into a five-membered ring without the introduction of substantial strain.

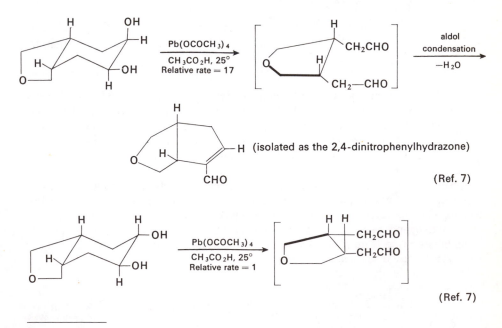

(isolated as the 2,4-dinitrophenylhydrazone)

(Ref. 7)

(Ref. 7)

14. F. J. Wolf and J. Weijlard, *Org. Syn.*, **Coll. Vol. 4**, 124 (1963).

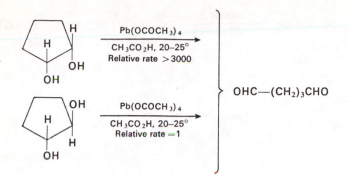

$$OHC{-}(CH_2)_3CHO$$

(Ref. 7, 15)

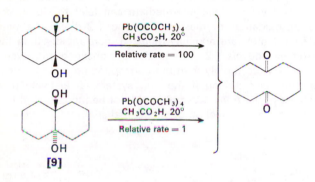

[9]

(Ref. 15a)

Certain examples of the cleavage of *trans*-1,2-diols, e.g. [9], are of special interest since the intervention of a cyclic intermediate prior to cleavage seems most unlikely. Although such *trans*-diols react more slowly than the corresponding *cis*-diols, it is of interest that they are cleaved at all. Furthermore, it has been noted that *trans*-1,2-diols, which are cleaved only slowly with lead tetraacetate in acetic acid, are more readily cleaved if pyridine is used as the reaction solvent.[16] These observations suggest that a second mechanism for glycol cleavage is possible which does not require a cyclic intermediate. The accompanying equations illustrate a possible pathway when the cyclic process is sterically unfavorable.[2b] An interesting application of this type of cleavage to the preparation of a 10-membered ring lactone is provided by the cleavage of the diol [10], a sequence similar to one discussed in Chapter 6.

15. (a) R. Criegee, E. Büchner, and W. Walther, *Ber.*, **73**, 571 (1940). Also see (b) S. J. Angyal and R. J. Young, *J. Am. Chem. Soc.*, **81**, 5467 (1959). (c) W. S. Trahanovsky, L. H. Young, and M. H. Bierman, *J. Org. Chem.*, **34**, 869 (1969).
16. H. R. Goldschmid and A. S. Perlin, *Can. J. Chem.*, **38**, 2280 (1960).

The oxidative cleavage of 1,2-diketones, α-hydroxy ketones, and α-keto acids with lead tetraacetate has been found to occur only when the reaction medium

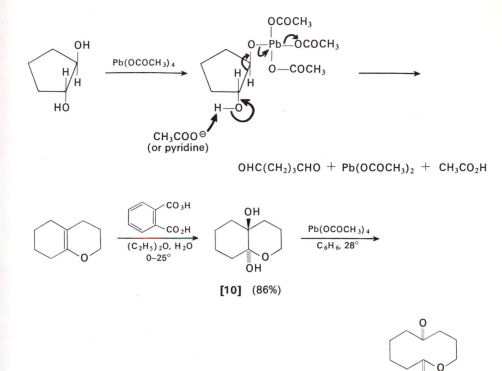

OHC(CH$_2$)$_3$CHO $+$ Pb(OCOCH$_3$)$_2$ $+$ CH$_3$CO$_2$H

[10]   (86%)

(83%)   (Ref. 17)

contains a hydroxylic component such as water or ethanol.[18] This observation, illustrated by the following examples, has been interpreted to mean that the hydrate, e.g. [11], or hemiketal derivative of the carbonyl function is the intermediate which reacts with lead tetraacetate. Presumably a cyclic intermediate such as [12] is involved in the subsequent cleavage reaction.

As noted earlier, monohydric alcohols react rapidly with lead tetraacetate to yield alkoxy lead(IV) intermediates, e.g. [13]. The rates of thermal decomposition for various lead(IV) derivatives have been found to be in the following order: acetates and benzoates < alkoxides ~ monoalkylacetates < dialkyl- and tri-alkylacetates.[11e] These decomposition rates are enhanced by adding donor ligands such as pyridine which can bond to the lead atom, increasing its electron

17. I. J. Borowitz, G. Gonis, R. Kelsey, R. Rapp, and G. J. Wiliams, *J. Org. Chem.,* **31**, 3032 (1966).
18. E. Baer, *J. Am. Chem. Soc.,* **62**, 1597 (1940).

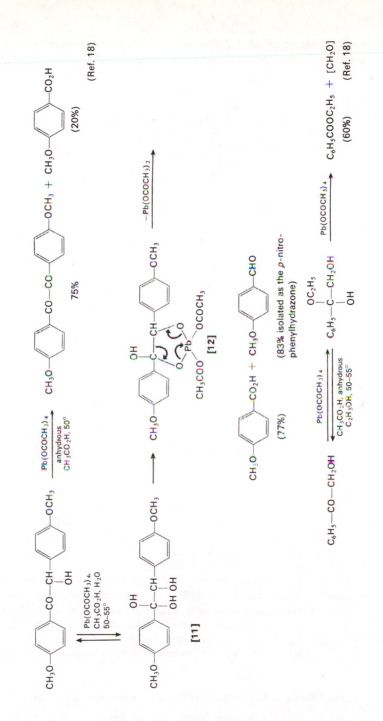

density and facilitating loss of an acetate anion.[11c,e,f,21j] The lead alkoxide intermediates have been found to decompose thermally or photolytically in a variety of ways as illustrated in the following equations.[11b,e,12b,19-21] Both the structural

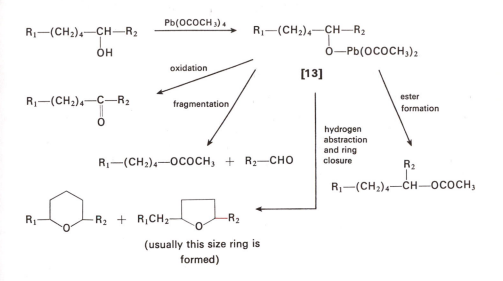

(usually this size ring is
formed)

features of the alcohol and the reaction conditions play a role in determining which of the decomposition paths is preferred. If the lead alkoxide from a primary or secondary alcohol, e.g. [14], is generated in the presence of a donor solvent, especially pyridine, oxidation to an aldehyde or ketone is the dominant mode of decomposition.[21a-c]

19. For a general review, see K. Heusler and J. Kalvoda, *Angew. Chem., Intern. Ed. Engl.,* **3,** 525 (1964).

20. (a) M. L. Mihailovic and co-workers, *Tetrahedron,* **20,** 2279, 2289 (1964); **21,** 1395, 2799, 2813 (1965); **22,** 723, 955 (1966); **23,** 3095 (1967); **24,** 4947, 6959 (1968); **25,** 2269, 3205 (1969); *Chem. Commun.,* **No. 5,** 236 (1969); **No. 14,** 854 (1970); *Helv. Chim. Acta,* **52,** 1146 (1969). (b) D. Hauser, K. Schaffner, and O. Jeger, *ibid.,* **47,** 1883 (1964). (c) D. Hauser, K. Heusler, J. Kalvoda, K. Schaffner, and O. Jeger, *ibid.,* **47,** 1961 (1964). (d) G. Ohloff, K. H. Schulte-Elte, and B. Willhalm, *ibid.,* **49,** 2135 (1966). (e) J. Lhomme and G. Ourisson, *Tetrahedron,* **24,** 3177 (1968). (f) A. C. Cope, M. A. McKervey, N. M. Weinshenker, and R. B. Kinnel, *J. Org. Chem.,* **35,** 2918 (1970).

21. (a) K. Heusler, *Tetrahedron Letters,* **No. 52,** 3975 (1964). (b) R. E. Partch, *ibid.,* **No. 41,** 3071 (1964). (c) R. E. Partch, *J. Org. Chem.,* **30,** 2498 (1965). (d) S. Moon and J. M. Lodge, *ibid.,* **29,** 3453 (1964). (e) S. Moon and P. R. Clifford, *ibid.,* **32,** 4017 (1967). (f) S. Moon and B. H. Waxman, *ibid.,* **34,** 288 (1969). (g) P. Morand and M. Kaufman, *ibid.,* **34,** 2175 (1969). (h) D. Rosenthal, C. F. Lefler and M. E. Wall, *Tetrahedron Letters,* **No. 36,** 3203 (1965). (i) P. Brun, M. Pally, and B. Waegell, *ibid.,* **No. 5,** 331 (1970). (j) W. H. Starnes, Jr., *J. Am. Chem. Soc.,* **90,** 1807 (1968); *J. Org. Chem.,* **33,** 2767 (1968). (k) R. O. C. Norman and R. A. Watson, *J. Chem. Soc.,* B, 184, 692 (1968). (l) I. G. Guest, J. G. L. Jones, B. A. Marples, and M. J. Harrington, *ibid.,* C, 2360 (1969). (m) C. W. Shoppee, J. C. Coll, and R. E. Lack, *ibid.,* C, 1893 (1970).

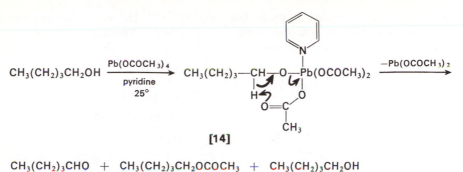

[14]

CH$_3$(CH$_2$)$_3$CHO + CH$_3$(CH$_2$)$_3$CH$_2$OCOCH$_3$ + CH$_3$(CH$_2$)$_3$CH$_2$OH

(75% of product)   (10% of product)   (15% of product)   (Ref. 21b, c)

The progress of this oxidation is easily followed because the red color resulting from interaction of pyridine with lead(IV) salts disappears when all of the lead tetraacetate has been consumed.[21b] The following equation illustrates the change in product composition when a nonbasic solvent such as acetic acid or benzene is

$$CH_3(CH_2)_3CH_2OH \xrightarrow[\substack{CH_3CO_2H, \\ reflux}]{Pb(OCOCH_3)_4} \quad CH_3-\text{[oxolane ring]}_O \quad +$$

(27% of the product)

CH$_3$(CH$_2$)$_3$CHO + CH$_3$(CH$_2$)$_3$CH$_2$OCOCH$_3$

(20% of the product)   (63% of the product)   (Ref. 21c)

employed. The fragmentation of lead alkoxide intermediates, e.g. [15], becomes more favorable with alkoxides from secondary and tertiary alcohols and is especially important if a relatively stable tertiary alkyl, benzyl, or allyl fragment may be lost.[19-21] Although this fragmentation usually involves the initial separation of a radical, the ultimate reactions of the fragment (coupling, elimination, reaction with nucleophilic solvents) may correspond to behavior expected of carbonium ions. Furthermore, in sterically favorable cases, the fragmentation process is reversible.[19,21g,22]

These observations would appear to be best accommodated by the indicated cleavage to a transient alkyl radical which then is oxidized either with the transient lead(III) acetate or with lead tetraacetate to form an alkyl lead intermediate e.g. [16], or a carbonium ion intermediate.[19-21]

This fragmentation may become the major reaction path for tertiary alcohols where a relatively stable alkyl radical may be lost and where competing abstraction of a nearby C—H bond is not sterically favorable. The use of calcium carbonate in the accompanying examples to neutralize the acetic acid formed during the reaction is also worthy of note.

---

22. For examples in which the initially formed radical or cation either rearranges or recombines with the carbonyl group in the reverse manner to form a new carbon-oxygen bond, see Refs. 21f, j, k.

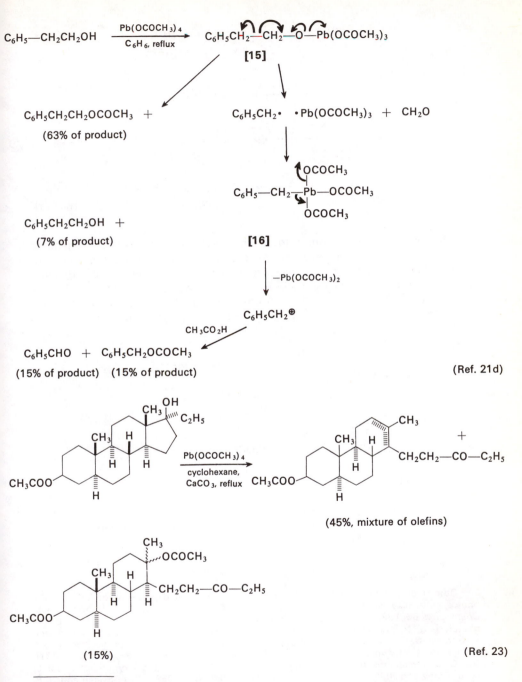

C_6H_5—CH_2CH_2OH $\xrightarrow[C_6H_6, \text{ reflux}]{Pb(OCOCH_3)_4}$ $C_6H_5CH_2$—CH_2—O—$Pb(OCOCH_3)_3$

[15]

$C_6H_5CH_2CH_2OCOCH_3$ +

(63% of product)

$C_6H_5CH_2\cdot$ $\cdot Pb(OCOCH_3)_3$ + CH_2O

$C_6H_5CH_2CH_2OH$ +

(7% of product)

C_6H_5—CH_2—Pb—$OCOCH_3$

[16]

$-Pb(OCOCH_3)_2$

$C_6H_5CH_2^\oplus$

CH_3CO_2H

C_6H_5CHO + $C_6H_5CH_2OCOCH_3$

(15% of product)   (15% of product)

(Ref. 21d)

$\xrightarrow[\substack{\text{cyclohexane,}\\ \text{CaCO}_3, \text{ reflux}}]{Pb(OCOCH_3)_4}$

(45%, mixture of olefins)

(15%)

(Ref. 23)

23. M. Amorosa, L. Caglioti, G. Cainelli, H. Immer, J. Keller, H. Wehrli, M. L. Mihailovic, K. Schaffner, D. Arigoni, and O. Jeger, Helv. Chim. Acta, 45, 2674 (1962).

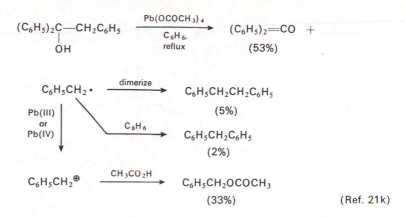

The conversion of alcohols to tetrahydrofurans or, to a lesser extent, to tetrahydropyran derivatives has been of considerable interest because this procedure offers a way to perform selective oxidations at unactivated C—H bonds which can approach sufficiently close to the hydroxyl oxygen atom so that the C—O distance is 2.5–2.8 Å.[11-21,24] A possible path for the reaction is illustrated by the following equations; it will be noted that the configuration at the original

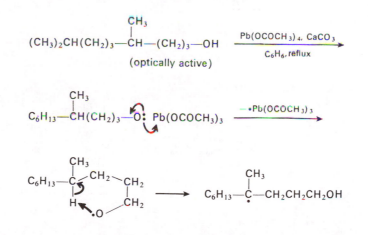

---

24. Other methods for introducing functionality at unactivated positions include (a) the photolysis of alkyl nitrites and alkyl hypoiodites [M. Akhtar and D. H. R. Barton, *J. Am. Chem. Soc.*, **86**, 1528 (1964) ; J. Kalvoda, *Chem. Commun.*, **No. 16**, 1002 (1970)]; (b) the photolysis of N-iodoamides to form γ-lactones [D. H. R. Barton, A. L. J. Beckwith, and A. Goosen, *J. Chem. Soc.*, 181 (1965)]; and (c) the decomposition of protonated N-halo amines, the Hofmann-Loeffler-Freytag reaction [M. E. Wolff, *Chem. Rev.*, **63**, 55 (1963)]. (d) Certain reactions have been suggested to involve a cationic oxygen intermediate [E. J. Corey and R. W. White, *J. Am. Chem. Soc.*, **80**, 6686 (1958) ; R. A. Sneen and N. P. Matheny, *ibid.*, **86**, 3905 (1964) ; A. Deluzarche, A. Maillard, P. Rimmelin, F. Schue, and J. M. Sommer, *Chem. Commun.*, **No. 16**, 976 (1970)].

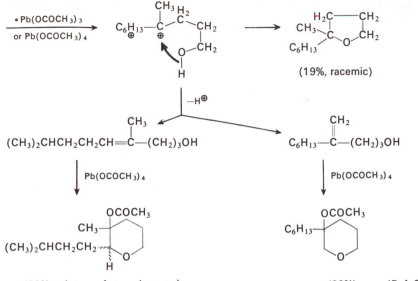

(33%, mixture of stereoisomers)   (20%)   (Ref. 20b)

C—H bond is lost during the ring closure reaction as would be expected with the indicated free radical and carbonium ion intermediates. The preference for abstraction of a hydrogen atom separated from the alkoxy radical oxygen atom by four intervening atoms is a general phenomenon. The following equations illustrate the fact that although activation of an adjacent CH bond by a phenyl group (to form a benzyl radical intermediate) is insufficient to favor formation of a six-membered cyclic ether, activation by an adjacent alkoxy group is effective in this respect.

$$C_6H_5(CH_2)_4CH_2OH \xrightarrow[\text{C}_6\text{H}_6,\text{ reflux}]{\text{Pb(OCOCH}_3)_4} C_6H_5-CH_2-\underset{\underset{H}{|}}{CH}\overset{CH_2-CH_2}{\underset{\cdot O}{\diagdown}}CH_2 \longrightarrow$$

+ $C_6H_5(CH_2)_4CH_2OCOCH_3$ +   other minor products

(50%)   (34%)   (Ref. 21e and 20a, f)

$$C_2H_5OCH_2(CH_2)_3CH_2OH \xrightarrow[\text{C}_6\text{H}_6,\text{ reflux}]{\text{Pb(OCOCH}_3)_4,\text{ CaCO}_3}$$

+   + $C_2H_5O(CH_2)_5OCOCH_3$

(46%)   (2%)   (12%)

(Ref. 20a)

By establishing the structures of ring closed products, this lead tetraacetate oxidation process can also serve to establish the stereochemistry of epimeric pairs of alcohols.[20e,21i] The following equations provide an example.

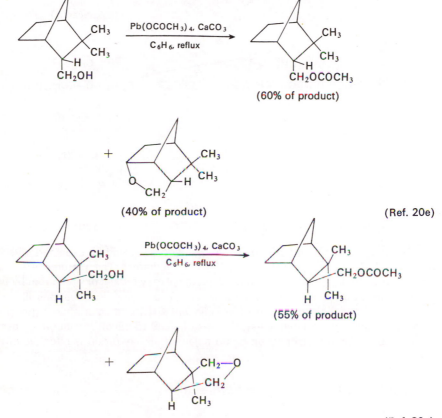

(60% of product)

(40% of product)   (Ref. 20e)

(55% of product)

(25% of product)   (Ref. 20e)

A procedure for achieving this oxidation at unactivated C—H bonds which provides an oxidation product with more synthetic utility than a tetrahydrofuran derivative consists of treatment of an alcohol, e.g. **[17]**, with a solution containing lead tetraacetate and iodine.[19,24a,25] The reaction path believed operative in this reaction is indicated in the accompanying equations. The iodine apparently traps the initially formed carbon radical **[19]** before it can be oxidized by a lead(III) or lead(IV) derivative to form a tetrahydrofuran and, consequently, allows oxidation to continue. The intermediate carbon radical has also been trapped by intra-molecular addition to an appropriately located cyano group.[24a] The alkoxy radical

25. (a) K. Heusler, P. Wieland, and C. Meystre, *Org. Syn.*, **45**, 57 (1965). Also see (b) E. Wenkert and B. L. Mylari, *J. Am. Chem. Soc.*, **89**, 174 (1967). (c) Y. Shalon, Y. Yanuka, and S. Sarel, *Tetrahedron Letters*, **No. 12**, 957, 961 (1969).

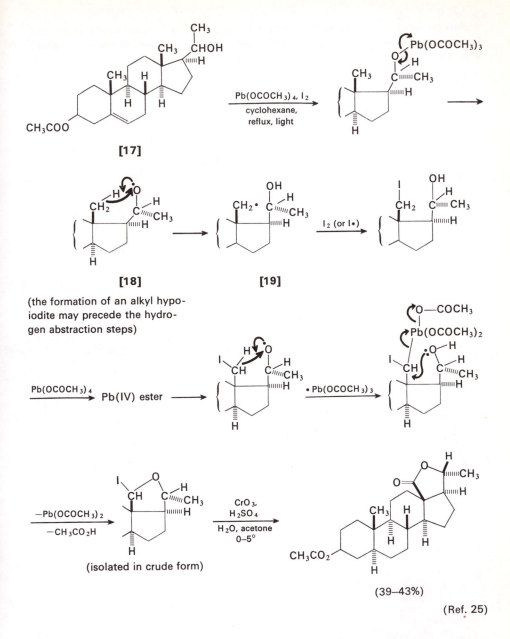

[17]

[18]   [19]

(the formation of an alkyl hypoiodite may precede the hydrogen abstraction steps)

(isolated in crude form)

(39–43%)

(Ref. 25)

intermediates, e.g. **[18]**, which abstract a hydrogen atom from a nearby C—H
bond, may be derived either from homolytic decomposition of the lead(IV) alkoxide
or from pyrolysis (or photolysis) or the I—O bond of an intermediate hypoiodite.[19,24a]

The ability of carboxylic acids to exchange rapidly with the acetate residues of
lead tetraacetate has provided a procedure for the oxidative degradation of

carboxylic acids.[26] The intermediate lead ester, e.g. [20], was originally suggested to undergo the heterolytic cleavage indicated to yield a carbonium ion intermediate which can either react with nearby nucleophilic reagents, lose a proton to form an olefin, or rearrange.

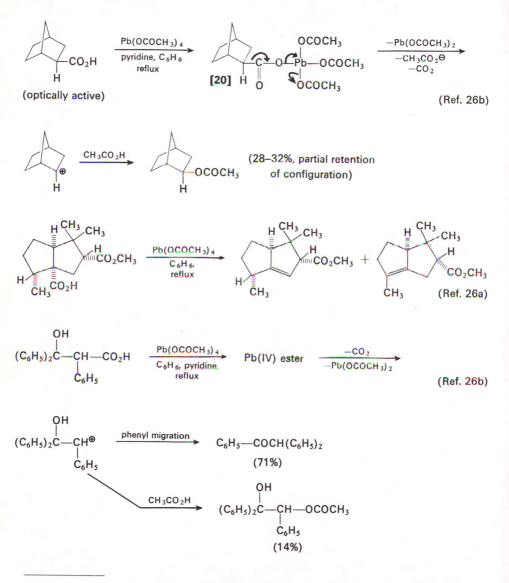

(Ref. 26b)

(28–32%, partial retention of configuration)

(Ref. 26a)

(Ref. 26b)

C₆H₅—COCH(C₆H₅)₂
(71%)

(14%)

26. (a) G. Büchi, R. E. Erickson, and N. Wakabayashi, *J. Am. Chem. Soc.*, **83**, 927 (1961). (b) E. J. Corey and J. Casanova, Jr., *ibid.*, **85**, 165 (1963). (c) N. A. LeBel and J. E. Huber, *ibid.*, **85**, 3193 (1963). (d) L. H. Zalkow and D. R. Brannon, *J. Chem. Soc.*, 5497 (1964). (e) C. R. Bennett and R. C. Cambie, *Tetrahedron*, **23**, 927 (1967).

More recent studies[11c,d,27] have indicated that the thermal or photolytic decomposition of a lead(IV) ester may lead to a transient acyloxy radical in a manner
analogous to the previously discussed fragmentation of lead(IV) alkoxides.
Rapid (or concerted) decarboxylation of this acyloxy radical forms an alkyl radical
e.g. [21], which may then be oxidized by a lead(III) [or a lead(IV)] derivative to
a carbonium ion. The alkyl radical [21] may also initiate a radical chain reaction
leading to decomposition of more lead(IV) ester. This radical chain reaction
may be inhibited by addition of oxygen to the reaction vessel to intercept the alkyl
radicals.[27a]

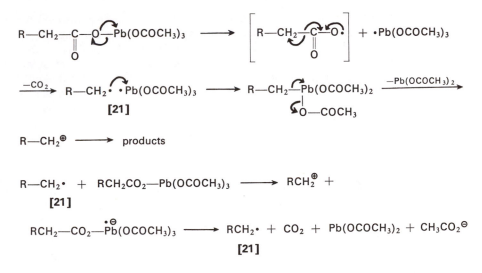

Although the decarboxylation reaction is retarded or inhibited by oxygen, it is
catalyzed by pryridine and by added copper(II) salts. The copper(II) salts are
believed to be more efficient than the lead salts in the oxidation of the alkyl radical
intermediates [21] as the following equation illustrates. The use of lead tetraacetate

$$R-CH_2\cdot + Cu(OCOCH_3)_2 \xrightarrow{-CH_3CO_2^{\ominus}} RCH_2^{\oplus} + CuOCOCH_3$$
[21]
                                           (or the corresponding
                                           acetate or olefin)

$$CuOCOCH_3 + Pb(OCOCH_3)_4 \longrightarrow Cu(OCOCH_3)_2 + \cdot Pb(OCOCH_3)_3$$

27. (a) J. K. Kochi, *J. Am. Chem. Soc.,* **87**, 3609 (1965). J. K. Kochi, J. D. Bacha, and
T. W. Bethea, *ibid.,* **89**, 6538 (1967); J. K. Kochi and J. D. Bacha, *J. Org. Chem.,* **33**, 83,
2746 (1968); *Tetrahedron,* **24**, 2215 (1968); J. K. Kochi, R. A. Sheldon, and S. S. Lande,
ibid., **25**, 1197 (1969). (b) K. Heusler, H. Labhart, and H. Loeliger, *Tetrahedron Letters,*
No. 32, 2847 (1965). (c) D. I. Davies and C. Waring, *J. Chem. Soc.,* C, 1639 (1967);
C, 2333, 2337 (1968). (d) G. E. Gream and D. Wege, *Tetrahedron Letters,* **No. 6**, 503
(1967). (e) R. O. C. Norman and M. Poustie, *J. Chem. Soc.,* B, 781 (1968).

with added pyridine and a copper(II) salt provides a useful method for converting alkylcarboxylic acids to terminal olefins.

$$HO_2C—(CH_2)_6—CO_2C_2H_5 \quad \xrightarrow[\substack{\text{pyridine, } C_6H_6, \\ \text{reflux}}]{Pb(OCOCH_3)_4, \; Cu(OCOCH_3)_2} \quad \xrightarrow[H_2O]{\substack{HNO_3 \text{ (to dissolve the} \\ \text{Pb(II) salts)}}}$$

$$CH_2{=}CH(CH_2)_4CO_2C_2H_5$$
$$(35\%)$$

(Ref. 27a)

As this reaction scheme would predict, the ease of decomposition of the lead(IV) esters increases as the radical [21] formed becomes progressively more stable.[11c,27a] The order is: $C_6H_5CH_2—CO_2H \sim R_3C—CO_2H > R_2CH—CO_2H > RCH_2—CO_2H > CH_3—CO_2H$. With substituted acetic acids which could form either stable radicals or stable carbonium ions, evidence has been presented[27c,d] that free alkyl radicals cannot be involved. In such cases either the alkyl radical is immediately trapped (and oxidized) by the lead(III) salt or a concerted process such as that illustrated in the following equation becomes important.

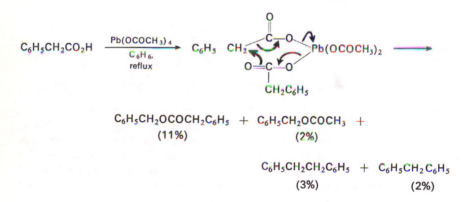

$$C_6H_5CH_2OCOCH_2C_6H_5 \; + \; C_6H_5CH_2OCOCH_3 \; +$$
$$(11\%) \qquad\qquad\qquad (2\%)$$

$$C_6H_5CH_2CH_2C_6H_5 \; + \; C_6H_5CH_2C_6H_5$$
$$(3\%) \qquad\qquad (2\%)$$

(Ref. 27c)

In one case [22], reaction of an acid with lead tetraacetate led to rearrangement rather than decarboxylation. The response of this rearrangement to the electrical effects of substituents suggested that the aryl group was migrating to a cationic center rather than to an acyloxy radical.[28]

The reaction of a carboxylic acid with lead tetraacetate and iodine in carbon tetrachloride with concurrent irradiation by light provides a method for the preparation of alkyl iodides from carboxylic acids.[29] This reaction, which is similar to the

28. W. H. Starnes, Jr., *J. Am. Chem. Soc.,* **86**, 5603 (1964).
29. (a) D. H. R. Barton, H. P. Faro, E. P. Serebryakov, and N. F. Woolsey, *J. Chem. Soc.,* 2438 (1965). (b) B. Weinstein, A. H. Fenselau, and J. G. Thoene, *ibid.,* 2281 (1965).

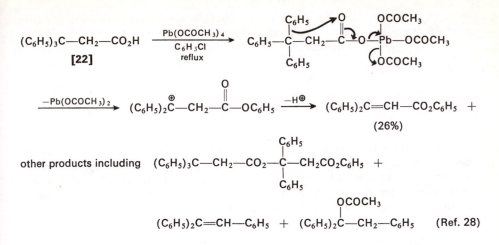

$$(C_6H_5)_2C=CH-CO_2C_6H_5 +$$

(26%)

other products including $(C_6H_5)_3C-CH_2-CO_2-\overset{\overset{\displaystyle C_6H_5}{|}}{\underset{\underset{\displaystyle C_6H_5}{|}}{C}}-CH_2CO_2C_6H_5 +$

$$(C_6H_5)_2C=CH-C_6H_5 + (C_6H_5)_2\overset{\overset{\displaystyle OCOCH_3}{|}}{C}-CH_2-C_6H_5 \qquad \text{(Ref. 28)}$$

Hunsdiecker degradation of carboxylic acids (see Chapter 8), is believed to involve the formation and subsequent homolytic fission of acyl hypoiodites.

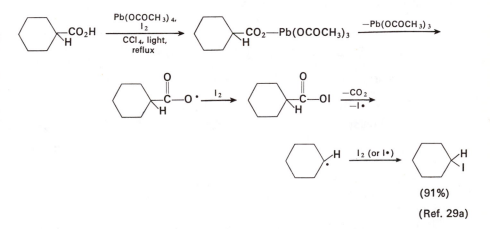

(91%)

(Ref. 29a)

The application of the lead tetraacetate degradation procedure to 1,2-dicarboxylic acids, e.g. [23], results in *bis*-decarboxylation to form olefins.[12b,26b,d,30] The rate of decarboxylation is enhanced by the addition of pyridine. Application of this decarboxylation procedure to either the *meso* or the racemic isomer of 2,3-diphenylsuccinic acid yielded only *trans*-stilbene indicating that loss of the

30. (a) C. A. Grob, M. Ohta, E. Renk, and A. Weiss, *Helv. Chim. Acta*, **41**, 1191 (1958); C. A. Grob and A. Weiss, *ibid.*, **43**, 1390 (1960). (b) N. B. Chapman, S. Sotheeswaran, and K. J. Toyne, *Chem. Commun.*, **No. 11**, 214 (1965). (c) J. J. Tufariello and W. J. Kissel, *Tetrahedron Letters*, **No. 49**, 6145 (1966). (d) T. D. Walsh and H. Bradley, *J. Org. Chem.*, **33**, 1276 (1968). (e) D. V. Hertzler, J. M. Berdahl, and E. J. Eisenbraun, *ibid.*, **33**, 2008 (1968).

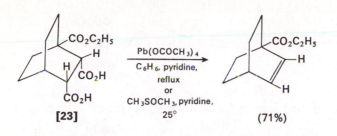

[23]   (71%)   (Ref. 12a, 30a, b)

two carboxyl functions is not a concerted process.[26b] Rather, the lead ester **[24]** is believed to yield a cationic intermediate, e.g. **[25]**, which loses its stereochemical integrity prior to loss of the second carboxyl group. It has recently been found that *bis*-decarboxylation can be effected more rapidly and at lower temperatures by the use of dioxane or dimethyl sulfoxide as the reaction solvent.[30b]

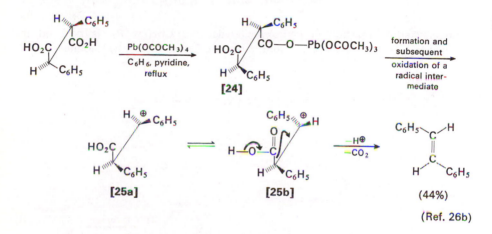

[25a]   **[25b]**   (44%)

(Ref. 26b)

The following equations indicate the use of this decarboxylation procedure for the degradation of disubstituted malonic acids[30c,d] to ketones and γ-keto acids[30e] to α,β-unsaturated ketones.

The reaction of olefins with lead tetraacetate has been suggested to involve the electrophilic attack by the cationic species $^{\oplus}Pb(OCOCH_3)_3$ or its equivalent followed by the addition of acetate anion,[12b] a sequence comparable to that suggested for the subsequently discussed reaction of olefins with mercuric

$$(n\text{-}C_4H_9)_2C(CO_2H)_2 \xrightarrow[\substack{\text{pyridine, } C_6H_6, \\ \text{reflux, } -CO_2}]{Pb(OCOCH_3)_4}$$

$$(n\text{-}C_4H_9)_2C(OCOCH_3)_2 \xrightarrow[\substack{H_2O, \text{ reflux}}]{KOH, \ CH_3OH} n\text{-}C_4H_9\text{—}CO\text{—}C_4H_9\text{-}n$$

(70%)   (Ref. 30c)

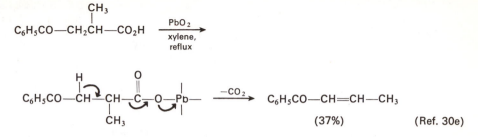

$$C_6H_5CO{-}CH{=}CH{-}CH_3$$

(37%)   (Ref. 30e)

acetate. The initial, symmetrical lead intermediate, e.g. **[26]**, is thought to undergo a variety of heterolytic cleavage processes as illustrated in the following equations.

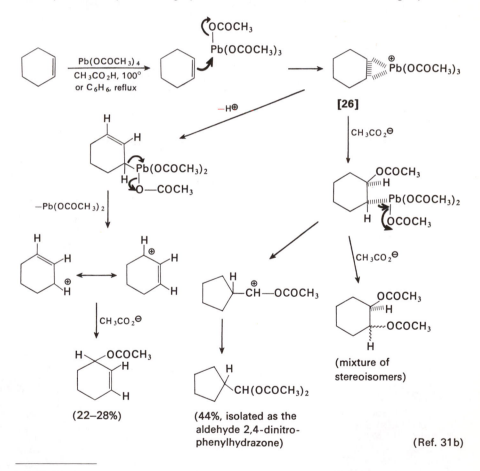

(22–28%)   (44%, isolated as the
                    aldehyde 2,4-dinitro-
                    phenylhydrazone)   (Ref. 31b)

31. (a) H. J. Kabbe, *Justus Liebigs Ann. Chem.*, **656**, 204 (1962). (b) K. B. Wiberg and S. D. Nielsen, *J. Org. Chem.*, **29**, 3353 (1964). (c) S. Wolfe, P. G. C. Campbell, and G. E. Palmer, *Tetrahedron Letters*, **No. 35**, 4203 (1966). (d) W. F. Erman, *J. Org. Chem.*, **32**, 765 (1967). (e) R. O. C. Norman and C. B. Thomas, *J. Chem. Soc.*, B, 604, 771 (1967);

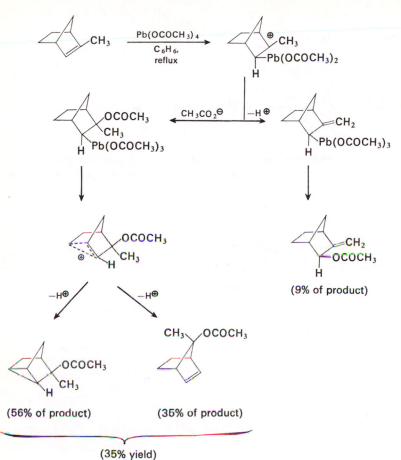

(9% of product)

(56% of product)   (35% of product)

(35% yield)   (Ref. 31d)

In at least some of the cases, the heterolytic cleavage of the lead bonds and nucleophilic attack (or proton loss) appear to be concerted processes so that no free carbonium ion is involved. Since the allylic acetates, e.g. [27], which may be produced in these reactions are especially prone to allylic rearrangement when formed in polar reaction solvents, the conditions used for a reaction may determine

ibid., B, 994 (1968); *ibid.,* B, 421 (1970). (f) D. R. Harvey and R. O. C. Norman, *ibid.,* 4860 (1964). (g) J. B. Aylward, *ibid.,* B, 1268 (1967). (h) J. M. Davidson and C. Triggs, *ibid.,* A, 1331 (1968). (i) R. E. Partch, *J. Am. Chem. Soc.,* **89**, 3662 (1967). (j) E. I. Heiba, R. M. Dessau, and W. J. Koehl, Jr., *ibid.,* **90**, 1082, 2706 (1968). (k) G. E. Gream and D. Wege, *Tetrahedron,* **22**, 2583 (1966). (l) H. Kropf, J. Gelbrich, and M. Ball, *Tetrahedron Letters,* **No. 39**, 3427 (1969). (m) For the oxidation of acetylene with lead tetraacetate, see S. Moon and W. J. Campbell, *Chem. Commun.,* **No. 14**, 470 (1966). (n) For the cleavage of cyclopropane derivatives with lead tetraacetate, see S. Moon, *J. Org. Chem.,* **29**, 3456 (1964); R. J. Ouellette, D. Miller, A. South, Jr., and R. D. Robins, *J. Am. Chem. Soc.,* **91**, 971 (1969).

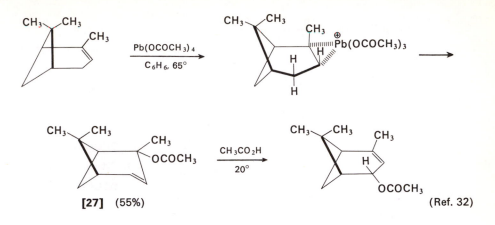

[27] (55%) (Ref. 32)

the structure of the product isolated.[12b,32] Reactive aromatic systems are also attacked by lead tetraacetate in a reaction which appears to be analogous to that described for olefins.

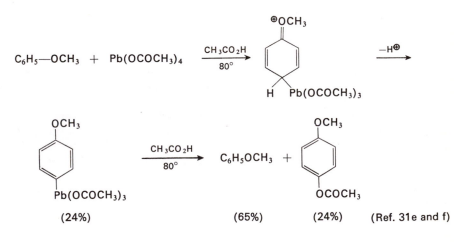

(24%) (65%) (24%) (Ref. 31e and f)

The decomposition of the intermediate lead derivative to form an aryl acetate is suggested to be a free-radical chain reaction.[31e] Even unactivated aromatic systems such as benzene are attacked if lead tetraacetate is used with a boron trifluoride catalyst[31g] or if the more reactive lead tetra(trifluoroacetate) is used.[31i] When vigorous reaction conditions or long reaction times are employed, the previously discussed reactions of olefins or aromatic systems may be accompanied by reactions in which the free radicals $CH_3\cdot$ (from $CH_3CO_2\cdot$) and $HO_2CCH_2\cdot$ have added to the unsaturated substrate.[31e,h,j−m] This type of reaction is illustrated in the following equations.

32. G. H. Whitham, *J. Chem. Soc.*, 2232 (1961).

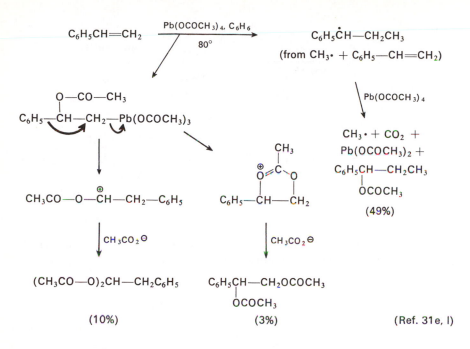

Other examples of this reaction include the methylation of quinones and aromatic nitro compounds by reaction of these substrates with the methyl radicals produced from lead tetraacetate.[11e]

If other electron-rich functional groups (e.g. hydroxyl, carboxyl) are present in the molecule containing the olefinic double bond, these functional groups may participate in the reaction of the olefin with lead tetraacetate.[33] A reaction of this type, which resembles the halolactonization reaction mechanistically (see Chapter 8), is illustrated by the following equations.

Enolizable ketones react with lead tetraacetate to form α-acetoxy ketones.[12b,34] This reaction, which is believed to proceed *via* a lead(IV) enol ether, e.g. [28], is markedly catalyzed by the presence of boron trifluoride[35] which presumably either catalyzes prior formation of an enol or increases the electrophilicity of the

33. (a) S. Moon and L. Haynes, *J. Org. Chem.,* **31**, 3067 (1966). (b) I. Tabushi and R. Oda, *Tetrahedron Letters,* **No. 22**, 2487 (1966). (c) R. M. Moriarty, H. G. Walsh, and H. Gopal, *ibid.,* **No. 36**, 4363 (1966). (d) R. M. Moriarty, H. Gopal, and H. G. Walsh, *ibid.,* **No. 36**, 4369 (1966). (e) T. Sasaki, S. Eguchi, and M. Ohno, *J. Org. Chem.,* **33**, 676 (1968).
34. (a) G. W. K. Cavill and D. H. Solomon, *J. Chem. Soc.,* 4426 (1955). (b) J. W. Ellis, *J. Org. Chem.,* **34**, 1154 (1969) ; *Chem. Commun.,* **No. 7**, 406 (1970). (c) For the reaction of α,β-epoxy ketones with lead tetraacetate, see M. L. Mihailovic, J. Forsek, L. Lorenc, Z. Maksimovic, H. Fuhrer, and J. Kalvoda, *Helv. Chim. Acta,* **52**, 459 (1969).
35. (a) H. B. Henbest, D. N. Jones, and G. P. Slater, *J. Chem. Soc.,* 4472 (1961). (b) J. D. Cocker, H. B. Henbest, G. H. Phillipps, G. P. Slater, and D. A. Thomas, *ibid.,* 6 (1965) ; H. B. Henbest, D. N. Jones, and G. P. Slater, *ibid.,* C, 756 (1967). (c) D. M. Piatak and E. Caspi, *Chem. Commun.,* **No. 15**, 501 (1966). (d) G. R. Pettit, C. L. Herald, and J. P. Yardley, *J. Org. Chem.,* **35**, 1389 (1970).

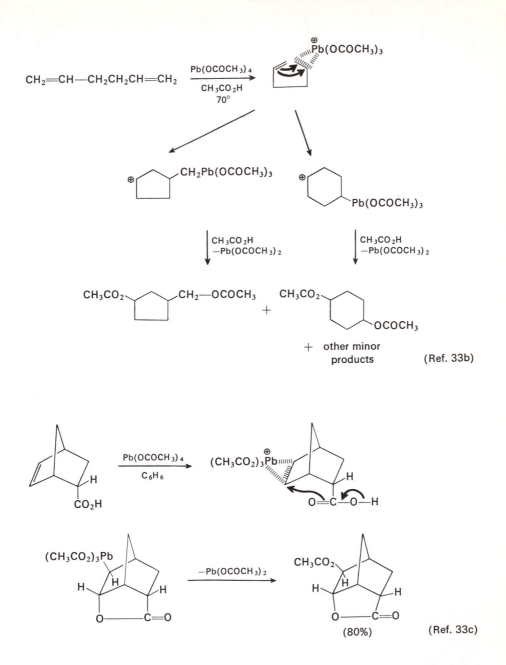

+ other minor
products (Ref. 33b)

(80%) (Ref. 33c)

lead tetraacetate. As the accompanying equation illustrates, reaction of an enolate anion with lead tetraacetate is even more effective. An alternative mechanism which may be operative in uncatalyzed reactions involves the attack of the lead(IV) species at the enol double bond to form an α-keto alkyllead triacetate intermediate.

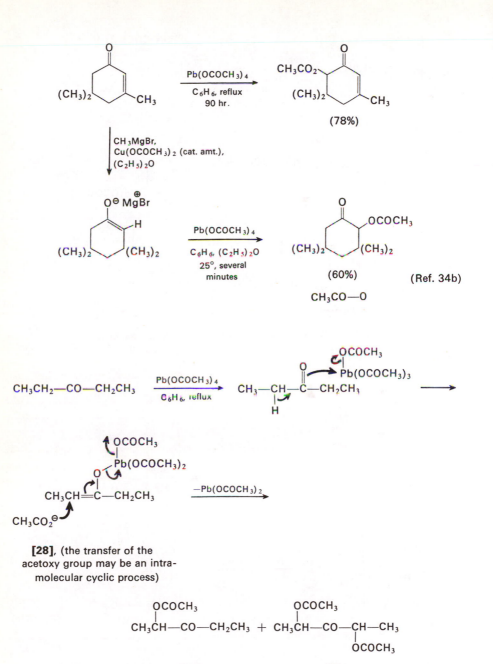

[28], (the transfer of the acetoxy group may be an intra-molecular cyclic process)

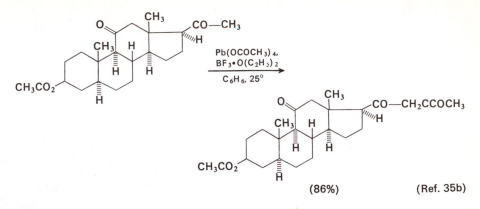

(86%)   (Ref. 35b)

In one instance **[29]** where attack at the enol double bond was sterically hindered, this acetoxylation procedure was accompanied by a concurrent rearrangement process as illustrated in the following equation.

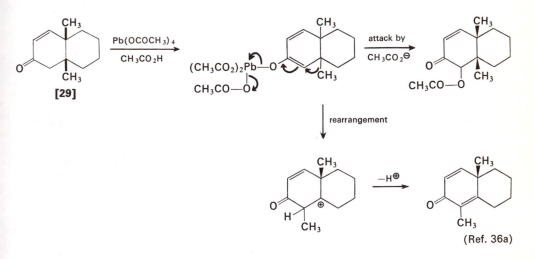

(Ref. 36a)

Lead tetraacetate undergoes a comparable reaction with phenols to form acetoxy cyclohexadienones; the indicated lead phenoxide is believed to be an intermediate.[36b] When ketones or β-dicarbonyl compounds are heated with lead(IV) dioxide, the observed dimeric products would appear to arise from the formation of an intermediate lead enolate which undergoes homolytic bond cleavage to form α-keto radicals which couple.[36c]

---

36. (a) J. A. Marshall and G. L. Bundy, *Chem. Commun.*, **No. 15**, 500 (1966). (b) M. J. Harrison and R. O. C. Norman, *J. Chem. Soc.*, C, 728 (1970). (c) R. Brettle and D. Seddon, *Chem. Commun.*, **No. 23**, 1546 (1968); R. Brettle, *ibid.*, **No. 6**, 342 (1970) and references therein.

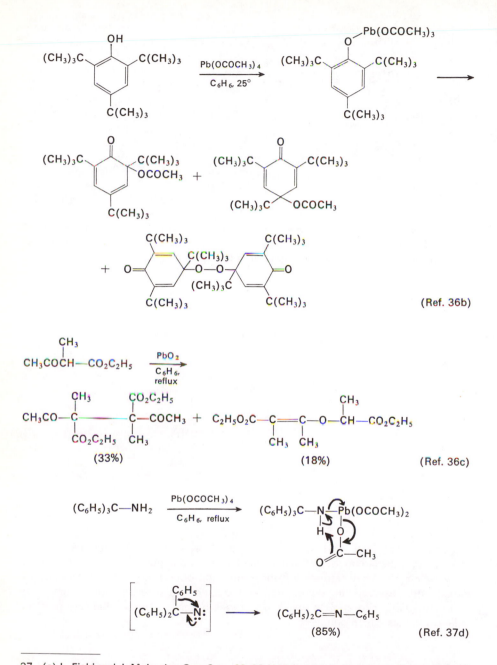

(Ref. 36b)

(33%)   (18%)   (Ref. 36c)

(85%)   (Ref. 37d)

37. (a) L. Field and J. M. Locke, *Org. Syn.*, **46**, 62 (1966). (b) A. Stojiljkovic, V. Andrejevic, and M. L. Mihailovic, *Tetrahedron,* **23**, 721 (1967). (c) K. Nakagawa and H. Onoue, *Chem. Commun.,* **No. 17**, 396 (1965). (d) A. J. Sisti, *ibid.,* **No. 21**, 1272 (1968). (e) W. Nagata, S. Hirai, K. Kawata, and T. Aoki, *J. Am. Chem. Soc.,* **89**, 5045 (1967).

A variety of other functional groupings may be oxidized with lead tetra-acetate.[11,12] Examples are the oxidation of disulfides[37a] and the oxidation of primary amines.[37b-e] The reaction of primary amines with lead tetraacetate appears to result in the initial production of a nitrene intermediate. This intermediate may either rearrange or, in sterically favorable cases, undergo intramolecular cyclo-addition to an olefinic double bond. A similar reaction is observed when trimethyl-silylazide or unsubstituted amides are oxidized with lead tetraacetate.[38a,b,j]

$$n\text{-}C_5H_{11}CH_2NH_2 \xrightarrow[C_6H_6, \text{ reflux}]{Pb(OCOCH_3)_4}$$

$$n\text{-}C_5H_{11}C{\equiv}N \;+\; n\text{-}C_5H_{11}\text{---}CO\text{---}NHCH_2C_5H_{11}\text{-}n \;+\; n\text{-}C_5H_{11}CH_2NHCOCH_3$$

(41%)   (5%)   (12%)   (Ref. 37b)

(80%)   (Ref. 37e)

(62%)   (Ref. 38a)

38. (a) H. E. Baumgarten and A. Staklis, *J. Am. Chem. Soc.,* **87**, 1141 (1965). (b) B. Acott and A. L. J. Beckwith, *Chem. Commun.,* **No. 8**, 161 (1965); A. L. J. Beckwith and R. J. Hickman, *J. Chem. Soc.,* C, 2756 (1968). (c) G. Just and K. Dahl, *Tetrahedron,* **24**, 5251 (1968); J. W. Lown, *J. Chem. Soc.,* B, 441, 644 (1966); B. C. Gilbert and R. O. C. Norman, *ibid.,* B, 86 (1966). (d) W. A. F. Gladstone and R. O. C. Norman, *ibid.,* C, 1527, 1531, 1536 (1966); M. J. Harrison, R. O. C. Norman, and W. A. F. Gladstone, *ibid.,* C, 735 (1967); W. A. F. Gladstone, J. B. Aylward, and R. O. C. Norman, *ibid.,* C, 2587 (1969); D. C. Iffland, L. Salisbury, and W. R. Schafer, *J. Am. Chem. Soc.,* **83**, 747 (1961). (e) B. T. Gillis and M. P. LaMontagne, *J. Org. Chem.,* **32**, 3318 (1967); *ibid.,* **33**, 762, 1294 (1968). (f) For the formation of an acetylene from a 1,2-bishydrazone, see A. Krebs, *Tetrahedron Letters,* **No. 43**, 4511 (1968). (g) D. H. R. Barton, P. L. Batten, and J. F. McGhie, *Chem. Commun.,* **No. 9**, 450 (1969); *J. Chem. Soc.,* C, 1033 (1970); A. Stojiljkovic, N. Orbovic, S. Sredojevic, and M. L. Mihailovic, *Tetrahedron,* **26**, 1101 (1970). (h) A. Bhati, *Chem. Commun.,* **No. 20**, 476 (1965); J. B. Aylward and R. O. C. Norman, *J. Chem. Soc.,* C, 2399 (1968). (i) J. B. Aylward, *ibid.,* C, 1663 (1969); *ibid.,* C, 1494 (1970). (j) K. Kischa and E. Zbiral, *Tetrahedron,* **26**, 1417 (1970); E. Zbiral, G. Nestler, and K. Kischa, *ibid.,* **26**, 1427 (1970). (k) W. J. Middleton and D. M. Gale, *Org. Syn.,* **50**, 6 (1970).

Other substrates which have been oxidized with lead tetraacetate include oximes,[38c] hydrazones,[38d−g,i] azines,[38e] N-acylhydrazines,[38h] and aryl hydrazines.[38i] The following equations provide some typical examples of these oxidation procedures; evidence has been provided for both the heterolytic and homolytic cleavages of the nitrogen-lead bond which are indicated and both reaction paths appear possible.

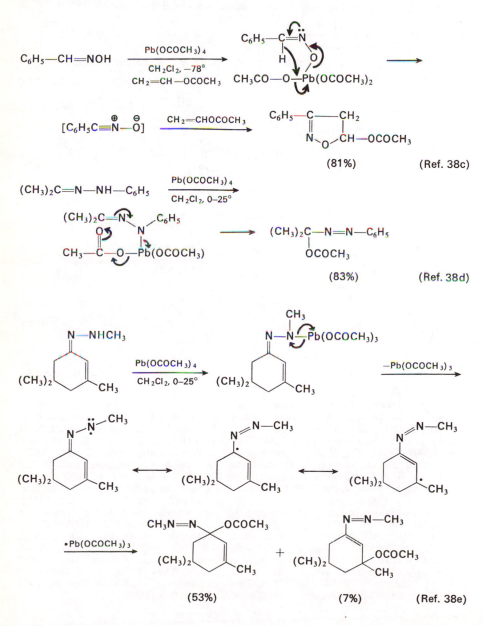

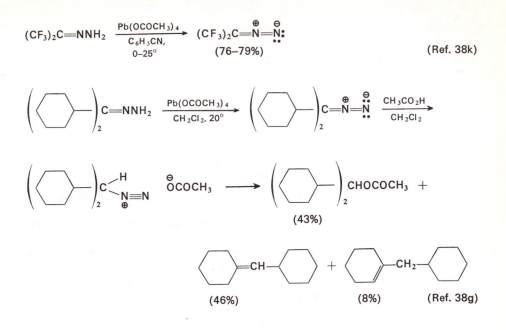

(Ref. 38k)

(43%)

(46%)   (8%)   (Ref. 38g)

OXIDATION WITH MERCURIC ACETATE

Mercuric acetate, a colorless, crystalline solid which is soluble in water, acetic acid, methanol, and ethanol, has been used for the oxidation of a wide variety of organic compounds. The reagent may be prepared in the reaction mixture by use of a mixture of mercuric oxide and acetic acid. In oxidations where an organomercury compound is not isolated, the mercury(II) salt is reduced either to the relatively insoluble mercurous acetate or to metallic mercury. The most common applications of mercuric acetate have been for the oxidation of olefins, ketones, and tertiary amines.

The addition of mercuric acetate to olefins[39] leads to the formation of 1-acetoxy-2-mercuriacetate derivatives. As illustrated in the following equations, the structural and stereochemical outcome of the addition is consistent with the formation of an intermediate acetoxymercurium cation [30] which is subsequently attacked by a nucleophile such as acetate ion. However, recent studies[31c] have suggested that in at least come cases two one-electron transfer steps may precede formation of this cation [30].

39. (a) J. Chatt, *Chem. Rev.,* **48**, 7 (1951). (b) W. Kitching, *Organometal. Chem. Rev.,* **3**, 35, 61 (1968). (c) T. G. Traylor, *Accts. Chem. Res.,* **2**, 152 (1969). (d) For the cleavage of cyclopropane derivatives with mercuric acetate, see R. J. Ouellett, R. D. Robins, and A. South, Jr., *J. Am. Chem. Soc.,* **90**, 1619 (1968); A. DeBoer and C. H. DePuy, *ibid.,* **92**, 4008 (1970). (e) Alkylboranes react with mercuric acetate to form alkylmercuric acetates. R. C. Larock and H. C. Brown, *ibid.,* **92**, 2467 (1970); J. J. Tufariello and M. M. Hovey, *ibid.,* **92**, 3221 (1970). (f) S. Bentham, P. Chamberlain, and G. H. Whitham, *Chem. Commun.,* **No. 22**, 1528 (1970).

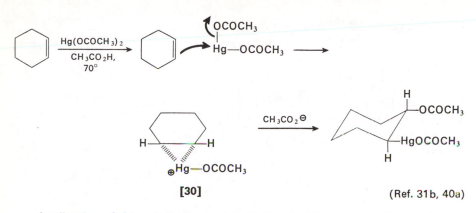

[30] (Ref. 31b, 40a)

Application of this reaction to certain hindered olefins such as the one in the following equation has resulted in the formation of vinylmercury acetates. Presumably, the intermediate acetoxymercurium ion formed in such cases loses a proton more rapidly than it is attacked by a nucleophile.

Mercuric trifluoracetate, prepared as indicated in the accompanying equation, is soluble in various aprotic solvents such as ether, tetrahydrofuran, dimethylformamide, and benzene. This reagent is more electrophilic than mercuric acetate; it reacts rapidly and reversibly with various olefins as the following example indicates.

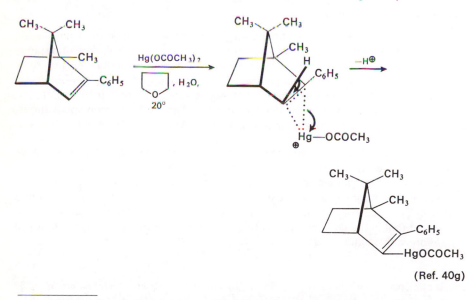

(Ref. 40g)

40. (a) S. Wolfe and P. G. C. Campbell, *Can. J. Chem.,* **43**, 1184 (1965). (b) H. C. Brown and M. H. Rei, *Chem. Commun.,* **No. 22**, 1296 (1969); *J. Am. Chem. Soc.,* **91**, 5646 (1969). (c) H. C. Brown, M. H. Rei, and K. T. Liu, *ibid.,* **92**, 1760 (1970). (d) H. C. Brown and J. T. Kurek, *ibid.,* **91**, 5647 (1969). (e) H. C. Brown, J. H. Kawakami, and S. Misumi, *J. Org. Chem.,* **35**, 1360 (1970). (f) S. Moon and B. H. Waxman, *ibid.,* **34**, 1157 (1969). (g) J. M. Coxon, M. P. Hartshorn, and A. J. Lewis, *Tetrahedron,* **26**, 3755 (1970).

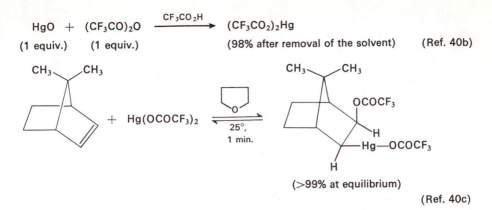

(>99% at equilibrium)

(Ref. 40c)

In the presence of other nucleophiles such as methanol the cyclic mercurium intermediate may react to form an alkoxy mercuriacetate in a reaction called oxymercuration.[41] The following equations illustrate the use of various nucleophiles

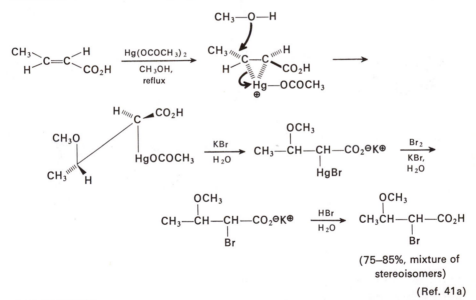

(75–85%, mixture of stereoisomers)

(Ref. 41a)

41. (a) H. E. Carter and H. D. West, *Org. Syn.,* **Coll. Vol. 3,** 813 (1965); also see H. E. Carter and H. D. West, *ibid.,* **Coll. Vol. 3,** 774 (1955). (b) H. C. Brown and P. J. Geoghegan, Jr., *J. Am. Chem. Soc.,* **89,** 1522 (1967); *J. Org. Chem.,* **35,** 1844 (1970). (c) H. C. Brown, and W. J. Hammar, *J. Am. Chem. Soc.,* **89,** 1524 (1967). (d) H. C. Brown, J. H. Kawakami, and S. Ikegami, *ibid.,* **89,** 1525 (1967). (e) D. H. Ballard, A. J. Bloodworth, and R. J. Bunce, *Chem. Commun.,* **No. 14,** 815 (1969); A. J. Bloodworth and R. J. Bunce, *ibid.,* **No. 12,** 753 (1970). (f) J. J. Perie and A. Lattes, *Tetrahedron Letters,* **No. 27,** 2289 (1969). (g) F. G. Bordwell and M. L. Douglass, *J. Am. Chem. Soc.,* **88,** 993 (1966). (h) D. J. Pasto and J. A. Gontarz, *ibid.,* **91,** 719 (1969); G. A. Gray and W. R. Jackson, *ibid.,* **91,** 6205 (1969); G. M. Whitesides and J. San Filippo, Jr., *ibid.,* **92,** 6611 (1970); V. M. A. Chambers, W. R. Jackson, and G. W. Young, *Chem. Commun.,* **No. 20,** 1275 (1970).

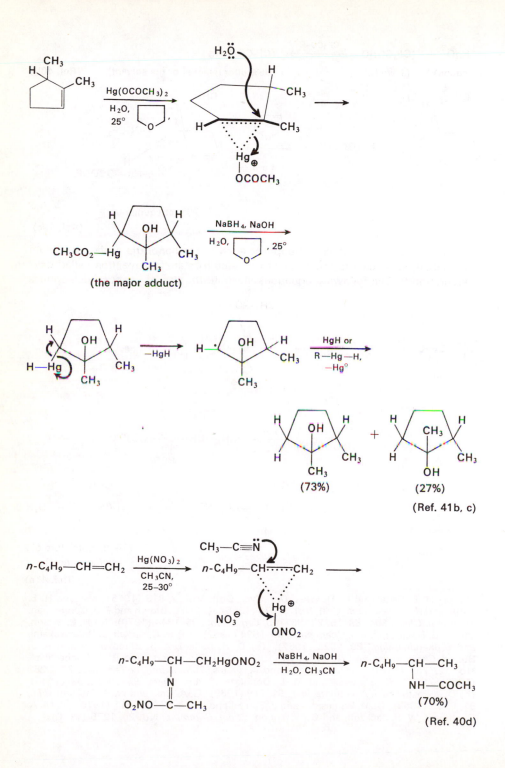

(the major adduct)

(73%)   (27%)

(Ref. 41b, c)

(70%)

(Ref. 40d)

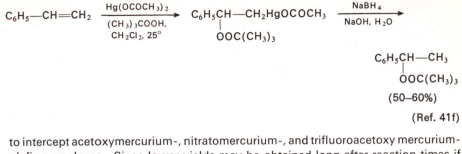

(Ref. 41f)

to intercept acetoxymercurium-, nitratomercurium-, and trifluoroacetoxy mercurium-olefin complexes. Since lower yields may be obtained long after reaction times if the mercury(II) nitrate or trifluoroacetate salts are used, mercury(II) acetate is normally the reagent of choice.[41b] It will be noted that the nucleophile normally attacks the more highly substituted olefinic carbon atom from the less hindered side of the double bond. The fact that the nucleophile and not the acetoxymercury group attacks from the less hindered side of the olefinic double bond in these oxymercuration reactions is in agreement with the belief that formation of the acetoxymercurium ion is a rapid and reversible process[39] which forms both stereo-isomeric acetoxymercurium ion intermediates. Reaction of the resulting substituted alkylmercury derivatives with sodium borohydride to cleave the carbon-mercury bond[40g,41b–h] is also illustrated. A number of these reductive cleavages have been found not to be stereospecific and the illustrated free radical pathway is believed operative in this carbon-mercury cleavage.[41h] Although it has been shown that the hydrogen atom transfered to the alkyl radical intermediate was originally derived from the metal hydride reducing agent, whether the actual hydrogen atom donor is a transient mercury(I) hydride or an alkylmercury hydride (in a radical chain reaction) is not presently known.[41h]

If the reacting olefin contains an internal nucleophile, e.g. [31], this group often participates in the addition reaction rather than the mercurium cation inter-mediate being attacked by a nucleophile from the reaction solvent.[42] Even in cases where a functional group in an olefin would not appear to be involved in the formation of the oxymercuration product, the functional group may influence the stereochemistry of the reaction at the olefin.[42e–g] For example, although the oxymercuration of α,β-unsaturated carbonyl compounds with no substituents at the β-position yields compounds with mercury bonded to the less highly sub-stituted β-position, analogous β-substituted-α,β-unsaturated carbonyl compounds give products with mercury bonded at the α-position.[41e] Such results are believed to arise from a combination of the inductive effect of the electronegative sub-stituent which influences the position of attack by the entering nucleophile

42. (a) F. R. Jensen and J. J. Miller, *Tetrahedron Letters*, **No. 40**, 4861 (1966). (b) G. V. Baddeley, R. A. Eade, J. Ellis, P. Harper, and J. J. H. Simes, *Tetrahedron*, **25**, 1643 (1969). (c) A. Factor and T. G. Traylor, *J. Org. Chem.*, **33**, 2607 (1968). (d) S. Moon, J. M. Takakis, and B. H. Waxman, *ibid.*, **34**, 2951 (1969). (e) M. R. Johnson and B. Rickborn, *ibid.*, **34**, 2781 (1969) ; *Chem. Commun.*, **No. 18**, 1073 (1968). (f) S. Moon and B. H. Waxman, *ibid.*, **No. 24**, 1283 (1967). (g) J. Klein and R. Levene, *Tetrahedron Letters*, **No. 54**, 4833 (1969). (h) P. Chamberlain and G. H. Whitham, *J. Chem. Soc.*, B, 1382 (1970). (i) F. D. Gunstone and R. P. Inglis, *Chem. Commun.*, **No. 14**, 877 (1970).

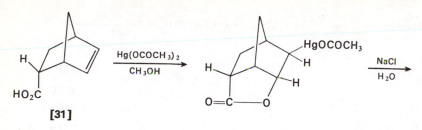

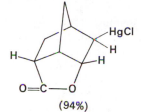

(94%)

(Ref. 42a)

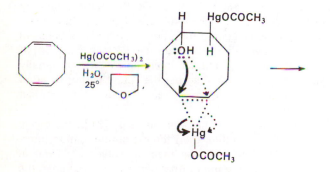

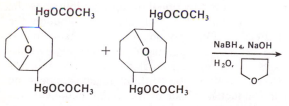

+

NaBH₄, NaOH

H₂O,

(54% of product)   +   (46% of product)

(63% yield)

(Ref. 42d)

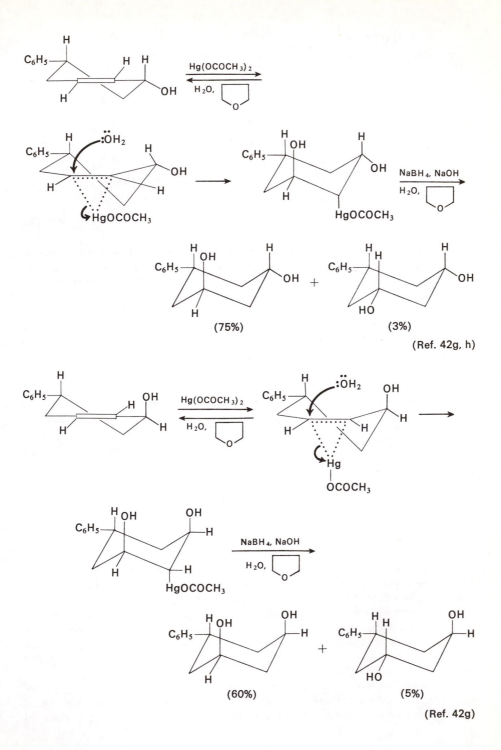

(75%) + (3%)

(Ref. 42g, h)

(60%) + (5%)

(Ref. 42g)

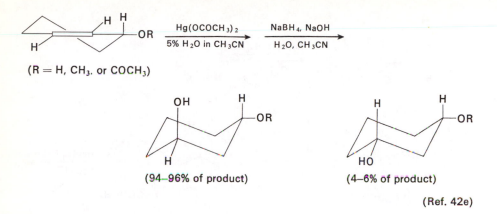

(R = H, CH₃, or COCH₃)

(94–96% of product)   (4–6% of product)

(Ref. 42e)

accompanied by the tendency of the mercurium ion-olefin complexes from cyclo-hexene derivatives to open in such a way that a *trans*-diaxial product will be formed. There is no compelling evidence for invoking complex formation with the hydroxyl group as a factor in controlling the course of the reaction of cyclohexenols with mercury acetate.[42h]

Both the initial formation of a cyclic acetoxymercurium cation[39b] and its subsequent acetoylsis in acetic acid are reversible processes as the following equation indicates.

C_6H_5—CH_2—CH—CH_2 ⎡OCOCH₃ / HgOCOCH₃⎤   + CD_3CO_2H ⇌ (70°) C_6H_5—CH_2CH—CH_2 Hg—OCOCH₃ ⇌

C_6H_5—CH_2—CH—CH_2 ⎡OCOCD₃ / Hg—OCOCH₃⎤   + CH_3CO_2H

(Ref. 31c)

In several of the accompanying examples it will be noted that the initially formed mercuriacetate has been converted to a mercurihalide to facilitate product isolation. The oxymercuration adducts may be reconverted to olefins by heating with a solution of lithium chloride in dimethylformamide.[40e] When the oxymercura-tion-demercuration reaction sequence was conducted in the presence of an equimolar amount of silver acetate, the expected alcohol product was accompanied by substantial amounts of the corresponding ketone; the mechanism of this transformation is uncertain.

If the olefin-mercury salt adducts are heated in refluxing acetic acid, they are usually converted to the corresponding allylic acetates accompanied by acetic acid and mercury (if no excess mercuric acetate is present) or mercurous acetate (if

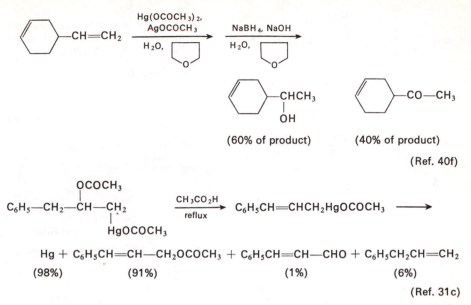

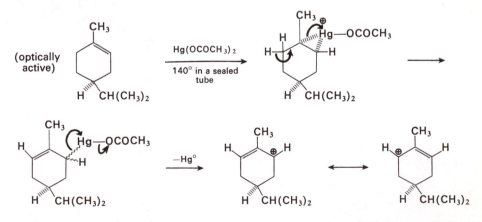

excess mercuric acetate is present).[31b,c,43] If formation of the allylic acetate is the desired result, the reaction is most commonly effected by heating the olefin with mercuric acetate either in acetic acid or in a sealed tube. The following equations illustrate these procedures and also indicate possible paths for the reactions.[31b,c,43] Frequently, an equilibrated mixture of allylic acetates is formed in this reaction because of a rapid mercury salt-catalyzed isomerization of the initially formed mixture of allyl esters.[43c]

43. (a) D. H. R. Barton and W. J. Rosenfelder, *J. Chem. Soc.*, 2381 (1951). (b) Z. Rappoport, P. D. Sleezer, S. Winstein, and W. G. Young, *Tetrahedron Letters,* **No. 42**, 3719 (1965). (c) Z. Rappoport, L. K. Dyall, S. Winstein, and W. G. Young, *ibid.,* **No. 40**, 3483 (1970). (d) For the formation of a diacetate by this procedure, see R. M. Carlson and R. K. Hill, *Org. Syn.,* **50**, 24 (1970).

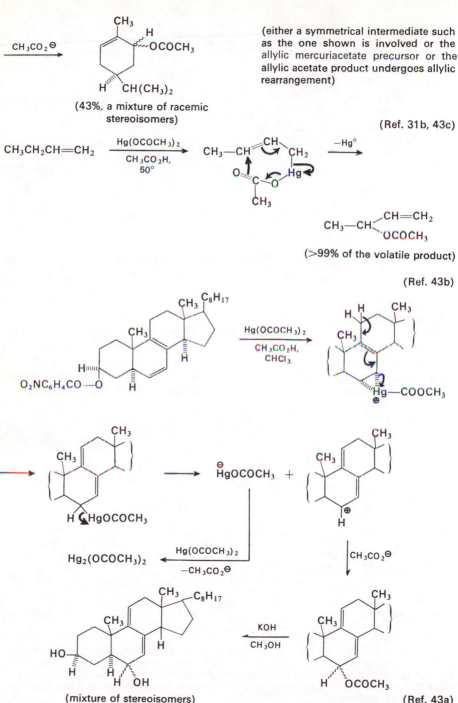

CH₃CO₂⁻ →

(43%, a mixture of racemic stereoisomers)

(either a symmetrical intermediate such as the one shown is involved or the allylic mercuriacetate precursor or the allylic acetate product undergoes allylic rearrangement)

(Ref. 31b, 43c)

$CH_3CH_2CH{=}CH_2$ $\xrightarrow[\substack{CH_3CO_2H,\\50°}]{Hg(OCOCH_3)_2}$

$-Hg°$ →

(>99% of the volatile product)

(Ref. 43b)

$O_2NC_6H_4CO{-}O$

$\xrightarrow[\substack{CH_3CO_2H,\\CHCl_3}]{Hg(OCOCH_3)_2}$

→

$\overset{\ominus}{Hg}OCOCH_3$ +

$Hg_2(OCOCH_3)_2$ $\xleftarrow[{-CH_3CO_2^\ominus}]{Hg(OCOCH_3)_2}$

$CH_3CO_2^\ominus$

HO—

(mixture of stereoisomers)

$\xleftarrow[CH_3OH]{KOH}$

(Ref. 43a)

In related reactions, mercury(II) salts have also been used to catalyze the exchange of the vinyl group of vinyl esters, e.g. **[32]**, and vinyl ethers with carboxylic acids and with alcohols. The following equations illustrate the reaction conditions and probable reaction path used for this vinyl exchange reaction. Mercury(II) salts have also been used frequently to catalyze the addition of nucleophiles to acetylenes and allenes (e.g., **[33]**).[47] The partial retention of optical activity after reduction of the allene **[33]** requires that at least some of the product was derived from attack of the nucleophile on the mercurium complex rather than the free allyl carbonium ion which is not asymmetric.[47c,d]

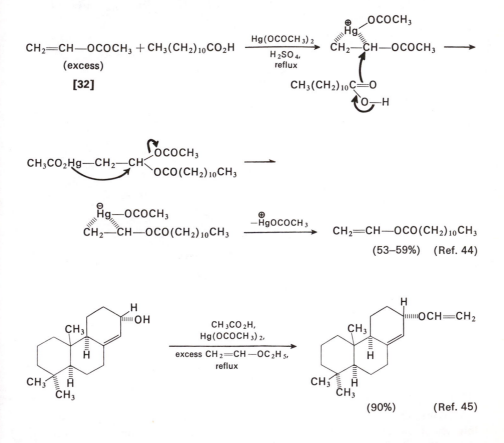

44. (a) D. Swern and E. F. Jordan, Jr., *Org. Syn.,* **Coll. Vol. 4**, 977 (1963). (b) H. Lüssi, *Helv. Chim. Acta,* **49**, 1681, 1684 (1966). (c) G. Slinckx and G. Smeis, *Tetrahedron,* **23**, 1395 (1967).
45. R. F. Church, R. E. Ireland, and J. A. Marshall, *J. Org. Chem.,* **31**, 2526 (1966). The preparation of allyl vinyl ethers by this procedure may require considerable experimentation to find the optimum reaction conditions.

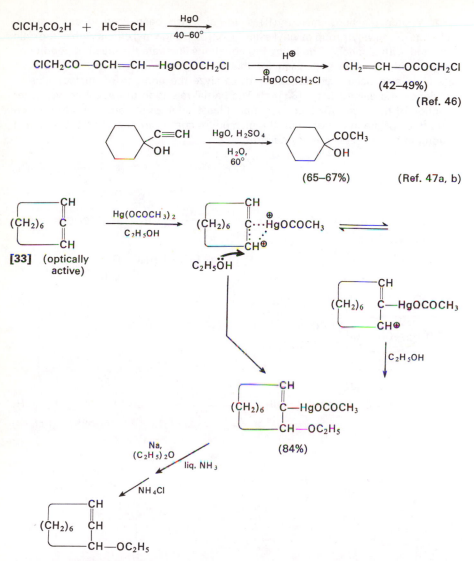

(part of optical activity retained)   (Ref. 47d)

Enolizable ketones, e.g. [34], react with mercuric acetate in acetic acid to form α-acetoxy derivatives in a process that appears to be mechanistically analogous to the reaction of mercuric acetate with olefins. However, an alternative mechanism

46. R. H. Wiley, *Org. Syn.*, **Coll. Vol. 3**, 853 (1955).
47. (a) G. W. Stacy and R. A. Mikulec, *Org. Syn.*, **Coll. Vol. 4**, 13 (1963). (b) M. S. Newman and W. R. Reichle, *ibid.*, **40**, 88 (1960). (c) W. L. Waters and E. F. Kiefer, *J. Am. Chem. Soc.*, **89**, 6261 (1967) ; W. S. Linn, W. L. Waters, and M. C. Caserio, *ibid.*, **92**, 4018 (1970). (d) R. D. Bach, *ibid.*, **91**, 1771 (1969).

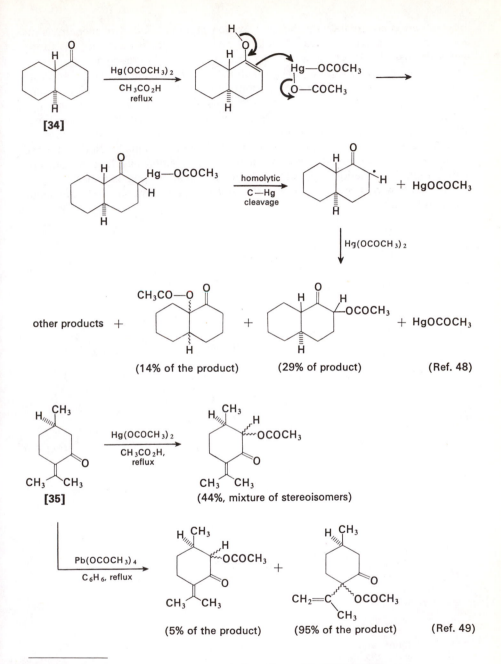

(14% of the product) (29% of product) (Ref. 48)

(44%, mixture of stereoisomers)

[35]

(5% of the product) (95% of the product) (Ref. 49)

48. H. O. House and H. W. Thompson, *J. Org. Chem.*, **26**, 3729 (1961).
49. (a) R. H. Reitsema, *J. Am. Chem. Soc.*, **79**, 4465 (1957). (b) L. H. Zalkow, J. W. Ellis, and M. R. Brennan, *J. Org. Chem.*, **28**, 1705 (1963). (c) L. H. Zalkow and J. W. Ellis, *ibid.*, **29**, 2626 (1964).

in which an intermediate with the enol oxygen atom is bonded to mercury is involved has not been excluded. This latter possibility would be analogous to the earlier discussed reactions of enolizable ketones with lead tetraacetate.

The behavior of pulegone [35] on acetoxylation with mercuric acetate and with lead tetraacetate would appear to be attributable to a change in the reaction solvent. As was discussed previously in connection with the dehydrogenation of α,β-unsaturated ketones with quinones (see Chapter 1), reaction in acetic acid may result in rapid formation and equilibration of the intermediate enols so that the product observed is derived from the more stable enol. The major product formed in benzene solution would appear to arise from the kinetically favored, but less stable, enol. The reaction of phenols with mercuric acetate appears to be analogous to the reaction of enols except that the aryl mercury compound is sufficiently stable to be isolated.

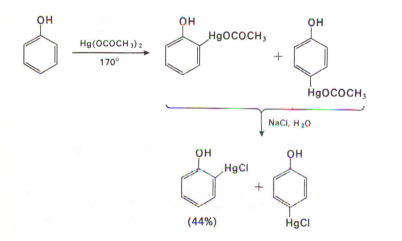

(44%)   (Ref. 50)

Although mercury(II) salts have been found to oxidize α-hydroxy ketones to 1,2-diketones,[51] this transformation is accomplished more satisfactorily by the use of copper(II) acetate or bismuth(III) acetate.

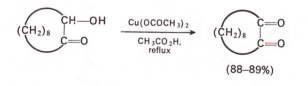

(88–89%)   (Ref. 52)

50. F. C. Whitmore and E. R. Hanson, *Org. Syn.*, **Coll. Vol. 1**, 161 (1944).
51. S. Patai and I. Shenfeld, *J. Chem. Soc.*, B, 366 (1966).
52. (a) A. T. Blomquist and A. Goldstein, *Org. Syn.*, **Coll. Vol. 4**, 838 (1963). (b) For the use of oxygen, copper sulfate, and pyridine to effect this type of oxidation, see H. T. Clarke and E. E. Dreger, *Org. Syn.*, **Coll. Vol. 1**, 87 (1955). (c) The mechanism of this oxidation is discussed by K. B. Wiberg and W. G. Nigh, *J. Am. Chem. Soc.*, **87**, 3849 (1965).

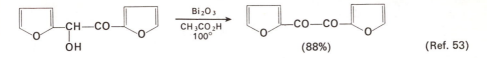

(88%)   (Ref. 53)

The oxidation of tertiary amines, e.g. [36], with mercuric acetate provides a preparative route to enamines (*cf.* Chapter 9). In certain cases the initially formed

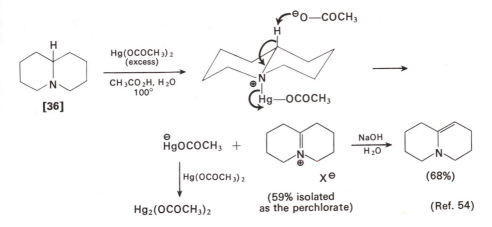

[36]

$\overset{\ominus}{HgOCOCH_3}$ +

Hg(OCOCH₃)₂

Hg₂(OCOCH₃)₂

(59% isolated as the perchlorate)

X⊖

NaOH H₂O

(68%)

(Ref. 54)

iminium salt [37], which is in equilibrium with the corresponding enamine in the weakly acid reaction medium, has been found to undergo further reaction with mercuric acetate to form, after saponification, a β-hydroxy enamine. As illustrated in the following equations, this process is apparently mechanistically analogous to the previously described α-acetoxylation of ketones with mercuric acetate.

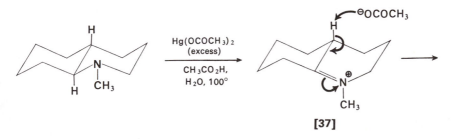

[37]

53. (a) W. Rigby, *J. Chem. Soc.*, 793 (1951). (b) also see B. Holden and W. Rigby, *ibid.,* 1924 (1951).
54. (a) N. J. Leonard, A. S. Hay, R. W. Fulmer, and V. W. Gash, *J. Am. Chem. Soc.*, **77**, 439 (1955). (b) N. J. Leonard and R. R. Sauers, *ibid.,* **79**, 6210 (1957). (c) also see N. J. Leonard, W. J. Middleton, P. D. Thomas, and D. Choudhury, *J. Org. Chem.,* **21**, 344 (1956).
55. (a) N. J. Leonard, L. A. Miller, and P. D. Thomas, *J. Am. Chem. Soc.,* **78**, 3463 (1956). (b) see also N. J. Leonard, R. W. Fulmer, and A. S. Hay, *ibid.,* **78**, 3457 (1956).
56. N. J. Leonard and F. P. Hauck, Jr., *J. Am. Chem. Soc.,* **79**, 5279 (1957).

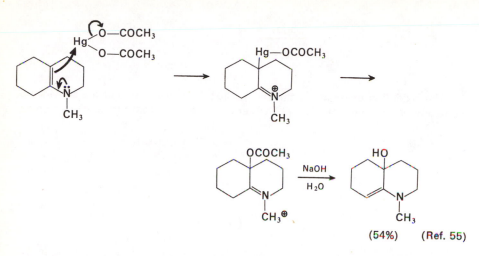

(54%)   (Ref. 55)

In the absence of subsequently discussed unfavorable conformation factors, the reaction of a tertiary amine with mercuric acetate normally yields the most highly substituted iminium salt. The following examples illustrate this generality as well as the tendency of monosubstituted enamines, e.g. **[38]**, to dimerize.

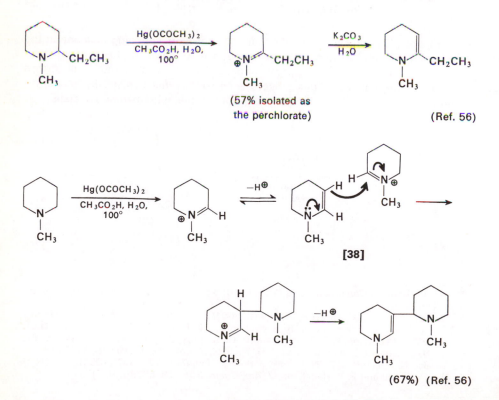

It has been shown that decomposition of the amine-mercuri-acetate complex, e.g. [39], occurs readily only with systems where the free electron pair on nitrogen and the α-hydrogen atom can readily adopt a *trans* coplanar arrangement. For example, under conditions where yohimbine [40] was readily oxidized, the epimeric pseudoyohimbine [41] failed to react. Also, the bicyclic amine [42] was oxidized at the methyl group rather than at one of the tertiary *alpha* C—H bonds because neither of the tertiary bonds can be arranged in a conformation which is *trans* to and coplanar with the electron pair on nitrogen.

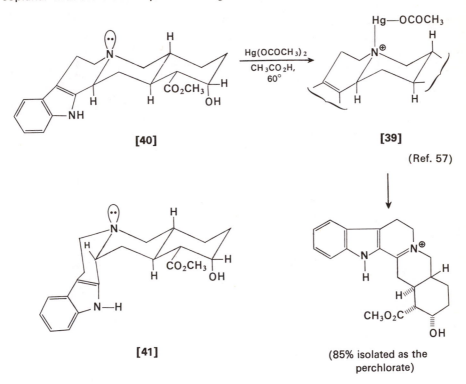

[40]

Hg(OCOCH₃)₂
CH₃CO₂H,
60°

[39]

(Ref. 57)

[41]

(85% isolated as the perchlorate)

The cyclic iminium salts, e.g. [43], produced in these oxidations may serve as synthetically useful electrophilic reagents. These intermediates, which reduced to the starting tertiary amines by catalytic hydrogenation (see Chapter 1), by the Clemmensen reduction procedure (see Chapter 3), by metal hydrides (see Chapter 2), or by formic acid,[54b] readily add nucleophiles such as cyanide ion and Grignard reagents to form substituted amines. The further transformation of the amine [44] to a medium-sized ring compound by a modified Hofmann degradation offers an interesting use of an iminium salt intermediate. Intramolecular additions of nucleophiles to iminium salt intermediates have also provided a useful method for ring closure, e.g. [45].

57. F. L. Weisenborn and P. A. Diassi, *J. Am. Chem. Soc.,* **78**, 2022 (1956).

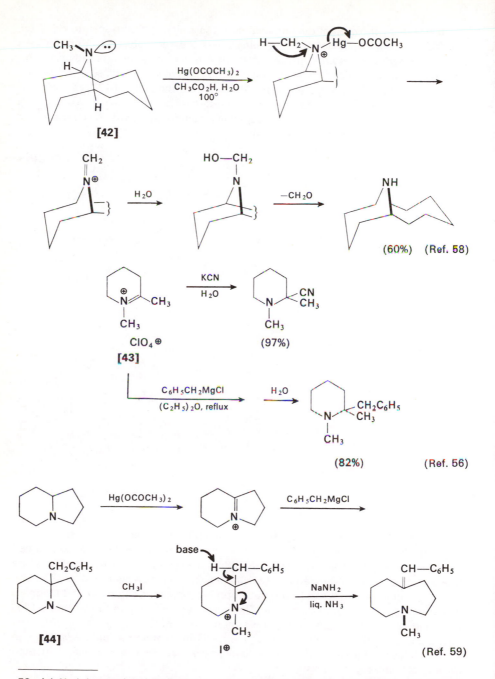

(60%)   (Ref. 58)

(97%)

(82%)   (Ref. 56)

(Ref. 59)

58. (a) N. J. Leonard and D. F. Morrow, *J. Am. Chem. Soc.,* **80**, 371 (1958). (b) also see F. Bohlman and P. Strehlke, *Tetrahedron Letters,* **No. 3**, 167 (1965).
59. M. G. Reinecke, L. R. Kray, and R. F. Francis, *Tetrahedron Letters,* **No. 40**, 3549 (1965).

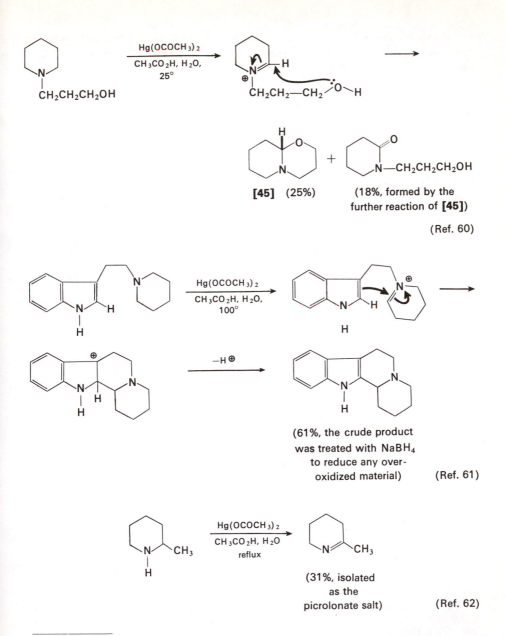

[45]   (25%)   (18%, formed by the
                    further reaction of [45])

(Ref. 60)

(61%, the crude product
was treated with NaBH₄
to reduce any over-
oxidized material)   (Ref. 61)

(31%, isolated
as the
picrolonate salt)   (Ref. 62)

60. (a) N. J. Leonard and W. K. Musker, *J. Am. Chem. Soc.,* **82**, 5148 (1960). (b) W. Schneider, *Angew. Chem., Intern. Ed. Engl.,* **4**, 536 (1965).
61. E. Wenkert and B. Wickberg, *J. Am. Chem. Soc.,* **84**, 4914 (1962).
62. (a) M. F. Grundon and B. E. Reynolds, *J. Chem. Soc.,* 2445 (1964). (b) For the reaction of secondary amines with mercuric acetate and carbon monoxide to form biscarbamoyl-mercury compounds, see U. Schöllkopf and F. Gerhart, *Angew. Chem., Intern. Ed. Engl.,* **5**, 664 (1966).

Although the oxidation of secondary amines with mercuric acetate is a less satisfactory procedure[54a,62] than oxidation of tertiary amines, moderate yields of cyclic imines have been obtained in certain cases. The oxidation of hydrazones with mercuric oxide to diazo compounds is also worthy of note.[63] Application of the same procedure to the bishydrazones, e.g. [46], constitutes a useful synthesis of acetylenes. The indicated cyclization of a bis-phenylhydrazone is also of interest.

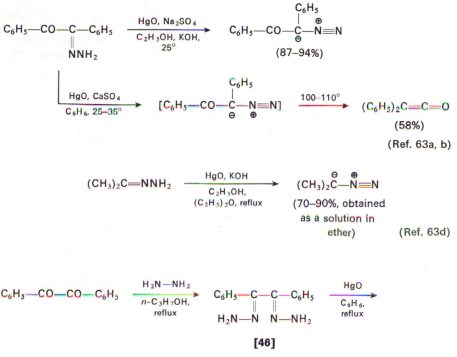

(87–94%)

(58%)

(Ref. 63a, b)

(70–90%, obtained as a solution in ether)   (Ref. 63d)

[46]

(83–89%)

$C_6H_5-C\equiv C-C_6H_5$

(67–63% overall)   (Ref. 64)

63. (a) C. D. Nenitzescu and E. Solomonica, *Org. Syn.,* **Coll. Vol. 2**, 496 (1943). (b) L. I. Smith and H. H. Hoehn, *ibid.,* **Coll. Vol. 3**, 356 (1955). (c) L. I. Smith and K. L. Howard, *ibid.,* **Coll. Vol. 3**, 351 (1955). (d) S. D. Andrews, A. C. Day, P. Raymond, and M. C. Whiting, *ibid.,* **50**, 27 (1970). (e) The use of nickel peroxide for the oxidation of hydrazones to diazo compounds has been recommended recently; K. Nakagawa, H. Onoue, and K. Minami, *Chem. Commun.,* **No. 20**, 730 (1966).
64. (a) A. C. Cope, D. S. Smith, and R. J. Cotter, *Org. Syn.,* **Coll. Vol. 4**, 377 (1963). (b) G. Wittig and H. Heyn, [*Chem. Ber.,* **97**, 1609 (1964)] have found manganese dioxide to be a more effective oxidant for the preparation of strained acetylene derivatives. (c) A. J. Bellamy, R. D. Guthrie, and G. J. F. Chittenden, *J. Chem. Soc.,* C, 1989 (1966).

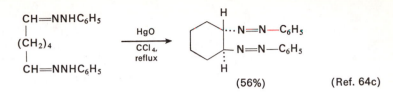

(56%)   (Ref. 64c)

OXIDATION WITH SELENIUM DIOXIDE

Selenium dioxide, a poisonous white crystalline solid which melts at 340° and may be sublimed at atmospheric pressure, has served as an oxidant for a variety of organic compounds.[65] It is most commonly used as a solution in dioxane, acetic acid, acetic anhydride, water, or ethanol. In aqueous or alcoholic solutions, the oxide is converted to selenious acid, $(HO)_2SeO$, or to the corresponding dialkyl-selenite ester. The reagent may be added to the reaction mixture as selenious acid. During the oxidation process, the selenium(IV) reactant is reduced to metallic selenium, a red to black insoluble solid.

The most common uses of this oxidant are the conversions of ketones or aldehydes to 1,2-dicarbonyl compounds,[65,66] the conversion of 1,4-diketones

$$C_6H_5-CO-CH_3 \xrightarrow[\substack{H_2O, \text{ dioxane,} \\ \text{reflux}}]{SeO_2} Se_8 + C_6H_5-CO-CHO$$

(69–72%)   (Ref. 68)

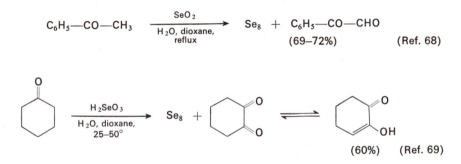

(60%)   (Ref. 69)

to 2,3-unsaturated-1,4-diketones,[67] and the oxidation of olefins to allylic alcohol derivatives.[31b,65] The accompanying equations illustrate the use of the reagent for preparing 1,2-dicarbonyl compounds. This oxidation, which is catalyzed by acid,[66] is believed to follow the reaction path indicated in the following equations.[70]

65. (a) N. Rabjohn, *Org. Reactions,* **5**, 331 (1959). (b) E. N. Trachtenberg in R. L. Augustine, ed., *Oxidation,* Marcel Dekker, New York, 1969, pp. 119–187.
66. (a) E. J. Corey and J. P. Schaefer, *J. Am. Chem. Soc.,* **82**, 918 (1960). (b) Also see J. P. Schaefer, *ibid.,* **84**, 717 (1962), for a discussion of the base-catalyzed reaction.
67. (a) J. C. Banerji, D. H. R. Barton, and R. C. Cookson, *J. Chem. Soc.,* 5041 (1957). (b) J. P. Schaefer, *J. Am. Chem. Soc.,* **84**, 713 (1962).
68. H. A. Riley and A. R. Gray, *Org. Syn., Coll. Vol. 2,* 509 (1943).
69. C. C. Hach, C. V. Banks, and H. Diehl, *Org. Syn., Coll. Vol. 4,* 229 (1963).
70. In certain cases, the selenium dioxide oxidation of cyclic ketones has resulted in dehydrogenation of the ketone (cf. Chapter 1) to form α,β-unsaturated ketones as well as (or instead of) 1,2-diketones. This reaction is discussed in Ref. 66a and by R. A. Jerussi and D. Speyer, *J. Org. Chem.,* **31**, 3199 (1966).

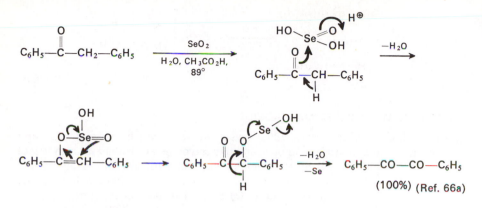

(100%) (Ref. 66a)

A similar intermediate may be involved in the dehydrogenation of 1,4-diketones, e.g. [47]. It has been noted with certain polycyclic diketones that this dehydro-

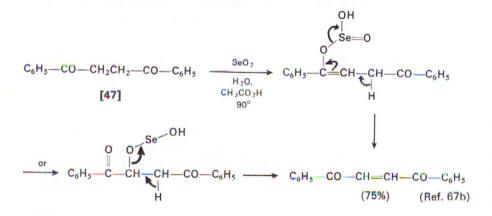

(75%)   (Ref. 67b)

genation process is more rapid when the two hydrogen atoms being removed bear a *cis*-relationship to one another, e.g. [48], than when the hydrogen atoms are *trans*, e.g. [49]. This observation was initially regarded as evidence for a mechanism in which the two hydrogen atoms are removed simultaneously by selenium dioxide.[67] However, these rate enhancements may be the result of relieving non-bonding interactions present in the starting diketone by forming the corresponding enol selenite ester intermediate.

The oxidation of olefins, e.g. [50], at allylic positions with selenium dioxide bears a formal resemblance to the previously discussed reactions of olefins with lead tetraacetate and with mercuric acetate. As indicated in the accompanying equations, the nature of the reaction product is influenced by the choice of reaction solvent, with carbonyl containing products predominating in aqueous media and allylic acetates being formed when an acetic acid-acetic anhydride mixture is used as solvent. This oxidation differs from the allylic oxidations with lead tetra-acetate and mercuric acetate in that partial retention of the double bond in its

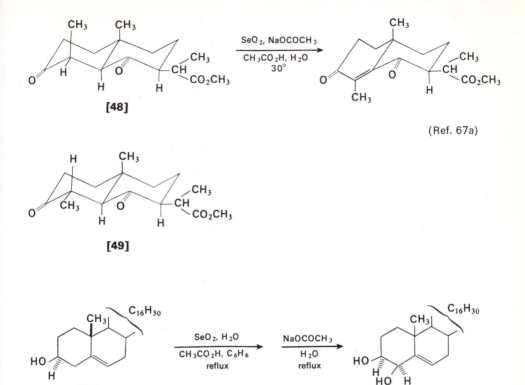

[48]

(Ref. 67a)

[49]

[50]

(39%)   (Ref. 65a)

original position is sometimes observed while a symmetrical intermediate is indicated in other cases. Possible mechanistic routes to account for these results are indicated in the accompanying equations. The attack of selenious acid at the double bond of cyclohexene derivatives is suggested to occur from the direction which will minimize 1,3-diaxial interactions between the selenium and other substituents.[71b]

71. (a) J. P. Schaefer, B. Horvath, and H. P. Klein, *J. Org. Chem.,* **33,** 2647 (1968). (b) E. N. Trachtenberg and J. R. Carver, *ibid.,* **35,** 1646 (1970) ; E. N. Trachtenberg, C. H. Nelson, and J. R. Carver, *ibid.,* **35,** 1653 (1970). (c) D. H. Olson, *Tetrahedron Letters,* **No. 19,** 2053 (1966). (d) K. A. Javaid, N. Sonoda, and S. Tsutsumi, *ibid.,* **No. 51,** 4439 (1969) ; *Ind. Eng. Chem., Prod. Res. Develop.,* **9,** 87 (1970). (e) For a discussion of the adducts from 1,3-dienes and selenium dioxide, see Ref. 71c and W. L. Mock and J. H. McCausland, *ibid.,* **No. 4,** 391 (1968). (f) For the preparation of unsaturated γ-lactones by the allylic oxidation of α,β-unsaturated esters with selenium dioxide, see N. Danieli, Y. Mazur, and F. Sondheimer, *Tetrahedron,* **22,** 3189 (1966) ; J. N. Marx and F. Sondheimer, *ibid.,* **Suppl. 8, Part 1,** 1 (1966). (g) J. J. Plattner, U. T. Bhalerao, and H. Rapoport, *J. Am. Chem. Soc.* **91,** 4933 (1969).

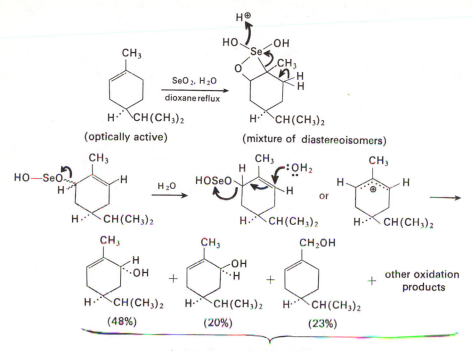

(optically active)   (mixture of diastereoisomers)

(48%)   (20%)   (23%)   + other oxidation products

(products optically active but of lower
optical purity than the starting olefin)   (Ref. 71b and 31b)

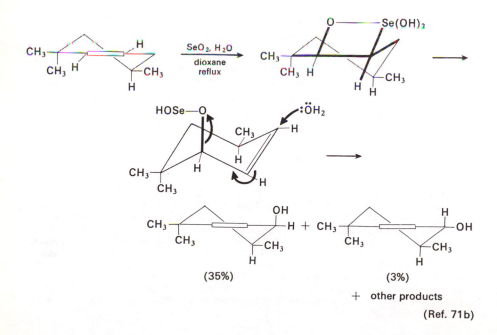

(35%)   (3%)

+ other products

(Ref. 71b)

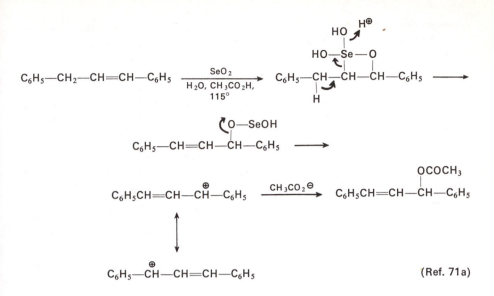

(Ref. 71a)

The following example of a remarkably selective oxidation with selenium dioxide in ethanol is worthy of special note. This procedure is reported to be general for the preparation of α-substituted-trans-α,β-unsaturated aldehydes.[71g]

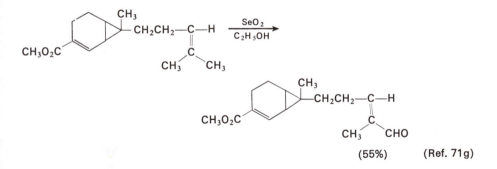

(55%) (Ref. 71g)

OTHER OXIDIZING AGENTS

Among the many other oxidation methods,[1] mention will be made only of several procedures which are of special value for water-soluble compounds or for compounds which are unusually sensitive to acidic or basic reaction media.

Among these procedures, catalytic oxidation over a platinum catalyst[72] offers the most convenience for water-soluble compounds since filtration of the catalyst after the oxidation is complete leaves a water solution of the product. The catalyst is normally prepared by reducing a platinum salt with hydrogen or formaldehyde in

72. K. Heyns and H. Paulsen in W. Foerst, ed., *Newer Methods of Preparative Organic Chemistry*, Vol. 2, Academic Press, New York, 1963, pp. 303–335.

the presence of finely powdered carbon to form a platinium-on-carbon catalyst or by hydrogenating platinum oxide to form a finely divided metallic platinum catalyst.[72] In the latter case, the suspension of catalyst in the reaction solvent is repeatedly flushed with air to remove adsorbed hydrogen from the catalyst. A mixture of the catalyst, solvent, and compound to be reduced are then shaken or stirred in contact with air; the course of the reaction may be followed by measuring the volume of oxygen (from the air) absorbed as the reaction proceeds. The following examples illustrate this method.

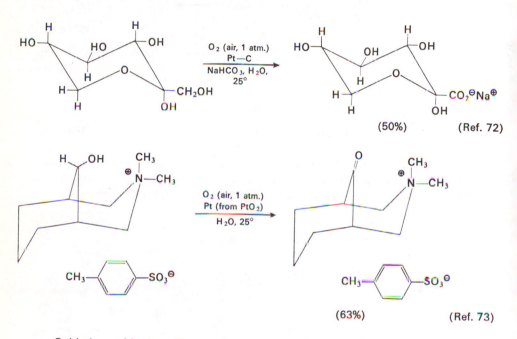

Oxidations of hydroxylic compounds can be accomplished much more rapidly with the very reactive oxidizing agent, ruthenium tetroxide.[74] Ruthenium tetroxide, a volatile yellow solid melting at 25°, can be used as a solution in water, carbon tetrachloride, or *alcohol-free* chloroform. The oxidant reacts, often violently, with most of the other common organic solvents. Ruthenium tetroxide is prepared either by the oxidation of ruthenium trichloride with a hot aqueous solution of sodium bromate[74b] or, more conveniently, by the oxidation of the water-insoluble ruthenium dioxide with an aqueous solution of sodium metaperiodate or sodium hypochlorite.[74d] The reagent is either distilled or extracted from these aqueous

73. (a) H. O. House and C. G. Pitt, *J. Org. Chem.*, **31**, 1062 (1966). (b) see also R. P. A. Sneedon and R. B. Turner, *J. Am. Chem. Soc.*, **77**, 130, 190 (1955).

74. (a) C. Djerassi and R. R. Engle, *J. Am. Chem. Soc.*, **75**, 3838 (1953). (b) L. M. Berkowitz and P. N. Rylander, *ibid.*, **80**, 6682 (1958). (c) H. Nakata, *Tetrahedron*, **19**, 1959 (1963). (d) S. Wolfe, S. K. Hasan, and J. R. Campbell, *Chem. Commun.*, **No. 21**, 1420 (1970).

solutions with carbon tetrachloride. When ruthenium tetroxide is used as an oxidizing agent, the oxidant is reduced to the insoluble ruthenium dioxide which may be filtered from the reaction mixture.

Although ruthenium tetroxide will effect a number of interesting oxidations including the conversion of ethers to esters,[74b] and the oxidative cleavage of olefinic

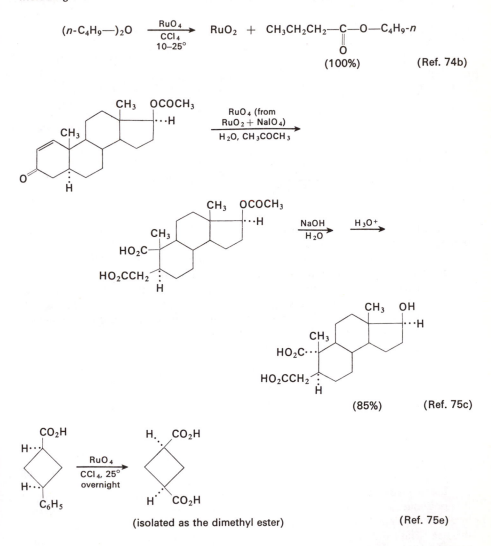

$$(n\text{-}C_4H_9\text{---})_2O \xrightarrow[\substack{CCl_4 \\ 10-25°}]{RuO_4} RuO_2 + CH_3CH_2CH_2\text{---}\overset{\displaystyle O}{\underset{\displaystyle \|}{C}}\text{---}O\text{---}C_4H_9\text{-}n$$

(100%)   (Ref. 74b)

(isolated as the dimethyl ester)   (Ref. 75e)

(85%)   (Ref. 75c)

75. (a) G. Snatzke and H. W. Fehlhaber, *Justus Liebigs Ann. Chem.*, **663**, 123 (1963). (b) F. M. Dean and J. C. Knight, *J. Chem. Soc.*, 4745 (1962). (c) D. M. Piatak, H. B. Bhat, and E. Caspi, *J. Org. Chem.*, **34**, 112 (1969). (d) D. M. Piatak, G. Herbst, J. Wicha, and E. Caspi, *ibid.*, **34**, 116 (1969). (e) J. A. Caputo and R. Fuchs, *Tetrahedron Letters*, **No. 47**, 4729 (1967).

and aromatic double bonds,[10h,74b,75] perhaps the most useful aspect of this reagent lies in the rapidity with which it oxidizes secondary alcohols to ketones.[74b,c,75e,76]

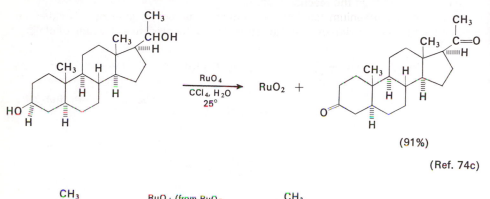

(91%)

(Ref. 74c)

$$C_6H_5-\underset{\underset{CH_3}{|}}{CH}-CH_2OH \xrightarrow[CCl_4,\ H_2O,\ 25°]{RuO_4\ (from\ RuO_2 + NaIO_4)} C_6H_5-\underset{\underset{CH_3}{|}}{CH}-CHO$$

(48%)

(Ref. 75e)

The oxidation of primary alcohols to aldehydes appears to be a less satisfactory procedure because of the rapid further oxidation of the aldehydes to carboxylic acids.[74b] It is possible to use only a catalytic amount of ruthenium tetroxide as the oxidant if the reaction is run as a two phase mixture with excess sodium periodate or sodium hypochlorite in the aqueous phase to reoxidize the ruthenium dioxide as it is formed.[74,75c-e] The following equation illustrates the use of this procedure with compounds which could undergo a slower reaction (cleavage of the vicinal diol or the α-hydroxy ketone) with sodium periodate.

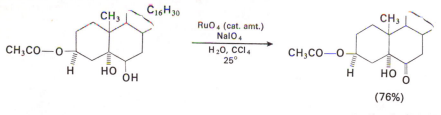

(76%)

(Ref. 74c)

Since the oxidant will distribute itself between carbon tetrachloride and water,[74a] it is possible to use this reagent for the oxidation of compounds which are soluble in water and insoluble in carbon tetrachloride. The following equations illustrate this use. The oxidation of the hydroxy acid salt [51] is of interest since acidification of this aqueous solution results in the spontaneous relactonization of the hydroxy acid and precludes the use of many common oxidation methods.

76. (a) P. J. Beynon, P. M. Collins, P. T. Doganges, and W. G. Overend, *J. Chem. Soc.,* C, 1131 (1966). (b) J. S. Brimacombe, *Angew. Chem., Intern. Ed. Engl.,* **8**, 401 (1969).

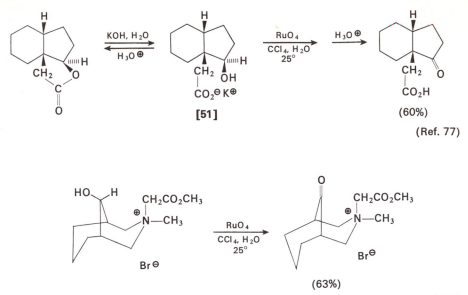

[51]

(60%)

(Ref. 77)

(63%)

(Ref. 78)

The reactions of alkyl halides and alkyl arenesulfonates with dimethyl sulfoxide[79] under various conditions have served as synthetic routes to aldehydes and ketones, especially when primary iodides or primary alkyl arenesulfonates are employed. Mechanistically similar oxidations may be accomplished by reaction of alkyl halides with amine oxides or by conversion of alkyl halides to pyridinium salts followed by reaction with nitroso compounds. When secondary alkyl arenesulfonates are heated with dimethyl sulfoxide, elimination to form olefinic products is usually the predominant reaction.[80] The direct oxidation of alcohols to aldehydes or

77. H. O. House and C. J. Blankley, *J. Org. Chem.*, **32**, 1741 (1967). Also see R. M. Moriarty, H. Gopal, and T. Adams, *Tetrahedron Letters*, **No. 46**, 4003 (1970).
78. H. O. House and B. A. Tefertiller, *J. Org. Chem.*, **31**, 1068 (1966).
79. For examples, see (a) N. Kornblum and co-workers, *J. Am. Chem. Soc.*, **79**, 6562 (1957). (b) N. Kornblum, W. J. Jones, and G. J. Anderson, *ibid.*, **81**, 4113 (1959). (c) H. R. Nace and J. J. Monagle, *J. Org. Chem.*, **24**, 1792 (1959). (d) R. N. Iacona, A. T. Rowland, and H. R. Nace, *ibid.*, **29**, 3495 (1964). (e) H. R. Nace and R. N. Iacona, *ibid.*, **29**, 3498 (1964). (f) A. P. Johnson and A. Pelter, *J. Chem. Soc.*, 520 (1964). (g) C. R. Johnson and W. G. Phillips, *Tetrahedron Letters*, **No. 25**, 2101 (1965). (h) K. Torssell, *ibid.*, **No. 37**, 4445 (1966). (i) A. Kalir, *Org. Syn.*, **46**, 81 (1966). (j) V. Franzen, *ibid.*, **47**, 96 (1967). (k) F. Kröhnke, *Angew. Chem., Intern. Ed. Engl.*, **2**, 380 (1963). (l) For examples in which the halogenation and oxidation of active methylene compounds is effected concurrently in dimethyl sulfoxide solution, see A. Markovac, C. L. Stevens, A. B. Ash, and B. E. Harkley, Jr., *J. Org. Chem.*, **35**, 841 (1970); E. Schipper, M. Cinnamon, L. Rascher, Y. H. Chiang, and W. Oroshnik, *Tetrahedron Letters*, **No. 59**, 6201 (1968). (m) M. W. Epstein and J. Ollinger, *Chem. Commun.*, **No. 20**, 1338 (1970).
80. For examples, see Ref. 79b and 79c and (a) D. N. Jones and M. A. Saeed, *J. Chem. Soc.*, 4657 (1963). (b) H. R. Nace, *J. Am. Chem. Soc.*, **81**, 5428 (1959).

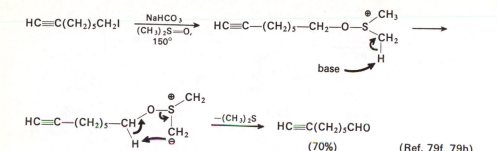

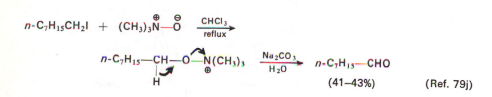

HC≡C—(CH₂)₅CHO

(70%)   (Ref. 79f, 79h)

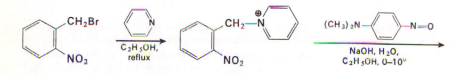

n-C₇H₁₅—CHO

(41–43%)   (Ref. 79j)

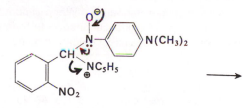

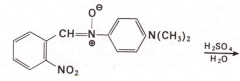

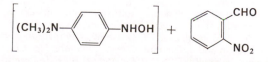

(47–53%)   (Ref. 79i, k)

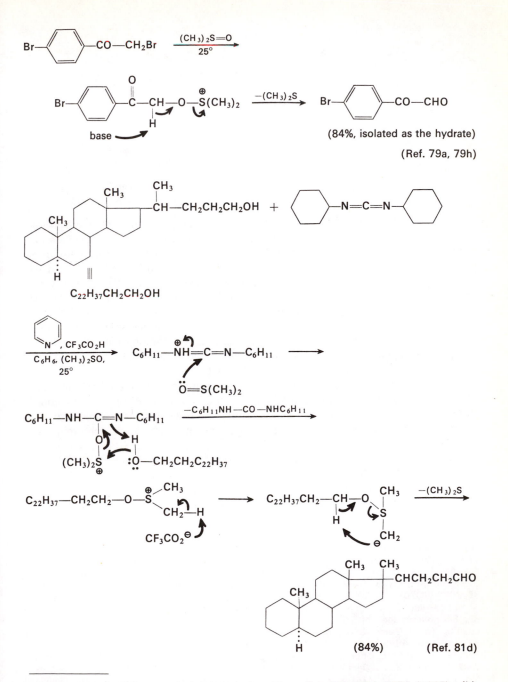

(84%, isolated as the hydrate)

(Ref. 79a, 79h)

(84%)   (Ref. 81d)

81. (a) K. E. Pfitzner and J. G. Moffatt, *J. Am. Chem. Soc.*, **87**, 5661, 5670 (1965). (b) A. H. Fenselau and J. G. Moffatt, *ibid.*, **88**, 1762 (1966). (c) A. F. Cook and J. G. Moffatt, *ibid.*, **89**, 2697 (1967); **90**, 740 (1968). (d) J. G. Moffatt, *Org. Syn.*, **47**, 25 (1967).

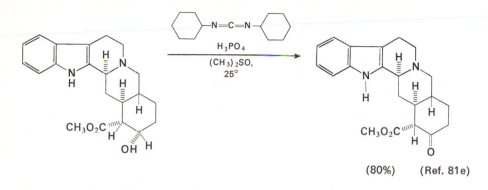

(80%)   (Ref. 81e)

ketones with mixtures of dimethyl sulfoxide, dicyclohexylcarbodiimide, and a proton donor[81] appears to be a more generally useful oxidation procedure.[82] This oxidation has been most commonly conducted by treatment of the alcohol with a solution of dicyclohexylcarbodiimide (3 molar equiv.) and pyridinium trifluoroacetate (0.5 molar equiv.) in dimethyl sulfoxide.[81] In certain cases, better results have been obtained by the use of phosphoric acid rather than pyridinium trifluoroacetate and by the addition of an inert cosolvent such as benzene or 1,2-dimethoxyethane. These conditions permit the oxidations to be performed under essentially neutral conditions to form aldehydes or ketones. The alkoxysulfonium ylid intermediate, e.g. [52], is also capable of undergoing an elimination reaction which can lead to the formation of a thiomethoxymethyl by-product.[81a,g]

Although the rates of oxidation of unhindered stereoisomeric alcohols are approximately equal, hindered equatorial alcohols are oxidized by this procedure more rapidly than the epimeric axial alcohols.[81a] This result is opposite to the effect of stereochemistry on the rates of oxidation of alcohols with chromic acid (see Chapter 5).

Since the use of carbodiimide as a condensing reagent in this oxidation procedure leads to the production of dicyclohexylurea which may be difficult to separate from the reaction product, other reagents have been examined. Among these are ketimines,[81h] ynamines,[81h] mercuric acetate,[83e] acetic anhydride or other

(e) J. D. Albright and L. Goldman, *J. Org. Chem.*, **30**, 1107 (1965). (f) F. W. Sweat and W. W. Epstein, *ibid.*, **32**, 835 (1967). (g) J. B. Jones and D. C. Wigfield, *Tetrahedron Letters*, **No. 46**, 4103 (1965). (h) R. E. Harmon, C. V. Zenarosa, and S. K. Gupta, *ibid.*, **No. 43**, 3781 (1969); *J. Org. Chem.*, **35**, 1936 (1970). (i) U. Brodbeck and J. G. Moffatt, *ibid.*, **35**, 3552 (1970). (j) For the oxidation of oximes and hydroxylamines by this procedure, see A. H. Fenselau, E. H. Hamamura, and J. G. Moffatt, *ibid.*, **35**, 3546 (1970).
82. For reviews see Ref. 76b and (a) W. W. Epstein and F. W. Sweat, *Chem. Rev.*, **67**, 247 (1967). (b) T. Durst, *Adv. Org. Chem.*, **6**, 285 (1969).
83. (a) J. D. Albright and L. Goldman, *J. Am. Chem. Soc.*, **89**, 2416 (1967). (b) J. R. Parikh and W. von E. Doering, *ibid.*, **89**, 5505 (1967). (c) M. VanDyke and N. D. Pritchard, *J. Org. Chem.*, **32**, 3204 (1967). (d) S. M. Ifzal and D. A. Wilson, *Tetrahedron Letters*, **No. 17**, 1577 (1967). (e) J. M. Tien, H. J. Tien, and J. S. Ting, *ibid.*, **No. 19**, 1483 (1969). (f) D. R. Dalton and D. G. Jones, *ibid.*, **No. 30**, 2875 (1967). (g) T. M. Santosusso and D. Swern, *ibid.*, **No. 40**, 4261 (1968). (h) D. H. R. Barton, B. J. Garner, and R. H. Wightman, *J. Chem. Soc.*, 1855 (1964).

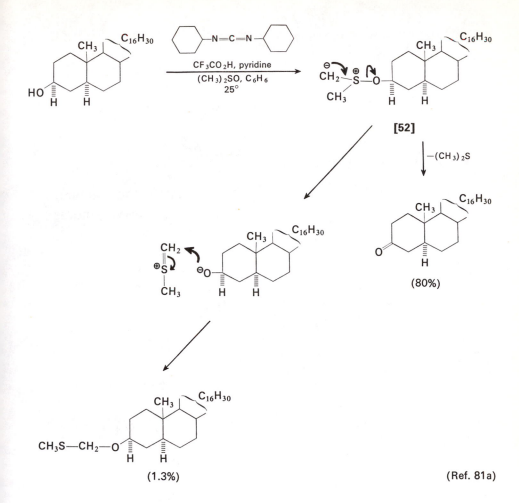

(Ref. 81a)

anhydrides,[83a,c,d] and the pyridine-sulfur trioxide complex.[83b] The use of acetic anhydride is restricted to oxidations of relatively hindered alcohols where competing acetate formation will be slow.[83a] Alcohol oxidation may also be accomplished by conversion of the hydroxy function to a chloroformate[83h] followed by reaction with dimethyl sulfoxide. Examples of certain of these procedures are provided in the following equations.

Other reactions in which dimethyl sulfoxide participates as a reactant[82b] include the conversion of active methylene compounds to sulfur ylids (see Chapter 10) by reaction with dimethyl sulfoxide and dicyclohexylcarbodiimide,[81c] the formation of bromohydrins (see Chapter 8) by reaction of olefins with bromine and water in dimethyl sulfoxide solution,[83f] and the reaction of epoxides with dimethyl sulfoxide to form α-hydroxy ketones.[83g] This latter reaction, illustrated by an accompanying equation, is comparable in mechanism to the previously discussed reaction of alkyl halides with dimethyl sulfoxide.

p-CH$_3$OC$_6$H$_4$—CO—CH—C$_6$H$_4$OCH$_3$-p $\xrightarrow[\text{(CH}_3)_2\text{SO, 25°}]{\text{(CH}_3\text{CO)}_2\text{O}}$

with OH below

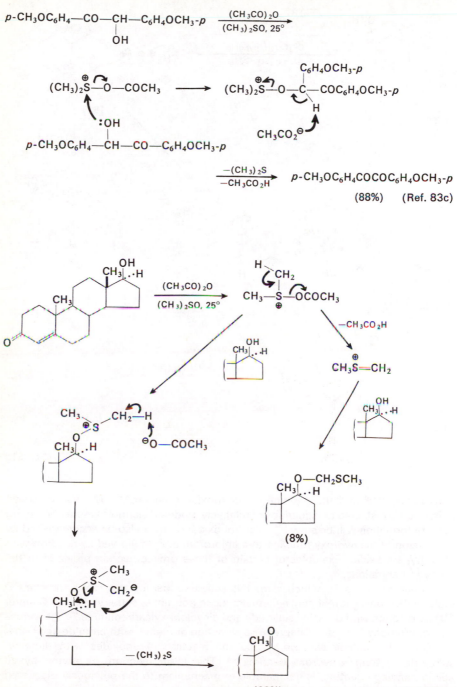

$\xrightarrow[-\text{CH}_3\text{CO}_2\text{H}]{-\text{(CH}_3)_2\text{S}}$ p-CH$_3$OC$_6$H$_4$COCOC$_6$H$_4$OCH$_3$-p

(88%)   (Ref. 83c)

(8%)

(68%)   (Ref. 83a)

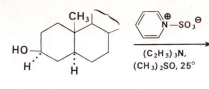 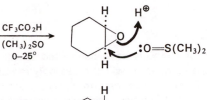

(83%)   (Ref. 83b)

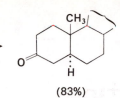

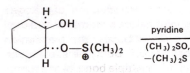

(Ref. 83g)

8

HALOGENATION

Although a variety of methods have been used to introduce halogen atoms into organic molecules, this chapter will discuss only reactions between halogens or halogen derivatives and carbon-carbon double bonds, introduction of halogen atoms alpha to a carbonyl function, and substitution of halogen atoms for hydroxyl groups or for hydrogen atoms at allylic or benzylic C—H bonds; the preparation of fluorine-containing compounds will not be considered.[1] Both the substitution of halogen for hydrogen and the addition of halogen to a multiple bond bear a formal resemblance to previously discussed oxidation reactions, In fact, many of the intermediates formed in halogenation reactions may be converted to products (e.g., 1,2-diols, epoxides, carbonyl compounds) also available from oxidation procedures.

REACTIONS WITH CARBON-CARBON DOUBLE BONDS

Solutions of bromine in carbon tetrachloride, chloroform, carbon disulfide, acetic acid, ether, or ethyl acetate are usually employed for reaction with olefins (e.g., [1] and [2]) to form 1,2-dibromides. The corresponding reactions with

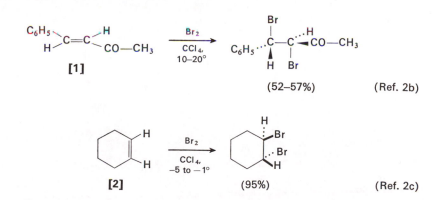

(52–57%) (Ref. 2b)

(95%) (Ref. 2c)

1. W. A. Sheppard and C. M. Sharts, *Organic Fluorine Chemistry* Benjamin, New York, 1969.
2. (a) C. F. H. Allen, R. D. Abell, and J. B. Normington, *Org. Syn.*, **Coll. Vol. 1**, 205 (1944). (b) N. H. Cromwell and R. Benson, *ibid.*, **Coll. Vol. 3**, 105 (1955). (c) H. R. Snyder and L. A. Brooks, *ibid.*, **Coll. Vol. 2**, 171 (1943).

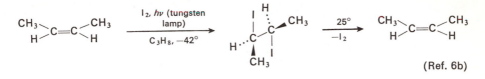

(Ref. 6b)

iodine to form thermally unstable 1,2-diiodides and with chlorine to form 1,2-dichlorides also occur but are less often used in the laboratory. The crystalline complex from pyridine hydrobromide and bromine (pyridinium hydrobromide perbromide)[3] has also found use[4] for the bromination of both olefins and ketones. The addition of bromine to an unhindered carbon-carbon double bond is often so rapid that the reaction may be performed as a titration, the bromine reacting with the olefin as quickly as it is introduced into the reaction mixture. Consequently, bromine will add to olefinic double bonds in the presence of aldehyde (e.g., [3]), ketone (e.g., [1], epoxide (e.g., [4]), ester (e.g., [5]), alcohol (e.g., [6]), carboxylic acid, or amide functions without serious competing reactions.

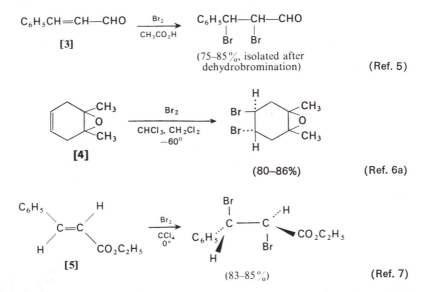

$C_6H_5CH=CH-CHO$ $\xrightarrow[CH_3CO_2H]{Br_2}$ $C_6H_5CH-CH-CHO$
                                                      Br Br

[3]

(75–85%, isolated after dehydrobromination)   (Ref. 5)

[4]

(80–86%)   (Ref. 6a)

[5]

(83–85%)   (Ref. 7)

3. Commercially available from Arapahoe Chemicals, Inc., Boulder, Colorado.
4. (a) C. Djerassi and C. R. Scholz, *J. Am. Chem. Soc.*, **70**, 417 (1948). (b) L. F. Fieser, *Experiments in Organic Chemistry*, Heath, Boston, 1955, pp. 65, 180. (c) L. F. Fieser and M. Fieser, *Reagents for Organic Synthesis*, Wiley, New York, 1967, pp. 967–970. (d) For a study of the halogenation of olefins and acetylenes with copper(II) halides in methanol solution, see C. E. Castro, E. J. Gaughan, and D. C. Owsley, *J. Org. Chem.*, **30**, 587 (1965).
5. C. F. H. Allen and C. O. Edens, Jr., *Org. Syn.*, **Coll. Vol. 3**, 731 (1955).
6. (a) L. A. Paquette and J. H. Barrett, *Org. Syn.*, **49**, 62 (1969). (b) P. S. Skell and R. R. Pavlis, *J. Am. Chem. Soc.*, **86**, 2956 (1964). (c) Vicinal dibromides react with iodide ion to form olefins. For examples and leading references, see I. M. Mathai and S. I. Miller, *J. Org. Chem.*, **35**, 3416 (1970) ; W. K. Kwok, I. M. Mathai, and S. I. Miller, *ibid.*, **35**, 3420 (1970).
7. (a) T. W. Abbott and D. Althousen, *Org. Syn.*, **Coll. Vol. 2**, 270 (1943). (b) S. M. McElvain and D. Kundiger, *ibid.*, **Coll. Vol. 3**, 123 (1955).

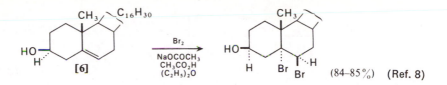

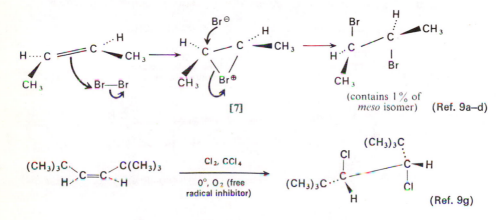

The reaction of bromine or chlorine with an olefin is thought to proceed by electrophilic attack of the halogen on the double bond to form an intermediate ion such as the bromonium ion [7]. Such an electrophilic attack, which, may involve the initial formation of a charge-transfer complex,[9e] is in accord with the observations[10] that the reaction is accelerated by electron-donating substituents in the olefin and by the use of polar solvents. Under special conditions, stable solutions of chloronium and bromonium ions such as [7] could be obtained for n.m.r. studies.[10e] The accompanying equations also illustrate the attack on the bromonium

8. (a) L. F. Fieser, *Org. Syn.,* **Coll. Vol. 4,** 195 (1963). (b) L. F. Fieser and M. Fieser, *Steroids,* Reinhold, New York, 1959, pp. 37–41.
9. (a) W. G. Young, R. T. Dillon, and H. J. Lucas, *J. Am. Chem. Soc.,* **51,** 2528 (1929). (b) S. Winstein and H. J. Lucas, *ibid.,* **61,** 1576 (1939). (c) H. O. House and R. S. Ro, *ibid.,* **80,** 182 (1958). (d) J. H. Rolston and K. Yates, *ibid.,* **91,** 1469, 1477, 1483 (1969). (e) J. E. Dubois and F. Garnier, *Chem. Commun.,* **No. 5,** 241 (1968). (f) R. C. Fahey and C. Schubert, *J. Am. Chem. Soc.,* **87,** 5172 (1965). (g) R. C. Fahey, *ibid.,* **88,** 4681 (1966). (h) M. L. Poutsma and J. L. Kartch, *ibid.,* **89,** 6595 (1967). (i) R. C. Fahey and H. J. Schneider, *ibid.,* **90,** 4429 (1968). (j) M. C. Cabaleiro and M. D. Johnson, *J. Chem. Soc.,* B, 565 (1967); M. D. Johnson and E. N. Trachtenberg, *ibid.,* B, 1018 (1968); M. C. Cabaleiro, M. D. Johnson, B. E. Swedlund, and J. G. Williams, *ibid.,* B, 1022 (1968); M. C. Cabaleiro, C. J. Cooksey, M. D. Johnson, B. E. Swedlund, and J. G. Williams, *ibid.,* B, 1026 (1968).
10. (a) K. Yates and W. V. Wright, *Tetrahedron Letters,* **No. 24,** 1927 (1965); J. E. Dubois and W. V. Wright, *ibid.,* **No. 32,** 3101 (1967). (b) J. E. Dubois and A. F. Hegarty, *J. Chem. Soc.,* B, 638 (1969). (c) G. Heublein and G. Agatha, *Tetrahedron,* **24,** 3799 (1968); G. Heublein and B. Rauscher, *ibid.,* **25,** 3999 (1969). (d) R. E. Buckles, J. L. Miller, and R. J. Thurmaier, *J. Org. Chem.,* **32,** 888 (1967). (e) G. A. Olah and J. M. Bollinger, *J. Am. Chem. Soc.,* **90,** 947 (1968).

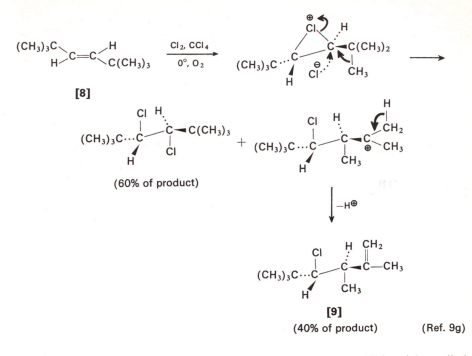

[8]

(60% of product)

+

↓ −H⊕

[9]
(40% of product)   (Ref. 9g)

or chloronium ion by halide ion; the overall result is a *trans* addition (also called *anti* addition) of two halogen atoms to the double bond. The reaction of the intermediate bromonium ion with a nucleophile is similar in many respects to the previously discussed (Chapter 6) acid-catalyzed ring opening of epoxides, which is believed to proceed by attack of the nucleophile on the conjugate acid of the epoxide. Not only is the reaction of bromine with alkyl-substituted olefins a *trans* addition, as several of the foregoing examples illustrate, but the predominant product from the addition of bromine to a conformationally rigid cyclohexene is a *trans* diaxial dibromide.[11a] This stereochemical result is illustrated, both by the aforementioned[8] bromination of cholesterol [6] to form the dibromide [10] via the bromonium ion [11], and by the bromination of 2-cholestene [12]. In each case the initial bromonium ion was produced by attack of the halogen from the less hindered side of the double bond. The initially formed *trans* diaxial dibromides [10] and [13] were found to rearrange on standing or heating to become the more stable *trans* diequatorial dibromides [14] and [15], respectively. This diaxial-to-diequatorial rearrangement, which presumably occurs[12,13] by formation and collapse

11. (a) D. H. R. Barton and R. C. Cookson, *Quart. Rev.,* **10**, 44 (1956). (b) D. H. R. Barton and E. Miller, *J. Am. Chem. Soc.,* **72**, 1066 (1950). (c) D. H. R. Barton, E. Miller, and H. T. Young, *J. Chem. Soc.,* 2598 (1951).
12. (a) G. H. Alt and D. H. R. Barton, *J. Chem. Soc.,* 4284 (1954). (b) C. A. Grob and S. Winstein, *Helv. Chim. Acta,* **35**, 782 (1952).
13. (a) D. H. R. Barton and J. F. King, *J. Chem. Soc.,* 4398 (1958). (b) J. F. King and R. G. Pews, *Can. J. Chem.,* **43** 847 (1965).

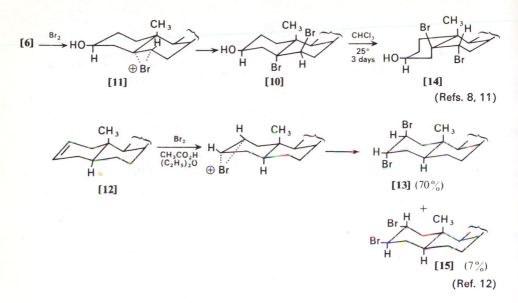

(Refs. 8, 11)

[13] (70%)

+

[15] (7%)

(Ref. 12)

of the ion pair [16], has been suggested to be the general behavior for compounds containing *trans*-diaxial substituents.[13]

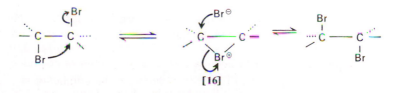

[16]

The previously described reaction of the olefin [8] with chlorine to form both the expected *meso*-dichloride and the rearranged chloro olefin [9] exemplifies the possibility of molecular rearrangements of the intermediate halonium ion which may accompany the halogenation of olefins.[14] In general, rearrangement during the halogenation of an olefin is enhanced by the use of polar solvents, by the presence of olefin substituents such as *t*-alkyl or triarylmethyl groups, and by the presence of neighboring groups which favor rearrangement.[14b] This rearrangement may sometimes be avoided by adjusting the reaction conditions to favor halogenation by a free radical rather than an ionic reaction path.[14h] Rearrangement

14. (a) M. J. Janssen, F. Wiegman, and H. J. Kooreman, *Tetrahedron Letters,* No. 51, 6375 (1966). (b) R. O. C. Norman and C. B. Thomas, *J. Chem. Soc.,* B, 598 (1967). (c) J. Wolinsky, R. W. Novak, and K. L. Erickson, *J. Org. Chem.,* 34, 490 (1969). (d) D. D. Tanner and G. C. Gidley, *ibid.,* 33, 38 (1968). (e) C. R. Johnson, C. J. Cheer, and D. J. Goldsmith *ibid.,* 29, 3320 (1964). (f) G. Berti and A. Marsili, *Tetrahedron,* 22, 2977 (1966); G. Bellucci, G. Berti, F. Marioni, and A. Marsili, *ibid.,* 26, 4627 (1970). (g) J. G. Traynham and D. B. Stone, Jr., *J. Org. Chem.,* 35, 2025 (1970). (h) A. Hassner and J. S. Teeter, *ibid.,* 35, 3397 (1970); A. Hassner and J. S. Teeter, to be published.

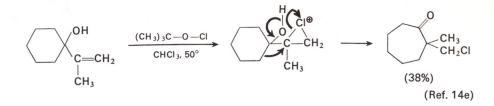

(38%)

(Ref. 14e)

may also result from catalytic amounts of hydrogen halide liberated during the course of the olefin halogenation; one possible source of hydrogen bromide is the subsequently discussed free-radical bromination at allylic positions. The hydrogen halide can catalyze *cis-trans* isomerization of olefins,[9g,14a] migration of the olefinic double bond,[14c] rearrangement of the carbon skeleton, or isomerization of the initially formed dihalide. The following equation illustrates the possibility of minimizing or eliminating completely acid catalyzed rearrangements by performing the bromination in the presence of pyridine;[14c] comparable results have been reported when pyridinium hydrobromide perbromide is used as the brominating reagent.[2c]

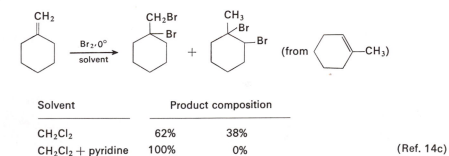

Solvent	Product composition	
CH₂Cl₂	62%	38%
CH₂Cl₂ + pyridine	100%	0%

(Ref. 14c)

The effect of added pyridine presumably is to trap any hydrogen halide liberated. However, the formation of optically acitve dibromides by reaction of bromine with olefins in the presence of optically active tertiary amines[14f] suggests that added amines may be bonded to the species which transfers one of the bromine atoms to the olefin.

Although the reaction of alkyl-substituted olefins with halogens appears to involve a covalent halonium ion intermediate [17a] which leads to a stereospecific *trans* (or *anti*) addition of the halogen atoms, the presence of substituents such as aryl groups which can stabilize an adjacent carbonium ion lead to increased ionic character (i.e., [17b] and [17c]) in the intermediate.[9,10,16] The tendency to form ionic intermediates [17b] and [17c] is also enhanced by the use of polar solvents and by the presence of a chlorine atom (X=Cl in [17]) rather than the large, less

15. S. J. Cristol, F. R. Stermitz, and P. S. Ramey, *J. Am. Chem. Soc.*, **78**, 4939 (1956).
16. (a) M. J. S. Dewar and R. C. Fahey, *J. Am. Chem. Soc.*, **85**, 2245, 2248, 3645 (1963).
(b) A. Hassner, F. P. Boerwinkle, and A. B. Levy, *ibid.*, **92**, 4879 (1970).

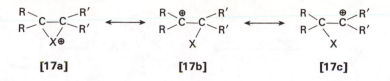

                [17a]   [17b]   [17c]

electronegative bromine or, especially, iodine atoms.[16b] As the reaction intermediate becomes more ionic in nature, the stereochemistry of the halogen addition has been found to change so that either *cis*-addition to the olefin or a non-stereospecific addition is observed. These stereochemical possibilities are illustrated in the following equations.

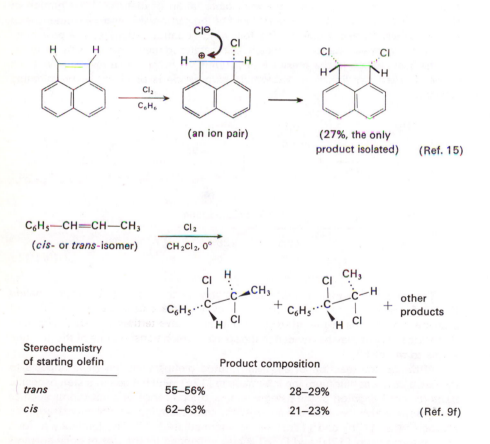

C_6H_5—CH=CH—CH_3

(*cis*- or *trans*-isomer)

(an ion pair)

(27%, the only product isolated)   (Ref. 15)

other products

+

Stereochemistry of starting olefin	Product composition	
trans	55–56%	28–29%
cis	62–63%	21–23%

(Ref. 9f)

It should be noted that a non-stereospecific addition may be observed if the halogenation follows a free-radical reaction path[16b] rather than the ionic pathway which is being discussed.

The reaction of bromine with an olefin to form a 1,2-dibromide does not differ fundamentally from its reaction with an aromatic compound (e.g., [18]) to

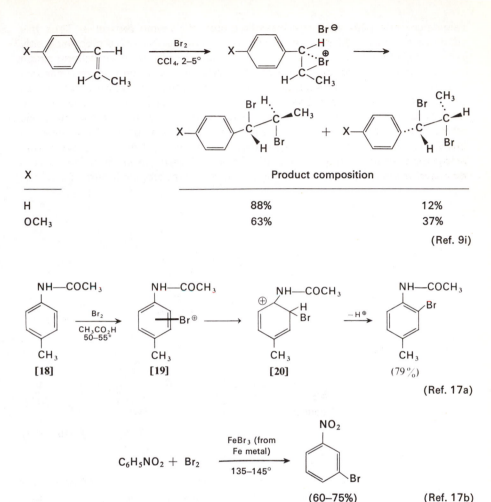

X	Product composition	
H	88%	12%
OCH₃	63%	37%
		(Ref. 9i)

[18] [19] [20] (79%)

(Ref. 17a)

$C_6H_5NO_2 + Br_2$ $\xrightarrow[\substack{135-145°}]{\substack{FeBr_3 \text{ (from} \\ \text{Fe metal)}}}$

(60–75%)   (Ref. 17b)

substitute a bromine atom for an aromatic hydrogen atom.[17] Aromatic bromination is usually regarded as an electrophilic attack of bromine (or a complex of bromine with a Lewis acid) on the aromatic ring to produce successively a pi complex, [19],

17. (a) J. R. Johnson and L. T. Sandborn, *Org. Syn.*, **Coll. Vol. 1**, 111 (1944). (b) J. R. Johnson and C. G. Gauerke, *ibid.*, **Coll. Vol. 1**, 123 (1944). (c) For reviews of halogenation and other electrophilic substitutions with heterocyclic aromatic compounds, see A. R. Katritzky and C. D. Johnson, *Angew. Chem., Intern. Ed. Engl.*, **6**, 608 (1967); J. L. Goldfarb, J. B. Volkenstein, and L. I. Belenkij, *ibid.*, **7**, 519 (1968). (d) D. E. Pearson, R. D. Wysong and C. V. Breder, *J. Org. Chem.*, **32**, 2358 (1967). (e) R. Adams and C. S. Marvel, *Org. Syn.*, **Coll. Vol. 1**, 128 (1944). (f) For the use of copper(II) halides to halogenate aromatic nuclei, see W. C. Baird, Jr., and J. H. Surridge, *J. Org. Chem.*, **35**, 3436 (1970); H. P. Crocker and R. Walser, *J. Chem. Soc.*, C, 1982 (1970); D. C. Nonhebel and J. A. Russell, *Tetrahedron*, **26**, 2781 (1970) and references therein.

a sigma complex [20], and, after loss of a proton, the bromo derivative. Thus, this substitution reaction differs from the conversion of an olefin to a dibromide only in the sense that loss of a proton from the intermediate bromine-containing cation is energetically favored over addition of a bromide ion to form another carbon-bromine bond. A similar situation exists in the reaction of bromine with enols, which will be considered subsequently. Although the bromination of substituted benzene derivatives can be expected to follow the usual rules for electrophilic substitution reactions, the following equations illustrate the possibility of selective bromination *ortho* to a phenolic hydroxyl function when the reaction is conducted at low temperatures in the presence of an amine. This *ortho* bromination process may involve the intramolecular rearrangement of a hypobromite intermediate.[17d]

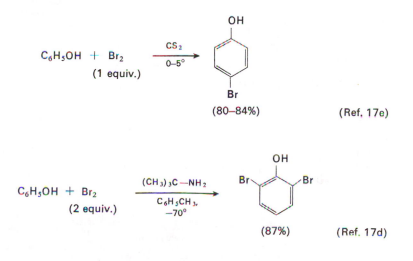

Allenes and conjugated dienes appear to react with bromine or chlorine by a process analogous to that described for olefins; however, more complex reaction mixtures are usually obtained.[18] Any allylic halides initially formed in these reactions may undergo allylic rearrangement if they are allowed to remain in the reaction mixture. The halogenation of acetylenes has not been studied extensively. The bromination reaction to form a *trans*-dibromide is catalyzed by bromide ion and has been suggested to follow the pathway indicated in an accompanying equation. Therefore, the amount of product formed by *trans* addition is enhanced by the presence of excess bromide ion in the reaction mixture.[18e]

18. (a) J. A. Pincock and K. Yates, *J. Am. Chem. Soc.,* **90**, 5643 (1968). (b) V. L. Heasley, C. L. Frye, R. T. Gore, Jr., and P. S. Wilday, *J. Org. Chem.,* **33**, 2342 (1968); V. L. Heasley, G. E. Heasley, S. K. Taylor, and C. L. Frye, *ibid.,* **35**, 2967 (1970). (c) M. L. Poutsma, *ibid.,* **33**, 4080 (1968). (d) For a review of the reactions of allenes, see K. Griesbaum, *Angew. Chem., Intern. Engl., Ed.* **5**, 933 (1966). (e) J. König and V. Wolf, *Tetrahedron Letters,* **No. 19**, 1629 (1970); R. Pettit and J. Henery, *Org. Syn.,* **50**, 36 (1970).

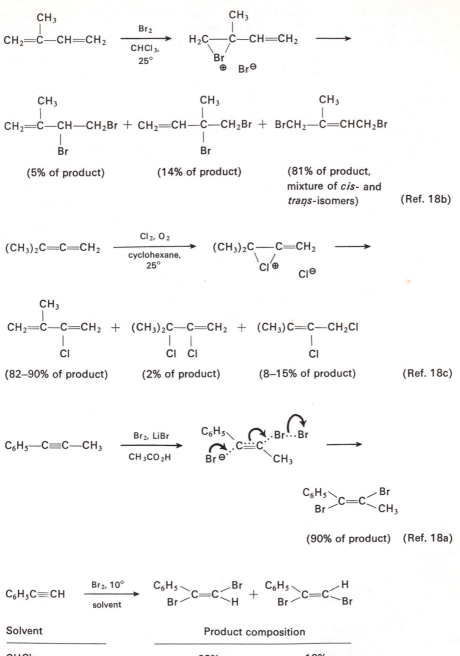

CH$_3$
CH$_2$=C—CH=CH$_2$ →(Br$_2$, CHCl$_3$, 25°)→ H$_2$C—C—CH=CH$_2$ →
with Br⊕ Br⊖ and CH$_3$

CH$_3$
CH$_2$=C—CH—CH$_2$Br + CH$_2$=CH—C—CH$_2$Br + BrCH$_2$—C=CHCH$_2$Br
         | |
         Br   Br
with CH$_3$

(5% of product)   (14% of product)   (81% of product, mixture of cis- and trans-isomers)   (Ref. 18b)

(CH$_3$)$_2$C=C=CH$_2$ →(Cl$_2$, O$_2$, cyclohexane, 25°)→ (CH$_3$)$_2$C—C=CH$_2$ →
with Cl⊕ Cl⊖

CH$_3$
CH$_2$=C—C=CH$_2$ + (CH$_3$)$_2$C—C=CH$_2$ + (CH$_3$)C=C—CH$_2$Cl
      | | | |
      Cl   Cl Cl   Cl

(82–90% of product)   (2% of product)   (8–15% of product)   (Ref. 18c)

C$_6$H$_5$—C≡C—CH$_3$ →(Br$_2$, LiBr, CH$_3$CO$_2$H)→ C$_6$H$_5$...Br⊖...C≡C...Br...Br with CH$_3$ →

C$_6$H$_5$, Br
C=C
Br, CH$_3$

(90% of product)   (Ref. 18a)

C$_6$H$_5$C≡CH →(Br$_2$, 10°, solvent)→ C$_6$H$_5$, Br / C=C \ Br, H + C$_6$H$_5$, H / C=C \ Br, Br

Solvent	Product composition	
CHCl$_3$	82%	18%
CH$_3$CO$_2$H	70%	30%
CH$_3$CO$_2$H + LiBr	97%	3%

(Ref. 18e)

As would be anticipated from the above discussion, the intermediate ion formed from bromine or chlorine and an olefin may react with any nucleophile in the reaction medium. For example, the olefin [21] will react with bromine in methanol solution to produce both a dibromide and a methoxybromide.[19] Also, the presence of excess halide ion during the halogenation of an olefin in a nucleophilic solvent will increase the fraction of dihalide in the product.[9d,h,j] The accompanying examples illustrate this effect and also the tendency of the nucleophile to attack that olefinic carbon atom which is better able to stabilize a positive charge. Reactions of this type have greater preparative value when an N-bromo amide (such as N-bromosuccinimide[3] or N-bromoacetamide[3,19d]) is used as the source of the

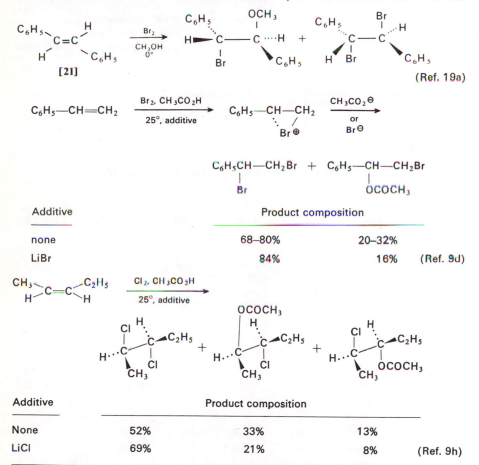

Additive	Product composition		
none	68–80%	20–32%	
LiBr	84%	16%	(Ref. 9d)

Additive	Product composition		
None	52%	33%	13%
LiCl	69%	21%	8%	(Ref. 9h)

19. (a) P. D. Bartlett and D. S. Tarbell, *J. Am. Chem. Soc.*, **58**, 466 (1936). (b) R. P. Bell and M. Pring, *J. Chem. Soc.*, B, 1119 (1966). (c) For studies of the reaction with halo olefins, see C. A. Clarke and D. L. H. Williams, *ibid.*, B, 1126 (1966); J. A. Hopwood and D. L. H. Williams, *ibid.*, B, 718 (1968); D. W. Pearson and D. L. H. Williams, *ibid.*, B, 436 (1970). (d) Directions for the preparation of this amide have been provided by E. P. Oliveto and C. Gerold, *Org. Syn., Coll. Vol. 4*, 104 (1963).

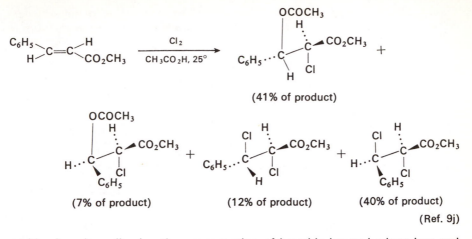

(41% of product)

(7% of product)   (12% of product)   (40% of product)

(Ref. 9j)

positive bromine, allowing the concentration of bromide ion to be kept low and minimizing dibromide formation. It is advantageous to use cosolvents such as dimethyl sulfoxide or dimethylformamide for the preparation of bromohydrins.[21b,c] The preparation of bromohydrin derivatives by this method, illustrated in the accompanying equations, is believed to involve an electrophilic attack on the olefin

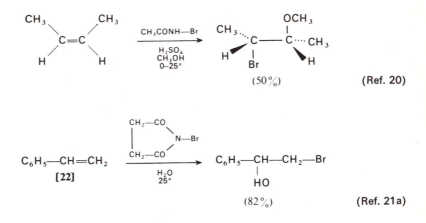

by the protonated bromoamide to form a bromonium ion. In the absence of the previously discussed conformational factors (favoring opening of the intermediate to give a *trans* diaxial product), the predominant product from subsequent reaction

20. (a) S. Winstein and R. B. Henderson, *J. Am. Chem. Soc.,* **65**, 2196 (1943). (b) For the analogous preparation of bromohydrins, see Ref. 9b.
21. (a) C. O. Guss and R. Rosenthal, *J. Am. Chem. Soc.,* **77**, 2549 (1955). (b) D. R. Dalton, V. P. Dutta, and D. C. Jones, *ibid.,* **90**, 5498 (1968). (c) D. R. Dalton, R. C. Smith, Jr., and D. G. Jones, *Tetrahedron,* **26**, 575 (1970). (d) For a discussion of the selectivity of this reaction with polyolefins, see E. E. van Tamelen and K. B. Sharpless, *Tetrahedron Letters,* **No. 28**, 2655 (1967). (e) A. J. Sisti, *ibid.,* **No. 38**, 3305 (1970). (f) F. Boerwinkle and A. Hassner, *ibid.,* **No. 36**, 3921 (1968).

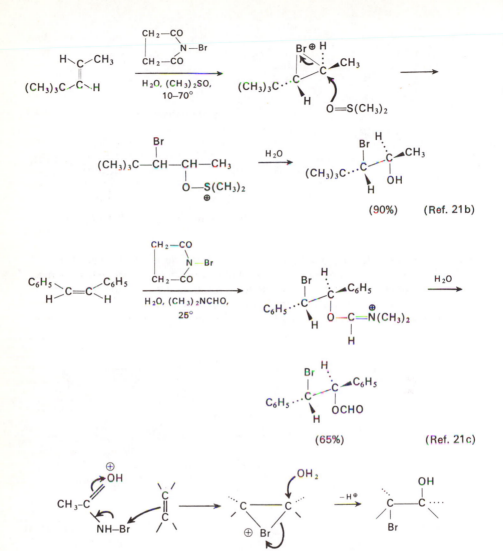

of the bromonium ion with the nucleophile is the isomer in which bromine is bonded to the less highly substituted carbon atom. As noted previously, this orientation, illustrated by the reaction of olefins [22] and [23], presumably reflects

$$(CH_3)_2C{=}CH_2 \xrightarrow[\substack{KBr \\ H_2O}]{Br_2} (CH_3)_2C{-}CH_2Br$$

[23]   OH   (Ref. 22a, b)

22. (a) C. M. Suter and H. D. Zook, *J. Am. Chem. Soc.*, **66**, 738 (1944). (b) H. O. House, *ibid.*, **77**, 5083 (1955). (c) S. Wolfe and D. V. C. Awang, *ibid.*, **89**, 5287 (1967).

the fact that the more important contributor to the bromonium ion is that structure [17b] or [17c] in which carbon is better able to tolerate a positive charge.[21] Solvents such as dimethylformamide, dimethyl sulfoxide, acetonitrile, and acetone may also serve as nucleophiles which attack the intermediate bromonium ion.[21f]

Since secondary alcohols are readily oxidized to ketones by reaction with N-bromoacetamide or N-bromosuccinimide[23] in aqueous acetone or aqueous t-butyl alcohol, the preparation of bromohydrins utilizing these reagents may be complicated by the further oxidation of the bromohydrin to a bromoketone. Apart from this possible side reaction, in the absence of a nucleophilic solvent, N-bromoacetamide was found to react with olefins to form imino ether derivatives[22c] as the following equation illustrates. The mechanism of this process remains to be established.

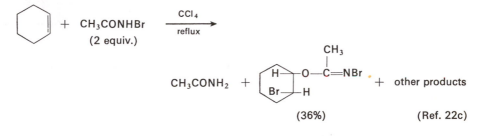

Although chlorohydrins may be similarly prepared by the reaction of olefins with N-chloroamides in aqueous acid[24] or with a preformed aqueous solution of hypochlorous acid.[25] perhaps the most convenient procedure utilizes hypochlorous acid generated from calcium hypochlorite in the reaction mixture. The reaction

$$CH_3-CH=CH-CH_3 \xrightarrow[\substack{CH_3CO_2H \\ H_2O}]{Ca(ClO)_2} CH_3-\underset{\underset{OH}{|}}{CH}-\underset{\underset{Cl}{|}}{CH}-CH_3 + CH_3-\underset{\underset{Cl}{|}}{CH}-\underset{\underset{Cl}{|}}{CH}-CH_3$$

(55%)   (Ref. 26)

of bromohydrins or chlorohydrins (e.g., [24]) with bases offers a useful route to epoxides and is often a more satisfactory method for the preparation of low-molecular-weight epoxides than is the direct reaction of the olefin with a peracid.

23. (a) L. F. Fieser and S. Rajagopalan, *J. Am. Chem. Soc.*, **71**, 3935, 3938 (1949); **72**, 5530 (1950); **73**, 118 (1951). (b) H. L. Herzog, M. A. Jevnik, and E. B. Hershberg, *ibid.*, **75**, 269 (1953). (c) R. Filler, *Chem. Rev.*, **63**, 21 (1963). (d) For a study of the oxidation of alcohols with N-chlorosuccinimide, see N. S. Srinivasan and N. Venkatasubramanian, *Tetrahedron Letters,* **No. 24**, 2039 (1970).
24. H. B. Donahoe and C. A. Vanderwerf, *Org. Syn.*, **Coll. Vol. 4**, 157 (1963).
25. (a) G. H. Coleman and H. F. Johnstone, *Org. Syn.*, **Coll. Vol. 1**, 158 (1944). (b) J. C. Richer and C. Freppel, *Tetrahedron Letters,* **No. 51**, 4411 (1969). (c) For a study of the addition of hypochlorous acid to allenes, see J. P. Bianchini and M. Cocordano, *Tetrahedron,* **26**, 3401 (1970).
26. C. E. Wilson and H. J. Lucas, *J. Am. Chem. Soc.*, **58**, 2396 (1936).

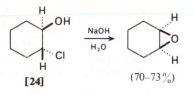

[24]   (70–73%)   (Ref. 27a)

The following equations provide examples of the preparation of iodohydrin derivatives by reaction of olefins with iodine in the presence of an oxidizing agent such as iodic acid or peracetic acid to form acyl hypoiodites or to convert the iodide ion formed to iodine.[27b,c] A comparable reaction occurs with olefins, bromine or chlorine, and peracetic acid forming esters of bromohydrins or chlorohydrins.[27c,d]

$$CH_3CH{=\!\!=}CH_2 + I_2 \xrightarrow[H_2O,\ 50°]{KIO_3,\ H_2SO_4} CH_3{-}\underset{OH}{\overset{\displaystyle |}{CH}}{-}CH_2I + CH_3\underset{I}{\overset{\displaystyle |}{CH}}{-}CH_2OH$$

(92% of product)   (8% of product)

(Ref. 27b)

$$CH_3CH{=\!\!=}CH_2 + I_2 + CH_3CO_3H \xrightarrow[\substack{(C_2H_5)_2O,\\16-19°}]{CH_3CO_2H}$$

$$CH_3C{-}O{-}I \longleftarrow CH_2{=}CH{-}CH_3 \longrightarrow ICH_2{-}\underset{OCOCH_3}{\overset{\displaystyle |}{CHCH_3}}$$
$$\overset{\|}{O}$$

(80%)   (Ref. 27c)

The formation of bromohydrin or iodohydrin esters as a result of reaction between a bromonium or iodonium ion and a carboxylate anion is most readily accomplished using an olefin (e.g., [25]), equimolar amounts of silver acetate or benzoate, and iodine or bromine[28] in an inert solvent such as carbon tetrachloride, chloroform, or ether. Under these circumstances the halogen and the silver

---

27. (a) A. E. Osterberg, *Org. Syn.*, **Coll. Vol. 1**, 185 (1944). (b) J. W. Cornforth and D. T. Green, *J. Chem. Soc.*, C, 846 (1970). (c) Y. Ogata and K. Aoki, *J. Org. Chem.*, **31**, 1625, 4181 (1966); **34**, 3974 (1969). (d) P. B. D. de la Mare, C. J. O'Connor, M. J. Rosser, and M. A. Wilson, *Chem. Commun.*, **No. 12**, 731 (1970).
28. (a) C. V. Wilson, *Org. Reactions*, **9**, 332 (1957); C. F. H. Allen and C. V. Wilson, *Org. Syn.*, **Coll. Vol. 3**, 578 (1955). (b) An improved procedure for the Hunsdiecker reaction utilizing bromine and mercuric oxide with the carboxylic acid has been described: S. J. Cristol and W. C. Firth, *J. Org. Chem.*, **26**, 280 (1961); J. A. Davis, J. Herynk, S. Carroll, J. Bunds, and D. Johnson, *ibid.*, **30**, 415 (1965); P. W. Jennings and T. D. Ziebarth, *ibid.*, **34**, 3216 (1969); J. S. Meek and D. T. Osuga, *Org. Syn.*, **43**, 9 (1963). (c) D. H. R. Barton, H. P. Faro, E. P. Serebryakov, and N. F. Woolsey, *J. Chem. Soc.*, 2438 (1965); J. K. Kochi, *J. Am. Chem., Soc.* **87**, 2500 (1965); *J. Org. Chem.*, **30**, 3265 (1965)l (d) F. D. Gunstone, *Adv. Org. Chem.*, **1**, 117 (1960).

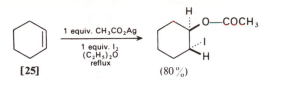

[25]   (80%)   (Ref. 28a)

carboxylate react to produce an acyl hypohalite (e.g., $CH_3CO-O-I$), which either can undergo thermal decomposition (presumably via free-radical chain mechanism) to form an alkyl or aryl halide (the Hunsdiecker reaction[28a]) or can add to an olefin. The following equations illustrate the Hunsdiecker reaction and other related procedures which have been used for the conversion of carboxylic acids to alkyl halides. The reaction utilizing lead tetraacetate may be used for alkyl iodides, bromides, or chlorides.[28c]

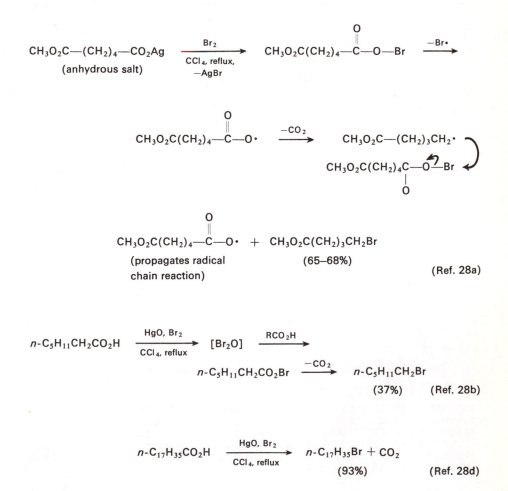

$$CH_3O_2C-(CH_2)_4-CO_2Ag \xrightarrow[\substack{CCl_4, \text{ reflux,} \\ -AgBr}]{Br_2} CH_3O_2C(CH_2)_4-\overset{O}{\overset{\|}{C}}-O-Br \xrightarrow{-Br\cdot}$$

(anhydrous salt)

$$CH_3O_2C(CH_2)_4-\overset{O}{\overset{\|}{C}}-O\cdot \xrightarrow{-CO_2} CH_3O_2C-(CH_2)_3CH_2\cdot$$

$$CH_3O_2C(CH_2)_4C-O-Br$$

$$CH_3O_2C(CH_2)_4-\overset{O}{\overset{\|}{C}}-O\cdot \; + \; CH_3O_2C(CH_2)_3CH_2Br$$

(propagates radical   (65–68%)
chain reaction)   (Ref. 28a)

$$n\text{-}C_5H_{11}CH_2CO_2H \xrightarrow[CCl_4, \text{ reflux}]{HgO, Br_2} [Br_2O] \xrightarrow{RCO_2H}$$

$$n\text{-}C_5H_{11}CH_2CO_2Br \xrightarrow{-CO_2} n\text{-}C_5H_{11}CH_2Br$$

(37%)   (Ref. 28b)

$$n\text{-}C_{17}H_{35}CO_2H \xrightarrow[CCl_4, \text{ reflux}]{HgO, Br_2} n\text{-}C_{17}H_{35}Br + CO_2$$

(93%)   (Ref. 28d)

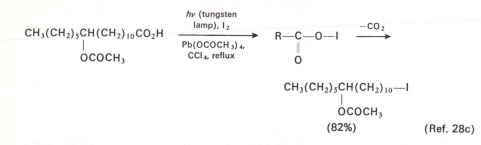

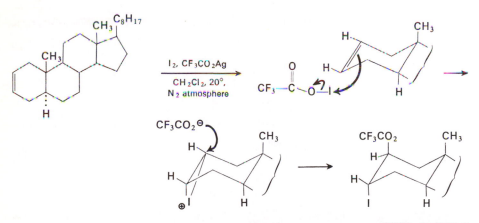

(72%) (Ref. 29d)

The addition of acyl hypoiodites to olefins is thought to be an ionic reaction in which the acyl hypoiodite attacks the olefin to form an iodonium ion that is in turn attacked by the carboxylate ion, resulting in overall *trans* addition. The accompanying example illustrates the preference for the formation of a *trans* diaxial product in the initial addition step. Two molar equivalents of the silver salt treated with one equivalent of iodine will produce a complex ($C_6H_5CO_2Ag \cdot C_6H_5CO_2I$, called a Simonini complex) which also adds the elements of the acyl hypoiodite to an olefin. If the resulting mixture is subsequently heated in an inert solvent (usually anhydrous benzene), the further changes indicated in the accompanying equations occur to form a *trans*-diol diester which may be saponified to form a *trans*-diol. This reaction sequence, called the Prevost reaction, involves two nucleophilic displacements with inversion of configuration and results in the overall *trans* addition of two oxygen functions to the double bond.

One of the accompanying equations shows a possible complication which

29. (a) G. E. McCasland and E. C. Horswill, *J. Am. Chem. Soc.,* **76**, 1654 (1954). (b) A. Ferretti and G. Tesi, *J. Chem. Soc.,* 5203 (1965). (c) P. S. Ellington, D. G. Hey, and G. D. Meakins, *ibid.,* C, 1327 (1966). (d) D. G. Hey, G. D. Meakins, and M. W. Pemberton, *ibid.,* C, 1331 (1966). (e) L. Mangoni and V. Dovinola, *Tetrahedron Letters,* **No. 60,** 5235 (1969). (f) V. L. Heasley, C. L. Frye, G. E. Heasley, K. A. Martin, D. A. Redfield, and P. S. Wilday, *ibid.,* **No. 18,** 1573 (1970).

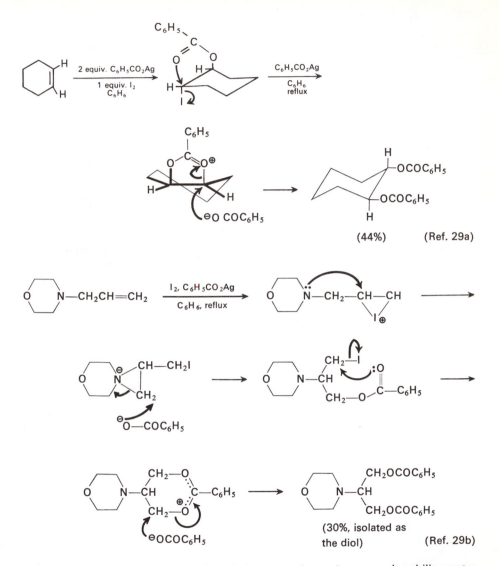

(44%)   (Ref. 29a)

(30%, isolated as
the diol)   (Ref. 29b)

may accompany the Prevost hydroxylation procedure when a nucleophilic center
(e.g. a tertiary amine) is located near the reacting double bond.

A useful modification of the Prevost reaction, the Woodward hydroxylation
procedure, utilizes an olefin with silver acetate and iodine in moist acetic acid. In
the second stage of this reaction, the oxonium ion intermediate [26] is produced in
the presence of at least one equivalent of water and under these conditions undergoes hydrolysis (rather than a second nucleophilic displacement) to form a hydroxy
acetate. Subsequent saponification leads to a diol [27] whose stereochemistry
corresponds to *cis* addition of the two hydroxyl groups from the more hindered
side of the double bond. The direction of this *cis* hydroxylation is of particular

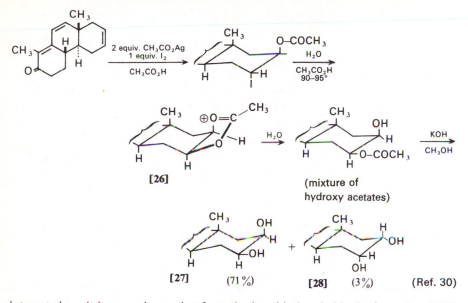

[26]   (mixture of
                        hydroxy acetates)

[27]   (71%)   [28]   (3%)   (Ref. 30)

interest since it is opposite to that from the less hindered side (to form diol [28])
which is observed with osmium tetroxide or potassium permanganate.

The following equations compare the stereochemical results when the Provost

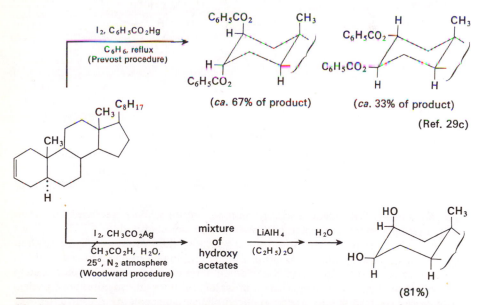

30. (a) R. B. Woodward and F. V. Brutcher, Jr., *J. Am. Chem. Soc.,* **80,** 209 (1958).
(b) C. B. Anderson, E. C. Friedrick, and S. Winstein, *Tetrahedron Letters,* **No. 29,** 2037
(1963). (c) J. F. King and A. D. Allbutt, *ibid.,* **No. 1,** 49 (1967); *Chem. Commun.,* **No. 1,** 14
(1966). (d) C. A. Bunton and M. D. Carr, *J. Chem. Soc.,* 770 (1963).

and Woodward hydroxylation procedures are applied to the same olefin. Attempts to use these procedures with trisubstituted olefins have indicated that the intervention of tertiary carbonium ion intermediates in these latter cases may lead both to the expected stereochemical results and to the formation of a number of by-products.[29e]

The reaction of an iodonium or bromonium ion intermediate with a carboxylate ion in the absence of silver ion occurs readily if an intramolecular reaction is sterically favorable. The process, called haloactonization,[31-33] is illustrated in the accompanying examples. This reaction, which is believed to proceed by intramolecular

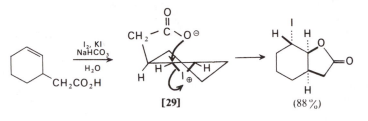

[29]   (88%)   (Ref. 31)

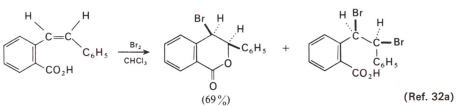

(69%)   (Ref. 32a)

attack of the carboxylate anion on the iodonium ion as in structure [29], provides a useful method for the conversion of unsaturated acids (e.g., [30]) to halolactones and, after hydrogenolysis,[34] to saturated lactones. It has also been useful for

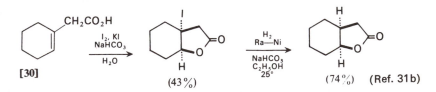

[30]   (43%)   (74%)   (Ref. 31b)

31. (a) E. E. van Tamelen and M. Shamma, *J. Am. Chem. Soc.,* **76**, 2315 (1954). (b) J. Klein, *ibid.,* **81**, 3611 (1959). (c) H. O. House, R. G. Carlson, and H. Babad, *J. Org. Chem.,* **28**, 3359 (1963).

32. (a) G. Berti, *Tetrahedron,* **4**, 393 (1958). (b) D. L. H. Williams, E. Beinvenüe-Goetz, and J. E. Dubois, *J. Chem. Soc.,* B, 517 (1969); E. Bienvenüe-Goetz, J. E. Dubois, D. W. Pearson, and D. L. H. Williams, *ibid.,* B, 1275 (1970).

33. A. W. Burgstahler and I. C. Nordin, *J. Am. Chem. Soc.,* **83**, 198 (1961).

34. For other hydrogenolysis procedures, see Ref. 35a. (a) R. Grewe, A. Heinke, and C. Sommer, *Chem. Ber.,* **89**, 1978 (1956). (b) H. W. Whitlock, Jr., *J. Am. Chem. Soc.,* **84**, 3412 (1962). (c) D. A. Denton, F. J. McQuillin, and P. L. Simpson, *J. Chem. Soc.,* 5535 (1964). (d) For the use of tri-*n*-butyltin hydride to effect such reductions, see H. O. House, S. G. Boots, and V. K. Jones, *J. Org. Chem.,* **30**, 2519 (1965).

separating and assigning stereochemistry to unsaturated acids (e.g., **[31]**) in cases where only one epimeric acid is capable of forming an iodolactone. As would be expected, other substituents such as hydroxyl groups and even halogen atoms can serve as intramolecular nucleophilic reactants.[32b]

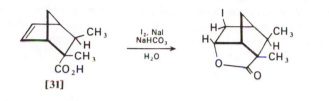

[31]   (Ref. 35)

In addition to water, alcohols, and carboxylic acids, a variety of other nucleophiles have been used to intercept halonium ions forming various β-substituted alkyl halides. The nitrogen containing nucleophiles such as the azide ion and the cyanate ion are particularly useful in synthesis. The accompanying equations illustrate the preparations and use of iodine isocyanate ($I—N\!=\!C\!=\!O$)[36] and the halogen azides ($I—N_3$, $Br—N_3$, $Cl—N_3$).[14h,16b,37] In the reactions with olefins, the nitrogen-containing nucleophiles tend to attack that carbon atom of the intermediate iodonium ion which is better able to bear a positive charge. In cases where similar substituents are present at each olefinic carbon atom and, consequently, the difference in charge density for the intermediate ion is slight, attack of the nucleophile at the less hindered atom is usually observed.

Iodine isocyanate fails to react with the electron-deficient double bonds of α,β-unsaturated carbonyl compounds, but these olefins readily add the more reactive iodine azide. Although the reaction of olefins with iodine azide is an ionic process, the reaction with bromine azide can proceed by either an ionic or a free radical pathway, and chlorine azide usually reacts with olefins by a free radical process.[37d,f]

35. (a) S. Beckmann and H. Geiger, *Chem. Ber.,* **92**, 2411 (1959). (b) J. Meinwald, S. S. Labana, and M. S. Chadha, *J. Am. Chem. Soc.,* **85**, 582 (1963). (c) J. A. Berson and A. Remanick, *ibid.,* **83**, 4947 (1961). (d) G. W. Oxer and D. Wege, *Tetrahedron Letters,* **No. 40**, 3513 (1969).
36. (a) C. G. Gebelein, G. Swift, and D. Swern, *J. Org. Chem.,* **32**, 3314 (1967). (b) C. G. Gebelein, S. Rosen, and D. Swern, *ibid.,* **34**, 1677 (1969). (c) A. Hassner, M. E. Lorber, and C. Heathcock, *ibid.,* **32** 540 (1967). (d) A. Hassner and C. Heathcock *ibid.,* **30**, 1748 (1964); *Tetrahedron,* **20**, 1037 (1964). (e) A. Hassner, R. P. Hoblitt, C. Heathcock, J. E. Kropp, and M. Lorber, *J. Am. Chem. Soc.,* **92**, 1326 (1970).
37. (a) F. W. Fowler, A. Hassner, and L. A. Levy, *J. Am. Chem. Soc.,* **89**, 2077 (1967). (b) A. Hassner, G. J. Matthews, and F. W. Fowler, *ibid.,* **91**, 5046 (1969). (c) A. Hassner and F. W. Fowler, *J. Org. Chem.,* **33**, 2686 (1968). (d) A. Hassner and F. Boerwinkle, *J. Am. Chem. Soc.,* **90**, 216 (1968); *Tetrahedron Letters,* **No. 36**, 3921 (1968); **No. 38**, 3309 (1969). (e) A. Hassner, R. J. Isbister, and A. Friederang, *ibid.,* **No. 34**, 2939 (1969). (f) A. Hassner, *Accts. Chem. Res.,* **4**, 9 (1971). (b) A. Hassner and J. E. Galle, *J. Am. Chem. Soc.,* **92**, 3733 (1970).

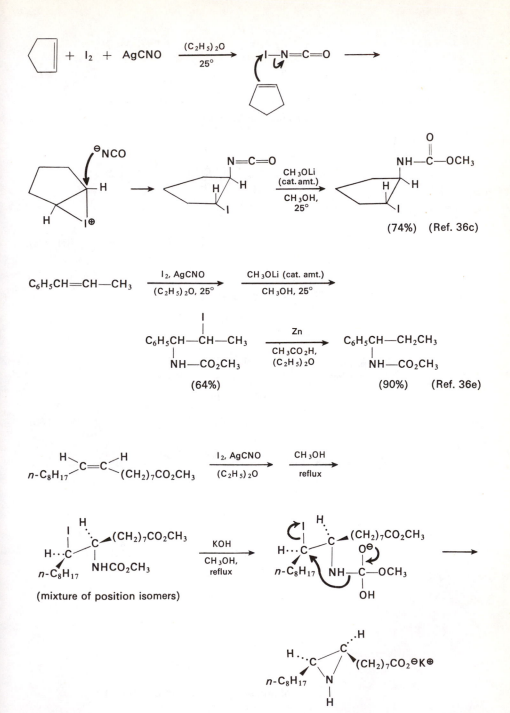

(74%)   (Ref. 36c)

$C_6H_5CH{=}CH{-}CH_3$ $\xrightarrow[\text{(C}_2\text{H}_5)_2\text{O, 25°}]{\text{I}_2,\ \text{AgCNO}}$ $\xrightarrow[\text{CH}_3\text{OH, 25°}]{\text{CH}_3\text{OLi (cat. amt.)}}$

(64%)   (90%)   (Ref. 36e)

(mixture of position isomers)

(97%)   (Ref. 36a)

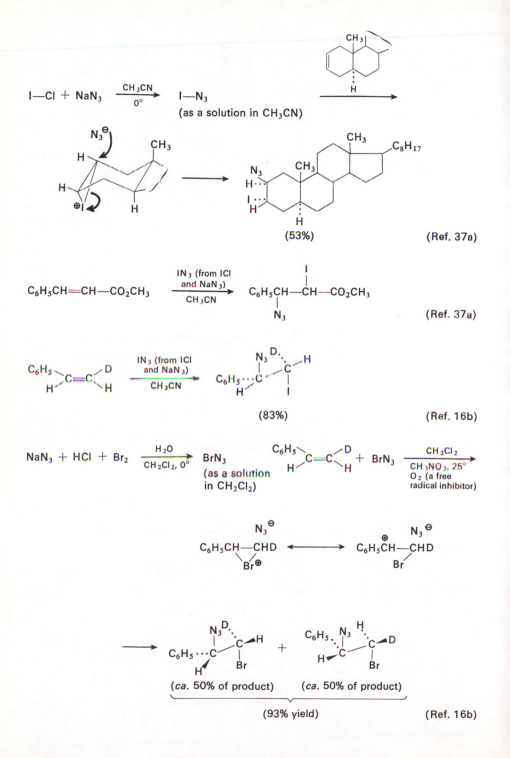

$I{-}Cl + NaN_3 \xrightarrow[0°]{CH_3CN} I{-}N_3$

(as a solution in CH_3CN)

(53%)   (Ref. 37a)

$C_6H_5CH{=}CH{-}CO_2CH_3 \xrightarrow[CH_3CN]{\substack{IN_3 \text{ (from ICl} \\ \text{and NaN}_3)}} C_6H_5CH{-}CH{-}CO_2CH_3$

(Ref. 37a)

$\xrightarrow[CH_3CN]{\substack{IN_3 \text{ (from ICl} \\ \text{and NaN}_3)}}$

(83%)   (Ref. 16b)

$NaN_3 + HCl + Br_2 \xrightarrow[CH_2Cl_2, 0°]{H_2O} BrN_3$

(as a solution in CH_2Cl_2)

$+ \, BrN_3 \xrightarrow[\substack{CH_3NO_2, 25° \\ O_2 \text{ (a free} \\ \text{radical inhibitor)}}]{CH_2Cl_2}$

$C_6H_5CH{-}CHD \longleftrightarrow C_6H_5CH{-}CHD$

(ca. 50% of product)   (ca. 50% of product)

(93% yield)   (Ref. 16b)

The reaction of iodine azide with acetylenes also appears to be an ionic process. Although the reason for the direction of addition observed remains to be established, it seems likely that the intermediate iodonium ion has the geometry indicated. In this event, the opening of the iodonium ion would be controlled by the inductive effects of the phenyl and methyl groups rather than by assistance from overlap of the benzene pi-orbital during the breaking of the carbon-iodine bond.

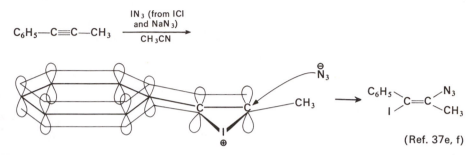

(Ref. 37e, f)

Other nucleophiles which have been observed to attack halonium ion inter-mediates include nitriles,[38a] acetylenes,[38b] and nitrate ion.[38c] Although the reactions of olefins with nitryl iodide $(I-NO_2)$[38d] and with nitrosyl chloride $(Cl-NO)$[38e-j] bear a superficial resemblance to the previously discussed halogenations, the orientation of the adducts in these latter cases corresponds to the initial addition of a nitrogen-containing species which is either electrophilic or free radical in nature. The following equations provide examples of the synthetic utility of olefin-nitrosyl chloride adducts.

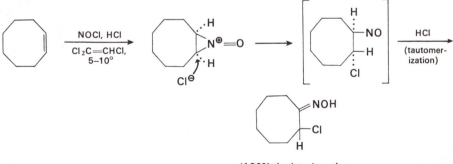

(100%, isolated as the
HCl salt)   (Ref. 38f)

38. (a) A. Hassner, L. A. Levy, and R. Gault, *Tetrahedron Letters*, **No. 27**, 3119 (1966); A. Hassner, R. A. Arnold, R. Gault, and A. Terada, *ibid.*, **No. 10**, 1241 (1968). (b) H. W. Whitlock, Jr., and P. E. Sandvick, *J. Am. Chem. Soc.*, **88**, 4525 (1966). (c) J. E. Kropp, A. Hassner, and G. J. Kent, *Chem. Commun.*, **No. 15**, 906 (1968). (d) A. Hassner, J. E. Kropp, and G. J. Kent, *J. Org. Chem.*, **34**, 2628 (1969). (e) J. Meinwald, Y. C. Meinwald, and T. N. Baker, *J. Am. Chem. Soc.*, **86**, 4074 (1964). (f) M. Ohno, N. Naruse, and I. Terasawa, *Org. Syn.*, **49**, 27 (1969). (g) B. W. Ponder and D. R. Walker, *J. Org. Chem.*, **32**, 4136 (1967). (h) M. Ohno, M. Okamoto, and K. Nukada, *Tetrahedron Letters*, **No. 45**, 4047 (1965). (i) A. Hassner and C. Heathcock, *J. Org. Chem.*, **29**, 1350 (1964). (j) For additions of nitrosyl fluoride to olefins see G. A. Boswell *ibid.*, **33** 3699 (1968).

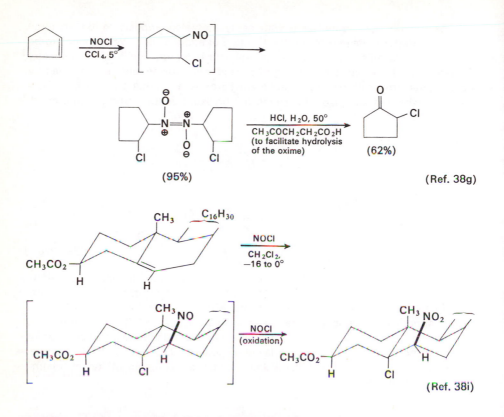

(95%)

HCl, H₂O, 50°
CH₃COCH₂CH₂CO₂H
(to facilitate hydrolysis
of the oxime)

(62%)

(Ref. 38g)

NOCl
CH₂Cl₂,
−16 to 0°

NOCl
(oxidation)

(Ref. 38i)

ALKYL HALIDES FROM OLEFINS OR ALCOHOLS

As would be expected from the reaction of halogens with olefins to produce 1,2-dihalides, the addition of hydrogen halides to olefins can be used to form alkyl halides. Recent studies of the additions of hydrogen chloride or hydrogen bromide to olefins[39,40] and acetylenes[41] have indicated the operation of two

39. (a) G. S. Hammond and C. H. Collins *J. Am. Chem. Soc.*, **82**, 4323 (1960). (b) R. C. Fahey and R. A. Smith, *ibid.*, **86**, 5035 (1964). (c) Y. Pocker, K. D. Stevens, and J. J. Champoux, *ibid.*, **91**, 4199 (1969); Y. Pocker and K. D. Stevens, *ibid.*, **91**, 4205 (1969). (d) R. C. Fahey and C. A. McPherson, *ibid.*, **91**, 3865 (1969). (e) R. C. Fahey, M. W. Monahan, and C. A. McPherson, *ibid.*, **92**, 2810 (1970). (f) R. C. Fahey, and M. W. Monahan, *ibid.*, **92**, 2816 (1970). (g) D. J. Pasto, G. R. Meyer, and S. Z. Kang, *ibid.*, **91**, 2163 (1969). (h) H. C. Brown and M. H. Rei, *J. Org. Chem.*, **31**, 1090 (1966). (i) P. K. Freeman, F. A. Raymond, and M. F. Grostic, *ibid.*, **32**, 24 (1967).
40. (a) B. A. Bohm and P. I. Abell, *Chem. Rev.*, **62**, 599 (1962). (b) P. S. Skell and P. K. Freeman, *J. Org. Chem.*, **29**, 2524 (1964). (c) P. D. Readio and P. S. Skell, *ibid.*, **31**, 753, 759 (1966); P. S. Skell and P. D. Readio, *J. Am. Chem. Soc.*, **86**, 3334 (1964). (d) N. A. LeBel, R. F. Czaja, and A. DeBoer, *J. Org. Chem.*, **34**, 3112 (1969). (e) K. D. Berlin, R. O. Lyerla, D. E. Gibbs, and J. P. Devlin, *Chem. Commun.*, **No. 19**, 1246 (1970).
41. (a) R. C. Fahey and D. J. Lee, *J. Am. Chem. Soc.*, **88**, 5555 (1966); **90**, 2124 (1968). (b) For a study of the addition of hydrogen chloride to allenes, see T. Okuyama, K. Izawa, and T. Fueno, *Tetrahedron Letters,* **No. 38**, 3295 (1970).

ionic pathways.[39,41] In addition, whenever hydrogen bromide is being added to an olefin a free-radical chain process may be an important competing reaction[39g,40] unless special precautions are taken, such as the use of free-radical inhibitors and protection of the reaction mixture from light, and other free-radical initiators. The ionic addition of hydrogen halides to olefins whose protonation can yield relatively stable carbonium ions (e.g. benzyl, allyl, or cyclopropylcarbinyl cations) is believed to proceed by the indicated proton transfer to form a carbonium ion intermediate. Reaction of these intermediate cations with the associated halide anion usually leads to a predominant *cis* addition of the hydrogen halide to the carbon-carbon double bond. Competing reactions of this cationic intermediate include attack by the reaction solvent and rearrangement.

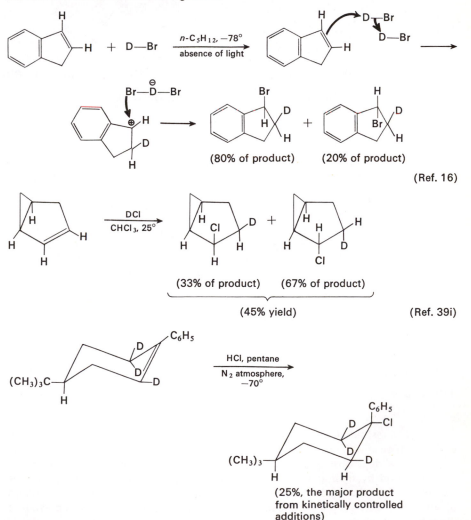

(80% of product)   (20% of product)

(Ref. 16)

(33% of product)   (67% of product)

(45% yield)

(Ref. 39i)

(25%, the major product from kinetically controlled additions)

(Ref. 40e)

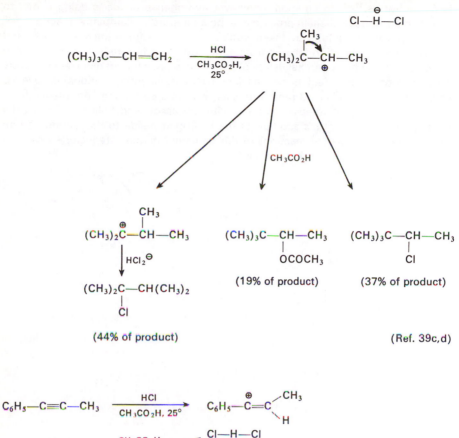

(44% of product)

(Ref. 39c,d)

(13% of product)

(Ref. 41)

Alkyl substituted acetylenes and relatively unhindered alkyl-substituted olefins exhibit a different stereochemical behavior in which *trans* addition of the hydrogen halide is observed. This result is attributed to a concerted attack on the multiple bond by hydrogen halide and halide anion as the following equations illustrate. Since these additions are catalyzed by halide anion, competing addition of solvent to the multiple bond can be minimized if excess halide ion is present in the reaction mixture.

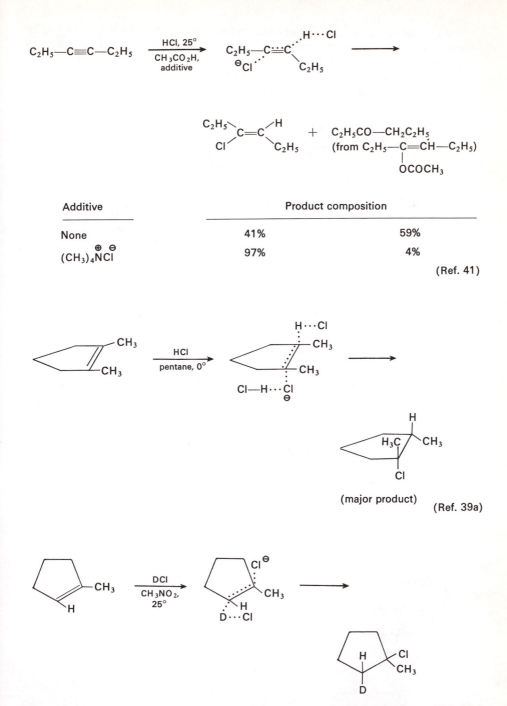

Additive	Product composition	
None	41%	59%
(CH$_3$)$_4$$\overset{\oplus}{N}$$\overset{\ominus}{Cl}$	97%	4%

(Ref. 41)

(major product)   (Ref. 39a)

(major product)   (Ref. 39c)

As noted previously, hydrogen bromide has a pronounced tendency to add to olefins in a free-radical chain process[40] unless special precautions are taken to avoid this reaction. The following equations illustrate this free radical reaction. It will be noted that whereas the ionic addition of hydrogen bromide to a terminal olefin yields mainly a 2-bromoalkane (addition following the Markownikoff rule),

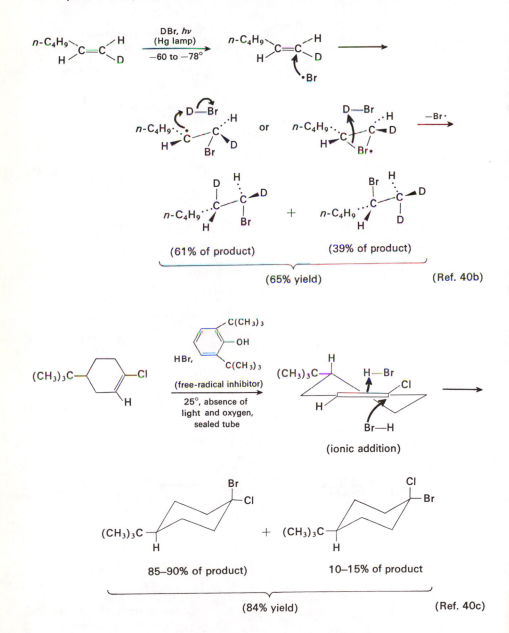

(61% of product)   (39% of product)

(65% yield)   (Ref. 40b)

(85–90% of product)   10–15% of product

(84% yield)   (Ref. 40c)

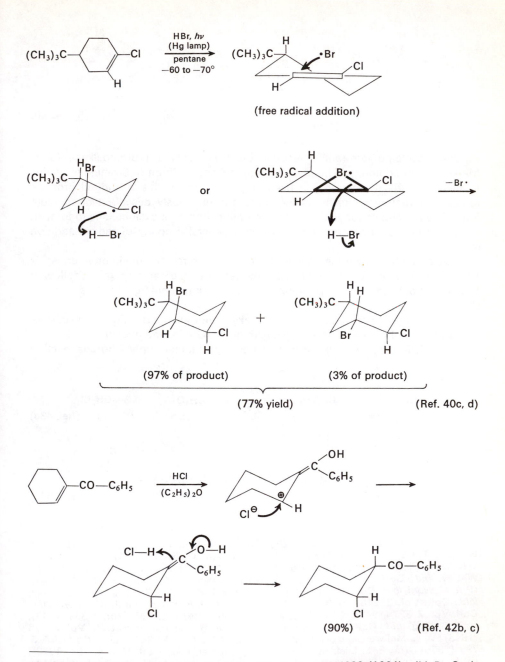

(free radical addition)

or

−Br·

(97% of product)   +   (3% of product)

(77% yield)   (Ref. 40c, d)

(90%)   (Ref. 42b, c)

42. (a) W. R. Vaughan and R. Caple, *J. Am. Chem. Soc.*, **86** 4928 (1964). (b) R. Caple and W. R. Vaughan, *Tetrahedron Letters,* **No. 34**, 4067 (1966). (c) C. Armstrong, J. A. Blair, and J. Homer, *Chem. Commun.,* **No. 3**, 103 (1969). (d) R. C. Fahey and H. J. Schneider, *J. Am. Chem. Soc.,* **92**, 6885 (1970). (e) K. Bowden and M. J. Price, *J. Chem. Soc.,* B, 1466, 1472 (1970).

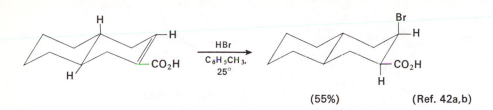

(55%)   (Ref. 42a,b)

the corresponding free-radical reaction gives primarily a 1-bromoalkane (anti-Markownikoff addition). The preference for a *trans*-addition of bromine to olefins in this free-radical reaction has been suggested to involve a bridged bromine radical, a structure which is analogous to the previously described bromonium ions. The preference for addition of the bromine atom to a cyclohexane to form an axial carbon-bromine bond appears to be a general property of radical addition reactions.[40c,d]

The addition of hydrogen halides to α,β-unsaturated carbonyl compounds[42] is believed to proceed by the indicated protonation of the carbonyl group followed by addition of halide ion and protonation of the resulting enol from the less hindered side.

The most common synthetic route to alkyl halides is the reaction of the corresponding alcohol with a hydrogen halide, thionyl chloride, or a phosphorus halide.[43] From the following examples, it will be noted that increasingly vigorous reaction

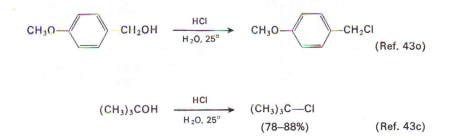

(Ref. 43o)

$(CH_3)_3COH$ $\xrightarrow[H_2O, 25°]{HCl}$ $(CH_3)_3C-Cl$

(78–88%)   (Ref. 43c)

43. (a) O. Kamm and C. S. Marvel, *Org. Syn.,* **Coll. Vol. 1**, 25–35 (1944). (b) R. H. Goshorn, T. Boyd, and E. F. Degering, *ibid.,* **Coll. Vol. 1**, 36–41 (1944). (c) J. F. Norris and A. W. Olmstead, *ibid.,* **Coll. Vol. 1**, 144 (1944). (d) J. E. Copenhaver and A. M. Whaley, *ibid.,* **Coll. Vol. 1**, 142 (1944). (e) F. Cortese, *ibid.,* **Coll. Vol. 2**, 91 (1943). (f) A. M. Ward, *ibid.,* **Coll. Vol. 2**, 159 (1943). (g) E. E. Reid, J. R. Ruhoff, and R. E. Burnett, *ibid.,* **Coll. Vol. 2**, 246 (1943). (h) W. W. Hartman, J. R. Byers, and J. B. Dickey, *ibid.,* **Coll. Vol. 2**, 322 (1943). (i) C. R. Noller and R. Dinsmore, *ibid.,* **Coll. Vol. 2**, 358 (1943). (j) H. S. King, *ibid.,* **Coll. Vol. 2**, 399 (1943). (k) H. B. Schurink, *ibid.,* **Coll. Vol. 2**, 476 (1943) ; W. L. McEwen, *ibid.,* **Coll. Vol. 3**, 227 (1955). (l) L. H. Smith, *ibid.,* **Coll. Vol. 3**, 793 (1955). (m) E. L. Eliel, M. T. Fisk, and T. Prosser, *ibid.,* **Coll. Vol. 4**, 169 (1963). (n) H. Stone and H. Shechter, *ibid.,* **Coll. Vol. 4**, 323 (1963). (o) K. Rorig, J. D. Johnston, R. W. Hamilton, and T. J. Telinski, *ibid.,* **Coll. Vol. 4**, 576 (1963). (P) S. Wawzonek, A. Matar, and C. H. Issidorides, *ibid.,* **Coll. Vol. 4**, 681 (1963). (q) L. A. R. Hall, V. C. Stephens, and J. H. Burckhalter, *ibid.,* **Coll. Vol. 4**, 333 (1963). (r) T. O. Soine and M. R. Buchdahl, *ibid.,* **Coll. Vol. 4**, 106 (1963).

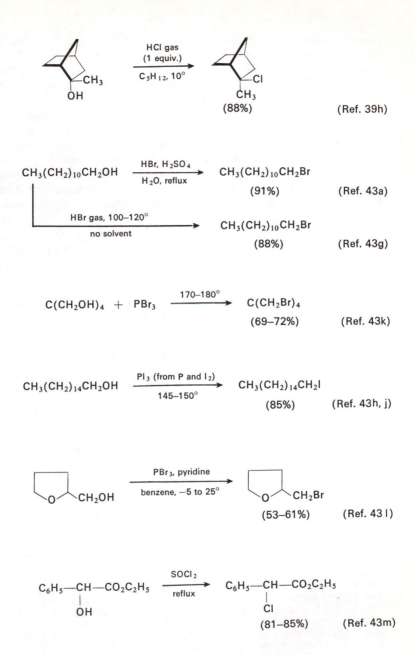

$$CH_3(CH_2)_{10}CH_2OH \xrightarrow[\text{H}_2\text{O, reflux}]{\text{HBr, H}_2\text{SO}_4} CH_3(CH_2)_{10}CH_2Br$$

(91%)   (Ref. 43a)

$$\xrightarrow[\text{no solvent}]{\text{HBr gas, 100–120°}} CH_3(CH_2)_{10}CH_2Br$$

(88%)   (Ref. 43g)

$$C(CH_2OH)_4 \; + \; PBr_3 \xrightarrow{170-180°} C(CH_2Br)_4$$

(69–72%)   (Ref. 43k)

$$CH_3(CH_2)_{14}CH_2OH \xrightarrow[\text{145–150°}]{\text{PI}_3 \text{ (from P and I}_2)} CH_3(CH_2)_{14}CH_2I$$

(85%)   (Ref. 43h, j)

(53–61%)   (Ref. 43 l)

$$\underset{\overset{|}{OH}}{C_6H_5-CH-CO_2C_2H_5} \xrightarrow[\text{reflux}]{\text{SOCl}_2} \underset{\overset{|}{Cl}}{C_6H_5-CH-CO_2C_2H_5}$$

(81–85%)   (Ref. 43m)

conditions are required as the reactant is changed from a tertiary, allylic, or benzylic alcohol (which can yield a relatively stable carbonium ion) to a secondary or primary alcohol. Recently, a detailed examination has been made of the ability of these methods to convert alcohols to alkyl halides with known stereochemical

consequences and without competing rearrangements.[44a-e] The reactions of primary alcohols with hydrogen halides at temperatures of 120° or less, with thionyl chloride, and with the phosphorus trihalides usually involve bimolecular nucleophilic displacements of some alcohol derivative (e.g. $-O^{\oplus}H_2$, $-OP^{\oplus}H(OR)_2$, or $-O-SO-Cl$) by a halide ion to form primary alkyl halides contaminated with little if any rearranged material.[44a-e] The formation of rearranged products from primary alcohols has generally been found only in cases where the alcohol was treated with a Lewis acid such as zinc chloride[44a] or a boron trihalide,[44b] where the chloride was formed by pyrolysis of the alkyl chloroformate,[44j] or where displacement reactions at the carbinol carbon of the alcohol were very hindered, as with neopentyl alcohol.[44d,g,h,k] In such hindered cases rearrangement is very common, but can sometimes be avoided by the use of a trialkyl- or triarylphosphine dihalide.[45]

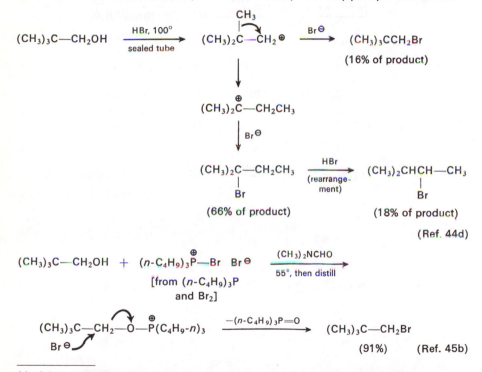

44. (a) W. Gerrard and H. R. Hudson, *J. Chem. Soc.,* 1059 (1963); 2310 (1964); *Chem. Rev.,* **65**, 697 (1965). (b) W. Gerrard, H. R. Hudson, and W. S. Murphy, *J. Chem. Soc.,* 2314 (1964). (c) E. J. Coulson, W. Gerrard, and H. R. Hudson, *ibid.,* 2364 (1965). (d) H. R. Hudson, *ibid.,* B, 664 (1968); *Synthesis,* **No. 3,** 112 (1969). (e) D. G. Goodwin and H. R. Hudson, *J. Chem. Soc.,* B, 1333 (1968). (f) C. W. Shoppee, T. E. Bellas, and R. Lack, *ibid.,* 6450 (1965); C. W. Shoppee and J. C. Coll, *ibid.,* C, 1121, 1124 (1970). (g) C. W. Shoppee, R. E. Lack, S. C. Sharma, and L. R. Smith, *ibid.,* C, 1155 (1967); C. W. Shoppee, R. E. Lack, and S. C. Sharma, *ibid.,* C, 2083 (1968). (h) R. A. Arain and M. K. Hargreaves, *ibid.,* C, 67 (1970). (i) G. Bellucci, F. Marioni, and A. Marsili, *Tetrahedron,* **25,** 4167 (1969). (j) P. W. Clinch and H. R. Hudson, *Chem. Commun.,* **No. 15,** 925 (1968). (k) D. Levy and R. Stevenson, *J. Org. Chem.,* **32,** 1265 (1967). (l) E. J. Corey and J. E. Anderson, *ibid.,* **32,** 4160 (1967).

The conversion of tertiary and benzylic alcohols to the corresponding alkyl halides by reaction with hydrogen halides is also frequently a satisfactory synthetic process if short reaction times, low reaction temperatures, and inert solvents are employed to minimize further transformation of the initial product.[39h] This conversion, which is believed to proceed by formation of a carbonium ion that is trapped by the halide ion, is frequently not useful for substituted allyl alcohols because the substituted allyl cation is not symmetrical and can give difficultly separable mixtures of halides.[45k] Methods which apparently avoid the formation of carbonium ion intermediates are illustrated in the following examples.

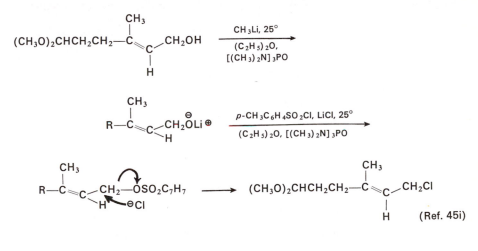

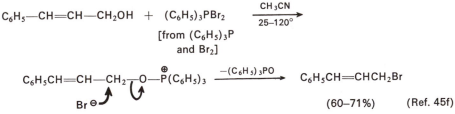

45. (a) L. Horner, H. Oediger, and H. Hoffmann, *Justus Liebigs Ann. Chem.*, **626**, 26 (1959). (b) G. A. Wiley, R. L. Hershkowitz, B. M. Rein, and B. C. Chung, *J. Am. Chem. Soc.*, **86**, 964 (1964); analogous results have been obtained by reaction of the alcohol with a mixture of triphenylphosphine and either carbon tetrachloride or carbon tetrabromide; R. G. Weiss and E. I. Snyder, *J. Org. Chem.*, **36**, 403 (1971). (c) L. Kaplan, *J. Org. Chem.*, **31**, 3454 (1966). (d) J. P. Schaefer and J. Higgins, *ibid.*, **32**, 1607 (1967). (e) J. P. Schaefer and D. S. Weinberg, *ibid.*, **30**, 2635, 2639 (1965). (f) J. P. Schaefer, J. G. Higgins, and P. K. Shenoy, *Org. Syn.*, **48**, 51 (1968). (g) *ibid.*, **49**, 6 (1969). (h) A. V. Bayless and H. Zimmer, *Tetrahedron Letters*, **No. 35**, 3811 (1968). (i) G. Stork, P. A. Grieco, and M. Gregson, *ibid.*, **No. 18**, 1393 (1969). (j) For a study of the iodide-ion catalyzed decomposition of chloroformates, see D. N. Kevill and F. L. Weitl, *J. Org. Chem.*, **32**, 2633 (1967). (k) For a review of the formation and rearrangement of allylic compounds, see R. H. DeWolfe and W. G. Young, *Chem. Rev.*, **56**, 753 (1956).

The alcohol-to-halide conversion is frequently most troublesome with secondary alcohols because a nucleophilic bimolecular displacement is retarded by steric hindrance, and the formation of carbonium ion intermediates (resulting in rearrangement and/or loss of stereospecificity) may become a serious competing reaction. Loss of stereospecificity may also result if the initially formed alkyl halide is allowed to remain in contact with excess halide ion under conditions favorable to bimolecular nucleophilic displacement.

$$(CH_3)_2CH-\underset{\underset{OH}{|}}{CH}-CH_3 \xrightarrow{HBr,\ H_2O} (CH_3)_2CH-\underset{\underset{Br}{|}}{CH}-CH_3 \ + \ (CH_3)_2C-CH_2CH_3$$

$$\underset{(3\%)}{} \qquad \qquad \underset{(54\%)}{\overset{\overset{Br}{|}}{}}$$

$$\xrightarrow[\substack{C_6H_6,\ pyridine, \\ 40-45°}]{(C_6H_5)_3PBr_2} (CH_3)_2CH-\underset{\underset{Br}{|}}{CH}-CH_3$$

(27%)   (Ref. 44h)

$$\underset{\substack{\\ \text{(optically active)}}}{HO-C\underset{H}{\overset{C_2H_5}{\cdots CH_3}}} \xrightarrow{\underset{-10°}{PBr_3}} \underset{\substack{\\ \text{(partial racemization)}}}{H_3C\underset{H}{\overset{C_2H_5}{\cdots}}C-Br \ + \ Br-C\underset{H}{\overset{C_2H_5}{\cdots CH_3}}}$$

$$\Big\downarrow \substack{(C_6H_5)_2PCl,\ (C_2H_5)_3N, \\ petroleum\ ether,\ 20°}$$

$$\underset{\substack{\\ (67\%,\ \text{optically} \\ \text{active})}}{(C_6H_5)_2P-O-C\underset{H}{\overset{C_2H_5}{\cdots CH_3}}} \xrightarrow{\underset{20°}{HBr}} \underset{\substack{\\ (79\%,\ \text{inversion})}}{CH_3\underset{H}{\overset{C_2H_5}{\cdots}}C-Br}$$

(Ref. 44e)

Although the initial reaction of alcohols with the phosphorus trihalides usually yields the illustrated phosphite esters without rearrangement, only the first step in the subsequent cleavage with a hydrogen halide occurs rapidly (and often stereospecifically with inversion and no rearrangement).[44c,e] Since only one-third of the alcohol is converted to the halide in such a stereospecific reaction with a phosphorus trihalide, use of the other phosphorus esters,[45] $R-O-P(C_6H_5)_2$ or $R-O-P\oplus-(C_6H_5)_3$, illustrated constitutes a better method for the stereospecific conversion of secondary alcohols to alkyl halides with a minimum amount of rearrangement.

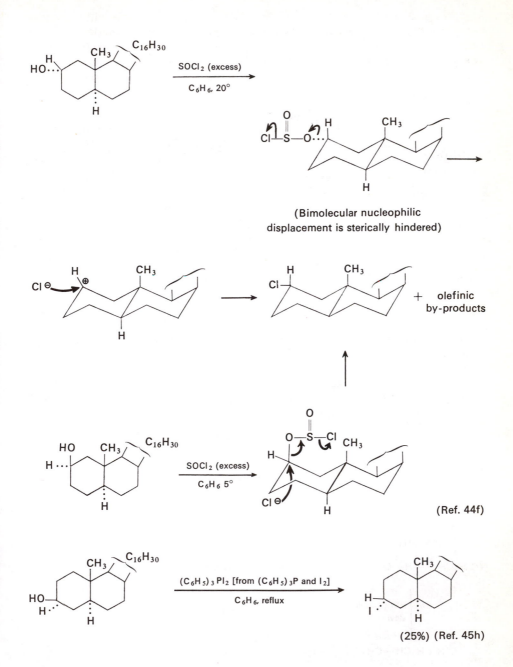

(Bimolecular nucleophilic
displacement is sterically hindered)

+ olefinic
by-products

(Ref. 44f)

(25%) (Ref. 45h)

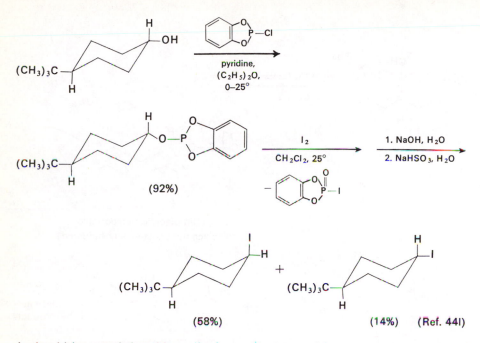

It should be noted that this method can also be used to convert phenols to aryl halides.[45b,d,g]

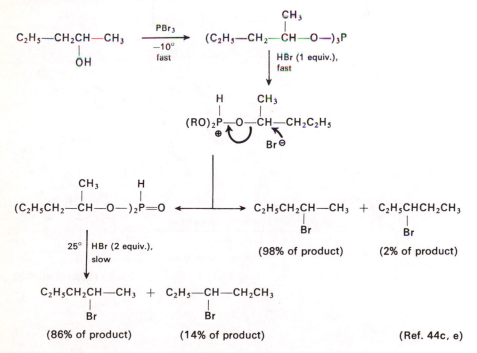

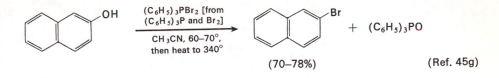

(70–78%) (Ref. 45g)

HALOGENATION OF CARBONYL COMPOUNDS

As the accompanying equations illustrate, the reaction conditions employed for the bromination or chlorination of ketones are often similar to those used for the addition of a halogen to olefins. Common solvents employed include carbon tetrachloride, chloroform, ether, and acetic acid. The reaction is catalyzed by acid and, if no acid is present initially (as when carbon tetrachloride or chloroform is used as solvent), often exhibits an induction period until some hydrogen halide

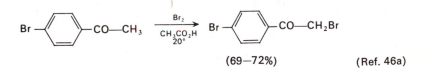

(69–72%) (Ref. 46a)

$$C_6H_5-CO-CH_3 \xrightarrow[\substack{CH_3CO_2H, \\ 60°}]{Cl_2\ (excess)} C_6H_5-CO-CHCl_2$$

(80–94%) (Ref. 46d)

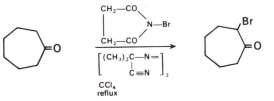

(Ref. 47a)

46. (a) W. D. Langley, *Org. Syn., Coll. Vol. 1*, 127 (1944). (b) R. Breslow and J. Posner, *ibid.*, **47**, 62 (1967). (c) J. J. Klingenberg, *ibid., Coll. Vol. 4*, 110 (1963). (d) J. G. Aston, J. D. Newkirk, D. M. Jenkins, and J. Dorsky, *ibid., Coll. Vol. 3*, 538 (1955).
47. (a) E. J. Corey, *J. Am. Chem. Soc.*, **75**, 2301, 3297, 4832 (1953); **76**, 175 (1954). (b) E. J. Corey, T. H. Topie, and W. A. Wozniak, *ibid.*, **77**, 5415 (1956). (c) E. J. Corey and H. J. Burke, *ibid.*, **77**, 5418 (1955). (d) E. J. Corey and R. A. Sneen, *ibid.*, **78**, 6269 (1956). (e) R. F. W. Cieciuch and F. H. Westheimer, *ibid.*, **85**, 2591 (1963). (f) J. E. Dubois and J. Toullec, *Chem. Commun., No. 6*, 292 (1969); *No. 9*, 478 (1969). (g) C. Rappe and W. H. Sachs, *J. Org. Chem.*, **32**, 3700 (1967). (h) C. Rappe, *Acta Chim. Scand.*, **22**, 219, 1359 (1968); and references therein. (i) C. Rappe, *Arkiv. Kemi.*, **23**, 81 (1965); **24**, 73 (1965). (j) M. D. Mehta, D. Miller, and D. J. D. Tidy, *J. Chem. Soc.*, 4614 (1963).

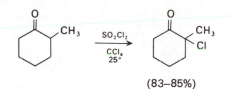

(83–85%)   (Ref. 48)

has been generated in the reaction mixture. Apart from adding acid, the bromination or chlorination of ketones may be initiated by holding a light bulb next to the reaction flask.

The reaction of a ketone with bromine proceeds by the acid-catalyzed[44c-h] enolation of the ketone and subsequent electrophilic attack on the enol by bromine; loss of a proton from the intermediate oxonium ion [32] leads to the bromoketone.

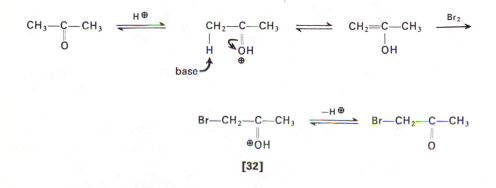

[32]

For the majority of ketones, which exist largely in the keto rather than the enol form, enolization is the rate-limiting step and the overall rate of halogenation is independent of the nature or concentration of the halogen except at low halogen concentrations.[47f,h] It will be noted that the formation of the enol requires the presence of both an acid and a base. In inert solvents such as carbon tetrachloride or chloroform, the only bases present are the unprotonated ketone and halide ion;[47e] a large excess of a Lewis acid such as aluminum chloride serves to convert these bases to their conjugate acids and to prevent enolization. The following equations illustrate the use of this technique to prevent bromination alpha to a carbonyl function during the bromination of an aromatic nucleus; this technique

48. (a) E. W. Warnhoff, D. G. Martin, and W. S. Johnson, *Org. Syn.*, **Coll. Vol. 4**, 162 (1963). (b) For further studies of the chlorination of ketones with sulfuryl chloride, see D. P. Wyman and P. R. Kaufman, *J. Org. Chem.*, **29**, 1956 (1964); D. P. Wyman, P. R. Kaufman, and W. R. Freeman, *ibid.*, **29**, 2706 (1964); F. Caujolle and D. Q. Quan, *Compt. rend.*, **265C**, 269 (1967); P. A. Peters, R. Ottinger, J. Reisse, and G. Chiurdoglu, *Bull. Soc., Chim. Belg.*, **77**, 407 (1968).

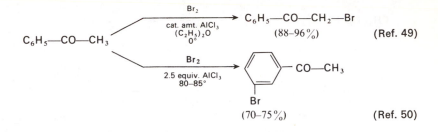

(Ref. 49)

(Ref. 50)

has also found utility in other types of reactions.[51a,b] Although halogenations of ketones with N-bromosuccinimide or with sulfuryl chloride (SO_2Cl_2) are frequently initiated by light, a peroxide, or some other free-radical initiator,[51h] it appears likely that once an appreciable concentration of halogen acid has accumulated in the reaction mixture, these halogenations, too, proceed by an ionic mechanism involving electrophilic attack on the enol. The use of N-bromosuccinimide for ketone bromination offers the advantage that little hydrogen bromide, which may catalyze aldol condensation of the ketone, is produced in the reaction mixture. Alternatively, as illustrated below, aqueous potassium chlorate may be used to remove hydrogen bromide during direct bromination of ketones. The reaction of carbonyl compounds with copper(II) chloride or bromide in various solvents[4d,51c-f] has also been used to prepare α-halo aldehydes and α-halo ketones. This reaction appears to involve a copper-halide-catalyzed enolization followed by transfer of a halogen atom from the copper(II) salt to the enolate. This procedure appears to offer a very selective method for the chlorination of an unsymmetrical ketone at the more highly substituted α-carbon atom (corresponding to the more stable enol).[51i]

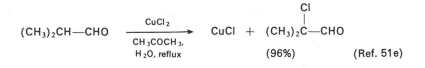

49. R. M. Cowper and L. H. Davidson, *Org. Syn.*, **Coll. Vol. 2**, 480 (1943).
50. (a) D. E. Pearson, H. W. Pope, and W. W. Hargrove, *Org. Syn.*, **40**, 7 (1960). (b) D. E. Pearson, H. W. Pope, W. W. Hargrove, and W. E. Stamper, *J. Org. Chem.*, **23**, 1412 (1958). (c) D. E. Pearson, W. E. Stamper, and B. R. Suthers, *ibid.*, **28**, 3147 (1963).
51. (a) H. O. House, V. Paragamian, R. S. Ro, and D. J. Wluka, *J. Am. Chem. Soc.*, **82**, 1457 (1960). (b) C. G. Swain and A. S. Rosenberg, *ibid.*, **83**, 2154 (1961). (c) L. C. King and G. K. Ostrum, *J. Org. Chem.*, **29**, 3459 (1964). (d) E. M. Kosower, W. J. Cole, G. S. Wu, D. E. Cardy, and G. Meisters, *ibid.*, **28**, 630 (1963); E. M. Kosower and G. S. Wu, *ibid.*, **28**, 633 (1963). (e) A. Lorenzini and C. Walling, *ibid.*, **32**, 4008 (1967). (f) For the use of copper(II) chloride to prepare α-chloro acids, see R. Louw, *Chem. Commun.*, **No. 15**, 544 (1966). (g) 2-Iodo-1,3-dicarbonyl compounds have been prepared by reaction of the 1,3-dicarbonyl compounds with periodic acid, A. J. Fatiadi, *ibid.*, **No. 1**, 11 (1970). (h) For a study of the gas-phase bromination of acetone, a free-radical chain reaction, see K. D. King, D. M. Golden, and S. W. Benson, *J. Am. Chem. Soc.*, **92**, 5541 (1970). (i) L. Werthemann and W. S. Johnson, to be published.

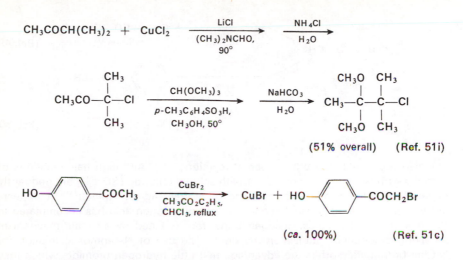

$$CH_3COCH(CH_3)_2 \ + \ CuCl_2 \ \xrightarrow[\substack{(CH_3)_2NCHO, \\ 90°}]{LiCl} \ \xrightarrow[H_2O]{NH_4Cl}$$

(51% overall)   (Ref. 51i)

(ca. 100%)   (Ref. 51c)

In the bromination of unsymmetrical ketones, the position taken by the entering bromine is determined by the relative ease of formation of the structurally isomeric enols. In general, this ease of formation is enhanced by the presence of an alpha alkyl substituent or other unsaturated alpha substituents that stabilize the enol by forming a conjugated system. Consequently, the predominant isomer produced on bromination of a methyl n-alkyl ketone or an α-alkylcycloalkanone is normally that in which the bromine atom has entered the more highly substituted alpha position,[47f–i,52a,b] as illustrated in the accompanying equation.

$$CH_3-CO-CH_2CH_2CH_3 \ \xrightarrow[\substack{H_2O \\ 40-45° \\ light}]{Br_2, \ KClO_3} \ CH_3-CO-\underset{\underset{\text{Br}}{|}}{CH}-CH_2CH_3 \ + \ BrCH_2-CO-CH_2CH_2CH_3$$

(53%)   (32%)

(Ref. 52a)

With methyl sec-alkyl ketones, the rate of acid-catalyzed enolization to form the less highly substituted (and less stable) enol is more rapid,[47f–h] possibly because of steric interactions that arise from eclipsed alkyl groups in the transition state leading to the more highly substituted enol [33]. Any studies of bromo ketone compositions in such cases can be complicated by the ready acid-catalyzed isomerization of the initially formed bromo ketones; consequently, conclusions based on product studies are always questionable unless the absence of isomeriza-

---

52. (a) J. R. Catch, D. H. Hey, E. R. H. Jones, and W. Wilson, *J. Chem. Soc.*, 276 (1948). (b) H. M. E. Cardwell and A. E. H. Kilner, *ibid.*, 2430 (1951); H. M. E. Cardwell, *ibid.*, 2442 (1951). (c) α-Fluoro and α-cyano substituents appear to have a similar effect. J. Canta-cuzene, M. Atlani, and J. Anibie, *Tetrahedron Letters*, **No. 19**, 2335 (1968). (d) The presence of methanol in the reaction mixture has been found to favor the bromination of methyl alkyl ketones at the methyl group; M. Gaudry and A. Marquet, *Tetrahedron*, **26**, 5611, 5617 (1970).

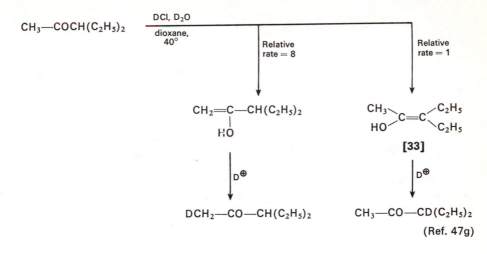

(Ref. 47g)

tion has been established. The probable cause of this isomerization, shown in the following equation, involves reconversion of the bromo ketone to bromine and the unhalogenated ketone.[47i,j]

$$Br—CH_2—CCH_2CH_2CH_3 \quad \xrightarrow[\substack{(C_2H_5)_2O, \\ 25°}]{HBr} \quad Br—CH_2—C—CH_2CH_2CH_3 \quad \xrightarrow{-Br_2}$$
$$\overset{\|}{O} \qquad\qquad\qquad Br^{\ominus} \nearrow \qquad \overset{+}{O}H$$

$$CH_2{=}C—CH_2CH_2CH_3 \rightleftharpoons CH_3—C{=}CHCH_2CH_3 \xrightarrow{Br_2}$$
$$\underset{OH}{|} \qquad\qquad\qquad \underset{OH}{|}$$

$$CH_3—C—CHCH_2CH_3 \;+\; BrCH_2COCH_2CH_2CH_3$$
$$\overset{\|}{O}\;\underset{Br}{|}$$

(58% of product)   (2% of product)

(Ref. 47j)

For either steric or electronic reasons, the presence of an α-halogen substituent usually retards the rate of acid-catalyzed enol formation.[52] As a result, the substitution of each successive halogen atom becomes more difficult, making it possible to introduce one, two, three, or more halogen atoms into a ketone in controlled fashion simply by limiting the amount of halogen added. A quite different order of reactivity is found for the base-catalyzed halogenations of ketones, where the presence of an α-halogen atom enhances the rate of further halogenation at this position. This result might be interpreted to mean the more readily formed α-halo enolate anion shown in the following equations is the intermediate which reacts with a halogen molecule; other data suggest that the mechanism(s) of base-catalyzed ketone halogenations are more complex.[47h] At the present time

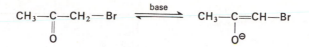

$$\text{CH}_3-\overset{\overset{\displaystyle O}{\|}}{\text{C}}-\text{CH}_2-\text{Br} \underset{\text{base}}{\rightleftharpoons} \text{CH}_3-\underset{\underset{\displaystyle O^\ominus}{|}}{\text{C}}=\text{CH}-\text{Br}$$

there is considerable uncertainty about whether the actual halogenating agent is a halogen molecule or some product resulting from reaction of the halogen with the basic medium. In any event base-catalyzed halogenation is usually of no value for the preparation of monohaloketones; however, it is used for the conversion of methyl ketones (e.g., [34] and [35]) to trihalomethyl ketones, which are usually cleaved in the basic reaction mixture to form acids and trihalomethanes (the haloform reaction[53]). The successful degradations of the unsaturated ketones [34] and [35] are interesting because halogenation alpha to the ketone is not complicated by halogenation of the carbon-carbon double bond of the ketones.

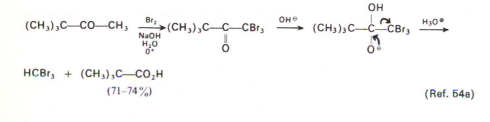

$$(\text{CH}_3)_3\text{C}-\text{CO}-\text{CH}_3 \xrightarrow[\substack{\text{NaOH} \\ \text{H}_2\text{O} \\ 0°}]{\text{Br}_2} (\text{CH}_3)_3\text{C}-\overset{\overset{\displaystyle O}{\|}}{\text{C}}-\text{CBr}_3 \xrightarrow{\text{OH}^\ominus} (\text{CH}_3)_3\text{C}-\overset{\overset{\displaystyle OH}{|}}{\underset{\underset{\displaystyle O^\ominus}{|}}{\text{C}}}-\text{CBr}_3 \xrightarrow{\text{H}_3\text{O}^\oplus}$$

HCBr$_3$ + (CH$_3$)$_3$C—CO$_2$H
      (71–74%)   (Ref. 54a)

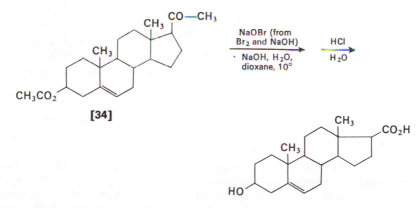

[34]

NaOBr (from Br$_2$ and NaOH) · NaOH, H$_2$O, dioxane, 10°   HCl H$_2$O

(91–95%)   (Ref. 54b)

53. R. C. Fuson and B. A. Bull, *Chem. Rev.*, **15**, 275 (1934).
54. (a) L. T. Sandborn and E. W. Bousquet, *Org. Syn.*, **Coll. Vol. 1**, 526 (1944). (b) J. Staunton and E. J. Eisenbraun, *ibid.*, **42**, 4 (1962). (c) For a study of the cleavage of enolizable ketones with base, iodine, and oxygen, see L. A. Freiberg, *J. Am. Chem. Soc.*, **89**, 5297 (1967).

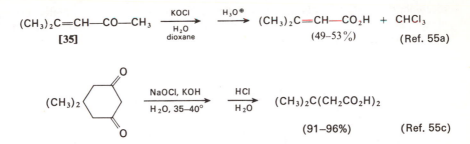

$$(CH_3)_2C=CH-CO-CH_3 \xrightarrow[\substack{H_2O \\ dioxane}]{KOCl} \xrightarrow{H_3O^{\oplus}} (CH_3)_2C=CH-CO_2H + CHCl_3$$

[35]   (49–53%)   (Ref. 55a)

$$(CH_3)_2 \quad \xrightarrow[\substack{H_2O, 35-40°}]{NaOCl, KOH} \xrightarrow[\substack{H_2O}]{HCl} (CH_3)_2C(CH_2CO_2H)_2$$

                                (91–96%)   (Ref. 55c)

The position of bromination of 3-ketosteroids (e.g., **[36]** and **[37]**) is determined not only by the presence of alpha substituents but also by the stereochemistry of the A-B ring junction. The tendency of the *trans*-fused system to brominate

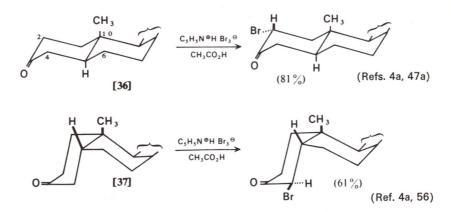

[36]   (81%)   (Refs. 4a, 47a)

[37]   (61%)   (Ref. 4a, 56)

selectively at position 2 has been attributed to the instability of a $\Delta^{3,4}$ double bond,[57] which results, at least in part, from repulsion between axial substituents at C-6 and C-10.[57b,58] The position and stereochemistry of a bromine atom in an α-bromoketone produced by ketone bromination may further depend on the reaction conditions employed, since the most rapidly formed bromoketone (kinetically controlled product) may be isomerized to a more stable bromoketone (thermodynamically controlled product) by the hydrogen bromide, also formed in

55. (a) L. I. Smith, W. W. Prichard, and L. J. Spillane, *Org. Syn.,* **Coll. Vol. 3,** 302 (1955). (b) M. S. Newman and H. L. Holmes, *ibid.,* **Coll. Vol. 2,** 428 (1943). (c) W. T. Smith and G. L. McLeod, *ibid.,* **Coll. Vol. 4,** 345 (1963). (d) For the cleavage of cyclobutanones with hypochlorous acid, see J. A. Horton, M. A. Laura, S. M. Kalbag, and R. C. Petterson, *J. Org. Chem.,* **34,** 3366 (1969).
56. (a) L. F. Fieser and R. Ettorre, *J. Am. Chem. Soc.,* **75,** 1700 (1953). (b) L. F. Fieser and X. A. Dominguez, *ibid.,* **75,** 1704 (1953).
57. (a) See Ref. 8b, pp. 276–279. (b) E. J. Corey and R. A. Sneen, *J. Am. Chem. Soc.,* **77,** 2505 (1955).
58. (a) C. Djerassi, N. Finch, R. C. Cookson, and C. W. Bird, *J. Am. Chem. Soc.,* **82,** 5488 (1960). (b) R. Mauli, H. J. Ringold, and C. Djerassi, *ibid.,* **82,** 5494 (1960). (c) R. Villotti, H. J. Ringold, and C. Djerassi, *ibid.,* **82,** 5693 (1960).

the reaction mixture. This acid-catalyzed isomerization,[47i,j,58,59] illustrated by the accompanying examples, may result either from enolization and reketonization of the bromoketone (as is probably the case for ketone **[38]**) or from the previously

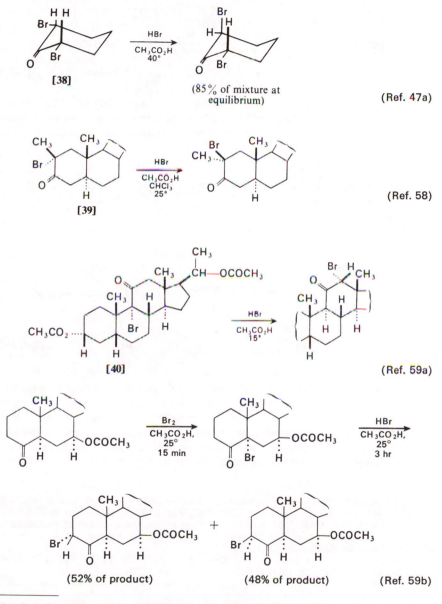

[38]

(85% of mixture at
equilibrium)   (Ref. 47a)

[39]   (Ref. 58)

[40]   (Ref. 59a)

(52% of product)   +   (48% of product)   (Ref. 59b)

59. (a) E. R. H. Jones and D. J. Wluka, *J. Chem. Soc.*, 907, 911 (1959). (b) A. R. Davies and G. H. R. Summers, *ibid.*, C, 1227 (1967). (c) M. P. Hartshorn and A. F. A. Wallis, *ibid.* 3839 (1962); *Tetrahedron*, **21**, 273 (1965).

discussed reaction of the protonated ketone with bromide ion or some other nucleophile to produce an enol that undergoes isomerization and rebromination. This latter sequence is clearly necessary to explain the isomerization of ketones [39] and [40]. To obtain the kinetically controlled product from a ketone bromination, the reaction is commonly run in acetic acid containing excess sodium acetate[58,60] or pyridine[58,59] to react with the hydrogen bromide as it is formed.

An alternative method for preparing a bromoketone of known structure consists of converting the ketone (e.g., [41]) to its enol acetate, which is then treated with

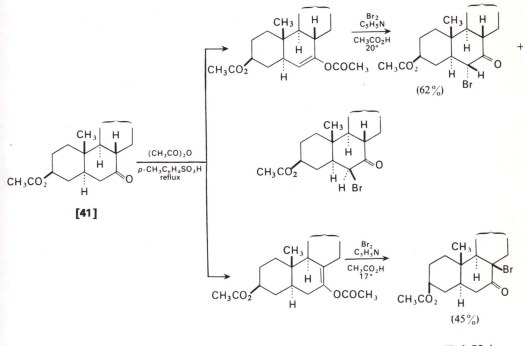

(Ref. 59a)

bromine and sodium acetate or pyridine in acetic acid to form the desired product.[59] Since the direct bromination of aldehydes is often complicated by a competing reaction with the aldehyde C—H bond, a similar procedure involving bromination of an enol acetate [42] is useful for the preparation of α-bromoaldehydes.

A related preparative method is the direct bromination of ketals with bromine, pyridinium hydrobromide perbromide, or phenyltrimethylammonium tribromide to

60. (a) C. W. Shoppee, G. A. R. Johnston, and R. E. Lack, *J. Chem. Soc.*, 3604 (1962). (b) C. W. Shoppee and T. E. Bellas, *ibid.*, 3366 (1963). (c) P. Z. Bedoukian, *Org. Syn.*, **Coll. Vol. 3**, 127 (1955). (d) For other preparations of α-halo aldehydes, see Y. Yanuka, R. Katz, and S. Sarel, *Chem. Commun.*, **No. 15**, 849 (1968) ; J. J. Riehl, A. Fougerousse, and F. Lamy, *Tetrahedron Letters*, **No. 42**, 4415 (1968).

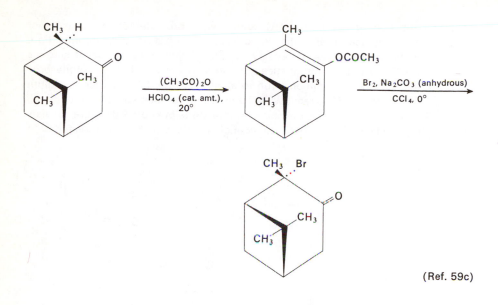

(Ref. 59c)

$$CH_3(CH_2)_4CH_2\text{---}CHO \xrightarrow[\substack{CH_3CO_2K \\ reflux}]{(CH_3CO)_2O} CH_3(CH_2)_4CH\text{=}CH\text{---}OCOCH_3 \xrightarrow[\substack{CCl_4 \\ 10°}]{Br_2} \left[CH_3(CH_2)_4\underset{\underset{Br}{|}}{CH}\text{---}\underset{\underset{Br}{|}}{CH}\text{---}OCOCH_3 \right]$$

[42]

$$\xrightarrow{CH_3OH} CH_3(CH_2)_4\underset{\underset{Br}{|}}{CH}\text{---}CH(OCH_3)_2 \xrightarrow[\substack{H_2O \\ reflux}]{HCl} CH_3(CH_2)_4\underset{\underset{Br}{|}}{CH}\text{---}CHO$$

(80–85%)   (90–95%)

(Ref. 60c)

form the ketal of the corresponding α-bromo ketone.[61] As indicated in the accompanying example, these brominations are believed to proceed by reaction of the ketal with a catalytic amount of acid generated in the reaction mixture to form an enol ether which is brominated. This procedure favors bromination at the less highly substituted α-carbon of the ketal,[61b] suggesting that formation of the less highly substituted enol ether intermediate is kinetically favored. A recent report[61d] indicating differing stereochemical results from the kinetically controlled bromination of an enol ether and the corresponding ketone (as its enol), suggests that the stereochemical outcome may also differ in bromination of ketones and the corresponding ketals.

Stereochemical studies[47,56,58–60] of the bromination of cyclohexanone derivatives have been aided by physical measurements, which serve to define the conformation and configuration of the alpha bromine atom. Of special value have

61. (a) P. E. Eaton, *J. Am. Chem. Soc.,* **84**, 2344 (1962). (b) W. S. Johnson, J. D. Bass, and K. L. Williamson, *Tetrahedron,* **19**, 861 (1963). (c) E. W. Garbisch, Jr., *J. Org. Chem.,* **30**, 2109 (1965). (d) M. Bettahar and M. Charpentier, *Chem. Commun.,* **No. 11**, 629 (1970).

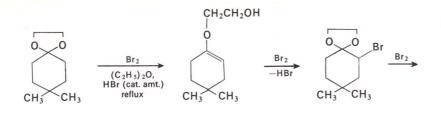

(47%)   (Ref. 61c)

been infrared spectrometry,[47,62,63] ultraviolet spectrometry,[64] nuclear magnetic resonance spectrometry,[65] dipole moment data,[63] and optical rotatory dispersion data.[58,66] The stereochemistry of bromination of enols and enol acetates appears to be controlled by two factors that may either oppose or reinforce one another. The stereoelectronic factor,[47] which is applicable to cyclohexanone derivatives, may be illustrated by the accompanying equations. The energetically most favorable transition states for removal of an alpha proton (e.g., [43] and [45]) to form the enol [47]) and for addition of a bromine atom (e.g., [44] and [46]) to form the product are those in which continuous overlap of the *p* orbitals involved is possible.

62. (a) R. N. Jones, D. A. Ramsay, F. Herling, and K. Dobriner, *J. Am. Chem. Soc.,* **74**, 2828 (1952). (b) R. N. Jones, *ibid.,* **75**, 4839 (1953). (c) J. Fishman, *J. Org. Chem.,* **27**, 1745 (1962). (d) J. Reisse, P. A. Peters, R. Ottinger, J. P. Bervelt, and G. Chiurdoglu, *Tetrahedron Letters,* No. 23, 2511 (1966).
63. (a) J. Allinger and N. L. Allinger, *Tetrahedron,* **2**, 64 (1958). (b) N. L. Allinger, J. Allinger, and N. A. Lebel, *J. Am. Chem. Soc.,* **82**, 2926 (1960). (c) N. L. Allinger and H. M. Blatter, *J. Org. Chem.,* **27**, 1523 (1962), and references therein. (d) M. J. Aroney, R. J. W. LeFevre, and A. N. Singh, *J. Chem. Soc.,* 564 (1965). (e) For a review of the use of various physical methods, see E. L. Eliel, N. L. Allinger, S. J. Angyal, and G. A. Morrison, *Conformational Analysis,* Wiley-Interscience, New York, 1966, pp. 129–188.
64. (a) R. C. Cookson, *J. Chem. Soc.,* 282 (1954). (b) R. C. Cookson and S. H. Dandegaonker, *ibid.,* 352 (1955). (c) C. Djerassi, H. Wolf, and E. Bunnenberg, *J. Am. Chem. Soc.,* **85**, 324 (1963). (d) N. L. Allinger, J. C. Tai, and M. A. Miller, *ibid.,* **88**, 4495 (1966). (e) R. M. Lynden-Bell and V. R. Saunders, *J. Chem. Soc.,* A, 2061 (1967).
65. (a) A. Nickon, M. A. Castle, R. Harada, C. E. Berkoff, and R. O. Williams, *J. Am. Chem. Soc.,* **85**, 2185 (1963). (b) R. J. Abraham and J. S. E. Holker, *J. Chem. Soc.,* 806 (1963). (c) See also K. L. Williamson and W. S. Johnson, *J. Am. Chem. Soc.,* **83**, 4623 (1961). (d) E. W. Garbisch, Jr., *ibid.,* **86**, 1780 (1964). (e) C. W. Shoppee, T. E. Bellas, R. E. Lack, and S. Sternhell, *J. Chem. Soc.,* 2483 (1965). (f) A. Baretta, J. P. Zahra, B. Waegell, and C. W. Jefford, *Tetrahedron,* **26**, 15 (1970).
66. (a) C. Djerassi, *Optical Rotatory Dispersion,* McGraw-Hill, New York, 1960, pp. 115–131. (b) G. Snatzke, D. M. Piatak, and E. Caspi, *Tetrahedron,* **24**, 2899 (1968). (c) G. Snatzke, *Angew. Chem., Intern. Ed. Engl.,* **7**, 14 (1968).

If only the usually[67] more stable chair forms **[48]** and **[49]** of the starting material and product are considered, it is clear that there should be a preference for the removal of the axial proton H_a and for the addition of the bromine at an axial bond. However, the importance of this preference should diminish in proportion to the degree to which the transition states **[43]** and **[44]** resemble the planar enol **[47]**

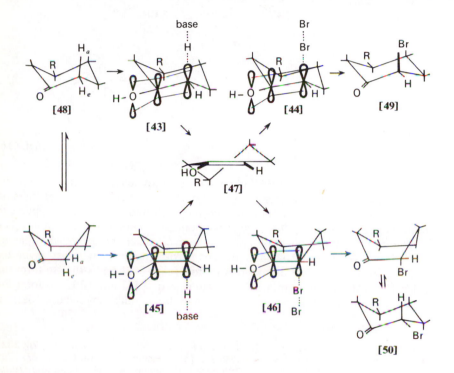

rather than the ketones **[48]** and **[49]**, and there is reason to believe that these transition states do resemble the enol rather than the ketones.[47d,68] The second factor of concern is the steric interference that exists in the transition states for proton removal (**[43]** and **[45]**) and for bromine addition (**[44]** and **[46]**). It is apparent that if serious steric interactions exist in the chairlike transition states **[43]** and **[44]**, the enolization and bromination may proceed via the boatlike transition states **[45]** and **[46]** and still allow continuous p-orbital overlap.

In general, the existence of one 1,3-diaxial interaction between bromine and a group larger than hydrogen in the chairlike transition state **[44]** appears to be sufficient cause for enols and enol acetates to react with bromine via the alternative

67. For a review of instances in which the boat form of a cyclohexane ring appears to be preferred, see M. Balasubramanian, *Chem. Rev.,* **62**, 591 (1962).

68. (a) H. E. Zimmerman in P. deMayo, ed., *Molecular Rearrangements,* Wiley-Interscience, New York, 1963, pp. 345–372. (b) H. Shechter, M. J. Collis, R. Dessy, Y. Okuzumi and A. Chen, *J. Am. Chem. Soc.,* **84**, 2905 (1962).

transition state [46], leading to an equatorial bromo ketone [50], in conformationally fixed systems.[58,60,69] In the absence of such 1,3-diaxial interactions, the
predominant product is usually the axial bromo ketone anticipated on stereoelectronic grounds. The brominations of ketones [51] and [52] illustrate these
generalities. The accompanying equations also illustrate the generality of these

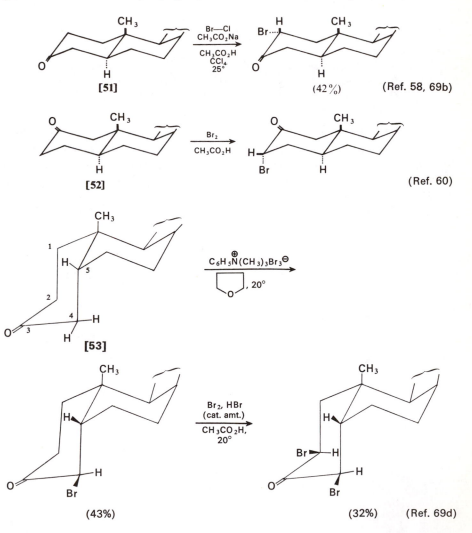

(Ref. 58, 69b)

(Ref. 60)

(43%)   (32%)   (Ref. 69d)

69. (a) E. W. Warnhoff, *J. Org. Chem.*, **28**, 887 (1963). (b) R. A. Jerussi, *ibid.*, **30**, 1650
(1965). (c) J. S. E. Holker, W. R. Jones, M. G. R. Leeming, G. M. Holder, and W. B. Whalley,
Chem. Commun., **No. 2**, 90 (1967). (d) C. W. Shoppee, A. B. Devine, and R. E. Lack, *J.
Chem. Soc.*, 6458 (1965). (e) C. W. Shoppee and S. C. Sharma, *ibid.*, C, 2385 (1967);
245 (1968). (f) For a study of the bromination of cyclobutanone derivatives, see J. M.
Conia and J. Gore, *Tetrahedron Letters*, **No. 21**, 1379 (1963).

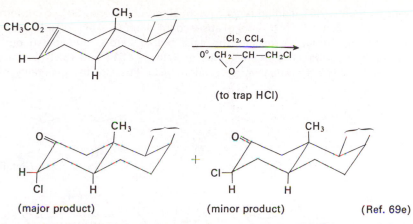

(to trap HCl)

(major product)   +   (minor product)   (Ref. 69e)

stereochemical considerations for chlorination reactions. The fact that 3-keto steroids with a *cis*-AB ring fusion (e.g., [53]) are brominated most rapidly at the 4 position, while the corresponding ketones with a *trans*-AB ring fusion (e.g., [51]) react at position 2 is attributable to the relative rates of enol formation. The quite different behavior of the stereoisomeric ketones [54a] and [54c] on bromina-

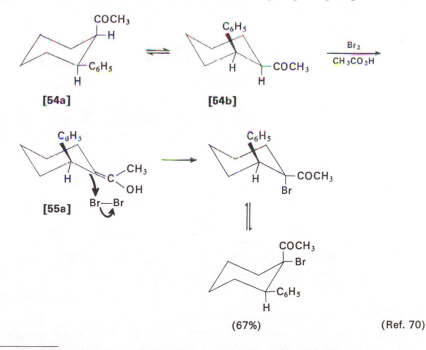

(67%)   (Ref. 70)

70. (a) H. E. Zimmerman, *J. Am. Chem. Soc.,* **79,** 6554 (1957). (b) F. Johnson, *Chem. Rev.,* **68,** 375 (1968). (c) In an analogous case the rate of base-catalyzed enolate anion formation is much slower for the *trans*-isomer, H. O. House and F. A. Richey, *J. Org. Chem.,* **32,** 2151 (1967).

tion illustrates a similar importance of steric factors on enolization rates in the reactions of ketones that are not cyclohexanone derivatives. The *cis*-ketone [54a] can adopt a conformation [54b] in which formation of a relatively unstrained enol [55a] is possible. However, the *trans*-ketone [54c] with two equatorial substituents would need to react in an energetically unfavorable conformation with both substituents in axial positions to form enol [55a]. The alternative enol [55b] which might be formed is destabilized by a serious steric interaction between substituents at the ends of the allylic system, a steric destablization called A^{1,3} strain.[70b] As a result, the *trans*-ketone reacts by way of the alternative enol [55c].

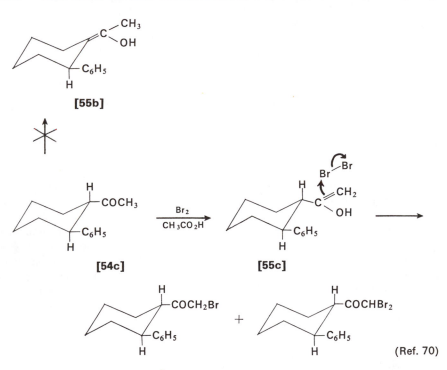

As the following example illustrates, no stereoselectivity is observed in the bromination of cyclohexyl ketones which do not possess a nearby substituent which interferes sterically with enol formation.

α-Halo ketones serve as useful synthetic intermediates, since they undergo rearrangement with bases (the Favorski rearrangement),[71a-c] dehydrohalogenation, or reduction with sodium borohydride (see Chapter 2) to form halohydrins. Although certain α-halo ketones have been dehydrohalogenated by reaction with

71. (a) D. W. Goheen and W. R. Vaughan, *Org. Syn.*, **Coll. Vol. 4**, 594 (1963). (b) A. S. Kende, *Org. Reactions*, **11**, 261 (1960). For recent summaries and leading references, see Ref. 70c; N. J. Turro, *Accts. Chem. Res.*, **2**, 25 (1969); F. G. Bordwell and R. G. Scamehorn, *J. Am. Chem. Soc.*, **90**, 6751 (1968); F. G. Bordwell, R. G. Scamehorn, and W. R. Springer, *ibid.*, **91**, 2087 (1966); F. G. Bordwell and M. W. Carlson, *ibid.*, **92**, 3370, 3377 (1970).

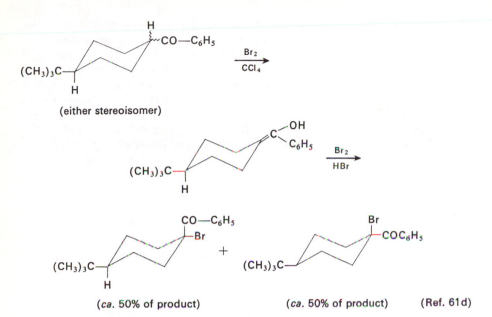

(either stereoisomer)

(*ca.* 50% of product)   (*ca.* 50% of product)   (Ref. 61d)

hindered nitrogenous bases such as γ-collidine,[48,72a-c] methods involving reaction of the α-halo ketone with lithium chloride[48,72b,d] or with lithium or calcium carbonate[59b,72d] in dimethylformamide are often better procedures in that fewer side reactions are encountered. Dehydrobromination has also been effected by heating α-bromo ketones with tetramethylammonium mesitoate in acetone solution.[72f] Alternatively, the α-halo ketones can be converted to the corresponding α-halo ketals, which may be dehydrohalogenated with bases such as sodium methoxide or potassium t-butoxide.[61b,c,72c] The bromohydrins formed by reduction of α-bromo ketones are often different stereoisomers from those available by the direct addition of hypobromous acid to olefins or by the opening of epoxides with hydrogen bromide, and they have been used to form epoxides (e.g., [56]: cf. Chapter 6) that are stereoisomeric with those obtained by direct reaction of the corresponding olefin with a peracid. Also, the fact that halohydrins (e.g., [57], regardless of their stereochemistry (see Chapter 3), react readily with zinc to form olefins offers a convenient synthesis for olefins of known structure.

Although bromine may be introduced alpha to a nitrile function (e.g., [58]) by direct bromination, such halogenation reactions are normally not a satisfactory procedure for monocarboxylic esters (which usually react at the carbon alpha to the

72. (a) For a review of base-catalyzed elimination reactions and recent leading references, see G. Biale, A. J. Parker, S. G. Smith, I. D. R. Stevens, and S. Winstein, *J. Am. Chem. Soc.,* **92**, 115 (1970); D. V. Banthorpe, *Elimination Reactions,* Elsevier, New York, 1963. (b) W. F. Johns, *J. Org. Chem.,* **28**, 1616 (1963). (c) E. W. Warnhoff and D. R. Marshall, *ibid.,* **32**, 2000 (1967). (d) H. O. House and R. W. Bashe, *ibid.,* **30**, 2942 (1965). (e) C. W. Shoppee and B. C. Newman, *J. Chem. Soc.,* C, 2767 (1969). (f) W. S. Johnson, J. F. W. Keana, and J. A. Marshall, *Tetrahedron Letters,* **No. 4**, 193 (1963).

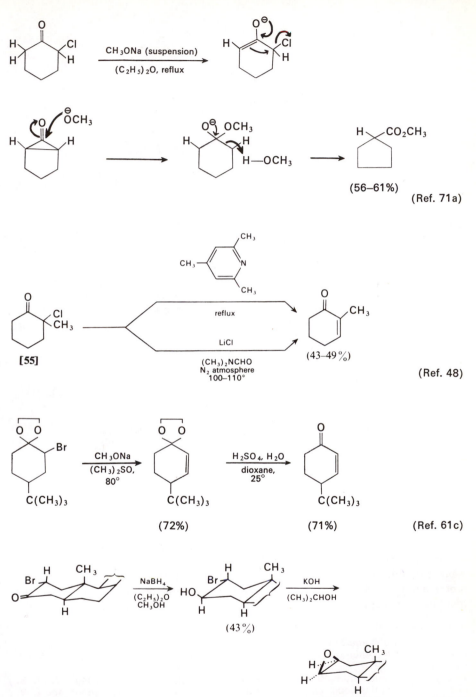

(56–61%)

(Ref. 71a)

[55]

(43–49%)

(Ref. 48)

(72%)   (71%)   (Ref. 61c)

(43%)

[56] (76%)   (Ref. 47a)

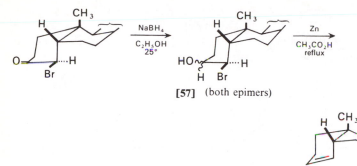

(80–85%)   (Ref. 56)

$$C_6H_5\text{—}CH_2\text{—}CN \xrightarrow[105\text{–}110^\circ]{Br_2} C_6H_5\text{—}CH\text{—}CN$$
[58] |
                                 Br

(Ref. 73a)

ethereal oxygen or at C—H bonds remote from the ester function)[73b] or for mono-carboxylic acids (which usually do not react).[51f] However, malonic esters (e.g., [59]) and malonic acids (e.g., [60]) will react with bromine. The latter reaction followed by decarboxylation serves as a convenient route to α-bromocarboxylic acids.

$$CH_2(CO_2C_2H_5)_2 \xrightarrow[\substack{CCl_4 \\ \text{(initiated with light)}}]{Br_2} BrCH(CO_2C_2H_5)_2$$
[59]   (73–75%)

(Ref. 74a)

$$(CH_3)_2CH\text{—}CH(CO_2C_2H_5)_2 \xrightarrow[\substack{H_2O \\ reflux}]{KOH} \xrightarrow[H_2O]{HCl} (CH_3)_2CH\text{—}CH(CO_2H)_2 \xrightarrow[\substack{(C_2H_5)_2O \\ reflux}]{Br_2}$$
[60]

$$(CH_3)_2CH\text{—}\underset{|}{\overset{}{C}}(CO_2H)_2 \xrightarrow{125\text{–}130^\circ} (CH_3)_2CH\text{—}\underset{|}{\overset{}{CH}}\text{—}CO_2H$$
                 Br   Br
                                                        (55–66%)

(Ref. 74b)

If carboxylic acids are first converted to their acid chloride, bromide, or an-hydride derivatives, bromination alpha to the carbonyl group occurs in satisfactory yield although the reaction is slow. The reaction is frequently effected by adding bromine to the carboxylic acid in the presence of a catalytic amount of phosphorus trichloride, phosphorus tribromide, or phosphorus (which is converted to the

73. (a) C. M. Robb and E. M. Schultz, *Org. Syn.*, **Coll. Vol. 3**, 347 (1955). (b) C. C. Price, C. D. Beard, and K. Akune, *J. Am. Chem. Soc.*, **92**, 5916 (1970); C. C. Price and C. D. Beard, *ibid.*, **92**, 5921 (1970).
74. (a) C. S. Palmer and P. W. McWherter, *Org. Syn.*, **Coll. Vol. 1**, 245 (1944). (b) C. S. Marvel and V. duVigneaud, *ibid.*, **Coll. Vol. 2**, 93 (1943). (c) L. A. Carpino and L. V. McAdams, *ibid.*, **50**, 31 (1970).

tribromide in the reaction mixture). This method, the Hell-Volhard-Zelinsky reaction, apparently relies upon an equilibrium between the small amount of brominated acid halide (or anhydride) present and the starting acid to permit complete bromination. Examples of the various procedures are provided in the following equations. The mechanism of this process has been thought to involve

$$CH_3(CH_2)_3CH_2CO_2H \xrightarrow[\substack{\text{cat. amt. } PCl_3 \\ 65-70°}]{Br_2} CH_3(CH_2)_3\underset{\underset{Br}{|}}{C}HCO_2H$$

(83–89%)   (Ref. 75a)

$$C_6H_5CH_2CO_2H + Br_2 \xrightarrow[\substack{C_6H_6, \\ \text{reflux}}]{PCl_3 \text{ (cat. amt.)}} C_6H_5\underset{\underset{Br}{|}}{C}H—CO_2H$$

(60–62%)   (Ref. 74c)

(94%)   (Ref. 75c)

$$CH_3CO_2H + (CH_3CO)_2O \xrightarrow[\substack{C_5H_5N \\ \text{reflux}}]{Br_2} \xrightarrow{H_2O} BrCH_2CO_2H$$

(80–85%)   (Ref. 76)

$$HO_2C(CH_2)_4CO_2H \xrightarrow[\text{reflux}]{SOCl_2} Cl—CO(CH_2)_4CO—Cl \xrightarrow{Br_2} Cl—CO—\underset{\underset{Br}{|}}{C}H(CH_2)_2\underset{\underset{Br}{|}}{C}H—CO—Cl$$

$$\xrightarrow{C_2H_5OH} C_2H_5O_2C—\underset{\underset{Br}{|}}{C}H(CH_2)_2\underset{\underset{Br}{|}}{C}H—CO_2C_2H_5$$

(91–97%)   (Ref. 77)

$$(CH_3)_2CH—CO_2H \xrightarrow[\substack{P \\ 100°}]{Br_2} (CH_3)_2\underset{\underset{Br}{|}}{C}—CO—Br$$

(75–83%)   (Ref. 78)

75. (a) H. T. Clarke and E. R. Taylor, *Org. Syn., Coll. Vol. 1*, 115 (1944). (b) H. Kwart and F. V. Scalzi, *J. Am. Chem. Soc., 86*, 5496 (1964). (c) J. C. Little, Y. L. C. Tong, and J. P. Heeschen, *ibid., 91*, 7090 (1969); J. C. Little, A. R. Sexton, Y. L. C. Tong, and T. E. Zurawic, *ibid., 91*, 7098 (1969). (d) J. G. Gleason and D. N. Harpp, *Tetrahedron Letters, No. 39*, 3431 (1970).
76. S. Natelson and S. Gottfried, *Org. Syn., Coll. Vol. 3*, 381 (1955).
77. P. C. Guha and D. K. Sankaran, *Org. Syn., Coll. Vol. 3*, 623 (1955).
78. C. W. Smith and D. G. Norton, *Org. Syn., Coll. Vol. 4*, 348 (1963).

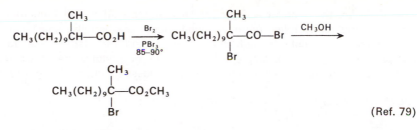

(Ref. 79)

the rate-limiting enolization of an acid chloride or acid anhydride intermediate, followed by reaction of the enol with the halogen. However, recent studies of these reactions[73b,c] are not consistent with such a reaction mechanism and suggest that some other reaction intermediate (possibly a ketene) is involved. A more satisfactory procedure for the bromination of acid chlorides involves their reaction with N-bromosuccinimide, as indicated below. This method does not appear to involve a free-radical chain reaction; however, which reaction path is being followed remains to be established.

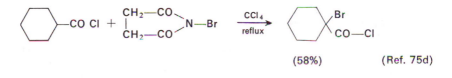

(58%)   (Ref. 75d)

SUBSTITUTION OF HALOGEN AT BENZYLIC AND ALLYLIC CARBON-HYDROGEN BONDS

Unlike the previously discussed halogenations, the majority of reactions that substitute bromine or chlorine at an allylic or benzylic position appear to be free-radical chain processes.[80] However, as the following example illustrates, either ionic or free-radical pathways may lead to addition products and allylic halides. The substitution reactions are frequently run at elevated temperatures or are promoted by free-radical initiators such as light, a peroxide [usually dibenzoyl peroxide, $C_6H_5CO—OO—COC_6H_5$, or di-t-butyl peroxide, $(CH_3)_3C—OO—C(CH_3)_3$], or an azo compound [usually azobisisobutyronitrile, $(CH_3)_2C(CN)—N=N—C(CN)CH_3)_2$]. The commonly employed chlorinating agents are molecular chlorine, sulfuryl chloride (SO_2Cl_2),[81b] trichloromethanesulfonyl chloride, and

79. C. F. Allen and M. J. Kalm, *Org. Syn.,* **Coll. Vol. 4**, 608 (1963).
80. (a) C. Walling, *Free Radicals in Solution,* Wiley, New York, 1957, pp. 347–396; W. A. Pryor, *Free Radicals,* McGraw-Hill, New York, 1966, pp. 179–213; M. L. Poutsma in E. S. Huyser, ed., *Methods in Free Radical Chemistry,* Vol. 1, Marcel Dekker, New York, 1969, pp. 79–193; W. A. Thaler in E. S. Huyser, ed., *Methods in Free Radical Chemistry,* Vol. 2, Marcel Dekker, New York, 1969, pp. 121–227. (b) E. S. Huyser, *Synthesis,* **2**, 7 (1970). (c) J. M. Tedder, *Quart. Rev.,* **14**, 336 (1960). (d) M. L. Poutsma, *J. Am. Chem. Soc.,* **87**, 2161, 2172 (1965).
81. (a) The preparation of this reagent is described by H. M. Teeter and E. W. Bell, *Org. Syn.,* **Coll. Vol. 4**, 125 (1963), and M. J. Mintz and C. Walling, *ibid.,* **49**, 9 (1969). (b) K. H. Lee, *Tetrahedron,* **25**, 4357, 4363 (1969); **26**, 2041 (1970).

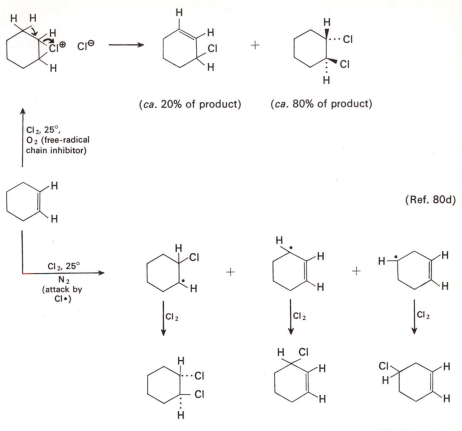

(*ca.* 20% of product)   (*ca.* 80% of product)

Cl₂, 25°,
O₂ (free-radical
chain inhibitor)

(Ref. 80d)

Cl₂, 25°
N₂
(attack by
Cl•)

(55% of product)   (28% of product)   (17% of product)

t-butyl hypochlorite,[81a,82] whereas either molecular bromine or N-bromosuccini-mide[3,83] is normally used for bromination.

In the absence of other reactive functional groups, halogenation at a benzylic position is most readily effected with molecular bromine or chlorine, as in the accompanying examples.

82. (a) C. Walling and B. B. Jacknow, *J. Am. Chem. Soc.,* **82**, 6108, 6113 (1960). (b) C. Walling and W. Thaler, *ibid.,* **83**, 3877 (1961). (c) C. Walling and P. S. Fredricks, *ibid.,* **84**, 3326 (1962). (d) C. Walling and P. J. Wagner, *ibid.,* **86**, 3368 (1964). (e) C. Walling and M. J. Mintz, *ibid.,* **89**, 1515 (1967). (f) C. Walling and J. A. McGuinness, *ibid.,* **91**, 2053 (1969). (g) D. J. Carlsson and K. U. Ingold, *ibid.,* **89**, 4885, 4891 (1967). (h) C. Walling and V. P. Kurkov, *ibid.,* **89**, 4895 (1967).
83. (a) C. Djerassi, *Chem. Rev.,* **43**, 271 (1948); L. Horner and E. H. Winkelmann, in W. Foerst, ed., *Newer Methods of Preparative Organic Chemistry,* **Vol. 3**, Academic Press, New York, 1964, pp. 151–198. (b) T. D. Waugh, *N-Bromosuccinimide: Its Reactions and Uses,* Arapahoe Chemicals, Inc., Boulder, Colorado, 1951. (c) H. J. Dauben and L. L. McCoy, *J. Am. Chem. Soc.,* **81**, 4863, 5404 (1959). (d) For the preparation of N-iodo-succinimide, see W. R. Benson, E. T. McBee, and L. Rand, *Org. Syn.,* **42**, 73 (1962). (e) For a review of the use of N-halo amides for oxidation reactions, see R. Filler, *Chem. Rev.,* **63**, 21 (1963) and Ref. 23d.

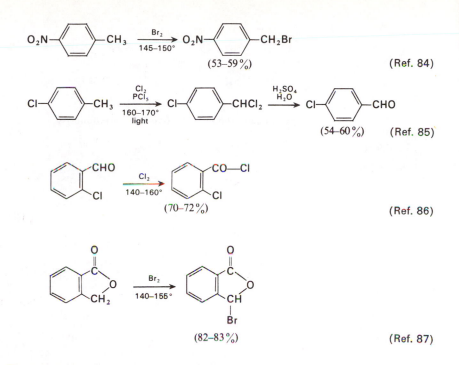

$$(53\text{--}59\%) \qquad \text{(Ref. 84)}$$

$$(54\text{--}60\%) \qquad \text{(Ref. 85)}$$

$$(70\text{--}72\%) \qquad \text{(Ref. 86)}$$

$$(82\text{--}83\%) \qquad \text{(Ref. 87)}$$

The course of these reactions is illustrated in the scheme below. The hydrogen-abstraction step in the reactions constituting the propagation stage is retarded by the presence of electron-withdrawing substituents.[80,81b] As a result, each successive

Initiation
$$R\cdot + Br_2 \longrightarrow R\text{---}Br + Br\cdot$$

Propagation
$$\begin{cases} Br\cdot + C_6H_5\text{---}CH_3 \longrightarrow HBr + C_6H_5\text{---}CH_2\cdot \\ C_6H_5\text{---}CH_2\cdot + Br_2 \longrightarrow C_6H_5\text{---}CH_2\text{---}Br + Br\cdot \end{cases}$$

Termination
$$\begin{cases} C_6H_5CH_2\cdot + Br\cdot \longrightarrow C_6H_5\text{---}CH_2\text{---}Br \\ 2\,C_6H_5\text{---}CH_2\cdot \longrightarrow C_6H_5\text{---}CH_2\text{---}CH_2\text{---}C_6H_5 \end{cases}$$

halogen atom substituted at a benzylic position makes more difficult the abstraction of other hydrogen atoms at that position, a situation that facilitates the stepwise substitution of halogen (e.g., [61] and [62]). This same retarding effect by electron-withdrawing substituents is found in halogenations with sulfuryl chloride, t-butyl hypochlorite, or N-bromosuccinimide.

84. G. H. Coleman and G. E. Honeywell, *Org. Syn., Coll. Vol. 2*, 443 (1943).
85. W. L. McEwen, *Org. Syn., Coll. Vol. 2*, 133 (1943).
86. H. T. Clarke and E. R. Taylor, *Org. Syn., Coll. Vol. 1*, 155 (1944).
87. R. L. Shriner and F. J. Wolf, *Org. Syn., Coll. Vol. 3*, 737 (1955).

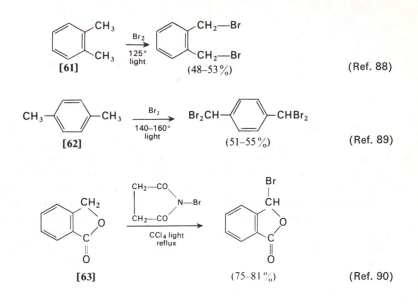

(Ref. 88)

(Ref. 89)

(Ref. 90)

Benzylic bromination with N-bromosuccinimide is often a more convenient procedure (e.g., [63]) than bromination with molecular bromine and is definitely the method of choice in the case of a reactive aromatic nucleus (e.g., [64]) or in the presence of another functional group that can react with bromine (e.g., [65])

(Ref. 91a)

$C_6H_5CH_2CH_2CH_2Br$ +

(Ref. 91c)

88. E. F. M. Stephenson, *Org. Syn.*, **Coll. Vol. 4**, 984 (1963).
89. J. M. Snell and A. Weissberger, *Org. Syn.*, **Coll. Vol. 3**, 788 (1955).
90. I. A. Koten and R. J. Sauer, *Org. Syn.*, **42**, 26 (1962).
91. (a) E. Campaigne and B. F. Tullar, *Org. Syn.*, **Coll. Vol. 4**, 921 (1963). (b) A. Kalir, *ibid.*, **46**, 81 (1966). (c) T. F. Corbin, R. C. Hahn, and H. Shechter, *ibid.*, **44**, 30 (1964).

$$C_6H_5-CH_2CH_2CH_2CH_2CO-C_6H_5 \xrightarrow[\text{reflux}]{\underset{\text{CCl}_4 \text{ light}}{\text{CH}_2-\text{CO}\diagdown N-\text{Br} \diagup \text{CH}_2-\text{CO}}} C_6H_5-\underset{\underset{\text{Br}}{|}}{CH}-CH_2CH_2CH_2CO-C_6H_5$$

[65]   (66%)   (Ref. 92a)

or hydrogen bromide. Allylic and benzylic brominations with N-bromosuccinimide had originally been suggested[80,83] to proceed by a free-radical chain mechanism, with hydrogen abstraction by the succinimide radical. However, recent studies[80,93] indicate that a bromine radical is the hydrogen-abstracting agent in benzylic bromination and strongly suggest a similar mechanism for allylic bromination. According to this scheme, the bromine that is produced slowly by reaction of the N-bromosuccinimide with hydrogen bromide, as indicated below, enters into the previously illustrated free-radical chain reaction. If N-bromosuccinimide as used in polar acid solvents such as acetic acid or aqueous sulfuric acid, this acid-catalyzed process predominates. In such reaction media, aromatic hydrocarbons are brominated at the aryl nucleus rather than at benzylic positions.[92b,c]

$$\underset{\text{CH}_2-\text{CO}}{\overset{\text{CH}_2-\text{CO}}{|}} \diagup N-\text{Br} + \text{HBr} \longrightarrow \underset{\text{CH}_2-\text{CO}}{\overset{\text{CH}_2-\text{CO}}{|}} \diagup N-\text{H} + \text{Br}_2$$

The halogenation of sterically hindered benzyl derivatives may lead to the rearrangement and/or elimination reaction of the radical intermediates, as illustrated in the following equation.

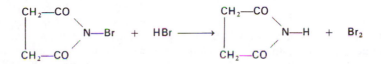

$$(C_6H_5)_3C-CH_2-C_6H_5 \xrightarrow[\text{CCl}_4, \text{ reflux}]{\underset{\text{CH}_2-\text{CO}}{\overset{\text{CH}_2-\text{CO}}{}} N-\text{Br, } h\nu} (C_6H_5)_2 \overset{C_6H_5}{\underset{\centerdot}{C-CH-C_6H_5}} \longrightarrow$$

$$(C_6H_5)_2\overset{\centerdot}{C}-\underset{\underset{\text{Br}-\text{Br}}{\overset{\downarrow}{H}}}{C(C_6H_5)_2} \xrightarrow[-\text{Br}\centerdot]{-\text{HBr}} (C_6H_5)_2C=C(C_6H_5)_2$$

                                                                          (43%)

(Ref. 93j)

92. (a) R. L. Huang and P. Williams, *J. Chem. Soc.*, 2637 (1958). (b) F. Dewhurst and P. K. J. Shah, *J. Chem. Soc.*, C, 1737 (1970). (c) For the use of iodine and peracetic acid in acetic acid to iodinate the nucleus of aromatic hydrocarbons, see Y. Ogata and I. Urasaki, *ibid.*, C, 1689 (1970).
93. (a) R. E. Pearson and J. C. Martin, *J. Am. Chem. Soc.*, **85**, 354, 3142 (1963); J. H. Incremona and J. C. Martin, *ibid.*, **92**, 627 (1970). (b) B. P. McGrath and J. M. Tedder, *Proc. Chem. Soc.*, 80 (1961). (c) P. S. Skell, D. L. Tuleen, and P. D. Readio, *J. Am. Chem. Soc.*, **85**, 2850 (1963). (d) E. Hedaya, R. L. Hinman, V. Schomaker, S. Theodoropulos, and L. M. Kyle, *ibid.*, **89**, 4875 (1967). (e) C. Walling, A. L. Rieger, and D. D. Tanner, *ibid.*, **85**, 3129 (1963). (f) G. A. Russell and K. M. Desmond, *ibid.*, **85**, 3139 (1963). (g) T. Koenig and W. Brewer, *ibid.*, **86**, 2728 (1964).   (h) R. L. Huang and K. H. Lee, *J. Chem.*

Bromine radicals generated photochemically[93m] or by radiolysis[91] will attack relatively unactivated C—H bonds. As the accompanying examples illustrate, tertiary alkyl radicals generated in this way in an excess of a hydrocarbon solvent

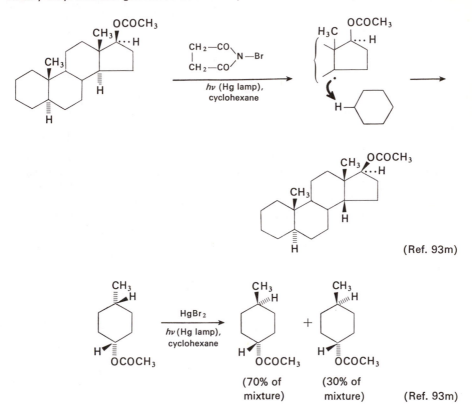

(Ref. 93m)

(70% of
mixture)

(30% of
mixture)

(Ref. 93m)

tend to abstract a hydrogen atom from the solvent. The overall reaction sequence serves to epimerize asymmetric centers which can be attacked by a bromine atom to form a relatively stable tertiary alkyl radical.

It has been shown that, at low concentration levels and in the absence of hydrogen bromide, bromine reacts with cyclohexene to form the allylic bromide [**66**] rather than the addition product, a 1,2-dibromide.[93b] Failure to observe addition of bromine to the double bond under these circumstances has been attributed to the reversibility of the first step in the addition reaction (by either a radical or an

Soc., C, 932, 935 (1966) ; T. P. Low and K. H. Lee, *ibid.,* B, 535 (1970). (i) S. S. Friedrich, E. C. Friedrich, L. J. Andrews, and R. M. Keefer, *J. Org. Chem.,* **34**, 900 (1969) ; I. Horman, S. S. Friedrich, R. M. Keefer, and L. J. Andrews, *ibid.,* **34**, 905 (1969). (j) H. Meislich, J. Costanza, and J. Strelitz, *ibid.,* **33**, 3221 (1968). (k) J. M. Landesberg and M. Siegel, *ibid.,* **35**, 1674 (1970). (l) D. H. Martin and F. Williams, *J. Am. Chem. Soc.,* **92**, 769 (1970). (m) M. Gorodetsky, D. Kogan, and Y. Mazur, *ibid.,* **92**, 1094 (1970) ; M. Gorodetsky and Y. Mazur, *ibid.,* **90**, 6540 (1968).

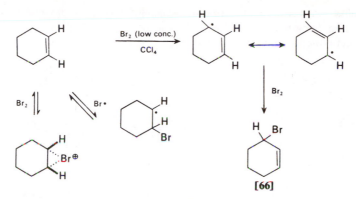

[66]

ionic mechanism). Since the concentrations of both bromine and hydrogen bromide (or bromide ion) are kept very low, little of the radical or ionic intermediate leading to addition is trapped to form the 1,2-dibromide. Under special circumstances, N-bromosuccinimide and its derivatives may be induced to undergo either free-radical or ionic additions to olefins. Thus, the reaction of N-bromosuccinimide with 3-sulfolene, an olefin with an electron-withdrawing sulfone function deactivating the allylic C—H bonds, yields an adduct which appears to

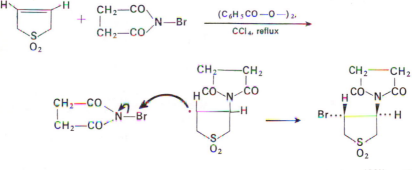

(8%)   (Ref. 93k)

arise from a radical addition reaction. The reaction of 1,1-diphenylpropene with N-bromotetrafluorosuccinimide yields a similar adduct by a process which appears to be ionic. The latter reaction path is also capable of yielding allylic halides if the succinimide anion abstracts a proton from the bromonium ion.[93a]

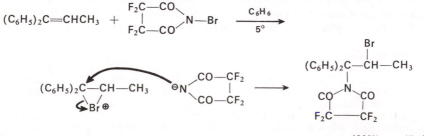

(60%)   (Ref. 93a)

The following equations illustrate the use of N-bromosuccinimide for allylic bromination. Since a free allylic carbon radical (e.g., **[67]** from olefin **[68]**) is

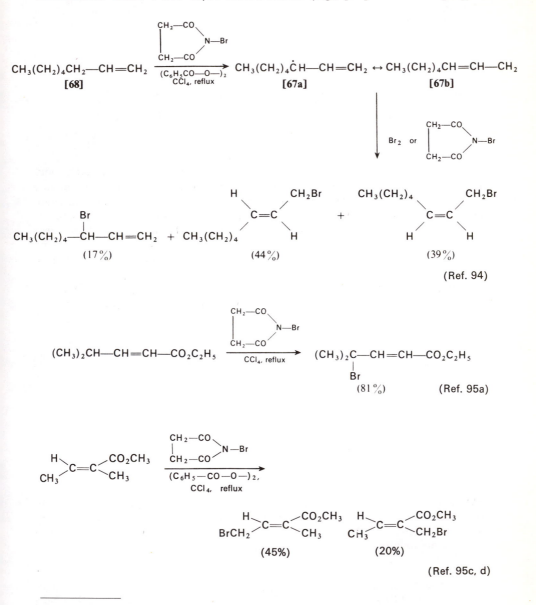

94. L. Bateman and J. I. Cunneen, *J. Chem. Soc.*, 941 (1950).
95. (a) H. J. Dauben, Jr., and L. L. McCoy, *J. Org. Chem.*, **24**, 1577 (1959). (b) I. Ahmad, R. N. Gedye, and A. Nechvatal, *J. Chem. Soc.*, C, 185 (1968). (c) A. Löffler, R. J. Pratt, H. P. Rüesch, and A. S. Dreiding, *Helv. Chim. Acta*, **53**, 383 (1970). (d) A. Löffler, F. Norris, W. Taub, K. L. Svanholt, and A. S. Dreiding, *ibid.*, **53**, 403 (1970).

presumably[80,83] an intermediate in these reactions no matter what the actual nature of the brominating agent, a mixture of allylic halides as products is to be expected.[80] The reported formation of only a single structural isomer from bromination of the olefin **[69]** therefore seems most curious; on the other hand, the absence of

$$CH_3(CH_2)_3—CH=CH—CH_3 \xrightarrow[\substack{(C_6H_5CO—O—)_2 \\ CCl_4, \text{ reflux}}]{} CH_3(CH_2)_2—CH—CH=CH—CH_3$$

[69]

Br

(58–64%)   (Ref. 92a, b)

terminal unsaturation or of a bromomethyl group in the product is in agreement with the generalization[83] that a hydrogen atom is abstracted from secondary allylic positions more readily than from primary allylic positions. A similar situation has been observed in the bromination of α,β-unsaturated esters. As the accompanying examples indicate, products with the bromine at the alpha carbon (adjacent to the deactivating carboalkoxyl group) have not been observed.[95] It should be noted that this result with α,β-unsaturated esters could also arise from an ionic bromination involving a dienol intermediate from the starting ester.

Application of this allylic bromination to enol acetates yields intermediates which can be converted to α,β-unsaturated aldehydes. Cyclic acetals are also

$$n\text{-}C_5H_{11}—CH_2—CH=CH—OCOCH_3 \xrightarrow[\substack{CCl_4, \text{ reflux} \\ \text{radical} \\ \text{initiator}}]{}$$

$$\left[n\text{-}C_5H_{11}—\underset{\underset{Br}{|}}{CH}—CH=CH—OCOCH_3 \right] \xrightarrow[\text{heat}]{\text{base or}} n\text{-}C_5H_{11}CH=CH—CHO$$

(50%)   (Ref. 96c)

attacked by bromine radicals to yield the fragmentation products indicated. In certain cases where olefins have been brominated in reaction media containing some water, the initially formed allylic halides have been hydrolyzed to alcohols which were oxidized to α,β-unsaturated ketones.[96e]

---

96. (a) F. L. Greenwood, M. D. Kellert, and J. Sedlak, *Org. Syn.,* **Coll. Vol. 4,** 108 (1963). (b) F. L. Greenwood and M. D. Kellert, *J. Am. Chem. Soc.,* **75,** 4842 (1953). (c) J. J. Riehl and F. Jung, *Tetrahedron Letters,* **No. 37,** 3139 (1969). (d) J. D. Prugh and W. C. McCarthy, *ibid.,* **No. 13,** 1351 (1966). It should be noted that the alpha C—H bonds of ethers are cleaved by aqueous bromine in a process which appears to involve ionic intermediates. N. C. Deno and N. H. Potter, *J. Am. Chem. Soc.,* **89,** 3550, 3555 (1967). (e) B. W. Finucane and J. B. Thompson, *Chem. Commun.,* **No. 20,** 1220 (1969).

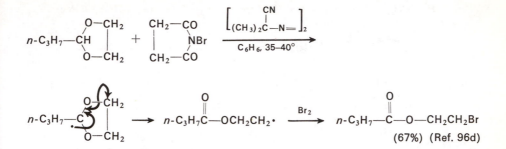

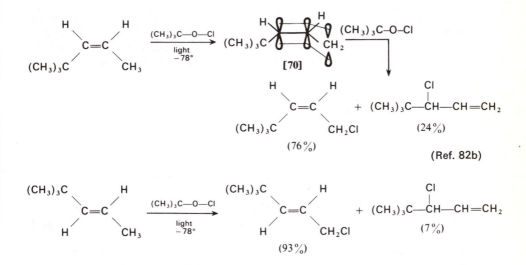

Examples of the allylic chlorination of olefins with *t*-butyl hypochlorite follow.

In these examples the formation of predominantly the 1-chloro-2-olefin rather than its allylic isomer is attributable to steric hindrance to attack at the 3 position. Of special interest is the fact that the intermediate allylic radicals (e.g., [70]) have a sufficient barrier to rotation about the C_2—C_3 bond to preserve their stereochemistry, at least at low temperatures.

This rotational barrier is presumably associated with the resonance energy (*ca*. 10 kcal/mole)[97] available to an allyl radical so long as it remains planar, allowing overlap of the orbitals at each of the three carbon atoms of the allyl system (as in structure [70]). Chlorinations at allylic and benzylic positions with *t*-butyl hypochlorite involve a radical-chain process in which a hydrogen atom is abstracted by a *t*-butoxy radical; the chain-propagation steps are illustrated in the following equation.[82] In cases in which the hydrogen abstraction step is slow, the *t*-butoxy radical can also undergo the fragmentation reaction indicated.

97. D. M. Golden, N. A. Gac, and S. W. Benson, *J. Am. Chem. Soc.*, **91**, 2136 (1969).

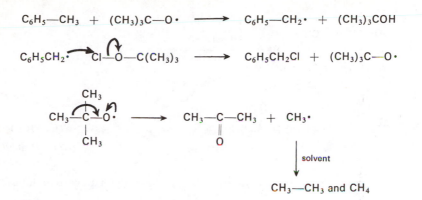

$$C_6H_5-CH_3 + (CH_3)_3C-O\cdot \longrightarrow C_6H_5-CH_2\cdot + (CH_3)_3COH$$

$$C_6H_5CH_2\cdot \quad Cl-O-C(CH_3)_3 \longrightarrow C_6H_5CH_2Cl + (CH_3)_3C-O\cdot$$

Acetylenes and allenes undergo a chlorination reaction similar to that observed with olefins. However, the intermediate propargylic radical [71] appears to react with chlorine or t-butyl hypochlorite to form acetylenic products containing little if any of the isomeric allenes.[98] t-Butyl hypochlorite has also been used to chlorinate

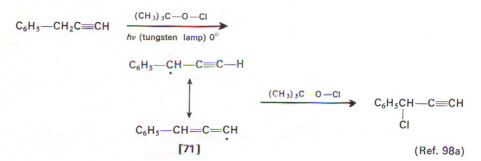

(Ref. 98a)

ethers, alcohols, aldehydes, ketones, and amines. The following equations provide examples of these uses.

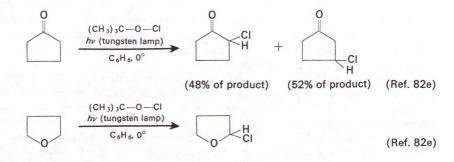

(48% of product)   (52% of product)   (Ref. 82e)

(Ref. 82e)

98. (a) M. C. Caserio and R. E. Pratt, *Tetrahedron Letters,* **No. 1**, 91 (1967). (b) M. L. Poutsma and J. L. Kartch, *Tetrahedron,* **22**, 2167 (1966). (c) R. M. Fantazier and M. L. Poutsma, *J. Am. Chem. Soc.,* **90**, 5490 (1968. (d) L. R. Byrd and M. C. Caserio, *ibid.,* **92**, 5422 (1970). (e) J. K. Kochi and P. J. Krusic, *ibid.,* **92**, 4110 (1970).

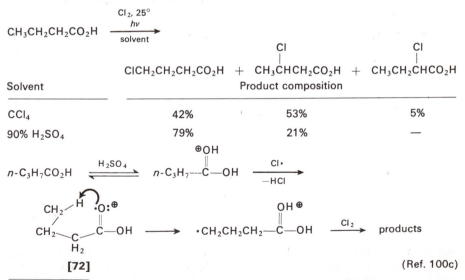

Free-radical halogenations of saturated compounds have been studied extensively[80] to establish the reactivity of various C—H bonds to the hydrogen atom abstraction reaction. These studies have established the reactivity order, tertiary CH > secondary CH > primary CH, and have also established the deactivating effect of nearby electron-withdrawing substituents.[80,82e,100] In spite of these selectivities in free-radical halogenation of alkanes, the halogenation of substrates without specially activated C—H bonds (e.g., benzyl or allyl CH bonds) frequently yields mixtures of halogenated products which are difficult to separate on a preparative scale. Notable exceptions are the previously discussed (see Chapter 7) *intramolecular* abstractions of a hydrogen atom from an unactivated C—H bond by a radical which is generated and held near a particular C—H bond. As indicated in Chapter 7, these radicals have been generated from alcohols and lead tetraacetate, from alkyl hypohalites, from N-halo amides, or from protonated N-halo amines. The following equations compare the product compositions from the free-radical

$$CH_3CH_2CH_2CO_2H \xrightarrow[\text{solvent}]{\substack{Cl_2, 25° \\ h\nu}}$$

$$ClCH_2CH_2CH_2CO_2H \; + \; CH_3\overset{\underset{\textstyle |}{Cl}}{C}HCH_2CO_2H \; + \; CH_3CH_2\overset{\underset{\textstyle |}{Cl}}{C}HCO_2H$$

Solvent	Product composition		
CCl$_4$	42%	53%	5%
90% H$_2$SO$_4$	79%	21%	—

$$n\text{-}C_3H_7CO_2H \underset{}{\overset{H_2SO_4}{\rightleftharpoons}} n\text{-}C_3H_7\text{---}\overset{\overset{\textstyle \oplus OH}{\|}}{C}\text{---}OH \xrightarrow[-HCl]{Cl\cdot}$$

$$ \cdot CH_2CH_2CH_2\text{---}\overset{\overset{\textstyle OH \oplus}{\|}}{C}\text{---}OH \xrightarrow{Cl_2} \text{products}$$

[72]   (Ref. 100c)

99. (a) H. E. Baumgarten and J. M. Petersen, *Org. Syn.,* **41**, 82 (1961). (b) G. H. Alt and W. S. Knowles, *ibid.,* **45**, 16 (1965). (c) For a related procedure employing N,N-dichlorobenzenesulfonamide as the halogenating agent, see T. Taguchi, Y. Shimizu, and Y. Kawazoe, *Tetrahedron Letters,* **No. 32**, 2853 (1970).
100. (a) D. S. Ashton and J. M. Tedder, *Chem. Commun.,* **No. 14**, 785 (1968). (b) F. Minisci, R. Galli, and R. Bernardi, *ibid.,* **No. 17**, 903 (1967); F. Minisci, R. Galli, A. Galli, and R. Bernardi, *Tetrahedron Letters,* **No. 23**, 2207 (1967). (c) N. C. Deno, R. Fishbein, and J. C. Wyckoff, *J. Am. Chem. Soc.,* **92**, 5274 (1970); J. Kollonitsch, G. A. Doldouras, and V. F. Verdi, *J. Chem. Soc.,* B, 1093 (1967).

chlorination of butyric acid in a nonpolar solvent (CCl_4) and in 90% sulfuric acid. In both cases, very little substitution is observed alpha to the electron-withdrawing carbonyl group. The enhanced tendency for the reaction in 90% sulfuric acid to form the γ-chloro acid is suggested to result from an intramolecular hydrogen atom abstraction, as illustrated in structure [72].

This selective halogenation of an unactivated position may also be accomplished by the thermal or photochemical decomposition of t-alkyl hypochlorites,[101] as indicated in the following example. The fragmentation of the t-alkyl hypochlorites

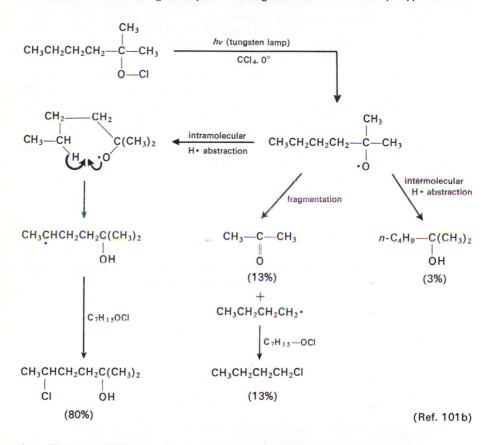

(Ref. 101b)

(e.g. [73] and [74]) can also serve as a useful procedure for the introduction of a halogen atom at a specific site with a free-radical chain process. As the accompanying equations indicate, the fragmentation of the alkoxyl radical intermediate occurs in such a way that a carbonyl compound and the most stable carbon radical are formed.

101. (a) F. D. Greene, M. L. Savitz, F. D. Osterholtz, H. H. Lau, W. N. Smith, and P. M. Zanet, *J. Org. Chem.*, **28**, 55 (1963). (b) C. Walling and A. Padwa, *J. Am. Chem. Soc.*, **85**, 1597 (1963). (c) M. Akhtar, P. Hunt, and P. B. Dewhurst, *ibid.*, **87**, 1807 (1965).

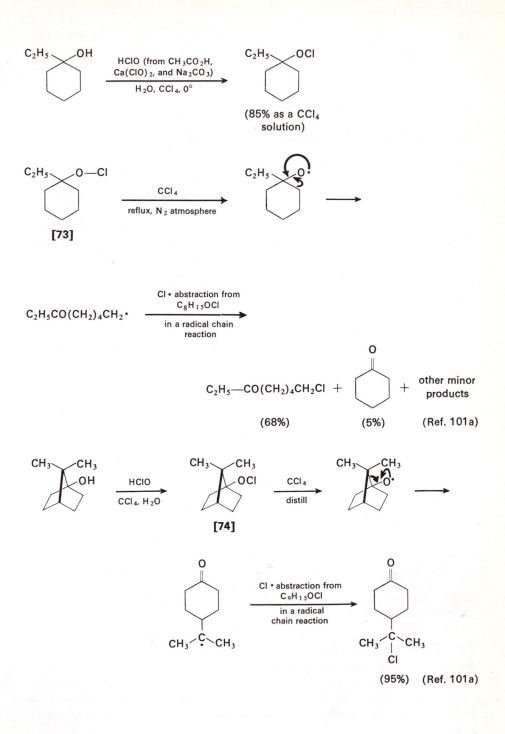

(85% as a CCl₄ solution)

[73]

Cl • abstraction from C₈H₁₅OCl

in a radical chain reaction

C_2H_5—$CO(CH_2)_4CH_2Cl$ +

(68%)   (5%)   (Ref. 101a)

other minor products

[74]

Cl • abstraction from C₉H₁₅OCl

in a radical chain reaction

(95%)   (Ref. 101a)

9

THE ALKYLATION OF ACTIVE
METHYLENE COMPOUNDS

The presence of certain unsaturated functions as substituents at a saturated carbon atom—nitro, carbonyl, cyano, sulfone, or phenyl groups, for example—renders any hydrogen atoms bonded to that carbon relatively acidic. Table 9–1 lists the pK_a values for representative compounds of this type and, for comparison, for some common solvents and reagents. The acidity of the C—H bond in these substances, often called active methylene compounds, is attributed to a combination of the inductive electron-withdrawing ability of the unsaturated substituents and the ability of these substituents to delocalize the negative charge remaining when a proton has been removed, as illustrated in the accompanying equations. It will be noted from Table 9–1 that the effectiveness of these unsaturated functions as activating groups follows the approximate order —NO$_2$ > —CO—R > —SO$_2$—R > —CO—OR and —C≡N > —SO—R > —C$_6$H$_5$. Also, the presence

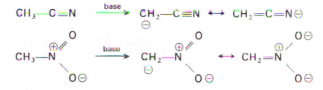

of two such unsaturated substituents further enhances the acidity of an active methylene compound. Acidity is increased (about 2 pK_a units)[1c,2d] by an electron-withdrawing chloro or bromo substituent[1j] and is decreased (about 1 to 2 pK_a units)[1c,2] by an alkyl substituent.

The values (or ranges of values) given in Table 9–1 for weakly acidic substances should be considered as only approximate. The values listed were obtained under a variety of conditions and are not corrected for variations in pK_a values which result when the measurements are made in different solvents with different bases and cations present.[1,2] A very different order of acidities is observed in the gas phase, where these solvation effects are not operative.[1n]

THE FORMATION OF ENOLS AND ENOLATE ANIONS

Procedures that involve the formation and subsequent reaction of anions derived from active methylene compounds constitute a very important and synthetically useful class of organic reactions. Perhaps the most common of this class are those reactions in which the anion has been derived by removal of a proton from the

492

carbon atom alpha to a carbonyl group. These anions, usually called enolate anions (e.g., [1]), are to be distinguished from enols (e.g., [2]), which are present in equilibrium with the carbonyl compounds in the presence of either acidic or basic catalysts.[3] In the case of most monoketones and esters the amount of enol

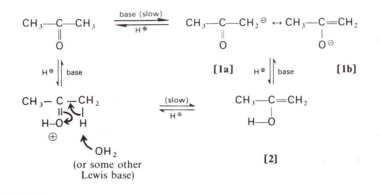

1. (a) J. B. Conant and G. W. Wheland, *J. Am. Chem. Soc.,* **54**, 1212 (1932). (b) W. K. McEwen, *ibid.,* **58**, 1124 (1936). (c) R. G. Pearson and R. L. Dillon, *ibid.,* **75**, 2439 (1953). (d) C. D. Ritchie and R. E. Uschold, *ibid.,* **89**, 1721, 2752, 2960 (1967); **90**, 2821, 3415 (1968); C. D. Ritchie, *ibid.,* **91**, 6749 (1969). (e) E. C. Steiner and J. M. Gilbert, *ibid.,* **87**, 382 (1965); E. C. Steiner and J. D. Starkey, *ibid.,* **89**, 2751 (1967). (f) A. Streitwieser, Jr., J. I. Brauman, J. H. Hammons, and A. H. Pudjaatmaka, *ibid.,* **87**, 384 (1965). (g) F. G. Bordwell, R. H. Imes, and E. C. Steiner, *ibid.,* **89**, 3905 (1967). (h) K. Bowden and A. F. Cockerill, *J. Chem. Soc.,* B, 173 (1970); K. Bowden A. F. Cockerill, and J. R. Gilbert, *ibid.,* B, 179 (1970). (i) D. J. Cram, *Fundamentals of Carbanion Chemistry,* Academic Press, New York, 1965, pp. 1–84; D. J. Cram, *Survey Progr. Chem.,* **4**, 45 (1968). (j) An α-fluoro substituent has been found to lower the acidity of an *alpha* C—H bond by 1–3 pK units: H. G. Adolph and M. J. Kamlet, *J. Am. Chem. Soc.,* **88**, 4761 (1966); J. Hine, L. G. Mahone, and C. L. Liotta, *ibid.,* **89**, 5911 (1967). (k) V. I. Slovetskii, A. I. Ivanov, S. A. Shevelev, A. A. Fainsilberg, and S. S. Novikov, *Tetrahedron Letters,* **No. 16**, 1745 (1966); M. E. Sitzmann, H. G. Adolph, and M. J. Kamlet, *J. Am. Chem. Soc.,* **90**, 2815 (1968). (l) S. J. Rhoads and A. W. Decora, *Tetrahedron,* **19**, 1645 (1963). (m) D. J. Schaeffer, *Chem. Commun.,* **No. 17**, 1043 (1970). (n) J. I. Brauman and L. K. Blair, *J. Am. Chem. Soc.,* **91**, 2126 (1969); **92**, 5986 (1970). (o) J. I. Brauman, J. A. Bryson, D. C. Kahl, and N. J. Nelson, *ibid.,* **92**, 6679 (1970).
2. (a) W. L. Rellahan, W. L. Gumby, and H. D. Zook, *J. Org. Chem.,* **24**, 709 (1959). (b) H. D. Zook, W. L. Kelly, and I. Y. Posey, *ibid.,* **33**, 3477 (1968). (c) H. D. Zook, T. J. Russo, E. F. Ferrand, and D. S. Stotz, *ibid.,* **33**, 2222 (1968). (b) R. P. Bell, G. R. Hillier, J. W. Mansfield, and D. G. Street, *J. Chem. Soc.,* B, 827 (1967). (e) R. P. Bell and B. G. Cox, *ibid.,* B, 194 (1970). (f) J. A. Feather and V. Gold, *ibid.,* 1752 (1965). (g) P. W. K. Flanagan, H. W. Amburn, H. W. Stone, J. G. Traynham, and H. Shechter, *J. Am. Chem. Soc.,* **91**, 2797 (1969); M. Fukuyama, P. W. K. Flanagan, F. T. Williams, Jr., L. Frainier, S. A. Miller, and H. Shechter, *ibid.,* **92**, 4689 (1970). (h) F. G. Bordwell, W. J. Boyle, Jr., and K. C. Yee, *ibid.,* **92**, 5926 (1970); F. G. Bordwell and K. C. Yee, *ibid.,* **92**, 5933, 5939 (1970).
3. (a) R. P. Bell, *The Proton in Chemistry,* Cornell University Press, Ithaca, N.Y., 1959. (b) V. J. Shiner, Jr., and B. Martin, *J. Am. Chem. Soc.,* **84**, 4824 (1962). (c) J. Stuehr, *ibid.,* **89**, 2826 (1967); M. Eigen, *Angew. Chem., Intern. Ed. Engl.,* **3**, 1 (1964). (d) G. E. Lienhard and T. C. Wang, *J. Am. Chem. Soc.,* **91**, 1146 (1969). (e) For Studies of amine-catalyzed enolization see M. L. Bender and A. Williams, *ibid.,* **88**, 2502 (1966); J. Hine, J. G. Houston, J. H. Jensen, and J. Mulders, *ibid.,* **87**, 5050 (1965); J. Hine, B. C. Menon, J. Mulders, and R. L. Flachskam, Jr., *J. Org. Chem.,* **34**, 4083 (1969) and references therein.

Table 9–1[1,2b] Approximate acidities of active methylene compounds and other common reagents[a]

Compound	pK_a	Compound	pK_a
$CH_2(NO_2)_2$	4	$(CH_3)_3COH$	19
CH_3CO_2H	5	$C_6H_5COCH_3$	19
$CH_2(CN)CO_2C_2H_5$	9	CH_3COCH_3	20
$CH_2(COCH_3)_2$	9		21
$CH_3CH_2NO_2$	9		
$(CH_3)_3NH^{\oplus}$	10	$(C_6H_5)_2NH$	21
C_6H_5OH	10	$(CH_3)_3C-CO-CH_3$	21
CH_3NO_2	10	$(CH_3)_3C-CO-CH(CH_3)_2$	23
$CH_3COCH_2CO_2C_2H_5$	11	$CH_3-SO_2-CH_3$	23–27
$CH_3CH(COCH_3)_2$	11	$CH_3CO_2C_2H_5$	25
$CH_2(CN)_2$	11	CH_3CN	25
$CH_2(CO_2C_2H_5)_2$	13	$HC\equiv CH$	25
	15	$C_6H_5NH_2$	27
H_2O	16	$(C_6H_5)_3CH$	28–33
	16	$(C_6H_5)_2CH_2$	33–35
$C_6H_5CH_2-CO-C_6H_5$	16	$CH_3-SO-CH_3$	35
CH_3OH	16–18	NH_3	35
CH_3COCH_2-Cl	17	$(C_2H_5)_2NH$	36
CH_3CH_2OH	18	$C_6H_5CH_3$	37
$(CH_3)_2CHOH$	18	$CH_2{=}CHCH_3$	38

[a] Acidic hydrogen atoms boldface.

present at equilibrium is small (less than 1%, see Table 9–2) ; on the other hand 1,2- and 1,3-dicarbonyl compounds often contain high percentages ($>$50 percent) of their enol tautomer at equilibrium. The following equation illustrates the general rule that an enol tautomer is more acidic (4–5 pK units) than a keto form. This acidity difference is attributable to the fact that the rate of proton abstraction from

Table 9–2[4]   Percentage of enol present in equilibrium with some typical carbonyl compounds

Compounds	% Enol (solvent)
CH_3COCH_3	$<$0.002% (H_2O)
	0.013% (H_2O)
	0.0004% (H_2O)
$CH_3COCH_2CO_2C_2H_5$	10–13% (C_2H_5OH)
	49% (C_6H_{14})
	59–60% (C_2H_5OH)
CH_3CO—CH_2—$COCH_3$	16% (H_2O)
	83% (C_2H_5OH)
	92% (C_6H_{14})
	3% (H_2O)
	31% (C_2H_5OH)
	59% (C_6H_{14})
	95% (H_2O)
	2–14% (cyclohexane, dependent on concentration)

a C—H bond is usually a slow process compared with the very fast (often diffusion-controlled) rate of proton removal from OH bonds or proton addition to either oxygen or carbon anions.[3]

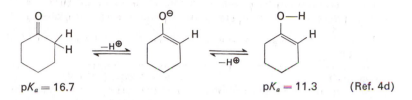

$$pK_a = 16.7 \qquad\qquad\qquad pK_a = 11.3 \qquad \text{(Ref. 4d)}$$

The extent of enolization and the structure and stereochemistry of the enols obtained from 1,3-dicarbonyl compounds are dependent both on the structure of the dicarbonyl compound and on the solvent in which the equilibria are measured.[4a,5] Recently, the positions and compositions of the equilibrium mixtures obtained from 1,3-dicarbonyl compounds have been determined from n.m.r. measurements which offer a distinct advantage over older chemical methods.[5]

As the data in Table 9–2 suggest, nonpolar, aprotic solvents usually favor the enol form which normally exists as a monomer with an intramolecular hydrogen bond as shown in the following example. Dimeric and polymeric hydrogen-

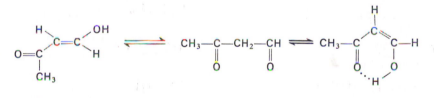

transoid conformation
favored in H_2O

cisoid conformation
favored in CCl_4

(Ref. 5a)

4. (a) G. W. Wheland, *Advanced Organic Chemistry*, Wiley, New York, 1960, pp. 663–730. (b) A. Gero, *J. Org. Chem.*, **19**, 1960 (1954); **26**, 3156 (1961). (c) N. L. Allinger, L. W. Chow, and R. A. Ford, *ibid.*, **32**, 1994 (1967). (d) R. P. Bell and P. W. Smith, *J. Chem. Soc.*, B, 241 (1966). (e) S. J. Rhoads, J. C. Gilbert, A. W. Decora, T. R. Garland, R. J. Spangler, and M. J. Urbigkit, *Tetrahedron*, **19**, 1625 (1963); S. J. Rhoads and C. Pryde, *J. Org. Chem.*, **30**, 3212 (1965). (f) A. Yogev and Y. Mazur, *ibid.*, **32**, 2162 (1967). (g) W. Hänsel and R. Haller, *Tetrahedron*, **26**, 2027, 2035 (1970).
5. (a) W. O. George and V. G. Mansell, *J. Chem. Soc.*, B, 132 (1968). (b) G. Allen and R. A. Dwek, *ibid.*, B, 161 (1966). (c) D. C. Nonhebel, *ibid.*, C, 1716 (1967); C, 676 (1968); *Tetrahedron*, **24**, 1869 (1968). (d) K. M. Baker, and J. P. Bartley, *ibid.*, **24**, 1651 (1968). (e) M. Gorodetsky, Z. Luz, and Y. Mazur, *J. Am. Chem. Soc.*, **89**, 1183 (1967). (f) E. W. Garbisch, Jr., *ibid.*, **85**, 1696 (1963); *ibid.*, **87**, 505 (1965); E. W. Garbisch, Jr., and J. G. Russell, *Tetrahedron Letters*, **No. 1**, 29 (1967). (g) S. T. Yoffe, P. V. Petrovskii, E. I. Fedin, K. V. Vatsuro, P. S. Burenko, and M. I. Kabachnick, *ibid.*, **No. 46**, 4525 (1967). (h) D. J. Sardella, D. H. Heinert, and B. L. Shapiro, *J. Org. Chem.*, **34**, 2817 (1969). (i) For a study of the enols from α-nitro ketones, see T. Simmons, R. F. Love, and K. L. Kreuz, *ibid.*, **31**, 2400 (1966). (j) For the stabilization of dienol tautomers as their tricarbonyliron complexes, see C. H. DePuy, R. N. Greene, and T. E. Schroer, *Chem. Commun.*, **No. 20**, 1225 (1968); C. H. DePuy and C. R. Jablonski, *Tetrahedron Letters*, **No. 45**, 3989 (1969).

bonded structures are formed from cyclic 1,3-diketones where intramolecular hydrogen bonding in the enol is precluded by the geometry of the molecule;[4f] the fraction of enol present in such cases is dependent on both the concentration and the temperature of the solution. In more polar solvents, and especially in protic solvents, the amount of the keto form is usually enhanced, since this tautomer is more prone to solvation (or intermolecular hydrogen bonding) with external solvent molecules. (Cyclic 1,3-diketones are an exception to this generalization.)

The presence of additional substituents, especially sterically bulky substituents, in the 1,3-dicarbonyl system usually decreases the amount of enol present at equilibrium, since formation of the previously discussed planar enol system with an intramolecular hydrogen bond forces adjacent substituents into a sterically unfavorable eclipsed conformation. The position of equilibrium between structurally isomeric enols (see the examples which follow) is apparently determined by the combined effects arising from conjugation with substituents and minimizing steric interactions and torsional strains.[5f,h]

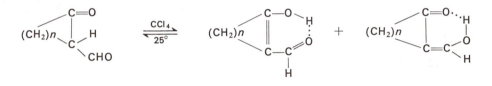

Value of n	Composition of enol mixture	
3	22%	78%
4	76%	24%
5	34%	66%
6	56%	44%
7	69%	31%
8	82%	18%

(Ref. 5f)

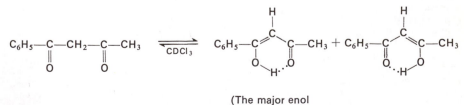

(The major enol
isomer present)   (Ref. 5e, k)

Data concerning the rates of proton removal from C—H bonds alpha to carbonyl groups are more plentiful[11,2,6] than measurements of the equilibrium dissociation constants for these weak acids.

Proton removal rates involving either acidic or basic catalysis are usually determined by measuring the rate of bromination (see Chapter 8), the rate of deuterium incorporation,[6-8] or the rate of racemization[6] (if the carbonyl compound is optically active). In the previous discussion of the relative rates of acid catalyzed enolization (see Chapter 8),[6o-q] methyl n-alkyl ketones and α-alkylcycloalkanones were noted to form the more highly substituted enols more rapidly; however, the presence of two α-alkyl substituents in an acyclic ketone provides sufficient steric interference so that formation of the less highly substituted enol becomes more rapid. This steric destabilization of the more highly substituted olefin is also seen

---

6. (a) W. G. Brown and K. Eberly, *J. Am. Chem. Soc.*, **62**, 113 (1940). (b) H. M. E. Cardwell and A. E. H. Kilner, *J. Chem. Soc.*, 2430 (1951); H. M. E. Cardwell, *ibid.*, 2442 (1951). (c) A. K. Mills, and A. E. Wildor Smith, *Helv. Chim. Acta*, **43**, 1915 (1960). (d) D. J. Cram, B. Rickborn, C. A. Kingsbury, and P. Haberfield, *J. Am. Chem. Soc.*, **83**, 3678 (1961); D. J. Cram and L. Gosser, *ibid.*, **86**, 5457 (1964). (e) R. E. Dessy, Y. Okuzumi, and A. Chen, *ibid.*, **84**, 2899 (1962). (f) H. Shechter, M. J. Collis, R. Dessy, Y. Okuzumi, and A. Chen, *ibid.*, **84**, 2905 (1962). (g) A. Schriesheim, R. J. Muller, and C. A. Rowe, Jr., *ibid.*, **84**, 3164 (1962). (h) J. Ringold and S. K. Malhotra, *Tetrahedron Letters*, **No. 15**, 669 (1962); *J. Am. Chem. Soc.*, **84**, 3402 (1962); S. K. Malhotra and H. J. Ringold, *ibid.*, **85**, 1538 (1963); **86**, 1997 (1964); **87**, 3228 (1965). (i) G. Subrahmanyam, S. K. Malhotra, and H. J. Ringold, *ibid.*, **88**, 1332 (1966). (j) R. Beugelmans, R. H. Shapiro, L. J. Durham, D. H. Williams, H. Budzikiewicz, and C. Djerassi, *ibid.*, **86**, 2832 (1964). (k) G. H. Whitman and J. A. F. Wickramasinghe, *J. Chem. Soc.*, C, 338 (1968). (l) G. Kruger, *J. Org. Chem.*, **33**, 1750 (1968). (m) J. Warkentin and R. A. Cox, *ibid.*, **33**, 1301 (1968). (n) A. A. Bothner-By and C. Sun, *ibid.*, **32**, 492 (1967). (o) C. Rappe and W. H. Sachs, *ibid.*, **32**, 3700, 4127 (1967); *Tetrahedron Letters*, **No. 19**, 2317 (1968); **No. 25**, 2921 (1968); *Tetrahedron*, **24**, 6287 (1968). (p) J. E. Dubois and J. Toullec, *Chem. Commun.*, **No. 6**, 292 (1969). (q) J. Hine, K. G. Hampton, and B. C. Menon, *J. Am. Chem. Soc.*, **89**, 2664 (1967). (r) J. Warkentin and C. Barnett, *ibid.*, **90**, 4629 (1968). (s) H. M. Walborsky and J. M. Motes, *ibid.*, **92**, 2445, 3697 (1970). (t) H. W. Amburn, K. C. Kauffman, and H. Shechter, *ibid.*, **91**, 530 (1969). (u) J. R. Jones and R. Stewart, *J. Chem. Soc.*, B, 1173 (1967). (v) For the use of gas chromatography columns with basic or acidic packing to effect hydrogen-deuterium exchange at positions alpha to carbonyl functions, see M. Senn, W. J. Richter and A. L. Burlingame, *J. Am. Chem. Soc.*, **87**, 680 (1965); *Tetrahedron Letters*, **No. 17**, 1235 (1965); G. J. Kallos and L. B. Westover, *ibid.*, **No. 13**, 1223 (1967); α-deuterio ketones may also be obtained by the decarboxylation of α-keto acids in deuterium oxide [J. Deutsch and A. Mandelbaum, *ibid.*, **No. 17**, 1351 (1969)] and from the acid-catalyzed hydrolysis of enol ethers or enamines [J. P. Schaefer and D. S. Weinberg, *ibid.*, **No. 23**, 1801 (1965), and Ref. 7b.] (w) For hydrogen-deuterium exchanges involving amides, see I. H. Klotz and P. L. Feidelseit, *J. Am. Chem. Soc.*, **88**, 5103 (1966); C. Y. S. Chen and C. A. Swenson, *ibid.*, **91**, 234 (1969). (x) Hydrogen-deuterium exchange alpha to carboxylate ions is reported by J. G. Atkinson, J. J. Csakvary, G. T. Herbert, and R. S. Stuart, *ibid.*, **90**, 499 (1968); P. Belanger, J. G. Atkinson, and R. S. Stuart, *Chem. Commun.*, **No. 18**, 1067 (1969). (y) Studies of hydrogen-deuterium exchanges of sulfones and sulfoxides are reviewed by C. D. Broaddus, *Accts. Chem. Res.*, **1**, 231 (1968); also see G. Maccagnani, F. Montanari, and F. Taddei, *J. Chem. Soc.*, B, 453 (1968). (z) For studies of hydrogen-deuterium exchange in nitroalkanes, see Ref. 2g; P. Jones, J. L. Longridge, and W. F. K. Wynne-Jones, *ibid.*, 3606 (1965); P. A. S. Smith, *The Chemistry of Open-Chain Organic Nitrogen Compounds*, Vol. 2, Benjamin, New York, 1966, pp. 391–398; I. T. Millar and H. D. Springall, *The Organic Chemistry of Nitrogen*, Clarendon Press, Oxford, 3rd ed., 1966, pp. 361–379.

in the compositions of enol acetate mixtures at equilibrium,[9] as the following equations show. In base-catalyzed enolization processes, alkyl substituents appear always to decrease the rate of proton removal from ketones at both alpha positions (see Table 9–3).[6,7]

This effect is attributable in part to an increased steric hindrance to approach of the base for proton abstraction.[7,8] Also, in the case of methylene groups activated

7. (a) H. O. House and V. Kramar, *J. Org. Chem.*, **28**, 3362 (1963). (b) H. O. House and B. M. Trost, *ibid.*, **30**, 1341, 2502 (1965); H. O. House, B. A. Tefertiller, and H. D. Olmstead, *ibid.*, **33**, 935 (1968); H. O. House and T. M. Bare, *ibid.*, **33**, 943 (1968). (c) H. O. House, L. J. Czuba, M. Gall, and H. D. Olmstead, *ibid.*, **34**, 2324 (1969). (d) H. O. House *Rec. Chem. Progr.*, **28**, 99 (1967). (e) D. H. R. Barton, R. H. Hesse, G. Tarzia, and M. M. Pechet, *Chem. Commun.*, **No. 24**, 1497 (1969); M. Tanabe and D. F. Crowe, *ibid.*, **No. 24**, 1498 (1969). (f) Various evidence for the planar nature of the anions derived from nitroalkanes is given by M. J. Kamlet, R. E. Oesterling, and H. G. Adolph, *J. Chem. Soc.*, 5838 (1965); M. J. Brookes and N. Jonathan, *ibid.*, A, 1529 (1968); A. A. Griswald and P. S. Starcher, *J. Org. Chem.*, **30**, 1687 (1965); J. R. Holden and C. Dickinson, *J. Am. Chem. Soc.*, **90**, 1975 (1968). (g) For studies of various metal complexes of β-diketones, see D. Gibson, J. Lewis, and C. Oldham, *J. Chem. Soc.*, A, 1453 (1966); J. Lewis and C. Oldham, *ibid.*, A, 1456 (1966); J. A. S. Smith and E. J. Wilkins, *ibid.* A, 1749 (1966); D. C. Luehrs, R. T. Iwamoto, and J. Kleinberg, *Inorg. Chem.* **4**, 1739 (1965). (h) For studies of the infrared and n.m.r. spectra of simple enolate anions, see Ref. 2b,c, 7c; G. Stork and P. F. Hudrlik, *J. Am. Chem. Soc.*, **90**, 4462, 4464 (1968); A. G. Pinkus, J. G. Lindberg, and A.-B. Wu, *Chem. Commun.*, **No. 22**, 1350 (1969); *ibid.*, **No. 14**, 859 (1970). (i) H. E. Zaugg and A. D. Schaefer, *J. Am. Chem. Soc.*, **87**, 1857 (1965). (j) B. J. L. Huff, F. N. Tuller, and D. Caine, *J. Org. Chem.*, **34**, 3070 (1969). (k) M. C. Caserio, W. Lauer, and T. Novinson, *J. Am. Chem. Soc.*, **92**, 6082 (1970). (l) A. Kergomard and M. F. Renard, *Tetrahedron Letters*, **No. 27**, 2319 (1970). (m) H. Morrison and S. R. Kurowsky, *Chem. Commun.*, **No. 21**, 1098 (1967).
8. (a) E. J. Corey and R. A. Sneen, *J. Am. Chem. Soc.*, **78**, 6269 (1956). (b) H. E. Zimmerman in P. deMayo, ed., *Molecular Rearrangements*, Vol. 1, Wiley-Interscience, New York, 1963, pp. 345–372. (c) F. Johnson and S. K. Malhotra, *J. Am. Chem. Soc.*, **87**, 5492, 5493, 5513 (1965); F. Johnson, *Chem. Rev.*, **68**, 375 (1968). (d) H. E. Zimmerman and P. S. Mariano, *J. Am. Chem. Soc.*, **90**, 6091 (1968). (e) F. G. Bordwell and R. G. Scamehorn, *ibid.*, **90**, 6749 (1968). (f) J. Fishman, *J. Org. Chem.*, **31**, 520 (1966). (g) For studies of the preferential *exo* hydrogen-deuterium exchange in bicyclo-[2.2.1]-heptan-2-one derivatives, see A. F. Thomas and B. Willhalm, *Tetrahedron Letters*, **No. 18**, 1309 (1965); J. M. Jerkunica, S. Borcic, and D. E. Sunko, *ibid.*, **No. 49**, 4465 (1965); P. Barraclough and D. W. Young, *ibid.*, **No. 26**, 2293 (1970); A. F. Thomas, R. A. Schneider, and J. Meinwald, *J. Am. Chem. Soc.*, **89**, 68 (1967); T. T. Tidwell, *ibid.*, **92**, 1448 (1970). (h) W. D. Emmons and M. F. Hawthorne, *ibid.*, **78**, 5593 (1956). (i) For a study of the amine-catalyzed epimerization of an α-methylcyclohexanone derivative, see T. A. Spencer and L. D. Eisenhauer, *J. Org. Chem.*, **35**, 2632 (1970). (j) F. G. Bordwell, *Accts. Chem. Res.*, **3**, 281 (1970). (k) E. Casadevall and P. Metzger, *Tetrahedron Letters*, **No. 48**, 4199 (1970).
9. For studies of the formation and equilibration of enol acetates, see Ref. 7a–d. (a) W. B. Smith and T. K. Chen, *J. Org. Chem.*, **30**, 3095 (1965). (b) W. M. Muir, P. D. Ritchie, and D. J. Lyman, *ibid.*, **31**, 3790 (1966). (c) B. E. Edwards and P. N. Rao, *ibid.*, **31**, 324 (1966). (d) O. R. Rodig and G. Zanati, *ibid.*, **32**, 1423 (1967). (e) A. J. Liston, *ibid.*, **31**, 2105 (1966); A. J. Liston and M. Howarth, *ibid.*, **32**, 1034 (1967); A. J. Liston and P. Toft, *ibid.*, **33**, 3109 (1968); **34**, 2288 (1969); *Chem. Commun.*, **No. 2**, 111 (1970). (f) H. Favre and A. J. Liston, *Can. J. Chem.*, **42**, 268 (1964); **47**, 3233 (1969). (g) J. Libman, M. Sprecher, and Y. Mazur, *Tetrahedron*, **25**, 1679 (1969); J. Libman and Y. Mazur, *ibid.*, **25**, 1699 (1969); D. Amar, V. Permutti, and Y. Mazur, *ibid.*, **25**, 1717 (1969). (h) G. C. Joshi, W. D. Chambers, and E. W. Warnhoff, *Tetrahedron Letters*, **No. 37**, 3613 (1967). (i) M. P. Hartshorn and E. R. H. Jones, *J. Chem. Soc.*, 1312 (1962). (j) B. Berkoz, E. P. Chavez, and C. Djerassi, *ibid.*, 1323 (1962). (k) A. J. Bellamy and G. H. Whitham, *ibid.*, C, 215 (1967).

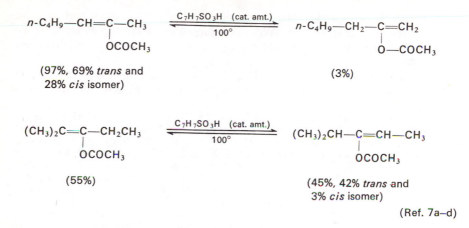

$n\text{-}C_4H_9\text{—}CH\!\!=\!\!C\text{—}CH_3$

$OCOCH_3$

$\xrightarrow[\;100°\;]{\;C_7H_7SO_3H\;(cat.\ amt.)\;}$

$n\text{-}C_4H_9\text{—}CH_2\text{—}C\!\!=\!\!CH_2$

$O\text{—}COCH_3$

(97%, 69% *trans* and
28% *cis* isomer)

(3%)

$(CH_3)_2C\!\!=\!\!C\text{—}CH_2CH_3$

$OCOCH_3$

$\xrightarrow[\;100°\;]{\;C_7H_7SO_3H\;(cat.\ amt.)\;}$

$(CH_3)_2CH\text{—}C\!\!=\!\!CH\text{—}CH_3$

$OCOCH_3$

(55%)

(45%, 42% *trans* and
3% *cis* isomer)

(Ref. 7a–d)

by carbonyl functions, the presence of alkyl substituents at either alpha carbon atom interferes sterically with attainment of a transition state for proton removal (e.g., [3]) that permits continuous overlap of the *p*-orbitals involved.[8]

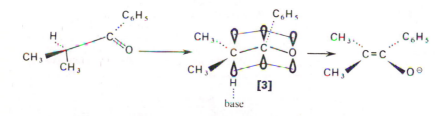

[3]

base

Weakly acidic monoketones such as those listed in Table 9–3 probably react with bases in protic solvents (e.g. water or an alcohol) by a process that involves

Table 9–3[60,r] Relative rates[a] of H–D exchange for a series of ketones R—CO—CH₃ with Na₂CO₃ or NaOD in D₂O or D₂O-dioxane mixtures

R	Exchange at CH_3	Exchange at an α-C—H bond of R
CH_3-	100	100
CH_3CH_2-	45–50	28
$CH_3CH_2CH_2$-	33–41	12–13
$(CH_3)_2CHCH_2$-	27	7
$(CH_3)_3CCH_2$-	5–21	0.3–0.7
$(CH_3)_2CH$-	33–45	0.1–0.9
$(CH_3)_3C$-	16–27	—

[a]   These values represent the total rate of exchange at each group and are not corrected for the number of hydrogen atoms at each alpha position.

addition of a proton to oxygen followed by removal of a proton from the alpha carbon (see structure [4a]).[3a,8h,j]

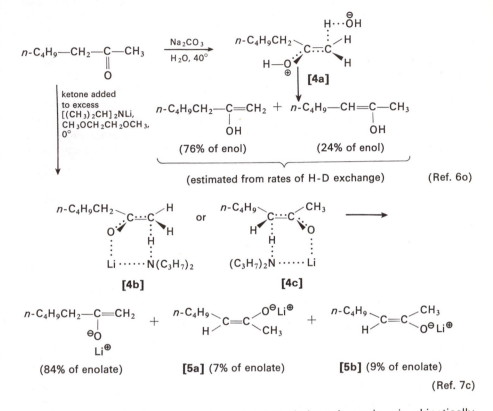

(estimated from rates of H-D exchange)   (Ref. 6o)

(Ref. 7c)

Ketones can also be converted completely to their enolate anions in a kinetically controlled process by adding the ketones slowly to an excess of a strong base in an aprotic solvent. As the accompanying example suggests, the relative rates of proton abstraction in each of these two ketone-base reactions are often surprisingly similar. The fact that the disubstituted enolates (e.g., [5]) formed in the kinetically controlled reactions of ketones with strong bases contain substantial amounts of the less stable *cis*-isomer (e.g., [5b])[7c,d] suggests the importance of transition states such as [4b] or [4c] in which steric interference at both the breaking carbon-hydrogen bond and the forming oxygen-metal bond is minimized.

A qualitative relationship exists between the rate of proton removal and the dissociation constant (pK_a values) for active methylene compounds in that proton removal is usually more rapid for more acidic compounds. Although this relationship provides a reasonable estimate (sometimes called the kinetic acidity) of the dissociation constant for very weak carbon acids,[1] the relationship is less satisfactory for relatively acidic carbon acids in which the negative charge in the anions is delocalized and resides mainly at atoms other than carbon. In particular, the rates of proton removal from nitroalkanes and from ketones do not provide a reliable measure of

the dissociation constants for these substances, since a number of examples exist in which proton removal is slower from the more acidic of a pair of nitroalkanes or ketones.[2] An important consequence of this fact is the ability of a ketone to give enolate anion mixtures of different compositions, depending on whether the enolates were formed under circumstances in which the composition was determined by the relative rates of proton abstraction (kinetic control) or by equilibration of the various enolate anions (equilibrium control).[7a-e,j] As the accompanying equation illustrates, rapid equilibration of enolate anions is achieved only when some proton donor such as a protic solvent or excess un-ionized ketone is present in the reaction mixture. Consequently, a kinetically controlled mixture of enolates is obtained by slowly adding a ketone to an excess of a strong base in an aprotic solvent, whereas the slow addition of a strong base to a ketone or the presence of excess ketone in a solution of enolate anions allows the formation of an equilibrium mixture of enolate anions.

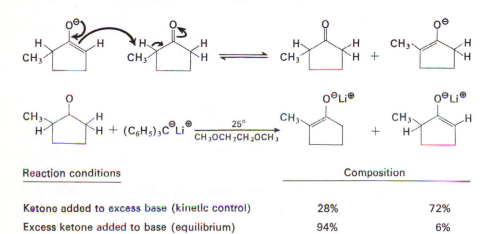

Reaction conditions	Composition	
Ketone added to excess base (kinetic control)	28%	72%
Excess ketone added to base (equilibrium)	94%	6%

(Ref. 7b, d)

The reaction of nitroalkane anions and enolate anions with proton donors (usually aqueous solutions of acetic acid or mineral acids) has also been examined.[2h,8c,d] The interpretation of the experimental results of these reactions is complicated by the facts that protonation of the anions is normally much faster (bimolecular rate constants 10^6 to 10^{11} M^{-1} sec^{-1})[3c] than the time required for mixing (typically 10^{-2} sec or more) and the anions are capable of being protonated at two or more sites. Anions possessing this property of being able to form new covalent bonds at two or more atoms are often called *ambident ions.*

Relatively stable anions such as those obtained from the illustrated nitroalkane and the dienone tend to undergo kinetically controlled protonation at the most electronegative atom, usually oxygen. The resulting *aci*-nitro compound and the enol isomerize more slowly to the more stable tautomeric forms. The application of this kinetically controlled protonation procedure to α,β-unsaturated ketones

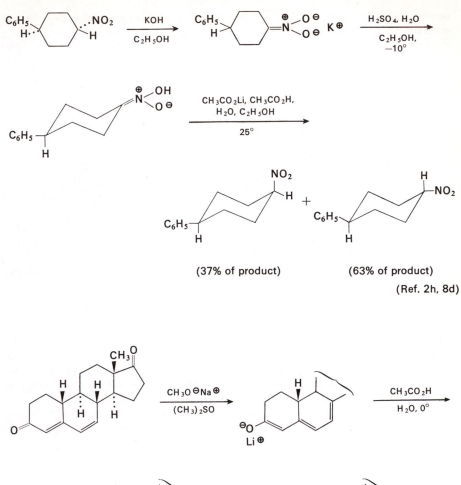

(37% of product)   (63% of product)

(Ref. 2h, 8d)

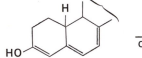

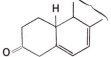

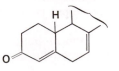

(Ref. 6l)

offers a useful method for obtaining the corresponding β,γ-unsaturated ketone.[6h,i,k,7m] This process, called deconjugation, is illustrated by the following example.

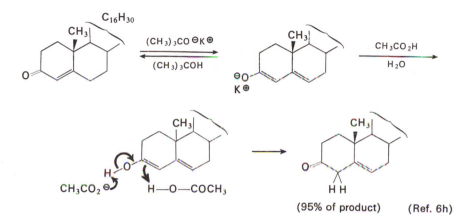

(95% of product)   (Ref. 6h)

It appears to involve kinetically controlled protonation at oxygen to form an enol which tautomerizes in neutral or weakly acidic media to the β,γ-unsaturated ketone. Although an alternative mechanism has been proposed,[6h,bb] in which the dienolate anion is suggested to undergo kinetically controlled protonation at the alpha carbon atom, whereas the dienol is protonated at the gamma position, the presently available data are more compatible with a reaction scheme in which the dienol is usually[6k,7k,l] tautomerized to a β,γ-unsaturated ketone in neutral or weakly acidic media and to an α,β-unsaturated ketone under more highly acidic conditions. The following equation provides an example in which the position of hydrogen-deuterium exchange does differ for reactions in acidic and neutral media, although an enol intermediate is believed to be formed[6h] in each case. The reaction of the enol in a relatively acidic medium is thought to be related to the acid-catalyzed hydrolysis of enol ethers, also illustrated.

The solvolysis of enol trichloracetates in methanol provides an additional example in which tautomerization of a dienol in neutral solution yields a β,γ-unsaturated ketone.

The course of protonation of unconjugated enolate anions is less clear, since protonation both at carbon (forming the keto tautomer) and at oxygen (forming the enol tautomer) may be occurring. The acid-catalyzed hydrolysis of cyclohexanone enol ethers (and enamines), like the previously illustrated dienol ether hydrolysis, favors the addition of an axial alpha proton. As the accompanying equations illustrate, reaction of the corresponding enolate anion with aqueous acetic acid is a less stereospecific process, yielding products in which the added proton (or deuteron) is in both the axial and the equatorial positions of the final product.

However, the proportion of the final mixture of ketones derived by direct addition of a proton (or deuteron) to the α-carbon of the enolate is not known.

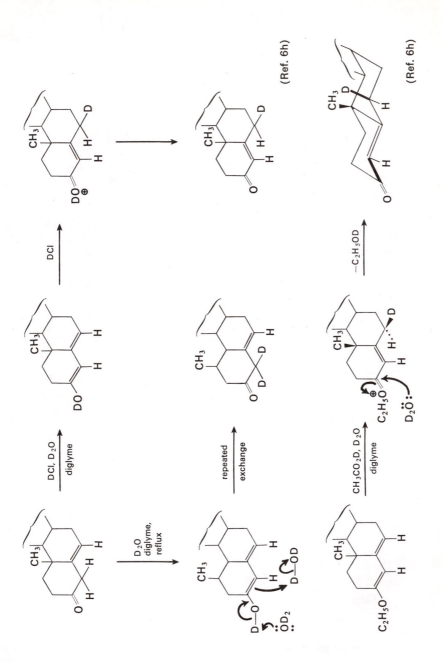

(Ref. 6h)

(Ref. 6h)

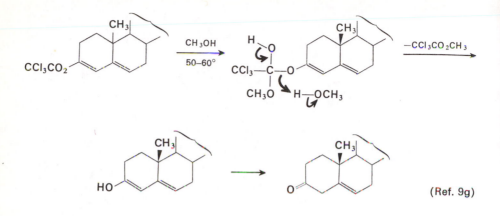

(Ref. 9g)

The following equations provide another comparison of stereochemical results from the protonation of an enolate anion and from a reduction reaction thought to involve an enol intermediate (see Chapter 3). Note that in each case the mixture obtained by this kinetically controlled protonation procedure contains a surplus of the less-stable isomer formed by adding a proton to the alpha carbon atom of the enol (or enolate) from the less-hindered side.[8b,d]

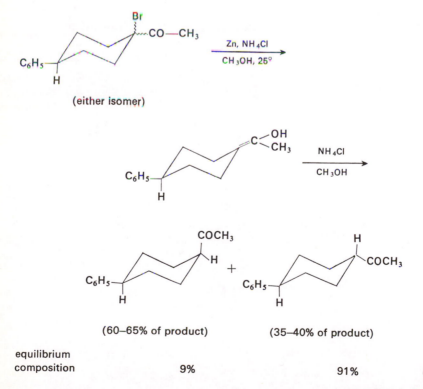

equilibrium
composition   9%   91%   (Ref. 8d)

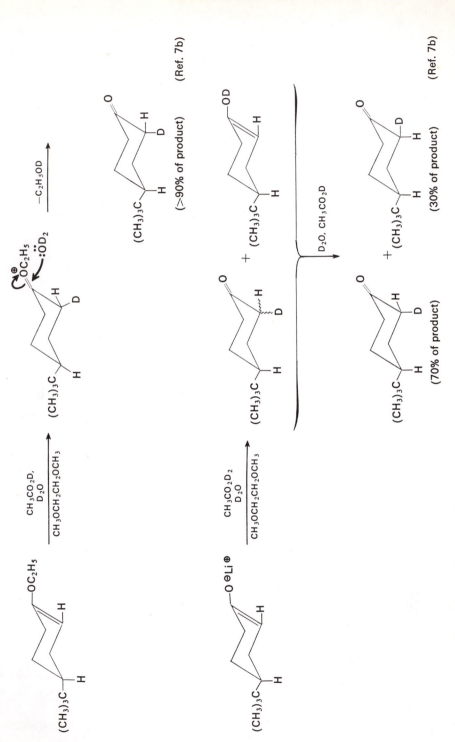

(>90% of product)   (Ref. 7b)

(70% of product)   +   (30% of product)   (Ref. 7b)

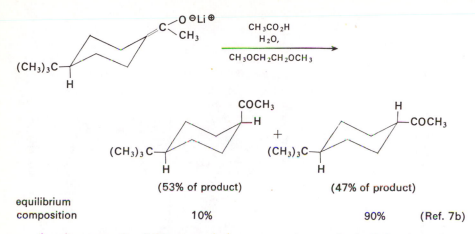

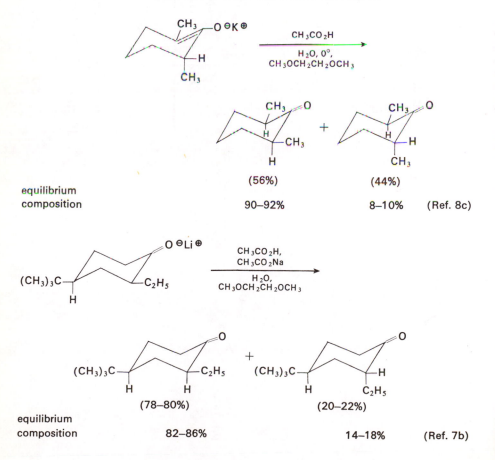

(53% of product)   (47% of product)

equilibrium
composition   10%   90%   (Ref. 7b)

Another example of this general phenomenon is seen in the following protonations of enolate anions derived from cyclohexanone derivatives.

(56%)   (44%)

equilibrium
composition   90–92%   8–10%   (Ref. 8c)

(78–80%)   (20–22%)

equilibrium
composition   82–86%   14–18%   (Ref. 7b)

In all these cases, there is the requirement that the proton be added to carbon from a direction perpendicular to the plane of the enol or enolate. As will be discussed later, this requirement does not preclude attack from either side of an enolate derived from a cyclohexanone derivative. This stereoelectronic requirement, also applicable to the formation of enols or enolates, arises from the need to maintain maximum overlap of the *p*-orbitals involved during the formation or breaking of the C—H bond. This stereoelectric requirement does not appear to be applicable to carbanions that are stabilized by adjacent sulfone or sulfoxide functions, presumably because overlap of the developing *p*-orbital on carbon with a *d*-orbital of sulfur is possible.[1i,6y,10] The synthetic utility of these carbon ions stabilized by various sulfur functions is discussed in Chapter 10.

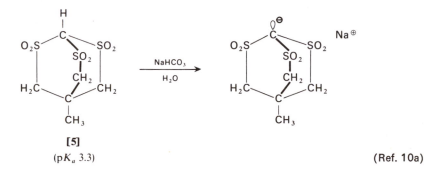

[5]

(pK_a 3.3)

(Ref. 10a)

Usually, reactions of active methylene compounds effected in the presence of bases involve enolate anions (or their analogs) as intermediates, whereas acid-catalyzed reactions involve enol intermediates. The majority of the alkylations to be discussed here will involve the formation of enolate anions, which then serve as nucleophiles in bimolecular nucleophilic (S_N2) displacements. An alternative mode of reaction, consisting of attack on the enol by an electrophilic reagent (cf. bromination, Chapter 8) leads to the formation of a new carbon-carbon bond and is occasionally useful for alkylation; more frequently, it is utilized in acid-catalyzed aldol condensations (Chapter 10) and acylations (Chapter 11).

10. (a) W. von E. Doering and L. K. Levy, *J. Am. Chem. Soc.,* **77,** 509 (1955). (b) W. von E. Doering and K. C. Schreiber, *ibid.,* **77,** 514 (1955). (c) W. von E. Doering and A. K. Hoffmann, *ibid.,* **77,** 521 (1955). (d) D. J. Cram, D. A. Scott, and W. D. Nielsen, *ibid.,* **83,** 3696 (1961). (e) H. L. Goering, D. L. Towns, and B. Dittmar, *J. Org. Chem.,* **27,** 736 (1962). (f) E. J. Corey, H. König, and T. H. Lowry, *Tetrahedron Letters,* **No. 12,** 515 (1962) ; E. J. Corey and T. H. Lowry, *ibid.,* **No. 13,** 793, 803 (1965). (g) D. J. Cram, R. D. Trepka, and P. St. Janiak, *J. Am. Chem. Soc.,* **86,** 2731 (1964) ; D. J. Cram and T. A. Whitney, *ibid.,* **89,** 4651 (1967). (h) E. J. Corey and M. Chaykovsky, *ibid.,* **87,** 1345 (1965). (i) L. H. Slaugh and E. Bergman, *J. Org. Chem.,* **26,** 3158 (1961). (j) S. Oae, W. Tagaki, and A. Ohno, *Tetrahedron,* **20,** 417, 427, 437, 443 (1964). (k) S. Wolfe and A. Rauk, *Chem. Commun.,* **No. 21,** 778 (1966) ; S. Wolfe, A. Rauk, L. M. Tel, and I. G. Csizmadia, *ibid.,* **No. 2,** 96 (1970). (l) K. C. Bank and D. L. Coffen, *ibid.,* **No. 1,** 8 (1969). (m) R. R. Fraser and F. J. Schuber, *ibid.,* **No. 24,** 1474 (1969). (n) D. Seebach, *Angew. Chem., Intern. Ed. Engl.,* **8,** 639 (1969). (o) T. Durst, *Adv. Org. Chem.,* **6,** 285 (1969).

ALKYLATION OF RELATIVELY ACIDIC ACTIVE METHYLENE COMPOUNDS

From the data in Table 9–1, recall that methylene groups activated by a single nitro group or by two or more carbonyl, ester, or cyano groups are more acidic than are the common aliphatic alcohols. As a result, compounds of these types (e.g., [6]) may be converted in large part to their enolate (or analogous) anions (e.g., [7]) by treatment either with relatively strong bases in aprotic solvents or with an anhydrous alcoholic solution of a metal alkoxide.[11] (If the product from alkylation of an ester is to be isolated prior to hydrolysis, the alcohol solvent and the metal alkoxide employed should correspond to the alkoxy group of the ester; otherwise the ester interchange that occurs will lead to a mixture of products.)

$$CH_2(CO_2C_2H_5)_2 \ + \ Na^{\oplus} \quad {}^{\ominus}OC_2H_5 \rightleftharpoons C_2H_5O—CO—CH=C—OC_2H_5 \ + \ C_2H_5OH$$

[6]

$$\underset{Na^{\oplus}}{\overset{|}{O}^{\ominus}}$$

[7]

The alcoholic solutions of enolate anions thus obtained are allowed to react with alkyl halides or other alkylating agents, as illustrated in the accompanying equations. β-Diketones (e.g., [8]) are often sufficiently acidic that their enolate

$$CH_3—CO—CH_2CO_2C_2H_5 \xrightarrow[\substack{C_2H_5OH \\ reflux}]{C_2H_5ONa} CH_3—\underset{\underset{Na^{\oplus}}{\overset{|}{O}^{\ominus}}}{C}=CHCO_2C_2H_5 \xrightarrow[\substack{C_2H_5OH \\ reflux}]{n\text{-}C_4H_9Br} n\text{-}C_4H_9—\underset{\underset{CO_2C_2H_5}{|}}{CH}—CO—CH_3$$

(69–72%)

(Ref. 12a)

$$CH_2(CO_2C_2H_5)_2 \xrightarrow[C_2H_5OH]{C_2H_5ONa} \xrightarrow[\substack{C_2H_5OH \\ reflux}]{n\text{-}C_4H_9Br} n\text{-}C_4H_9—CH(CO_2C_2H_5)_2$$

(80–90%)

(Ref. 12b)

$$CH_2(CO_2C_2H_5)_2 \xrightarrow[C_2H_5OH]{C_2H_5ONa} \xrightarrow[\substack{C_2H_5OH \\ reflux}]{\overset{CH_3}{\overset{|}{C_2H_5—CH—Br}}} \overset{CH_3}{\overset{|}{C_2H_5—CH}}—CH(CO_2C_2H_5)_2$$

(83–84%)

(Ref. 13)

$$CH_3—CO—CH_2—CO—CH_3 \xrightarrow[\substack{CH_3COCH_3 \\ reflux}]{\overset{CH_3I}{K_2CO_3}} CH_3—CO—\overset{CH_3}{\overset{|}{CH}}—CO—CH_3$$

[8]

(75–77%)

(Ref. 14a)

11. (a) A. C. Cope, H. L. Holmes, and H. O. House, *Org. Reactions,* **9**, 107 (1957). (b) F. Freeman, *Chem. Rev.,* **69**, 591 (1969).
12. (a) C. S. Marvel and F. D. Hager, *Org. Syn., Coll. Vol. 1*, 248 (1944). (b) R. Adams and R. M. Kamm, *ibid., Coll. Vol. 1*, 250 (1954).
13. (a) C. S. Marvel, *Org. Syn., Coll. Vol. 3*, 495 (1955). (b) E. B. Vliet, C. S. Marvel, and C. M. Hsueh, *ibid., Coll. Vol. 2*, 416 (1943).
14. (a) A. W. Johnson, E. Markham, and R. Price, *Org. Syn., 42*, 75 (1962). (b) S. Boatman and C. R. Hauser, *ibid., 47*, 87 (1967). (c) J. J. Bloomfield, *J. Org. Chem., 26*, 4112 (1961). The alkylation of malononitrile in alcoholic solutions is complicated by imino ester formation. See Ref. 11.

$$CH_2(CN)_2 + NaH \xrightarrow[\substack{(CH_3)_2SO}]{-H_2} \xrightarrow[\substack{(CH_3)_2SO,\ 25°}]{\substack{C_6H_5CH_2Cl \\ (2\ equiv.)}} (C_6H_5CH_2)_2C(CN)_2$$

$$(75\%) \qquad (Ref.\ 14c)$$

anies may be formed with alkali metal hydroxides or alkali metal carbonates in water, aqueous alcohol, or acetone. The alkylation products of malonic esters (e.g., [6]) and β-keto esters may be hydrolyzed and decarboxylated, as illustrated in the equations below, to yield acids and ketones, respectively. These decarboxylations involve the free acids (e.g., [9]) and are known to proceed by a six-center

$$\underset{\substack{| \\ CH_3}}{C_2H_5-CH-CH(CO_2C_2H_5)_2} \xrightarrow[\substack{H_2O \\ reflux}]{KOH} \underset{\substack{| \\ CH_3}}{C_2H_5-CH-CH(CO_2K)_2} \xrightarrow[\substack{H_2O \\ reflux}]{H_2SO_4}$$

$$\underset{\substack{| \\ CH_3}}{C_2H_5-CH-CH_2-CO_2H}$$

$$(62\text{--}65\%) \qquad (Ref.\ 13)$$

$$\underset{\substack{| \\ CO_2C_2H_5}}{n\text{-}C_4H_9-CH-CO-CH_3} \xrightarrow[\substack{H_2O \\ 25°}]{NaOH(5\%\ soln.)} \underset{\substack{| \\ CO_2Na}}{n\text{-}C_4H_9-CH-CO-CH_3} \xrightarrow[\substack{H_2O \\ reflux}]{H_2SO_4}$$

$$n\text{-}C_4H_9-CH_2-CO-CH_3$$

$$(52\text{--}61\%) \qquad (Ref.\ 15)$$

transition state (e.g., [10])[16a] which initially forms the enol; the decarboxylation of β,γ-unsaturated acids follows an analogous reaction path, although a significantly higher temperature is required.[16b]

If the structure of the keto acid (e.g., [11]) is such that the enol (e.g., [12] with a bridgehead double bond which violates Bredt's rule[17a]) would be excessively

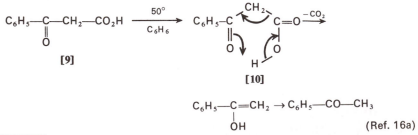

$$\underset{\substack{\| \\ O}}{C_6H_5-C-CH_2-CO_2H} \xrightarrow[\substack{C_6H_6}]{50°} \quad [10] \xrightarrow{-CO_2}$$

$$[9]$$

$$\underset{\substack{| \\ OH}}{C_6H_5-C=CH_2} \rightarrow C_6H_5-CO-CH_3$$

$$(Ref.\ 16a)$$

15. J. R. Johnson and F. D. Hager, *Org. Syn.,* **Coll. Vol. 1,** 315 (1944).

16. (a) C. G. Swain, R. F. W. Bader, R. M. Esteve, Jr., and R. N. Griffin, *J. Am. Chem., Soc.* **83,** 1951 (1961). (b) D. B. Bigley, *J. Chem. Soc.,* 3897 (1964); D. B. Bigley and J. C. Thurman, *ibid.,* 6202 (1965); B, 1076 (1966); B, 941 (1967); B, 436 (1968); D. B Bigley and R. W. May, *ibid.,* B, 557 (1967).

17. (a) F. S. Fawcett, *Chem. Rev.,* **47,** 219 (1950). (b) H. O. House and H. C. Müller, *J. Org. Chem.,* **27,** 4436 (1962). (c) J. P. Ferris and N. C. Miller, *J. Am. Chem. Soc.,* **85,** 1325 (1963); **88,** 3522 (1966). (d) P. I. Abell and R. Tien, *J. Org. Chem.,* **30,** 4212 (1965). (e) D. B. Bigley and J. C. Thurman, *Tetrahedron Letters,* **No. 51,** 4687 (1965). (f) J. Bus, H. Steinberg, and Th. J. deBoer, *ibid.,* **No. 18,** 1979 (1966).

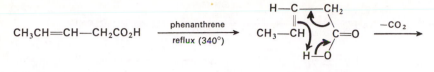

CH₃CH=CH—CH₂CO₂H $\xrightarrow[\text{reflux (340°)}]{\text{phenanthrene}}$

$$CH_3CH_2CH=CH_2$$

(72%, distilled from the
reaction mixture as
formed)   (Ref. 17b)

strained, the decarboxylation either fails or takes a different reaction path.[17] In intermediate cases, such as those illustrated, in which the formation of a strained enol is possible, the success of decarboxylation appears to be related to the ability of the keto acid to adopt a conformation in which the carboxyl group is perpendicular to the plane of the carbonyl group; a discrete enol intermediate may not be formed in such decarboxylations.[17c]

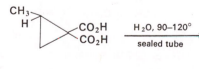

[11]   [12]

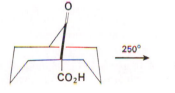

250°   no reaction

(Ref. 17c)

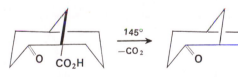

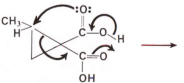

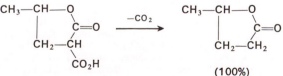

(100%)   (Ref. 17f)

The decarboxylation of cyclopropane carboxylic acid derivatives in which a strained enol would be involved is found to proceed with rearrangement as the accompanying example indicates.[17d-f] Other decarboxylation procedures often used include isolation and subsequent heating of the malonic acid, or heating of the malonic ester or β-keto ester with aqueous constant-boiling hydrochloric acid (20 per cent solution) or hydrobromic acid (48 per cent solution) to effect hydrolysis and decarboxylation in the same reaction.[18] The following equations illustrate the use of these procedures for the preparation of α-substituted carboxylic acids.

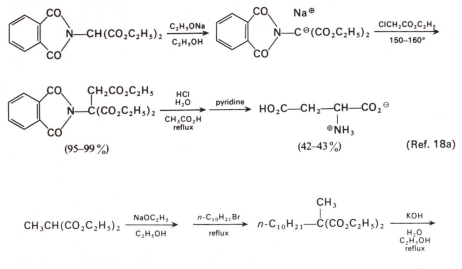

An alternative synthesis of α-substituted acids is based on the fact that even the C—H bonds alpha to carboxylate anions are sufficiently acidic to be removed by bases.[6x,19b,c] The following equations provide examples in which use of the strong base, lithium diisopropylamide, permits the formation of high concentrations of dianions which can be alkylated at carbon. (A related reaction, in which strong bases are used to form the enolate anions of monoesters, will be discussed later in this chapter.)

18. (a) M. S. Dunn and B. W. Smart, *Org. Syn.*, **Coll. Vol. 4**, 55 (1963). (b) G. Barger and T. E. Weichselbaum, *ibid.*, **Coll. Vol. 2**, 384 (1943).
19. (a) C. F. Allen and M. J. Kalm, *Org. Syn.*, **Coll. Vol. 4**, 616 (1963). (b) P. L. Creger, *ibid.*, **50**, 58 (1970). (c) P. L. Creger, *J. Am. Chem. Soc.*, **89**, 2500 (1967) ; **92**, 1396, 1397 (1970).

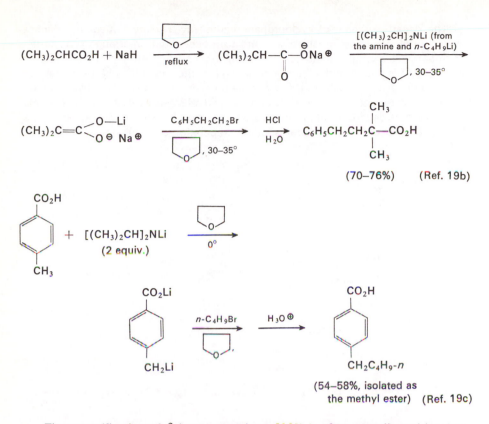

(70–76%)   (Ref. 19b)

(54–58%, isolated as
the methyl ester)   (Ref. 19c)

The saponification of β-keto esters (e.g., [13]) is often complicated by competing attack of the hydroxide anion at the ketone function, leading to cleavage rather than saponification, especially in cases in which the alpha position is

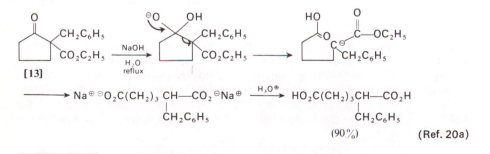

(90%)   (Ref. 20a)

20. (a) R. Mayer in W. Foerst, ed., *Newer Methods of Preparative Organic Chemistry*, Vol. 2, Academic, New York, 1963, pp. 101–131. (b) W. B. Renfrow and G. B. Walker, *J. Am. Chem. Soc.*, **70**, 3957 (1948). (c) For a discussion of the structure and alkylation of cyclic β-keto esters as a function of ring size, see Refs. 1l, 4e, and S. J. Rhoads and R. W. Hasbrouck, *Tetrahedron*, **22**, 3557 (1966); S. J. Rhoads and R. W. Holder, *ibid.*, **25**, 5443 (1969); J. P. Ferris, C. E. Sullivan, and B. G. Wright, *J. Org. Chem.*, **29**, 87 (1964).

disubstituted. For this reason hydrolysis and decarboxylation are best accomplished with aqueous acid, as illustrated in the following equation. Other methods that

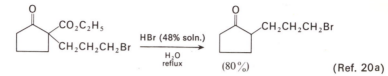

(80%)   (Ref. 20a)

have been used for cleavage of the ester function in β-keto esters to permit de-carboxylation include hydrogenolysis of benzyl esters[21] (see Chapter 1) and acid-catalyzed cleavage of t-butyl esters (e.g., [14])[20b,22] or 2-tetrahydropyranyl esters (e.g., [15]).[23a,b] Even ethyl esters undergo an analogous acid-catalyzed cleavage

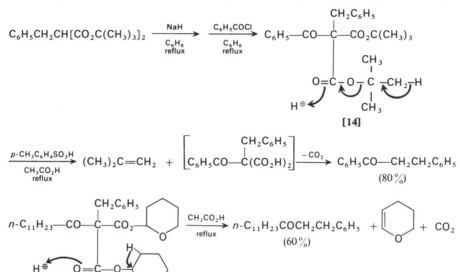

(Ref. 23a, b)

21. (a) R. E. Bowman, *J. Chem. Soc.*, 325 (1950). (b) R. E. Bowman and W. D. Fordham, *ibid.*, 2758 (1951).

22. (a) G. S. Fonken and W. S. Johnson, *J. Am. Chem. Soc.*, **74**, 831 (1952). (b) For the preparation of di-t-butyl malonate, see A. L. McCloskey, G. S. Fonken, R. W. Kluiber, and W. S. Johnson, *Org. Syn.*, **Coll. Vol. 4**, 261 (1963). (c) For the preparation of t-butyl cyanoacetate, see R. E. Ireland and M. Chaykovsky, *ibid.*, **41**, 5 (1961). (d) For the prepara-tion of t-butyl acetoacetate, see S. O. Lawesson, S. Gronwall, and R. Sandberg, *ibid.*, **42**, 28 (1962). (e) B. Riegel and W. M. Lilienfeld, *J. Am. Chem. Soc.*, **67**, 1273 (1945). (f) J. M. Lalancette and A. Lachance, *Tetrahedron Letters*, **No. 45**, 3903 (1970).

23. (a) R. E. Bowman and W. D. Fordham, *J. Chem. Soc.*, 3945 (1952). (b) The 4-(4-methoxytetrahydropyranyl) group has been recommended as an acid-labile blocking group, since use of this blocking group does not introduce an additional asymmetric center. C. B. Reese, R. Saffhill, and J. E. Sulston, *Tetrahedron*, **26**, 1023 (1970). (c) F. Elsinger, *Org. Syn.*, **45**, 7 (1965) ; F. Elsinger, J. Schreiber, and A. Eschenmoser, *Helv. Chim. Acta*, **43**, 113 (1960). (d) D. Y. Chang, S. H. Lee, and H. C. Lee, *J. Org. Chem.*, **32**, 3716 (1967). (e) A. P. Krapcho, G. A. Glynn, and B. J. Grenon, *Tetrahedron Letters*, **No. 3**, 215 (1967). (f) W. S. Johnson, C. A. Harbert, and R. D. Stipanovic, *J. Am. Chem. Soc.*, **90**, 5279 (1968).

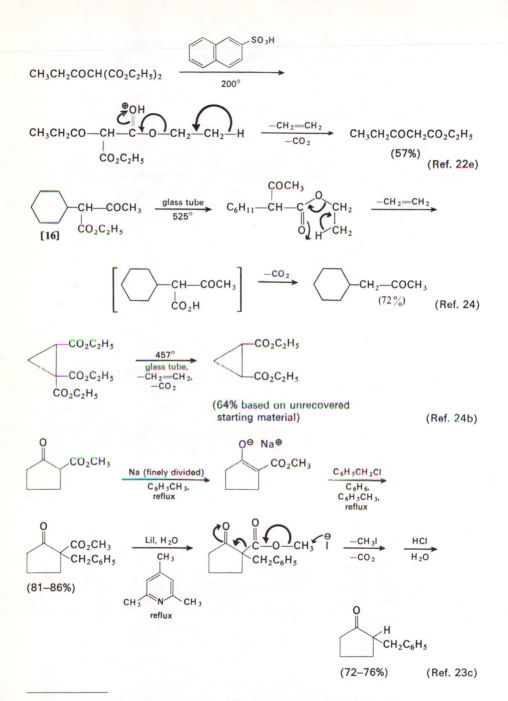

24. (a) W. J. Bailey and J. J. Daly, Jr., *J. Org. Chem.*, **22**, 1189 (1957). (b) *ibid.*, **29**, 1249 (1964). (c) W. J. Bailey and W. G. Carpenter, *ibid.*, **29**, 1252 (1964).

$$CH_3CH_2CH(CO_2C_2H_5)_2 \xrightarrow[\substack{(CH_3)_2SO, \\ 160°}]{NaCN \text{ (excess)}} CH_3CH_2CH_2CO_2C_2H_5$$
$$(80\%) \qquad \text{(Ref. 23e)}$$

if they are heated with an arenesulfonic acid[22e] or with the anhydride of boric acid.[22f] Alternatively, cleavage and subsequent decarboxylation of β-keto esters and malonic esters have been achieved by passing the materials (e.g., [16]) through a glass tube heated to 460–560°,[24] or by heating the ester with lithium iodide in refluxing γ-collidine,[23c] with hydrated sodium iodide or calcium iodide,[23d] or with sodium cyanide in dimethylsulfoxide.[23e,f] Each of these nonhydrolytic methods enjoys the advantage of minimizing cleavage at the ketone group. The successful application of the ester pyrolysis to a cyclopropane derivative is noteworthy, in view of the rearrangement observed in conventional procedures.

The cleavage of β-keto esters at the ketone function has also been observed during the alkylation reactions leading to their preparation. As was noted earlier, such cleavage is most pronounced when the alkylation product (e.g., [17]) has two alpha substituents. With only one alpha substituent, the product exists in

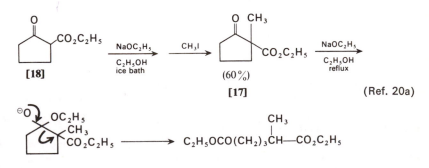

the basic solution as a stable enolate anion in which the carbonyl group is relatively resistant to attack by alkoxide ion. The cleavage has been minimized by use of low reaction temperatures with sodium ethoxide; of sterically hindered bases such as potassium t-butoxide in t-butyl alcohol;[25] or of sodium hydride[20b,22] as a base in a nonhydroxylic solvent such as dioxane, benzene, dimethylformamide, or 1,2-dimethoxyethane. Interestingly, it is possible to prepare the potassium enolate

$$n\text{-}C_4H_9-\underset{\underset{CO_2C_2H_5}{|}}{CH}-COCH_3 \xrightarrow[(CH_3)_3COH]{(CH_3)_3COK} \xrightarrow[(CH_3)_3COH]{n\text{-}C_4H_9I} (n\text{-}C_4H_9)_2\underset{\underset{CO_2C_2H_5}{|}}{C}-COCH_3 \qquad \text{(Ref. 25)}$$
$$(80\%)$$

of 2-carboethoxycyclopentanone [18] by reaction of this relatively acidic[11] keto ester with potassium hydroxide in cold aqueous alcohol.[20a] Isolation of the pure

25. (a) W. B. Renfrow, Jr., J. Am. Chem. Soc., 66, 144 (1944). (b) W. B. Renfrow and A. Renfrow, ibid., 68, 1801 (1946). (c) W. G. Dauben, J. W. McFarland, and J. B. Rogan, J. Org. Chem., 26, 297 (1961). (d) D. M. Pond and R. L. Cargill, ibid., 32, 4064 (1967).

potassium enolate, followed by reaction with an alkyl halide in an inert solvent, or preferably, in dimethyl sulfoxide[25d] constitutes the best method for alkylating this keto ester. However, alkali metal hydroxides (or other bases in partially aqueous media) normally cannot be used for the alkylation of malonic esters or less acidic[11] β-keto esters because the rapid competing saponification of the ester function lowers the yield of alkylated product.[12]

A cleavage mechanistically similar to that previously discussed for β-keto esters, called decarbethoxylation, is almost always a significant side reaction when malonic esters (e.g., [19]) or cyanoacetic esters are heated with alcoholic sodium

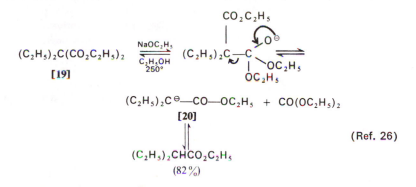

$$(C_2H_5)_2C^{\ominus}-CO-OC_2H_5 \;+\; CO(OC_2H_5)_2$$
$$[20]$$

(Ref. 26)

$$(C_2H_5)_2CHCO_2C_2H_5$$
$$(82\%)$$

ethoxide for long periods of time (e.g. 18 to 24 hr).[11,26] The rate of cleavage is enhanced by aryl or vinyl substituents, which further stabilize the intermediate anion [20]. Since the reaction is reversible, it may be suppressed by the use of diethyl carbonate as the reaction solvent.[27]

Comparable cleavage reactions have also been observed during the alkylation of α-substituted β-diketones (e.g., [21]). This ready cleavage serves as the basis

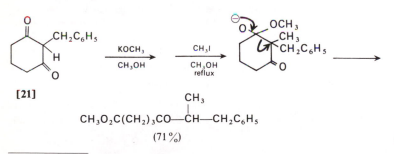

[21]

$$CH_3O_2C(CH_2)_3CO-\overset{\overset{\displaystyle CH_3}{|}}{C}H-CH_2C_6H_5$$
$$(71\%)$$

(Ref. 28a, b)

26. A. C. Cope and S. M. McElvain, *J. Am. Chem., Soc.* **54**, 4319 (1932).
27. (a) V. H. Wallingford, A. H. Homeyer, and D. M. Jones, *J. Am. Chem. Soc.,* **63**, 2056 (1941). (b) V. H. Wallingford, M. A. Thorpe, and A. H. Homeyer, *ibid.,* **64**, 580 (1942).
28. (a) H. Stetter and W. Dierichs, *Chem. Ber.,* **85**, 1061 (1952). (b) H. Stetter in W. Foerst, ed., *Newer Methods of Preparative Organic Chemistry,* Vol. 2, Academic, New York, 1963, pp. 51–99. (c) S. Boatman, T. M. Harris, and C. R. Hauser, *J. Org. Chem.,* **30**, 3321 (1965). (d) P. G. Gassman and G. D. Richmond, *ibid.,* **31**, 2355 (1966). (e) H. O. House and J. K. Larson, *ibid.,* **33**, 61 (1968). (f) E. J. Corey and M. Chaykovsky, *J. Am. Chem. Soc.,* **86**, 1639 (1964). (g) G. H. Russell and G. J. Mikol, *ibid.,* **88**, 5498 (1966).

for a ketone synthesis in which 2,4-pentanedione is alkylated in the presence of potassium carbonate and boiling ethanol; the alkylated product is cleaved in the reaction mixture.[14b,28c] A similar ketone synthesis involves the alkylation of the

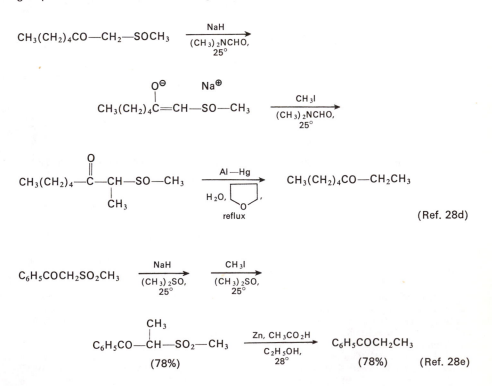

acidic β-ketosulfones or β-ketosulfoxides (see Chapter 10 for preparations) followed by reductive cleavage (see Chapter 3) of the sulfur-containing functional group with zinc or aluminum amalgam.[9h,o,28d−g]

The base-promoted cleavage of α-substituted β-diketones (effected with hydroxide ion) has also been used to advantage in the scheme for the synthesis of certain acids, as the following equations illustrate.

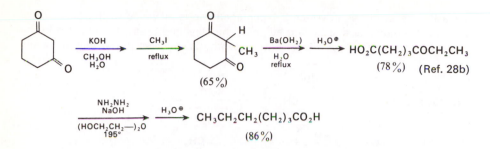

$$\xrightarrow[\substack{(HOCH_2CH_2—)_2O \\ 195°}]{\substack{NH_2NH_2 \\ NaOH}} \xrightarrow{H_3O^{\oplus}} CH_3CH_2CH_2(CH_2)_3CO_2H$$

$$(86\%)$$

A difficulty that frequently arises in the alkylation of 1,3-dicarbonyl compounds (including the diketone [22]) is the concurrent formation of both C-alkylated and O-alkylated products.[29] This possibility becomes apparent when the charge distribution in the intermediate enolate anion [23] is examined. A more common

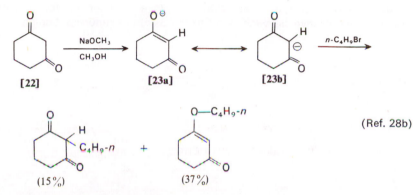

(Ref. 28b)

example of usually predominant O-alkylation is the alkylation of phenols (e.g., [24]). In general, O-alkylation competes significantly with C-alkylation only when the active methylene compounds involved are relatively acidic (e.g. nitro-alkanes) ;[29d,30b–d] in such cases the equilibrium concentration of the enol tautomer

29. For reviews and general discussion of the reactions of ambient ions, see : (a) S. Hünig, *Angew. Chem., Intern. Ed. Engl.,* **3,** 548 (1964). (b) R. Gompper, *ibid.,* **3,** 560 (1964). (c) W. J. LeNoble, *Synthesis,* **2,** 1 (1970). (d) N. Kornblum, R. A. Smiley, R. K. Blackwood, and D. C. Iffland, *J. Am. Chem. Soc.,* **77,** 6269 (1955). (e) R. G. Pearson and J. Songstad, *ibid.,* **89,** 1827 (1967).
30. (a) J. S. Buck, *Org. Syn.,* **Coll. Vol. 2,** 619 (1943). (b) S. J. Etheredge, *Tetrahedron Letters,* **No. 50,** 4527 (1965). (c) N. Kornblum and R. A. Brown, *J. Am. Chem. Soc.,* **86,** 2681 (1964). (d) R. C. Kerber, G. W. Urry, and N. Kornblum, *ibid.,* **87,** 4521 (1965); N. Kornblum, R. E. Michel, and R. C. Kerber, *ibid.,* **88,** 5660, 5662 (1966) ; N. Kornblum and co-workers, *ibid.,* **89,** 5714 (1967) ; **90,** 6219, 6221 (1968) ; **92,** 5513, 5783, 5784 (1970) ; G. A. Russell and W. C. Danen, *ibid.,* **88,** 5663 (1966) ; (e) W. J. leNoble and H. F. Morris, *J. Org. Chem.,* **34,** 1969 (1969) ; W. J. leNoble and J. E. Puerta, *Tetrahedron Letters,* **No. 10,** 1087 (1966). (f) G. Brieger and W. M. Pelletier, *ibid.,* **No. 40,** 3555 (1965). (g) A. Chatterjee, D. Banerjee, and S. Banerjee, *ibid.,* **No. 43,** 3851 (1965). (h) F. H. Bottom and F. J. McQuillin, *ibid.,* **No. 21,** 1975 (1967). (i) A. L. Kurz, I. P. Beletskaya, A. Macias, and O. A. Reutov, *ibid.,* **No. 33,** 3679 (1968). (j) R. M. Coates and J. E. Shaw, *J. Org. Chem.,* **35,** 2597, 2601 (1970) ; J. A. Marshall, G. L. Bundy, and W. I. Fanta, *ibid.,* **33,** 3913 (1968).

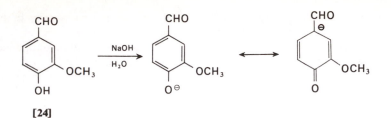

(82–87%)   (Ref. 30a)

is often relatively high (e.g. β-keto esters, 1,3-dicarbonyl compounds, and phenols). Alkylation at the more electronegative atom of the ambident anion (at oxygen rather than at carbon for an enolate anion) is usually favored by the use of polar, aprotic solvents,[31] especially hexamethylphosphoramide,[2c,20c,29,30e,f,i,j,32d,e] and by the presence of large (R⁴N⊕ > K⊕ > Na⊕ > Li⊕) cations which have a tendency to dissociate from the anion.[29,30e–i,32e]

Various observations[29,30,32] suggest that the reaction *in solution* of a particular enolate anion with a particular alkylating agent will give the greatest proportion of O-alkylation when reaction conditions (solvent, cation, temperature) are chosen which allow the maximum amount of the free enolate anion to be present. Presumably the greatest fraction of the negative charge in the enolate anion is located on oxygen, the most electronegative atom present, and the maximum opportunity for O-alkylation exists when the oxygen atom is not shielded by association with a metal cation or a hydrogen-bonding solvent.[32d]

In cases when the metal enolate is not in solution, any heterogeneous reaction of the alkylating agent with the solid enolate will necessarily involve reaction

31. For reviews of the properties and uses of dipolar, aprotic solvents, see: (a) A. J. Parker, *Quart. Rev.* (London), **16**, 163 (1962); *Adv. Org. Chem.*, **5**, 1 (1965); *Chem. Rev.*, **69**, 1 (1969). (b) H. Normant, *Bull. Soc. Chim. France*, 791 (1968); *Angew. Chem., Intern. Ed. Engl.*, **6**, 1046 (1967). (c) H. G. Hertz, *ibid.*, **9**, 124 (1970). (d) D. Martin, A. Weise, and H. J. Niclas, *ibid.*, **6**, 318 (1967). (e) M. Szwarc, *Progr. Phys. Org. Chem.*, **6**, 323 (1968); *Accts. Chem. Res.*, **2**, 87 (1969).

32. (a) D. Y. Curtin and A. R. Stein, *Org. Syn.*, **46**, 115 (1966); D. Y. Curtin, R. J. Crawford, and M. Wilhelm, *J. Am. Chem. Soc.*, **80**, 1391 (1958). (b) D. Y. Curtin and D. H. Dybvig, *ibid.*, **84**, 225 (1962). (c) N. Kornblum and A. Lurie, *ibid.*, **81**, 2705 (1959). (d) N. Kornblum, P. J. Berrigan, and W. J. leNoble, *ibid.*, **85**, 1141 (1963). (e) N. Kornblum, R. Seltzer, and P. Haberfield, *ibid.*, **85**, 1148 (1963); (f) S. Masamune, *ibid.*, **83**, 1009 (1961); **86**, 288–291 (1964). (g) For other synthetic applications, see E. Wenkert, R. D. Youssefyeh, and R. G. Lewis, *ibid.*, **82**, 4675 (1960); R. S. Atkinson and A. S. Dreiding, *Helv. Chim. Acta*, **50**, 23 (1967); J. A. Marshall and S. F. Brady, *Tetrahedron Letters*, 1387 (1969). (h) E. C. Taylor, G. H. Hawkes, and A. McKillop, *J. Am. Chem. Soc.*, **90**, 2421 (1968). (i) For reviews, see A. G. Lee, *Quart. Rev.*, **24**, 310 (1970); E. C. Taylor and A. McKillop, *Accts. Chem. Res.*, **3**, 338 (1970).

with an anion in which the oxygen is shielded by the associated metal atom in the crystal lattice; in such cases, C-alkylation is favored.[32c]

The following equations provide examples in which the C-alkylation of phenols is favored by the use of heterogeneous reaction conditions (e.g., [25]) or by the use of a protic solvent which can hydrogen-bond with the oxygen atom of the anion (e.g., [26]).

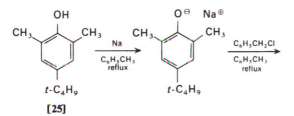

[25]

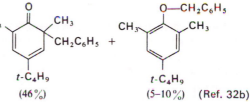

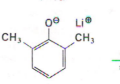

(46%)   (5–10%)   (Ref. 32b)

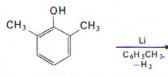

[25]   (98–100%)

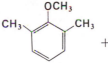   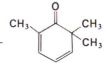

(51–55%)   (18–19% isolated as the dimer)   (Ref. 32a)

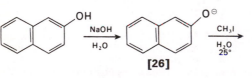

[26]

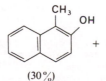   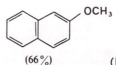

(30%)   (66%)   (Ref. 32e)

An interesting illustration of the C-alkylation of phenols is the indicated intramolecular cyclization of a phenol to form the tricyclic system. In this case intramolecular O-alkylation is not possible. This cyclization has provided a useful synthetic route to comparable ring systems present in certain diterpenes and diterpene alkaloids.[32f,g]

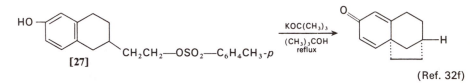

(Ref. 32f)

Another procedure for effecting the C-alkylation of β-dicarbonyl compounds utilizes the thallium(I) enolates[32h,i] prepared as indicated in the following equation. The thallium(I) enolates (extremely toxic substances) are isolated and subsequently heated with the alkyl halide to achieve alkylation. Since these reactions are both heterogeneous and involve a cation which is tightly associated with the enolate oxygen,[32h,i] the previously discussed criteria favoring C-alkylation are provided. These thallium enolates, like the subsequently discussed magnesium and zinc enolates of monoketones, apparently exist as tightly associated ion pairs which favor C-alkylation and retard the equilibration of isomeric enolate structures. However, these advantages are accompanied by a distinct loss of reactivity of these enolates as nucleophiles in reactions with alkyl halides.

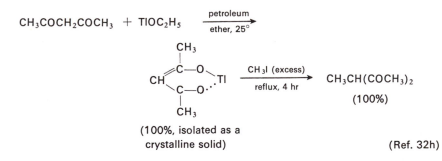

(Ref. 32h)

Although the previously discussed reaction conditions (heterogeneous or homogeneous reaction, solvent polarity, choice of cation) clearly influence the proportions of O-alkylated and C-alkylated products formed, other factors of comparable importance are the nature of the alkylating agent and, especially, the nature of the ambident ion. If the transition states [27] for O- and C-alkylation (which localize the negative charge at either oxygen or carbon) are considered, it will be seen that the difference in activation energy between them will be some fraction of the energy difference between the alkylated products ([28], R = alkyl) or the corresponding keto-enol tautomers ([28], R = H). This consideration serves as a basis for the previously mentioned observation that O-alkylation (e.g. via [27a]) is most favorable for those active methylene compounds which are highly enolic (relative stabilities; [28a] > [28b]).

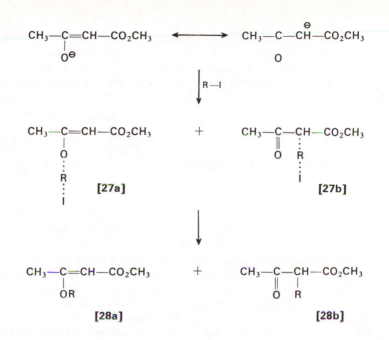

The following examples illustrate typical proportions of O- and C-alkylated products which may be observed in the alkylation of carbonyl compounds.

CH₃COCH₂CO₂C₂H₅ + n-C₄H₉-X $\xrightarrow[\text{solvent}]{\text{K}_2\text{CO}_3,\ 100°}$

CH₃COCHCO₂C₂H₅ + CH₃—C=CHCO₂C₂H₅
        | |
     n-C₄H₉   O—C₄H₉-n

Solvent	X	Product composition	
CH₃COCH₃	Cl	90%	10%
CH₃CN	Cl	81%	19%
CH₃SOCH₃	Cl	53%	47%
(CH₃)₂NCHO	Cl	54%	46%
(CH₃)₂NCHO	Br	67%	33%
(CH₃)₂NCHO	I	>99%	<1%

(Ref. 30e, f)

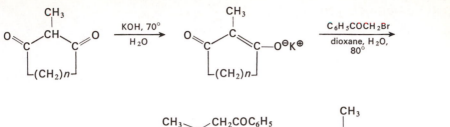

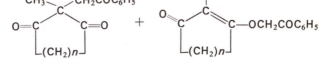

n	Total yield	Product composition	
2	41%	20%	80%
3	71%	52%	48%

(Ref. 33a)

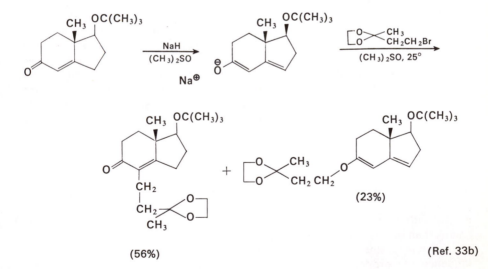

(56%)

(23%)

(Ref. 33b)

33. (a) D. Rosenthal and K. H. Davis, Jr., *J. Chem. Soc., C*, 1973 (1966). (b) Z. G. Hajos, R. A. Micheli, D. R. Parrish, and E. P. Oliveto, *J. Org. Chem.,* **32**, 3008 (1967). (c) B. Miller and H. Margulies, *ibid.,* **30**, 3895 (1965). (d) S. T. Yoffe, K. V. Vatsuro, E. E. Kugutcheva, and M. I. Kabachnik, *Tetrahedron Letters,* **No. 10**, 593 (1965). (e) R. Chong and P. S. Clezy, *ibid.,* **No. 7**, 741 (1966). (f) G. J. Heiszwolf and H. Kloosterziel, *Chem. Commun.,* **No. 2**, 51 (1966). (g) H. E. Zaugg, R. J. Michaels, and E. J. Baker, *J. Am. Chem. Soc.,* **90**, 3800 (1968).

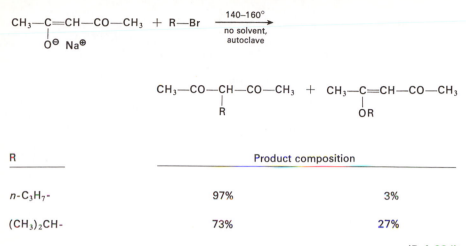

R	Product composition	
$n\text{-}C_3H_7\text{-}$	97%	3%
$(CH_3)_2CH\text{-}$	73%	27%

(Ref. 33d)

As certain of the preceding examples suggest, the choice of the alkylating agent may exert a significant influence on the proportions of C- and O-alkylated products. Variations in the leaving group of the alkylating agent are usually found to favor C-alkylation in the order: $R\!-\!I > R\!-\!Br > R\!-\!Cl > R\!-\!O\!-\!SO_2\!-\!O\!-\!R$ and $R\!-\!OSO_2Ar > R_3O^{\oplus}BF_4^{\ominus}$.[2o,7b,25d,30c,e,f,h,i,31,33e,f] These observations have been correlated by the principle of hard and soft acids and bases which states that hard Lewis acids prefer to coordinate (i.e. bond) to hard Lewis bases and soft acids prefer to coordinate to soft bases.[29e] Soft Lewis acids and bases are characterized as being highly polarizable; usually the donor atoms in soft bases are large and in a low oxidation state with a relatively low electronegativity. Hard Lewis acids and bases have a low polarizability; the donor atom of hard bases is usually small and highly electronegative. Thus the common leaving groups of alkylating agents, listed in order of increasing hardness, would be: $-I < -Br < -Cl < -O\!-\!SO_2R$. The relatively small, electronegative oxygen atom of enolate anions and other ambident anions is harder than the larger, less electronegative, and more polarizable carbon atom. This principle correlates both the effect of variations in the leaving group of the alkylating agent with the proportion of C- and O-alkylation and also the fact that, at least in some solvents,[31a] sulfates and sulfonates react much more rapidly with the hard oxygen donor atom of alkoxide, phenoxide, and carboxylate anions than do the corresponding alkyl bromides and iodides.[29e]

Changes in the structure of the alkyl group of an alkylating agent may also influence the proportions of C- and O-alkylated products. Usually more O-alkylation is obtained with secondary alkyl halides than with primary alkyl halides;[30c,g,33d] allyl and benzyl halides tend to give less O-alkylation than the corresponding less polarizable (harder) saturated alkyl halides.[30e,33c] The accompanying equation illustrates the unusual orientation observed when the highly polarizable triphenylmethyl cation is allowed to react with a sterically hindered enolate anion.

$$(CH_3)_3C\text{—}CO\text{—}CH_2\text{—}CO\text{—}C(CH_3)_3 \xrightarrow[\substack{(CH_3)_2NCHO,\\ C_6H_6}]{NaH}$$

$$(CH_3)_3C\text{—}\underset{\underset{O^{\ominus}\ \ Na^{\oplus}}{|}}{C}\text{=}CH\text{—}CO\text{—}C(CH_3)_3 \xrightarrow[\substack{(CH_3)_2NCHO,\\ C_6H_6,\ 29\text{–}36°}]{(C_6H_5)_3C^{\oplus}BF_4^{\ominus}}$$

$$(C_6H_5)_2C=\!\!\left\langle\!\!\!\begin{array}{c} H \\ CH[COC(CH_3)_3]_2 \end{array}\!\!\!\right. \xrightarrow{\text{isomerization}}$$

$$(C_6H_5)_2CH\text{—}\!\left\langle\!\!\!\right\rangle\!\!\text{—}CH[COC(CH_3)_3]_2$$

(72%)   (Ref. 33g)

An unexpectedly large tendency to form C-alkylated products has been shown to result from a change in the reaction mechanism when p-nitrobenzyl chloride is used as an alkylating agent.[30d] As the following equation indicates, there is superimposed upon the usual bimolecular nucleophilic displacement (S_N2) process a much faster electron-transfer reaction (see Chapter 3). Unless the p-nitrobenzyl radical formed in this way is intercepted by some additive such as copper(II) chloride or p-dinitrobenzene, it attacks more enolate anion to propagate the indicated chain reaction.

Studies[11,2a–c,11a,20c,30e,31,34] of alkylations in dipolar, aprotic solvents such as hexamethylphosphoramide, dimethylformamide, dimethyl sulfoxide, and 1,2-dimethoxyethane have demonstrated that very substantial increases in the rates of reaction of enolate (or analogous) anions with alkylating agents result from their use in preference to alcohols or inert solvents. Their advantage over protonic solvents (e.g., ethanol) lies in the fact that they presumably do not solvate the enolate anion and, consequently, do not diminish its reactivity as a nucleophile.[31] On the other hand, these aprotic solvents do have the ability to solvate the cation, separating it from the cation–enolate anion pair [29a] and leaving a relatively free anion [29b–29d] in the reaction mixture.

Conformationally mobile enolate anions derived from 1,3-dicarbonyl compounds can, in principle, adopt any one of the three planar conformations [29b], [29c], or [29d], termed U-, W-, and sickle-shaped, respectively.[7i,20c] Of these

34. (a) H. E. Zaugg, B. W. Horrom, and S. Borgwardt, J. Am. Chem. Soc., 82, 2895 (1960). (b) H. E. Zaugg, ibid., 82, 2603 (1960); 83, 837 (1961). (c) H. E. Zaugg, D. A. Dunnigan, R. J. Michaels, L. R. Swett, T. S. Wang, A. H. Sommers, and R. W. Denet, J. Org. Chem., 26, 644 (1961). (d) J. J. Bloomfield, ibid., 26, 4112 (1961). (e) R. Fuchs, G. E. McCrary, and J. J. Bloomfield, J. Am. Chem. Soc., 83, 4281 (1961). (f) H. D. Zook and T. J. Russo, ibid., 82, 1258 (1960). (g) H. D. Zook and W. L. Gumby, ibid., 82, 1386 (1960). (h) K. Bowden and R. S. Cook, J. Chem. Soc., B, 1529 (1968). (i) For a study of the effect of added salts on the rate and position of enolate alkylation, see G. Bram, F. Guibe, and M. F. Mollet, Tetrahedron Letters, No. 34, 2951 (1970).

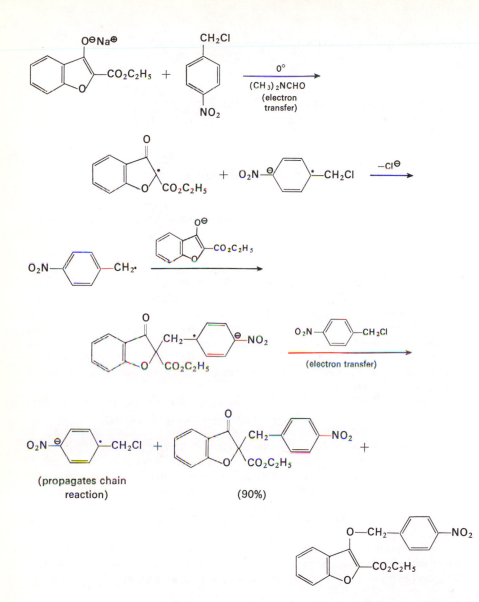

(90%)

(2%)   (Ref. 30d)

conformations only the U-shaped enolate **[29b]** is capable of serving as a bidentate ligand for association with a metal ion as indicated in structure **[29a]**. However, the most stable form of the free enolate anion is expected to be the W-shaped conformer **[29c]** in which dipole-dipole repulsion is minimized.[20c] In both inert solvents and protic solvents such as alcohols, such conformationally mobile enolate anions appear to exist primarily in the associated U-shaped form **[29a]**.

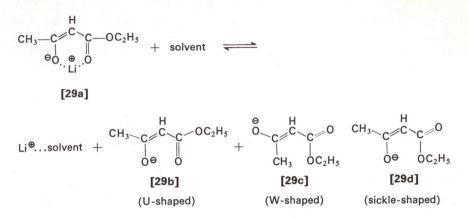

[29a]

[29b]   [29c]   [29d]

(U-shaped)   (W-shaped)   (sickle-shaped)

In very polar aprotic solvents, such as dimethyl sulfoxide or hexamethylphosphoramide, which can strongly solvate metal cations, the W-shaped conformer [29c] appears to be preferred.[35d] With compounds such as cyclic β-keto esters, in which only the U-shaped [29b] and sickle-shaped [29c] conformers are geometrically possible, the associated U-shaped ion pairs [29a] appear to be the predominant conformers in all solvents.[7i,20c]

Support for the idea that enolate anions exist as ion pairs with cations is also found in observations[2a−c,20c,30e,31,34g,35] that the type and degree of reactivity of an enolate ion is often influenced by the nature of the cation present; the lithium cation forms more tightly associated ion pairs than sodium or potassium. Usually the free enolate anions are more reactive as nucleophiles than are the associated ion pairs (e.g., [29a]);[31,34] however, as noted earlier, this increased nucleophilic reactivity is accompanied by an increased tendency to form O-alkylated products. Consequently, the selection of conditions which favor the presence of unassociated enolate anions may be a poor choice for preparative reactions if O-alkylation is apt to be a significant side reaction.

The reaction of the associated ion pairs with alkylating agents to form C-alkylated products has been suggested to incorporate electrophilic catalysis by the metal cation in a six-membered cyclic transition state.[35a] Although this idea could account for the variations in reaction rate observed with different cations, such a cyclic process can be criticized for requiring a energetically unfavorable nonlinear arrangement of the entering nucleophile, the carbon atom where displacement is occurring, and the leaving halide ion. This criticism might be avoided if the reacting enolate were considered to be an aggregate of two, three, or more ion pairs, as seems to be true for the metal enolates of monoketones.[2b,c,7h,34g]

The variations in reactivity of enolates with an alkylating agent which are observed with changes in the structure of the enolate anions are difficult to

35. (a) A. Brändström, *Acta Chem. Scand.,* **7**, 223 (1953) ; A. Brändström, *Arkiv. Kemi,* **6**, 155 (1954) ; **7**, 81 (1954) ; **11**, 567 (1957) ; **13**, 51 (1958). (b) K. G. Hampton, T. M. Harris, and C. R. Hauser, *J. Org. Chem.,* **28**, 1946 (1963) ; **31**, 1035 (1966). (c) D. Caine and B. J. L. Huff, *Tetrahedron Letters,* **No. 39**, 4695 (1966) ; **No. 35**, 3399 (1967). (d) B. Miller, H. Margulies, T. Drabb, Jr., and R. Wayne, *ibid.,* **No. 43**, 3801, 3805 (1970).

disentangle from effects arising from differences in the extent of dissociation of ion pairs and differences in the degree of aggregation.[11,2b,c,20c] However, it is usually true that C-alkylation increases with significant decreases in the acidity of the active methylene compound.[11] Presumably in the strongly acidic compounds the negative charge is extensively delocalized on atoms other than carbon, so that the localization of the electron pair at carbon required to form a new carbon-carbon bond is associated with the loss of a significant amount of resonance energy.

Although the substitution of alkyl groups at the α-carbon of enolates might be expected to retard the rate of reaction of enolate anions with alkylating agents for steric reasons, this effect may[2b] or may not[35b,c] be observed. In some cases, the less highly substituted enolate anions appear also to be more highly associated[7h] with the overall result that the less highly substituted enolates react more slowly than analogs having more alkyl substituents.

The alkylating agent frequently employed for reaction with an enolate (or analogous) anion is an alkyl halide. As has been illustrated in previous equations, both primary and secondary halides may be used successfully, as may primary and secondary allylic and benzylic halides. However, tertiary alkyl halides having at least one beta hydrogen atom (e.g., [30]) are usually of little value as alkylating

$$(CH_3)_3C-Br \; + \; CH_2(CO_2C_2H_5)_2 \; \xrightarrow[\substack{C_2H_5OH \\ 5-25°}]{NaOC_2H_5} \; (CH_3)_3C-CH(CO_2C_2H_5)_2 \; + \; (CH_3)_2C{=}CH_2$$

[30]   (6%)   (Ref. 36a)

agents because the major reaction that occurs when these materials are treated with enolate anions is a bimolecular elimination ($E2$), illustrated in the accompanying equation, leading to formation of an olefin. Dehydrohalogenation may also be a serious side reaction with certain secondary alkyl halides (e.g., [31]) and leads to diminished yields of alkylated products. The proportion of dehydrohalogenation to alkylation appears to increase with increasing basicity of the enolate anion (i.e. decreasing acidity of the active methylene compound); with strongly basic enolates, elimination may become the major reaction, even with primary and secondary alkyl halides.[2b]

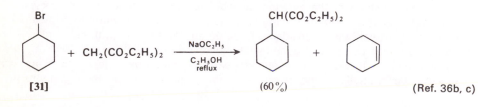

[31]   (60%)   (Ref. 36b, c)

---

36. (a) A. W. Dox and W. G. Bywater, *J. Am. Chem. Soc.,* **58**, 731 (1936). (b) M. Kopp and B. Tchoubar, *Bull. Soc. Chim. France,* 30 (1951). (c) J. F. Eykman, *Chem. Weekblad,* **6**, 699 (1909) ; G. S. Fonken and W. S. Johnson, *J. Am. Chem. Soc.,* **74**, 831 (1952). (d) E. Bowden, *ibid.,* **60**, 131 (1938).

$$(CH_3)_3C-\underset{\underset{O^{\ominus}\quad Na^{\oplus}}{|}}{C}=CH_2 \;+\; n\text{-}C_4H_9-Br \;\xrightarrow[\text{diglyme}]{25°}$$

$$(CH_3)_3C-\underset{\underset{O}{\|}}{C}CH_2C_4H_9 \;+\; (CH_3)_3CCOCH_3 \;+\; CH_3CH_2CH=CH_2$$

(24% of product)   (76% of product)   (Ref. 2b)

In certain cases, where competing O-alkylation is not a serious problem, sulfates and arenesulfonates are more useful alkylating agents than the corresponding alkyl halides.[37] For example, the commercially available methyl p-toluenesulfonate, dimethyl sulfate, and diethyl sulfate, being less volatile than the methyl and ethyl halides, can be used to advantage in alkylation reactions that require elevated temperatures, eliminating the need for sealed reaction vessels or pressure equipment to prevent the loss of the alkylating agent.[36d] Use of the p-toluenesulfonate rather than the halide also has a very real advantage when the alkylating agent is not commercially available and must be prepared from the corresponding alcohol. As discussed in Chapter 8, conversion of an alcohol to a halide may be complicated by molecular rearrangement, and the stereochemical relationship of the alcohol to the halide is frequently uncertain.

The following equations exemplify the preparation and use of p-toluenesulfonate esters to avoid these difficulties. The reaction of arenesulfonic ester [32] illustrates the general observation[11a,37d,38a] that alkylation occurs with inversion of the configuration of the alkylating agent, as expected for an S_N2 reaction. Arenesulfonic esters offer the additional advantage that they are frequently crystalline solids, which may be readily purified before use. The relative reactivities of the common alkylating agents follow approximately the order $R-Cl < RO-SO_2-$

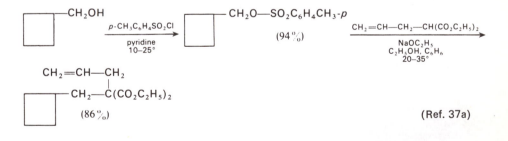

37. (a) W. Braker, E. J. Pribyl, and W. A. Lott, *J. Am. Chem. Soc.*, **69**, 866 (1947). (b) Procedures for the preparation of very unstable p-toluenesulfonates such as those derived from tertiary alcohols have been described by H. M. R. Hoffmann, *J. Chem. Soc.*, 6748, 6753, 6762 (1965); R. M. Coates and J. P. Chen, *Tetrahedron Letters*, **No. 32**, 2705 (1969). (c) For the preparation of various water-soluble alkanesulfonic esters as alkylating agents, see P. Blumbergs and co-workers, *J. Org. Chem.*, **34**, 4065 (1969); A. B. Ash and co-workers, *ibid.*, **34**, 4070 (1969). (d) J. A. Marshall and R. D. Carroll, *ibid.*, **30**, 2748 (1965); J. A. Marshall, M. T. Pike, and R. D. Carroll, *ibid.*, **31**, 2933 (1966).

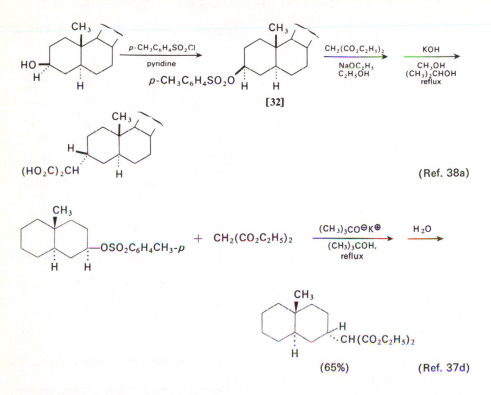

(Ref. 38a)

(65%)   (Ref. 37d)

$C_6H_4CH_3$-p < R—Br < R—I < R—O—SO$_2$—O—R (for displacement of the first alkyl group).[7b,31a,37b,39]

In circumstances in which highly reactive alkylating agents are required, the trialkyloxonium salts [e.g. $(CH_3)_3O^{\oplus}BF_4^{\ominus}$ or $(C_2H_5)_3O^{\oplus}BF_4^{\ominus}$][38b–d] or the dialkoxycarbonium ion salts[38e] should be considered. The use of these reagents to alkylate enolate anions is not an entirely satisfactory procedure, however, both because of the occurrence of competing O-alkylation, even with simple ketones,[7b,c,33f] and because these oxonium salts rapidly alkylate many of the common reaction solvents, including 1,2-dimethoxyethane, dimethylformamide, dimethyl sulfoxide, and hexamethylphosphoramide.

38. (a) C. W. Shoppee and R. J. Stephenson, *J. Chem. Soc.*, 2230 (1954). (b) For reviews of the preparation and use of trialkyloxonium salts, see H. Meerwein in E. Muller, ed., *Methoden der organischen Chemie* (Houben-Weyl), Vol. 6, Pt. 3, Georg Thieme Verlag, Stuttgart, Germany, 1965, p. 325. (c) Preparations of trimethyloxonium fluoroborate and triethyloxonium fluoroborate are described in Refs. 30c, 38b, and by H. Meerwein, *Org. Syn.*, **46**, 113, 120 (1966). (d) The preparation of trimethyloxonium 2,4,6-trinitrobenzene-sulfonate is described by G. K. Helmkamp and D. J. Pettitt, *ibid.*, **46**, 122 (1966); *J. Org. Chem.*, **28**, 2932 (1963); **29**, 2702 (1964). (e) For the preparation and use of dialkoxy-carbonium ions as alkylating agents, see S. Kabuss, *Angew. Chem., Intern. Ed. Engl.*, **5**, 675 (1966); K. Dimroth and P. Heinrich, *ibid.*, **5**, 676 (1966).
39. (a) J. M. Conia, *Record Chem. Progr.*, **24**, 43 (1963). (b) J. M. Conia, *Bull. Soc. Chim. France*, 533, 537 (1950). (c) J. M. Conia, *ibid.*, 1040 (1956).

Since the unsuitability of tertiary halides and tertiary sulfonates[37b] as alkylating agents for enolate anions results from a competing dehydrohalogenation reaction, it is possible to solve this problem with an acid-catalyzed procedure in which the enol (not enolate) is attacked by a potential tertiary carbonium ion. Such a reaction has been realized in the alkylation of β-keto esters and β-diketones with secondary and tertiary alcohols and ethers in the presence of boron trifluoride.[40a−c] Although this procedure gave satisfactory yields with secondary alcohols, the yields were poor (6–14%) with tertiary alcohols. By modifying the reaction procedure so

$$CH_3\text{—}CO\text{—}CH_2\text{—}CO_2C_2H_5 \ + \ (CH_3)_2CHOH \ \xrightarrow[0-7°]{BF_3} \ CH_3\text{—}C\underset{\underset{\text{CH}}{\diagdown}}{\overset{\overset{O\text{—}BF_2\cdots O}{|}}{\quad}}C\text{—}OC_2H_5 \ \xrightarrow[H_2O]{NaOCOCH_3}$$

$$(CH_3)_2CH\text{—}O\overset{\oplus}{\diagup}\underset{BF_3}{\overset{\diagup H}{\diagdown}{\ominus}}$$

$$\underset{\begin{array}{c}(CH_3)_2CH\\|\\CH_3\text{—}CO\text{—}CH\text{—}CO_2C_2H_5\\(60\text{–}67\%)\end{array}}{}$$

(Ref. 40a–c)

that the t-alkyl carbonium ion is generated in nitromethane solution from an alkyl halide and boron trifluoride, aluminum chloride, or silver perchlorate or from an olefin and perchloric acid, satisfactory yields of t-alkyl derivatives have been obtained with malononitrile, ethyl cyanoacetate, diethyl malonate, ethyl acetoacetate and pentane-1,3-dione.[40e]

An alternative route to t-alkyl derivatives of malonic and cyanoacetic esters involves the subsequently discussed aldol condensation (see Chapter 10) of

$$CH_3COCH_2CO_2C_2H_5 \ + \ (CH_3)_3CBr \ \xrightarrow[CH_3NO_2, 0°]{AgClO_4}$$

$$CH_3\text{—}C\underset{\underset{H}{\curvearrowleft}}{\overset{\overset{O\text{—}H\cdots O}{|}}{\quad}}C\text{—}OC_2H_5 \ \xrightarrow[H_2O]{NaHCO_3} \ (CH_3)_3C\text{—}\overset{\overset{COCH_3}{|}}{C}HCO_2C_2H_5$$

$$(CH_3)_3C^{\oplus}ClO_4^{\ominus}$$

(68%) (Ref. 40e)

40. (a) J. T. Adams, R. Levine, and C. R. Hauser, *Org. Syn.*, **Coll. Vol. 3**, 405 (1955). (b) J. T. Adams, B. Abramovitch, and C. R. Hauser, *J. Am. Chem., Soc.* **65**, 552 (1943). (c) T. F. Crimmins and C. R. Hauser, *J. Org. Chem.*, **32**, 2615 (1967). For further study of the reactions of borofluoride complexes of 1,3-dicarbonyl compounds, see R. A. J. Smith and T. A. Spencer, *ibid.*, **35**, 3220 (1970). (d) N. C. Deno, W. E. Billups, R. E. DiStefano, K. M. McDonald, and S. Schneider, *ibid.*, **35**, 278 (1970). (e) P. Boldt and W. Thielecke, *Angew. Chem., Intern. Ed. Engl.*, **5**, 1044 (1966); P. Boldt and H. Militzer, *Tetrahedron Letters*, **No. 30**, 3599 (1966); P. Boldt, H. Militzer, W. Thielecke, and L. Schulz, *Justus Liebigs Ann. Chem.*, **718**, 101 (1968); P. Boldt, A. Ludwieg, and H. Militzer, *Chem. Ber.*, **103**, 1312 (1970). (f) E. S. Prout, E. P. Y. Huang, R. J. Hartman, and C. J. Korpics, *J. Am. Chem. Soc.*, **76**, 1911 (1954). (g) E. L. Eliel, R. O. Hutchins, and Sr. M. Knoeber, *Org. Syn.*, **50**, 38 (1970).

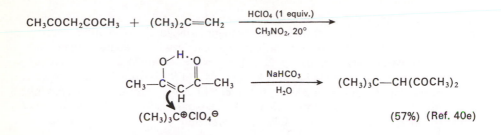

these active methylene compounds with ketones to form alkylidene derivatives such as [**33**]. Reaction of these α,β-unsaturated esters with Grignard reagents, particularly in the presence of copper(I) salts, or with lithium dialkylcuprates (R_2CuLi), results in the indicated conjugated addition to form the enolates of t-alkylmalonates or cyanoacetates.[11a,40f,g]

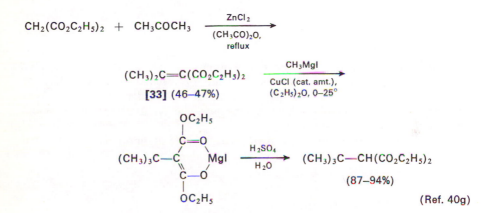

As noted previously (Chapter 6), epoxides may also serve as alkylating agents for active methylene compounds.[10a] When an ester function is present (e.g., [**34**]),

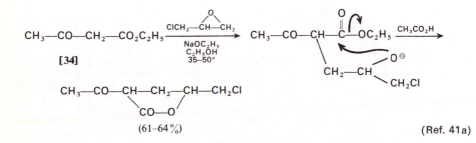

---

41. (a) G. D. Zuidema, E. van Tamelen, and G. Van Zyl, *Org. Syn.*, **Coll. Vol. 4**, 10 (1963).
(b) C. H. DePuy, F. W. Breitbeil, and K. L. Eilers, *J. Org. Chem.*, **29**, 2810 (1964).

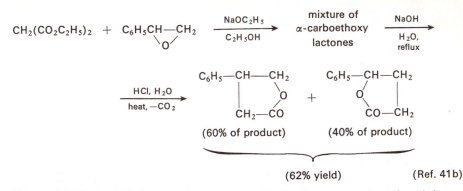

the product isolated is frequently a lactone. The further conversion of such keto-lactones to γ-haloketones is illustrated below.

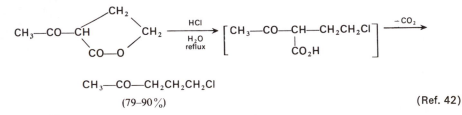

$$CH_3-CO-CH_2CH_2CH_2Cl$$

(79–90%)   (Ref. 42)

In reactions with polarizable nucleophiles (soft Lewis bases) such as enolate anions, epoxides are less reactive alkylating agents than the corresponding alkyl bromides. This reactivity order is illustrated by the following reaction of epoxyalkyl

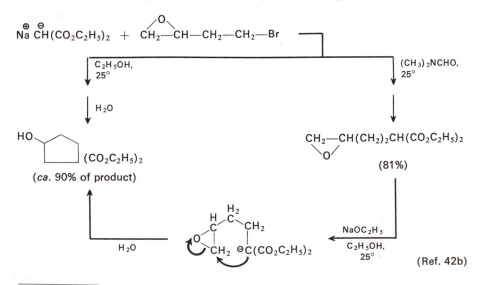

42. (a) G. W. Cannon, R. C. Ellis, and J. R. Leal, *Org. Syn.,* **Coll. Vol. 4,** 597 (1963). (b) P. A. Cruickshank and M. Fishman, *J. Org. Chem.,* **34,** 4060 (1969).

bromides with the enolate of diethyl malonate in which displacement of bromide ion is the more rapid process.

Under the normal conditions employed for the alkylation of active methylene compounds, vinyl and aryl halides usually fail to react, a result that would be anticipated from the failure of these halides to undergo other bimolecular nucleo-philic displacements. However, aryl halides that are activated by the presence of electron-withdrawing groups at the *ortho* and/or *para* positions (e.g., [**35**]) will

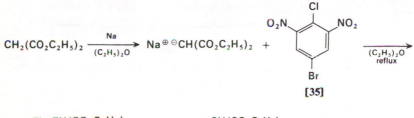

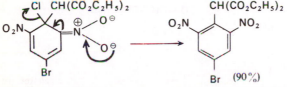

act as alkylating agents. Such alkylations are believed to proceed by the initial addition of the enolate anion to the aromatic system and subsequent elimination of halide ion as illustrated;[44] a mechanistically similar reaction is possible with vinyl halides that have an electron-withdrawing group at the beta carbon atom. A variant of the usual alkylation procedure permits the formation of active methylene compounds having vinyl substituents. An alkylidene derivative (e.g., [**36**]) of

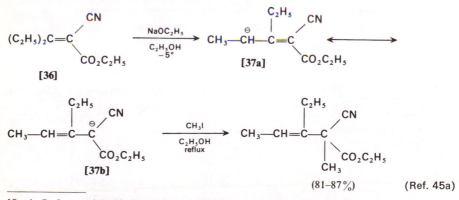

---

43. A. B. Sen and P. M. Bhargava, *J. Indian Chem. Soc.*, **24**, 371 (1947).
44. J. F. Bunnett, *Quart. Rev.*, **12**, 1 (1958).
45. (a) E. M. Hancock and A. C. Cope, *Org. Syn.*, **Coll. Vol. 3**, 397 (1955). (b) D. Nasipuri, G. Sarkar, M. Guha, and R. Roy, *Tetrahedron Letters*, **No. 9**, 927 (1966). See also L. I. Smith and G. F. Rouault, *J. Am. Chem. Soc.*, **65**, 631 (1943); A. J. B. Edgar, S. H. Harper, and M. A. Kazi, *J. Chem. Soc.*, 1083 (1957).

the active methylene compound is treated with a metal alkoxide to form the corresponding enolate anion; reaction of this anion with an alkylating agent usually introduces an alkyl substituent at the alpha position, a result reminiscent of the previous discussion (Chapter 3) concerning the site of kinetically controlled protonation of pentadienyl anions. A comparable result will also be found in the subsequently discussed alkylations of α,β-unsaturated ketones. In the alkylation of enolate anions such as that derived from Hagemann's ester [37], alkylation occurs alpha to each of the activating groups.

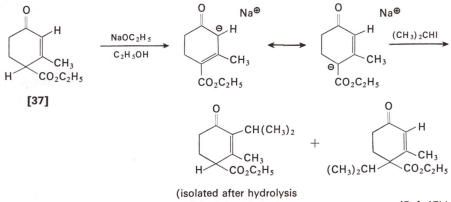

[37]

(isolated after hydrolysis
and decarboxylation)   (Ref. 45b)

By using different reaction conditions, namely treatment of an aryl halide (e.g., [38]) with a solution of an enolate anion and an excess of a strong base such

$$CH_2(CO_2C_2H_5)_2 \xrightarrow[\text{liq. NH}_3]{\text{4 equiv. NaNH}_2} {}^{\ominus}CH(CO_2C_2H_5)_2 +$$

(2 equiv.)   Na${}^{\oplus}$

[38]

(1 equiv.)

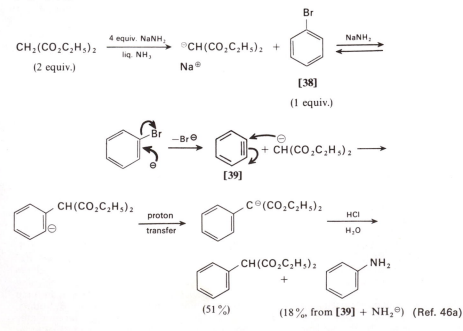

[39]

proton
transfer

HCl
H$_2$O

(51%)   (18%, from [39] + NH$_2{}^{\ominus}$) (Ref. 46a)

as sodium amide in liquid ammonia, it is possible to obtain fair yields of arylated products. This reaction is not a nucleophilic displacement but rather follows an elimination-addition sequence in which an intermediate [39], called a benzyne or a 1,2-dehydrobenzene[44,47] is involved. Analogous intermediates, called hetarynes, may also be formed from heterocyclic aryl halides.[47d] These highly reactive intermediates are capable not only of reacting with various nucleophiles but also of undergoing various cycloaddition reactions including dimerization. The following equations illustrate some of these possibilities as well as other preparative routes

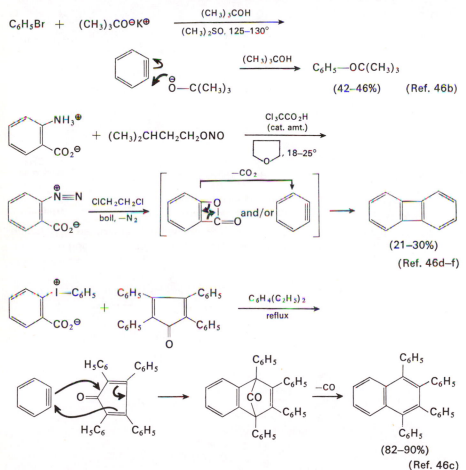

46. (a) W. W. Leake and R. Levine, *J. Am. Chem. Soc.,* **81**, 1169, 1627 (1959). (b) M. R. V. Sahyun and D. J. Cram, *Org. Syn.,* **45**, 89 (1965); J. I. G. Cadogan, J. K. A. Hall, and J. T. Sharp, *J. Chem. Soc.,* C, 1860 (1967). (c) L. F. Fieser and M. J. Haddadin, *Org. Syn.,* **46**, 107 (1966). (d) F. M. Logullo, A. H. Seitz, and L. Friedman, *ibid.,* **48**, 12 (1968); L. Friedman and F. M. Logullo, *J. Org. Chem.,* **34**, 3089 (1969). (e) R. Gompper, G. Seybold, and B. Schmolke, *Angew. Chem., Intern. Ed. Engl.,* **7**, 389 (1968). (f) Even tetrahydrofuran adds to benzyne in the absence of other more reactive nucleophiles. E. Wolthuis, B. Bouma, J. Modderman, and L. Sytsma, *Tetrahedron Letters,* **No. 6**, 407 (1970). (g) Benzyne may also be generated by the oxidation of l-aminobenzotriazole with lead tetraacetate or other

which have been used to generate arynes.[46] It will be noted that mixtures of structural isomers will usually be obtained if a substituted aryl halide is employed.

The following examples illustrate the generalization[47g,h] that nucleophilic anions tend to add to the triple bond in such a way that the resulting negative

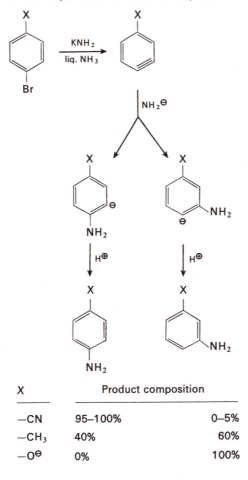

X	Product composition	
—CN	95–100%	0–5%
—CH$_3$	40%	60%
—O$^{\ominus}$	0%	100%

(Ref. 47h)

oxidants. C. D. Campbell and C. W. Rees, *J. Chem. Soc.*, C, 742, 748, 752 (1969). A similar procedure has been used to generate 1,8-dehydronaphthalene. C. W. Rees and R. C. Storr, *ibid.*, C, 756, 760, 765 (1969); R. W. Hoffmann, G. Guhn, M. Preiss, and B. Dittrich, *ibid.*, C, 769 (1969). (h) For the generation of benzyne from o-carboxyphenyl-triazene derivatives, see J. Nakayama, O. Simamura, and M. Yoshida, *Chem. Commun.*, No. 18, 1222 (1970).
47. (a) R. Huisgen and J. Sauer, *Angew. Chem.*, **72**, 91 (1960). (b) H. Heaney, *Chem. Rev.*, **62**, 81 (1962). (c) G. Wittig, *Angew. Chem., Intern. Ed. Engl.*, **4**, 731 (1965). (d) Th. Kauffmann, *ibid.*, **4**, 543 (1965). (e) R. S. Berry, J. Clardy, and M. E. Schafer, *J. Am. Chem. Soc.*, **86**, 2738 (1964). (f) B. H. Klanderman and T. R. Criswell, *ibid.*, **91**, 510 (1969). (g) J. D. Roberts, C. W. Vaughan, L. A. Carlsmith, and D. A. Semenow, *ibid.*, **78**, 611 (1956). (h) G. B. R. deGraaff, H. J. denHertog, and W. Ch. Melger, *Tetrahedron Letters*, No. 15, 963 (1965).

charge is nearer to electron-attracting substituents and further from electron-donating substituents.

A useful synthetic application of this reaction is the intramolecular addition of an anion to a benzyne intermediate as shown in the following equations.

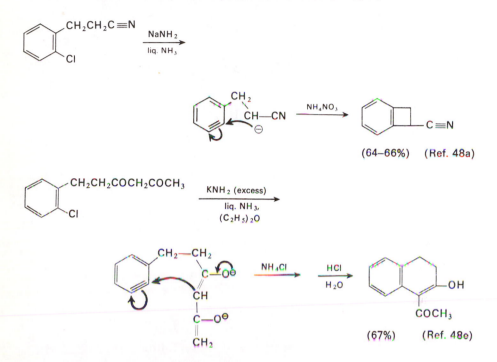

(64–66%)   (Ref. 48a)

(67%)   (Ref. 48e)

Aryl groups may also be introduced at enolate carbon atoms by reaction with diaryliodonium salts.[49] The following equations indicate one preparative route to these salts and an example of their use as arylating agents. The suggested mechanism[49c-e] for this process, an electron transfer followed by coupling of the resultant radicals, is similar to the previously discussed reaction path believed operative in alkylations with p-nitrobenzyl chloride.

48. (a) J. A. Skorcz and F. E. Kaminski, *Org. Syn.*, **48**, 53 (1968); J. F. Bunnett and J. A. Skorcz, *J. Org. Chem.*, **27**, 3836 (1962). (b) J. F. Bunnett, T. Kato, R. R. Flynn, and J. A. Skorcz, *ibid.*, **28**, 1 (1963). (c) See also J. F. Bunnett, B. F. Hrutfiord, and S. M. Williamson, *Org. Syn.*, **40**, 1 (1960). (d) T. M. Harris and C. R. Hauser, *J. Org. Chem.*, **29**, 1391 (1964). (e) C. F. Beam, R. L. Bissell, and C. R. Hauser, *ibid.*, **35**, 2083 (1970).
49. (a) F. M. Beringer and R. A. Nathan, *J. Org. Chem.*, **34**, 685 (1969). (b) D. J. LeCount and J. A. W. Reid, *J. Chem. Soc.*, C, 1298 (1967). (c) F. M. Beringer and R. A. Falk, *ibid.*, 4442 (1964). (d) F. M. Beringer and P. S. Forgione, *Tetrahedron*, **19**, 739 (1963); *J. Org. Chem.*, **28**, 714 (1963). (e) F. M. Beringer, W. J. Daniel, S. A. Galton, and G. Rubin, *ibid.*, **31**, 4315 (1966). (f) K. G. Hampton, T. M. Harris, and C. R. Hauser, *ibid.*, **29**, 3511 (1964). (g) D. F. Banks, *Chem. Rev.*, **66**, 243 (1966).

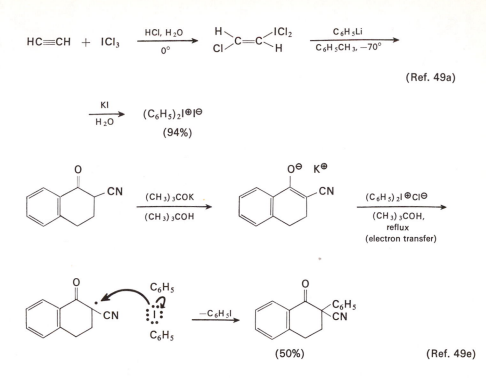

(Ref. 49a)

(Ref. 49e)

The formation of cyclic compounds by the alkylation of active methylene compounds, illustrated in the following equations, has found widespread use.[11a,50a]

$$Br(CH_2)_4Br + CH_3COCH_2CO_2C_2H_5 \xrightarrow[\substack{C_2H_5OH \\ 100°}]{NaOC_2H_5}$$ [cyclopentane ring with COCH$_3$ and CO$_2$C$_2$H$_5$]   (Ref. 50b)

$$\underset{\substack{| \\ (CH_2)_3 \\ |}}{CH(CO_2C_2H_5)_2} \quad + \quad CH_2I_2 \xrightarrow[\substack{C_2H_5OH \\ 100°}]{NaOC_2H_5}$$ [cyclohexane ring with (CO$_2$C$_2$H$_5$)$_2$ and (CO$_2$C$_2$H$_5$)$_2$]   (Ref. 51a)

$$CH(CO_2C_2H_5)_2$$

50. (a) R. C. Fuson in H. Gilman, ed., *Organic Chemistry*, Vol. 1, Wiley, New York, 1943, pp. 82–88. (b) L. J. Goldsworthy, *J. Chem. Soc.*, 377 (1934). (c) A. C. Knipe and C. J. M. Stirling, *ibid.*, B, 67 (1968); B, 808 (1967); R. Bird and C. J. M. Stirling, *ibid.*, B, 111 (1968).
51. (a) W. H. Perkin, *J. Chem. Soc.*, **59**, 798 (1891). (b) R. A. Bartsch and D. M. Cook, *J. Org. Chem.*, **35**, 1714 (1970). (c) S. J. Etheredge, *ibid.*, **31**, 1990 (1966). (d) C. F. Wilcox, Jr., and G. C. Whitney, *ibid.*, **32**, 2933 (1967). (e) C. H. Heathcock, *Tetrahedron Letters*, **No. 18**, 2043 (1966). (f) For other examples of ring closures to form bicyclic ketones, see E. N. Marvell, D. Sturmer, and C. Rowell, *Tetrahedron*, **22**, 861 (1966); R. D. Sands, *J. Org. Chem.*, **29**, 2488 (1964); **32**, 3681 (1967); **34**, 2794 (1969).

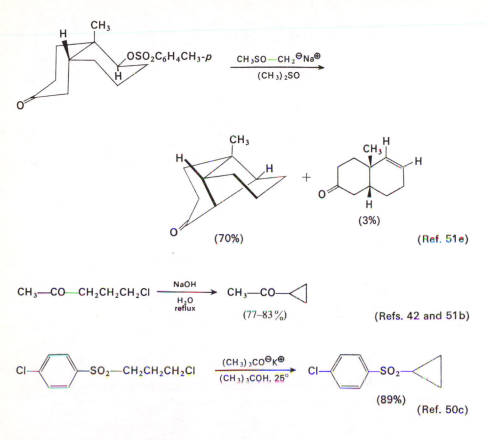

$$CH_3—CO—CH_2CH_2CH_2Cl \xrightarrow[\substack{H_2O \\ reflux}]{NaOH} CH_3—CO—\triangleleft$$

(77–83%)   (Refs. 42 and 51b)

$$Cl—\langle\ \rangle—SO_2—CH_2CH_2CH_2Cl \xrightarrow[(CH_3)_3COH,\ 25°]{(CH_3)_3CO^{\ominus}K^{\oplus}} Cl—\langle\ \rangle—SO_2—\triangleleft$$

(89%)

(Ref. 50c)

In general, the relative rates of closure for rings of varying size follow the order: $3 > 5 > 6 > 7 > 4$, 8 and larger, or intermolecular reactions.[50c,52] The accompanying examples indicate the very large differences in reaction rates which are observed in the formation of various ring sizes; the very high rate of closure of the three-membered ring accounts for the successful formation of the cyclopropyl derivatives illustrated under mild conditions from relatively weakly acidic active methylene compounds. The fact that four-membered rings are produced at a rate comparable to that of intermolecular alkylation reactions requires that they be prepared under carefully controlled conditions[53] to minimize the formation of acylic products.

---

52. (a) E. L. Eliel in M. S. Newman, ed., *Steric Effects in Organic Chemistry,* Wiley, New York, 1956, pp. 114–120. (b) E. L. Eliel, *Stereochemistry of Carbon Compounds,* McGraw-Hill, New York, 1962, pp. 198–202.
53. (a) J. Cason and H. Rapoport, *Laboratory Text in Organic Chemistry,* 2d ed., Prentice-Hall, Englewood Cliffs, N.J., 1962, pp. 401–408. (b) For a less satisfactory procedure, see G. B. Heisig and F. H. Stodola, *Org. Syn., Coll. Vol. 3*, 213 (1955). (c) Note that alkylation of acetoacetic ester with 1,3-dibromopropane does not yield a four-membered carbocyclic ring but rather a six-membered dihydropyran derivative that results from C-alkylation followed by O-alkylation.

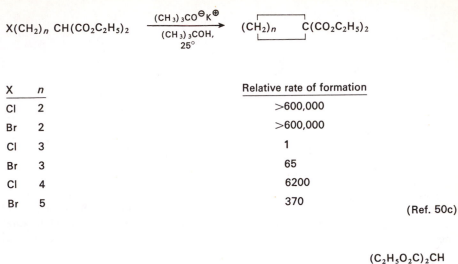

X	n	Relative rate of formation
Cl	2	>600,000
Br	2	>600,000
Cl	3	1
Br	3	65
Cl	4	6200
Br	5	370

(Ref. 50c)

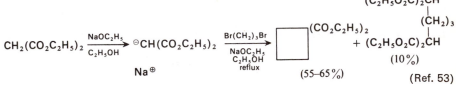

(Ref. 53)

The following examples illustrate the fact that if alternative modes of cyclization by either C- or O-alkylation exist, these reactions often occur more rapidly than the formation of cyclobutane derivatives.

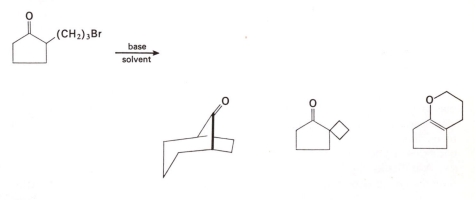

Reaction conditions	Product yields		
$(CH_3)_3COK$, C_6H_6, reflux	19%	30%	22%
KOH, H_2O, reflux	6%	13%	15%

(Ref. 51c, d)

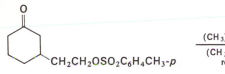

$$\xrightarrow[\substack{(CH_3)_3COH, \\ reflux}]{(CH_3)_3CO^{\ominus}K^{\oplus}}$$

(73%, isolated as the
2,4-dinitrophenylhydrazone)

(Ref. 51c)

Although the preparation of three-membered rings by alkylation is a kinetically favored process, the initial products are capable of reacting with an additional equivalent of the anion derived from the active methylene compound to form acyclic compounds, as illustrated below. The most successful preparations of

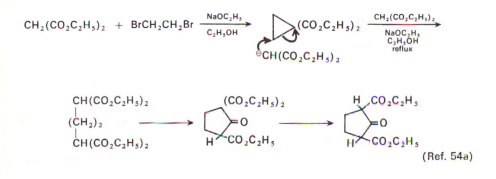

(Ref. 54a)

cyclopropane derivatives[42,55] utilize the intramolecular alkylation of weakly acidic active methylene compounds so that the concentration of enolate anion in solution is kept low and little attack on the cyclopropane ring is observed.

The reaction of anions from active methylene compounds with 1,2-dihalides may also be complicated by a competing elimination reaction which ultimately yields coupled products rather than cyclopropane derivatives.[11a,54b] This possibility, illustrated by the following example, is particularly favorable when 1,2-diiodides or ditertiary 1,2-dihalides are employed.

A difficulty always encountered in the alkylation of active methylene compounds having two or more acidic hydrogen atoms is the possibility of dialkylation.

54. (a) R. W. Kierstead, R. P. Linstead, and B. C. L. Weedon, *J. Chem. Soc.*, 3610, 3616 (1952). (b) W. G. Kofron and C. R. Hauser, *J. Org. Chem.*, **35**, 2085 (1970).
55. (a) C. M. McCloskey and G. H. Coleman, *Org. Syn.*, **Coll. Vol. 3**, 221 (1955). (b) M. J. Schlatter, *ibid.*, **Coll. Vol. 3**, 223 (1955).

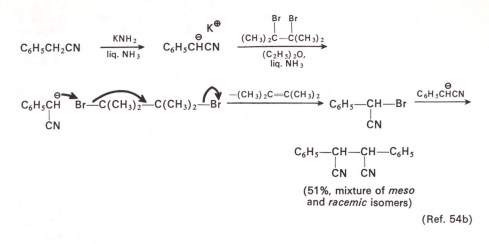

$$C_6H_5—CH—CH—C_6H_5$$
$$\quad\quad | \quad\quad |$$
$$\quad\quad CN \quad CN$$

(51%, mixture of *meso*
and *racemic* isomers)

(Ref. 54b)

This problem, illustrated in the accompanying equations, arises because the monoalkylated product is also acidic and will be in equilibrium with its enolate anion [**39**]. The possibility is usually not serious for the relatively acidic active

$$CH_2(CO_2C_2H_5)_2 + C_6H_5CH_2Cl \xrightarrow[\substack{C_2H_5OH \\ reflux}]{NaOC_2H_5} C_6H_5CH_2—CH(CO_2C_2H_5)_2 + (C_6H_5CH_2)_2C(CO_2C_2H_5)_2$$

[40]   (51–57%)

$$Na^{\oplus \ominus}CH(CO_2C_2H_5)_2 + C_2H_5OH \rightleftharpoons CH_2(CO_2C_2H_5)_2 + C_2H_5O^{\ominus}Na^{\oplus}$$

$$C_6H_5CH_2—CH(CO_2C_2H_5)_2 + C_2H_5O^{\ominus}Na^{\oplus} \rightleftharpoons C_6H_5CH_2—C^{\ominus}(CO_2C_2H_5)_2 + C_2H_5OH$$

[39] $Na^{\oplus}$

(Ref. 11a, 56a)

methylene compounds being discussed, since the solvent (ethanol) and the monoalkylated product are of similar acidic strength and both are weaker acids than the starting compound; as a result, the concentration of enolate anion derived from the monoalkylated product is relatively low. However, dialkylation does become a significant side reaction in the alkylation of active methylene compounds with benzyl halides, allyl halides, or α-haloketones, a fact possibly attributable to the greater acidity of the monoalkylated products as compared to that of the corresponding compounds derived from simple alkyl halides. For alkylations carried out with strong bases and nonhydroxylic solvents—such as the subsequently discussed alkylations of ketones and nitriles—dialkylation is almost always a significant side reaction. As would be predicted from the above equilibria, it is

56. (a) C. S. Marvel, *Org. Syn.,* **Coll. Vol. 3,** 705 (1955). (b) For a general discussion of the factors determining the composition of polyalkylated products, see J. E. Dubois and A. Panaye, *Tetrahedron Letters,* **No. 19,** 1501 (1969); **No. 38,** 3275 (1969).

possible to reduce the concentration of the monoalkylated enolate anion [39] (and consequently the amount of dialkylation) by adding an excess of the starting active methylene compound to the reaction mixture. For example, in the case cited, the yield of the monoalkylated product [40] can be increased from 57 to 85 per cent by using two equivalents of diethyl malonate with one equivalent of the halide and one equivalent of base.[11a] This procedure is desirable if the starting active methylene compound is readily available and easily separable from the product.

ALKYLATION OF KETONES AND NITRILES

From reference to Table 9–1 it will be apparent that a stronger base than sodium ethoxide and a less acidic solvent than ethanol must be used in order to obtain an appreciable concentration of an anion from a methylene group activated only by a nitrile or ketone function. The metal alkoxides derived from tertiary alcohols such as *t*-butyl alcohol or *t*-amyl alcohol are usually sufficiently basic to provide appreciable equilibrium concentrations of enolate anions from ketones, especially in cases where the anion is stabilized both by the carbonyl group and by an alpha vinyl or phenyl substituent. Alternatively, much stronger bases will convert either ketones or nitriles quantitatively to their anions. Table 9–4 lists some of the common bases and solvents that have been employed for the conversion of weakly acidic active methylene compounds to their anions. The accompanying equations illustrate the use of certain of these bases for alkylations.

57. Commercially available from the MSA Research Corp., Callery, Pennsylvania.
58. For preparations see (a) W. S. Johnson and W. P. Schneider, *Org. Syn.,* **Coll. Vol. 4**, 132 (1963). (b) W. S. Johnson and G. H. Daub, *Org. Reactions,* **6**, 41 (1951). (c) Ref. 6d.
59. (a) H. Rinderknecht, *J. Am. Chem. Soc.,* **73**, 5770 (1951). (b) F. Sondheimer and Y. Mazur, *ibid.,* **79**, 2906 (1957). (c) Y. Mazur and F. Sondheimer, *ibid.,* **80**, 5220, 6296 (1958). (d) H. J. Ringold and G. Rosenkrantz, *J. Org. Chem.,* **22**, 602 (1957). (e) M. Yanagita, M. Hirakura, and F. Seki, *ibid.,* **23**, 841 (1958). (f) N. W. Atwater, *J. Am. Chem. Soc.,* **82**, 2847 (1960). (g) Crystalline potassium *t*-butoxide has a tetrameric structure with the potassium and oxygen atoms located at alternate corners of a cube; E. Weiss, H. Alsdorf, and H. Kühr, *Angew. Chem., Intern. Ed. Engl.,* **6**, 801 (1967).
60. (a) J. M. Conia and P. Gosselin, *Bull. Soc. Chim. France,* 836 (1961). (b) J. M. Conia and F. Rouessac, *Tetrahedron,* **16**, 45 (1961). (c) J. M. Conia and A. Sandre-Le Craz, *Tetrahedron Letters,* **No. 12**, 505 (1962). (d) J. M. Conia and F. Rouessac, *Bull. Soc. Chim. France,* 1925, 1930 (1963).
61. Commercially available from Roberts Chemicals, Inc., Nitro, West Virginia.
62. (a) F. W. Bergstrom, *Org. Syn.,* **Coll. Vol. 3**, 778 (1955). (b) C. A. Vanderwerf and L. F. Lemmerman, *ibid.,* **Coll. Vol. 3**, 44 (1955). (c) E. M. Hancock and A. C. Cope, *ibid.,* **Coll. Vol. 3**, 219 (1955). (d) C. R. Hauser and W. R. Dunnavant, *ibid.,* **40**, 38 (1960). (e) G. Wash, B. Shive, and H. L. Lochte, *J. Am. Chem. Soc.,* **63**, 2975 (1941). (f) A. Haller and E. Bauer, *Ann. Chim. (Paris),* **(8)28**, 373 (1913). (g) The properties of alkali metal amides have been reviewed by R. Juza, *Angew. Chem., Intern. Ed. Engl.,* **3**, 471 (1964). (h) E. M. Kaiser, W. G. Kenyon, and C. R. Hauser, *Org. Syn.,* **47**, 72 (1967); W. G. Kenyon, E. M. Kaiser, and C. R. Hauser, *J. Org. Chem.,* **30**, 2937, 4135 (1965). (i) K. G. Hampton, T. M. Harris, and C. R. Hauser, *Org. Syn.,* **47**, 92 (1967). (j) W. S. Murphy, P. J. Hamrick, and C. R. Hauser, *ibid.,* **48**, 80 (1968). (k) P. G. Gassman and B. L. Fox, *J. Org. Chem.,* **31**, 982 (1966); H. L. Needles and R. E. Whitfield, *ibid.,* **31**, 989 (1966).
63. (a) C. R. Hauser and W. R. Dunnavant, *Org. Syn.,* **39**, 73 (1959). (b) S. Boatman, T. M. Harris, and C. R. Hauser, *ibid.,* **48**, 40 (1968).

Table 9-4 Commonly used strong bases

Base	References to preparation and use	Frequently used solvents
$(CH_3)_3C\overset{\ominus\ \oplus}{O}K$	6d, g–i, 46b, 57, 58, 59, 66a	t-Butyl alcohol, dimethyl sulfoxide, ether,[a] 1,2-dimethoxyethane,[a] benzene,[a] tetrahydrofuran
$(CH_3)_2\underset{\underset{C_2H_5}{\mid}}{\overset{\ominus\ \oplus}{C}}\!-\!ONa$	39, 60	Ether, benzene, toluene
$Na\overset{\oplus}{}\overset{\ominus}{N}H_2$	48a, 61, 62, 66a	Liquid ammonia, ether,[a] benzene,[a] toluene,[a] 1,2-dimethoxyethane[a]
$K\overset{\oplus}{}\overset{\ominus}{N}H_2$	35b, 62g, 63, 64	Liquid ammonia, ether[a]
$(C_2H_5)_2\overset{\ominus\ \oplus}{N}Li$	7b, 65, 66a	1,2-Dimethoxyethane, ether
$[(CH_3)_2CH]_2\overset{\ominus\ \oplus}{N}Li$	7c, 19b,c, 65h, 66a	1,2-Dimethoxyethane, ether, tetrahydrofuran
$[(CH_3)_3Si]_2\overset{\ominus\ \oplus}{N}Na$	7e, 65d	Tetrahydrofuran, ether, benzene
NaH	2a–c, 7a,b, 22, 34a,c,g, 66, 67	Ether,[a] benzene,[a] toluene,[a] xylene,[a] 1,2-dimethoxyethane,[a] dimethylformamide[a]
LiH	2a–c, 34g, 66a	Ether[a]
$CH_3SO\!-\!\overset{\ominus\ \oplus}{C}H_2Na$	10h,o, 34d, 68	Dimethyl sulfoxide (prepared by reaction of sodium hydride with excess dimethyl sulfoxide)
$(C_6H_5)_3\overset{\ominus\ \oplus}{C}Li$	7a,b, 69	Ether, 1,2-dimethoxyethane
$(C_6H_5)_3\overset{\ominus\ \oplus}{C}Na$	2a, 70, 71b	Ether, benzene, toluene, liquid ammonia
$(C_6H_5)_3\overset{\ominus\ \oplus}{C}K$	7a,b, 71	1,2-Dimethoxyethane
$\left[\text{naphthalene}^{\cdot}\right]^{\ominus}Na^{\oplus}$	72	1,2-Dimethoxyethane

[a] Solvents in which the base either is insoluble or is only slightly soluble.

Among these base-solvent combinations, sodium amide in liquid ammonia or in an inert solvent (usually ether or toluene) and the metal *t*-alkoxides (especially potassium *t*-butoxide in *t*-butyl alcohol) have enjoyed the most widespread use. The *t*-alkoxide bases suffer one major disadvantage: at equilibrium, most ketones are only partially converted to their enolate anions; as a result, an aldol condensation (see Chapter 10) may occur between the free ketone and its enolate anion,[39] especially if the intermediate aldol condensation product can be dehydrated by base to form an α,β-unsaturated ketone. Although the other bases listed in Table 9–4 are capable of converting ketones essentially quantitatively to their enolate anions, competing aldol condensation may also be a problem with sodium

64. (a) C. R. Hauser and T. M. Harris, *J. Am. Chem. Soc.*, **80**, 6360 (1958); **81**, 1154, 1160 (1959). (b) R. B. Meyer and C. R. Hauser, *J. Org. Chem.*, **25**, 158 (1960). (c) W. G. Kofron, W. R. Dunnavant, and C. R. Hauser, *ibid.*, **27**, 2737 (1962). (d) S. Boatman, T. M. Harris, and C. R. Hauser, *J. Am. Chem. Soc.*, **87**, 82 (1965).

65. (a) K. Ziegler and H. Ohlinger, *Justus Liebigs Ann. Chem.*, **495**, 84 (1932). (b) J. Cason, G. Sumrell, and R. S. Mitchell, *J. Org. Chem.*, **15**, 850 (1950). (c) A. C. Cope and B. D. Tiffany, *J. Am. Chem. Soc.*, **73**, 4158 (1951). (d) C. R. Krüger and E. G. Rochow, *J. Organometal. Chem.*, **1**, 476 (1964); M. W. Rathke, *J. Am. Chem. Soc.*, **92**, 3222 (1970). The precursor for this sodium derivative is commercially available (Ref. 66a). (e) H. Gilman, N. N. Crounse, S. P. Massie, Jr., R. A. Benkeser, and S. M. Spatz, *ibid.*, **67**, 2106 (1945). (f) C. H. Horning and F. W. Bergstrom, *ibid.*, **67**, 2110 (1945). (g) G. Wittig and H. D. Frommeld, *Chem. Ber.*, **97**, 3548 (1964). (h) G. Wittig and A. Hesse, *Org. Syn.*, **50**, 66 (1970).

66. (a) Commercially available from Alfa Inorganics, Inc., Beverly, Massachusetts. (b) See Ref. 58b for precautions in handling dry sodium hydride. The currently available sodium hydride is sold as a dry solid or as a dispersion in mineral oil. It is almost always desirable to wash this dispersion with ether or pentane prior to use in order to remove the mineral oil. For specific directions, see Ref. 34g: J. P. Schaefer and J. J. Bloomfield, *Org. Reactions*, **15**, 1 (1967); A. P. Krapcho, J. Diamanti, C. Cayen, and R. Bingham, *Org. Syn.*, **47**, 20 (1967).

67. (a) M. D. Soffer, R. A. Stewart, J. C. Cavagnol, H. E. Gellerson, and E. A. Bowler, *J. Am. Chem. Soc.*, **72**, 3704 (1950). (b) J. H. Fried, G. E. Arth, and L. H. Sarett, *ibid.*, **82**, 1684 (1960). (c) J. H. Fried, A. N. Nutile, and G. E. Arth, *ibid.*, **82**, 5704 (1960). (d) For the preparation of a very reactive form of sodium hydride by reaction of hydrogen with sodium naphthalenide in tetrahydrofuran solution, see S. Bank and T. A. Lois, *ibid.*, **90**, 4505 (1968). (e) J. S. McConaghy, Jr., and J. J. Bloomfield, *J. Org. Chem.*, **33**, 3425 (1968).

68. (a) E. J. Corey and M. Chaykovsky, *J. Am. Chem. Soc.*, **84**, 866 (1962). (b) For a study of the thermal decomposition of dimethylsulfoxide, see V. J. Traynelis and W. L. Hergenrother, *J. Org. Chem.*, **29**, 221 (1964). (c) A. Ledwith and N. McFarlane, *Proc. Chem. Soc.*, 108 (1964).

69. (a) P. Tomboulian, *J. Org. Chem.*, **24**, 229 (1959). (b) H. Gilman and B. J. Gaj, *ibid.*, **28**, 1725 (1963). (c) For the use of an alkyllithium reagent to generate a cyclohexadienyl anion for alkylation, see G. Brieger and D. W. Anderson, *Chem. Commun.*, **No. 20**, 1325 (1970).

70. (a) H. D. Zook and W. L. Rellahan, *J. Am. Chem. Soc.*, **79**, 881 (1957). (b) C. R. Hauser and W. B. Renfrew, Jr., *Org. Syn.*, **Coll. Vol. 2**, 268 (1943). (c) H. Adkins and W. Zartman, *ibid.*, **Coll. Vol. 2**, 607 (1943).

71. (a) H. O. House and V. Kramar, *J. Org. Chem.*, **27**, 4146 (1962). (b) D. F. Thompson, P. L. Bayless, and C. R. Hauser, *ibid.*, **19**, 1490 (1954); C. R. Hauser, D. S. Hoffenberg, W. H. Puterbaugh, and F. C. Frostick, Jr., *ibid.*, **20**, 1531 (1955).

72. (a) N. D. Scott, J. F. Walker, and V. L. Hansley, *J. Am. Chem. Soc.*, **58**, 2442 (1936). (b) H. Normant and B. Angelo, *Bull. Soc. Chim. France*, 354 (1960). See also (c) J. J. Eisch and W. C. Kaska, *J. Org. Chem.*, **27**, 3745 (1962).

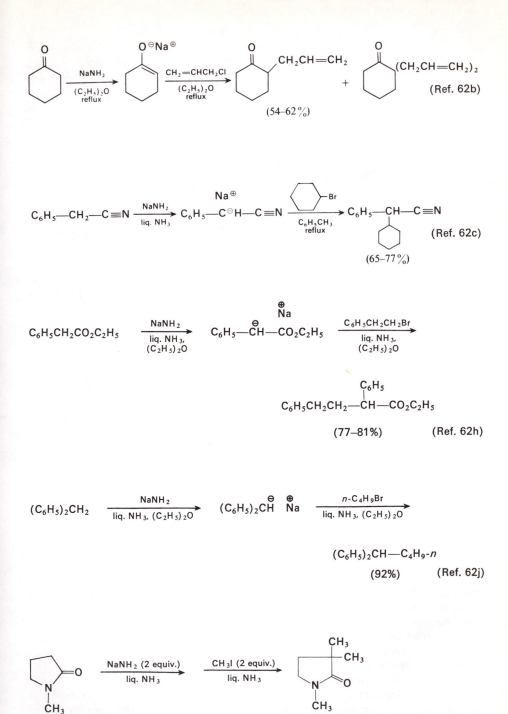

(54–62%)

(Ref. 62b)

(65–77%)

(Ref. 62c)

(77–81%)   (Ref. 62h)

(92%)   (Ref. 62j)

(45%)   (Ref. 62k)

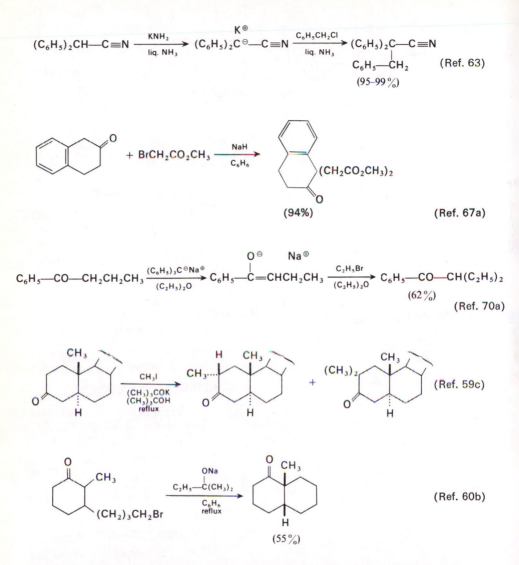

(Ref. 63)

(94%)   (Ref. 67a)

(62%)
(Ref. 70a)

(Ref. 59c)

(Ref. 60b)

(55%)

hydride[7a−d] or sodium amide[39] in an inert solvent. When either of these bases is used, the enolate is formed relatively slowly in the heterogeneous reaction mixture, with the result that appreciable concentrations of both the free ketone and its enolate anion are present at some point. This difficulty may be mitigated, at least in part, by allowing a mixture of the ketone and the alkylating agent to react with sodium hydride in 1,2-dimethoxyethane,[7a,67e] instead of attempting to form the enolate anion before the alkylating agent is added.

Another less common side reaction which has been noted in the alkylation of several strained bicyclic ketones is the competing reduction of the carbonyl group by sodium hydride indicated in the following equations.

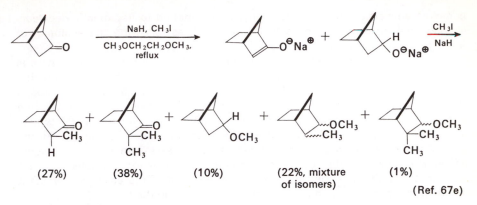

(27%)   (38%)   (10%)   (22%, mixture   (1%)
                                           of isomers)
                                                          (Ref. 67e)

The foregoing complications are most easily avoided by the slow addition of the ketone (or other active methylene compound) to a *solution* of a strong base in a suitable aprotic solvent. In this way, the ketone is converted to its enolate as it is added to the reaction solution and no excess ketone is present to react with the enolate in an aldol condensation.

The various alkali metal derivatives of triphenylmethane offer the special advantage of being dissociated into a solvated metal cation and the intensely red-colored triphenylmethyl anion in a variety of dipolar, aprotic solvents including 1,2-dimethoxyethane, tetrahydrofuran, and diethyl ether. [One exception is triphenylmethyllithium which is dissociated in 1,2-dimethoxyethane and in tetrahydrofuran but is colorless in diethyl ether solution.] Because of the intense color of this anion, ketones or other active methylene compounds may be added dropwise to solutions of the triphenylmethyl anions until the red color is just discharged. In this way, ketones may be converted quantitatively to their enolate anions in the absence of excess base or excess ketone.

The major objection to the use of metal triphenylmethides as bases is the fact that the reaction product must be separated from relatively large quantities of triphenylmethane. However, this objection is easily overcome by the use of only a few milligrams of triphenylmethane as indicator in a solution of some other strong base. By use of this titration procedure with a triphenylmethane indicator, active methylene compounds may be converted to their enolates employing solutions of lithium dialkylamides in 1,2-dimethoxyethane,[7b,c] of sodium amide in liquid ammonia, or of the dimethyl sulfoxide anion in dimethyl sulfoxide.

A related technique makes use of the fact that either 2,2-bipyridyl or 1,10-phenanthroline gives colored charge-transfer complexes not only with organolithium and organomagnesium reagents,[73d] but also gives an intense purple to red color when mixed with solutions of lithium dialkylamides in 1,2-dimethoxyethane, tetrahydrofuran, or diethyl ether.[73e] This titration procedure for establishing the

73. (a) G. A. Russell, E. G. Janzen, H. D. Becker, and F. J. Smentowski, *J. Am. Chem. Soc.,* **84**, 2652 (1962). (b) M. Chaykovsky and E. J. Corey, *J. Org. Chem.* **28**, 254 (1963). (c) C. Walling and L. Bollyky, *ibid.,* **28**, 256 (1963). (d) S. C. Watson and J. F. Eastham, *J. Organometal. Chem.,* **9**, 165 (1967); W. Voskuil and J. F. Arens, *Org. Syn.,* **48**, 47 (1968). (e) H. O. House, M. Gall, and H. D. Olmstead, *J. Org. Chem.,* **36**, 2361 (1971).

complete conversion of an active methylene compound to its anion is decidedly simpler than others which have been used such as following the evolution of hydrogen when sodium hydride is employed as the base.[7a,67a]

Utilization of the dimethyl sulfoxide anion in dimethyl sulfoxide as a base is frequently unsatisfactory because this anion may add to the carbonyl function of a ketone or ester (see Chapter 10) instead of abstracting an alpha hydrogen atom.[10h,68,73a-c] Furthermore, the facts that dimethyl sulfoxide reacts slowly with alkylating agents and is sometimes difficult to separate from the products may make this compound an undesirable solvent for alkylation reactions. At the present time, 1,2-dimethoxyethane, dimethylformamide, tetrahydrofuran, or liquid ammonia appear to be the best solvents to employ in alkylation reactions; of these, 1,2-dimethoxyethane is the most convenient because of the ease with which it can be purified, handled, and separated from reaction products.

Among the strong bases, the lithium dialkylamides (some of which are commercially available[66a]) are particularly convenient to prepare and use. As noted earlier, these amide bases can be employed with a 2,2-bipyridyl indicator in various ethereal solvents to permit quantitative conversion of active methylene compounds to their enolates by a titration procedure.

Although the lithium amides, formed from a variety of secondary amines including diethylamine (b.p. 55°), diisopropylamine (b.p. 84°), piperidine (b.p. 106°), and hexamethyldisilazine[$(CH_3)_3SiNHSi(CH_3)_3$, b.p. 125°)] and various alkyllithium reagents, have been used as bases, perhaps the most convenient materials are the lithium amides formed from methyllithium and either diethylamine or diisopropylamine. The latter amine offers the advantage of reacting only very slowly even with reactive alkyl halides. Also, as will be discussed later, lithium diisopropylamide is a very hindered base which exhibits considerable positional selectivity in forming enolates under kinetically controlled conditions.[7c,73e]

The following equations illustrate the use of a lithium dialkylamide in 1,2-dimethoxyethane solution for the alkylation of an active methylene compound. It should be noted that these amides slowly attack ethereal solvents (presumably in a β-elimination reaction) such as 1,2-dimethoxyethane and tetrahydrofuran, especially at elevated temperatures; consequently the ethereal solutions of lithium dialkylamides should be prepared at 25° or less and used promptly. Alkyllithium reagents (especially the commercially available[66a] reagents CH_3Li and n-C_4H_9—Li) may be used as strong bases for certain compounds with acid C—H bonds[69c] (see Chapter 10). However, these compounds are usually not satisfactory bases for forming enolate anions because of other rapid competing reactions including addition to carbonyl groups.

The alkylation of nitriles, of symmetrical ketones, and of ketones that can enolize in only one direction usually presents no problem other than the separation of the monoalkylated product from any unchanged starting material or dialkylated product. The formation of substantial amounts of O-alkylated products from ketone enolates is only rarely observed.[2c,7b,59a,62e,73e] The alkylation of nitriles (e.g., [41]) and subsequent hydrolysis often provides a good synthetic route to α-substituted acetic acids in spite of the rather vigorous conditions required for the hydrolysis.

The alkylation of ketones that can form structurally isomeric enolate anions

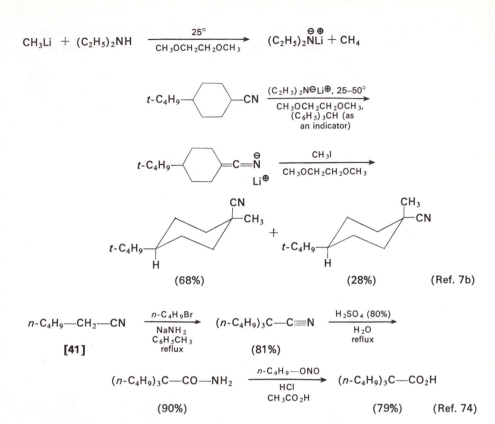

$$CH_3Li \;+\; (C_2H_5)_2NH \xrightarrow[CH_3OCH_2CH_2OCH_3]{25°} (C_2H_5)_2\overset{\ominus}{N}\overset{\oplus}{Li} + CH_4$$

$$t\text{-}C_4H_9\text{—}\bigcirc\text{—}CN \xrightarrow[\substack{CH_3OCH_2CH_2OCH_3, \\ (C_6H_5)_3CH \text{ (as} \\ \text{an indicator)}}]{(C_2H_5)_2\overset{\ominus}{N}Li^{\oplus},\; 25\text{–}50°}$$

raises two questions: which enolate will predominate in the equilibrium mixture and which monoalkylated structure will predominate in the product. As illustrated previously in this chapter, if only one of the structurally isomeric enolate anions is stabilized by an alpha substituent such as a carboalkoxy, cyano, or carbonyl group, then essentially only this highly stabilized enolate is formed and alkylation occurs at the position activated by both functional groups. Interestingly, if such stabilized enolates are treated with an excess of a strong base such as sodium amide or potassium amide in liquid ammonia, dianions may be formed;[35b,48e,62i,64,75] as the

74. (a) N. Sperber, D. Papa, and E. Schwenk, *J. Am. Chem. Soc.*, **70**, 3091 (1948). (b) The N-alkylation of very hindered nitriles to form ketimines has been observed: see M. S. Newman, T. Fukunaga, and T. Miwa, *ibid.*, **82**, 873 (1960). (c) For a study of the factors influencing the proportions of mono- and dialkylated products obtained from nitriles, see M. Makosza, *Tetrahedron*, **24**, 175 (1968).

75. (a) T. M. Harris and C. M. Harris, *Org. Reactions*, **17**, 155 (1969). (b) C. Mao, C. R. Hauser, and M. L. Miles, *J. Am. Chem. Soc.*, **89**, 5303 (1967). (c) K. G. Hampton, T. M. Harris, and C. R. Hauser, *J. Org. Chem.*, **30**, 61 (1965); **31**, 663 (1966). (d) J. F. Wolfe, T. M. Harris, and C. R. Hauser, *ibid.*, **29**, 3249 (1964). (e) N. M. Carroll and W. I. O'Sullivan, *ibid.*, **30**, 2830 (1965). (f) S. D. Work, D. R. Bryant, and C. R. Hauser, *ibid.*, **29**, 722 (1964); *J. Am. Chem. Soc.*, **86**, 872 (1964). (g) E. M. Kaiser, R. L. Vaulx, and C. R. Hauser, *Tetrahedron Letters*, **No. 40**, 4833 (1966). (h) J. F. Wolfe and T. G. Rogers, *J. Org. Chem.*, **35**, 3600 (1970). (i) L. Weiler, *J. Am. Chem. Soc.*, **92**, 6702 (1970).

accompanying equations show, reaction of these dianions with only one equivalent of an alkylating agent results in the predominant alkylation of the less stable of the two enolate anions present. This result is in accord with the previously stated generalization that the more basic (and less stable) enolates usually react more readily with alkylating agents. (It will be recalled that these more basic enolates may also lead to an increased amount of dehydrohalogenation of the alkyl halide.)

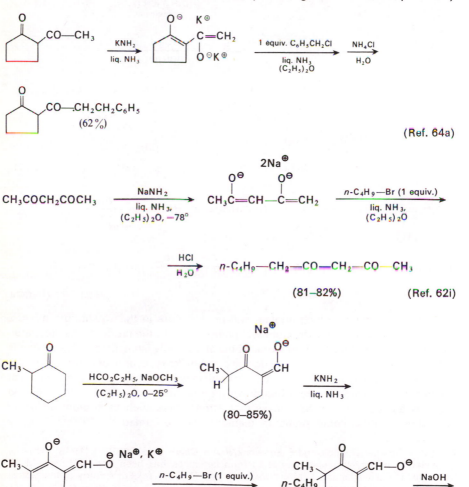

(Ref. 64a)

(81–82%)   (Ref. 62i)

(80–85%)

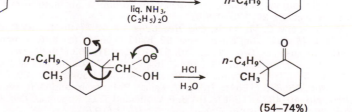

(54–74%)

(Ref. 63b, 64d)

   Although the alkylation of dipotassio dianions is complicated by equilibration of the alkylated product with starting dianion to form dialkylated by-products, this equilibration is slower with disodio and dilithio salts and dialkylation is not a serious side reaction.[75c] The relatively slow equilibration of sodium, and especially lithium enolates, is a general phenomenon which will be discussed later.

   Comparable dianions may also be obtained from β-keto esters,[75d] β-keto sulfones,[75e] and various amides and imides;[75f−h] the accompanying equation

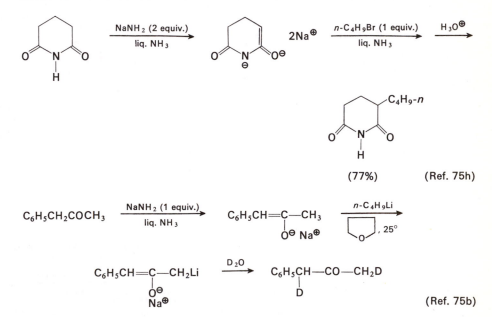

(77%)   (Ref. 75h)

(Ref. 75b)

illustrates the monoalkylation of one of these derivatives. The activation provided by one phenyl group and one ketone function is sufficient to permit dianion formation if the monoanion is treated with a sufficiently strong base such as n-butyllithium. The position of alkylation of such dianions has not yet been reported.

   Not only an α-keto or an α-cyano substituent, but also an α-phenyl group[7a,39] or an α-vinyl group[6h,39,60c] provides sufficient stabilization to determine the structure of an enolate anion derived from an unsymmetrical ketone. This selectivity is illustrated by the accompanying examples. As noted previously, the anion [44] derived from an α,β-unsaturated ketone [43] by abstraction of a gamma proton is alkylated at the alpha position to form a β,γ-unsaturated ketone [45]. This initial product [45] may be isomerized to an α-alkyl-α,β-unsaturated ketone [46] or may undergo further alkylation ([48]). Dialkylation has often been the major reaction in such cases[6h,59b,59d−59f,76] because a proton is abstracted more readily from the intermediate β,γ-unsaturated ketone [45] than from the starting material [43] or the alkylated α,β-unsaturated ketone [46].[6h] Dialkylation may be diminished either by the slow addition of the alkylating agent[59f] or by the use of a less reactive alkylating agent (e.g., methyl chloride).[6h] Such procedures permit the isomerization of the β,γ-unsaturated ketone [45] (to the less acidic α,β-unsaturated ketone

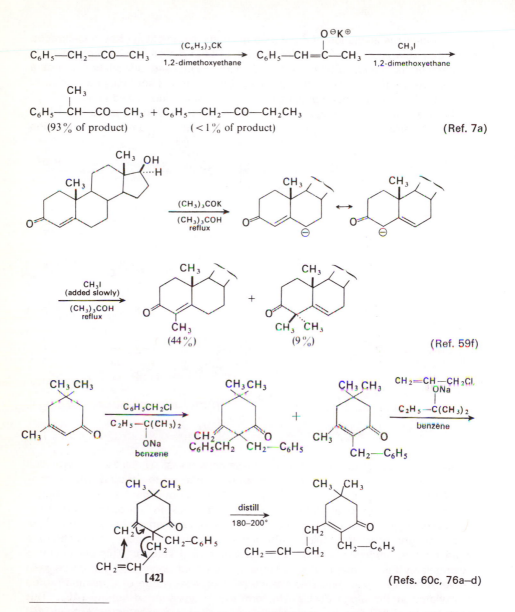

$$C_6H_5-CH_2-CO-CH_3 \xrightarrow[\text{1,2-dimethoxyethane}]{(C_6H_5)_3CK} C_6H_5-CH=\overset{\overset{\displaystyle O^{\ominus}K^{\oplus}}{|}}{C}-CH_3 \xrightarrow[\text{1,2-dimethoxyethane}]{CH_3I}$$

$$\overset{\overset{\displaystyle CH_3}{|}}{C_6H_5-CH-CO-CH_3} + C_6H_5-CH_2-CO-CH_2CH_3$$

(93% of product)   (<1% of product)   (Ref. 7a)

76. (a) J. M. Conia, *Bull. Soc. Chim. France,* 690 (1954). (b) J. M. Conia and C. Nevot, *ibid.,* 493 (1959). (c) J. M. Conia and A. LeCraz, *ibid.,* 1929, 1934 (1960). (d) J. M. Conia and P. LePerchec, *Tetrahedron Letters,* **No. 39**, 2791 (1964). (e) R. B. Woodward, A. A. Patchett, D. H. R. Barton, D. A. J. Ives, and R. B. Kelly, *J. Chem. Soc.,* 1131 (1957). (f) M. S. Newman, V. DeVries, and R. Darlak, *J. Org. Chem.,* **31**, 2171 (1966). (g) P. S. Wharton and C. E. Sundin, *ibid.,* **33**, 4255 (1968). (h) S. J. Rhoads in P. deMayo, ed., *Molecular Rearrangements,* Vol. 1, Wiley-Interscience, New York, 1963, pp. 655–706; G. B. Gill, *Quart. Rev.,* **22**, 338 (1968); R. B. Woodward and R. Hoffmann, *Angew. Chem., Intern. Ed. Engl.,* **8**, 781 (1969).

[46]) to occur more rapidly than alkylation of the intermediate anion [47]. The ready thermal isomerization of the α-allyl-β,γ-unsaturated ketone [42], an example of the Cope rearrangement,[76h] is also noteworthy as a possible complication in the alkylation of an α,β-unsaturated ketone with an allyl halide.

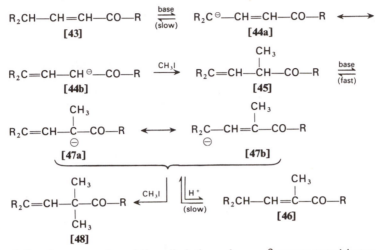

The following examples of the alkylation of an α,β-unsaturated ketone (e.g., [49]) with dihalides illustrate the fact that initial alkylation at the alpha carbon atom is also observed in these cases.[76f] In this example [49], there are two different γ-positions from which a proton may be abstracted by the base and products from each of the possible dienolates are observed. The variable proportions of double bond isomers obtained in the final product suggests that the extent of equilibration of the intermediate enolate anions varied in the different runs. The complexity possible in such a circumstance is illustrated in a detailed study of the products obtained in the methylation of 4-methyl-$\Delta^{4,10}$-octal-1-one.[76g]

Unsymmetrical ketones with only alkyl groups as alpha substituents usually yield mixtures of the two possible structurally isomeric enolate anions when treated with base. The various potential effects of α-alkyl substituents, including destabilization of resonance structure [50a] (for electrostatic reasons) and stabilization of resonance structure [50b] (by hyperconjugation) appear to be approximately balanced. As a result, the equilibrium between structurally isomeric enolates often may be influenced not only by the structure of the ketone but also by factors sometimes considered of lesser importance, among them the nature of the cation and the solvent.[2b,c,6b,7a–7d,j,35c,77d] These factors may influence the position of the

77. (a) H. Sobotka and J. D. Chanley, *J. Am. Chem. Soc.*, **71**, 4136 (1949). (b) W. J. Bailey and M. Madoff, *ibid.*, **76**, 2707 (1954). (c) F. E. King, T. J. King and J. G. Topliss, *J. Chem. Soc.*, 919 (1957). (d) H. O. House, W. L. Roelofs, and B. M. Trost, *J. Org. Chem.*, **31**, 646 (1966). (e) D. Caine, *ibid.*, **29**, 1868 (1964). (f) G. Stork, P. Rosen, N. Goldman, R. V. Coombs, and J. Tsuji, *J. Am. Chem. Soc.*, **87**, 275 (1965). (g) H. A. Smith, B. J. L. Huff, W. J. Powers, III, and D. Caine, *J. Org. Chem.*, **32**, 2851 (1967). (h) L. E. Hightower, L. R. Glasgow, K. M. Stone, D. A. Albertson, and H. A. Smith, *ibid.*, **35**, 1881 (1970). (i) For the formation of specific enolate anions by the reduction of *n*-butylthiomethylene derivatives of ketones, see R. M. Coates and R. L. Sowerby, *J. Am. Chem. Soc.*, **93**, 1027 (1971).

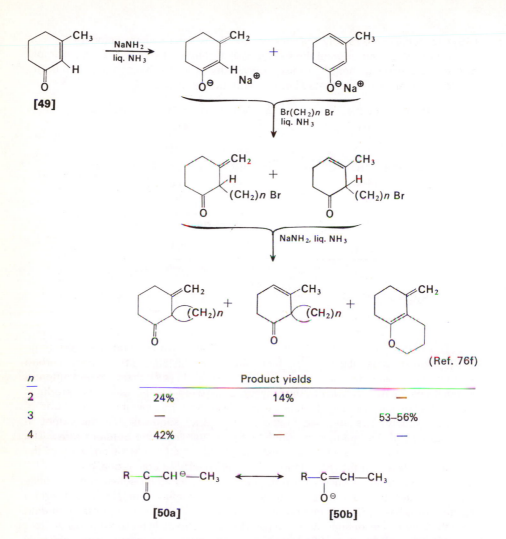

(Ref. 76f)

n	Product yields		
2	24%	14%	—
3	—	—	53–56%
4	42%	—	—

equilibrium among metal enolates by changing the degree of dissociation or the degree of aggregation of the various enolate isomers since metal enolates of monoketones appear to exist as aggregated structures (e.g., dimers,[7h] trimers,[2c,34g] tetramers) in solution.

Table 9–5 summarizes the equilibrium concentrations for a number of metal enolates as solutions in 1,2-dimethoxyethane. Usually, changing the metal cation from lithium to potassium shifts the equilibrium toward the less highly substituted enolate anion.[7a–d] This result is explicable in terms of the lithium-oxygen bond of the lithium enolate having more covalent character than the corresponding potassium-oxygen bond. Consequently, the lithium enolate would resemble an olefin (i.e. [50b]) more closely and would derive more hyperconjugative stabilization from an alkyl substituent. An exception to this generalization about cation

Table 9–5 Equilibrium positions for solutions of metal enolates in
1,2-dimethoxyethane[7a–d,j,35c,77d]

$n\text{-}C_4H_9\text{—}CH\text{=}C\text{—}CH_3$
$\quad\quad\quad\quad\quad\overset{|}{O^{\ominus}}\ M^{\oplus}$

58% (M = K)
ca. 87% (M = Li)

$\rightleftharpoons$

$n\text{-}C_4H_9\text{—}CH_2\text{—}C\text{=}CH_2$
$\quad\quad\quad\quad\quad\overset{|}{O^{\ominus}}\ M^{\oplus}$

42% (M = K)
ca. 13% (M = Li)

$(CH_3)_2CH\text{—}CH\text{=}C\text{—}CH_3$
$\quad\quad\quad\quad\quad\quad\overset{|}{O^{\ominus}}M^{\oplus}$

25% (M = K)

$\rightleftharpoons$

$(CH_3)_2CH\text{—}CH_2\text{—}C\text{=}CH_2$
$\quad\quad\quad\quad\quad\quad\quad\overset{|}{O^{\ominus}}\ M^{\oplus}$

75% (M = K)

$(CH_3)_2C\text{=}C\text{—}CH_2CH_3$
$\quad\quad\quad\overset{|}{O^{\ominus}}M^{\oplus}$

12% (M = K)
1% (M = Li)

$\rightleftharpoons$

$(CH_3)_2CH\text{—}C\text{=}CHCH_3$
$\quad\quad\quad\quad\quad\overset{|}{O^{\ominus}}M^{\oplus}$

88% (M = K)
99% (M = Li)

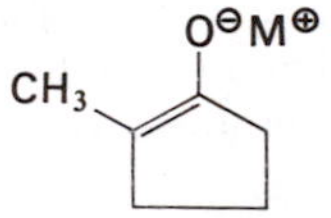

78% (M = K)
94% (M = Li)

$\rightleftharpoons$

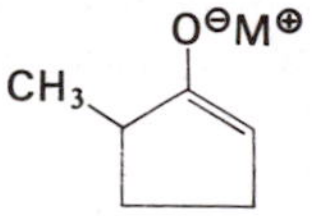

22% (M = K)
6% (M = Li)

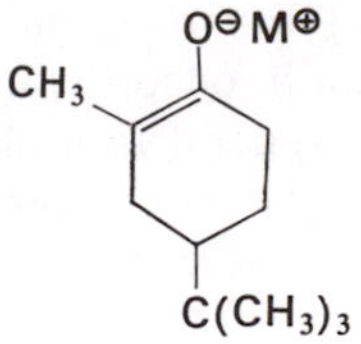

89% (M = Li)

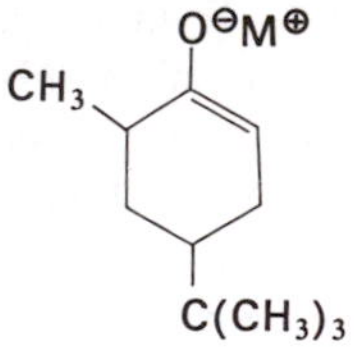

$\rightleftharpoons$

11% (M = Li)

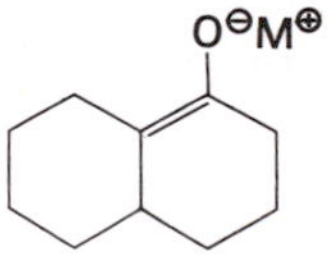

66% (M = Li)

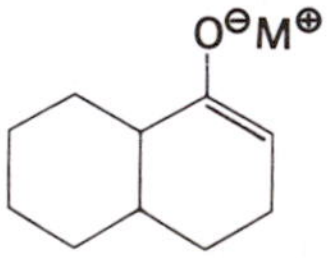

$\rightleftharpoons$

34% (M = Li)

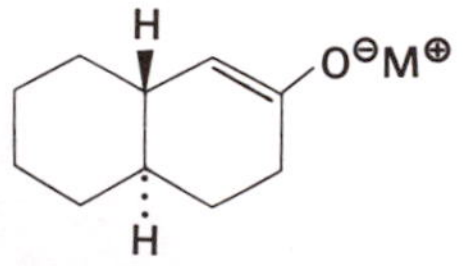

53% (M = Li)

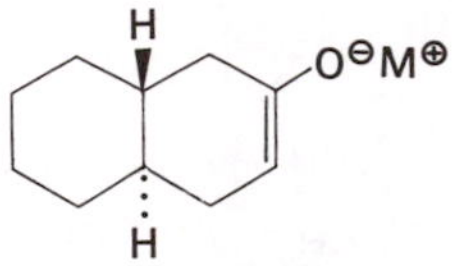

$\rightleftharpoons$

47% (M = Li)

effects is found with *acyclic* enolates with tetrasubstituted double bonds. In these cases the more highly substituted enolate is relatively unstable and is destabilized by changing the cation from potassium to lithium, possibly because of steric interference with solvation or with the formation of enolate aggregates. The rate of attaining equilibrium among isomeric metal enolates (in the presence of excess ketone or some other proton donor) is increased substantially as the cation is changed from lithium to sodium or potassium.[7a-7d,j,75c,77e,f] Increasing the polarity of the solvent also appears to increase the rate of equilibration.[77f]

As noted previously, the rate of reaction of metal enolates with alkylating agents is increased substantially as the metal cation is changed from lithium to sodium to potassium and as the polarity and solvating ability of the solvent are increased.[2c,34g] Both effects presumably arise from an increased amount of dissociated and/or non-aggregated enolate anion in the reaction solution. The rate of alkylation is also influenced, but to a lesser extent, by the structure of the enolate ion; the usual order of reactivity appears to be:

$$R_2C{=}\overset{\underset{|}{O^{\ominus}}}{C}{-}R \;\geqslant\; RCH{=}\overset{\underset{|}{O^{\ominus}}}{C}{-}R \;>\; CH_2{=}\overset{\underset{|}{O^{\ominus}}}{C}{-}R.^{[2a,7j,35b,c,73c]}$$

This variation in reactivity with structure such that the less highly substituted enolates are often less reactive would appear to result from differences in the degree of metal enolate aggregation. The more highly substituted (and more reactive) enolates offer more steric interference to the formation of large, tightly associated aggregates.[7h]

The proportions of isomeric monoalkylated products formed in ketone alkylations may also be affected by further alkylation to form dialkylated and polyalkylated products.[7a,76g,77d] If the initially formed monoalkylated products undergo further alkylation at different rates, the proportions of monoalkylated product in the reaction mixture will change as further alkylation occurs. An indication of the extent to which polyalkylation may occur in a preparative-scale reaction is provided in the methylation of ketone [51]. As a result, rather complex mixtures may be formed in the alkylation of unsymmetrical ketones; and if the physical properties of the mono- and dialkylated products are similar (as is often the case when a methyl or ethyl group is being introduced), isolation of pure compounds from the mixture frequently requires either a very efficient physical separation (e.g., gas chromatography) or a chemical transformation of some of the components prior to separation.[77a-c]

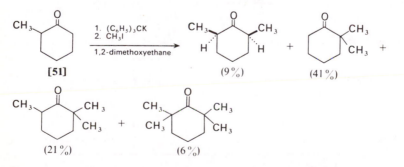

(Ref. 7a)

Three general procedures have been developed to avoid this problem in the alkylation of unsymmetrical ketones. One approach involves the introduction of an activating functional group alpha to the carbonyl group to stabilize the desired enolate anion. Aside from the previously illustrated α-aryl substituents or α-vinyl substituents (an α,β-unsaturated ketone with a gamma proton is equivalent), the commonly used activating groups are carbethoxy, formyl,[78] and ethoxyoxalyl.[59c] As in the example below, these activating groups may be introduced by acylation

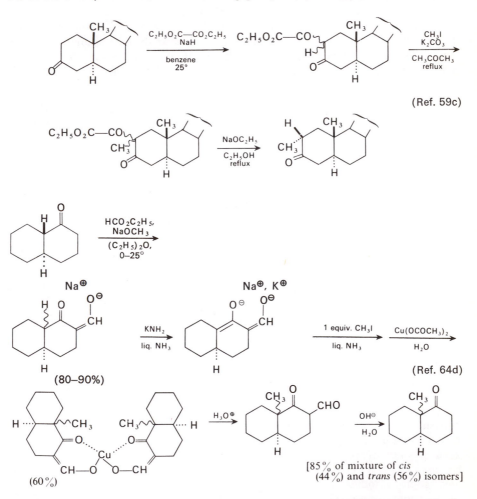

(Ref. 59c)

(Ref. 64d)

[85% of mixture of cis (44%) and trans (56%) isomers]

of the enolate anion (see Chapter 11). This acylation is usually a thermodynamically controlled process in which the major product is formed by introduction of the acyl group at the less hindered alpha carbon. A modification of this method is the

78. R. B. Woodward, F. Sondheimer, D. Taub, K. Heusler, and W. M. McLamore, *J. Am. Chem. Soc.*, **74**, 4223 (1952).

previously illustrated conversion of the intermediate β-dicarbonyl compound to its dianion which is then monoalkylated at the more reactive position.[63b,64d] In either case, after alkylation the activating group is removed by base-catalyzed cleavage. The previously described (see Chapter 2) reaction of the enolates of α-halo ketones with trialkylboranes to form α-alkyl ketones could be considered as a special case of the use of an activating group (an α-halogen substituent) to achieve a selective alkylation.

An alternative approach to selective alkylation of an unsymmetrical ketone involves the introduction of a blocking group at a methyl or methylene group alpha to the carbonyl function, thus preventing the formation of the corresponding enolate. One such type of blocking is achieved by an aldol condensation (Chapter 10) between the ketone and an aromatic aldehyde (usually benzaldehyde or furfural).[79]

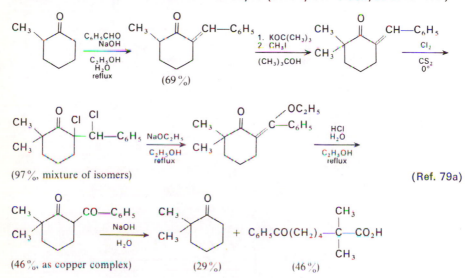

(Ref. 79a)

Although this procedure has the advantage of often producing easily crystallized ketone derivatives, it suffers from the difficulty that the arylidene blocking group is frequently difficult to remove after alkylation. A second type of blocking involves the initial acylation of a methyl or methylene group with ethyl formate and the subsequent transformation of the resulting formyl (or hydroxymethylene) substituent to a grouping that is stable to base. Three different procedures have been used: reaction of the formyl group with N-methylaniline (or some other secondary amine) to form an enamine (e.g., [52])[80a]; O-alkylation of the formyl substituent with isopropyl iodide to form an isopropoxy ether (e.g., [53])[80b]; and reaction of the formyl group with n-butyl mercaptan to form a thioether (e.g., [54]).[81]

79. (a) W. S. Johnson, *J. Am. Chem. Soc.*, **65**, 1317 (1943); **66**, 215 (1944). (b) W. S. Johnson, D. S. Allen, Jr., R. R. Hindersinn, G. H. Sausen, and R. Pappo, *ibid.*, **84**, 2181 (1962).
80. (a) A. J. Birch and R. Robinson, *J. Chem. Soc.*, 501 (1944). (b) W. S. Johnson and H. Posvic, *J. Am. Chem. Soc.*, **69**, 1361 (1947).
81. R. E. Ireland and J. A. Marshall, *J. Org. Chem.*, **27**, 1615, 1620 (1962).

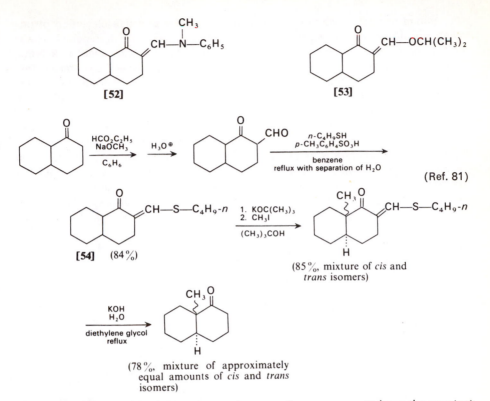

[52]

[53]

(Ref. 81)

[54]   (84%)

(85%, mixture of *cis* and *trans* isomers)

(78%, mixture of approximately equal amounts of *cis* and *trans* isomers)

The latter, illustrated in the accompanying equations, appears to have the greatest synthetic utility of the three. The thioether blocking group has also proved useful by virtue of its reducibility with Raney nickel or with lithium in liquid ammonia to form an α-methyl ketone. Reaction of an α-formylketone with 1,3-propanedithiol

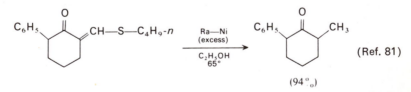

(Ref. 81)

(94%)

di-*p*-toluenesulfonate in alcoholic potassium acetate yields a keto dithioketal (e.g., [55]), in which the dithioketal function serves to block a methylene group during alkylation;[76e] it is removed by hydrogenolysis over Raney nickel (see Chapter 1).

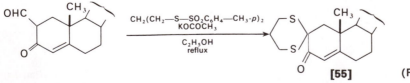

[55]   (Ref. 76e)

A third general procedure which is useful for the selective alkylation of un-symmetrical ketones makes use of the previously discussed fact that metal enolates equilibrate readily only in the presence of some proton donor such as a protic solvent or excess un-ionized ketone. Consequently, if a specific metal enolate can be generated in an aprotic, oxygen-free[82] medium and in the absence of un-ionized ketone, the enolate will retain both its structural and its stereochemical integrity for periods of at least several hours.[7a–d] In principle, this method avoids the need for either blocking groups or additional activating groups and constitutes the most efficient procedure for the selective alkylation of a ketone.[7a–d,73e,77e–g] In practice, this method is complicated by the fact that enolate equilibration can occur once some monoalkylated product (an un-ionized ketone such as [56a]) is produced in the reaction mixture. If this equilibration rate is comparable to the rate of reaction of the initial enolate with the alkylating agent, appreciable amounts of the isomeric monoalkylated product (e.g., [56b]) may be formed. This compli-

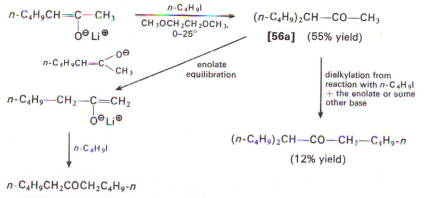

$$n\text{-}C_4H_9CH=\!\!\overset{\overset{\displaystyle |}{O^\ominus Li^\oplus}}{C}\!\!-CH_3 \xrightarrow[\substack{CH_3OCH_2CH_2OCH_3,\\0-25^\circ}]{n\text{-}C_4H_9I} (n\text{-}C_4H_9)_2CH\!\!-\!\!CO\!\!-\!\!CH_3$$

[56a]   (55% yield)

$$n\text{-}C_4H_9CH=C\!\!\overset{\displaystyle O^\ominus}{\underset{\displaystyle CH_3}{}}$$ enolate equilibration

$$n\text{-}C_4H_9\!\!-\!\!CH_2\!\!-\!\!\overset{\overset{\displaystyle |}{O^\ominus Li^\oplus}}{C}\!\!=\!\!CH_2$$

dialkylation from reaction with $n\text{-}C_4H_9I$ + the enolate or some other base

$$(n\text{-}C_4H_9)_2CH\!\!-\!\!CO\!\!-\!\!CH_2\!\!-\!\!C_4H_9\text{-}n$$

(12% yield)

$$n\text{-}C_4H_9I$$

$$n\text{-}C_4H_9CH_2COCH_2C_4H_9\text{-}n$$

[56b]   (6% yield)

(Ref. 7b)

---

82. Metal enolates, like allylic carbanions [see D. H. Hunter and D. J. Cram, *J. Am. Chem. Soc.,* **86**, 5478 (1964) ; H. H. Freedman, V. R. Sandel, and B. P. Thill, *ibid.,* **89**, 1762 (1967)], have an appreciable energy barrier to rotation. Consequently, solutions of *cis-* or *trans-*isomers of metal enolates do not interconvert in the absence of proton donors.[7a–d] However, small amounts of oxygen catalyze the interconversion of geometrically isomeric enolates, possibly by the reversible formation of an α-keto radical, as indicated in the accompanying equation.

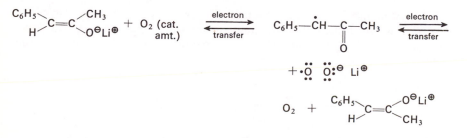

$$\underset{H}{\overset{C_6H_5}{}}\!\!C\!\!=\!\!C\!\!\overset{\overset{\displaystyle CH_3}{}}{\underset{\displaystyle O^\ominus Li^\oplus}{}} + O_2 \text{ (cat. amt.)} \underset{\text{transfer}}{\overset{\text{electron}}{\rightleftarrows}} C_6H_5\!\!-\!\!\overset{\bullet}{C}H\!\!-\!\!\overset{\overset{\displaystyle |}{\underset{\displaystyle O}{\|}}}{C}\!\!-\!\!CH_3 \underset{\text{transfer}}{\overset{\text{electron}}{\rightleftarrows}}$$

$$+ \cdot\overset{..}{\underset{..}{O}} \quad \overset{..}{\underset{..}{O}}{:}^\ominus \text{ Li}^\oplus$$

$$O_2 \quad + \quad \underset{H}{\overset{C_6H_5}{}}\!\!C\!\!=\!\!C\!\!\overset{\overset{\displaystyle O^\ominus Li^\oplus}{}}{\underset{\displaystyle CH_3}{}}$$

(Ref. 7c)

cation can be largely avoided by the correct choice of a cation in the metal enolate accompanied by the use of reactive alkylating agents in relatively high concentrations to increase the rate of alkylation of the initially formed enolate.

The rate of enolate equilibration is retarded by choosing a metal cation such as lithium,[7a–d,77e–g] calcium,[83a–c] zinc,[83d] or magnesium[83e–g] which tends to form covalent bonds or tightly associated ion pairs with the enolate anion.[83] Although the zinc and magnesium enolates can be used with highly reactive alkylating agents (e.g., methyl iodide), these alkaline-earth-metal enolates are relatively unreactive toward many alkylating agents unless very polar solvents such as hexamethylphosphoramide are used.[83f,h,i] At the present time the lithium enolates offer the best compromise to obtain reasonable reaction rates with alkylating agents, and yet to have relatively slow equilibration rates among enolates. Two general methods have been employed to generate specific metal enolates for alkylation: One method involves the dissolving metal reduction (see Chapter 3) of an α,β-unsaturated ketone[77f–i] or an α-halo or α-acyloxy ketone,[83a–e] whereas the other uses the reaction of an enol ester,[7b,c,73e,83e,g,k,l] an enol silyl ether,[7c,h,65d,73e,83h–j] or some other enol derivative[83i] with an organolithium or organomagnesium reagent. Examples of these procedures are illustrated in the following equations. The use

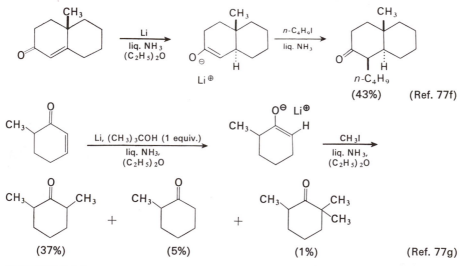

(43%)   (Ref. 77f)

(37%)   +   (5%)   +   (1%)   (Ref. 77g)

83. (a) R. E. Schaub and M. J. Weiss, *Chem. Ind.* (*London*), 2003 (1961). (b) M. J. Weiss, R. E. Schaub, J. F. Poletto, G. R. Allen, Jr., and C. J. Coscia, *ibid.,* 118 (1963). (c) M. J. Weiss and co-workers, *Tetrahedron,* **20**, 357 (1964); R. E. Schaub, W. Fulmor, and M. J. Weiss, *ibid.,* **20**, 373 (1964). (d) T. A. Spencer, R. W. Britton, and D. S. Watt, *J. Am. Chem. Soc.,* **89**, 5727 (1967). (e) E. R. H. Jones and D. A. Wilson, *J. Chem. Soc.,* 2933 (1965). (f) J. Fauvarque and J. F. Fauvarque, *Bull. Soc. Chim. France,* 160 (1969). (g) C. J. R. Adderley, G. V. Baddeley, and F. R. Hewgill, *Tetrahedron,* **23**, 4143 (1967). (h) P. A. Tardella, *Tetrahedron Letters,* **No. 14**, 1117 (1969). (i) For the use of trialkyltin enol ethers, see M. Pereyre and Y. Odic, *ibid.,* **No. 7**, 505 (1969). (j) R. Bourhis and E. Frainnet, *Bull. Soc. Chim. France,* 3552 (1967). (k) H. O. House and C. J. Blankley, *J. Org. Chem.,* **32**, 1741 (1967). (l) H. W. Whitlock and L. E. Overman, *ibid.,* **34**, 1962 (1969). For examples of the reaction of enolates with dihalocarbenes, see T. D. J. D'Silva and H. J. Ringold, *Tetrahedron Letters,* **No. 50**, 4487 (1965).

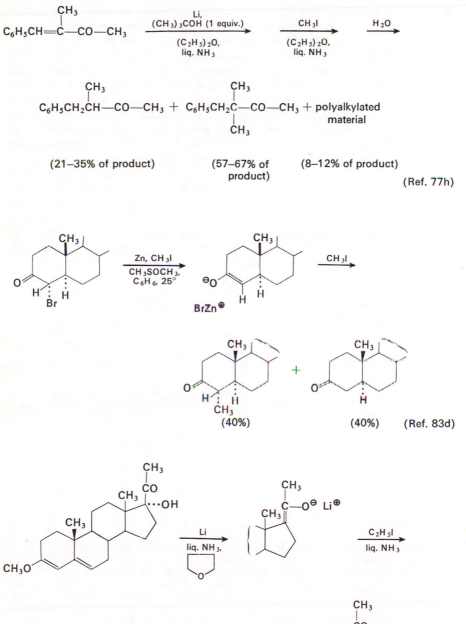

$$\underset{CH_3}{\overset{|}{C_6H_5CH{=}C{-}CO{-}CH_3}} \xrightarrow[\substack{(C_2H_5)_2O,\\ liq.\ NH_3}]{\substack{Li,\\ (CH_3)_3COH\ (1\ equiv.)}} \xrightarrow[\substack{(C_2H_5)_2O,\\ liq.\ NH_3}]{CH_3I} \xrightarrow{H_2O}$$

$$\underset{CH_3}{\overset{CH_3}{C_6H_5CH_2CH{-}CO{-}CH_3}} + \underset{CH_3}{\overset{CH_3}{C_6H_5CH_2C{-}CO{-}CH_3}} + \text{polyalkylated material}$$

(21–35% of product)   (57–67% of product)   (8–12% of product)

(Ref. 77h)

Zn, CH₃I
CH₃SOCH₃,
C₆H₆, 25°

CH₃I

(40%)   (40%)   (Ref. 83d)

Li
liq. NH₃,

C₂H₅I
liq. NH₃

(45%)   (Ref. 83c)

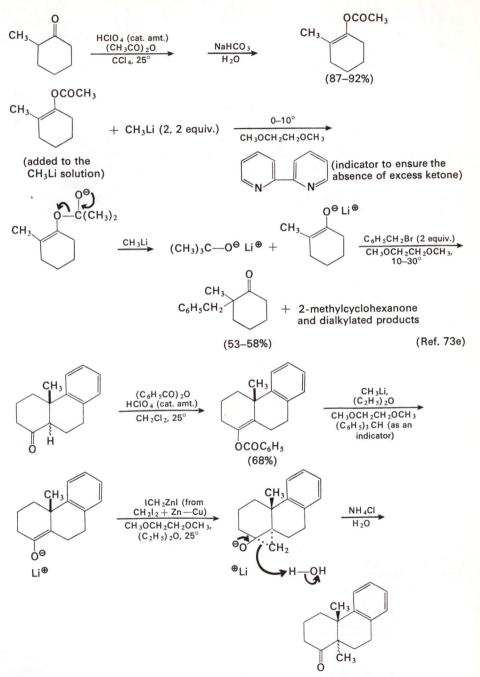

(87–92%)

+ CH₃Li (2, 2 equiv.) $\xrightarrow[\text{CH}_3\text{OCH}_2\text{CH}_2\text{OCH}_3]{0-10°}$

(added to the CH₃Li solution)

(indicator to ensure the absence of excess ketone)

$\xrightarrow{\text{CH}_3\text{Li}}$ (CH₃)₃C—O⁻ Li⁺ + $\xrightarrow[\substack{\text{CH}_3\text{OCH}_2\text{CH}_2\text{OCH}_3, \\ 10-30°}]{\text{C}_6\text{H}_5\text{CH}_2\text{Br (2 equiv.)}}$

+ 2-methylcyclohexanone and dialkylated products

(53–58%)   (Ref. 73e)

$\xrightarrow[\text{CH}_2\text{Cl}_2, 25°]{\substack{(\text{C}_6\text{H}_5\text{CO})_2\text{O} \\ \text{HClO}_4 \text{ (cat. amt.)}}}$ OCOC₆H₅ (68%) $\xrightarrow[\substack{\text{CH}_3\text{OCH}_2\text{CH}_2\text{OCH}_3 \\ (\text{C}_6\text{H}_5)_3\text{CH (as an} \\ \text{indicator)}}]{\substack{\text{CH}_3\text{Li}, \\ (\text{C}_2\text{H}_5)_2\text{O}}}$

$\xrightarrow[\substack{\text{CH}_3\text{OCH}_2\text{CH}_2\text{OCH}_3, \\ (\text{C}_2\text{H}_5)_2\text{O}, 25°}]{\substack{\text{ICH}_2\text{ZnI (from} \\ \text{CH}_2\text{I}_2 + \text{Zn}-\text{Cu})}}$ $\xrightarrow[\text{H}_2\text{O}]{\text{NH}_4\text{Cl}}$

(64%, mixture of 70% *cis* and 30% *trans* isomers)

(Ref. 83l)

of iodomethylzinc iodide (the Simmons-Smith reagent) rather than a conventional methylating agent to convert an enolate to a monomethyl ketone uncontaminated with polyalkylation products is worthy of special note.[831]

Among these methods for generating specific metal enolates, the reduction of enones with lithium and one equivalent of a proton donor in liquid ammonia and the reaction of enol acetates with methyllithium appear to offer the most synthetic utility.   Both methods are limited by the availability of the necessary starting materials (i.e. the correct α,β-unsaturated ketone isomer or the correct enol acetate isomer). As noted previously, it is usually possible to obtain the more highly substituted (and more stable) enol acetate isomers by the O-acetylation of ketones under equilibrating conditions.[7a,c,d,9]   Thus, the preparation of the more highly substituted metal enolate from an unsymmetrical ketone is readily accomplished by the enol acetate-methyllithium procedure. The use of silyl enol ethers as precursors to metal enolates[7c,h,83h] was introduced in an effort to avoid the

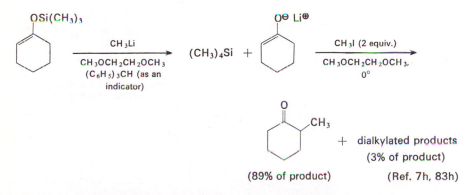

(89% of product)   (Ref. 7h, 83h)

production of an equivalent amount of base (usually lithium t-butoxide) which is formed when specific metal enolate isomers are formed from enol acetates or by the reduction of enones. The presence of this additional base has been found to enhance the formation of di- and polyalkylated products from the initially formed monoalkylated material. Although the use of silyl ethers as enolate precursors offers some advantage in this respect,[7h,73e,83h] this is offset by the difficulty of obtaining a single isomer of a silyl enol ether from an unsymmetrical ketone.[7c] This difficulty is illustrated by the conversion of 2-methylcyclohexanone to a mixture of silyl enol ethers under equilibrating conditions. As noted previously, the corresponding more highly substituted enol acetate may be obtained directly from an acetylation under equilibrating conditions.

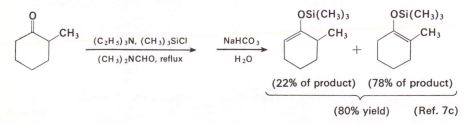

(22% of product)   (78% of product)

(80% yield)   (Ref. 7c)

In certain cases, the kinetic preference for a strong base to abstract an alpha proton from the less hindered position of an unsymmetrical ketone is sufficiently great that this method can be used to generate a specific enolate anion. For this purpose the hindered base, lithium diisopropylamide, appears to be significantly more selective than the other commonly used strong bases. The following equations illustrate the use of this kinetically controlled process to generate specific enolate

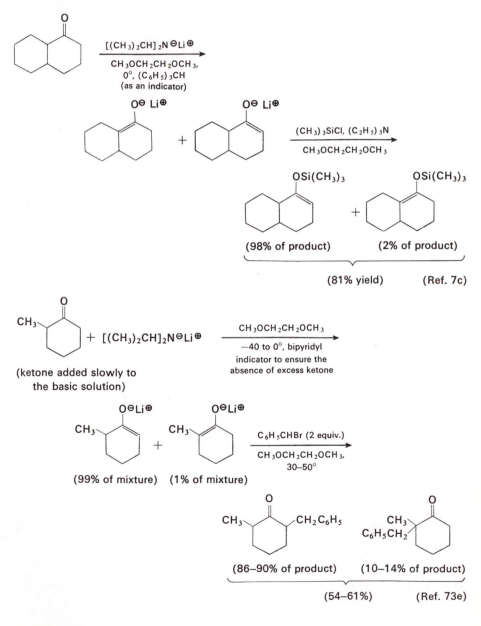

(98% of product)   (2% of product)

(81% yield)   (Ref. 7c)

(99% of mixture)   (1% of mixture)

(86–90% of product)   (10–14% of product)

(54–61%)   (Ref. 73e)

anions. The sodium or lithium derivative of hexamethyldisilizane may offer similar selectivity.[7e]

THE FORMATION AND ALKYLATION OF ENAMINES

An indirect, but widely used, procedure for the selective alkylation (or acylation: see Chapter 11) of an aldehyde or ketone involves the initial reaction of the carbonyl compound with a secondary amine to form an intermediate enamine.[84] Typical enamine preparations are outlined in the accompanying equations. It will be noted that the less highly substituted enamine is usually the predominant product unless the enamine function can be stabilized by conjugation with an adjacent function such as a carbonyl group.[85c-l] It will also be noted from one of the accompanying examples that the composition of equilibrium mixtures of enamine isomers is

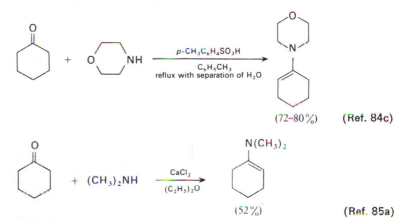

(72–80%)   (Ref. 84c)

(52%)   (Ref. 85a)

84. (a) G. Stork, A. Brizzolara, H. Landesman, J. Szmuszkovicz, and R. Terrell, *J. Am. Chem. Soc.*, **85**, 207 (1963). (b) J. Szmuszkovicz, *Adv. Org. Chem.*, **4**, 1 (1963). (c) S. Hünig, E. Lücke, and W. Brenninger, *Org. Syn.*, **41**, 65 (1961). (d) A. G. Cook, ed., *Enamines: Their Synthesis, Structure, and Reactions*, Marcel Dekker, New York, 1969. (e) For reviews of the properties of tetraaminoethylene derivatives and other electron-rich olefins, see R. W. Hoffmann, *Angew. Chem., Intern. Ed. Engl.*, **7**, 754 (1968); N. Wiberg, *ibid.*, **7**, 766 (1968). (f) M. E. Kuehne, *Synthesis*, **2**, 510 (1970).
85. (a) E. P. Blanchard, Jr., *J. Org. Chem.*, **28**, 1397 (1963). (b) W. A. White and H. Weingarten, *J. Org. Chem.*, **32**, 213 (1967); **31**, 4041 (1966); H. Weingarten, J. P. Chupp, and W. A. White, *ibid.*, **32**, 3246 (1967); H. Weingarten and M. G. Miles, *ibid.*, **33**, 1506 (1968). (c) A. I. Meyers, A. H. Reine, and R. Gault, *ibid.*, **34**, 698 (1969). (d) N. J. Leonard and J. A. Adamcik, *J. Am. Chem. Soc.*, **81**, 595 (1959). (e) Y. Shvo and H. Shanan-Atidi, *ibid.*, **91**, 6683, 6689 (1969). (f) P. M. Vay, *Chem. Commun., No. 15*, 861 (1969). (g) Y. Shvo and I. Belsky, *Tetrahedron*, **25**, 4649 (1969). (h) A. Mannschreck and U. Koelle, *Tetrahedron Letters, No. 10*, 863 (1967). (i) J. V. Greenhill, *J. Chem. Soc.*, B, 299 (1969). (j) For studies of cyano enamines, see T. Sasaki, T. Yoshioka, and K. Shoji, *ibid.*, C, 1086 (1969); T. Sasaki and A. Kojima, *ibid.*, C, 476 (1970). (k) For studies of nitro enamines, see F. W. Lichtenthaler and N. Majer, *Tetrahedron Letters, No. 6*, 411 (1969). (l) For studies of halo enamines see S. J. Huang and M. V. Lessard, *J. Am. Chem. Soc.*, **90**, 2432 (1968); R. Buyle and H. G. Viehe, *Tetrahedron*, **25**, 3447, 3453 (1969). (m) H. G. Viehe, *Angew. Chem., Intern. Ed. Engl.*, **6**, 767 (1967). (n) M. E. Kuehne and P. J. Sheeran, *J. Org. Chem.*, **33**, 4406 (1968). (o) For a study of the use of aziridine to form enamines from ketones, see S. C. Kuo and W. H. Daly, *ibid.*, **35**, 1861 (1970).

influenced not only by the structure of the ketone but also by the structure of secondary amine used to form the enamine.[8c,86c-f] Mixtures of double bond isomers also frequently result when α,β-unsaturated ketones are converted to dien-amines;[86g-i] the isomeric dienamine in which both double bonds are not contained is a single six-membered ring appears to be favored.[86g] Usually the reaction of

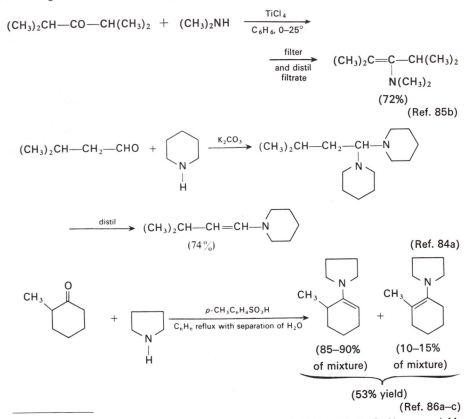

86. (a) M. E. Kuehne, *J. Am. Chem. Soc.*, **81**, 5400 (1959). (b) H. O. House and M. Schellenbaum, *J. Org. Chem.*, **28**, 34 (1963). (c) W. D. Gurowitz and M. A. Joseph, *ibid.*, **32**, 3289 (1967). (d) F. P. Colonna, M. Forchiassin, A. Risaliti, and E. Valentin, *Tetrahedron Letters*, **No. 8**, 571 (1970). (e) K. Nagarajan and S. Rajappa, *ibid.*, **No. 27**, 2293 (1969). (f) A. L. Ham and P. R. Leeming, *J. Chem. Soc.*, C, 2017 (1969). (g) N. F. Firrell and P. W. Hickmott, *ibid.*, B, 293 (1969) ; C, 716 (1970). (h) H. Nozaki, T. Yamaguti, S. Ueda, and K. Kondo, *Tetrahedron*, **24**, 1445 (1968.) (i) H. O. House, B. M. Trost, R. W. Magin, R. G. Carlson, R. W. Franck, and G. H. Rasmusson, *J. Org. Chem.*, **30**, 2513 (1965). (j) H. Ahlbrecht, J. Belcher, and F. Kröhnke, *Tetrahedron Letters*, **No. 6**, 439 (1969) ; H. Ahlbrecht, *ibid.*, **No. 42**, 4421 (1968). (k) G. Bianchetti, P. D. Croce, and D. Pocar, *ibid.*, **25**, 2043 (1965). (l) D. A. Nelson and J. J. Worman, *Chem. Commun.*, **No. 14**, 487 (1966). (m) The enamine tautomer predominates in the product from primary amines and β-dicarbonyl compounds. See G. O. Dudek, *J. Org. Chem.*, **30**, 548 (1965) and references therein. (n) S. K. Malhotra, D. F. Moakley, and F. Johnson, *J. Am. Chem. Soc.*, **89**, 2794 (1967). (o) J. Sauer and H. Prahl, *Tetrahedron Letters*, **No. 25**, 2863 (1966). (p) A. J. Hubert, *J. Chem. Soc.*, C, 2048 (1968). (q) H. Ahlbrecht and S. Fisher, *Tetrahedron*, **26**, 2837 (1970).

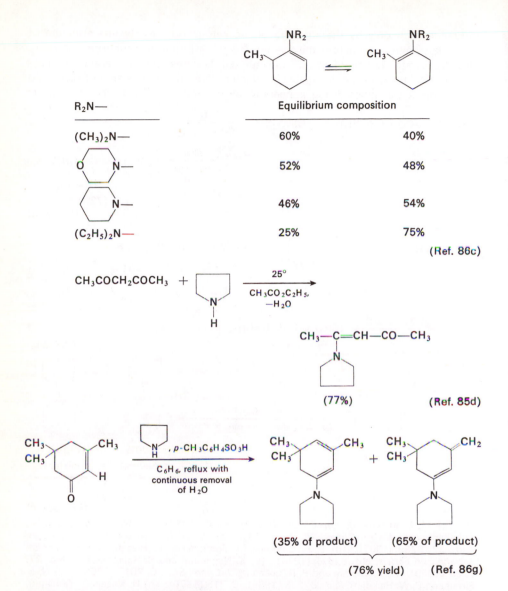

R_2N—	Equilibrium composition	
$(CH_3)_2N$—	60%	40%
O‿N— (morpholine)	52%	48%
‿N— (piperidine)	46%	54%
$(C_2H_5)_2N$—	25%	75%

(Ref. 86c)

$CH_3COCH_2COCH_3$ + (pyrrolidine) →[25°, $CH_3CO_2C_2H_5$, $-H_2O$]

$$CH_3—C=CH—CO—CH_3$$

(77%)   (Ref. 85d)

→[, p-$CH_3C_6H_4SO_3H$, C_6H_6, reflux with continuous removal of H_2O]

(35% of product)   (65% of product)

(76% yield)   (Ref. 86g)

primary amines with aldehydes and ketones affords products which contain primarily the imino compounds (Schiff bases) rather than the enamine tautomers.[86j−p] The following examples illustrate this fact and also an interesting base-catalyzed isomerization of an imino compound derived from benzylamine and an α,β-unsaturated ketone which serves to convert an α,β-unsaturated ketone to the corresponding saturated carbonyl compound.[86n] A similar base-catalyzed isomerization has been used to convert allylamine derivatives to enamines.[86o,p] The acetylenic analogues of enamines, called ynamines or alkynylamines, have also been prepared and studied.[85m,n]

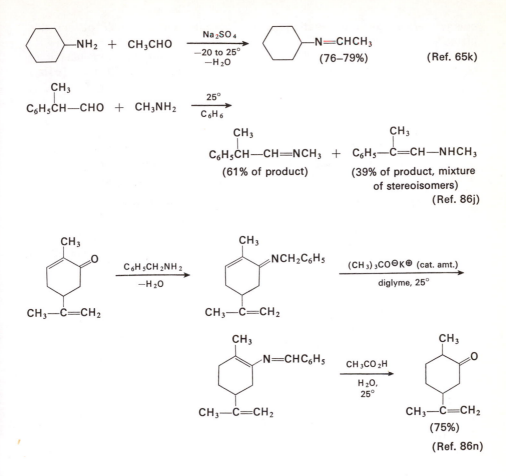

The electronic distribution in enamines as exemplified by the resonance structures [57] and [58] is such that the beta carbon atom bears an appreciable negative charge and may serve as a nucleophile. As the charge-separated structures [57b] and [58b] imply, for electron delocalization to occur it is necessary that the unshared electron pairs on nitrogen lie in the same plane as the pi orbital of the carbon-carbon double bond (see structures [59] and [61]) and an appreciable

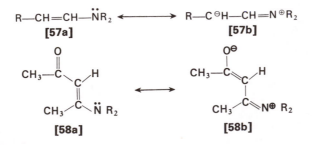

energy barrier exists to rotation of the nitrogen electron pair out of this plane.[8c,84a,b,85e−k,86c−f,87a−h]

Any structural features which interfere sterically with this requirement of coplanarity of certain enamine isomers (see arrows in structures [59b] and [61b])

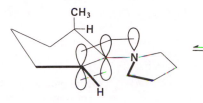

[59a] (more stable)   **[59b]** (less stable)

$CH_3CO_2D,$
D_2O, diglyme, 25°

(Ref. 87b, c)

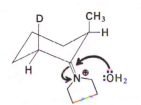

[60]

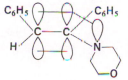

[61a] (88%)

CH₃OH
25°

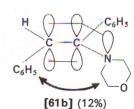

[61b] (12%)   (Ref. 87a)

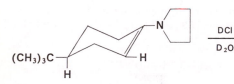

DCl
D₂O

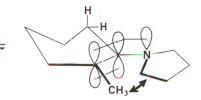

H₂O
pentane

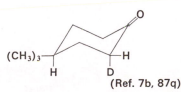

(Ref. 7b, 87q)

will destabilize these isomers. Also note that the more stable form [59a] of the cyclohexenyl enamine [59] has the alpha substituent in a pseudoaxial conformational to minimize steric interference. Enamines are ambident nucleophiles that may be protonated either at nitrogen or at the beta carbon atom; usually the iminium salts (e.g., [60]) are the more stable protonated species.[87f-n] These iminium salts, whose synthetic utility and synthesis by alternative methods has been discussed elsewhere (see Chapter 7), are relatively stable in strongly acidic media and have been used as protecting groups[87k] and as basic derivatives for the resolution of asymmetric ketones.[87l] However, the iminium salts react with weak acids (e.g., aqueous acetic acid) to form hydrolysis products.[7b,87b,c,i-n,q] The ease of hydrolysis of enamines varies with the amine; among the common secondary amines used the order is:

$$-N\langle\square\rangle \;>\; -N(CH_3)_2 \;>\; -N\langle\hexagon\rangle \;>\; -N\langle\hexagon\rangle O.^{[87m,n]}$$

The proton is usually added to the beta carbon of the enamine system from the less hindered side;[87j] with cyclohexenyl enamines (e.g., [60]) the proton is added to the beta carbon from an axial direction for stereoelectronic reasons[7b,8c,87b,c,q] so that an axial monodeuterio ketone is the initially formed hydrolysis product. It is this rapid and reversible formation of iminium salts (e.g., [60]) which is believed responsible for the ready equilibration of structurally isomeric enamines.

Another reaction of iminium salts is their reduction when they are heated in the presence of amines.[87o,p] This reaction, believed to occur by the hydrogen transfer illustrated in the following equation, can sometimes constitute a significant side reaction during the hydrolysis of enamines with aqueous acid.[86i]

When enamines are treated with alkylating agents, reaction paths leading to C-alkylation or N-alkylation are possible. Subsequent hydrolysis of the C-alkylated

87. (a) M. E. Munk and Y. K. Kim, *J. Org. Chem.,* **30**, 3705 (1965). (b) F. Johnson and A. Whitehead, *Tetrahedron Letters,* **No. 50**, 3825 (1964). (c) S. K. Malhotra and F. Johnson, *ibid.,* **No. 45**, 4027 (1965). (d) S. Danishefsky and M. Feldman, *ibid.,* **No. 16**, 1131 (1965). (e) S. K. Malhotra, D. F. Moakley, and F. Johnson, *Chem. Commun.,* **No. 9**, 448 (1967). (f) R. L. Hinman, *Tetrahedron,* **24**, 185 (1968). (g) J. Elguero, R. Jacquier, and G. Tarrago, *Tetrahedron Letters,* **No. 51**, 4719 (1965). (h) Stabilized enamines (β-amino-α,β-unsaturated ketones) are protonated predominantly at oxygen (see structure [58b]); H. E. A. Kramer and R. Gompper, *ibid.,* **No. 15**, 969 (1963). (i) G. Schroll, P. Klemmensen, and S. O. Lawesson, *ibid.,* **No. 33**, 2869 (1965). (j) J. Ficini and A. Krief, *ibid.,* **No. 17**, 1397 (1970). (k) B. Gadsby and M. R. G. Leeming, *Chem. Commun.,* **No. 11**, 596 (1968). (l) W. R. Adams, O. L. Chapman, J. B. Sieja, and W. J. Welstead, Jr., *J. Am. Chem. Soc.,* **88**, 162 (1966). (m) P. Y. Sollenberger and R. B. Martin, *ibid.,* **92**, 4261 (1970). (n) E. J. Stamhuis and W. Maas, *J. Org. Chem.,* **30**, 2156 (1965); E. J. Stamhuis, W. Maas, and H. Wynberg, *ibid.,* **30**, 2160 (1965); W. Maas, M. J. Janssen, E. J. Stamhuis, and H. Wynberg, *ibid.,* **32**, 1111 (1967). (o) A. G. Cook, W. C. Meyer, K. E. Ungrodt, and R. H. Mueller, *ibid.,* **31**, 14 (1966); A. G. Cook and C. R. Schulz, *ibid.,* **32**, 473 (1967). (p) E. L. Patmore and H. Chafetz, *ibid.,* **32**, 1254 (1967). (q) J. P. Schaefer and D. S. Weinberg, *Tetrahedron Letters,* **No. 23**, 1801 (1965). (r) M. G. Reinecke and L. R. Kray, *J. Org. Chem.,* **30**, 3671 (1965); **31**, 4215 (1966). (s) J. M. McEuen, R. P. Nelson, and R. G. Lawton, *ibid.,* **35**, 690 (1970).

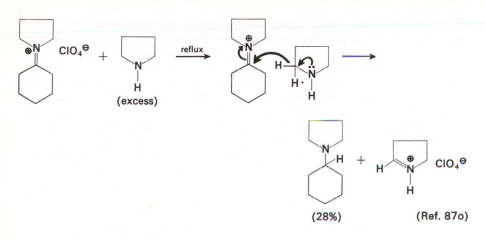

(28%)   (Ref. 87o)

iminium salt [62] yields an alkylated ketone, whereas the N-alkylated product [63] is usually water soluble, and relatively inert to hydrolysis. This alkylation procedure does not preclude the possibility of dialkylation since the iminium salt can react

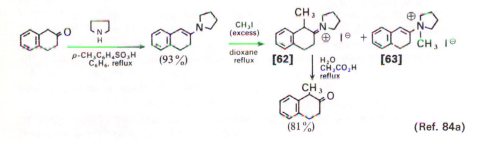

(81%)   (Ref. 84a)

with the original enamine to form a new enamine (e.g., [64]), which can react with the alkylating agent. In such cases, hydrolysis of the reaction mixture yields, as neutral products, a mixture of the starting ketone (from the salt of the starting

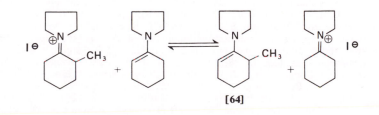

[64]

enamine), the monoalkylated product, and the dialkylated product(s). The following equations provide examples of enamine alkylation in which dialkylation has been observed; in the second example, the dialkylation is used to advantage in forming a new ring.

Frequently, the alkylation of enamines with simple alkylating agents is not a good preparative method because the major reaction is N-alkylation rather than

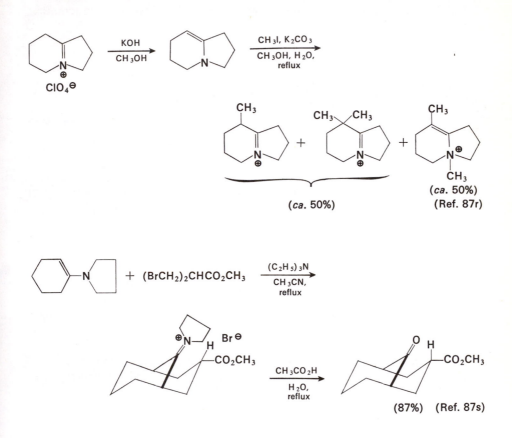

C-alkylation.[84,87a−e] The situation is similar to the previously discussed C- or O-alkylation of enolate anions; however, the enamine being uncharged, has no cation to shield the nitrogen from attack by the alkylating agent. Also, the nitrogen atom of enamines is softer (more polarizable) than the corresponding enolate oxygen atom. Both of these factors apparently contribute to making N-alkylation a serious competing reaction in attempts to alkylate enamines at carbon. The stabilized enamines (e.g., **[58]**) derived from β-dicarbonyl compounds react with alkylating agents to form either O-alkylated products (favored by protic solvents) or mixtures of O- and C-alkylated products (favored by aprotic solvents).[85c,d] However, good yields of C-alkylated products have been obtained from enamines with such very reactive alkylating agents as methyl halides, allyl halides, benzyl halides, or α-halocarbonyl compounds. The successful use of these reagents (e.g., **[67]**) may be attributable to the ability of initially formed

N-alkylated products to undergo either an intramolecular (e.g., [65]) or an inter-molecular (e.g., [66]) transfer of the alkyl group to carbon.[88a] The amount of C-alkylation obtained with aldehyde enamines has also been enhanced by the use of the relatively hindered secondary amine, n-butylisobutylamine, to form the enamine.[88b]

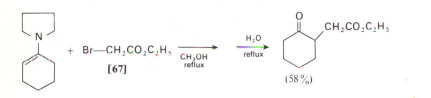

(Ref. 84b)

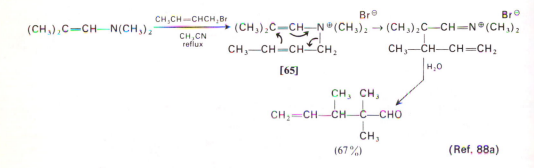

(Ref. 88a)

New carbon-carbon bonds have also been formed by the reaction of enamines with dichlorocarbene[88j-l] or by reaction of the corresponding iminium salts

88. (a) K. C. Brannock and R. D. Burpitt, *J. Org. Chem.*, **26**, 3576 (1961). (b) T. J. Curphey and J. C. Hung, *Chem. Commun.*, **No. 10**, 510 (1967). (c) F. E. Ziegler and P. A. Zoretic, *Tetrahedron Letters*, **No. 22**, 2639 (1968). (d) C. F. Hobbs and H. Weingarten, *J. Org. Chem.*, **33**, 2385 (1968). (e) M. E. Kuehne and T. Garbacik, *ibid.*, **35**, 1555 (1970). (f) For the use of epoxides to alkylate enamines, see P. Jakobsen and S. O. Lawesson, *Tetrahedron*, **24**, 3671 (1968); A. Z. Britten, W. S. Owen, and C. W. Went, *ibid.*, **25**, 3157 (1969). (g) For the alkylation of an enamine with N-carboethoxyaziridine, see J. E. Dolfini and J. D. Simpson, *J. Am. Chem. Soc.*, **87**, 4381 (1965). (h) G. H. Alt and A. J. Speziale, *J. Org. Chem.*, **31**, 1340, 2073 (1966). (i) A. Lukasiewicz and J. Lesinka, *Tetrahedron*, **24**, 7 (1968). (j) M. Ohno, *Tetrahedron Letters*, **No. 25**, 1753 (1963). (k) U. K. Pandit and S. A. G. deGraaf, *Chem. Commun.*, **No. 6**, 381 (1970). (l) J. Wolinsky and D. Chan, *ibid.*, **No. 16**, 567 (1966). (m) G. Stork and S. R. Dowd, *J. Am. Chem. Soc.*, **85**, 2178 (1963). (n) W. E. Harvey and D. S. Tarbell, *J. Org. Chem.*, **32**, 1679 (1967). (o) A. I. Meyers, A. Nabeya, H. W. Adickes, and I. R. Politzer, *J. Am. Chem. Soc.*, **91**, 763 (1969); A. I. Meyers, H. W. Adickes, I. R. Politzer, and W. N. Beverung, *ibid.*, **91**, 765 (1969); H. W. Adickes, I. R. Politzer, and A. I. Meyers, *ibid.*, **91**, 2155 (1969); A. I. Meyers, G. R. Malone, and H. W. Adickes, *Tetrahedron Letters*, **No. 42**, 3715 (1970). (p) J. M. Conia and P. Briet, *Bull. Soc. Chim. France*, 3881, 3888 (1966). (q) E. J. Corey and D. E. Cane, *J. Org. Chem.*, **35**, 3405 (1970).

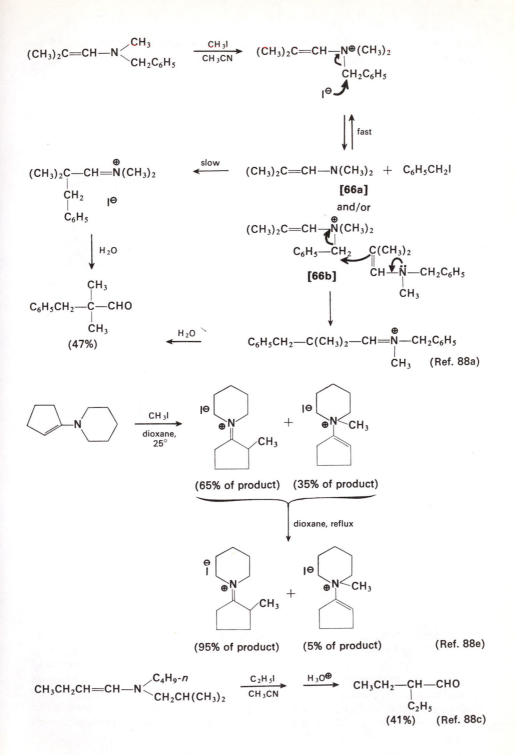

with the trichloromethyl anion.[88h,i] The following examples illustrate two of the several types of reaction which have been observed.

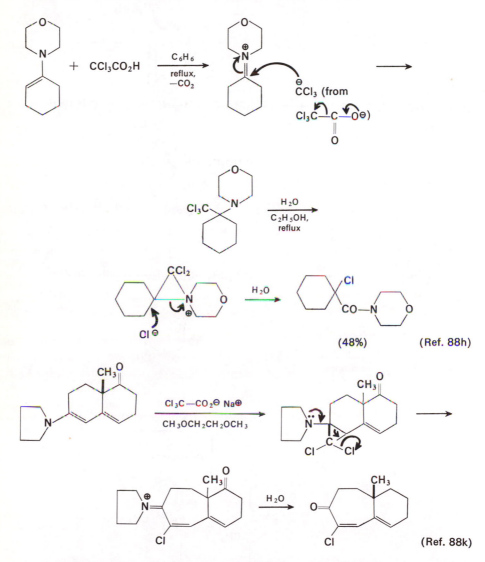

(48%) (Ref. 88h)

(Ref. 88k)

In a modification of the enamine alkylation procedure,[88m,n] an aldehyde (e.g., [68]) or ketone is allowed to react with an aliphatic amine (*t*-butylamine and cyclohexylamine have been used) to form an imine (e.g., [69]). Subsequent reaction with ethylmagnesium bromide forms a magnesium salt that undergoes C-alkylation when treated with alkylating agents. This modification appears to be distinctly more versatile than the original enamine procedure for the alkylation of aldehydes and ketones. Related procedures include the use of lithio derivatives

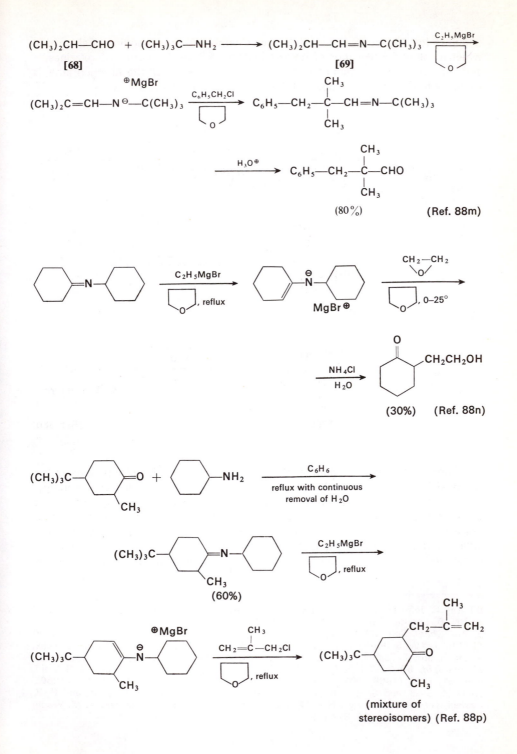

$$(CH_3)_2CH-CHO \; + \; (CH_3)_3C-NH_2 \longrightarrow (CH_3)_2CH-CH=N-C(CH_3)_3 \xrightarrow{C_2H_5MgBr}$$

[68]   **[69]**

$$(CH_3)_2C=CH-\overset{\ominus}{N}\overset{\oplus MgBr}{-}C(CH_3)_3 \xrightarrow[\text{O}]{C_6H_5CH_2Cl} C_6H_5-CH_2-\overset{\overset{\displaystyle CH_3}{|}}{\underset{\underset{\displaystyle CH_3}{|}}{C}}-CH=N-C(CH_3)_3$$

$$\xrightarrow{H_3O^{\oplus}} C_6H_5-CH_2-\overset{\overset{\displaystyle CH_3}{|}}{\underset{\underset{\displaystyle CH_3}{|}}{C}}-CHO$$

(80%)   (Ref. 88m)

$$\xrightarrow[\text{O}, \text{ reflux}]{C_2H_5MgBr}$$

$$\xrightarrow[\text{O}, 0-25°]{CH_2-CH_2 \diagdown O}$$

$$\xrightarrow[H_2O]{NH_4Cl}$$

(30%)   (Ref. 88n)

$$(CH_3)_3C- =O \; + \; -NH_2 \xrightarrow[\text{reflux with continuous removal of } H_2O]{C_6H_6}$$

$$(CH_3)_3C- =N- \xrightarrow[\text{O}, \text{ reflux}]{C_2H_5MgBr}$$

(60%)

$$(CH_3)_3C- \xrightarrow[\text{O}, \text{ reflux}]{CH_2=\overset{\overset{\displaystyle CH_3}{|}}{C}-CH_2Cl} (CH_3)_3C-$$

(mixture of stereoisomers) (Ref. 88p)

of imines for aldol condensations[65g,h] (see Chapter 10), a preparation of aldehydes which involves the alkylation of lithio derivatives of 2-alkyldihydro—1,3 oxazines,[88o] and the alkylation of lithio derivatives of ynamines in which the nitrogen atom is sterically hindered.[88q] Examples of the latter two procedures follow.

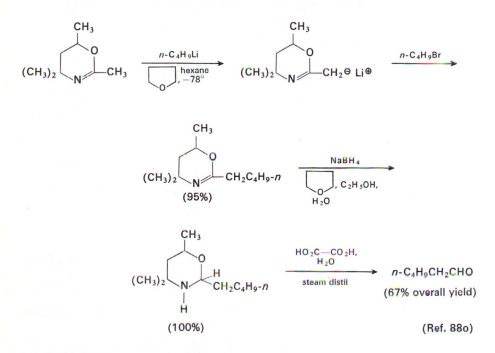

(Ref. 88o)

Another indirect procedure for the introduction of an alkyl group alpha to the carbonyl function involves the Claisen rearrangement[76h,89] of allyl and propargyl enol ethers. This reaction appears to be particularly useful for the allyl enol ethers of esters[89e] or amides[89f] which are generated and rearranged in the same reaction vessel without isolation of the enol ether intermediates. As the following examples indicate, this rearrangement is a concerted process involving a six-membered transition state; a chair conformation of this six-membered transition state is

89. (a) A. Jefferson and F. Scheinmann, *Quart. Rev.*, **22**, 391 (1968). (b) W. L. Howard and N. B. Lorette, *Org. Syn.*, **42**, 14, 34 (1962). (c) A. W. Burgstahler and I. C. Nordin, *J. Am. Chem. Soc.*, **83**, 198 (1961). (d) G. Buchi and J. E. Powell, Jr., *ibid.*, **92**, 3126 (1970). (e) W. S. Johnson and co-workers, *ibid.*, **92**, 741 (1970). (f) A. E. Wick, D. Felix, K. Steen, and A. Eschenmoser, *Helv. Chim. Acta*, **47**, 2425 (1964). (g) G. Saucy and R. Marbet, *ibid.*, **50**, 2091 (1967). (h) P. Vittorelli, T. Winkler, H. J. Hansen, and H. Schmid, *ibid.*, **51**, 1457 (1968). (i) H. M. Frey and B. M. Pope, *J. Chem. Soc.*, B, 209 (1966). (j) D. K. Black and S. R. Landor, *ibid.*, 6784 (1965). (k) R. Gardi, R. Vitali, and P. P. Castelli, *Tetrahedron Letters*, **No. 27**, 3203 (1966). (l) D. St. C. Black and A. M. Wade, *Chem. Commun.*, **No. 14**, 871 (1970).

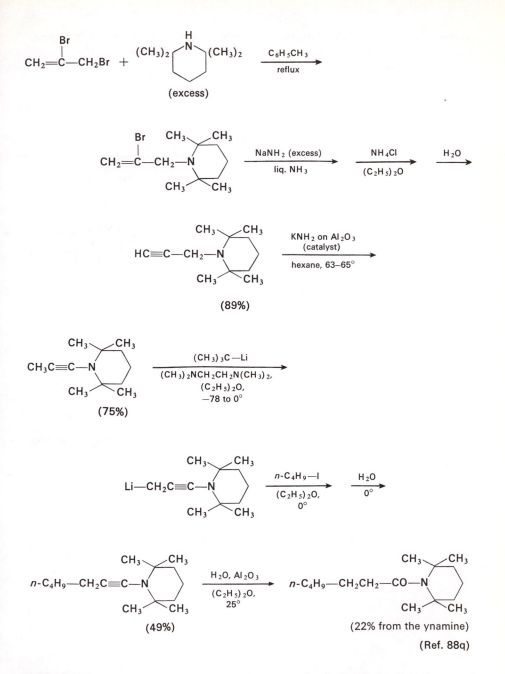

(Ref. 88q)

preferred.[89h]   Several of the accompanying examples indicate the high degree of stereospecificity which may be attained in these rearrangements. The third example illustrates how this stereospecificity may be rationalized in terms of chair

conformations for the transition states; the transition state leading to the minor isomer of the product is destabilized by a sterically unfavorable 1,3-diaxial inter-action (see heavy arrow).[89e]

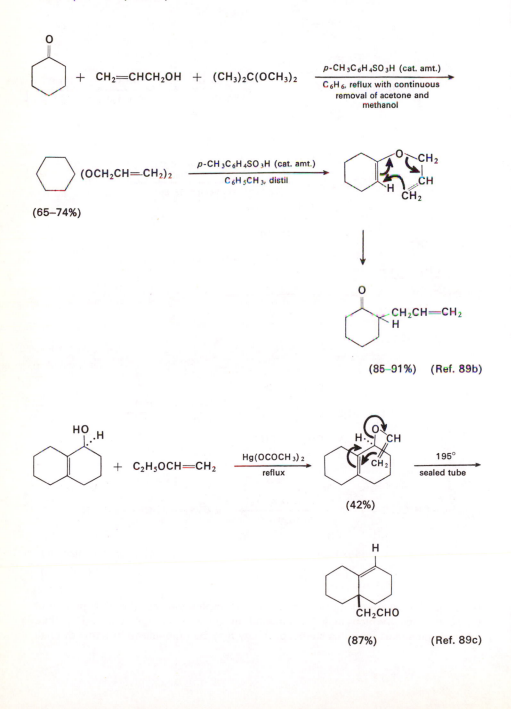

(65–74%)

(85–91%)   (Ref. 89b)

(42%)

(87%)   (Ref. 89c)

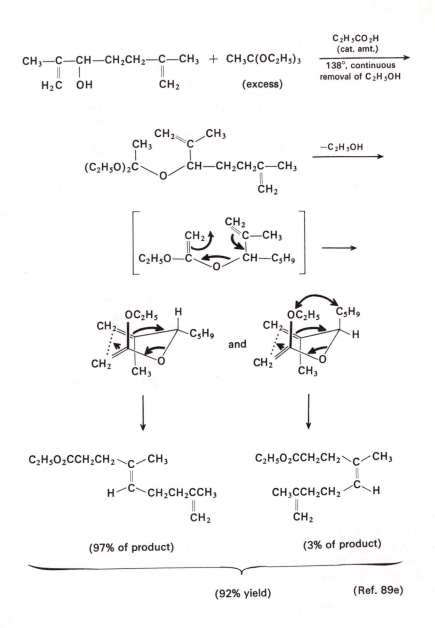

(97% of product)

(3% of product)

(92% yield)

(Ref. 89e)

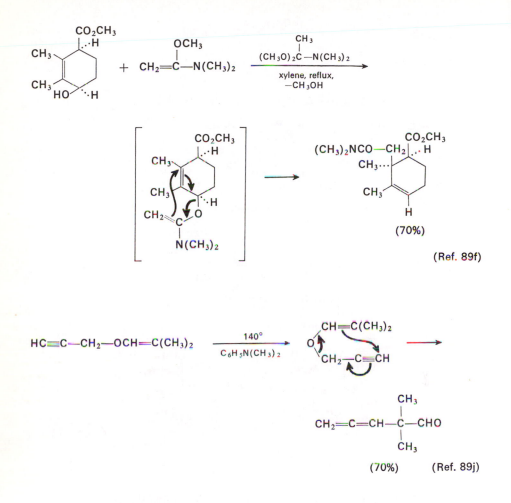

(70%)

(Ref. 89f)

(70%)   (Ref. 89j)

THE STEREOCHEMISTRY OF ALKYLATION

As noted earlier in this chapter, the reaction of an alkylating agent with the anion of an active methylene compound is usually a typical bimolecular nucleophilic displacement (S_N2) reaction, which results in inversion of configuration at the carbon atom of the alkylating agent where displacement occurs. The stereochemistry at the carbon atom alpha to the carbonyl function in the alkylated product cannot be predicted with such certainty.[90] If the alkylated alpha position still retains one hydrogen atom, the alkaline reaction conditions used will often permit isomerization at this position and the more stable epimer will predominate, irrespective of the initial direction of attack on the enolate anion. A case in point is the previously described methylation of 3-cholestanone[59e] to form 2α-methyl-3-cholestanone, the epimer in which the methyl group occupies the more stable

equatorial position.[90a] If the alkylated product has no hydrogen atom at the alkylated position, the stereochemical problem presented is similar to that encountered in the halogenation of ketones (see Chapter 8). In other words, the direction of attack on the enolate anion by the alkylating agent will be perpendicular to the plane of the enolate anion and, if one side of the enolate anion (e.g., [70]) is clearly less hindered, the predominant product will usually be the one resulting from introduction of the alkyl group from the less hindered side.

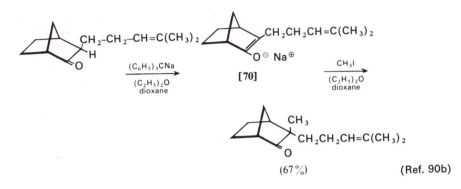

(67%)   (Ref. 90b)

If an enolate anion derived from a cyclohexanone is being alkylated, two factors need to be considered: steric hindrance to approach of the alkylating agent and the effect of stereoelectronic control. The following equation illustrates the possibility of stereoelectronic control in the alkylation of a cyclohexanone enolate.[7b,j,90c]

Attack by the alkylating agent from the direction labelled "Path A" can proceed directly by a chair-like transition state [71A] to the more stable chair conformation of the product with an axial alkyl group. The alternative attack, labelled Path E, leads via a boat-like transition state [71E] to an initial twist-boat conformation of

90. (a) For cases where part of the less stable isomer was isolated before isomerization occurred, see Refs. 7b and 7j and J. L. Beton, T. G. Halsall, E. R. H. Jones, and P. C. Phillips, J. Chem. Soc., 753 (1957). (b) E. J. Corey, R. Hartmann, and P. A. Vatakencherry, J. Am. Chem. Soc., 84, 2611 (1962). (c) L. Velluz, J. Valls, and G. Nomine, Angew. Chem., Intern. Ed. Engl., 4, 181 (1965); L. Velluz, J. Valls, and J. Mathieu, ibid., 6, 778 (1967). (d) E. Wenkert, A. Afonso, J. B. Bredenberg, C. Kaneko, and A. Tahara, J. Am. Chem. Soc., 86, 2038 (1964). (e) V. Permutti and Y. Mazur, J. Org. Chem., 31, 705 (1966). (f) R. E. Ireland and R. C. Kierstead, ibid., 31, 2543 (1966). (g) C. Djerassi, J. Osiecki, and E. J. Eisenbraun, J. Am. Chem. Soc., 83, 4433 (1961); M. V. Kulkarni, E. J. Eisenbraun and M. M. Marsh, J. Org. Chem., 33, 1661 (1968). (h) T. A. Spencer and co-workers, ibid., 33, 712, 719 (1968). (i) M. E. Kuehne and J. A. Nelson, ibid., 35, 161 (1970); M. E. Kuehne, ibid., 35, 171 (1970). (j) A. Afonso, ibid., 35, 1949 (1970). (k) P. Beak and T. L. Chaffin, ibid., 35, 2275 (1970). (l) C. L. Graham and F. J. McQuillin, J. Chem. Soc., 4634 (1963); F. J. McQuillin and P. L. Simpson, ibid., 4726 (1963); F. J. McQuillin and R. B. Yeats, ibid., 4273 (1965); F. H. Bottom and F. J. McQuillin, Tetrahedron Letters, No. 4, 459 (1968). (m) R. S. Matthews, P. K. Hyer, and E. A. Folkers, Chem. Commun., No. 1, 38 (1970); R. S. Matthews, S. J. Girgenti, and E. A. Folkers, ibid., No. 11, 708 (1970). (n) W. T. Pike, G. H. R. Summers, and W. Klyne, J. Chem. Soc., 5064, 7199 (1965). (o) D. K. Banerjee, S. N. Balasubrahmanyam, and R. Ranganathan, ibid., C, 1458 (1966). (p) M. Sharma, U. R. Ghatak, and P. C. Dutta, Tetrahedron, 19, 985 (1963).

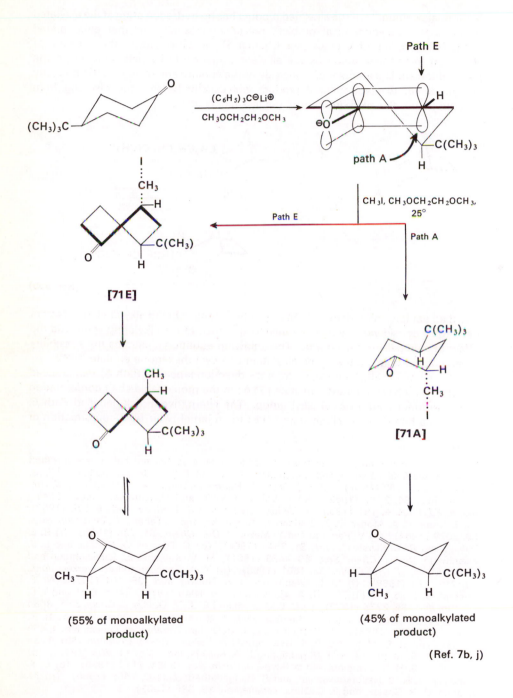

[71E]

(55% of monoalkylated
product)

[71A]

(45% of monoalkylated
product)

(Ref. 7b, j)

the alkylated product; conformational interconversion gives the more stable chair conformer of the cyclohexane product with an equatorial alkyl group. If the geometries of the transition states [71] resemble the initial products, then the chair-like transition state [71A] is expected to be of lower energy and the product with an axial alkyl group should predominate. On the other hand, if the transition state resembles the planar starting enolate, then very little difference is to be expected between the transition states [71].

As the example indicates, the latter circumstance is in better agreement with the experimental results for the alkylation of unhindered cyclohexanes. The stereochemical results for the alkylation of a variety of cyclohexanone derivatives have been examined.[7b,j,89p,90c−p] In view of the comparable activation energies for forming equatorial and axial alkylated products from a simple unhindered cyclohexanone, it is not surprising that changes in the structure of the starting metal enolate have led to results ranging from practically complete formation of axial alkylated products to the formation of essentially pure equatorial alkylated products. Although no studies have yet been reported it also seems possible that changes in factors (solvent, cation identity) which influence the degree of association and/or aggregation of the enolate may also influence the stereochemistry of these alkylation reactions.

Among the empirical correlations which may be drawn from published alkylation studies, two appear to have a reasonable generality: (1) if a cyclohexanone derivative already has one alkyl substituent at the alpha position, the proportion the second alkyl group introduced at this alpha position from an axial direction is enhanced;[7j,90f,g] (2) if the introduction of the alkyl group from an axial direction will produce a 1,3-diaxial interaction between alkyl groups, the product with an equatorial alkyl group is frequently favored.[90i−n] Of course, if one side of a cyclohexanone enolate anion is substantially more hindered than the other side, alkylation will occur from the less hindered side.[67b,90b] The following equations provide examples of stereochemical results which have been obtained.

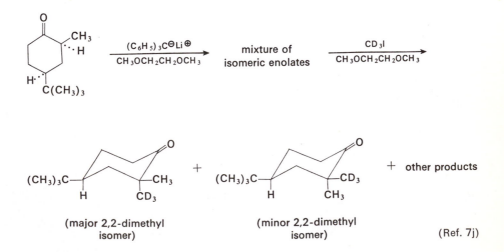

(major 2,2-dimethyl
isomer)

(minor 2,2-dimethyl
isomer)

(Ref. 7j)

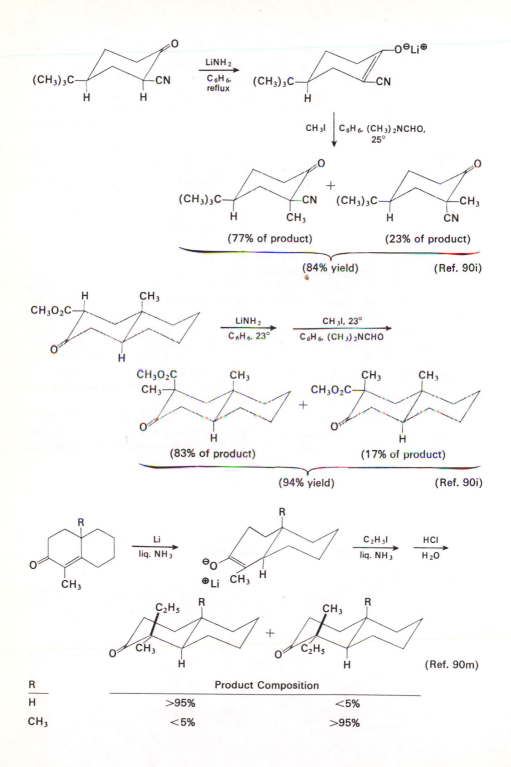

(77% of product)   (23% of product)

(84% yield)   (Ref. 90i)

(83% of product)   (17% of product)

(94% yield)   (Ref. 90i)

(Ref. 90m)

R	Product Composition	
H	>95%	<5%
CH₃	<5%	>95%

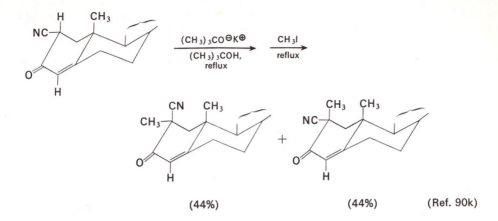

(44%)   (44%)   (Ref. 90k)

The presence of an additional carbon-carbon double bond in the six-membered ring of a cyclohexanone enolate usually appears to decrease the interference of a 1,3-diaxial interaction between an axial alkyl group and an alkylating agent entering from an axial direction.[901,p] When polar axial substituents such as cyano or carboethoxy groups rather than axial alkyl groups are present, the 1,3-diaxial interaction with an entering alkyl group does not prevent predominant alkylation from an axial direction as the following equation illustrates. At the present time it is not clear whether such results are best attributable to a diminished steric interaction or to the polar axial substituent favoring the attack of the alkylating agent from an axial direction.

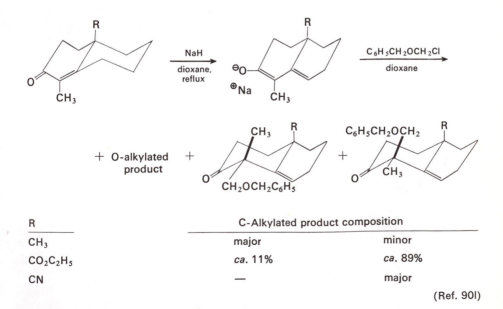

R	C-Alkylated product composition	
CH_3	major	minor
$CO_2C_2H_5$	ca. 11%	ca. 89%
CN	—	major

(Ref. 90l)

Even the alkylation of relatively simple 1-decalone (e.g., **[54]** and **[71]**) and perhydroindanone (e.g., **[72]**) systems at the bridgehead position leads to mixtures of stereoisomers in which the *cis*- and *trans*-isomers are formed in comparable amounts or in which the *cis*-isomer predominates. These results suggest that

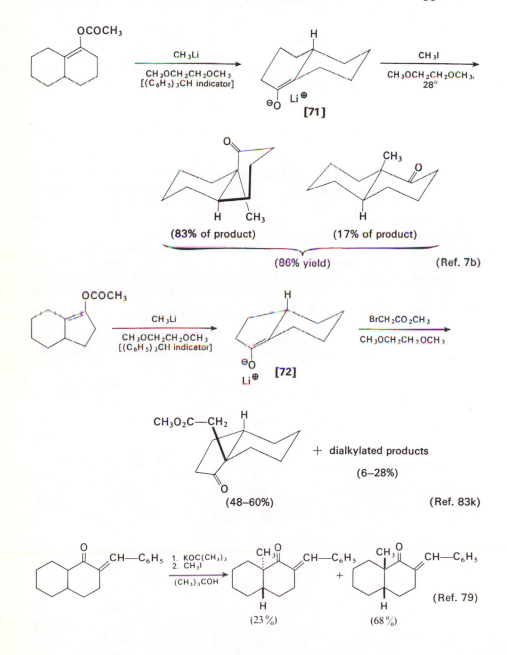

(83% of product)   (17% of product)

(86% yield)   (Ref. 7b)

(48–60%)   (Ref. 83k)

+ dialkylated products

(6–28%)

(23%)   (68%)   (Ref. 79)

stereoelectronic control is less important than steric hindrance in determining the direction of attack on an enolate anion. The presence of a 6,7-double bond in a 1-decalone system (as in [73] and [74]) favors the formation of a *trans*-fused ring

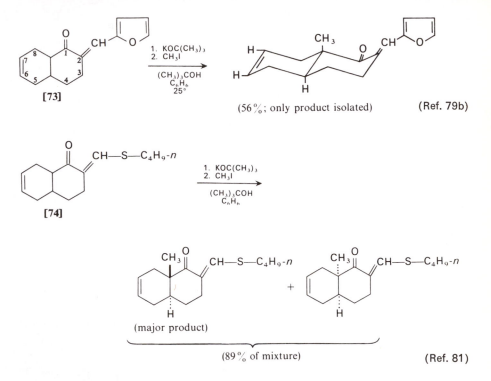

(56%; only product isolated)   (Ref. 79b)

(major product)

(89% of mixture)   (Ref. 81)

system on alkylation. This could be considered evidence for stereoelectronic control, which becomes more important as one of the 1,3-diaxial interactions (that between the axial hydrogen atom at C-7 and the methyl group) in a 9-methyl-*trans*-1-decalone is removed. However, the force of this argument is diminished by the fact that introduction of double bonds elsewhere in the 1-decalone system (which also diminishes the number of 1,3-diaxial interactions in the alkylated product) does not favor the formation of a *trans*-fused product with an axial methyl group.[79b]

Although the intramolecular alkylation with the bromoketone [75] to form 9-methyl-*cis*-1-decalone has been interpreted as evidence for a preferential axial attack on the enolate anion (i.e. [76])[39a,60b,60d] the geometry of the enolate anion [77] favors the formation of the *cis*-fused decalone in any event. The stereochemistry of alkylation of cyclohexyl ketones (e.g., [78]) to form predominantly products in which the alkylating agent has been introduced from an equatorial direction also seems best explained in terms of steric effects controlling the direction of reaction.[7b] However, the alkylation of enolate anions such as [78] would appear to be subject to the same influences from steric and torsional strain as the reduction of cyclohexanone derivatives with metal hydrides (see Chapter 2) to

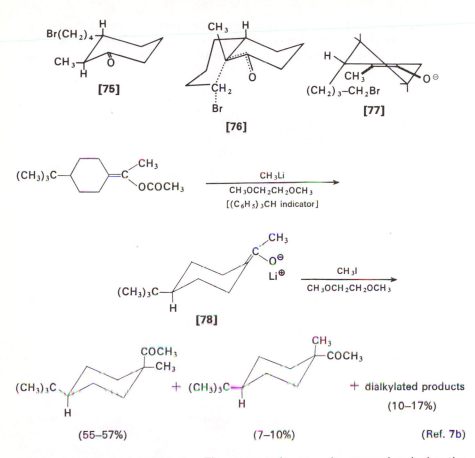

91. (a) W. G. Kenyon, R. B. Meyer, and C. R. Hauser, *J. Org. Chem.*, **28**, 3108 (1963). (b) D. M. von Schriltz, K. G. Hampton, and C. R. Hauser, *ibid.*, **34**, 2509 (1969). (c) S. Karady, M. Lenfant, and R. E. Wolff, *Bull. Soc. Chim. France*, 2472 (1965).

form mainly equatorial alcohols. The reason why opposite stereochemical paths are preferred in these cases is not apparent.

Although investigations of the stereochemical results of alkylating acyclic enolates have not been extensive, several examples have been reported in which there is a clear preference for the formation of the *erythro*-isomer.[91a,b] The following equation exemplifies these results; the illustrated transition state leading to the *erythro*-isomer is suggested to have less non-bonded steric interaction than the corresponding transition state leading to the *threo*-isomer.[91a,b]

In spite of the fact that the alkylation of enamines at carbon is often complicated by competing N-alkylation, the C-alkylated product obtained is often formed very stereoselectively.[91c] Thus, both the C-alkylation and the protonation of enamines frequently show a much higher tendency to occur from an axial direction than is the case for the corresponding reactions with enolate anions.[8c,88p]

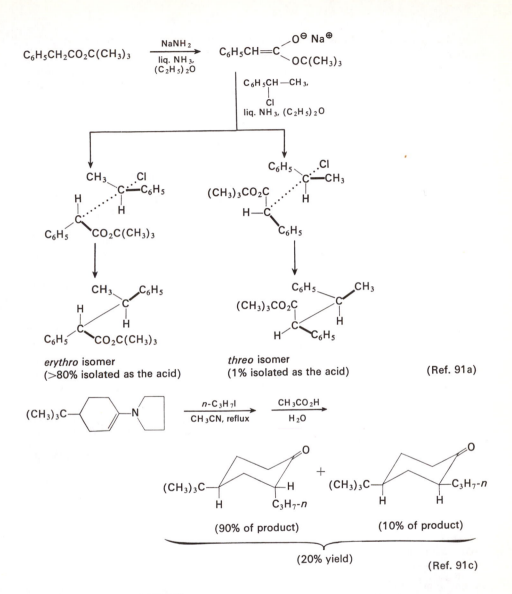

erythro isomer
(>80% isolated as the acid)

threo isomer
(1% isolated as the acid)

(Ref. 91a)

(90% of product)

(10% of product)

(20% yield)

(Ref. 91c)

THE MICHAEL REACTION

The nucleophilic addition of enolate (or analogous) anions (e.g., [79]) to the carbon-carbon double bond of α,β-unsaturated ketones, aldehydes, nitriles, or carboxylic acid derivatives, a process known as the Michael reaction,[92] also

92. (a) E. D. Bergmann, D. Ginsburg, and R. Pappo, *Org. Reactions,* **10**, 179 (1959). (b) H. A. Bruson, *ibid.,* **5**, 79 (1949). (c) F. Kröhnke and W. Zecher, *Angew. Chem., Intern. Ed. Engl.,* **1**, 626 (1962). (d) L. A. Kaplan and D. J. Glover, *J. Am. Chem. Soc.,* **88**, 84 (1966). (e) J. A. Markisz and J. D. Gettler, *Can. J. Chem.,* **47**, 1965 (1969).

constitutes a method for the alkylation of active methylene compounds. The unsaturated compounds used in the reaction, often called Michael acceptors, may include any unsaturated system having a functional group capable of stabilizing the carbanionic intermediate (e.g., **[80]**). The Michael reaction differs from

$$CH_2(CO_2C_2H_5)_2 \xrightarrow[\substack{C_2H_5OH \\ 25-35°}]{NaOC_2H_5} {}^{\ominus}CH(CO_2C_2H_5)_2 \xrightarrow{CH_2=CH-CN}$$

[79]

$$\left.\begin{array}{c} N\equiv C-CH-CH_2-CH(CO_2C_2H_5)_2 \\ {}^{\ominus}\quad \textbf{[80]} \\ \Updownarrow \\ N\equiv C-CH_2-CH_2-C^{\ominus}(CO_2C_2H_5)_2 \end{array}\right\} \xrightarrow{H_2O} N\equiv C-CH_2-CH_2-CH(CO_2C_2H_5)_2$$

$$(57-63\%)$$

(Ref. 93a)

previously discussed alkylation reactions in that the base which generates the enolate anion is regenerated, so that usually only a catalytic amount is required. Furthermore, the rate-limiting reaction step (i.e. **[79]** → **[80]**) which forms a new carbon-carbon bond is reversible,[93b-e] and the product, usually a 1,5-dicarbonyl compound, is frequently capable of further transformation in the presence of base. The Michael acceptors may also add a variety of nucleophiles other than enolate anions such as alcohols, thiols, or amines.[93b,f-p] In addition, many α,β-unsaturated ketones which might be used as Michael acceptors undergo self-condensation reactions (either two successive Michael reactions or a Michael reaction followed by an aldol condensation) in the presence of bases.[94b-j] These many possibilities

93. (a) N. F. Albertson and J. F. Fillman, *J. Am. Chem. Soc.*, **71**, 2818 (1949). (b) S. Patai and Z. Rappoport, *J. Chem. Soc.*, 377, 383, 392, 396 (1962); S. Patai, S. Weinstein, and Z. Rappoport, *ibid.*, 1741 (1962); Z. Grünbaum, S. Patai, and Z. Rappoport, *ibid.*, B, 1133 (1966); H. Shenhav, Z. Rappoport, and S. Patai, *ibid.*, B, 469 (1970). (c) J. H. Burckhalter and B. A. Brown, *J. Org. Chem.*, **30**, 1291 (1965). (d) P. Beak and B. M. Monroe, *ibid.*, **32**, 2778 (1967). (e) H. Junek and H. Sterk, *Tetrahedron Letters*, **No. 40**, 4309 (1968). (f) C. H. McMullen and C. J. M. Stirling, *J. Chem. Soc.*, B, 1217, 1221 (1966); S. T. McDowell and C. J. M. Stirling, *ibid.*, B, 343, 348, 351 (1967). (g) E. Winterfeldt, *Angew. Chem., Intern. Ed. Engl.*, **6**, 423 (1967). (h) R. Huisgen, B. Giese, and H. Huber, *Tetrahedron Letters*, **No. 20**, 1883 (1967); B. Giese and R. Huisgen, *ibid.*, **No. 20**, 1889 (1967). (i) F. M. Menger and J. H. Smith, *J. Am. Chem. Soc.*, **91**, 4211 (1969). (j) P. Chamberlain and G. H. Whitham, *J. Chem. Soc.*, B, 1131 (1969). (k) B. A. Feit and A. Zilkha, *J. Org. Chem.*, **28**, 406 (1963); B. A. Feit, J. Sinnreich, and A. Zilkha, *ibid.*, **32**, 2570 (1967); B. A. Feit and Z. Bigon, *ibid.*, **34**, 3942 (1969). (l) R. N. Ring, G. C. Tesoro, and D. R. Moore, *ibid.*, **32**, 1091 (1967). (m) C. M. Wynn and W. R. Vaughan, *ibid.*, **33**, 2371 (1968). (n) W. E. Truce and D. G. Brady, *ibid.*, **31**, 3543 (1966). (o) J. B. Hendrickson, R. Rees, and J. F. Templeton, *J. Am. Chem. Soc.*, **86**, 107 (1964). (p) For the use of a Michael adduct from *n*-butyl mercaptan as a protecting group for an α-methylene lactone, see S. M. Kupchan, T. J. Giacobbe, and I. S. Krull, *Tetrahedron Letters*, **No. 33**, 2859 (1970).
94. (a) R. Connor and D. B. Andrews, *J. Am. Chem. Soc.*, **56**, 2713 (1934). (b) D. W. Theobald, *Tetrahedron*, **23**, 2767 (1967). (c) W. A. Ayer and W. I. Taylor, *J. Chem. Soc.*, 2227 (1955). (d) J. Grimshaw and W. B. Jennings, *ibid.*, C, 817 (1970). (e) A. T. Nielsen, *J. Org. Chem.*, **28**, 2115 (1963). (f) A. T. Nielsen and H. J. Dubin, *ibid.*, **28**, 2120 (1963). (g) A. T. Nielsen and S. Haseltine, *ibid.*, **33**, 3264 (1968). (h) A. T. Nielsen and D. W. Moore, *ibid.*, **34**, 444 (1969). (i) A. T. Nielsen, H. Dubin, and K. Hise, *ibid.*, **32**, 3407 (1967). (j) G. Kabas and H. C. Rutz, *Tetrahedron*, **22**, 1219 (1966).

permit a variety of side reactions[92a] not encountered in the previously discussed alkylations. Consequently, it is generally desirable to use the mildest reaction conditions possible for effecting any given Michael reaction.

It will be noted that many Michael products may, in principle, be obtained from either of two different pairs of reactants. For example, a combination either of benzalacetophenone and diethyl malonate or of diethyl benzalmalonate ([81]) and acetophenone ([82]) could theoretically effect preparation of the adduct [83]. Of these possibilities, diethyl malonate rather than acetophenone as the active

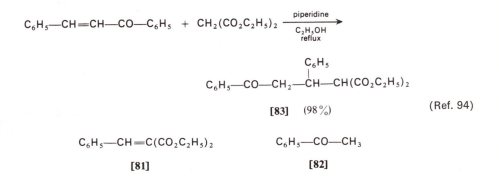

$$C_6H_5-CH=CH-CO-C_6H_5 \ + \ CH_2(CO_2C_2H_5)_2 \ \xrightarrow[\substack{C_2H_5OH \\ reflux}]{piperidine}$$

$$C_6H_5-CO-CH_2-\overset{\displaystyle \overset{C_6H_5}{|}}{C}H-CH(CO_2C_2H_5)_2$$

[83]   (98%)   (Ref. 94)

$$C_6H_5-CH=C(CO_2C_2H_5)_2 \qquad\qquad C_6H_5-CO-CH_3$$

[81]   [82]

methylene component is the proper choice because formation of its anion can be accomplished with a weaker base and less vigorous reaction conditions. When possible, relatively weak basic catalysts such as piperidine, pyridine, triethylamine, benzyltrimethylammonium hydroxide (Triton B), or potassium hydroxide should be selected. If stronger bases (e.g., sodium ethoxide, potassium t-butoxide, sodium hydride, or a metal amide) are required, it is normally appropriate to use only 0.1 to 0.3 equivalent of the base and to employ low reaction temperatures (25° or less) and short reaction times in order to minimize side reactions.[95] It is normally best to perform the reaction in the presence of a protic solvent (e.g., water or an alcohol) if further transformation of the initially formed Michael adduct is to be minimized.[77b,95d−i] Also, use of an excess of the active methylene compound is recommended if dialkylation and reverse Michael reactions are to be minimized. Examples of these various procedures are illustrated in the previous and following equations.

95. (a) R. Connor and W. R. McClellan, *J. Org. Chem.*, **3**, 570 (1939). (b) J. A. Gardner and H. N. Rydon, *J. Chem. Soc.*, 45 (1938). (c) H. Wachs and O. F. Hedenburg, *J. Am. Chem. Soc.*, **70**, 2695 (1948). (d) D. J. Goldsmith and J. A. Hartman, *J. Org. Chem.*, **29**, 3520, 3524 (1964). (e) T. A. Spencer, M. D. Newton, and S. W. Baldwin, *ibid.*, **29**, 787 (1964); T. A. Spencer and M. D. Newton, *Tetrahedron Letters*, **No. 22**, 1019 (1962). (f) R. A. Abramovitch and D. L. Newton, *Tetrahedron*, **24**, 357 (1968). R. A. Abramovitch, M. M. Rogić, S. S. Singer, and N. Venkateswaran, *J. Am. Chem. Soc.*, **91**, 1571 (1969). (g) N. C. Ross and R. Levine, *J. Org. Chem.*, **29**, 2341, 2346 (1964). (h) J. J. Beereboom, *ibid.*, **31**, 2026 (1966). (i) R. Selvarajan, J. P. John, K. V. Narayanan, and S. Swaminathan, *Tetrahedron*, **22**, 949 (1966).

$(CH_3)_2CH—NO_2$ + $CH_2\!\!=\!\!CH—CO_2CH_3$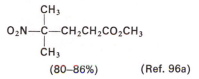

$$O_2N—\underset{\underset{CH_3}{|}}{\overset{\overset{CH_3}{|}}{C}}—CH_2CH_2CO_2CH_3$$

(80–86%)   (Ref. 96a)

$CH_2(CO_2C_2H_5)_2$ + $CH_3COCH\!\!=\!\!CH_2$ $\xrightarrow[\substack{C_2H_5OH, \\ -10 \text{ to } 25°}]{\substack{NaOC_2H_5 \\ (cat.\ amt.)}}$ $CH_3COCH_2CH_2CH(CO_2C_2H_5)_2$

(71%)

(Ref. 95e, g)

$$C_6H_5—\underset{\underset{CO_2C_2H_5}{}}{\overset{\overset{C\equiv N}{|}}{CH}}—CO_2C_2H_5$$ + $CH_2\!\!=\!\!CH—C\!\!\equiv\!\!N$ $\xrightarrow[\substack{(CH_3)_3COH \\ 40-45°}]{KOH}$ $C_6H_5—\underset{\underset{CO_2C_2H_5}{|}}{\overset{\overset{C\equiv N}{|}}{C}}—CH_2CH_2C\!\!\equiv\!\!N$

(69–83%)

(Ref. 97)

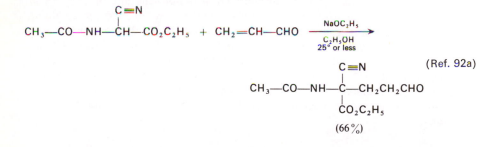

$$CH_3—CO—NH—\underset{\underset{CO_2C_2H_5}{|}}{\overset{\overset{C\equiv N}{|}}{CH}}—CO_2C_2H_5$$ + $CH_2\!\!=\!\!CH—CHO$ $\xrightarrow[\substack{C_2H_5OH \\ 25° \text{ or less}}]{NaOC_2H_5}$

(Ref. 92a)

$$CH_3—CO—NH—\underset{\underset{CO_2C_2H_5}{|}}{\overset{\overset{C\equiv N}{|}}{C}}—CH_2CH_2CHO$$

(66%)

96. (a) R. B. Moffett, *Org. Syn., Coll. Vol. 4*, 652 (1963). (b) For other examples of the use of nitro compounds in the Michael reaction, see G. N. Walker, *J. Org. Chem.*, **30**, 1416 (1965). (c) For the use of unsaturated sulfones as Michael acceptors, see C. S. Argyle, S. C. Goadby, K. G. Mason, R. A. Reed, M. A. Smith, and E. S. Stern, *J. Chem. Soc.*, C, 2156 (1967). (d) Both C- and O-alkylated products were obtained from the cyanoethylation of 2-naphthol; K. H. Takemura, K. E. Dreesen, and M. A. Petersen, *J. Org. Chem.*, **32**, 3412 (1967). (e) γ-Pyrones [T. M. Harris and C. M. Harris, *ibid.*, **32**, 970 (1967)] and cyclo-hexadienones [E. Wenkert, F. Haviv, and A. Zeitlin, *J. Am. Chem. Soc.*, **91**, 2299 (1969)] have also been used as Michael acceptors.
97. E. C. Horning and A. F. Finelli, *Org. Syn., Coll. Vol. 4*, 776 (1963).

$(CH_3\!-\!CO)_2CH_2 \;+\; CH_2\!=\!CH\!-\!C\!\equiv\!N \xrightarrow[\substack{(CH_3)_3COH,\; H_2O \\ 25°}]{(C_2H_5)_3N} (CH_3\!-\!CO)_2C(CH_2CH_2C\!\equiv\!N)_2$

(77%) (Ref. 98)

$C_6H_5COCH_2COCH_3 \xrightarrow[\text{liq. NH}_3]{KNH_2 \;(2\;\text{equiv.})} C_6H_5\overset{\overset{\displaystyle\ominus}{O}}{C}\!=\!CH\!-\!\overset{\overset{\displaystyle\ominus}{O}}{C}\!=\!CH_2 \;\; 2K^{\oplus} \xrightarrow[\substack{\text{liq. NH}_3, \\ (C_2H_5)_2O}]{C_6H_5CH=CHCO_2C(CH_3)_3}$

$\xrightarrow[\text{H}_2O]{NH_4Cl} C_6H_5\!-\!\underset{\underset{\displaystyle CH_2COCH_2COC_6H_5}{|}}{CH}\!-\!CH_2CO_2C(CH_3)_3$

(71%) (Ref. 98b)

In certain cases, it has also been possible to effect acid-catalyzed Michael reactions;[98c-e] two examples, shown in the following equations, both involve further transformation of the intermediate 1,5-dicarbonyl compounds.

$(CH_3)_2C\!=\!CHCOCH_3 \;+\; CH_3COCH_2CO_2C_2H_5 \xrightarrow[\substack{C_7H_{16},\; C_6H_6, \\ \text{reflux with continuous} \\ \text{separation of water}}]{ZnCl_2}$

$$\left[\begin{array}{l} (CH_3)_2C\!-\!\!-\!CH\!-\!CO_2C_2H_5 \\ \quad\;\; | \qquad\quad | \\ \quad\; CH_2 \quad CO \\ \quad\;\; | \qquad\quad | \\ \quad\; CO \quad\;\; CH_3 \\ \quad\;\; | \\ \quad\; CH_3 \end{array} \right] \xrightarrow{-H_2O}$$

(19%)   +   (32%)   +   (8%)   (Ref. 98c)

98. (a) J. A. Adamcik and E. J. Miklasiewicz, *J. Org. Chem.*, **28**, 336 (1963). (b) F. B. Kirby, T. M. Harris, and C. R. Hauser, *ibid.*, **28**, 2266 (1963). (c) J. D. Surmatis, A. Walser, J. Gibas, and R. Thommen, *ibid.*, **35**, 1053 (1970). (d) J. A. VanAllan and G. A. Reynolds, *ibid.*, **33**, 1102 (1968). (e) K. Dimroth, C. Reichardt, and K. Vogel, *Org. Syn.*, **49**, 121 (1969).

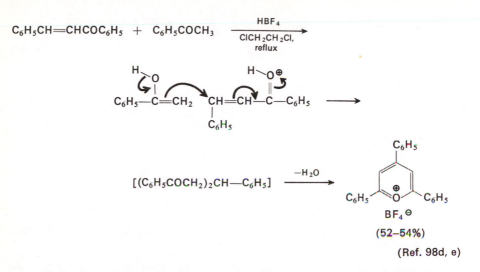

(52–54%)

(Ref. 98d, e)

The use of a full equivalent of base, elevated reaction temperatures, and long reaction times frequently promotes reversal of the Michael reaction (called a retrograde Michael reaction) or further transformation of the initial product. Since a retrograde Michael reaction may lead to compounds other than the original starting materials (e.g., [81] and [82] from [83]), complex mixtures may result. Examples of further transformations of Michael products are provided in the accompanying equations; it will be noted that these transformations usually lead to a product more acidic (i.e. the corresponding anion is more stable) than the initial Michael adduct.

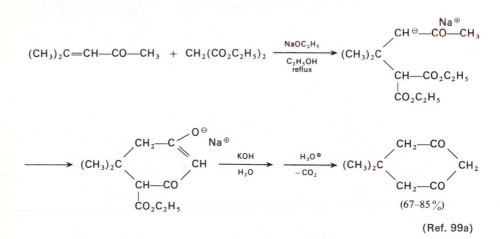

(67–85%)

(Ref. 99a)

99. (a) R. L. Shriner and H. R. Todd, *Org. Syn.,* **Coll. Vol. 2**, 200 (1943). For other examples of this type of cyclization of Michael adducts, see Refs. 95e and 95h. (b) P. Bladon and T. Sleigh, *J. Chem. Soc.*, 3264 (1962).

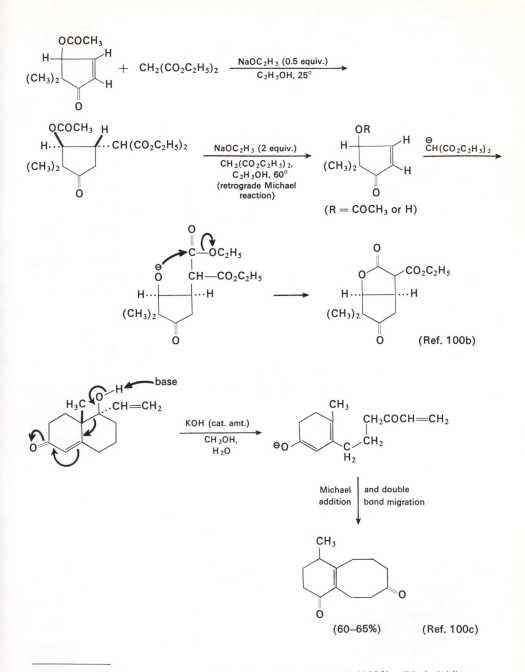

(Ref. 100b)

(60–65%)   (Ref. 100c)

100. (a) A. Michael and J. Ross, *J. Am. Chem. Soc.,* **52**, 4598 (1930). (b) A. Ichihara, J. Morita, K. Kobayashi, S. Kagawa, H. Shirahama, and T. Matsumoto, *Tetrahedron,* **26**, 1331 (1970). (c) S. Swaminathan, J. P. John, and S. Ramachandran, *Tetrahedron Letters,* **No. 16**, 729 (1962).

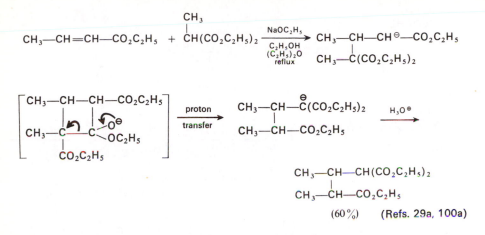

A variant of the usual procedure for the Michael reaction consists of treating the active methylene compound with a β-halocarbonyl compound, a β-dialkyl-aminocarbonyl compound, or the quaternary salt from a β-dialkylaminocarbonyl compound. These reactants are rapidly converted to α,β-unsaturated carbonyl compounds in the reaction mixture by base-catalyzed elimination,[101,102a] and a normal Michael reaction then occurs. The β-halocarbonyl compounds and corresponding quaternary ammonium salts consume a full equivalent of base during the elimination reaction; consequently, Michael reactions using these materials are often run like conventional alkylation reactions. Perhaps the most useful precursors of Michael acceptors are the β-dialkylaminocarbonyl compounds and corresponding quaternary salts,[102a] which are readily available from Mannich reactions (see Chapter 10) involving a ketone, formaldehyde, and a secondary amine. Since vinyl ketones are frequently unstable liquids that tend to dimerize or polymerize on standing, the generation of these materials from β-aminoketones in the reaction mixture is advantageous.

β-Chlorovinyl ketones as well as the corresponding nitriles and esters have also been studied as Michael acceptors for both enolate anions and, especially, for alkoxide ions, thiolate ions, and amines.[77d,102b-f] The geometrical isomers of

101. (a) N. Ferry and F. J. McQuillin, *J. Chem. Soc.,* 103 (1962). (b) G. L. Buchanan, A. C. W. Curran, J. M. McCrae, and G. W. McLay, *Tetrahedron,* **23,** 4729 (1967); H. L. Brown, G. L. Buchanan, A. C. W. Curran, and G. W. McLay, *ibid.,* **24,** 4565 (1968); G. L. Buchanan, A. C. W. Curran, and R. T. Wall, *ibid.,* **25,** 5503 (1969); E. M. Austin, H. L. Brown, and G. L. Buchanan, *ibid.,* **25,** 5509 (1969); E. M. Austin, H. L. Brown, G. L. Buchanan, and R. A. Raphael, Jr., *ibid.,* **25,** 5517 (1969). (c) L. B. Barkley and R. Levine, *J. Am. Chem. Soc.,* **72,** 3699 (1950); (d) S. Danishefsky and B. H. Migdalof, *ibid.,* **91,** 2806 (1969); *Tetrahedron Letters,* **No. 50,** 4331 (1969).
102. (a) J. H. Brewster and E. L. Eliel, *Org. Reactions,* **7,** 99 (1953). (b) A. E. Pohland and W. R. Benson, *Chem. Rev.,* **66,** 161 (1966). (c) D. E. Jones, R. O. Morris, C. A. Vernon, and R. F. M. White, *J. Chem. Soc.,* 2349 (1960). (d) D. Landini, F. Montanari, G. Modena, and F. Naso, *ibid.,* B, 243 (1969); B. Cavalchi, D. Landini, and F. Montanari, *ibid.,* C, 1204 (1969). (e) F. Scotti and E. J. Frazza, *J. Org. Chem.,* **29,** 1800 (1964). (f) W. E. Truce and M. L. Gorbaty, *ibid.,* **35,** 2113 (1970).

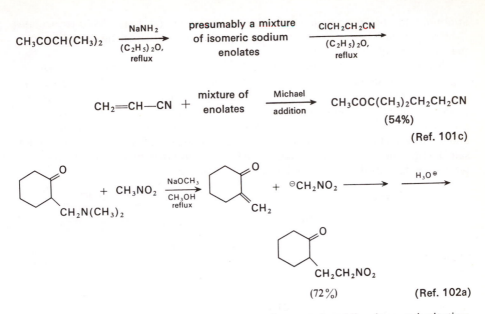

$$CH_3COCH(CH_3)_2 \xrightarrow[\substack{(C_2H_5)_2O, \\ reflux}]{NaNH_2} \begin{array}{c} \text{presumably a mixture} \\ \text{of isomeric sodium} \\ \text{enolates} \end{array} \xrightarrow[\substack{(C_2H_5)_2O, \\ reflux}]{ClCH_2CH_2CN}$$

$$CH_2{=}CH{-}CN + \begin{array}{c} \text{mixture of} \\ \text{enolates} \end{array} \xrightarrow[\text{addition}]{\text{Michael}} CH_3COC(CH_3)_2CH_2CH_2CN$$

(54%)

(Ref. 101c)

these chlorovinyl compounds react with various nucleophiles in a substitution reaction which occurs with retention of configuration.

A reaction mechanism involving the addition-elimination sequence indicated in the following equation has been suggested; the stereospecificity has been

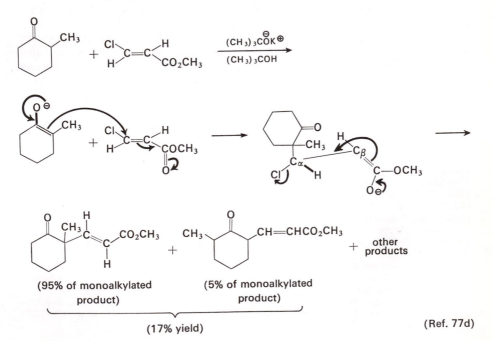

(95% of monoalkylated product)

+

(5% of monoalkylated product)

+ other products

(17% yield)

(Ref. 77d)

attributed to the immediate elimination of chloride ion from the intermediate β-chloro enolate before conformational equilibrium is attained by rotation about the C_α—C_β single bond.[102c,e]

An interesting and commercially valuable synthetic procedure which would appear to be related to the previously illustrated acid-catalyzed Michael reaction, is the condensation of a 6-hydroxy- or 6-alkoxy-1-vinyl-1-tetralol (or the corresponding amine[103c] or isothiouronium salt[103e,f]) with a 1,3-dicarbonyl compound.[103] As illustrated below, the reaction is believed to involve proton transfer from the enolic diketone to the allylic alcohol. Subsequent loss of water affords the oxonium salt [84] (or the corresponding ketone if a 6-hydroxytetralin derivative is employed) which reacts with the enolate anion to form the product.

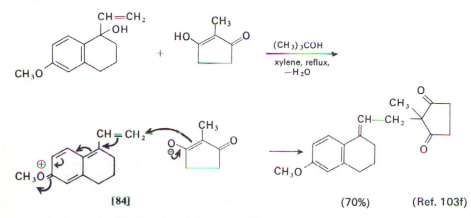

[84]   (70%)   (Ref. 103f)

Although monoketones may be used as active methylene compounds in the Michael reaction with very reactive Michael acceptors such as acrylonitrile [85], or acrylic acid esters, the reaction conditions required with less reactive unsaturated systems are often sufficiently vigorous that further transformations of the initial Michael adducts are observed. From several of the accompanying examples it will be noted that the Michael acceptor is normally introduced largely, if not exclusively, at the more highly substituted position of unsymmetrical ketones.[77d,92a,95g,101c,104,105]

103. (a) S. N. Ananchenko, V. Y. Limanov, V. N. Leonov, V. N. Rzhezniko, and I. V. Torgov, *Tetrahedron,* **18**, 1355 (1962). (b) A. B. Zakharichev, S. N. Ananchenko, and I. V. Torgov, *Tetrahedron Letters,* **No. 3,** 171 (1964); S. N. Ananchenko and I. V. Torgov, *ibid.,* **No. 23,** 1553 (1963); A. V. Zakharychev, D. R. Lagidze, and S. N. Ananchenko, *ibid.,* **No. 9,** 803 (1967). (c) U. K. Pandit, F. A. van der Vlugt, and A. C. van Dalen, *ibid.,* **No. 42,** 3697 (1969). (d) T. B. Windholz, J. H. Fried, and A. A. Patchett, *J. Org. Chem.,* **28,** 1092 (1963); G. Modena, *Accts. Chem. Res.,* **4,** 73 (1971). (e) R. D. Hoffsommer, D. Taub, and N. L. Wendler, *J. Org. Chem.,* **32,** 3074 (1967). (f) C. H. Kuo, D. Taub, and N. L. Wendler, *ibid.,* **33,** 3126 (1968).
104. (a) H. A. Bruson and T. W. Riener, *J. Am. Chem. Soc.,* **64,** 2850 (1942); R. Bertocchio and J. Dreux, *Bull. Soc. Chim. France,* 823, 1809 (1962). (b) J. J. Miller and P. L. deBenneville, *J. Org. Chem.,* **22,** 1268 (1957).
105. (a) R. L. Frank and R. C. Pierle, *J. Am. Chem. Soc.,* **73,** 726 (1951). (b) See also V. Boekelheide, *ibid.,* **69,** 790 (1947); W. E. Bachmann and L. B. Wick, *ibid.,* **72,** 3388 (1950).

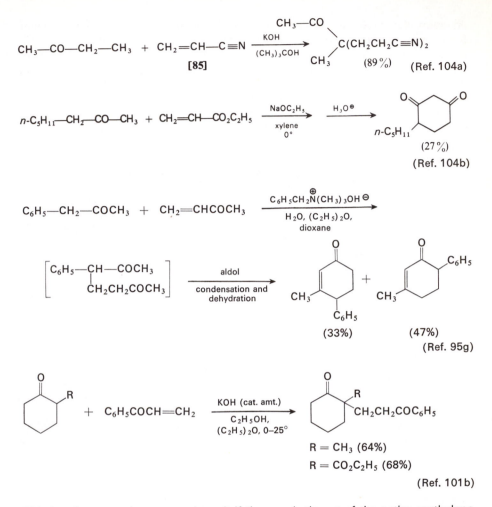

$$CH_3—CO—CH_2—CH_3 \ + \ CH_2{=}CH—C{\equiv}N \xrightarrow[\text{(CH}_3)_3\text{COH}]{\text{KOH}}$$

(89%)   (Ref. 104a)

$$n\text{-}C_5H_{11}—CH_2—CO—CH_3 \ + \ CH_2{=}CH—CO_2C_2H_5 \xrightarrow[\text{xylene} \atop 0°]{\text{NaOC}_2\text{H}_5} \xrightarrow{\text{H}_3\text{O}^{\oplus}}$$

(27%)
(Ref. 104b)

$$C_6H_5—CH_2—COCH_3 \ + \ CH_2{=}CHCOCH_3 \xrightarrow[\text{H}_2\text{O, (C}_2\text{H}_5)_2\text{O,} \atop \text{dioxane}]{C_6H_5CH_2\overset{\oplus}{N}(CH_3)_3OH^{\ominus}}$$

$$\left[\begin{array}{c} C_6H_5—CH—COCH_3 \\ CH_2CH_2COCH_3 \end{array} \right] \xrightarrow[\text{condensation and} \atop \text{dehydration}]{\text{aldol}}$$

(33%)   (47%)
(Ref. 95g)

$$+ \ C_6H_5COCH{=}CH_2 \xrightarrow[\text{C}_2\text{H}_5\text{OH,} \atop \text{(C}_2\text{H}_5)_2\text{O, 0–25°}]{\text{KOH (cat. amt.)}}$$

R = CH$_3$ (64%)
R = CO$_2$C$_2$H$_5$ (68%)

(Ref. 101b)

This is, of course, the expected result if the α-substituent of the active methylene component is a function which can stabilize the enolate anion (e.g., a formyl, carboalkoxyl, phenyl, or vinyl group). The reaction of enolizable α,β-unsaturated carbonyl compounds (as the active methylene component) with Michael acceptors has also usually resulted in the expected reaction at the alpha position;[95e,h,106a] however, reaction at the gamma position has also been observed.[106b] Substitution at the latter position, which is not observed in normal alkylations, may possibly be a consequence of the fact that the Michael reaction is reversible.

The reasons for the product orientation observed in reactions of unsymmetrical ketones which lack additional activating groups is not entirely clear. The following

106. (a) H. A. Bruson and T. W. Riener, *J. Am. Chem. Soc.*, **75**, 3585 (1953) : correction of the structural assignments is made in Ref. 92a. (b) C. R. Engel and J. Lessard, *ibid.*, **85**, 638 (1963).

equation illustrates the fact that the Michael reaction of 2-methylcyclohexanone forms both structurally isomeric adducts rather than the single 2,2-isomer originally reported. Furthermore, when precautions are taken to minimize removal of the more reactive 2,6-isomer from the reaction mixture by dialkylation, the proportions of the two adducts correspond to the composition of the equilibrium mixture of enolate ions from the starting ketone. Also, the initially formed products were not interconverted (by a retrograde Michael reaction) under the conditions of the study.[77d] However, a number of Michael reactions with unsymmetrical *acyclic*

Reaction conditions	Solvent	Monoalkylated product composition	
1 equiv. of ketone	$(CH_3)_3COH$	93%	7%
	$CH_3OCH_2CH_2OCH_3$	88%	12%
excess ketone to	$(CH_3)_3COH$	88%	12%
minimize dialkylation	$CH_3OCH_2CH_2OCH_3$	67%	33%

(Refs. 77d, 86b, 105)

ketones have been reported to give substantially greater amounts of the adduct from reaction at the more highly substituted position than can be accounted for by the equilibrium composition of the starting enolate ions. In such cases, one or more of the following circumstances seems probable: (1) the reactivity of the enolate anion toward a Michael acceptor is significantly enhanced by alkyl substitution; (2) the adduct formed by reaction at the less highly substituted position is being selectively destroyed by further transformation or by dialkylation; (3) the study was performed under conditions where the Michael reaction was reversible and the adduct from reaction at the less highly substituted position is being destroyed by a retrograde Michael reaction.

One of the most useful synthetic applications of the Michael reaction has been the Robinson annelation reaction,[92a,95g,i,102a,107−112,113a−c] illustrated by the accompanying equations. The formation of the new six-membered ring involves the

107. (a) E. C. duFeu, F. J. McQuillin, and R. Robinson, *J. Chem. Soc.*, 53 (1937). (b) F. J. McQuillin and R. Robinson, *ibid.*, 1097 (1938). (c) J. W. Cornforth and R. Robinson, *ibid.*, 1855 (1949).
108. (a) A. L. Wilds and C. H. Shunk, *J. Am. Chem. Soc.*, **65**, 469 (1943). (b) A. V. Logan, E. N. Marvell, R. LaPore, and D. C. Bush, *ibid.*, **76**, 4127 (1954). (c) The enolate anion initially formed in Michael reactions has also been intercepted by an intermolecular aldol condensation with added benzaldehyde [T. Henshall and E. W. Parnell, *J. Chem. Soc.*, 3040

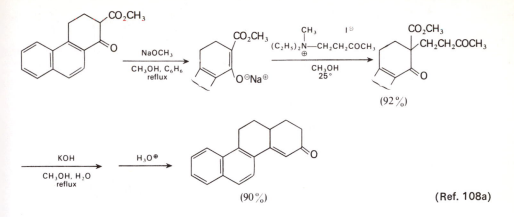

(92%)

(90%)   (Ref. 108a)

(1962)] or by alkylation with an alkyl halide [E. M. Kaiser, C. L. Mao, C. F. Hauser, and C. R. Hauser, *J. Org. Chem.*, **35**, 410 (1970)].

109. (a) J. A. Marshall, H. Faubl, and T. M. Warne, Jr., *Chem. Commun.*, **No. 15**, 753 (1967) ; J. A. Marshall and T. M. Warne, Jr., *J. Org. Chem.*, **36**, 178 (1971). (b) R. M. Coates and J. E. Shaw, *Chem. Commun.*, **No. 1**, 47 (1968). (c) R. L. Hale and L. H. Zalkow, *ibid.*, **No. 20**, 1249 (1968). (d) H. C. Odom and A. R. Pinder, *ibid.*, **No. 1**, 26 (1969). See also J. A. Marshall and R. A. Ruden, *Tetrahedron Letters*, **No. 15**, 1239 (1970). (e) R. Howe and F. J. McQuillin, *J. Chem. Soc.*, 1194 (1958). (f) D. J. Baisted and J. S. Whitehurst, *ibid.*, 2340 (1965). (g) J. N. Gardner, B. A. Anderson, and E. P. Oliveto, *J. Org. Chem.*, **34**, 107 (1969). (h) J. A. Marshall and W. I. Fanta, *ibid.*, **29**, 2501 (1964). (i) J. A. Marshall and D. J. Schaeffer, *ibid.*, **30**, 3642 (1965) and references therein. (j) G. Stork, S. Danish-efsky, and M. Ohashi, *J. Am. Chem. Soc.*, **89**, 5459 (1967) ; G. Stork and J. E. McMurry, *ibid.*, **89**, 5461, 5463, 5464 (1967).

110. (a) W. G. Dauben and J. W. McFarland, *J. Am. Chem. Soc.*, **82**, 4245 (1960). (b) For a modification of the Wichterle procedure, see D. Caine and N. F. Tuller, *J. Org. Chem.*, **34**, 222 (1969). (c) P. T. Lansbury, E. J. Nienhouse, D. J. Scharf, and F. R. Hilfiker, *J. Am. Chem. Soc.*, **92**, 5649 (1970). (d) S. Danishefsky and R. Cavanaugh, *ibid.*, **92**, 520 (1968).

111. (a) R. B. Woodward and T. Singh, *J. Am. Chem. Soc.*, **72**, 494 (1950) ; For other examples of the use or attempted use of acetylenic ketones and propiolic esters as Michael acceptors, see Ref. 77d. (b) G. Stork and M. Tomasz, *ibid.*, **86**, 471 (1964). (c) T. Metler, A. Uchida, and S. I. Miller, *Tetrahedron*, **24**, 4285 (1968). (d) H. Kappeler and E. Renk, *Helv. Chim. Acta*, **44**, 1541 (1961). (e) C. A. Grob and A. Kaiser, *ibid.*, **50**, 1599 (1967).

112. S. Ramachandran and M. S. Newman, *Org. Syn.*, **41**, 38 (1961) ; A. B. Mekler, S. Ramachandran, S. Swaminathan, and M. S. Newman, *ibid.*, **41**, 56 (1961) ; M. S. Newman and A. B. Mekler, *J. Am. Chem. Soc.*, **82**, 4039 (1960).

113. (a) W. S. Johnson, J. Ackerman, J. F. Eastham, and H. A. DeWalt, Jr., *J. Am. Chem. Soc.*, **78**, 6302 (1956). (b) W. S. Johnson, J. J. Korst, R. A. Clement, and J. Dutta, *ibid.*, **82**, 614 (1960). (c) T. A. Spencer and K. K. Schmiegel, *Chem. Ind. (London)*, 1765 (1963) ; T. A. Spencer, K. K. Schmiegel, and K. L. Williamson, *J. Am. Chem. Soc.*, **85**, 3785 (1963) ; T. A. Spencer, H. S. Neel, T. W. Flechtner, and R. A. Zayle, *Tetrahedron Letters*, **No. 43**, 3889 (1965) ; T. A. Spencer, H. S. Neel, D. C. Ward, and K. L. Williamson, *J. Org. Chem.*, **31**, 434 (1966). (d) L. L. McCoy, *ibid.*, **25**, 2078 (1960) ; **29**, 240 (1964) ; *J. Am. Chem. Soc.*, **84**, 2246 (1962). (e) Y. C. Kim and H. Hart, *J. Chem. Soc.*, C, 2409 (1969). (f) Y. Inouye, S. Inamasu, M. Horiike, M. Ohno, and H. M. Walborsky, *Tetrahedron*, **24**, 2907 (1968). (g) W. S. Johnson, S. Shulman, K. L. Williamson, and R. Pappo, *J. Org. Chem.*, **27**, 2015 (1962). (h) A. M. Baradel, R. Longeray, and J. Dreux, *Bull. Soc. Chim. France*, 252, 255, 258 (1970). (i) D. W. Theobald, *Tetrahedron*, **25**, 3139 (1969). (j) G. S. Abernethy, Jr., and M. E. Wall, *J. Org. Chem.*, **34**, 1606 (1969). (k) N. B. Haynes and C. J. Timmons, *J. Chem. Soc.*, C, 224 (1966).

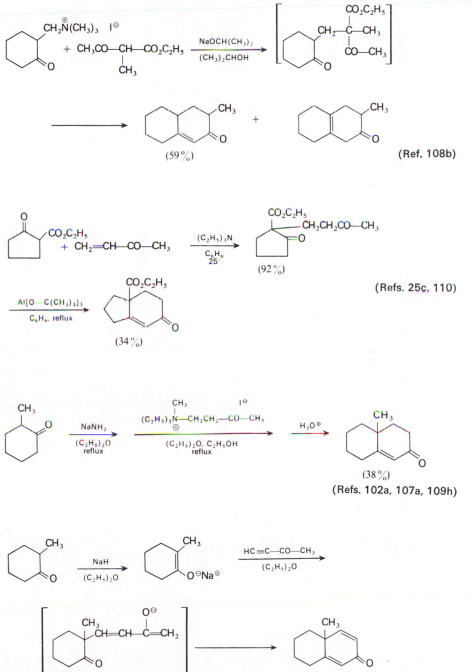

(59 %)   +   (Ref. 108b)

(92 %)   (Refs. 25c, 110)

(34 %)

(38 %)
(Refs. 102a, 107a, 109h)

(Ref. 111a)

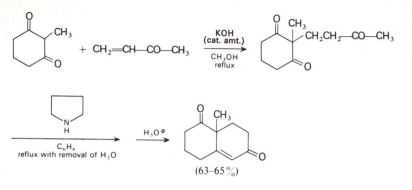

(63–65%)   (Ref. 112)

intramolecular aldol condensation (see Chapter 10) of the initial Michael adduct
and subsequent dehydration. Although a wide variety of experimental procedures
have been used to accomplish the Robinson annelation reaction, two appear to be
most reliable. In one procedure,[95i,112,113c] the uncyclized Michael adduct is
formed from methyl vinyl ketone and a relatively acidic active methylene compound
under mild conditions (triethylamine or a catalytic amount of sodium hydroxide
in an alcohol solvent). The adduct is then converted to the aldol product(s) by
treatment with pyrrolidine and acetic acid and finally dehydrated by brief treatment
with acid. The second procedure[109h] involves the slow addition of methyl vinyl
ketone to a cold, ethereal solution of the active methylene compound and a
catalytic amount of sodium ethoxide to form the aldol product. This aldol product
is separated from the reaction mixture and dehydrated by subsequent treatment with
acid or base. Examples of these methods are found in the preceding and following
equations.

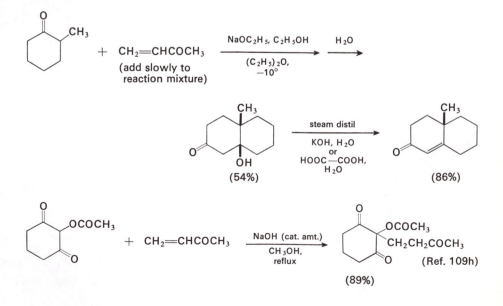

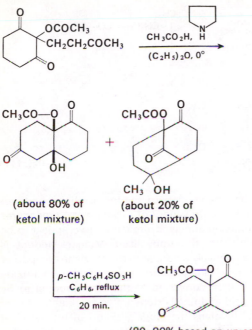

(about 80% of ketol mixture)   (about 20% of ketol mixture)

(80–90% based on unrecovered starting material)   (Ref. 113c)

As indicated in the accompanying equations, two different modes of intra-molecular aldol condensation are often found. The bicyclic ketols such as **[86]**

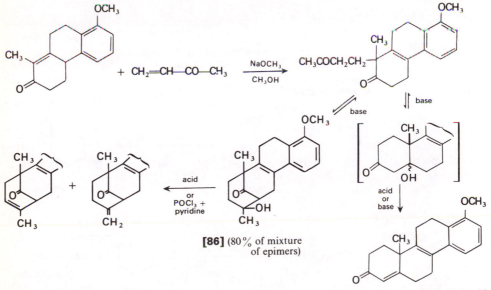

[86] (80% of mixture of epimers)

(Ref. 113a, b)

are reconverted to the initial Michael adducts upon base treatment; with acids they may either revert to the Michael adduct or undergo dehydration.[113]

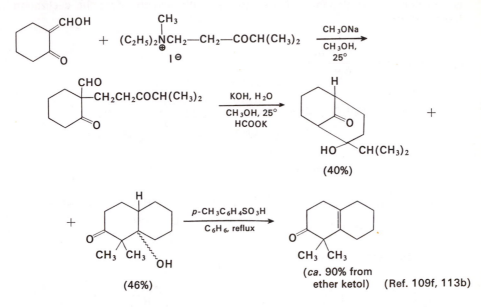

(40%)

(46%)

(ca. 90% from
ether ketol)   (Ref. 109f, 113b)

Several other general annelation procedures have been developed. The Wichterle reaction[109i,110a,b] introduces the four-carbon chain destined to become a new six-member ring by alkylation of a ketone enolate with 1,3-dichloro-*cis*-2-butene. The resulting vinyl chloride is hydrolyzed with concentrated sulfuric acid to form a 1,5-diketone which can cyclize by the pathways previously discussed.

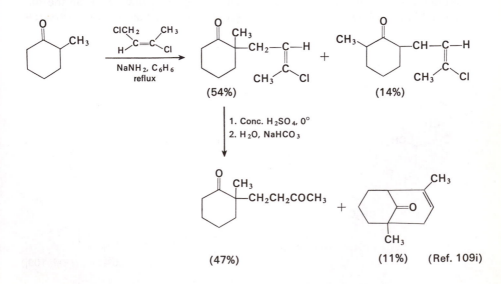

(54%)   (14%)

1. Conc. H_2SO_4, 0°
2. H_2O, $NaHCO_3$

(47%)   (11%)   (Ref. 109i)

A related reaction for forming five-membered rings involves the alkylation of an active methylene compound with 2,3-dichloropropene. The alkylated product is converted to an alcohol and then treated with acid to effect the cyclization indicated.[110c]

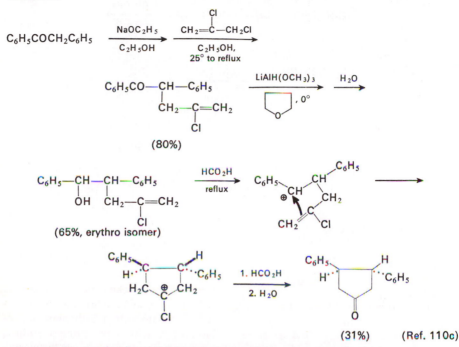

(80%)

(65, erythro isomer)

(31%)   (Ref. 110c)

In another procedure, the isoxazole annelation reaction, the active methylene compound is alkylated with a chloromethylisoxazole derivative such as the one indicated in the following equations. Subsequent hydrogenolysis of the isoxazole

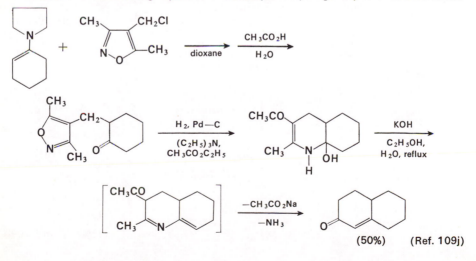

(50%)   (Ref. 109j)

ring followed by the base-catalyzed processes indicated completes the annelation sequence. A procedure which utilizes 2-vinyl-6-methylpyridine as a Michael acceptor provides a synthetic method for forming two new six-membered rings.[110d]

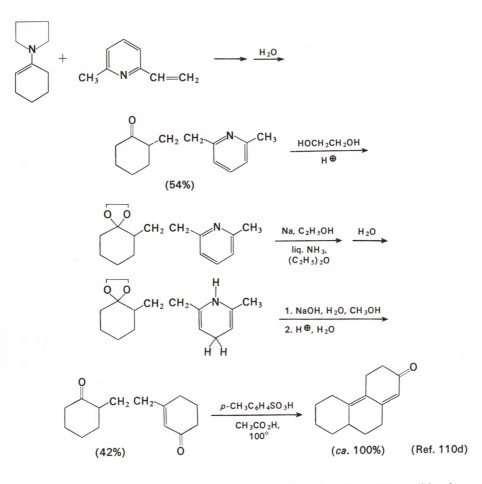

The use of a reaction sequence involving a Michael addition followed by ring closure also provides a useful route to cyclopropane derivatives.[113d-f] The stereo-chemistry of the ring-closure process exhibits an interesting dependence on the solvent and the metal cation.[113d,f] In non-polar media the sterically unfavorable *cis* isomer is formed whereas in polar solvents the *trans* product, expected if steric interactions are minimized,[113e] becomes the favored product. The results in non-polar solvents may be rationalized by the assumption that the initial Michael adduct is formed in a conformation which allows the monoenolate of the 1,5-dicarbonyl compound to serve as a bidentate ligand for the metal cation. Interest-ingly, formation of the same type of cation-bidentate anion intermediate can provide a rational for the stereochemical results obtained in certain other Michael

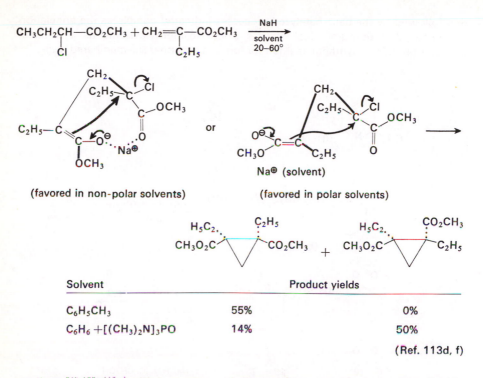

Solvent	Product yields	
$C_6H_5CH_3$	55%	0%
$C_6H_6 + [(CH_3)_2N]_3PO$	14%	50%

(Ref. 113d, f)

reactions.[94h,109a,113g,h] The accompanying equations provide two examples in which the stereochemical outcome of Michael reactions is explicable in terms of forming the initial adduct in such a way that the two carbonyl formations remain

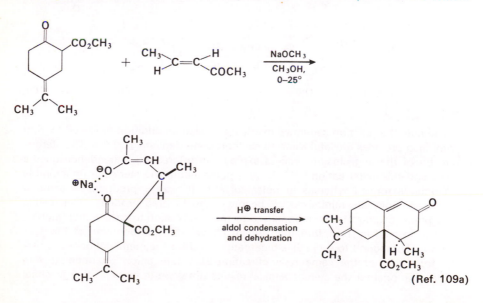

(Ref. 109a)

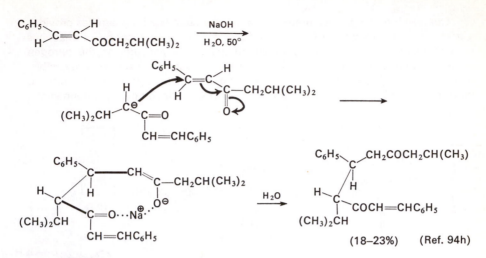

(18–23%)   (Ref. 94h)

near one another and other steric interactions are minimized. Other stereochemical questions related to the Michael condensation have been explored but no clear-cut generalizations are evident. Thus, in reactions of cyclohexanone derivatives with Michael acceptors examples of both axial[109e] and equatorial[109d,g] attack are reported. In cases where cyclic unsaturated ketones have served as Michael acceptors, many of the stereochemical results obtained are explicable as attack of the enolate from the less hindered side of the carbon-carbon double bond.[96e,99b,100b,113i,j] The prediction of the stereochemical outcome of Michael reactions is inevitably complicated by the possible intervention of retrograde Michael reactions and by further transformations of the initial Michael adduct. The following examples, believed to be kinetically controlled processes, illustrate the stereochemical change which may accompany a modification in reaction conditions.

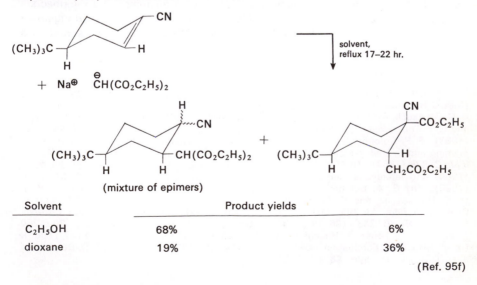

Solvent	Product yields	
C_2H_5OH	68%	6%
dioxane	19%	36%

(Ref. 95f)

The reaction of α,β-unsaturated carbonyl compounds with enamines provides a useful alternative route to Michael adducts.[84] Unlike the reaction of enamines with alkylating agents, the introduction of a Michael acceptor at the nitrogen atom of an enamine is not a serious problem because this reaction is reversible. The initial adduct [88] from reaction at carbon can undergo ring closure[114] to form either [89] or the subsequently discussed dihydropyran intermediate.

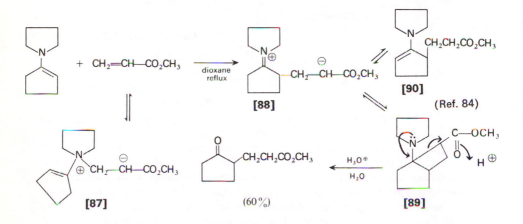

Alternatively, proton transfer may occur to form [90]. In either case, acid-catalyzed hydrolysis leads to the Michael adduct. Enamines have also been found to react as 1,3-dipolarophiles (e.g., azides and nitrile oxides add to enamines)[114f] and to undergo stepwise cycloaddition reactions with other unsaturated materials such as cyclopropenones[114g] and dienoic acid derivatives.[114h,i] If acetylenic esters (e.g., [91]) are employed as Michael acceptors with enamines, the intermediate cyclic product [92] is capable of thermal isomerization. Should this rearrangement occur prior to hydrolysis, the resulting product [92] does not correspond to a Michael adduct.

114. (a) K. C. Brannock, A. Bell, R. D. Burpitt, and C. A. Kelly, *J. Org. Chem.,* **29**, 801 (1964). (b) K. C. Brannock, R. D. Burpitt, V. W. Goodlett, and J. G. Thweatt, *ibid.,* **29**, 813, 818 (1964); K. C. Brannock, R. D. Burpitt, and J. G. Thweatt, *ibid.,* **29**, 940 (1964). (c) G. A. Berchtold and G. F. Uhlig, *ibid.,* **28**, 1459 (1963). (d) C. F. Huebner and co-workers, *ibid.,* **28**, 3134 (1963); F. P. Colonna, S. Fatutta, A. Risaliti, and C. Russo, *ibid.,* C, 2377 (1970). (e) I. Fleming and J. Harley-Mason, *J. Chem. Soc.,* 2165 (1964); I. Fleming and M. H. Karger, *ibid.,* C, 226 (1967). (f) M. E. Kuehne, S. J. Weaver, and P. Franz, *J. Org. Chem.,* **29**, 1582 (1964) and references therein. (g) J. Ciabattoni and G. A. Berchtold, *ibid.,* **31**, 1336 (1966). (h) G. A. Berchtold, J. Ciabattoni, and A. A. Tunick, *ibid.,* **30**, 3679 (1965). (i) S. Danishefsky and R. Cunningham, *ibid.,* **30**, 3676 (1965); S. Danishefsky and R. Cavanaugh, *ibid.,* **33**, 2959 (1965). (j) A. Risaliti, E. Valentin, and M. Forchiassin, *Chem. Commun.,* No. 5, 233 (1969). (k) M. E. Kuehne and L. Foley, *J. Org. Chem.,* **30**, 4280 (1965). (l) H. Feuer, A. Hirschfeld, and E. D. Bergman, *Tetrahedron,* **24**, 1187 (1968). (m) A. Risaliti, M. Forchiassin, and E. Valentin, *ibid.,* **24**, 1889 (1968). (n) R. D. Burpitt and J. G. Thweatt, *Org. Syn.,* **48**, 56 (1968).

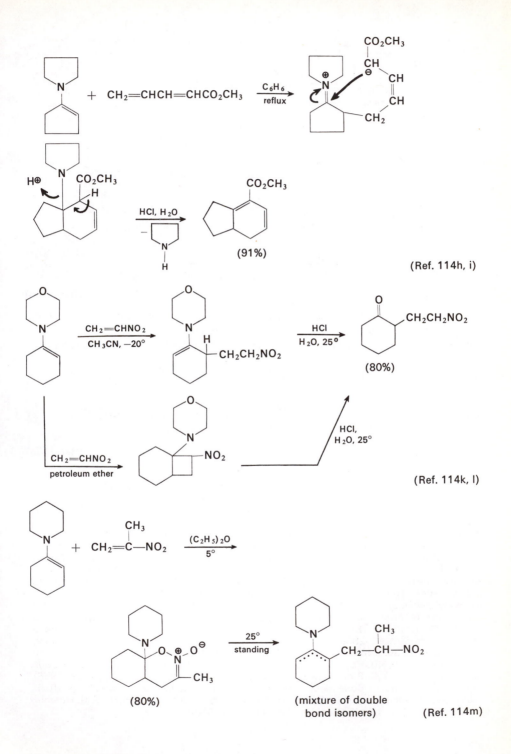

(Ref. 114h, i)

(80%)

(Ref. 114k, l)

(80%)

(mixture of double
bond isomers)   (Ref. 114m)

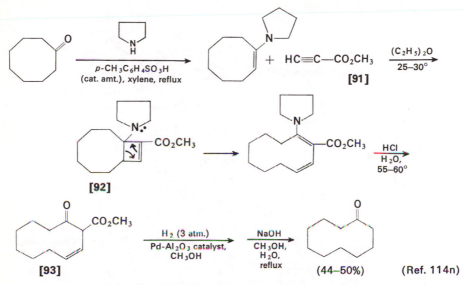

[91]

[92]

[93]   (44–50%)   (Ref. 114n)

The preparation of Michael adducts by the enamine procedure rather than by the older base-catalyzed method offers advantages in several circumstances. As the following equations show, monoalkylated Michael adducts can be obtained

$$CH_3(CH_2)_4CH=CH-N\bigcirc \quad + \quad CH_2=CHCO_2CH_3 \xrightarrow[\text{reflux}]{CH_3CN}$$

$$\xrightarrow[\substack{H_2O, \\ CH_3CN, \\ \text{reflux}}]{CH_3CO_2H} CH_3(CH_2)_4\underset{\underset{CHO}{|}}{CH}CH_2CH_2CO_2CH_3$$

(75%)   (Ref. 115a)

115. (a) G. R. Pettit, D. C. Fessler, K. D. Paull, P. Hofer, and J. C. Knight, *J. Org. Chem.*, **35**, 1398 (1970). (b) T. L. Westman, R. Paredes, and W. S. Brey, Jr., *ibid.*, **28**, 3512 (1963) ; T. L. Westman and A. E. Kober, *ibid.*, **29**, 2448 (1964). (c) L. Mandell, B. A. Hall, and K. P. Singh, *ibid.*, **29**, 3067 (1964). (d) M. vonStrandtmann, M. P. Cohen, and J. Shavel, Jr., *ibid.*, **30**, 3240 (1965). (e) L. H. Hellberg, R. J. Milligan, and R. N. Wilke, *J. Chem. Soc.*, C, 35 (1970). (f) N. F. Firrell and P. W. Hickmott, *Chem. Commun.*, **No. 10**, 544 (1969). (g) For the use of cyclopropyl derivatives as Michael acceptors with enamines, see S. Danishefsky, G. Rovnyak, and R. Cavanaugh, *ibid.*, **No. 12**, 636 (1969) ; J. E. Dolfini, K. Menich, and P. Corliss, *Tetrahedron Letters*, **No. 37**, 4421 (1966). (h) For the use of vinylpyridines as Michael acceptors, see G. Singerman and S. Danishefsky, *ibid.*, **No. 33**, 2249 (1964). (i) U. K. Pandit and H. O. Huisman, *ibid.*, **No. 40**, 3901 (1967). (j) T. J. Curphey and H. L. Kim, *ibid.*, **No. 12**, 1441 (1968). (k) R. L. Augustine and H. V. Cortez, *Chem. Ind. (London)*, 490 (1963). (l) R. L. Augustine, and J. A. Caputo, *Org. Syn.*, **45**, 80 (1965). (m) H. O. House, R. W. Giese, K. Kronberger, J. P. Kaplan, and J. F. Simeone, *J. Am. Chem. Soc.*, **92**, 2800 (1970). (n) H. A. P. deJongh, F. J. Gerhartl, and H. Wynberg, *J. Org. Chem.*, **30**, 1409 (1965). (o) R. P. Nelson, J. M. McEuen, and R. G. Lawton, *ibid.*, **34**, 1225 (1969). (p) H. Stetter and H. G. Thomas, *Angew. Chem., Intern. Ed. Engl.*, **6**, 554 (1969). (q) G. Stork and H. K. Landesman, *J. Am. Chem. Soc.*, **78**, 5129 (1956). (r) V. Dressler and K. Bodendorf, *Tetrahedron Letters*, **No. 43**, 4243 (1967). (s) R. D. Allan, B. G. Cordiner, and R. J. Wells, *ibid.*, **No. 58**, 6055 (1968) ; A. J. Birch, E. G. Hutchinson, and G. S. Rao, *Chem. Commun.*, **No. 11**, 657 (1970).

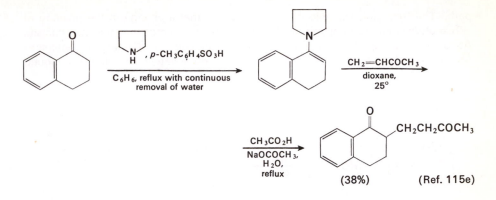

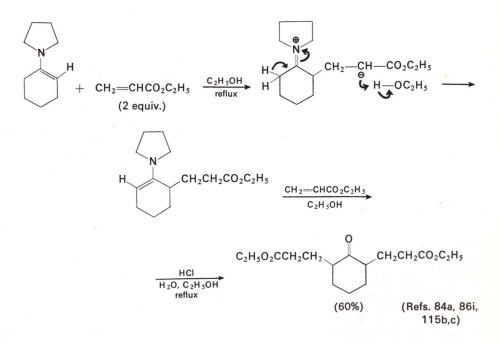

from either aldehydes or ketones under relatively mild reaction conditions. Of equal importance is the change in the favored position of substitution with unsymmetrical ketones. Whereas a conventional, base-catalyzed Michael reaction favors reaction of the Michael acceptor at the more highly substituted alpha carbon, the reverse orientation is found in the reaction of enamines (where the

(Ref. 115e)

(38%)

(2 equiv.)

(60%)   (Refs. 84a, 86i, 115b,c)

less highly substituted double bond isomer is usually more stable) with Michael acceptors *in protic solvents*.[84a,115f] In non-polar, aprotic solvents where the indicated intermolecular protonation of the zwitterionic intermediate is not favorable, then a slower intramolecular proton transfer[115f,i] becomes the dominant reaction

path. The following example illustrates the fact that a change in the position of substitution may accompany this change in the proton-transfer mechanism. An alternative procedure which has been developed for effecting an enamine-Michael reaction consists of heating a ketone or other active methylene compound

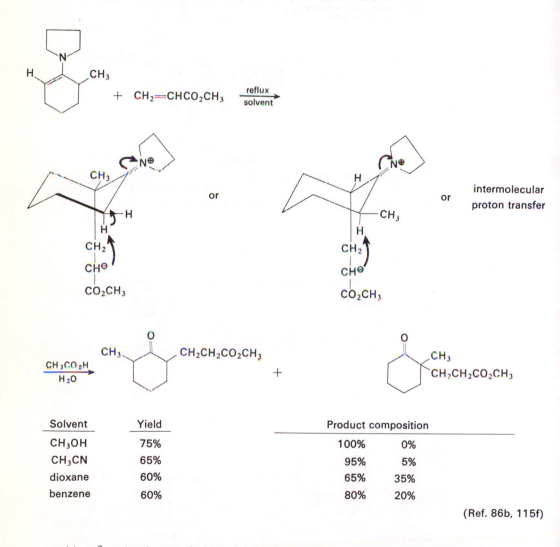

Solvent	Yield	Product composition	
CH_3OH	75%	100%	0%
CH_3CN	65%	95%	5%
dioxane	60%	65%	35%
benzene	60%	80%	20%

(Ref. 86b, 115f)

with a β-amino ketone (a Mannich base, see Chapter 10). As the accompanying equation indicates, this process, called a thermal Michael reaction, is believed to involve the formation of an enamine and an α,β-unsaturated ketone which then undergo further reaction.[101b] This thermal Michael reaction exhibits the same orientation seen in other enamine-Michael reactions with attack at the less substituted alpha carbon being favored.

A useful variant of the Robinson annelation reaction makes use of enamine percursors.[84a,b,113c,115j–m] (In some cases the enamine intermediate may be formed in the reaction mixture when a primary or secondary amine is used to catalyze the reaction of an active methylene compound with a Michael acceptor.)

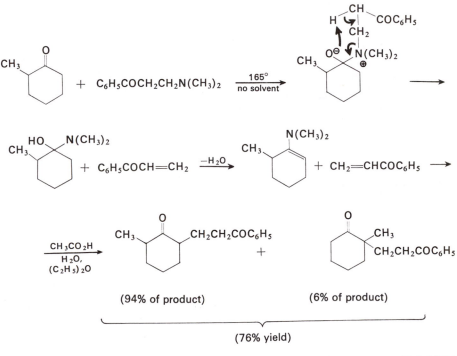

(94% of product)   (6% of product)

(76% yield)

(Ref. 101b)

The accompanying equations indicate two of the reaction paths which have been proposed[84c,86g,113c,114e] for the process. Whether both pathways are involved is not entirely clear; however, the isolation of ketol intermediates has established that at least this path is followed in some cases.[113c] The initial rapid formation of an unstable dihydropyran intermediate appears to be general for the reaction of enamines with unhindered Michael acceptors such as acrolein and methyl vinyl ketone.[114e] A number of by-products have been noted in this annelation method;[86i,115k] the most satisfactory procedure for suppressing these side reactions appears to be the slow addition of methyl vinyl ketone to a pyrrolidine enamine[115m] dissolved in an aprotic solvent (to minimize protonation of the intermediate and subsequent dialkylation).[86i] It also appears to be important not to heat the resulting enamine with acid if the previously discussed reduction of an iminium salt is to be avoided. Other ring-closure reactions which utilize an enamine-Michael reaction as one step in the synthesis sequence have also been described.[115n–s] The reaction

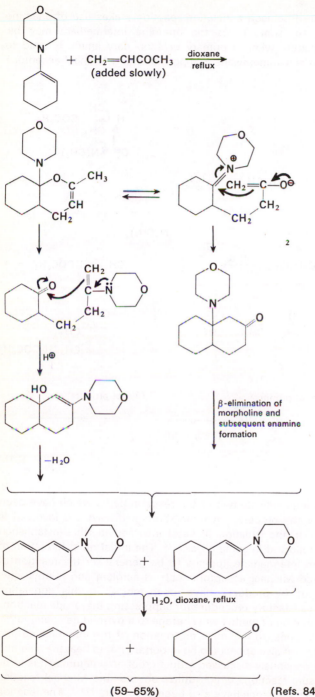

(59–65%)

(Refs. 84c, 86g, 113c, 114e, 115k, l)

of acrolein with the enamines of cyclic ketones, illustrated below, is of particular interest as a useful route to bridged bicyclic systems.[115q-s]

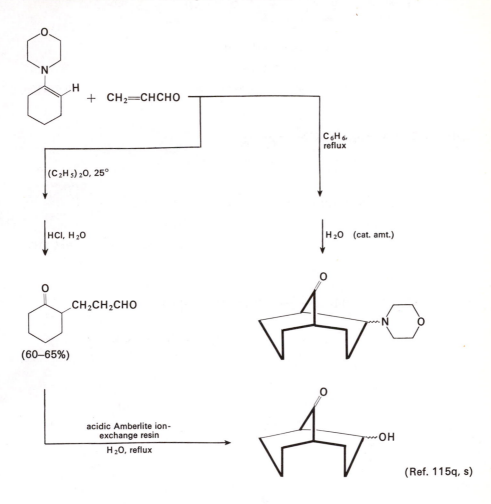

(Ref. 115q, s)

THE HYDROCYANATION REACTION

A reaction closely related to the Michael reaction in mechanism is the conjugate addition of cyanide ion to various unsaturated systems, a procedure called hydrocyanation.[116-118] Some typical examples of this reaction are provided in the accompanying equations.

116. (a) J. A. McRae and R. A. B. Bannard, *Org. Syn.,* **Coll. Vol. 4**, 393 (1963). (b) G. B Brown, *ibid.,* **Coll. Vol. 3**, 615 (1955). (c) C. F. H. Allen and H. B. Johnson, *ibid.,* **Coll. Vol. 4**, 804 (1963). See also (d) A. K. Kundu, N. G. Kundu, and P. C. Dutta, *J. Chem. Soc.,* 2749 (1965).

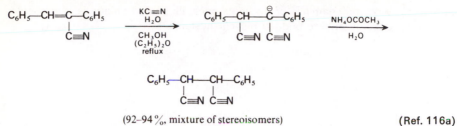

$$C_6H_5-CH-CH-C_6H_5$$
$$\quad\quad\quad | \quad\quad |$$
$$\quad\quad C\equiv N \quad C\equiv N$$

(92–94 %, mixture of stereoisomers)   (Ref. 116a)

$C_6H_5CH{=}C(CO_2C_2H_5)_2$ + KCN $\xrightarrow[\substack{C_2H_5OH, \\ 65-75° \\ -C_2H_5OH \\ -KHCO_3}]{H_2O}$

$$C_6H_5CH-CH_2CO_2C_2H_5 \xrightarrow[\substack{H_2O, \\ reflux}]{HCl} C_6H_5CH-CH_2CO_2H$$
$$\quad\quad |\quad\quad\quad\quad\quad\quad\quad\quad\quad\quad\quad\quad\quad | $$
$$\quad\quad CN \quad\quad\quad\quad\quad\quad\quad\quad\quad\quad CO_2H$$

(67–70%)   (Ref. 116c)

$$C_6H_5-CO-CH_2CH_2-N(CH_3)_2 \xrightarrow[\substack{H_2O \\ reflux}]{KCN} [C_6H_5COCH{=}CH_2] \longrightarrow$$

$$C_6H_5-CO-CH_2CH_2-C\equiv N$$

(67%)   (Ref. 102a)

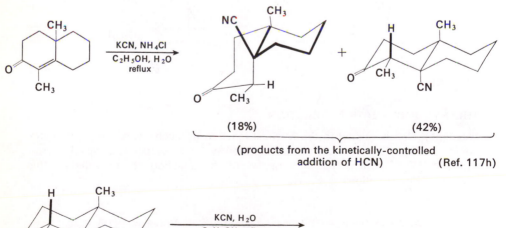

(18%)   (42%)

(products from the kinetically-controlled
                          addition of HCN)   (Ref. 117h)

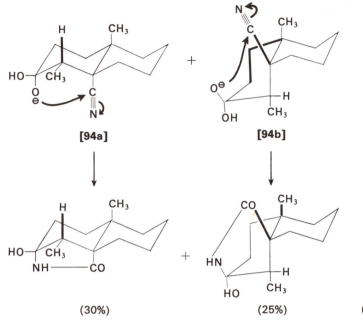

[94a]   [94b]

(30%)   +   (25%)   (Ref. 117h)

Early procedures for the addition of hydrogen cyanide to α,β-unsaturated ketones and related Michael acceptors employed a reaction with sodium or potassium cyanide in aqueous or aqueous ethanolic solution. This procedure suffers from the disadvantage that a full equivalent of a strong base (e.g., potassium hydroxide) is formed if the addition of cyanide ion is complete. This strong base promotes reversal of the original addition and may also lead to a variety of hydrolysis products.[116c,d,117a,c,d,h,i,l] If the keto nitrile produced has suitable geometry, the adduct of hydroxide ion to the carbonyl group may participate in the partial hydrolysis of the nitrile to form a lactam (see structure [94]). These difficulties arising from base-catalyzed reversal or hydrolysis may be largely avoided by buffering the reaction mixture with a weak acid such as acetic acid[117e,g] or ammonium chloride[117a,b,d,f,h−l,n,p] or by reaction of the unsaturated system with anhydrous

117. (a) W. Nagata, S. Hirai, H. Itazaki, and K. Takeda, *J. Org. Chem., 26*, 2413 (1961). See also (b) W. Nagata, *Tetrahedron, 13*, 278 (1961). (c) A. Bowers, *J. Org. Chem., 26*, 2043 (1961). (d) W. L. Meyer and N. G. Schnautz, *ibid., 27*, 2011 (1962). (e) C. Djerassi, R. A. Schneider, H. Vorbrueggen, and N. L. Allinger, *ibid., 28*, 1632 (1963). (f) W. L. Meyer and J. F. Wolfe, *ibid., 29*, 170 (1964). (g) P. N. Rao and J. E. Burdett, Jr., *ibid., 34*, 1090 (1969). (h) O. R. Rodig and N. J. Johnston, *ibid., 34*, 1942, 1949 (1969). (i) E. Wenkert, R. L. Johnson, and L. L. Smith, *ibid., 32*, 3224 (1969). (j) D. W. Theobald, *Tetrahedron, 21*, 791 (1965). (k) J. McKenna, J. M. McKenna, and P. B. Smith, *ibid., 21*, 2983 (1965). (l) M. Torigoe and J. Fishman, *ibid., 21*, 3669 (1965). (m) R. D. Haworth, B. G. Hutley, R. G. Leach, and G. Rodgers, *J. Chem. Soc.*, 2720 (1962). (n) A. T. Glen, W. Lawrie, and J. McLean, *ibid., C*, 661 (1966). (o) H. B. Henbest and W. R. Jackson, *ibid., C*, 2465 (1967). (p) W. L. Meyer and K. K. Maheshwari, *Tetrahedron Letters*, No. 32, 2175 (1964).

hydrogen cyanide[117n] or with a solution of calcium cyanide in N-methylpyrroli-
done.[117o] Examples of these various procedures follow. The use of dimethyl-
formamide-water mixtures as the reaction solvent has been especially recommended.[117a,b]

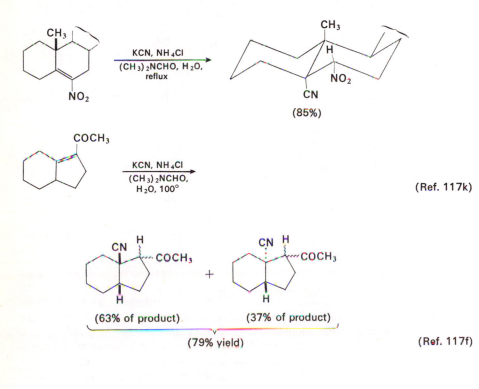

KCN, NH$_4$Cl
(CH$_3$)$_2$NCHO, H$_2$O,
reflux

(85%)

KCN, NH$_4$Cl
(CH$_3$)$_2$NCHO,
H$_2$O, 100°

(Ref. 117k)

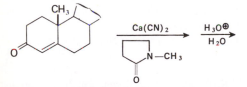

(63% of product)   +   (37% of product)

(79% yield)   (Ref. 117f)

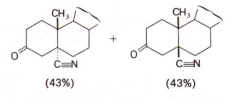

Ca(CN)$_2$ H$_3$O$^{\oplus}$
              H$_2$O

(43%)   +   (43%)

(Ref. 117o)

As the accompanying examples suggest, the kinetically-controlled addition of cyanide ion to a cyclohexenone derivative usually yields a mixture of products in which the product with an axial cyano group predominates. This has been suggested to be the result of the previously discussed stereoelectronic control.[117d,f,j,k,p] In cases where an axial alkyl substituent in the starting enone will bear a 1,3-diaxial relationship to the entering cyano group, the addition of cyanide ion is either slow or fails to occur.[117j]

An alternative set of reaction conditions—the use of hydrogen cyanide and triethylaluminum or a preformed dialkylaluminum cyanide in an inert solvent such as ether, tetrahydrofuran, or benzene—has been found to increase markedly the stereoselectivity of this addition,[118] apparently favoring the product with the cyano group in an axial position. The two procedures illustrated below differ in that the reaction of an enone with a preformed dialkylaluminum cyanide is a reversible process. The corresponding reaction with a mixture of a trialkylaluminum and anhydrous hydrogen cyanide, although much slower than the reaction with preformed dialkylaluminum cyanide, is apparently irreversible because the initial dialkylaluminum enolate formed is protonated by the free hydrogen cyanide in the reaction mixture.[118c] In both procedures a dialkylaluminum cyanide would appear to be the reactive intermediate which donates a cyanide ion to the enone system. One of the accompanying examples illustrates the successful addition of diethylaluminum cyanide to an enone even when the entering cyano group bears a 1,3-diaxial relationship to alkyl substituents.

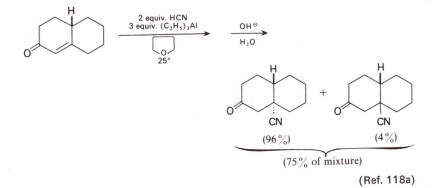

(Ref. 118a)

118. (a) W. Nagata, M. Yoshioka, and S. Hirai, *Tetrahedron Letters,* **No. 11**, 461 (1962). (b) W. Nagata, T. Terasawa, and T. Aoki, *ibid.,* **No. 14**, 865 (1963). (c) W. Nagata and M. Yoshioka, *ibid.,* **No. 18**, 1913 (1966). (d) H. Minato and T. Nagasaki, *J. Chem. Soc.,* C, 1866 (1966). (e) W. Nagata, M. Narisada, and T. Sugasawa, *ibid.,* C, 648 (1967). (f) A. C. Campbell, W. Lawrie, and J. McLean, *ibid.,* C, 554 (1969). (g) W. S. Johnson, J. A. Marshall, J. F. W. Keana, R. W. Franck, D. G. Martin, and V. J. Bauer, *Tetrahedron,* **Suppl. 8, Pt. II,** 541 (1966). (h) W. Nagata, T. Okumura, and M. Yoshioka, *J. Chem. Soc.,* C, 2347, 2365 (1970) ; W. Nagata, M. Yoshioka, T. Okumura, and M. Murakami, *ibid.,* C, 2355 (1970).

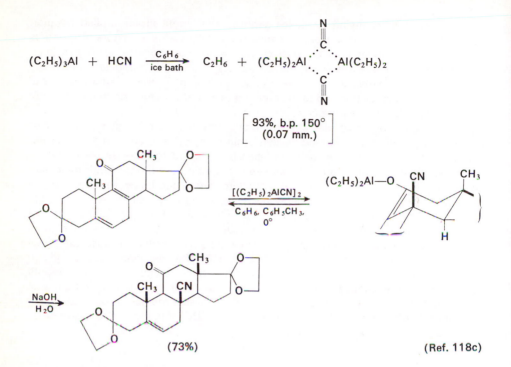

$$(C_2H_5)_3Al \ + \ HCN \ \xrightarrow[\text{ice bath}]{C_6H_6} \ C_2H_6 \ + \ (C_2H_5)_2Al \quad Al(C_2H_5)_2$$

$$\left[\begin{array}{c} 93\%, \text{ b.p. } 150° \\ (0.07 \text{ mm.}) \end{array} \right]$$

$$\xrightarrow[\substack{C_6H_6, \ C_6H_5CH_3, \\ 0°}]{[(C_2H_5)_2AlCN]_2}$$

$\xrightarrow[\text{H}_2\text{O}]{\text{NaOH}}$

(73%)

(Ref. 118c)

10

THE ALDOL CONDENSATION
AND RELATED REACTIONS

Together with the previously discussed (Chapter 9) reactions of enolate anions with alkylating agents and additions of enolate anions to the carbon-carbon double bonds of conjugated systems (the Michael reaction), additions of enolate anions to carbonyl functions constitute an important group of organic reactions. If the carbonyl function is part of a carboxylic acid derivative, such an addition leads to the introduction of an acyl group (Chapter 11), whereas addition of an enolate anion to the carbonyl group of an aldehyde or ketone followed by protonation constitutes a reaction known as aldol condensation.[1a] The series of equilibria involved in base-catalyzed aldol condensation are illustrated in the accompanying equations.

Kinetically, either the first step (proton abstraction) or the second step (carbon-carbon bond formation) may be rate-limiting.[1a,c] In cases where a reactive carbonyl component (e.g., an unhindered aldehyde such as [2]) serves as the acceptor, the

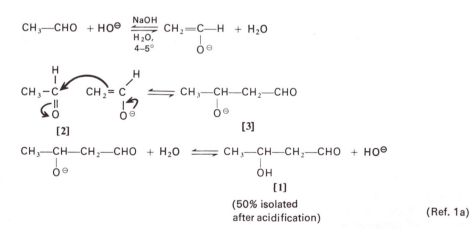

(50% isolated
after acidification)

(Ref. 1a)

1. (a) A. T. Nielsen and W. J. Houlihan, *Org. Reactions,* **16**, 1 (1968). (b) J. B. Conant and N. Tuttle, *Org. Syn.,* **Coll. Vol. 1**, 199 (1944). (c) J. Hine, J. G. Houston, J. H. Jensen, and J. Mulders, *J. Am. Chem. Soc.,* **87**, 5050 (1965); J. Hine, J. G. Houston, and J. H. Jensen, *J. Org. Chem.,* **30**, 1184 (1965) and references therein. (d) A. T. Nielsen, *ibid.,* **30**, 3650 (1965). (e) M. Stiles, R. R. Winkler, Y. Chang, and L. Traynor, *J. Am. Chem. Soc.,* **86**, 3337 (1964). (f) H. E. Zimmerman and M. D. Traxler, *ibid.,* **79**, 1920 (1957). (g) J. E. Dubois and M. Dubois, *Tetrahedron Letters,* **No. 43,** 4215 (1967); *Chem. Commun.,* **No. 24**, 1567 (1968).

first proton-abstraction step becomes rate-limiting; this is especially true at the relatively high reactant concentrations normally employed in preparative reactions. If catalysts no more basic than hydroxide ion or an alkoxide anion are employed, the first equilibrium is usually unfavorable and the last is usually favorable to the formation of the aldol product (e.g., [1]). A successful reaction in the presence of bases no stronger than metal alkoxides in polar solvents is therefore normally dependent on a favorable position for the second equilibrium, that forming the alkoxide (e.g., [3]); this equilibrium is usually unfavorable if the carbonyl compound to which the enolate anion adds (e.g., [2]) is a ketone rather than an aldehyde. In spite of this fact reasonable yields of aldol products may occasionally be obtained from ketones (e.g., [4]) if the reaction conditions employed permit the continuous separation of the product (e.g., [5]) from the catalyst as it is formed.

$$CH_3—CO—CH_3 \xrightarrow[\text{refluxing acetone}]{\substack{\text{Ba(OH)}_2 \text{ contained in the} \\ \text{thimble of a Soxhlet extractor}}} (CH_3)_2C—CH_2—CO—CH_3$$

[4]

with OH below the central carbon.

[5]   (71%)   (Ref. 1b)

As mentioned previously (Chapter 9), aldol condensations may also be catalyzed by acids, in which case the reaction is believed to follow the path indicated below. The enol (e.g., [6]) undergoes an electrophilic attack by the conjugate acid (e.g.,

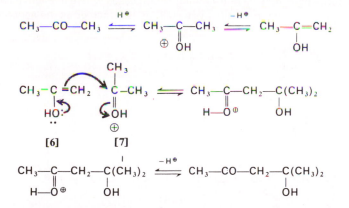

[7]) of the carbonyl component. It will become apparent in the subsequent discussion that the success of most of the preparatively useful reactions involving aldol condensation is attributable either to the selection of appropriate reaction conditions or to further transformations (e.g., dehydration) of the initial aldol product, which serve to displace the various prior equilibria in favor of the final product.

In many cases, the initial aldol condensation may lead to diastereoisomeric hydroxy ketone products.[1a,d–g] The ready reversibility of the aldol condensation in polar, protic solvents often leads to an equilibrium mixture of aldol products unless care is taken (e.g. low temperatures and short reaction times) to intercept

the initial product mixture formed under conditions of kinetic control. The following example illustrates the results of a kinetically controlled aldol condensation. The favored product of the less stable *threo*-isomer[1e,g] from condensations performed in the presence of a metal cation in nonpolar media has been suggested[1f,g] to involve the formation of an intermediate [8] in which the hydroxy ketone anion serves as a bidentate ligand for the metal cation; in this intermediate, steric interactions are minimized in formation of the *threo*-isomer. This point will be considered further in the discussion of the Reformatsky reaction and related condensations with preformed metal enolates.

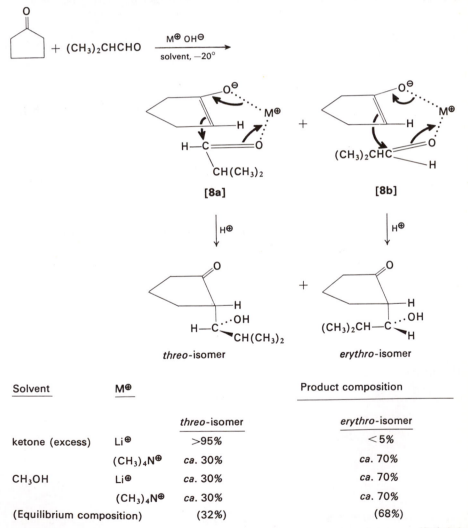

Solvent	M⊕	Product composition	
		threo-isomer	*erythro*-isomer
ketone (excess)	Li⊕	>95%	<5%
	(CH₃)₄N⊕	*ca.* 30%	*ca.* 70%
CH₃OH	Li⊕	*ca.* 30%	*ca.* 70%
	(CH₃)₄N⊕	*ca.* 30%	*ca.* 70%
(Equilibrium composition)		(32%)	(68%)

(Ref. 1g)

CONDENSATIONS IN WHICH EQUILIBRIUM IS FAVORED BY DEHYDRATION OF THE INTERMEDIATE ALDOL PRODUCTS: THE CLAISEN-SCHMIDT AND KNOEVENAGEL REACTIONS

In the presence of relatively strong acidic or basic catalysts (e.g., metal hydroxides) at elevated temperatures, the β-hydroxycarbonyl compound resulting from aldol condensation may often be dehydrated to form α,β-unsaturated carbonyl compounds. Although the base-catalyzed dehydration step is often reversible, it is usually energetically favorable to the production of the conjugated product and may serve to displace preceding unfavorable equilibria. The base-catalyzed dehydration reaction is believed to proceed by formation of an enolate anion (e.g., [9]) and subsequent elimination of hydroxide ion as illustrated. The use of a

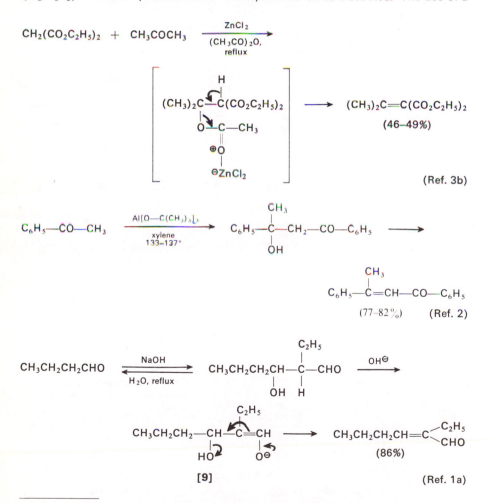

2. (a) W. Wayne and H. Adkins, *Org. Syn.*, **Coll. Vol. 3**, 367 (1955). (b) W. Wayne and H. Adkins, *J. Am. Chem. Soc.*, **62**, 3401 (1940).

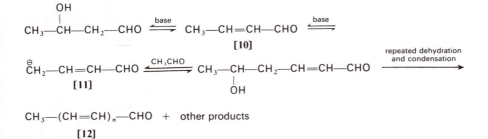

relatively strong base at elevated temperatures as catalyst to effect dehydration and force an aldol condensation to completion is not universally successful because α,β-unsaturated carbonyl compounds having gamma hydrogen atoms (e.g., [10]) may be converted to enolate anions (e.g., [11]), which can undergo further condensation and lead ultimately to polymeric products (e.g., [12]).

$$CH_3—CH—CH_2—CHO \overset{base}{\rightleftharpoons} CH_3—CH=CH—CHO \overset{base}{\rightleftharpoons}$$
$$\qquad\quad \underset{\text{OH}}{|}$$
$$\qquad\qquad\qquad\qquad\qquad [10]$$

repeated dehydration and condensation

$$\overset{\ominus}{C}H_2—CH=CH—CHO \overset{CH_3CHO}{\rightleftharpoons} CH_3—CH—CH_2—CH=CH—CHO \longrightarrow$$
$$\quad [11] \qquad\qquad\qquad\qquad\qquad \underset{\text{OH}}{|}$$

$$CH_3—(CH=CH)_n—CHO \quad + \quad \text{other products}$$
$$\quad [12]$$

However, this method for forcing aldol condensations to completion is more often useful when one of the carbonyl components is an aromatic aldehyde because the resulting α,β-unsaturated product (e.g., [13]) has no gamma hydrogen atoms and cannot undergo further condensation. The condensation of an aromatic aldehyde with an aliphatic aldehyde or ketone in the presence of a relatively strong base (hydroxide or alkoxide ion) to form an α,β-unsaturated aldehyde or ketone is

$$C_6H_5—CHO + CH_3—CO—C(CH_3)_3 \xrightarrow[\substack{H_2O \\ C_2H_5OH \\ 25°}]{NaOH} C_6H_5—CH=CH—CO—C(CH_3)_3$$
$$\qquad\qquad\qquad\qquad\qquad\qquad\qquad\qquad [13] \quad (88–93\%)$$

(Ref. 4)

known as the Claisen-Schmidt reaction and has been widely used in synthesis. The following examples are illustrative. Some workers prefer the use of methanol rather than ethanol as a co-solvent in these reactions to avoid the discoloration and the aldol polymers that may result if any of the ethanol is oxidized to acetaldehyde during the course of the reaction. This discoloration and polymer formation is most

3. (a) A. Russell and R. L. Kenyon, *Org. Syn.*, **Coll. Vol. 3**, 747 (1955). (b) E. L. Eliel, R. O. Hutchins, and Sr. M. Knoeber, *ibid.*, **50**, 38 (1970).
4. (a) G. A. Hill and G. M. Bramann, *Org. Syn.*, **Coll. Vol. 1**, 81 (1944). (b) C. S. Marvel and W. B. King, *ibid.*, **Coll. Vol. 1**, 252 (1944).

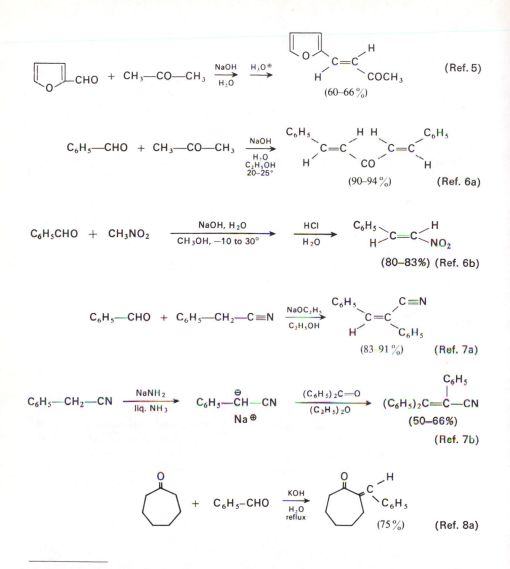

5. (a) G. J. Leuck and L. Cejka, *Org. Syn.*, **Coll. Vol. 1**, 283 (1944). (b) N. L. Drake and P. Allen, Jr., *ibid.*, **Coll. Vol. 1**, 77 (1944). (c) When the arylidene group is being introduced as a blocking group (see Chapter 9), the use of furfural rather than benzaldehyde is desirable because the Claisen-Schmidt condensation is more rapid: cf. W. S. Johnson, B. Bannister, and R. Pappo, *J. Am. Chem. Soc.*, **78**, 6331 (1956).
6. (a) C. R. Conard and M. A. Dolliver, *Org. Syn.*, **Coll. Vol. 2**, 167 (1943). (b) D. E. Worrall, *ibid.*, **Coll. Vol. 1**, 413 (1944).
7. (a) S. Wawzonek and E. M. Smolin, *Org. Syn.*, **Coll. Vol. 3**, 715 (1955). (b) S. Wawzonek and E. M. Smolin, *ibid.*, **Coll. Vol. 4**, 387 (1963).
8. (a) R. Baltzly, E. Lorz, P. B. Russell, and F. M. Smith, *J. Am. Chem. Soc.*, **77**, 624 (1955). (b) G. Kabas, *Tetrahedron*, **22**, 1213 (1966). (c) S. G. Powell and W. J. Wasserman, *J. Am. Chem. Soc.*, **79**, 1934 (1957).

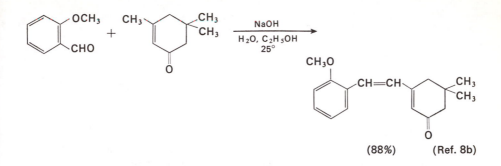

(88%)   (Ref. 8b)

apt to occur when solutions of bases in ethanol are heated or are allowed to stand for relatively long periods of time.

The stereochemistry of the Claisen-Schmidt reaction is of interest in that the unsaturated product normally has the carbonyl function *trans* to the larger group at the beta carbon atom.[1a,9b-e,10] This stereoselectivity appears to be attributable to the preferential dehydration of the enolate anion in conformation [14] rather than in conformation [15]. Conformation [14], in which there is less steric interference between the planar enolate anion system and the substituent at the beta carbon atom, is energetically favored.[10] Furthermore, the transition state [16] leading to

$$C_6H_5\text{—CHO} + C_6H_5\text{—CO—CH}_3 \xrightarrow[\substack{H_2O \\ C_2H_5OH \\ 15-30°}]{NaOH} C_6H_5\text{—CH—CH}_2\text{—CO—C}_6H_5 \underset{}{\overset{OH^{\ominus}}{\rightleftharpoons}}$$
$$\underset{OH}{|}$$

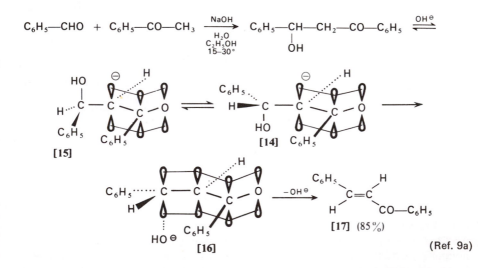

[17] (85%)

(Ref. 9a)

9. (a) E. P. Kohler and H. M. Chadwell, *Org. Syn.*, **Coll. Vol. 1**, 78 (1944). (b) M. Brink, *Tetrahedron*, **25**, 995 (1969). (c) A. Hassner and T. C. Mead, *ibid.*, **20**, 2201 (1964). (d) J. Zabicky, *J. Chem. Soc.*, 683 (1961). (e) H. O. House and R. S. Ro, *J. Am. Chem. Soc.*, **80**, 2428 (1958).
10. (a) H. E. Zimmerman, L. Singer, and B. S. Thyagarajan, *J. Am. Chem. Soc.*, **81**, 108 (1959). (b) H. E. Zimmerman and L. Ahramjian, *ibid.*, **81**, 2086 (1959). (c) H. E. Zimmerman and L. Ahramjian, *ibid.*, **82**, 5459 (1960). (d) H. E. Zimmerman in P. deMayo, ed., *Molecular Rearrangements*, Vol. 1, Wiley-Interscience, New York, 1963, pp. 345–406.

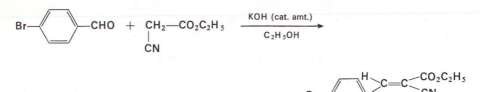

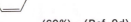

(60%)   (Ref. 9d)

the *trans* product [17] is favored, since in the transition state two large substituents are not eclipsed[11] and there is no interference with coplanarity of the enolate system. Of the two, the latter factor is in general the more important.[10]

Reaction of the intermediate aldol condensation products with base may lead either to the previously discussed dehydration reaction or to reversal of the aldol condensation (called a retrograde aldol or retroaldol reaction). Usually dehydration (i.e. the Claisen-Schmidt reaction) predominates when aromatic aldehydes are condensed with methyl ketones or with the methylene group of cyclic ketones. However, condensation at the methylene group of acyclic ketones having alpha substituents often fails because the intermediate aldol product (e.g., [18]) under-

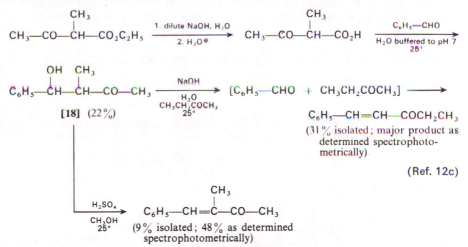

goes a retrograde aldol reaction more rapidly than it dehydrates.[12] For this reason the Claisen-Schmidt condensation of aromatic aldehydes with methyl alkyl ketones normally yields the product of condensation at the methyl group, as is illustrated by the accompanying examples. In acid solution both dehydration and retrograde aldol condensation of the hydroxy ketone [18] have been observed, but under the reaction conditions used the retrograde aldol product did not recondense.

11. D. Y. Curtin, *Rec. Chem. Progr.,* **15**, 111 (1954).
12. (a) J. D. Gettler and L. P. Hammett, *J. Am. Chem. Soc.,* **65**, 1824 (1943). (b) D. S. Noyce and W. L. Reed, *ibid.,* **81**, 624 (1959). (c) M. Stiles, D. Wolf, and G. V. Hudson, *ibid.,* **81**, 628 (1959). (d) A. T. Nielsen, H. Dubin, and K. Hise, *J. Org. Chem.,* **32**, 3407 (1967).

$$C_6H_5\text{—CHO} + CH_3\text{—CO—CH}_2\text{—CH}_3 \xrightarrow[H_2O]{NaOH} C_6H_5\text{—CH}=CH\text{—CO—CH}_2\text{—CH}_3$$

$$(99\%) \qquad \text{(Ref. 12a)}$$

$$C_6H_5\text{—CHO} + C_6H_5\text{—CH}_2\text{—CO—CH}_3 \xrightarrow[\substack{H_2O \\ 55°}]{KOH} C_6H_5\text{—CH}=CH\text{—CO—CH}_2\text{—C}_6H_5$$

$$(48\%) \qquad \text{(Ref. 13a)}$$

The reaction sequence shown for the preparation of the β-hydroxy ketone [18] in which a β-keto acid undergoes concurrent decarboxylation and aldol condensation is known as the Schöpf procedure.[13b,c] It is not presently clear whether this procedure for a selective aldol condensation involves the initial decarboxylation of the acid to form an enol which reacts with the aldehyde or whether condensation precedes or accompanies decarboxylation as illustrated in the following equation.

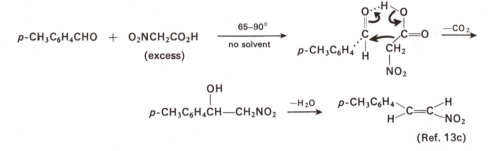

$$\text{(Ref. 13c)}$$

Condensations to form aldol products which are catalyzed by relatively weak bases under equilibrating conditions also usually result in condensation at the methyl group of methyl alkyl ketones. The ready reversibility of the aldol condensation and the possible reversibility of the subsequent base-catalyzed dehydration need

$$CCl_3CHO + (CH_3)_2CHCH_2COCH_3 \xrightarrow[\substack{CH_3CH_2CO_2H, \\ 100°}]{CH_3CH_2CO_2{}^{\ominus}Na^{\oplus}}$$

$$(CH_3)_2CHCH_2COCH_2\text{—CH—CCl}_3$$
$$|$$
$$OH$$

$$(59\%) \quad \text{(Ref. 13d)}$$

13. (a) P. L. Southwick, L. A. Pursglove, and P. Numerof, *J. Am. Chem. Soc.*, **72**, 1604 (1950). (b) C. Schöpf and K. Thierfelder, *Justus Liebigs Ann. Chem.*, **518**, 127 (1935). (c) W. L. F. Armarego, *J. Chem. Soc.*, C, 986 (1969). (d) W. Reeve and E. Kiehlmann, *J. Org. Chem.*, **31**, 2164 (1966). (e) P. Kurath, *ibid.*, **32**, 3626 (1967). See also R. E. Ireland, P. S. Grand, R. E. Dickerson, J. Bordner, and D. R. Rydjeski, *ibid.*, **35**, 570 (1970). (f) B. L. Yates and J. Quijano, *ibid.*, **34**, 2506 (1969); *ibid.*, **35**, 1239 (1970). (g) M. H. Benn and R. Shaw, *Chem. Commun.*, No. 6, 327 (1970). (h) R. E. Kent and S. M. McElvain, *Org. Syn.*, **Coll. Vol. 3**, 591 (1955). (i) P. Margaretha and O. E. Polansky, *Tetrahedron Letters*, **No. 57**, 4983 (1969). (j) C. A. Kingsbury, R. S. Egan, and T. J. Perun, *J. Org. Chem.*, **35**, 2913 (1970).

to be borne in mind whenever β-hydroxy ketones are to be subject to basic reaction conditions. The accompanying equations provide examples where these reverse reactions have been observed, including an example of a reaction claimed to involve a thermal rather than a base-catalyzed retrograde aldol reaction.[13f] The reverse reaction of α,β-unsaturated ketones involves the previously discussed (Chapter 9) Michael addition of hydroxide ion to the enone followed by a retrograde aldol reaction.

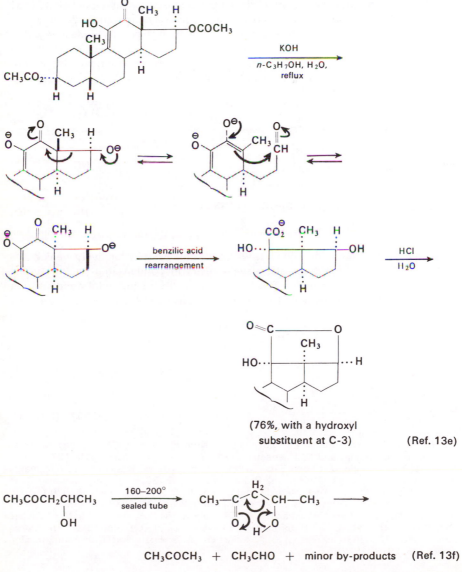

(76%, with a hydroxyl
substituent at C-3)   (Ref. 13e)

(Ref. 13f)

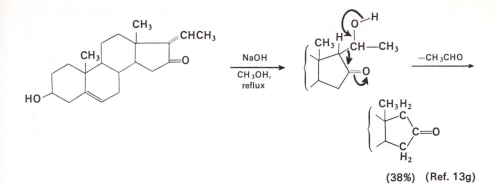

(38%) (Ref. 13g)

A very common side reaction accompanying the base-catalyzed aldol con-
densation-dehydration sequence is a further Michael reaction (see Chapter 9)
of the initially formed α,β-unsaturated carbonyl compound with an additional
equivalent of the active methylene compound.[1a,d,12d,13h−j]

Examples follow in which this aldol-Michael sequence is used as a method of
synthesis for a substituted glutaric acid[13h] and in which the Michael reaction of the
active methylene compound is prevented by performing the aldol condensation in
methanol with the strongly acidic isopropylidene malonate (Meldrum's acid).

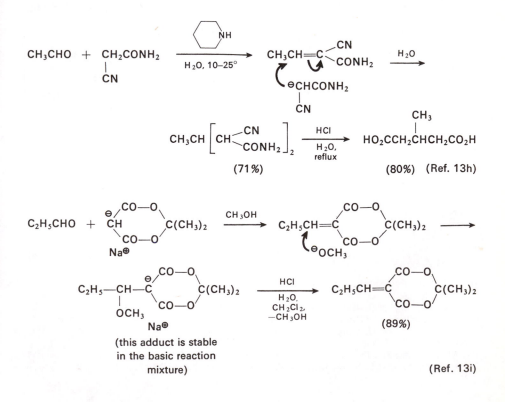

(71%)   (80%) (Ref. 13h)

(this adduct is stable
in the basic reaction
mixture)

(89%)

(Ref. 13i)

The successful condensation of aldehydes at the methylene group of methyl alkyl ketones is most readily accomplished with acid catalysis. This process involves the intermediate enol,[12,14] and, as noted previously (Chapters 8 and 9), the more highly substituted enol derived from an unsymmetrical ketone is usually the more stable. As a result, condensation in the presence of acid catalysts occurs predominantly at the methylene group.[12] Although the intermediate aldol condensation product may be dehydrated by the acid present in the original reaction mixture, a more common method for effecting this condensation involves reaction of the carbonyl compound with dry hydrogen chloride to form a β-chloroketone; subsequent heating or treatment with base produces the α,β-unsaturated ketone. These procedures are illustrated below.

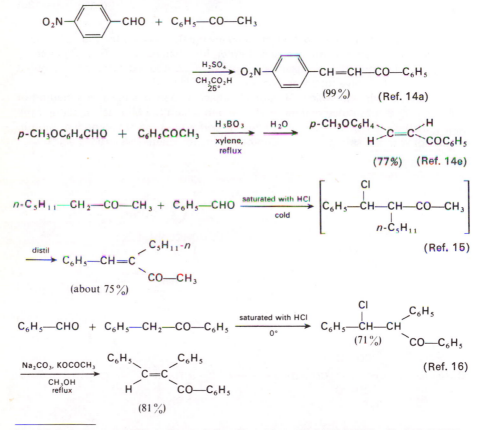

14. (a) D. S. Noyce and W. A. Pryor, *J. Am. Chem. Soc.,* **77**, 1397 (1955); **81**, 618 (1959). (b) D. S. Noyce, W. A. Pryor, and A. H. Bottini, *ibid.,* **77**, 1402 (1955). (c) D. S. Noyce and L. R. Snyder, *ibid.,* **80**, 4033, 4324 (1958); **81**, 620 (1959). (d) D. S. Noyce and W. L. Reed, *ibid.,* **80**, 5539 (1958). (e) R. D. Offenhauer and S. F. Nelsen, *J. Org. Chem.,* **33**, 775 (1968). (f) G. N. Dorofeenko, J. A. Shdanow, G. I. Shungijetu, and S. W. Kriwun, *Tetrahedron,* **22**, 1821 (1966).
15. M. T. Bogert and D. Davidson, *J. Am. Chem. Soc.,* **54**, 334 (1932).
16. E. P. Kohler and E. M. Nygaard, *J. Am. Chem. Soc.,* **52**, 4128 (1930).

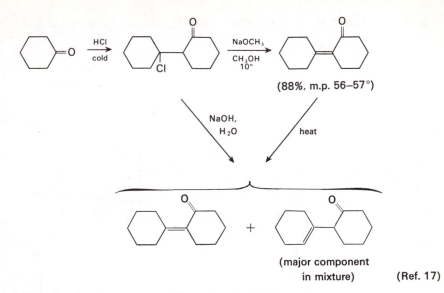

(88%, m.p. 56–57°)

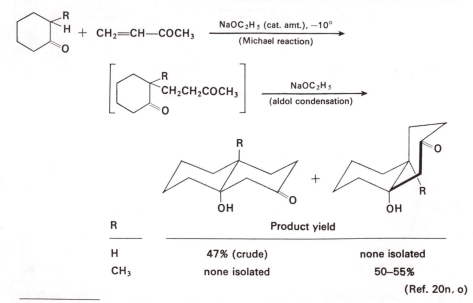

(major component
in mixture)   (Ref. 17)

Intramolecular base- or acid-catalyzed aldol condensation, followed by dehydration, for the preparation of cyclic α,β-unsaturated ketones is a widely used synthetic sequence, especially in cases where a five- or six-membered cyclic ketone is to be formed. The second or ring-closure step in the Robinson annelation reaction (see Chapter 9) provides a number of examples of the preparation of cyclohexenone derivatives. Others, are illustrated below. The first of the accom-

R	Product yield	
H	47% (crude)	none isolated
CH$_3$	none isolated	50–55%

(Ref. 20n, o)

17. (a) J. Reese, *Ber.* **75**, 384 (1942).   (b) E. Wenkert, S. K. Bhattacharya, and E. M. Wilson, *J. Chem. Soc.*, 5617 (1964).

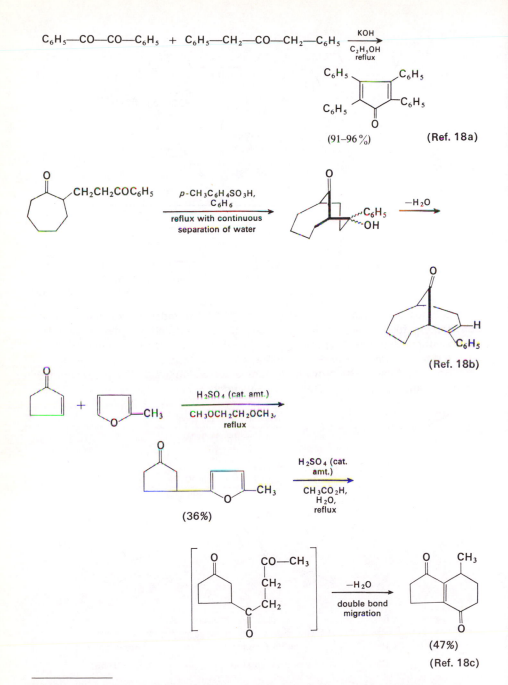

$$C_6H_5\text{—}CO\text{—}CO\text{—}C_6H_5 \ + \ C_6H_5\text{—}CH_2\text{—}CO\text{—}CH_2\text{—}C_6H_5 \ \xrightarrow[\substack{C_2H_5OH \\ \text{reflux}}]{KOH}$$

(91–96 %)   (Ref. 18a)

p-CH$_3$C$_6$H$_4$SO$_3$H, C$_6$H$_6$

reflux with continuous separation of water

—H$_2$O

(Ref. 18b)

H$_2$SO$_4$ (cat. amt.)

CH$_3$OCH$_2$CH$_2$OCH$_3$, reflux

H$_2$SO$_4$ (cat. amt.)

CH$_3$CO$_2$H, H$_2$O, reflux

(36%)

—H$_2$O

double bond migration

(47%)

(Ref. 18c)

18. (a) J. R. Johnson and O. Grummitt, *Org. Syn.*, **Coll. Vol. 3**, 806 (1955). (b) G. L. Buchanan, C. Maxwell, and W. Henderson, *Tetrahedron*, **21**, 3273 (1965). (c) M. A. Tobias, *J. Org. Chem.*, **35**, 267 (1970).

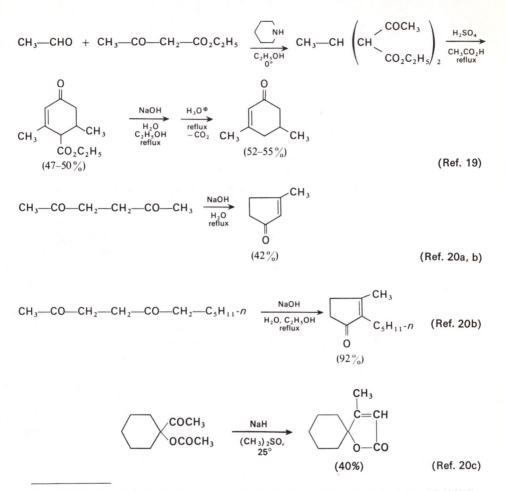

$$CH_3\text{—}CHO \;+\; CH_3\text{—}CO\text{—}CH_2\text{—}CO_2C_2H_5 \;\xrightarrow[\substack{C_2H_5OH \\ 0°}]{} \; CH_3\text{—}CH\left(CH\begin{array}{l} COCH_3 \\ CO_2C_2H_5 \end{array}\right)_2 \xrightarrow[\substack{CH_3CO_2H \\ reflux}]{H_2SO_4}$$

(47–50%)

(52–55%)

(Ref. 19)

$$CH_3\text{—}CO\text{—}CH_2\text{—}CH_2\text{—}CO\text{—}CH_3 \;\xrightarrow[\substack{H_2O \\ reflux}]{NaOH}\;$$

(42%)

(Ref. 20a, b)

$$CH_3\text{—}CO\text{—}CH_2\text{—}CH_2\text{—}CO\text{—}CH_2\text{—}C_5H_{11}\text{-}n \;\xrightarrow[\substack{H_2O,\ C_2H_5OH \\ reflux}]{NaOH}\;$$

(92%) (Ref. 20b)

(40%) (Ref. 20c)

19. (a) E. C. Horning, M. O. Denekas, and R. E. Field, *Org. Syn.,* **Coll. Vol. 3**, 317 (1955). (b) E. C. Horning, M. O. Denekas, and R. E. Field, *J. Org. Chem.,* **9**, 547 (1944). (c) E. C. Horning and R. E. Field, *J. Am. Chem. Soc.,* **68**, 384 (1946).

20. (a) R. M. Acheson and R. Robinson, *J. Chem. Soc.,* 1127 (1952). (b) H. Hunsdiecker, *Ber,* **75**, 447, 455, 460 (1942). (c) H. G. Lehmann, *Angew. Chem., Intern. Ed. Engl.,* **4**, 783 (1965). (d) G. R. Pettit, B. Green, and G. L. Dunn, *J. Org. Chem.,* **35**, 1367 (1970). (e) For other aldol condensations with glyoxylic acid, see M. Debono, R. M. Molloy, and L. E. Patterson, *ibid.,* **34**, 3032 (1969). (f) B. G. Cordiner, M. R. Vegar, and R. J. Wells, *Tetrahedron Letters,* **No. 26**, 2285 (1970). (g) K. G. Lewis and G. J. Williams, *ibid.,* **No. 50**, 4573 (1965). (h) M. D. Soffer and A. C. Williston, *J. Org. Chem.,* **22**, 1254 (1957). (i) L. J. Dolby and G. N. Riddle, *ibid.,* **32**, 3481 (1967). (j) P. M. Taylor and G. Fuller, *ibid.,* **34**, 3627 (1969). (k) D. E. O'Connor and W. I. Lyness, *J. Am. Chem. Soc.,* **86**, 3840 (1964). (l) A. B. Galun and A. Kalir, *Org. Syn.,* **48**, 27 (1968). (m) E. L. Compere, Jr., *J. Org. Chem.,* **33**, 2565 (1968) and references therein. (n) J. A. Marshall and W. J. Fanta, *ibid.,* **29**, 2501 (1964). (o) T. A. Spencer, H. S. Neel, D. C. Ward, and K. L. Williamson, *ibid.,* **31**, 434 (1966). See also Ref. 37a; (p) S. J. Rhoads, J. K. Chattopadhyay, and E. E. Waali, *J. Org. Chem.,* **35**, 3352 (1970); S. J. Rhoads and E. E. Waali, *ibid.,* **35**, 3358 (1970).

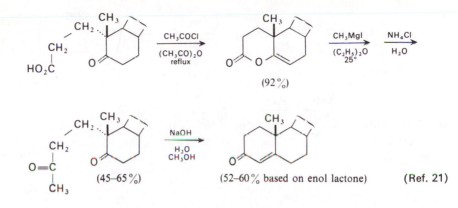

(92%)

(45–65%)   (52–60% based on enol lactone)   (Ref. 21)

panying examples illustrates the fact that the stereochemistry of the aldol condensation leading to ring closure is influenced by the nature of the α-substituent present. As was noted previously (Chapter 9), this type of intramolecular condensation is often more efficiently performed by the subsequently discussed Knoevenagel reaction employing pyrrolidine and acetic acid as catalysts.[20o]

Although the acid- and base-catalyzed aldol condensations and dehydrations which have been described usually form α,β-unsaturated carbonyl compounds as the major products, it is appropriate to note that the reaction conditions employed are capable of equilibrating enolizable α,β- and β,γ-unsaturated carbonyl compounds. In cases where the double bond has more alkyl substituents when it is located in the β,γ- rather than the α,β-position, the difference in stability between the conjugated and unconjugated isomers is small as the following examples indicate.[20f–k,p] In such cases, the product from an aldol condensation will normally be a mixture of double bond isomers from which separation of the lower boiling nonconjugated isomer by distillation may be difficult because of the ease with which the isomers interconvert on heating. Such mixtures of isomers can frequently be separated by chromatography on neutral adsorbents. With acid catalysis the position of a conjugated C—C double bond in an α-substituted ketone may be

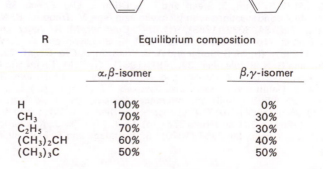

R	Equilibrium composition	
	α,β-isomer	β,γ-isomer
H	100%	0%
CH_3	70%	30%
C_2H_5	70%	30%
$(CH_3)_2CH$	60%	40%
$(CH_3)_3C$	50%	50%

(Ref. 20g, h)

$$R_1-\overset{\underset{\displaystyle R_2}{|}}{CH}-CH=\overset{\underset{\displaystyle CH_3}{|}}{C}-CO_2C_2H_5 \;\rightleftharpoons\; R_1-\overset{\underset{\displaystyle R_2}{|}}{C}=CH-\overset{\underset{\displaystyle CH_3}{|}}{CH}-CO_2C_2H_5$$

R_1	R_2	Equilibrium composition	
		α,β-isomer	β,γ-isomer
C_2H_5	CH_3	45%	55%
C_2H_5	C_2H_5	68%	32%
C_2H_5	$(CH_3)_3C$	86%	14%
C_6H_5	H	59%	41%

$$CH_3CH_2CH_2CH_2CH=CH-R \;\rightleftharpoons\; CH_3CH_2CH_2CH=CHCH_2-R$$

(Ref. 20i)

R	Equilibrium composition	
	α,β-isomer	β,γ-isomer
$-SO_2-CH_3$	<1%	>99%
$-SO-CH_3$	4%	96%
$-S-CH_3$	67%	33% (Ref. 20k)

isomerized to form a different conjugated isomer. The following example is illustrative.

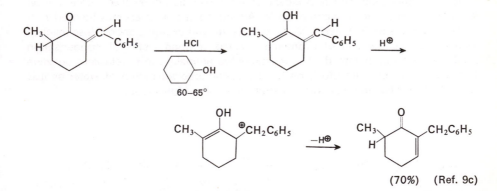

(70%)   (Ref. 9c)

Other reactions which closely resemble the base-catalyzed aldol condensations include certain of the subsequently discussed reactions of carbanions stabilized by sulfur-containing functional groups and the reaction with haloforms.[201,m] Examples of the latter condensations follow.

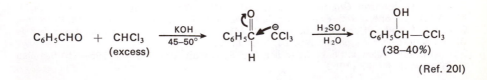

(Ref. 20l)

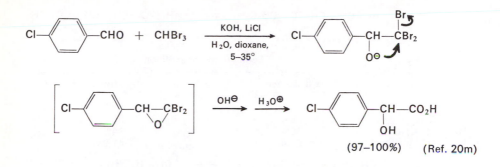

(97–100%)   (Ref. 20m)

A reaction related to the Claisen-Schmidt reaction, in the sense that condensation is forced to completion by dehydration of the intermediate aldol product, is the Knoevenagel reaction.[22] It is effected by treating an aldehyde or ketone with an active methylene compound in the presence of a catalytic amount of ammonia or a primary or secondary amine and at least a catalytic amount of a carboxylic acid.[23] Various ammonium salts (e.g. ammonium acetate, piperidinium acetate) are frequently used to supply the necessary catalysts. Since only small amounts of acid are required, a number of reports of successful Knoevenagel condensations, even in cases where no acid was added, are probably attributable to the presence of acidic impurities in the reactants or the solvent.

The active methylene compounds commonly employed include esters of malonic, cyanoacetic, and acetoacetic acid, as well as phenylacetonitrile, benzyl ketones, and aliphatic nitro compounds. Although malonic esters may be used with aldehydes and such reactive ketones as acetone and cyclohexanone, less reactive ketones require the use of cyanoacetic esters if good yields of condensation products are to be obtained. The reactions are normally run in refluxing benzene or toluene solution with provision for the continuous separation of water as it is formed. The following examples illustrate the utility of this reaction.

$$C_6H_5\text{—CHO} + CH_2(CO_2C_2H_5)_2 \xrightarrow[\substack{C_6H_6 \\ \text{reflux with separation} \\ \text{of } H_2O}]{\substack{\overset{\displaystyle NH}{\big|} \\ C_6H_5CO_2H}} C_6H_5\text{—CH}\text{=}C(CO_2C_2H_5)_2$$
(89–91%)   (Ref. 24)

21. (a) G. I. Fujimoto, *J. Am. Chem. Soc.,* **73**, 1856 (1951). (b) R. B. Turner, *ibid.,* **72**, 579 (1950). (c) R. B. Woodward, F. Sondheimer, D. Taub, K. Heusler, and W. M. McLamore, *ibid.,* **74**, 4223 (1952). (d) C. F. H. Allen, J. W. Gates, Jr., and J. A. VanAllan, *Org. Syn.,* **Coll. Vol. 3**, 353 (1955).
22. (a) J. R. Johnson, *Org. Reactions,* **1**, 210 (1942). (b) G. Jones, *ibid.,* **15**, 204 (1967).
23. (a) A. C. Cope, *J. Am. Chem. Soc.,* **59**, 2327 (1937). (b) A. C. Cope, C. M. Hofmann, C. Wyckoff, and E. Hardenbergh, *ibid.,* **63**, 3452 (1941). (c) J. L. van der Baan and F. Bickelhaupt, *Chem. Commun.,* **No. 24**, 1661 (1968).
24. (a) C. F. H. Allen and F. W. Spangler, *Org. Syn.,* **Coll. Vol. 3**, 377 (1955). (b) See also G. Billek, *ibid.,* **43**, 49 (1963).

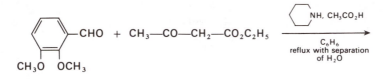

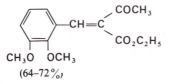

(64–72%)   (Ref. 25)

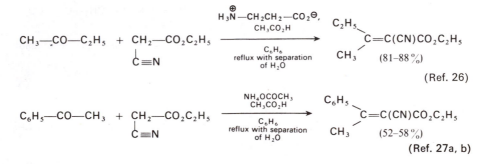

(Ref. 26)

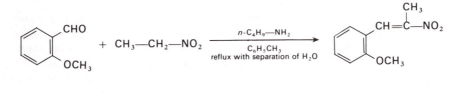

(Ref. 27a, b)

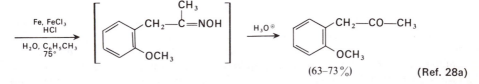

(63–73%)   (Ref. 28a)

25. E. C. Horning, J. Koo, M. S. Fish, and G. N. Walker, *Org. Syn.,* **Coll. Vol. 4**, 408 (1963).
26. F. S. Prout, R. J. Hartman, E. P. Y. Huang, C. J. Korpics, and G. R. Tichelaar, *Org. Syn.,* **Coll. Vol. 4**, 93 (1963). See also F. S. Prout, *J. Org. Chem.,* **18**, 928 (1953).
27. (a) S. M. McElvain and D. H. Clemens, *Org. Syn.,* **Coll. Vol. 4**, 463 (1963). (b) A. C. Cope and E. M. Hancock, *ibid.,* **Coll. Vol. 3**, 399 (1955); A. Sakurai and H. Midorikawa, *J. Org. Chem.,* **34**, 3612 (1969). (c) T. Hayashi, *ibid.,* **31**, 3253 (1966); T. Hayashi, I. Hori, H. Baba, and H. Midorikawa, *ibid.,* **30**, 695 (1965). (d) G. Jones and W. J. Rae, *Tetrahedron,* **22**, 3021 (1966). (e) M. Schwarz, *Chem. Commun.,* **No. 5**, 212 (1969).
28. (a) R. V. Heinzelman, *Org. Syn.,* **Coll. Vol. 4**, 573 (1963). (b) G. Billek, *ibid.,* **43**, 49 (1963).

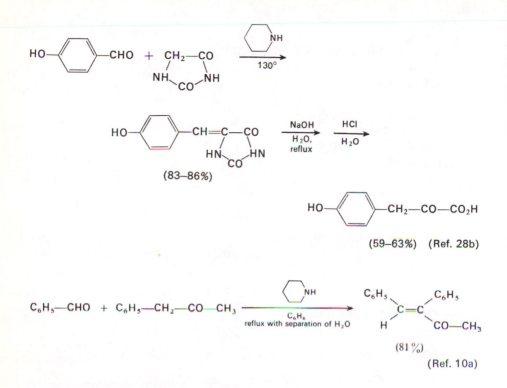

(83–86%)

(59–63%)   (Ref. 28b)

(81 %)

(Ref. 10a)

It should be noted that condensations of ketones with cyanoacetic esters are often not stereoselective but rather yield mixtures of geometrically isomeric alkylidenecyanoacetates.[27c,d]   However, the corresponding condensations with aldehydes are stereoselective yielding the stereoisomer in which the β-alkyl or β-aryl group is *cis* to the smaller cyano function.[9d,27e] Similar condensations have been effected employing either an ion-exchange resin,[29] certain metal fluorides,[30a] or potassium cyanide[30b] as a reaction catalyst.

These condensations were formerly considered to occur by a mechanism comparable to that previously discussed for the Claisen-Schmidt reaction in which the amine served only as a base to generate an enolate.   However, the facts that catalytic amounts of acid are often required and that ammonium salts and primary and secondary amines are usually much more effective catalysts than tertiary amines have led to the view that the Knoevenagel condensations often involve the inter-mediate formation of an imine or iminium salt (e.g., [19]) from the ketone or

29. (a) M. J. Astle and W. C. Gergel, *J. Org. Chem.,* **21**, 493 (1956). (b) R. W. Hein, M. J. Astle, and J. R. Shelton, *ibid.,* **26**, 4874 (1961).
30. (a) L. Rand, J. V. Swisher, and C. J. Cronin, *J. Org. Chem.,* **27**, 3505 (1962) ; P. M. Lelean and J. A. Morris, *Chem. Commun.,* **No. 5**, 239 (1968). (b) F. S. Prout, V. N. Aguilar, F. H. Girard, D. D. Lee, and J. P. Shoffner, *Org. Syn.,* **44**, 59 (1964). In this case, the initially formed conjugated system undergoes a further conjugate addition of cyanide ion (see Chapter 9).

aldehyde and the amine.[22b,31] Subsequent reaction of this imine (or iminium salt) with an enolate anion, an enol, or an enamine derived from the active methylene compound produces an intermediate amino compound (e.g., **[20]**), which in turn forms the unsaturated product by elimination of the amine. It will be noted that the position of condensation of an aldehyde with an unsymmetrical ketone (e.g., benzaldehyde with benzyl methyl ketone) is sometimes not the same in the Knoevenagel reaction as in the Claisen-Schmidt reaction.

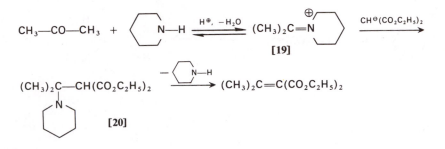

Malonic acid, the monoesters of malonic acid, and cyanoacetic acid are sufficiently reactive as active methylene compounds to permit successful Knoevenagel condensations with aldehydes and very reactive ketones (e.g., **[21]**). As indicated in the accompanying equations, thermal decarboxylation of the condensation product (e.g., **[22]**) leads predominantly to the formation of a β,γ-unsaturated acid derivative (e.g., **[23]**), presumably by the initial isomerization of the conjugated acid (e.g., **[22]**) to the unconjugated acid (e.g., **[24]**) and the

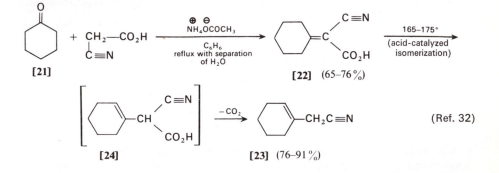

31. (a) T. I. Crowell and D. W. Peck, *J. Am. Chem. Soc.,* **75**, 1075 (1953). (b) E. H. Cordes and W. P. Jencks, *ibid.,* **84**, 826 (1962). (c) D. N. Robertson, *J. Org. Chem.,* **25**, 47 (1960). (d) C. Schroeder, S. Preis, and K. P. Link, *Tetrahedron Letters,* **No. 13**, 23 (1960). (e) G. Charles, *Bull. Soc. Chim. France,* 1559, 1566, 1573, 1576 (1963). (f) A. H. Blatt and N. Gross, *J. Org. Chem.,* **29**, 3306 (1964). (g) For examples in which the iminium salt intermediates may not be involved, see H. Dressler and J. E. Graham, *ibid.,* **32**, 985 (1967), and Ref. 37.
32. A. C. Cope, A. A. D'Addieco, D. E. Whyte, and S. A. Glickman, *Org. Syn.,* **Coll. Vol. 4,** 234 (1963).

usual subsequent decarboxylation of the nonconjugated cyanoacetic acid.[33] If condensations of this type are run in pyridine solution (the Doebner modification), decarboxylation usually occurs in the reaction mixture, leading to formation of α,β-unsaturated carboxylic acid derivatives (usually the stereoisomer with the carboxyl group *trans* to the larger beta substituent) as shown in the following equations. This procedure is generally superior to the subsequently discussed

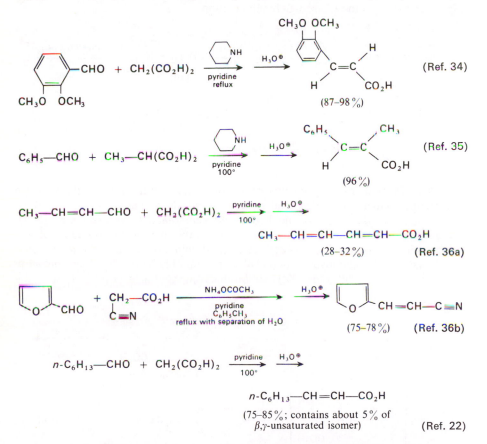

Perkin reaction because it is applicable to aliphatic aldehydes that would be polymerized by more vigorous reaction conditions and because it usually results in

33. (a) E. J. Corey, *J. Am. Chem. Soc.*, **74**, 5897 (1952). (b) E. J. Corey, *ibid.*, **75**, 1163 (1953). (c) E. J. Corey and G. Fraenkel, *ibid.*, **75**, 1168 (1953). (d) J. Klein and A. Y. Meyer, *J. Org. Chem.*, **29**, 1038 (1964).
34. (a) J. Koo, M. S. Fish, G. N. Walker, and J. Blake, *Org. Syn.*, **Coll. Vol. 4**, 327 (1963). (b) S. Rajagopalan and P. V. A. Raman, *ibid.*, **Coll. Vol. 3**, 425 (1955). (c) R. H. Wiley and N. R. Smith, *ibid.*, **Coll. Vol. 4**, 731 (1963).
35. W. J. Gensler and E. Berman, *J. Am. Chem. Soc.*, **80**, 4949 (1958).
36. (a) C. F. H. Allen and J. VanAllan, *Org. Syn.*, **Coll. Vol. 3**, 783 (1955). (b) J. M. Patterson, *ibid.*, **40**, 46 (1960).

better yields for aromatic aldehyde reactions. An investigation[33] of the decarboxylation that accompanies this condensation indicates that decarboxylation must either precede or occur concurrently with the introduction of the carbon-carbon double bond in condensations involving aliphatic aldehydes. The reaction is believed to proceed by the decarboxylative elimination outlined below. Although

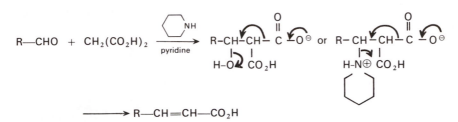

$$\longrightarrow R—CH{=}CH—CO_2H$$

the same sequence may be applicable to reactions involving aromatic aldehydes, an alternative addition-elimination process has been suggested for the decarboxylation of arylidenemalonic acid derivatives (e.g., [25]) in pyridine solution.[33c] In a related study of the decarboxylation of arylidenecyanoacetic acids,[33d] the stereochemistry of the decarboxylation process was found to change as the solvent was changed from pyridine to quinoline suggesting that a direct decarboxylation of the α,β-unsaturated cyanoacetic acid may also occur.

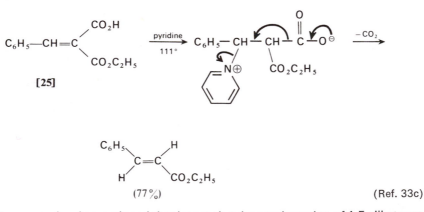

(77%)   (Ref. 33c)

It was previously mentioned that intramolecular condensation of 1,5-diketones and related substances is very efficiently accomplished by the use of secondary amines (especially pyrrolidine) and acetic acid as catalysts.[20n,37a,b] Although it might appear that these intramolecular aldol condensations also involve the previously discussed formation of an electrophilic iminium salt which is attacked by an enolate anion, studies of the reaction illustrated in the following equation

37. (a) T. A. Spencer, K. K. Schmiegel, and K. L. Williamson, *J. Am. Chem. Soc.*, **85**, 3785 (1963). (b) T. A. Spencer, H. S. Neel, T. W. Flechtner, and R. A. Zayle, *Tetrahedron Letters*, **No. 43**, 3889 (1965). (c) S. Tomoda, Y. Takeuchi, and Y. Nomura, *ibid.*, **No. 40**, 3549 (1969). (d) L. A. Paquette, *ibid.*, **No. 18**, 1291 (1965). (e) J. W. Lewis, P. L. Meyers, and M. J. Readhead, *J. Chem. Soc.*, C, 771 (1970). (f) K. C. Brannock, R. D. Burpitt, H. E. Davis, H. S. Pridgen, and J. G. Thweatt, *J. Org. Chem.*, **29**, 2579 (1964).

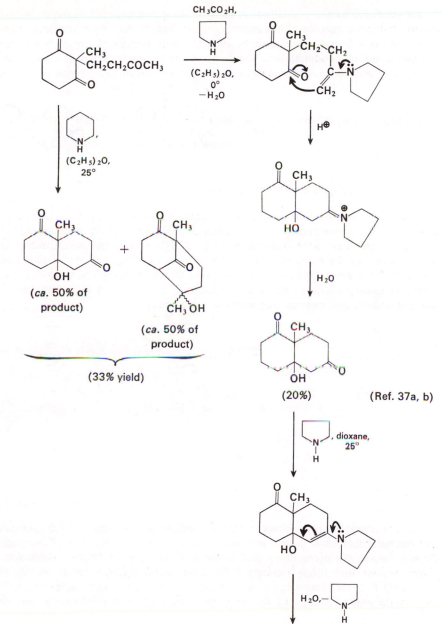

(ca. 50% of
product)

+

(ca. 50% of
product)

(33% yield)

(20%)   (Ref. 37a, b)

suggest that these Knoevenagel reactions follow a different course. It is believed that an enamine is formed and serves as a nucleophile in an attack on the second carbonyl group; hydrolysis yields the aldol product. Further reaction of this aldol product with a secondary amine is also believed to involve the formation of an enamine which undergoes the indicated elimination prior to hydrolysis. Support for this hypothesis is found in the ability of enamines to add to other double bonds of both Schiff bases (imines)[37c] and aldehydes[37d-f] in condensations analogous to the aldol condensation. The initial products of these enamine analogs of aldol condensations· frequently undergo further transformation such as that shown in the following example. Thus, it seems likely that at least two different

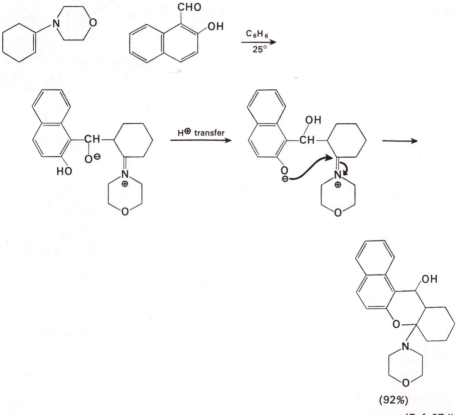

(92%)

(Ref. 37d)

reaction mechanisms may be operative in the Knoevenagel reaction.[22b] The formation of iminium salts which are attacked by the enolate anions of active methylene compounds appears applicable to the condensation of aldehydes or ketones with derivatives of malonic acid. Although this pathway may also be followed in some condensations involving a ketone as the active methylene component, a second reaction path involving the initial conversion of the ketone to an enamine appears to be more likely in many cases.

THE MANNICH REACTION

The aldol condensation of active methylene compounds (e.g., [26]), with formaldehyde (called Tollen's condensation when basic catalysis is used) occurs readily

$$CH_3{-}NO_2 + CH_2O \xrightarrow[\substack{CH_3OH \\ CH_3NO_2 \\ 25-35°}]{KOH} \xrightarrow{H_2SO_4} HO{-}CH_2{-}CH_2{-}NO_2$$

[26]   (introduced as   (46–49%)   (Ref. 38a)
the trimer)

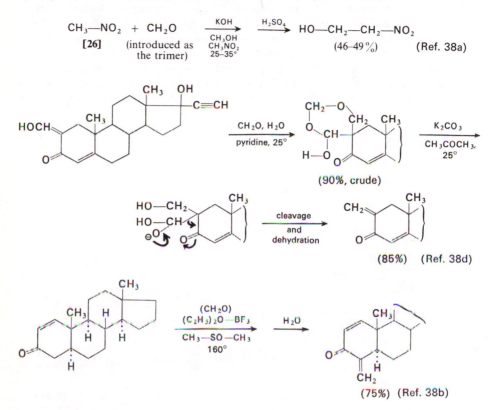

(90%, crude)

(85%)   (Ref. 38d)

(75%) (Ref. 38b)

but the reaction is often difficult to control in routine laboratory preparations, and
polycondensation results.[41]   A further complication in base-catalyzed reactions
arises from the fact that formaldehyde may reduce other carbonyl groups present
(the Cannizzaro reaction[39]).   Examples of the reaction of carbonyl compounds
with excess formaldehyde are provided in the following equations.

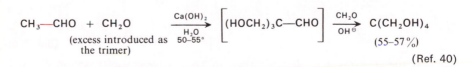

$$CH_3{-}CHO + CH_2O \xrightarrow[\substack{H_2O \\ 50-55°}]{Ca(OH)_2} \Big[(HOCH_2)_3C{-}CHO\Big] \xrightarrow[OH^\ominus]{CH_2O} C(CH_2OH)_4$$

(excess introduced as   (55–57%)
the trimer)
(Ref. 40)

38. (a) W. E. Noland, *Org. Syn.,* **41**, 67 (1961). (b) W. H. W. Lunn, *J. Org. Chem.,* **30**,
2925 (1965). (c) I. Laos, *ibid.,* **32**, 1409 (1967); (d) A. J. Manson and D. Wood, *ibid.,* **32**,
3434 (1967). (e) B. Wesslen, *Acta Chem. Scand.,* **21**, 713, 718 (1967); **22**, 2085 (1968);
23, 1017 (1969).
39. T. A. Geissman, *Org. Reactions,* **2**, 94 (1944).
40. H. B. J. Schurink, *Org. Syn.,* **Coll. Vol. 1**, 425 (1944).

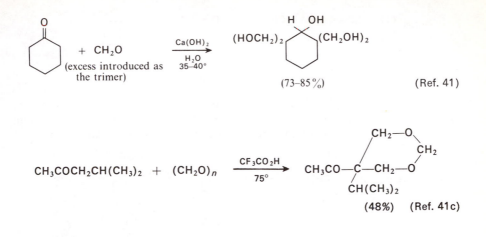

(73–85%)   (Ref. 41)

$$CH_3COCH_2CH(CH_3)_2 + (CH_2O)_n \xrightarrow[75°]{CF_3CO_2H}$$

(48%)   (Ref. 41c)

A more satisfactory method for the introduction of a single carbon atom is the reaction of an active methylene compound with formaldehyde and an amine to form a β-aminocarbonyl compound. This reaction, known as the Mannich reaction,[42] is usually run in water, methanol, ethanol, or acetic acid. The formaldehyde is introduced as an aqueous solution (20 to 40 per cent formaldehyde by weight), as the solid trimer (trioxymethylene), or as the solid polymer (polyoxymethylene); the latter two reagents are converted to formaldehyde by acid added to the reaction mixture. The amine is normally introduced as its hydrochloride, and several drops of hydrochloric acid are frequently added to ensure that the reaction mixture is not basic. Usually a secondary amine (e.g., dimethylamine, diethylamine, piperidine, morpholine, pyrrolidine) is employed to avoid side reactions, such as might occur between the initially formed Mannich base and additional formaldehyde and active methylene compound if a primary amine or ammonia were used. Typical examples of the Mannich are provided in the following equations.

$$C_6H_5-CO-CH_3 + (CH_2O)_3 + (CH_3)_2NH_2^{\oplus}Cl^{\ominus} \xrightarrow[\substack{C_2H_5OH \\ reflux}]{HCl}$$

$$C_6H_5-CO-CH_2-CH_2-NH^{\oplus}(CH_3)_2Cl^{\ominus}$$
(68–72%)   (Ref. 43)

41. (a) H. Wittcoff, *Org. Syn., Coll. Vol. 4*, 907 (1963). (b) D. R. Moore and A. Oroslan, *J. Org. Chem.*, **31**, 2620 (1966). (c) W. C. Lumma, Jr., and O. H. Ma, *ibid.*, **35**, 2391 (1970). (d) T. B. H. McMurry and M. T. Richardson, *J. Chem. Soc.*, C, 1804 (1967); W. A. Kennedy and T. B. H. McMurry, *ibid.*, C, 879 (1969).
42. (a) F. F. Blicke, *Org. Reactions*, **1**, 303 (1942). (b) B. Reichert, *Die Mannich Reaktion*, Springer Verlag, Berlin, 1959. (c) H. Hellmann and G. Opitz, *Angew. Chem.*, **68**, 265 (1956).
43. C. E. Maxwell, *Org. Syn., Coll. Vol. 3*, 305 (1955).

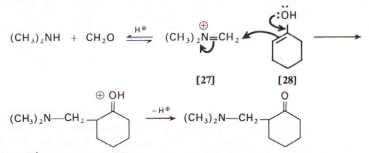

$$CH_3-CO-CH_3 + (CH_2O)_3 + (C_2H_5)_2NH_2^{\oplus}Cl^{\ominus} \xrightarrow[\substack{CH_3OH \\ CH_3COCH_3 \\ reflux}]{HCl} \xrightarrow[H_2O]{NaOH}$$

$$CH_3-CO-CH_2-CH_2-N(C_2H_5)_2$$
$$(62-70\%)$$

(Ref. 44)

Under the usual slightly acidic reaction conditions, the mechanism of the Mannich reaction is believed[42c,45] to involve electrophilic attack by an iminium salt (e.g., [27]) on the enol (e.g., [28]) of the active methylene compound.

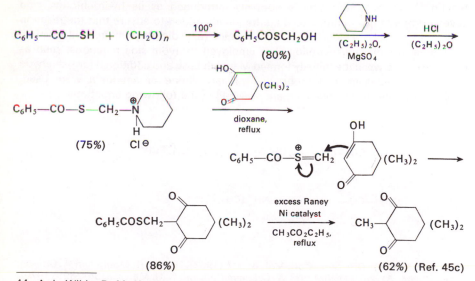

$$(CH_3)_2NH + CH_2O \xrightleftharpoons{H^{\oplus}} (CH_3)_2\overset{\oplus}{N}{=}CH_2$$

[27]   [28]

An interesting process, which is related to the Mannich reaction, introduces an acylthiomethyl group rather than an aminomethyl group.[45c] This procedure is illustrated below along with catalytic desulfurization (see Chapter 1) of the thiol ester to form a methyl group.

44. A. L. Wilds, R. M. Nowak, and K. E. McCaleb, *Org. Syn.*, **Coll. Vol. 4**, 281 (1963).
45. (a) T. F. Cummings and J. R. Shelton, *J. Org. Chem.*, **25**, 419 (1960). (b) J. E. Fernandez and J. S. Fowler, *ibid.*, **29**, 402 (1964) ; J. E. Fernandez, J. S. Fowler, and S. J. Glaros, *ibid.*, **30**, 2787 (1965). (c) E. E. Smissman, J. R. J. Sorenson, W. A. Albrecht, and M. W. Creese, *ibid.*, **35**, 1357 (1970). See also D. N. Kirk and V. Petrow, *J. Chem. Soc.*, 1091 (1962). (d) M. Masui, K. Fujita, and H. Ohmori, *Chem. Commun.*, **No. 13**, 182 (1970). (e) H. Volz and H. H. Kiltz, *Tetrahedron Letters*, **No. 22**, 1917 (1970).

A consequence of the foregoing mechanistic scheme is the expectation that unsymmetrical ketones will react predominantly at the more highly substituted alpha position, corresponding to the more stable enol (see Chapters 8 and 9). Although early studies[46] provided examples that appeared contrary to this expectation, reinvestigation[47] has shown the Mannich reaction products from unsymmetrical ketones such as [29] and [30] to be predominantly those that result from attack at the more highly substituted position.

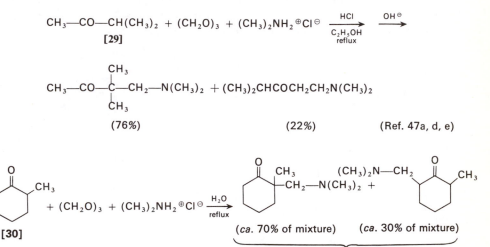

Since many of the initially formed Mannich bases appear to be capable of isomerization (presumably by reversal of the Mannich reaction) when they are heated to temperatures above 100°,[47e] it is advisable to make structural assignments to the isomer(s) from Mannich reactions by n.m.r. spectrometry rather than by chemical transformation. Although the Mannich reactions with unsymmetrical ketones would appear to involve a kinetically controlled attack at the more highly substituted (and more stable) enol, examples (e.g., [31] exist in which steric hindrance in an unsymmetrical ketone leads to the formation of a Mannich base

$(CH_3)_2CHCH_2COCH_3$ + $(CH_2O)_n$ + $(C_2H_5)_2\overset{\oplus}{N}H_2$ $Cl^\ominus$ $\xrightarrow[\substack{CH_3OH, \\ reflux}]{HCl, H_2O}$

[31]

$(CH_3)_2CHCH_2COCH_2CH_2N(C_2H_5)_2$

(63%)   (Ref. 47c, e)

46. (a) E. C. duFeu, F. J. McQuillin, and R. Robinson, *J. Chem. Soc.*, 53, (1937). (b) R. Jaquier, M. Mousseron, and S. Boyer, *Bull. Soc. Chim. France*, 1653 (1956).
47. (a) M. Brown and W. S. Johnson, *J. Org. Chem.*, **27**, 4706 (1962). (b) H. O. House and B. M. Trost, *ibid.*, **29**, 1339 (1964). (c) T. A. Spencer, D. S. Watt, and R. J. Friary, *ibid.*, **32**, 1234 (1967). (d) N. B. Haynes and C. J. Timmons, *J. Chem. Soc.*, C, 224 (1966). (e) G. L. Buchanan, A. C. W. Curran, and R. T. Wall, *Tetrahedron*, **25**, 5503 (1969); H. L. Brown, G. L. Buchanan, A. C. W. Curran, and G. W. McLay, *ibid.*, **24**, 4565 (1968).

which does not correspond to the more stable enol of the starting ketone. In such cases, it appears likely that the reaction product is determined by thermodynamic control so that the product which would result from attack at the more highly substituted enol is destabilized by steric interaction and reverts to the starting materials.

Although the use of primary amines in the Mannich reaction is undesirable when a monocondensation product is required, a number of cyclic products have been prepared via double Mannich condensations with primary amines. The second of the two cases cited below, an example of the Robinson-Schöpf reaction, also illustrates the use of an aldehyde other than formaldehyde in the Mannich reaction.

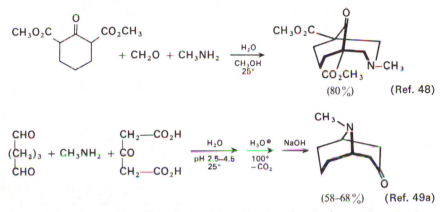

(80%)   (Ref. 48)

(58–68%)   (Ref. 49a)

Apart from the applicability of the Mannich reaction to ketones and other active methylene compounds,[42,45e] the intermediate iminium salt [27] is a sufficiently reactive electrophilic agent to attack such relatively reactive aromatic nuclei as phenol and indole ([32]).

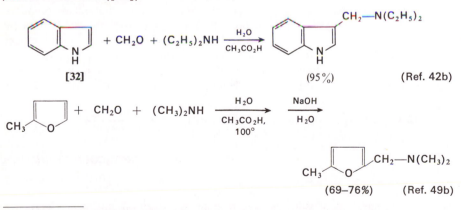

[32]

(95%)   (Ref. 42b)

(69–76%)   (Ref. 49b)

48. (a) F. F. Blicke and F. J. McCarty, *J. Org. Chem.*, **24**, 1379 (1959). (b) W. Schneider and H. Götz, *Arch. Pharm.*, **294**, 506 (1961). (c) H. O. House and H. C. Müller, *J. Org. Chem.*, **27**, 4436 (1962).
49. (a) A. C. Cope, H. L. Dryden, Jr., and C. F. Howell, *Org. Syn.*, **Coll. Vol. 4**, 816 (1963). (b) E. L. Eliel and M. T. Fisk, *ibid.*, **Coll. Vol. 4**, 626 (1963).

Mannich bases are used primarily as synthetic intermediates in which the amino function is subsequently replaced by some other group.[42,50] The most common reactions of this type involve elimination of the amine function from a β-aminoketone (e.g., [33]) or from the corresponding quaternary ammonium salt (e.g., [34]) to produce an α,β-unsaturated ketone (e.g., [35]). If the un-

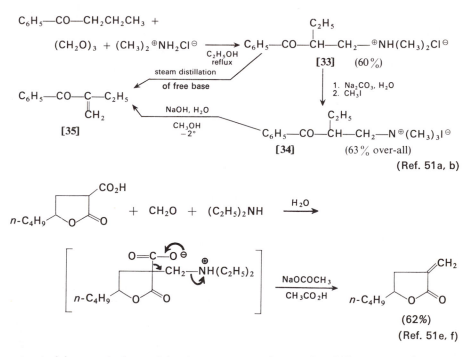

saturated ketone is formed in the presence of a nucleophilic reagent (e.g., an enolate anion or cyanide anion), conjugate addition of the nucleophile to the unsaturated system occurs in the reaction mixture. Several examples are cited in the Chapter 9 discussion of the Michael reaction; another is given below, along with an illustration of the use of a Mannich base derived from an aromatic system. As indicated, the latter reaction occurs[51d] by the same type of elimination-addition sequence that is operative with β-amino ketones.

50. (a) J. H. Brewster and E. L. Eliel, Org. Reactions, 7, 99 (1953). (b) A. F. Casy and J. L. Myers, J. Chem. Soc., 4639 (1964). (c) R. Andrisano, A. S. Angeloni, P. DeMaria, and M. Tramontini, ibid., C, 2307 (1967). (d) J. C. Craig, M. Moyle, and L. F. Johnson, J. Org. Chem., 29, 410 (1964).
51. (a) J. H. Burckhalter and R. C. Fuson, J. Am. Chem. Soc., 70, 4184 (1948). (b) H. O. House, D. J. Reif, and R. L. Wasson, ibid., 79, 2490 (1957). (c) The same procedure has been used to prepare aldehydes with an α-methylene substituent. L. Lardicci, F. Navari, and R. Rossi, Tetrahedron, 22, 1991 (1966). (d) J. D. Albright and H. R. Synder J. Am. Chem. Soc., 81, 2239 (1959). (e) J. Martin, P. C. Watts, and F. Johnson, Chem. Commun., No. 1, 27 (1970). (f) E. S. Behare and R. B. Miller, ibid., No. 7, 402 (1970). (g) N. R. Unde, S. V. Hiremath, G. H. Kulkarni, and G. R. Kelkar, Tetrahedron Letters, No. 47, 4861 (1968).

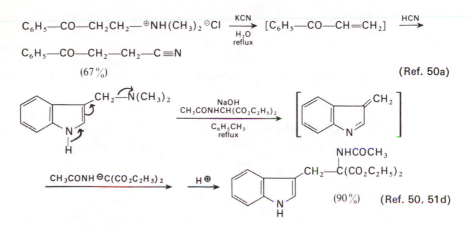

$$C_6H_5\text{—}CO\text{—}CH_2CH_2\text{—}^{\oplus}NH(CH_3)_2\ ^{\ominus}Cl \xrightarrow[\substack{H_2O \\ reflux}]{KCN} [C_6H_5\text{—}CO\text{—}CH=CH_2] \xrightarrow{HCN}$$

$$C_6H_5\text{—}CO\text{—}CH_2\text{—}CH_2\text{—}C\equiv N$$

(67%)   (Ref. 50a)

(90%)   (Ref. 50, 51d)

CONDENSATIONS IN WHICH PRODUCT FORMATION IS FAVORED BY RING CLOSURE INVOLVING THE INTERMEDIATE ALDOL PRODUCT: THE PERKIN, STOBBE, AND DARZENS REACTIONS

These modifications of the aldol condensation have as a common feature the presence of a functional group that can react intramolecularly with the alkoxide anion (analogous to [3]) initially formed, thus favoring the formation of the condensation product. The Perkin reaction,[22a] which is normally applicable only to aromatic aldehydes, is usually effected by heating a mixture of the aldehyde, an acid anhydride (the active methylene component), and a weak base (such as the sodium or potassium salt of the acid or triethylamine) to relatively high temperatures (150–200°), as illustrated in the accompanying equations. Note in the second

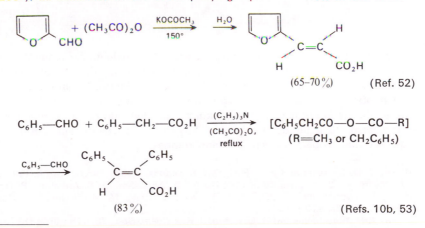

(65–70%)   (Ref. 52)

$$C_6H_5\text{—}CHO + C_6H_5\text{—}CH_2\text{—}CO_2H \xrightarrow[\substack{(CH_3CO)_2O, \\ reflux}]{(C_2H_5)_3N} [C_6H_5CH_2CO\text{—}O\text{—}CO\text{—}R]$$
$$(R=CH_3 \text{ or } CH_2C_6H_5)$$

(83%)   (Refs. 10b, 53)

52. (a) J. R. Johnson, *Org. Syn.,* **Coll. Vol. 3**, 426 (1955). (b) F. K. Thayer, *ibid.,* **Coll. Vol. 1**, 398 (1944).
53. (a) L. F. Fieser, *Experiments in Organic Chemistry,* 3d ed., Heath, Boston, 1955, pp. 182–185. (b) R. E. Buckles and J. A. Cooper, *J. Org. Chem.,* **30**, 1588 (1965). (c) R. E. Buckles and K. Bremer, *Org. Syn.,* **Coll. Vol. 4**, 777 (1963). See also D. F. DeTar, *ibid.,* **Coll. Vol. 4**, 730 (1963). (d) B. J. Kurtev and C. G. Kratchanov, *J. Chem. Soc.,* B, 649 (1969).

of the accompanying examples that acids and anhydrides present in the reaction mixture are rapidly equilibrated.

The Perkin reaction is believed to proceed by formation of an enolate anion (e.g., [36]) from the acid anhydride and subsequent reaction with the aldehyde to form an alkoxide (e.g., [37]). The ensuing intramolecular acylation affords a β-acyloxy derivative (e.g., [38]), which can either undergo elimination to form a cinnamic acid or decarboxylative elimination (e.g., [39]) to form an olefin. As in previously discussed condensations, the kinetically favored stereoisomer formed in the elimination reaction is the one with the carboxyl group *trans* to the larger group at the beta carbon atom.[10b] Decarboxylative elimination (e.g., [39]) to form an olefin is usually minor in the Perkin reaction; however, higher temperatures

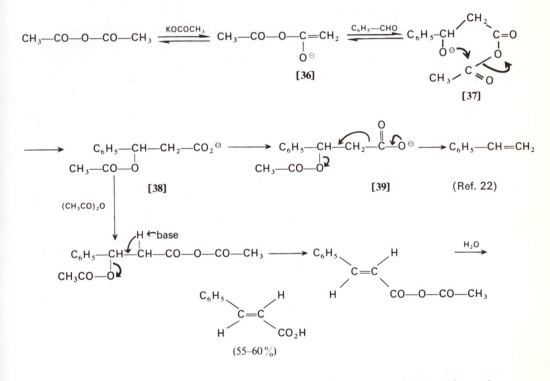

or an adjacent activating substituent may cause this process to become the major reaction, as the following equations demonstrate.

$$C_6H_5-CH=CH-CHO + C_6H_5-CH_2-CO_2H \xrightarrow[\substack{(CH_3CO)_2O \\ reflux}]{PbO} C_6H_5-CH=CH-CH=CH-C_6H_5$$

$$(27-29\%)$$

(Ref. 54)

54. B. B. Corson, *Org. Syn.*, **Coll. Vol. 2**, 229 (1943).

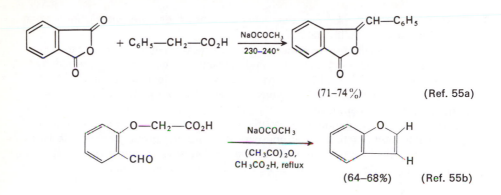

(71–74%)   (Ref. 55a)

(64–68%)   (Ref. 55b)

The use of α-acylamido acids in the Perkin reaction leads to the formation of arylidene derivatives of oxazolone called azlactones.[56] These substances, whose preparations are illustrated by the accompanying equations, serve as useful

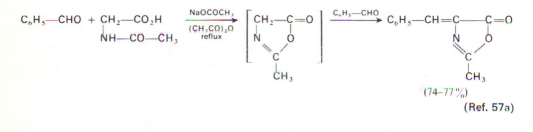

(74–77%)

(Ref. 57a)

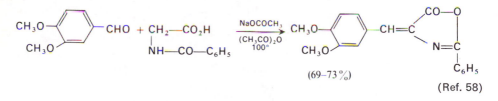

(69–73%)

(Ref. 58)

intermediates for the synthesis of α-amino acids (e.g., [40]) and α-keto acids (e.g., [41]). Similar intermediates may be obtained by the Knoevenagel condensa-

55. (a) R. Weiss, *Org. Syn.,* **Coll. Vol. 2**, 61 (1943). (b) A. W. Worden, *ibid.,* **46**, 28 (1966).

56. H. E. Carter, *Org. Reactions,* **3**, 198 (1946).

57. (a) R. M. Herbst and D. Shemin, *Org. Syn.,* **Coll. Vol. 2**, 1 (1943). (b) R. Filler, E. J. Piasek, and H. A. Leipold, *ibid.,* **43**, 3 (1963).

58. (a) J. S. Buck and W. S. Ide, *Org. Syn.,* **Coll. Vol. 2**, 55 (1943). (b) G. E. VandenBerg, J. B. Harrison, H. E. Carter, and B. J. Magerlein, *ibid.,* **47**, 101 (1967).

59. (a) R. M. Herbst and D. Shemin, *Org. Syn.,* **Coll. Vol. 2**, 491 (1943). (b) H. B. Gillespie and H. R. Snyder, *Org. Syn.,* **Coll. Vol. 2**, 489 (1943).

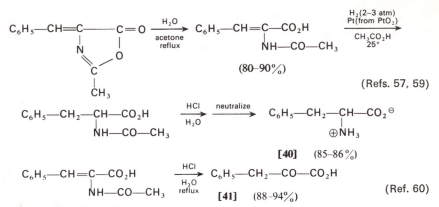

$$C_6H_5-CH=C-CO_2H \xrightarrow[\substack{H_2O \\ reflux}]{HCl} C_6H_5-CH_2-CO-CO_2H$$

$$\underset{NH-CO-CH_3}{|}$$

[41]   (88–94%)   (Ref. 60)

tion of an aldehyde with hydantoin, shown earlier in this chapter, and by the condensation of an enol lactone with an aldehyde as shown in the following equation.

$$C_6H_5-COCH_2CH_2CO_2H + C_6H_5-CHO \xrightarrow[\substack{(CH_3CO)_2O, \\ 95-100°}]{NaOCOCH_3}$$

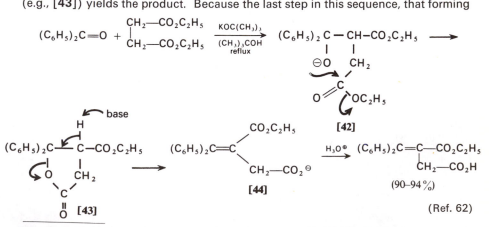

(45–50%) (Ref. 57b)

Ketones or aldehydes react with diethyl succinate in the presence of a strong base such as sodium hydride or potassium *t*-butoxide to form monoesters of an α-alkylidene- (or arylidene) succinic acid. The reaction, known as the Stobbe condensation,[61] is thought to proceed by initial aldol condensation and subsequent intramolecular lactone formation (i.e. **[42]**); ensuing base-catalyzed elimination (e.g., **[43]**) yields the product. Because the last step in this sequence, that forming

60. (a) R. M. Herbst and D. Shemin, *ibid.*, **Coll. Vol. 2**, 519 (1943). For a related synthetic procedure, see Ref. 24b.
61. W. S. Johnson and G. H. Daub, *Org. Reactions*, **6**, 1 (1951).
62. W. S. Johnson and W. P. Schneider, *Org. Syn.*, **Coll. Vol. 4**, 132 (1963).

the carboxylate ion (e.g., [44]), is essentially irreversible, this condensation may be applied successfully even to relatively hindered ketones, where other types of aldol condensation fail. Additional Stobbe reactions are outlined in the following equations. It will be noted that the unsaturated product may consist of mixtures, both of stereoisomers and of structural isomers in which the double bond can occupy several positions.[61,63]

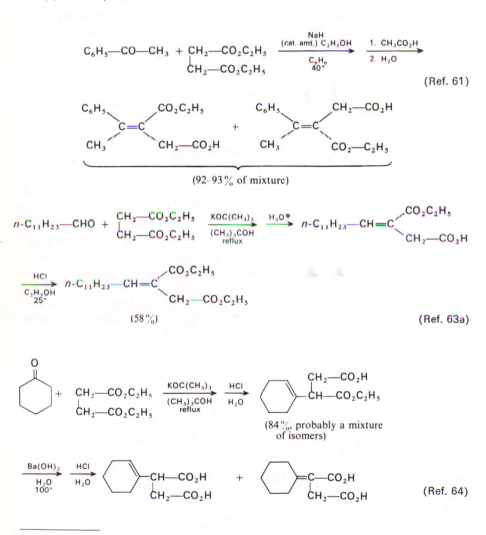

(Ref. 61)

(92–93% of mixture)

(58%)   (Ref. 63a)

(84%, probably a mixture of isomers)

(Ref. 64)

63. (a) C. G. Overberger and C. W. Roberts, *J. Am. Chem. Soc.*, **71**, 3618 (1949). (b) H. G. Heller and B. Swinney, *J. Chem. Soc.*, C, 2452 (1967). (c) H. O. House and J. K. Larson, *J. Org. Chem.*, **33**, 448 (1968).
64. W. S. Johnson, C. E. Davis, R. H. Hunt, and G. Stork, *J. Am. Chem. Soc.*, **70**, 3021 (1948).

The prime synthetic utility of the Stobbe condensation arises from the fact that the initial condensation product (e.g., [45]) may be decarboxylated as illustrated below. The propionic acid derivatives formed by this reaction sequence

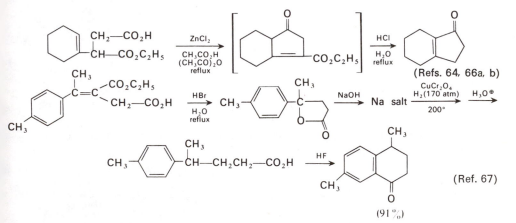

serve as intermediates for the preparation of cyclic ketones, as shown in the following examples.

The initial unsaturated Stobbe product may also be cyclized before reduction of the olefinic double bond to form either a phenol or an indenone derivative as the following examples illustrate. Although the direction of these cyclizations has been useful for assigning stereochemistry to the substituents about the olefinic double bond,[61,67b-d] it is always advisable to study the cyclization of both geometrical isomers before a stereochemical assignment is made. In some cases,

65. W. S. Johnson, J. W. Petersen, and W. P. Schneider, J. Am. Chem. Soc., 69, 74 (1947).
66. (a) D. W. Mathieson, J. Chem. Soc., 3248 (1953). (b) M. M. Coombs, S. B. Jaitly, and F. E. H. Crawley, ibid., C, 1266 (1970). (c) A. Takeda, K. Takahasi, S. Torii, and T. Moriwake, J. Org. Chem., 31, 616 (1966).
67. (a) W. S. Johnson and A. R. Jones, J. Am. Chem. Soc., 69, 792 (1947). (b) F. G. Baddar, V. B. Baghos, and A. Habashi, J. Chem. Soc., C, 603 (1966). (c) W. I. Awad, F. G. Baddar, F. A. Fouli, S. M. A. Omran, and M. I. B. Selim, ibid., C, 507 (1968). (d) S. M. Abdel-Wahhab, L. S. El-Assal, N. Ramses, and A. H. Shehab, ibid., C, 863 (1968) ; S. M. Abdel-Wahhab and L. S. El-Assal, ibid., C, 867 (1968) ; F. G. Baddar, M. F. El-Neweihy, and R. O. Loutfy, ibid., C, 620 (1970). (e) Although γ- and δ-keto esters have been used successfully as the carbonyl component in Stobbe condensations, attempts to use α- and β-keto esters have failed ; see R. L. Augustine and L. P. Calbo, J. Org. Chem., 33, 838 (1968) and references therein.

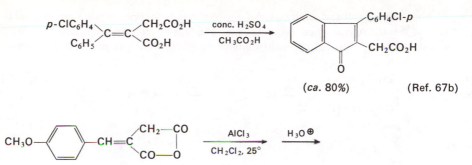

(ca. 80%)   (Ref. 67b)

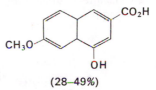

(either *cis* or *trans* isomer)

(28–49%)

(Ref. 63c)

the conditions required for the acid-catalyzed cyclization (see Chapter 11) also interconvert the *cis-* and *trans-*isomers of the starting diacid so that neither isomer affords the strained indenone system.[63c,67d]

The Stobbe condensation products have also been used to prepare γ-keto acids as illustrated in the following equations.[66c]

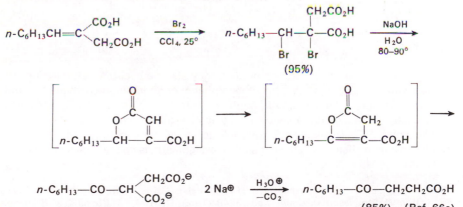

(85%)   (Ref. 66c)

The Darzens glycidic ester condensation[68] involves the reaction of an α-halo ester with a ketone or an aromatic aldehyde in the presence of a strong base such as potassium *t*-butoxide or sodium amide. The intermediate aldol condensation

68. (a) M. S. Newman and B. J. Magerlein, *Org. Reactions,* **5**, 413 (1949). (b) M. Ballester, *Chem. Rev.,* **55**, 283 (1955). (c) An analogous reaction has also been effected in the presence of acid catalysts. G. Sipos, G. Schöbel, and L. Balaspiri, *J. Chem. Soc.,* C, 1154 (1970).

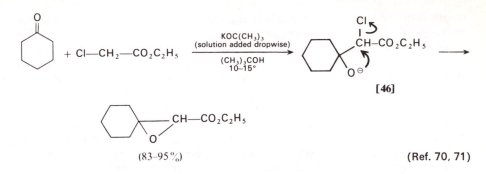

(83–95%) (Ref. 70, 71)

product (e.g., [46])[10c,68,69] undergoes intramolecular ring closure to form an α,β-epoxy ester called a glycidic ester. The predominant stereoisomer produced in this condensation appears to depend upon whether the aldol-condensation step or the subsequent ring-closure step is rate-limiting.[10c,69c−e] In cases such as the one shown in the following example the rate-limiting ring closure step allows reversal (and consequent equilibration) of the preceding aldol intermediates. The favored production of the stereoisomer with the ester function *trans* to the larger group at the beta carbon atom in such circumstances has been attributed to assistance by the carboalkoxyl function in the ring-closure step (e.g., [47]).[10c,69c]

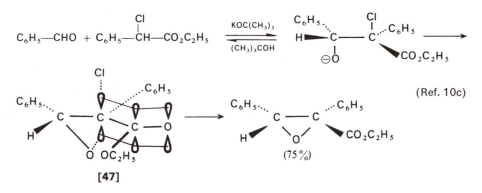

(Ref. 10c)

[47]

The assistance is suggested to involve overlap of the pi-orbital of the ester with the forming and breaking bonds at the α-carbon. Such orbital overlap is only possible when the forming C—O bond and the breaking C—Cl bond are both perpendicular to the plane of the ester function. This geometrical arrangement is energetically

69. (a) M. Ballester and P. D. Bartlett, *J. Am. Chem. Soc.*, **75**, 2042 (1953). (b) M. Ballester and D. Perez-Blanco, *J. Org. Chem.*, **23**, 652 (1958). (c) C. C. Tung, A. J. Speziale, and H. W. Frazier, *ibid.*, **28**, 1514 (1963). (d) F. W. Bachelor and R. K. Bansal, *ibid.*, **34**, 3600 (1969). (e) J. Seyden-Penne, M. C. Roux-Schmitt, and A. Roux, *Tetrahedron*, **26**, 2649, 2657 (1970).
70. R. H. Hunt, L. J. Chinn, and W. S. Johnson, *Org. Syn.*, **Coll. Vol. 4**, 459 (1963).
71. (a) W. S. Johnson, J. S. Belew, L. J. Chinn, and R. H. Hunt, *J. Am. Chem. Soc.*, **75**, 4995 (1953). (b) H. O. House and J. W. Blaker, *ibid.*, **80**, 6389 (1958). (c) C. F. H. Allen and J. VanAllan, *Org. Syn.*, **Coll. Vol. 3**, 727 (1955).

favorable only if no large *cis*-substituent is present at the β-carbon. As was true for the base-catalyzed dehydration of aldol products, maintaining this orbital overlap is apparently the dominant factor in determining the favored transition state for closure of the epoxide ring. This variation in the identity of the rate-limiting step may result either from changing the structure of the reactants or from altering the reaction conditions. A change from a non-polar solvent (e.g., benzene) or a protic solvent (e.g., ethanol) to a dipolar aprotic solvent (e.g., hexamethylphosphoramide) increases the rate of the displacement step and tends to make the aldol condensation rate limiting.[63e] When the initial aldol condensation is rate-limiting, equilibration of the aldol intermediate is slower than the ring closure step and the stereochemistry of the Darzens product is established during the initial aldol condensation.[69d,e,72c]

The following equations illustrate the fact that mixtures of stereoisomers are formed in such cases and the predominant glycidic ester product may have the ester function *cis* to the larger β-substituent. It is not clear which factors are primarily responsible for the formation of the major diastereoisomeric aldol product (e.g., [**48**]).

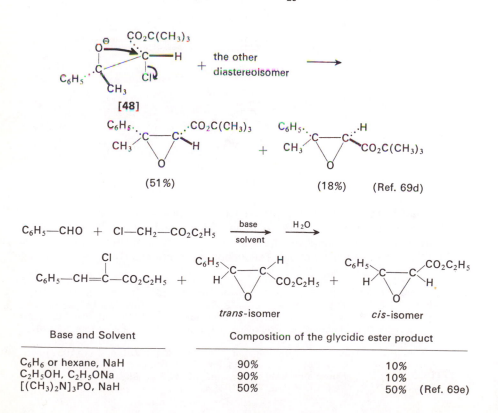

(51%)   (18%)   (Ref. 69d)

Base and Solvent	Composition of the glycidic ester product	
	trans-isomer	*cis*-isomer
C_6H_6 or hexane, NaH	90%	10%
C_2H_5OH, C_2H_5ONa	90%	10%
$[(CH_3)_2N]_3PO$, NaH	50%	50% (Ref. 69e)

Other Darzens condensations, including that with an α-haloketone (e.g., [49]) and with a *p*-nitrobenzyl halide, are exemplified by the accompanying equations. An analogous reaction has also been observed when the acceptor in the aldol condensation is a Schiff base (an imine) rather than an aldehyde or a ketone.[72c]

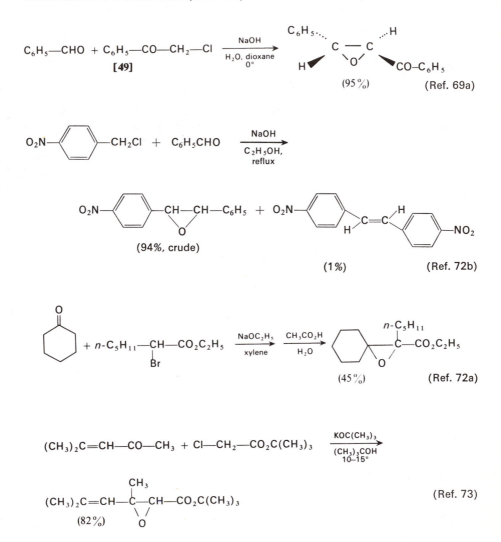

72. (a) H. H. Morris and M. L. Lusth, *J. Am. Chem. Soc.*, **76**, 1237 (1954). (b) W. L. Dilling, R. A. Hickner, and H. A. Farber, *J. Org. Chem.*, **32**, 3489 (1967). (c) J. A. Deyrup, *ibid.*, **34**, 2724 (1969). (d) The use of methyl dichloroacetate in the Darzens reaction yields a β-chloro-α-keto ester, possibly by way of an α-chloroglycidic ester which rearranges (see Chapter 6). R. N. McDonald and P. A. Schwab, *ibid.*, **29**, 2459 (1964).
73. E. P. Blanchard, Jr. and G. Büchi, *J. Am. Chem. Soc.*, **85**, 955 (1963).

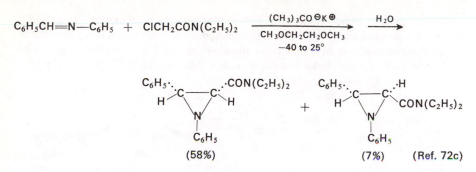

$$C_6H_5CH{=}N{-}C_6H_5 \ + \ ClCH_2CON(C_2H_5)_2$$

(CH₃)₃CO⊖K⊕ → CH₃OCH₂CH₂OCH₃ / −40 to 25° → H₂O →

(58%) + (7%)   (Ref. 72c)

Careful saponification and subsequent acidification of glycidic esters (e.g., [50]) produce glycidic acids (e.g., [51]) that are thermally unstable, decomposing to the enols of aldehydes (e.g., [52]) or ketones when warmed. Other examples

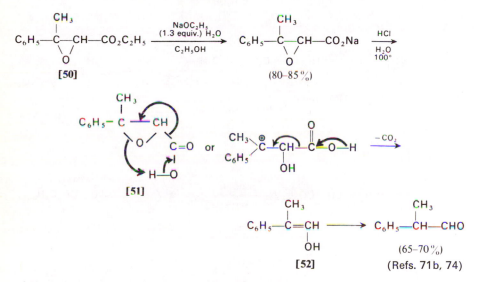

[50]   NaOC₂H₅ (1.3 equiv.) H₂O / C₂H₅OH →   (80–85%)   HCl H₂O 100° →

[51]   or

− CO₂ →

[52]   (65–70%)   (Refs. 71b, 74)

of the preparation and decomposition of glycidic acids, including the pyrolytic decomposition of a *t*-butyl ester, are outlined below.

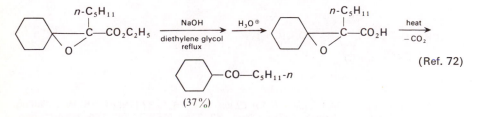

NaOH diethylene glycol reflux → H₃O⊕ → heat − CO₂ →   (Ref. 72)

(37%)

74. (a) C. F. H. Allen and J. VanAllan, *Org. Syn.*, **Coll. Vol. 3**, 733 (1955). (b) V. J. Shiner, Jr., and B. Martin, *J. Am. Chem. Soc.*, **84**, 4824 (1962). (c) S. P. Singh and J. Kagan, *J. Org. Chem.*, **35**, 2203 (1970).

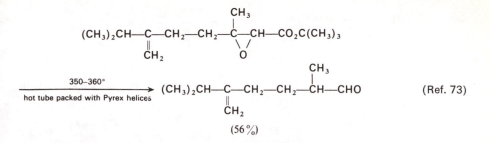

The presence of an α-phenyl substituent in glycidic esters has recently been found to lead to an abnormal decarboxylation as illustrated below. This result is ascribed to opening of the protonated oxide ring to form a carbonium ion intermediate alpha rather than beta to the carboxyl function.[74c]

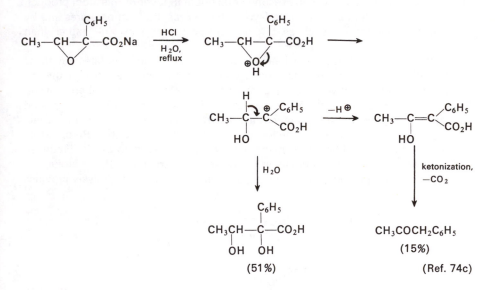

THE REFORMATSKY REACTION AND RELATED PROCESSES EMPLOYING PREFORMED METAL ENOLATES

An alternative method for forcing an aldol condensation to completion utilizes a base sufficiently strong to convert an active methylene compound completely to its enolate anion; this enolate anion is then allowed to react with an electrophilic carbonyl compound to form the alkoxide salt (e.g., [53]) of an aldol product. It is instructive to compare this process with the series of three equilibria described at the beginning of this chapter for the formation of an aldol product in protic media. At first sight, the major difference between the two reactions would appear to be the selection of conditions which cause the first equilibrium (often unfavorable in protic solvents) to lie far on the side of the enolate ion. However, it will be recalled that the success of an aldol condensation is dependent on a favorable position for the second equilibrium, a circumstance that is not assured even if the active methylene

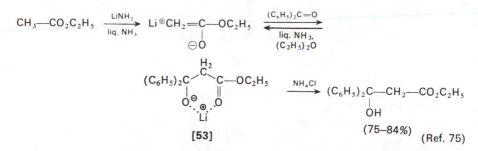

[53]

(75–84%)

(Ref. 75)

component is converted completely to its enolate anion. There is increasing reason to believe that the aldol condensations discussed in this section owe at least part of their success to the ability of the initially formed aldol product to serve as a bidentate ligand for the metal cation present. The formation of such metal chelates (e.g., [53]) displaces the second equilibrium in favor of this aldol product. This metal chelate formation may be favored by low reaction temperatures and by the use of metal cations such as Li$^\oplus$ or, especially, Mg$^{2\oplus}$ or Zn$^{2\oplus}$ rather than Na$^\oplus$ or K$^\oplus$.

Two general methods have been used to prepare metal enolates for these condensations. The first, illustrated above, consists of treating the active methylene component with a relatively strong base (often a metal amide) in inert solvent. The second method consists of the reduction of an α-halo carbonyl compound with zinc or magnesium (see Chapter 3) in an inert solvent such as ether or benzene. Among the latter procedures, the oldest and best known is the Reformatsky reaction;[1c,76] in this reaction an α-halo ester (usually an α-bromo ester) is reduced with zinc to form a halozinc enolate in the presence of an aldehyde or ketone. As indicated in the equation below, further reaction followed by hydrolysis provides

$$C_6H_5\text{—CHO} + Br\text{—}CH_2\text{—}CO_2C_2H_5 \xrightarrow[\substack{C_6H_6, (C_2H_5)_2O \\ reflux}]{Zn} BrZn^\oplus \quad CH_2\text{=}C\text{—}OC_2H_5 \xrightarrow{C_6H_5\text{—CHO}}$$

$$\underset{\oplus ZnBr}{\overset{\ominus O}{|}} \qquad \text{(Ref. 76b)}$$

$$\underset{C_6H_5\text{—CH}}{\overset{\ominus}{O}} \overset{\oplus}{\underset{CH_2}{ZnBr}} \overset{O}{\underset{}{\parallel}} C\text{—}OC_2H_5 \xrightarrow[cold]{H_3O^\oplus} C_6H_5\text{—CH—CH}_2\text{—CO}_2C_2H_5$$

$$\underset{OH}{|}$$

(61–64%)

75. (a) W. R. Dunnavant and C. R. Hauser, *Org. Syn.*, **44**, 56 (1964). (b) W. R. Dunnavant and C. R. Hauser, *J. Org. Chem.*, **25**, 503 (1960). (c) C. R. Hauser and W. R. Dunnavant, *ibid.*, **25**, 1296 (1960). (d) β-hydroxy esters and/or α,β-unsaturated esters are also available from the reaction of ketones with the lithium derivative of ethoxyacetylene and subsequent reactions of the resultant acetylenic ethers with dilute aqueous acid: see J. F. Arens, *Adv. Org. Chem.*, **2**, 117 (1960).
76. (a) R. L. Shriner, *Org. Reactions*, **1**, 1 (1942). (b) C. R. Hauser and D. S. Breslow, *Org. Syn.*, **Coll. Vol. 3**, 408 (1955). See also K. L. Rinehart, Jr. and E. G. Perkins, *ibid.*, **Coll. Vol. 4**, 444 (1963). (c) W. E. Bachmann, W. Cole, and A. L. Wilds, *J. Am. Chem. Soc.*, **62**, 824 (1940). (d) For the use of α-bromo lactones in the Reformatsky reaction,

a useful preparative route to β-hydroxy esters. Reactions of this type, involving condensation with a preformed metal enolate, are unique among aldol condensations in that isolation of the initial aldol product is possible even with relatively hindered ketones as electrophilic carbonyl components in the condensation. The following equations illustrate the formation of β-hydroxy esters from hindered ketones, the subsequent dehydration of this initial aldol product, and the reversal of the aldol condensation when a sterically congested β-hydroxy ester is treated with base under conditions where the formation of a metal chelate is not favorable.

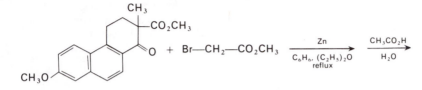

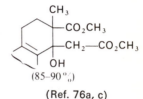

(85–90 %)

(Ref. 76a, c)

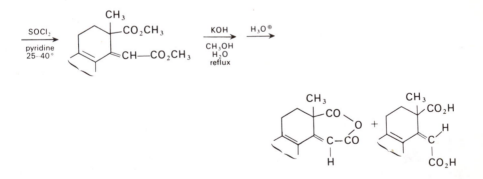

The most common side reactions encountered in the Reformatsky reaction are the previously illustrated condensation of the α-bromo ester with itself during the formation of the zinc enolate to form, after hydrolysis, a β-keto ester[76a,e,f,77a,c] and reaction of the ester enolate anion with the ketone to abstract a proton and

see H. Torabi, R. L. Evans, and H. E. Stavely, *J. Org. Chem.*, **34**, 3792, 3796 (1969). (e) W. R. Vaughan, S. C. Bernstein, and M. E. Lorber, *ibid.*, **30**, 1790 (1965). (f) W. R. Vaughan and H. P. Knoess, *ibid.*, **35**, 2394 (1970).

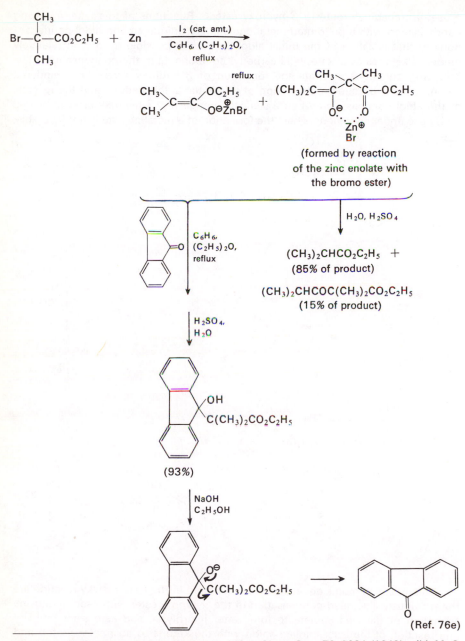

(93%)

(Ref. 76e)

77. (a) A. S. Hussey and M. S. Newman, *J. Am. Chem. Soc.,* **70,** 3024 (1948). (b) M. S. Newman, *ibid.,* **62,** 870 (1940) ; *ibid.,* **64,** 2131 (1942). (c) D. A. Cornforth, A. E. Opara, and G. Read, *J. Chem. Soc.,* C, 2799 (1969). (d) J. W. Frankenfeld and J. J. Werner, *J. Org. Chem.,* **34,** 3689 (1969). (e) E. H. Charlesworth and P. Charleson, *Can. J. Chem.,* **46,** 1843 (1968). (f) K. H. Fung, K. J. Schmalzl, and R. N. Mirrington, *Tetrahedron Letters,* **No. 57,** 5017 (1969). (g) F. Gaudemar-Bardone and M. Gaudemar, *Bull. Soc. Chim. France,* 2088

produce the ketone enolate anion. The latter process, illustrated in the accompanying equation, may lead to aldol condensation products derived from the ketone.[76a,77b] The β-hydroxy ester formed after hydrolysis may also undergo further transformations; dehydration or retrograde aldol condensation are the most common side reactions.[76a,77c-f]

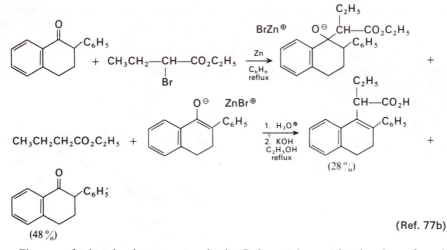

CH₃CH₂CH₂CO₂C₂H₅ +

(Ref. 77b)

The use of t-butyl α-bromo esters in the Reformatsky reaction has been found to suppress the side reaction leading to β-keto ester derivatives. Also, as shown below, the ready acid-catalyzed cleavage of t-butyl esters allows the initial

BrCH₂CO₂C(CH₃)₃ + n-C₃H₇CHO

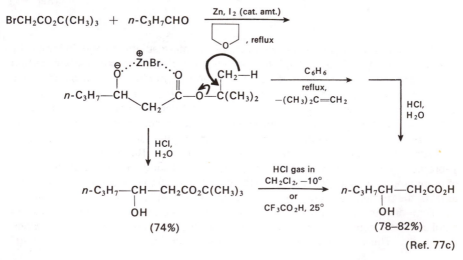

(Ref. 77c)

(1969). (h) Y. Beziat and M. Mousseron-Canet, *ibid.*, 1187, 2572 (1968). (i) J. Canceill, J. Gabard, and J. Jacques, *ibid.*, 2653 (1966); J. Canceill, J. J. Basselier, and J. Jacques, *ibid.*, 1024 (1967). (j) For a review of earlier work, see M. Mousseron, M. Mousseron, J. Neyrolles, and Y. Beziat, *ibid.*, 1483 (1963). (k) F. Lauria, V. Vecchietti, W. Logemann, G. Tosolini, and E. Dradi, *Tetrahedron*, **25**, 3989 (1969). (l) M. W. Rathke and A. Lindert, *J. Org. Chem.*, **35**, 3966 (1970).

Reformatsky products to be converted to β-hydroxy acids. Other modifications in reaction conditions have also been recommended to improve yields[77d,e,l] or to modify the proportion of diastereoisomers formed.[77g]

The stereochemistry of the diastereoisomeric products obtained from the Reformatsky reaction has been studied by a number of workers.[1e,77g-k] The major product isolated usually corresponds to attack of the zinc enolate at the less hindered side of the carbonyl group[77h,k] When the reaction is performed in non-polar solvents (e.g., benzene or ether), the predominant diastereoisomer formed is usually that which corresponds to the formation of an intermediate metal chelate with the smaller number of unfavorable steric interactions. The following example illustrates these generalities.

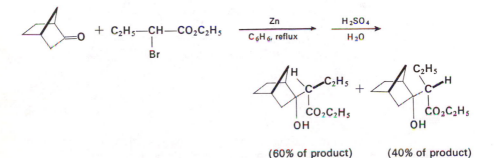

(60% of product)   (40% of product)

(Ref. 77k)

The zinc enolates of esters which serve as intermediates in the Reformatsky reaction have also been added to α-chloro ketones,[78a] to the carbon-carbon double bond of α,β-unsaturated ketones,[78c] and to nitriles.[78d] γ-Bromocrotonic esters have been found to give moderate yields in the Reformatsky reaction. As indicated in the accompanying equations,[78b,e] an interesting change in the nature of the condensation product was observed when the reaction solvent was changed from benzene or 1,2-dimethoxyethane to ether. In ether solution, the proportions of structurally analogous products obtained from the Reformatsky reaction of γ-bromocrotonic esters with aldehydes have been found to vary with the extent of branching in the alkyl chain of the aldehyde. Increasing steric hindrance appears to favor the product resulting from reaction at the γ-position of the crotonic ester.[78b]

Bromomagnesium and bromozinc enolates of ketones and esters for use in aldol condensations[79] have been prepared by both of the general methods described previously. The reductions of α-bromo carbonyl compounds with zinc or mag-

78. (a) W. W. Epstein and A. C. Sonntag, *Tetrahedron Letters,* **No. 8,** 791 (1966). (b) J. Colonge and J. P. Cayrel, *Bull. Soc. Chim. France,* 3596 (1965). (c) J. C. Dubois, J. P. Guette, and H. B. Kagan, *ibid.,* 3008 (1966). (d) H. B. Kagan and Y. H. Suen, *ibid.,* 1819 (1966). (e) A. S. Dreiding and R. J. Pratt, *J. Am. Chem. Soc.,* **75,** 3717 (1953).

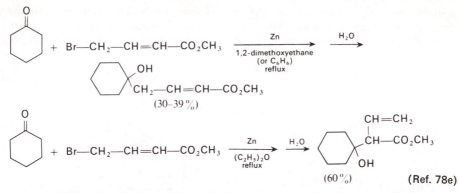

nesium are illustrated below. Since the bromometal enolate is generated slowly, it is desirable to have the other carbonyl component also in the reaction mixture to react with the enolate faster than the enolate can condense with the starting

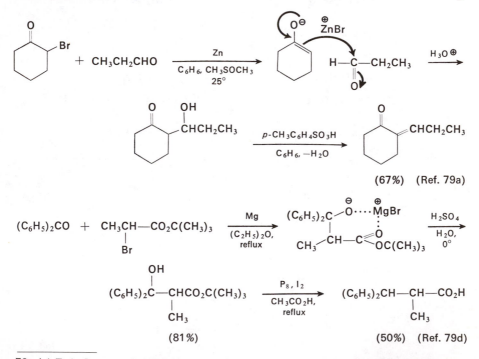

79. (a) T. A. Spencer, R. W. Britton, and D. S. Watt, *J. Am. Chem. Soc.*, **89**, 5727 (1967). (b) K. Sisido, H. Nozaki, and O. Kurihara, *ibid.*, **74**, 6254 (1952). (c) A. T. Nielsen, C. Gibbons, and C. A. Zimmerman, *ibid.*, **73**, 4696 (1951). (d) T. Moriwake, *J. Org. Chem.*, **31**, 983 (1966). (e) J. E. Dubois and M. Chastrette, *Tetrahedron Letters*, **No. 32**, 2229 (1964). (f) J. A. Miller, M. H. Durand, and J. E. Dubois, *ibid.*, **No. 32**, 2831 (1965); J. E. Dubois and J. Itzkowitch, *ibid.*, **No. 32**, 2839 (1965). (g) S. Mitsui and Y. Kudo, *Tetrahedron*, **23**, 4271 (1967). (h) A. G. Pinkus, J. G. Lindberg, and A. B. Wu, *Chem. Commun.*, **No. 22**, 1350 (1969); *ibid.*, **No. 14**, 859 (1970). (i) J. E. Dubois, G. Schutz, and J. M. Normant, *Bull. Soc. Chim. France*, 3578 (1966).

α-halo carbonyl compound. In the absence of the second reactive carbonyl component, condensation with starting α-bromo ketone can lead to the production of the furan derivatives and other products[79a,f] as the following equations indicate.

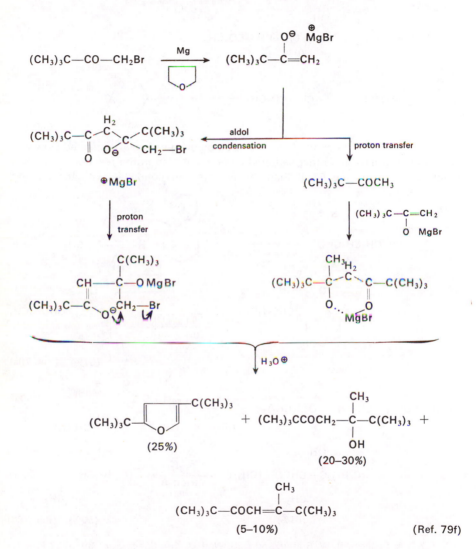

An alternative route to bromomagnesium enolates for aldol condensations involves treatment of the active methylene compound with either methylanilino-magnesium bromide[1a,79c] or diethylaminomagnesium bromide;[79b,g] hindered ketones may also be converted to the halomagnesium enolates by treatment with Grignard reagents.[79e,h] The methylene group of arylacetic acids is also sufficiently acidic to be converted to an enolate by reaction with excess Grignard reagent. The

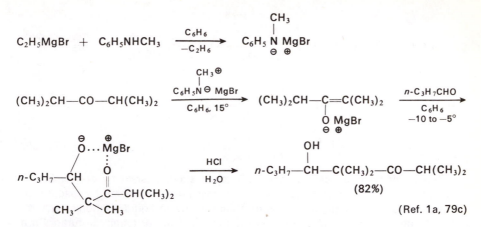

(Ref. 1a, 79c)

resulting halomagnesium enolate, called an Ivanov reagent, may be used in aldol condensations as shown below.

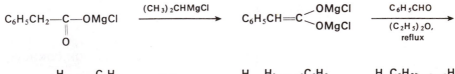

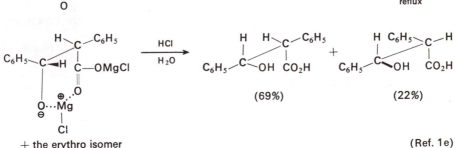

(69%)   (22%)

+ the erythro isomer   (Ref. 1e)

As indicated previously, the lithium enolates, prepared from esters and lithium amide in mixtures of ether and liquid ammonia,[75,80a] can be used to prepare β-hydroxy esters in a reaction which resembles the Reformatsky reaction. A comparable reaction can be effected by forming the lithium enolate of the ester with strong base, lithium bis(trimethylsilyl)amide.[80i]

80. (a) C. R. Hauser and W. H. Puterbaugh, *J. Am. Chem. Soc.*, **75**, 1068 (1953). (b) R. L. Gay and C. R. Hauser, *ibid.*, **89**, 1647 (1967). (c) E. M. Kaiser and C. R. Hauser, *ibid.*, **89**, 4566 (1967). (d) C. Mao, C. R. Hauser, and M. L. Miles, *ibid.*, **89**, 5303 (1967). (e) N. S. Narasimhan and A. C. Ranade, *Tetrahedron Letters*, **No. 46**, 4145 (1965). (f) E. M. Kaiser and C. R. Hauser, *J. Org. Chem.*, **31**, 3317 (1966). (g) D. M. von Schriltz, E. M. Kaiser, and C. R. Hauser, *ibid.*, **32**, 2610 (1967); *ibid.*, **33**, 4275 (1968). (h) H. O. House and A. Y. Teranishi. To be published. (i) M. W. Rathke, *J. Am. Chem. Soc.*, **92**, 3222 (1970).

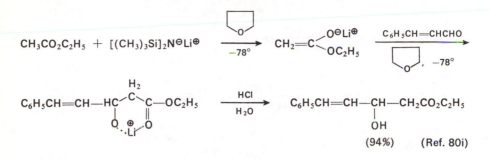

Similar aldol condensations have also been accomplished by conversion of amides,[80b,e-g] nitriles,[80d] and ketones[80d,h] to their lithium enolates by reaction of the active methylene compounds with lithium amide in liquid ammonia, or with either n-butyllithium or lithium diisopropylamide in ether or tetrahydrofuran. In a number of instances (especially when the active methylene component possesses an α-phenyl substituent), the aldol product is less stable than the starting material

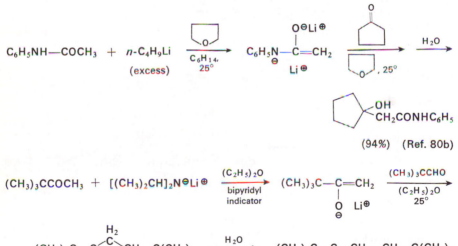

and undergoes a retrograde aldol condensation when treated with weak bases in protic solvents.[80c,f,g] This decomposition may be avoided by using an inverse hydrolysis procedure in which the reaction mixture is *added to* aqueous ammonium chloride.

The success of these aldol condensations is attributable to the previously discussed formation of an intermediate metal chelate which permits the equilibrium leading to the condensation product to be favorable. In certain cases, this equilibrium is more effectively displaced toward the condensation product by adding

di- or trivalent metal salts such as zinc chloride, magnesium bromide, or aluminum chloride;[80a,c,h] these di- and trivalent metal cations presumably form a more stable metal chelate than the lithium cation initially present. The stereochemistry

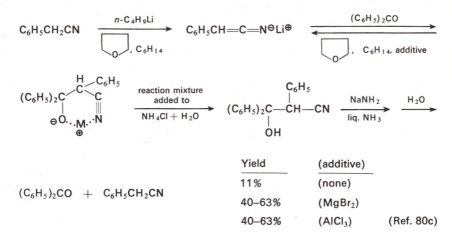

observed in these condensations is also explicable in terms of forming a metal chelate such as that illustrated below in which steric interactions are minimized. In the following example, the reversible condensation step initially leads to a mixture of diastereoisomeric sodium alkoxides which equilibrate relatively slowly to form the *threo*-isomer.[80g] Consequently, the composition of the mixture of diastereoisomeric hydroxy amide products was found to vary with time.

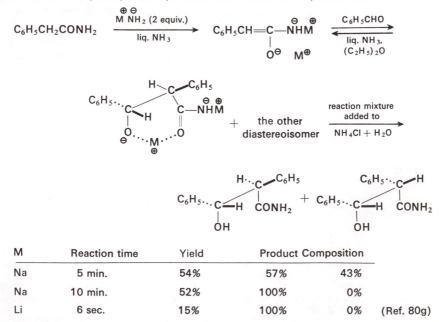

M	Reaction time	Yield	Product Composition	
Na	5 min.	54%	57%	43%
Na	10 min.	52%	100%	0%
Li	6 sec.	15%	100%	0% (Ref. 80g)

A related method for controlling the direction of an aldol condensation involves the initial conversion of the active methylene component (an aldehyde or ketone) to the corresponding imine (Schiff base) which is then converted to its lithio derivative by reaction with lithium diisopropylamide.[81] The example of this procedure presented in the following equation is one in which conventional aldol conditions (a weak base in a protic solvent) could not be used because the highly reactive acetaldehyde would self-condense faster than it would react with benzophenone.

This procedure for effecting directed aldol condensations, like those previously discussed in this section, owes its success at least in part to the formation of the illustrated metal chelate intermediate. It seems likely that other procedures for

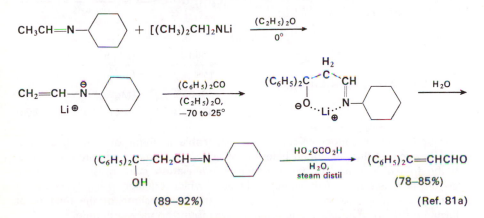

obtaining structurally specific aldol condensations can be developed by preforming single structural isomers of metal enolates as described in Chapter 9 and then allowing these metal enolates to react with other electrophilic carbonyl functions under conditions where the initial aldol product will be stabilized as a metal chelate. For successful condensations of this type it will be necessary to avoid proton-transfer reactions which can give rise both to equilibration among the enolates of the starting material and to dehydration (and subsequent Michael reactions)[80h] of the initially formed aldol product.

THE WITTIG REACTION

The reaction of a tertiary phosphine (usually triphenylphosphine) with an alkyl halide yields a phosphonium salt (e.g., [54]) in which the alpha C—H bonds are sufficiently acidic to be removed by a strong base (e.g., an organolithium compound, sodium hydride, or sodium amide). The resulting ylid (e.g., [55]) is believed to be stabilized by overlap between the p-orbital at carbon and one of the d-orbitals

81. (a) G. Wittig and A. Hesse, *Org. Syn.*, **50**, 66 (1970). (b) G. Wittig and H. D. Frommeld, *Chem. Ber.*, **97**, 3548 (1964). (c) G. Wittig, H. D. Frommeld, and P. Suchanek, *Angew. Chem., Intern. Ed. Engl.*, **2**, 683 (1963). (d) G. Wittig and H. Reiff, *ibid.*, **7**, 7 (1968). (e) G. Wittig, *Rec. Chem. Progr.*, **28**, 45 (1967). (f) G. Wittig and P. Suchanek, *Tetrahedron, Suppl. 8, Pt. I*, 347 (1966).

of the phosphorus atom.[82] Subsequent reaction of these ylids (e.g., [**55**]) with aldehydes or ketones offers the very useful synthesis for olefins known as the Wittig reaction.[83] The initial nucleophilic attack by the ylid on the carbonyl function usually forms a betaine intermediate (e.g., [**56**]). However, the inter-

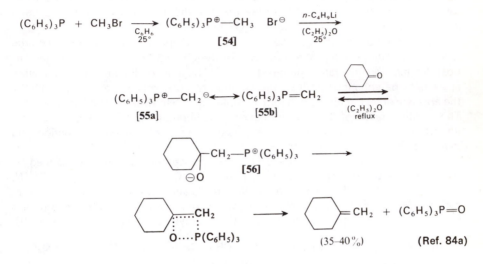

mediate derived from the highly electrophilic ketone, hexafluoroacetone, has a covalent four-membered ring structure.[82d] This attack is accelerated by the presence of electron-withdrawing groups in the component containing the carbonyl

82. (a) W. von E. Doering and A. K. Hoffmann, *J. Am. Chem. Soc.,* **77**, 521 (1955). (b) For a discussion of the mechanism of proton removal, see D. Seyferth, W. B. Hughes, and J. K. Heeren, *ibid.,* **87**, 2847 (1965); H. Schmidbaur and W. Tronich, *Angew. Chem., Intern. Ed. Engl.,* **6**, 448 (1967); S. E. Cremer and R. J. Chorvat, *Tetrahedron Letters,* No. **4**, 419 (1966). (c) R. F. Hudson, *Structure and Mechanism in Organophosphorus Chemistry,* Academic Press, New York, 1965. (d) C. N. Matthews and G. H. Birum, *Accts. Chem. Res.,* **2**, 373 (1969); G. Chioccola and J. J. Daly, *J. Chem. Soc.,* A, 568 (1968). (e) J. C. J. Bart, *ibid.,* B, 350 (1969). (f) F. S. Stephens, *ibid.,* 5640, 5658 (1965). (g) A. J. Speziale and K. W. Ratts, *J. Am. Chem. Soc.,* **87**, 5603 (1965). (h) S. O. Grim and J. H. Ambrus, *J. Org. Chem.,* **33**, 2993 (1968).

83. (a) G. Wittig and U. Schöllkopf, *Chem. Ber.,* **87**, 1318 (1954). (b) G. Wittig and W. Haag, *ibid.,* **88**, 1654 (1955). (c) G. Wittig, H. D. Weigmann, and M. Schlosser, *ibid.,* **94**, 676 (1961). (d) U. Schöllkopf, *Angew. Chem.,* **71**, 260 (1959). (e) S. Trippett, *Adv. Org. Chem.,* **1**, 83 (1960). (f) S. Trippett, *Quart. Rev.,* **17**, 406 (1963). (g) L. D. Bergelson and M. M. Shemyakin, *Angew. Chem., Intern. Ed. Engl.,* **3**, 250 (1964); L. D. Bergelson and M. M. Shemyakin, in W. Foerst, ed., *Newer Methods of Preparative Organic Chemistry,* Vol. 5, Academic Press, New York, 1968, pp. 154–175. (h) A. Maercker, *Org. Reactions,* **14**, 270 (1965). (i) A. W. Johnson, *Ylid Chemistry,* Academic Press, New York, 1966, pp. 5–247. (j) H. J. Bestmann, *Angew. Chem., Intern. Ed. Engl.,* **4**, 583, 645, 830 (1965); H. J. Bestmann, in W. Foerst, ed., *Newer Methods of Preparative Organic Chemistry,* Vol. 5, Academic Press, New York, 1968, pp. 1–60.

84. (a) G. Wittig and U. Schöllkopf, *Org. Syn.,* **40**, 66 (1960). (b) One example of the reverse process, reaction of triphenylphosphine oxide with dicyanoacetylene to form a Wittig reagent, is known; E. Ciganek, *J. Org. Chem.,* **35**, 1725 (1970).

function, by the presence of alkyl groups rather than phenyl groups bonded to the phosphorus atom of the ylid, and by the use of polar reaction solvents.[83,85] The formation of the betaine appears to be reversible in most, if not all, cases.[85a,d,i,j] Subsequent decomposition of the betaine intermediate (e.g., [56]) to produce an olefin and a tertiary phosphine oxide is believed to occur by the indicated cyclic four-center transition state.[83,84b] At least part of the driving force for this reaction is the formation of the very stable $P^{\oplus}$—$O^{\ominus}$ bond.

Depending on the structures of the carbonyl compound and the ylid and on the reaction conditions, either the formation of the betaine or its subsequent decomposition may be the rate-limiting step.[83,85] The rate of betaine decomposition appears to be enhanced both by the addition of polar solvents and, especially by the absence of metal salts which form complexes with the intermediate betaine.[85i,k,l] The relative rates of betaine formation, equilibration, and decomposition are important factors in determining the stereochemistry of the Wittig reaction and will be discussed later in this connection.

The acidity of the alpha C—H bonds of phosphonium salts is enhanced by the presence of an alpha substituent (e.g., —$CO_2C_2H_5$, —CO—C_6H_5, —C≡N, —C_6H_5, —CH=CH_2), which aids the stabilization of a negative charge at carbon.[82f,g,83,85,86] From a comparison of the pK_a values cited for the formulas below[85a,d] with the acidities of other active methylene compounds (Chapter 9),

$(C_6H_5)_3P^{\oplus}$—CH_2—CO—C_6H_5 $(C_6H_5)_3P^{\oplus}$—CH_2—C≡N $(C_6H_5)_3P^{\oplus}$—CH_2—$CO_2C_2H_5$

pK_a 6.0 pK_a 7.5 pK_a 9.0–9.2

one may conclude that a triphenylphosphonium substituent will stabilize an adjacent carbanion to a slightly greater extent than will an adjacent carbonyl function. As a result, the various ylids that have an additional stabilizing substituent may be prepared by reaction of the corresponding phosphonium salts with relatively weak bases such as metal alkoxides, metal hydroxides, and, in some cases, even

85. (a) S. Fliszar, R. F. Hudson, and G. Salvadori, *Helv. Chim. Acta,* **46**, 1580 (1963). (b) H. Goetz, F. Nerdel, and H. Michaelis, *Naturwissenschaften,* **50**, 496 (1963). (c) A. W. Johnson and R. B. LaCount, *Tetrahedron,* **9**, 130 (1960). (d) A. J. Speziale and K. W. Ratts, *J. Am. Chem. Soc.,* **85**, 2790 (1963). (e) A. J. Speziale and D. E. Bissing, *ibid.,* **85**, 1888, 3878 (1963). (f) R. Greenwald, M. Chaykovsky, and E. J. Corey, *J. Org. Chem.,* **28**, 1128 (1963). (g) C. Rüchardt, S. Eichler, and P. Panse, *Angew. Chem., Intern. Ed. Engl.,* **2**, 619 (1963); *Chem. Ber.,* **100**, 1144 (1967). (h) S. Fliszar, R. F. Hudson, and G. Salvadori, *Helv. Chim. Acta,* **47**, 159 (1964). (i) M. Schlosser and K. F. Christmann, *Angew. Chem., Intern. Ed. Engl.,* **3**, 636 (1964); **4**, 689 (1965); **5**, 126 (1966). (j) M. Schlosser, G. Müller, and K. F. Christmann, *ibid.,* **5**, 667 (1966). (k) M. Schlosser, *ibid.,* **7**, 650 (1968). (l) M. Schlosser and K. F. Christmann, *Justus Liebigs Ann. Chem.,* **708**, 1 (1967). (m) M. E. Jones and S. Trippett. *J. Chem. Soc.,* C, 1090 (1966). (n) A. K. Bose, M. S. Manhas, and R. M. Ramer, *ibid.,* C, 2728 (1969).
86. (a) F. Ramirez and S. Dershowitz, *J. Org. Chem.,* **22**, 41 (1957). (b) S. Trippett and D. M. Walker, *J. Chem. Soc.,* 1266 (1961). (c) P. Crews, *J. Am. Chem. Soc.,* **90**, 2961 (1968). (d) H. J. Bestmann, H. G. Liberda, and J. P. Snyder, *ibid.,* **90**, 2963 (1968); H. I. Zeliger, J. P. Snyder, and H. J. Bestmann, *Tetrahedron Letters,* No. 26, 2199 (1969); *ibid.,* **No. 38,** 3313 (1970); J. P. Snyder and H. J. Bestmann, *ibid.,* **No. 38,** 3317 (1970); I. F. Wilson and J. C. Tebby, *ibid.,* **No. 43,** 3769 (1970). (e) D. M. Crouse, A. T. Wehman, and E. E. Schweizer, *Chem. Commun.,* **No. 15,** 866 (1968).

metal carbonates. The ylids stabilized by an adjacent carbonyl function are readily isolated stable crystalline solids, as illustrated by the following examples.

$$C_6H_5-CO-CH_2-Br \ + \ (C_6H_5)_3P \ \xrightarrow{CHCl_3} \ C_6H_5-CO-CH_2-P^{\oplus}(C_6H_5)_3 \ Br^{\ominus} \ \xrightarrow[H_2O]{Na_2CO_3}$$

$$(79\%)$$

$$\underset{O}{\overset{\overset{O}{\parallel}}{C_6H_5-C}}-CH=P(C_6H_5)_3 \ \longleftrightarrow \ \underset{O^{\ominus}}{\overset{O^{\ominus}}{C_6H_5-C}}=CH-P^{\oplus}(C_6H_5)_3 \qquad \text{(Ref. 86a)}$$

$$(96\%; \ \text{m.p.} \ 178–180°)$$

$$\underset{Br}{\overset{}{CH_3-CH-CO_2CH_3}} \ + \ (C_6H_5)_3P \ \xrightarrow[70°]{C_6H_6} \ (C_6H_5)_3P^{\oplus}-\overset{CH_3}{\underset{}{CH}}-CO_2CH_3 \ Br^{\ominus} \ \xrightarrow[H_2O]{NaOH}$$

$$(C_6H_5)_3P=\overset{CH_3}{\underset{}{C}}-CO_2CH_3 \qquad \text{(Ref. 87)}$$

$$(44\%; \ \text{m.p.} \ 152–154.5°)$$

The additional electron delocalization present in these stabilized ylids requires a planar arrangement of the substituents at the carbon atoms alpha and beta to the phosphorus atom of the ylids.[82f,g,86] As was the case with enolate ions and enamines (see Chapter 9), there is a substantial energy barrier to rotation about this carbon-carbon bond;[86] the proportions of geometrical isomers present in solutions of stabilized ylids are dependent, among other factors, on the steric requirements and conjugating ability of the substituents as the following examples show.[82g,86c-e]

$$\underset{(C_6H_5)_3\overset{\oplus}{P}}{\overset{R}{\diagdown}}C=C\underset{O^{\ominus}}{\overset{OCH_3}{\diagup}} \quad \underset{\substack{\text{(free from} \\ \text{acid impurities)}}}{\overset{CDCl_3}{\rightleftharpoons}} \quad \underset{(C_6H_5)_3\overset{\oplus}{P}}{\overset{R}{\diagdown}}C=C\underset{OCH_3}{\overset{O^{\ominus}}{\diagup}}$$

R	Equilibrium concentration	
H	82%	18%
CH₃	50%	50%
(CH₃)₂CH	28%	72%
C₆H₅	79%	21%

$$\text{(Ref. 86d)}$$

The ylids that do not contain stabilizing alpha substituents react rapidly with oxygen,[88] a condition requiring their preparation and use in an inert atmosphere. Ylids having stabilizing alpha carbonyl substituents are not readily attacked by molecular oxygen but may be oxidized by peracids or potassium permanganate.[89]

---

87. (a) O. Isler, H. Gutmann, M. Montavon, R. Rüegg, G. Ryser, and P. Zeller, *Helv. Chim. Acta*, **40**, 1242 (1957). (b) H. O. House and G. Rasmusson, *J. Org. Chem.*, **26**, 4278 (1961).
88. (a) H. J. Bestmann and O. Kratzer, *Angew. Chem., Intern. Ed. Engl.*, **1**, 512 (1962). (b) H. J. Bestmann and O. Kratzer, *Chem. Ber.*, **96**, 1899 (1963).
89. (a) D. B. Denney, L. C. Smith, J. Song, C. J. Rossi, and C. D. Hall, *J. Org. Chem.*, **28**, 778 (1963). (b) H. O. House and H. Babad, *ibid.*, **28**, 90 (1963). (c) E. Zbiral and M. Rasberger, *Tetrahedron*, **24**, 2419 (1968); **25**, 1871 (1969). (d) H. Gross and B. Costisella, *Angew. Chem., Intern. Ed. Engl.*, **7**, 391 (1968).

All ylids are hydrolyzed by reaction with water, though the stabilized ylids require rather vigorous conditions. The probable course of this hydrolysis[90] is outlined below. It will be noted that the ease of cleaving groups from the phosphorus atom

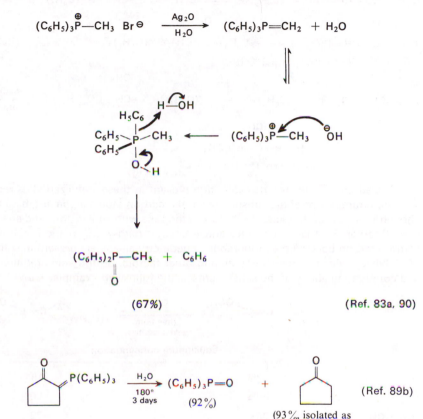

(67%)   (Ref. 83a, 90)

(92%)

(Ref. 89b)

(93%, isolated as
2,4-dinitrophenylhydrazone)

(i.e. $-CH_2-CO-R > -C_6H_5 >$ alkyl) is in the same order as the stabilities of the corresponding carbanions. Most of these substituents probably acquire some negative charge during the hydrolytic cleavage but are protonated before cleavage of the carbon-phosphorus bond is complete.[90e] One of the accompanying equations also illustrates the currently held view[90f] that both the entering group

90. (a) R. F. Hudson and B. A. Chopard, *Helv. Chim. Acta,* **45**, 1137 (1962); (b) R. F. Hudson and M. Green, *Angew. Chem., Intern. Ed. Engl.,* **2**, 11 (1963). (c) C. T. Eyles and S. Trippett, *J. Chem. Soc.,* C, 67 (1966). (d) S. Trippett and B. J. Walker, *ibid.,* C, 887 (1966). (e) J. R. Corfield and S. Trippett, *Chem. Commun.,* **No. 19**, 1267 (1970). (f) P. Haake and G. W. Allen, *Tetrahedron Letters,* **No. 35**, 3113 (1970); F. H. Westheimer, *Accts. Chem. Res.,* **1**, 70 (1968). (g) E. H. Axelrod, G. M. Milne, and E. E. van Tamelen, *J. Am. Chem. Soc.,* **92**, 2139 (1970).

and the leaving group occupy apical positions in the pentacoordinate phosphorus intermediate. The preparation and subsequent hydrolysis of appropriately substituted phosphoranes has served as a synthesis for ketones, hydrocarbons[90g] α-chloro ketones,[89c] and carboxylic acids.[89d] A more satisfactory method for the cleavage of α-ketoalkyl groups from phosphonium salts (e.g., **[57]**) is reduction with zinc and acetic acid[86b] or with lithium in ethylamine;[89g] this type of reductive cleavage has been discussed in Chapter 3.

$$
\underset{\textbf{[57]}}{(C_6H_5)_3P^{\oplus}\!\!-\!CH_2\!-\!CO\!-\!C_6H_5} \;\; \overset{Br^{\ominus}}{} \;\; \xrightarrow[\substack{CHCl_3\\CH_3CO_2H\\reflux}]{Zn} \;\; \underset{(56\%)}{(C_6H_5)_3P} \; + \; \underset{\substack{(95\%,\ isolated\ as\\2,4\text{-}dinitrophenylhydrazone)}}{C_6H_5\!-\!CO\!-\!CH_3} \qquad \text{(Ref. 86b)}
$$

Although ylids that do not contain good stabilizing substituents in the alpha position are slowly cleaved by reaction with alcohols, it is nonetheless possible to generate these ylids in relatively low concentrations with metal alkoxides in alcohol solution. The ylid (e.g., **[58]**), which is apparently in equilibrium with the phosphonium alkoxide (e.g., **[59]**)[83] under such conditions, will react with a carbonyl

$$
C_6H_5\!-\!CH\!=\!CH\!-\!CH_2\!-\!Cl \; + \; (C_6H_5)_3P \;\; \xrightarrow[reflux]{xylene} \;\; \underset{(91\text{-}93\%)}{C_6H_5\!-\!CH\!=\!CH\!-\!CH_2\!-\!P^{\oplus}(C_6H_5)_3} \;\; Cl^{\ominus}
$$

$$
\xrightarrow[\substack{C_2H_5OH\\25^\circ}]{LiOC_2H_5} \;\; \underset{\textbf{[58]}}{C_6H_5\!-\!CH\!=\!CH\!-\!CH\!=\!P(C_6H_5)_3} \;\; \xrightarrow{C_6H_5CHO} \;\; \underset{(60\text{-}67\%)}{C_6H_5\!-\!CH\!=\!CH\!-\!CH\!=\!CH\!-\!C_6H_5}
$$

$$
\Updownarrow \; C_2H_5OH
$$

$$
\underset{\textbf{[59]}}{C_6H_5\!-\!CH\!=\!CH\!-\!CH_2\!-\!P^{\oplus}(C_6H_5)_3\,^{\ominus}OC_2H_5} \qquad\qquad \text{(Ref. 91)}
$$

compound present in the reaction mixture. Although the procedure has been used to prepare olefins from some highly reactive alkylidenephosphoranes, the major product obtained from a methyltriphenylphosphonium salt was the indicated rearranged material. The mechanism of this rearrangement is presently unknown.

The prime utility of the Wittig reaction lies in the ease with which the reaction occurs under mild conditions and in the fact that no ambiguity exists concerning the location of the double bond in the initially formed product. The following equations provide additional examples of its use for the preparation of olefins. It should be noted that the alkylidenephosphoranes normally react with α,β-unsaturated ketones at the carbonyl carbon atom; only when the carbonyl function is

91. (a) R. N. McDonald and T. W. Campbell, *Org. Syn.,* **40**, 36 (1960). (b) T. W. Campbell and R. N. McDonald, *ibid.,* **40**, 85 (1960).

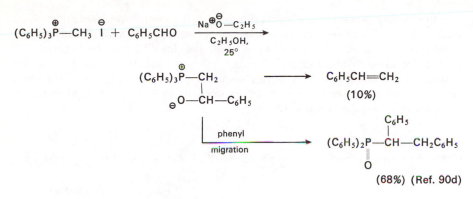

sterically hindered from attack by nucleophiles is conjugate addition of the Wittig reagent observed.[94b] The ylids that are stabilized by the presence of an alpha carbonyl substituent are sufficiently less reactive than other ylids that it is difficult to effect their successful reaction with ketones, especially in the absence of

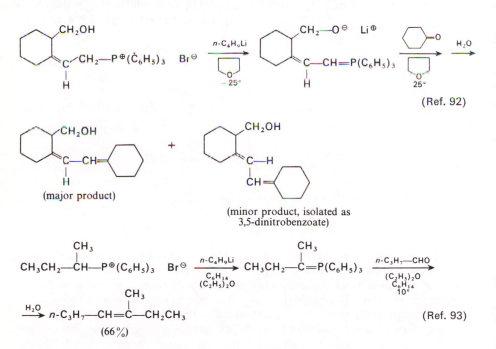

92. I. T. Harrison and B. Lythgoe, *J. Chem. Soc.*, 843 (1958).
93. C. F. Hauser, T. W. Brooks, M. L. Miles, M. A. Raymond, and G. B. Butler, *J. Org. Chem.*, **28**, 372 (1963).
94. (a) F. Sondheimer and R. Mechoulam, *J. Am. Chem. Soc.*, **79**, 5029 (1957). (b) J. P. Freeman, *J. Org. Chem.*, **31**, 538 (1966). (c) H. O. House and T. H. Cronin, *ibid.*, **30**, 1061 (1965). (d) J. C. Stowell, *ibid.*, **35**, 244 (1970).

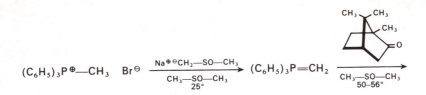

$(C_6H_5)_3\overset{\oplus}{P}—CH_3$　$Br^{\ominus}$ $\xrightarrow[\substack{CH_3—SO—CH_3 \\ 25°}]{Na^{\oplus \ominus}CH_2—SO—CH_3}$ $(C_6H_5)_3P=CH_2$ $\xrightarrow[\substack{CH_3—SO—CH_3 \\ 50–56°}]{}$

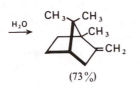

$\xrightarrow{H_2O}$

(73%)

(Ref. 85f)

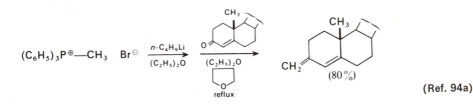

$(C_6H_5)_3\overset{\oplus}{P}—CH_3$　$Br^{\ominus}$ $\xrightarrow[\substack{(C_2H_5)_2O}]{n\text{-}C_4H_9Li}$ $\xrightarrow[\substack{(C_2H_5)_2O \\ \text{reflux}}]{}$

(80%)

(Ref. 94a)

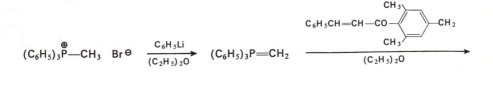

$(C_6H_5)_3\overset{\oplus}{P}—CH_3$　$Br^{\ominus}$ $\xrightarrow[\substack{(C_2H_5)_2O}]{C_6H_5Li}$ $(C_6H_5)_3P=CH_2$ $\xrightarrow[\substack{(C_2H_5)_2O}]{}$

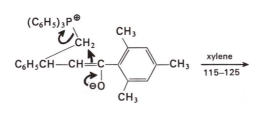

$\xrightarrow[\substack{115–125}]{\text{xylene}}$

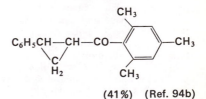

(41%)　(Ref. 94b)

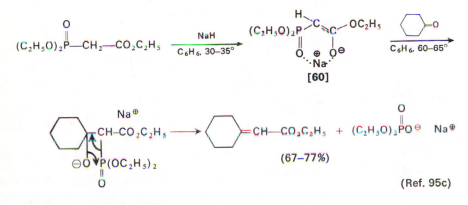

$$
\underset{\substack{|\\ \text{CH}_3}}{(C_6H_5)_3P{=}C{-}CO_2CH_3} + CH_2{=}CH{-}CHO \xrightarrow[\text{reflux}]{CH_2Cl_2} CH_2{=}CH\underset{H}{\diagup}\overset{}{C}{=}C\underset{CO_2CH_3}{\diagup}\overset{CH_3}{} \quad \text{(Ref. 87b)}
$$

(60% yield, >96% trans isomer)

benzoic acid as a catalyst.[8sg,h,n] A modification of the Wittig reaction, utilizing a phosphonate ylid (e.g., **[60]**) with the carbonyl compound,[95] has proved of value in such cases. This modification also circumvents the problems often encountered in separating the product from the triphenylphosphine oxide in the usual Wittig procedure.[95j]

(67–77%)

(Ref. 95c)

The following example is a variant of this reaction which has been developed for the synthesis of α,β-unsaturated aldehydes.

95. (a) W. S. Wadsworth and W. D. Emmons, *J. Am. Chem. Soc.*, **83**, 1733 (1961); F. A. Cotton and R. A. Schunn, *ibid.*, **85**, 2394 (1963); A. V. Dombrovskii and V. A. Dombrovskii, *Russ. Chem. Rev.*, **35**, 733 (1966). (b) A. K. Bose and R. T. Dahill, Jr., *J. Chem. Org.*, **30**, 505 (1965). (c) W. S. Wadsworth, Jr. and W. D. Emmons, *Org. Syn.*, **45**, 44 (1965). (d) M. J. Jorgenson and A. F. Thacher, *ibid.*, **48**, 75 (1968). (e) F. F. Blicke and S. Raines, *J. Org. Chem.*, **29**, 2036 (1964). (f) I. C. Popoff, J. L. Dever, and G. R. Leader, *ibid.*, **34**, 1128 (1969). (g) W. Nagata and Y. Hayase, *J. Chem. Soc.*, C, 460 (1969). (h) R. K. Huff, C. E. Moppett, and J. K. Sutherland, *ibid.*, C, 2725 (1968). (i) J. L. Baas, A. Davies-Fidder, F. R. Visser, and H. O. Huisman, *Tetrahedron*, **22**, 265 (1966); P. J. van den Tempel and H. O. Huisman, *ibid.*, **22**, 293 (1966). (j) The use of ylids derived from diphenyl-*p*-carboxyphenylphosphine has been recommended because the phosphine oxide produced is soluble in aqueous base and can be easily removed from the neutral reaction products; G. P. Schiemenz and J. Thobe, *Chem. Ber.*, **99**, 2663 (1966). (k) The anions derived from alkyl phosphonates, thiophosphonates, and phosphonamides may be prepared at −78°. The latter two systems offer useful modifications of the usual Wittig reaction. E. J. Corey and G. T. Kwiatkowski, *J. Am. Chem. Soc.*, **88**, 5652, 5653, 5654 (1966); **90**, 6816 (1968); E. J. Corey and D. E. Cane, *J. Org. Chem.*, **34**, 3053 (1969). (l) G. Lefebvre and J. Seyden-Penne, *Chem. Commun.*, **No. 20**, 1308 (1970).

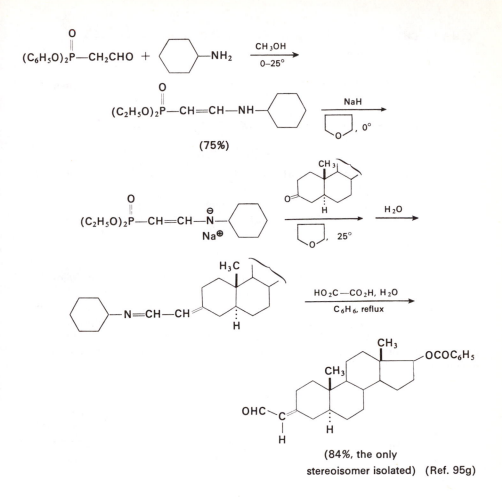

(84%, the only
stereoisomer isolated)   (Ref. 95g)

The reaction between phosphorus ylids and carboxylic acid derivatives is mechanistically related to the acylation of enolate anions (Chapter 11).[86,96] This reaction provides an alternative preparative route to stabilized ylids and, when accompanied by cleavage of the resulting acylated phosphonium salts, constitutes a synthesis for ketones, as illustrated in the accompanying equations. Stabilized phosphoranes are acylated at carbon (more stable product) by acid anhydrides but at oxygen (kinetically controlled product) by acid chlorides.[96e]

96. (a) H. J. Bestmann and B. Arnason, *Chem. Ber.,* **95**, 1513 (1962). (b) A. J. Speziale and K. W. Ratts, *J. Org. Chem.,* **28**, 465 (1963). (c) G. Märkl, *Tetrahedron Letters,* **No. 22**, 1027 (1962). (d) H. J. Bestmann, G. Graf, and H. Hartung, *Angew. Chem., Intern. Ed. Engl.,* **4**, 596 (1965). (e) P. A. Chopard, R. J. G. Searle, and F. H. Devitt, *J. Org. Chem.,* **30**, 1015 (1965). (f) R. G. Barnhardt, Jr., and W. E. McEwen, *J. Am. Chem. Soc.,* **89**, 7009 (1967). (g) C. A. Henrick, E. Böhme, J. A. Edwards, and J. H. Fried, *ibid.,* **90**, 5926 (1968). (h) W. Flitsch and H. Peters, *Tetrahedron Letters,* **No. 15**, 1161 (1969). (i) A. P. Gara, R. A. Massy-Westropp, and G. D. Reynolds, *ibid.,* **No. 48**, 4171 (1969).

$(C_6H_5)_3P{=}CH_2 \xrightarrow[\substack{C_6H_6 \\ reflux}]{C_6H_5{-}CO{-}Cl} [(C_6H_5)_3P^{\oplus}{-}CH_2{-}CO{-}C_6H_5] \; Cl^{\ominus} \xrightarrow{(C_6H_5)_3P{=}CH_2}$

$(C_6H_5)_3P{=}CH{-}CO{-}C_6H_5 \; + \; (C_6H_5)_3P^{\oplus}{-}CH_3 \;\; Cl^{\ominus}$ (Ref. 96a)
(71%)

$(C_6H_5)_3P{=}CH{-}CH_3 \; + \; C_6H_5{-}CH{=}CH{-}CO{-}S{-}C_2H_5 \xrightarrow[reflux]{C_6H_5CH_3}$

$\overset{\displaystyle CH_3}{\underset{\vert}{(C_6H_5)_3P^{\oplus}{-}CH}}{-}CO{-}CH{=}CH{-}C_6H_5 \;\; S^{\ominus}{-}C_2H_5$ (Ref. 96a)

$\xrightarrow[reflux]{-C_2H_5SH} \overset{\displaystyle CH_3}{\underset{\vert}{(C_6H_5)_3P{=}C}}{-}CO{-}CH{=}CH{-}C_6H_5 \xrightarrow[\substack{CH_3OH \\ reflux}]{\substack{NaOH \\ H_2O}} C_6H_5{-}CH{=}CH{-}CO{-}CH_2CH_3$
(70%) (74%)

The acylation of phosphorus ylids with nitriles, exemplified in the following equation, is catalyzed by the lithium cation in the reaction solution.[96f] For this reason, the reactions were successful when the ylids were prepared from the corresponding phosphonium iodides which generated a solution of lithium iodide in the reaction mixture; only very poor yields were obtained starting with the phosphonium bromides or chlorides because the corresponding lithium salts are relatively insoluble.

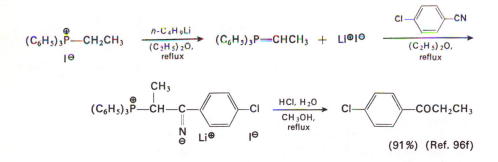

Cyclic anhydrides have been found to react with stabilized ylids to form the corresponding enol lactones.[96h,i] With more reactive ylids, enol lactones react to form cyclic unsaturated ketones such as that shown.[96g] This transformation is similar to the previously illustrated reaction of enol lactones with Grignard reagents to form cyclic enones.[21]

There is a sufficient difference between the reaction rates of Wittig reagents with ketones and with esters to make possible the selective reaction of an ylid with a keto ester (e.g., [61]) only at its ketone function. The presence of an ester function in the phosphonium salt at a position other than alpha or beta to the

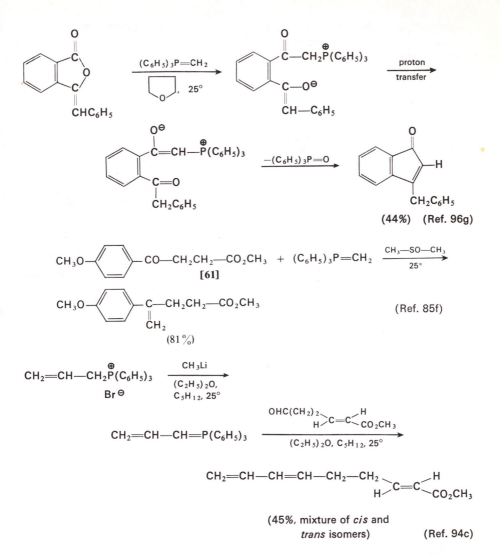

(44%) (Ref. 96g)

(81%) (Ref. 85f)

(45%, mixture of *cis* and *trans* isomers) (Ref. 94c)

phosphorus atom (e.g., **[62]**) may also result in either intramolecular or intermolecular acylation upon generation of the ylid; however, this acylation is apparently slow enough to permit the use of ylids containing ester functions in a normal Wittig

$(C_6H_5)_3P^{\oplus}\!-\!CH_2\!-\!(CH_2)_3\!-\!CO_2C_2H_5$ $\xrightarrow[\substack{(CH_3)_3COH \\ reflux}]{KOC(CH_3)_3}$ $(C_6H_5)_3P\!=\!CH\!-\!(CH_2)_3\!-\!CO_2C_2H_5$

$I^{\ominus}$ **[62]**

(Ref. 89b)

$\xrightarrow[\text{2. KOC(CH}_3)_3]{\text{1. cyclization}}$

(84%)

reaction provided that the very favorable intramolecular acylation to form a five- or six-membered ring is not possible.[83g,97] Although it does not appear possible to prepare stable ylids from phosphonium salts which have a beta carboalkoxy function because of a competing elimination of triphenylphosphine (c.f. the quaternary ammonium salts derived from Mannich bases), these ylids are formed as transient intermediates which can be trapped by carbonyl compounds.[97c−f]

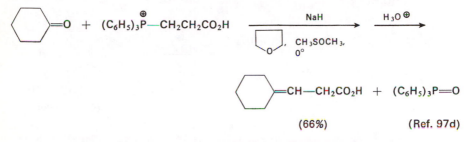

(66%)   (Ref. 97d)

The alkylation of phosphorus ylids with alkyl halides has also been observed.[86a,98] An intramolecular alkylation reaction has been used to prepare cyclobutylidene and cyclopropylidene phosphoranes[98c−f] as illustrated by the following example.

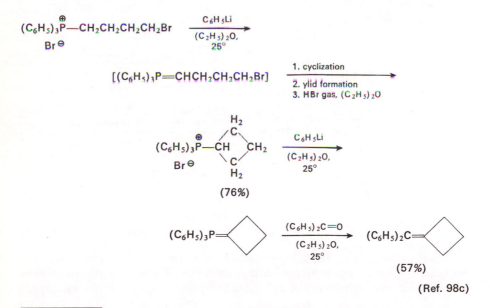

(76%)

(57%)

(Ref. 98c)

97. (a) L. D. Bergelson, V. A. Vaver, V. Yu. Kovtun, L. B. Senyavina, and M. M. Shemyakin, *J. Gen. Chem., U.S.S.R.*, **32**, 1785 (1962). (b) L. D. Bergelson, V. A. Vaver, A. A. Bezzubov, and M. M. Shemyakin, *ibid.*, **32**, 1790 (1962). (c) H. J. Bestmann, H. Häberlein, and I. Pils, *Tetrahedron*, **20**, 2079 (1964). (d) H. S. Corey, Jr., J. R. D. McCormick, and W. E. Swensen, *J. Am. Chem. Soc.*, **86**, 1884 (1964). (e) A. R. Hands and A. J. H. Mercer, *J. Chem. Soc.*, C, 2448 (1968). (f) The corresponding phosphonate derivative undergoes a reaction analogous to the Stobbe condensation; D. J. Martin, M. Gordon, and C. E. Griffin, *Tetrahedron*, **23**, 1831 (1967).

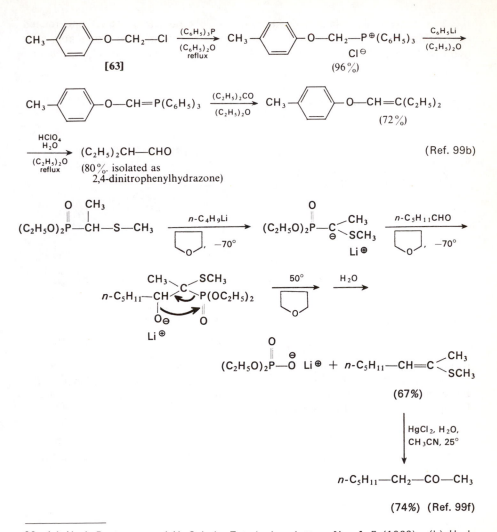

(Ref. 99b)

98. (a) H. J. Bestmann and H. Schulz, *Tetrahedron Letters,* **No. 4,** 5 (1960). (b) H. J. Bestmann and H. Häberlein *Z. Naturforsch.,* **17B,** 787 (1962); H. J. Bestmann and E. Kranz, *Angew. Chem., Intern. Ed. Engl.,* **6,** 81 (1967). (c) K. V. Scherer, Jr. and R. S. Lunt, *J. Org. Chem.,* **30,** 3215 (1965). (d) K. Sisido and K. Utimoto, *Tetrahedron Letters,* **No. 28,** 3267 (1966). (e) H. J. Bestmann and T. Denzel, *ibid.,* **No. 30,** 3591 (1966); H. J. Bestmann, T. Denzel, R. Kunstmann, and J. Lengyel, *ibid.,* **No. 24,** 2895 (1968). (f) E. E. Schweizer, C. J. Berninger, and J. G. Thompson, *J. Org. Chem.,* **33,** 336 (1968). (g) For a preparation of the cyclopropylidene ylid from cyclopropyllithium, see D. T. Longone and R. R. Doyle, *Chem. Commun.,* **No. 6,** 300 (1967). (h) The ylids derived from cyclopropylcarbinyl-phosphonium salts exhibit no tendency to isomerize to the corresponding allylcarbinyl derivatives; A. Maercker, *Angew. Chem., Intern. Ed. Engl.,* **6,** 557 (1967); E. E. Schweizer, J. G. Thompson, and T. A. Ulrich, *J. Org. Chem.,* **33,** 3082 (1968). (i) For examples of the alkylation of phosphorus ylids with epoxides, see R. M. Gerkin and B. Rickborn, *J. Am. Chem. Soc.,* **89,** 5850 (1967); W. E. McEwen, A. P. Wolf, C. A. VanderWerf, A. Blade-Font, and J. W. Wolfe, *ibid.,* **89,** 6685 (1967) and references therein.

The phosphonium salts derived from α-halo ethers (e.g., [63]) and related compounds have been used in the Wittig reaction to form vinyl ethers and, after hydrolysis, aldehydes and ketones.[99] Other modifications of the Wittig reaction

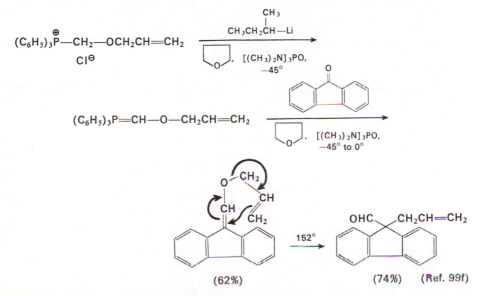

(62%)   (74%)   (Ref. 99f)

include addition of phosphorus ylids to conjugated double bonds, a process analogous to the Michael reaction (Chapter 9) and preparation of haloolefins by reaction of carbonyl compounds with phosphorus ylids that have one or two halogen substituents at the alpha carbon atom.[101]

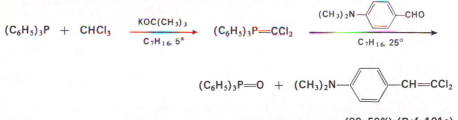

(39–56%) (Ref. 101c)

99. (a) S. G. Levine, *J. Am. Chem. Soc.,* **80**, 6150 (1958). (b) G. Wittig, W. Böll, and K. H. Krück, *Chem. Ber.,* **95**, 2514 (1962). (c) D. R. Coulson, *Tetrahedron Letters,* **No. 45**, 3323 (1964). (d) E. Zbiral, *ibid.,* **No. 20**, 1483 (1965). (e) G. R. Pettit, B. Green, A. K. Das Gupta, P. A. Whitehouse, and J. P. Yardley, *J. Org. Chem.,* **35**, 1381 (1970) ; G. R. Pettit, B. Green, G. L. Dunn, and P. Sunder-Plassmann, *ibid.,* **35**, 1385 (1970). (f) E. J. Corey and J. I. Shulman, *ibid.,* **35**, 777 (1970) ; *J. Am. Chem. Soc.,* **92**, 5522 (1970).
100. (a) R. Mechoulam and F. Sondheimer, *J. Am. Chem. Soc.,* **80**, 4386 (1958). (b) S. Trippett, *J. Chem. Soc.,* 4733 (1962). (c) H. J. Bestmann and F. Seng, *Angew. Chem., Intern. Ed. Engl.,* **1**, 116 (1962). (d) P. T. Keough and M. Grayson, *J. Org. Chem.,* **29**, 631 (1964). (e) M. I. Kabachnik, *Tetrahedron,* **20**, 655 (1964). (f) E. E. Schweizer and R. D. Bach, *Org. Syn.,* **48**, 129 (1968) ; E. E. Schweizer and G. J. O'Neill, *J. Org. Chem.,* **30**, 2082

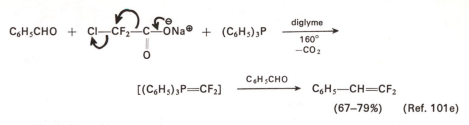

$$[(C_6H_5)_3P{=}CF_2] \xrightarrow{C_6H_5CHO} C_6H_5{-}CH{=}CF_2$$

(67–79%)   (Ref. 101e)

The ability of nucleophiles to add to the C—C double bond of vinylphosphonium salts[100d,e] has been utilized to generate intermediates capable of undergoing an intramolecular Wittig reaction to form carbocyclic or heterocyclic systems.[100f–i] The conjugate addition of phosphines or phosphite esters to α,β-unsaturated esters, ketones, or nitriles followed by proton transfer has also served to generate ylid intermediates which possess a beta electron-withdrawing group.[100j–n] The

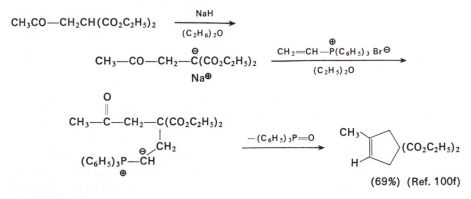

(69%) (Ref. 100f)

generation of ylids which can undergo intramolecular reactions has been used by a number of workers to prepare cyclic compounds such as those shown.

(1965). (g) E. E. Schweizer and K. K. Light, *ibid.*, **31**, 870, 2912 (1966). (h) E. E. Schweizer, L. D. Smucker, and R. J. Votral, *ibid.*, **31**, 467 (1966); E. E. Schweizer, E. T. Schaffer, C. T. Hughes, and C. J. Berninger, *ibid.*, **31**, 2907 (1966). (i) E. E. Schweizer and J. G. Liehr, *ibid.*, **33**, 583 (1968); E. E. Schweizer, J. Liehr, and D. J. Monaco, *ibid.*, **33**, 2416 (1968); E. E. Schweizer and A. T. Wehman, *J. Chem. Soc.*, C, 1901 (1970). (j) R. Oda, T. Kawabota, and S. Tanimoto, *Tetrahedron Letters*, **No. 25**, 1653 (1964). (k) S. Trippett, *Chem. Commun.*, **No. 14**, 468 (1966). (l) R. G. Harvey, *Tetrahedron*, **22**, 2561 (1966). (m) J. D. McClure, *Tetrahedron Letters*, **No. 25**, 2407 (1967). (n) J. Asunskis and H. Shechter, *J. Org. Chem.*, **33**, 1164 (1968); J. D. McClure, *ibid.*, **35**, 3045 (1970). (o) C. E. Griffin and G. Witschard, *ibid.*, **29**, 1001 (1964). (p) E. E. Schweizer, C. J. Berninger, D. M. Crouse, R. A. Davis, and R. S. Logothetis, *ibid.*, **34**, 207 (1969). (q) H. J. Bestmann and H. Morper, *Angew. Chem., Intern. Ed. Engl.*, **6**, 561 (1967). (r) C. Brown and M. V. Sargent, *J. Chem. Soc.*, C, 1818 (1969).
101. (a) D. Seyferth, S. O. Grim, and T. O. Read, *J. Am. Chem. Soc.*, **83**, 1617 (1961). (b) A. J. Speziale and K. W. Ratts, *ibid.*, **84**, 854 (1962). (c) A. J. Speziale, K. W. Ratts, and D. E. Bissing, *Org. Syn.*, **45**, 33 (1965). (d) H. Zimmer, P. J. Bercz, O. J. Maltenieks, and M. W. Moore, *J. Am. Chem. Soc.*, **87**, 2777 (1965). (e) S. A. Fuqua, W. G. Duncan, and R. M. Silverstein, *Org. Syn.*, **47**, 49 (1967). (f) F. E. Herkes and D. J. Burton, *ibid.*, **48**, 116 (1968).

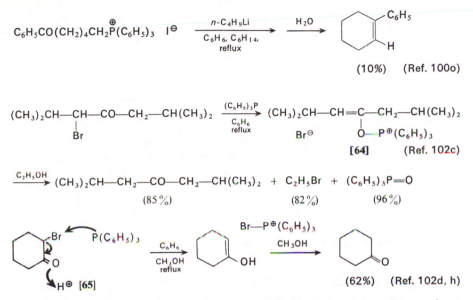

(10%)   (Ref. 100o)

[64]   (Ref. 102c)

(85%)   (82%)   (96%)

[65]

(62%)   (Ref. 102d, h)

The reaction of α-halocarbonyl compounds with triphenylphosphine to form phosphonium salts that may be used to generate α-keto ylids is sometimes complicated by the formation of enol phosphonium salts (e.g., [64]), which react with alcohol, water, or other protonic solvents to yield a dehalogenated ketone.[82c,102] This complication appears to result from an acid-catalyzed attack (e.g., [65]) by the triphenylphosphine at the bromine atom of α-bromo ketones to form an enol which is converted to the ketone in the protic medium. Reactions run in aprotic media may also be complicated by side reactions such as dehydrohalogenation of the halo ketone[102d,i,j] or the formation of an enol phosphonium salt.[102e,j] In

102. (a) F. W. Lichtenthaler, *Chem. Rev.,* **61**, 607 (1961). (b) H. Hoffmann and H. J. Diehr, *Angew. Chem., Intern. Ed. Engl.,* **3**, 737 (1964). (c) S. Trippett, *J. Chem. Soc.,* 2337 (1962); F. Hampson and S. Trippett, *ibid.,* 5129 (1965). (d) P. A. Chopard, R. F. Hudson, and G. Klopman, *ibid.,* 1379 (1965); P. A. Chopard and R. F. Hudson, *ibid.,* B, 1089 (1966); R. F. Hudson and G. Salvadori, *Helv. Chim. Acta,* **49**, 96 (1966). (e) A. J. Speziale and L. J. Taylor, *J. Org. Chem.,* **31**, 2450 (1966). (f) J. E. Thompson, *ibid.,* **30**, 4276 (1965). (g) K. Fukui, R. Sudo, M. Masaki, and M. Ohta, *ibid.,* **33**, 3504 (1968). (h) I. J. Borowitz and R. Virkhaus, *J. Am. Chem. Soc.,* **85**, 2183 (1963); I. J. Borowitz, K. C. Kirby, Jr., and R. Virkhaus, *J. Org. Chem.,* **31**, 4031 (1966); I. J. Borowitz and H. Parnes, *ibid.,* **32**, 3560 (1967). (i) I. J. Borowtiz, K. C. Kirby, Jr., P. E. Rusek, and R. Virkhaus, *ibid.,* **33**, 3686 (1968). (j) I. J. Borowitz, P. E. Rusek, and R. Virkhaus, *ibid.,* **34**, 1595 (1969); I. J. Borowitz, K. C. Kirby, Jr., P. E. Rusek, and E. Lord, *ibid.,* **34**, 2687 (1969). (k) I. J. Borowitz, M. Anschel, and S. Firstenberg, *ibid.,* **32**, 1723 (1967); I. J. Borowitz, K. C. Kirby, Jr., P. E. Rusek, and E. W. R. Casper, *ibid.,* **36**, 88 (1971). (l) D. B. Denney, N. Gershman, and J. Giacin, *ibid.,* **31**, 2833 (1966). (m) H. Machleidt and G. U. Strehlke, *Angew. Chem., Intern. Ed. Engl.,* **3**, 443 (1964). (n) P. A. Chopard, V. M. Clark, R. F. Hudson, and A. J. Kirby, *Tetrahedron,* **21**, 1961 (1965). (o) H. Tomioka, Y. Izawa, and Y. Ogata, *ibid.,* **24**, 5739 (1968). (p) J. C. Craig, M. D. Bergenthal, I. Fleming, and J. Harley-Mason, *Angew. Chem., Intern. Ed. Engl.,* **8**, 429 (1969). (q) M. Fetizon, M. Jurion, and N. T. Anh, *Chem. Commun.,* **No. 3**, 112 (1969).

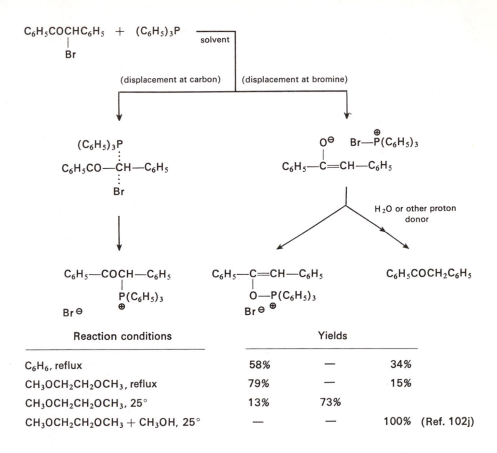

Reaction conditions	Yields		
C_6H_6, reflux	58%	—	34%
$CH_3OCH_2CH_2OCH_3$, reflux	79%	—	15%
$CH_3OCH_2CH_2OCH_3$, 25°	13%	73%	
$CH_3OCH_2CH_2OCH_3 + CH_3OH$, 25°	—	—	100% (Ref. 102j)

general, the formation of an α-keto phosphonium salt by the indicated S_N2 displacement is favored by the use of α-chloro rather than α-bromo ketones, by the use of a primary rather than a secondary halide (dehydrohalogenation is usually observed with tertiary halides), by the use of a dipolar, aprotic reaction solvent at elevated temperatures with high reactant concentrations, and by the presence of a small amount of a base such as triethylamine to suppress the acid-catalyzed debromination.[102g,j]

A similar set of side reactions is possible in the reaction of α-halo ketones with trialkyl phosphites to form α-ketoalkyl phosphonates.[82c,102a,b,k−o] However, the principal side reaction in these cases, formation of an enol phosphate (the Perkow reaction), is believed not to arise from attack of the phosphite at the halogen but rather from attack at the carbon of the carbonyl group as illustrated in the following scheme.[102k,l,n] This side reaction, to form an enol phosphate, is catalyzed by acid;[102k] strongly acidic conditions also catalyze the isomerization of an α-keto phosphonate to an enol phosphate.[102m]

The enol phosphates obtained by the Perkow reaction can also serve as useful synthetic intermediates.[102a] Among other uses, they have served as precursors

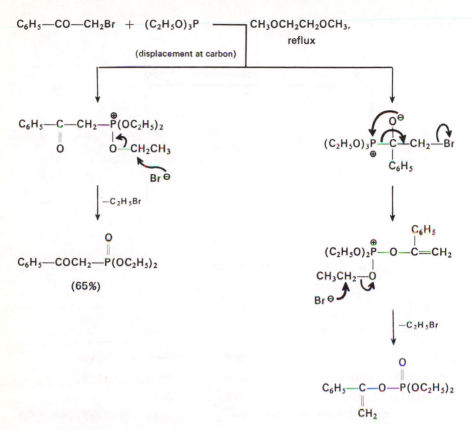

for acetylenes[102p] and as intermediates in the conversion of α-bromo ketones to olefins by the reaction sequence indicated below.

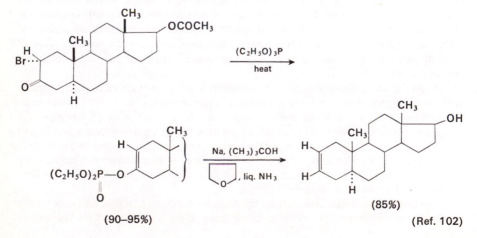

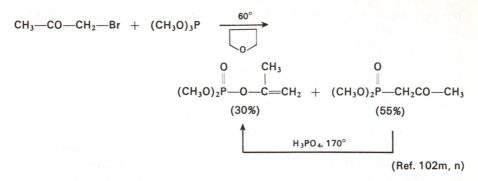

(Ref. 102m, n)

Although the position of the carbon–carbon double bond in an olefin formed by a Wittig reaction may be predicted with certainty, the stereochemistry of the olefin product is sometimes less predictable.[103b] The following examples illustrate the fact that stereoisomeric mixtures are often produced when reactive ylids (i.e. ylids not stabilized by an adjacent carbonyl function) are employed. However,

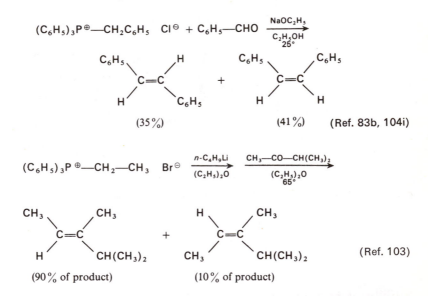

103. (a) J. P. Dusza, *J. Org. Chem.,* **25**, 93 (1960). (b) For a recent review, see M. Schlosser, *Topics in Stereochemistry,* **5**, 1 (1970).
104. (a) R. Ketcham, D. Jambotkar, and L. Martinelli, *J. Org. Chem.,* **27**, 4666 (1962). (b) H. J. Bestmann and O. Kratzer, *Chem. Ber.,* **95**, 1894 (1962). (c) H. Heitman, U. K. Pandit, and H. O. Huisman, *Tetrahedron Letters,* **No. 14**, 915 (1963). (d) H. O. House, V. K. Jones, and G. Frank, *J. Org. Chem.,* **29**, 3327 (1964). (e) D. E. Bissing, *ibid.,* **30**, 1296 (1965). (f) A. W. Johnson and V. L. Kyllingstad, *ibid.,* **31**, 334 (1966). (g) D. E. Bissing and A. J. Speziale, *J. Am. Chem. Soc.,* **87**, 2683 (1965). (h) D. H. Wadsworth, O. E. Schupp, E. J. Seus, and J. A. Ford, Jr., *J. Org. Chem.,* **30**, 680 (1965). (i) O. H. Wheeler and H. N. Batlle de Pabon, *ibid.,* **30**, 1473 (1965). (j) L. Horner and W. Klink, *Tetrahedron Letters,* **No. 36**, 2467 (1964). (k) G. Jones and R. F. Maisey, *Chem. Commun.,* **No. 10**, 543 (1968). (l) T. H. Kinstle and B. Y. Mandanas, *ibid.,* **No. 24**, 1699 (1968).

reactions employing stabilized ylids frequently yield as the predominant product the stereoisomer in which the carbonyl function is *trans* to the larger group at the beta carbon atom, a circumstance reminiscent of the stereoselectivity observed in the aldol condensation. This selectivity, illustrated in the equations below, is believed to result, at least in part, from the more rapid decomposition of the betaine diastereoisomer (e.g., [66]) in which there is less steric interference with overlap between the pi-orbital of the carbonyl function and the pi-orbital of the developing double bond. The degree of stereoselectivity observed may be influenced by the nature of the substituents both on the ylid and on the carbonyl component, as indicated in the accompanying examples.[85,104] The proportion of the olefin isomer with the two largest substituents *cis* is normally increased by the use of more electrophilic carbonyl reactants, by the use of ylids derived from triarylphosphines

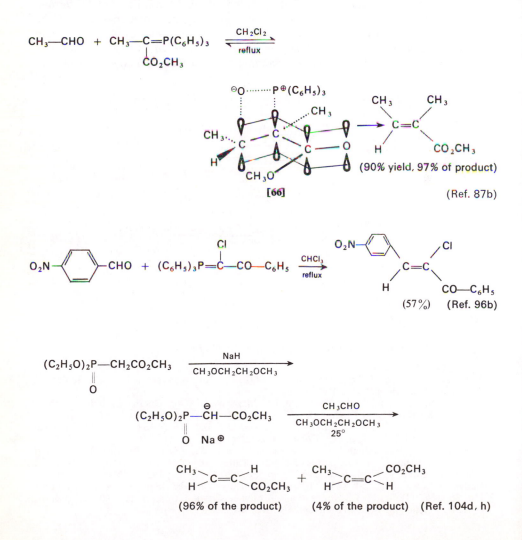

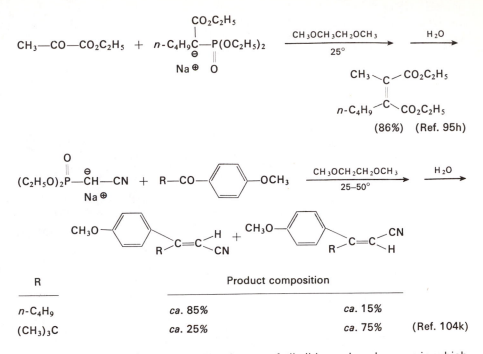

R	Product composition	
n-C_4H_9	ca. 85%	ca. 15%
$(CH_3)_3C$	ca. 25%	ca. 75% (Ref. 104k)

rather than trialkylphosphines, and by the use of alkylidene phosphoranes in which the negative charge is not stabilized by an electron-withdrawing alpha substituent. The olefin-forming reactions which use phosphonate anions (stabilized by either α-aryl or α-acyl substituents) rather than the corresponding Wittig reagents are often more stereoselective in leading to formation of the stereoisomer with the activating group at the α-carbon of the olefin *trans* to the larger group at the β-carbon.[95,104d,h,i,k] However, where steric interactions in the product become sufficiently large, even the condensation with a phosphonate anion loses its stereoselectivity.[104l]

R	Product composition	
CH_3—	82%	18%
C_2H_5—	59%	41%
$(CH_3)_2CH$—	10%	84%
$(CH_3)_3C$—	45%	55% (Ref. 104l)

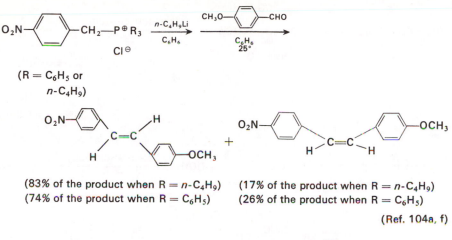

(R = C_6H_5 or n-C_4H_9)

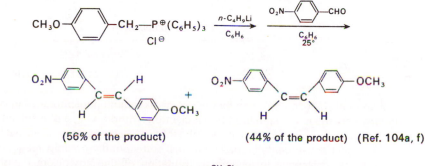

(83% of the product when R = n-C_4H_9)
(74% of the product when R = C_6H_5)

(17% of the product when R = n-C_4H_9)
(26% of the product when R = C_6H_5)

(Ref. 104a, f)

(56% of the product)

(44% of the product) (Ref. 104a, f)

X—CH_2—CHO + (C_6H_5)_3P=CH—CO_2CH_3 →(CH_2Cl_2) X—CH_2—CH=CH—CO_2CH_3

(X = H: 96% *trans* isomer in product)
(X = Cl: 71% *trans* isomer in product) (Ref. 104d)

The nature of the reaction medium has also been found to influence the proportions of stereoisomers[83b,i,104d,105] as illustrated below. In general the formation of *cis*-olefins from ylids stabilized by α-aryl and α-acyl substituents is favored by the use of either protic solvents or dipolar, aprotic solvents; in some cases the presence of dissolved lithium salts or proton donors may also enhance the proportion of *cis*-isomer. Since lithium iodide (and its complexes with nucleophiles) is

---

105. (a) L. D. Bergel'son and M. M. Shemyakin, *Tetrahedron,* **19,** 149 (1963). (b) G. Drefahl, D. Lorenz, and G. Schnitt, *J. prakt. Chem.,* [4] **23,** 143 (1964). (c) L. D. Bergel'son, V. A. Vaver, L. I. Barsukov, and M. M. Shemyakin, *Tetrahedron Letters,* **No. 38,** 2669 (1964). (d) L. D. Bergel'son, L. I. Barsukov, and M. M. Shemyakin, *Tetrahedron,* **23,** 2709 (1967). (e) W. P. Schneider, *Chem. Commun.,* **No. 14,** 785 (1969). (f) G. Pattenden and B. C. L. Weedon, *J. Chem. Soc.,* C, 1984, 1997 (1968). (g) C. F. Garbers, D. F. Schneider, and J. P. van der Merwe, *ibid.,* C, 1982 (1968). (h) L. Crombie, P. Hemesley, and G. Pattenden, *ibid.,* C, 1016, 1024 (1969). (i) E. J. Corey, J. I. Shulman, and H. Yamamoto, *Tetrahedron Letters,* **No. 6,** 447 (1970); E. J. Corey and H. Yamamoto, *J. Am. Chem. Soc.,* **92,** 226 (1970).

much more soluble in hydrocarbon solvents than many other lithium salts (e.g., LiBr or LiCl)[95i] the stereochemistry of certain Wittig reactions has been found[105a,c,d] to be influenced by the nature of the anion present. That this influence should be ascribed to solubility differences is indicated by the fact that when the reactions are studied under circumstances where the various lithium salts are in solution, the anion effect disappears;[98i,104d,105b,d,f] suspensions of *undissolved* lithium salts appear to have little or no influence on the stereochemistry of the reaction.[104d,105f]

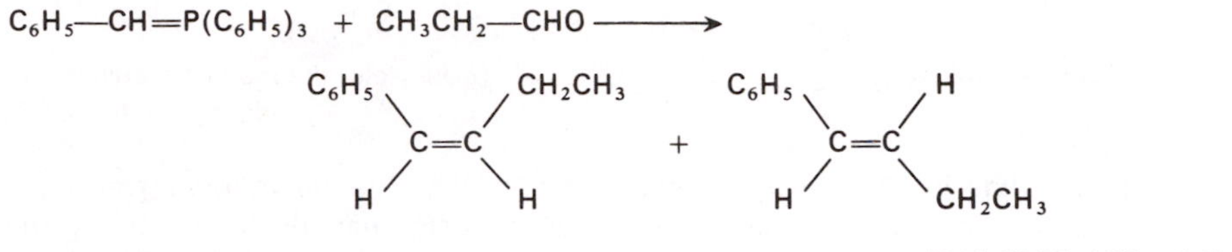
C6H5—CH=P(C6H5)3 + CH3CH2—CHO
C6H5
CH2CH3
C=C
H
H
+
C6H5
H
C=C
H
CH2CH3

(Ref. 104d, 105a–d, f)

Reaction medium	% of product	% of product
C_6H_6	18–27	73–82
C_6H_6 + suspended LiBr	23–27	73–77
C_2H_5OH	47–52	48–53
$(CH_3)_2NCHO$	39–46	54–61
$(CH_3)_2NCHO$ + dissolved LiBr	41–46	54–59

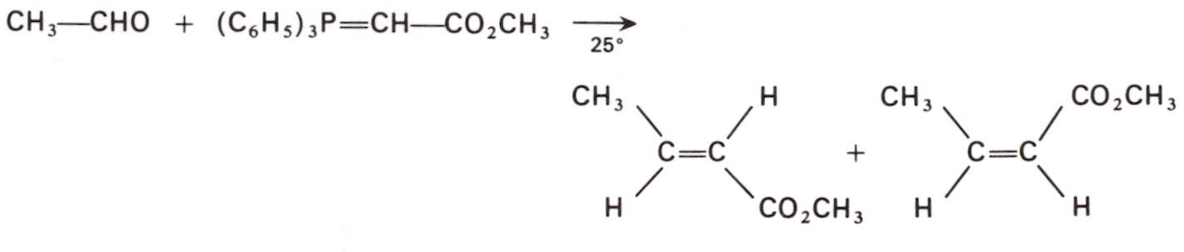
CH3—CHO + (C6H5)3P=CH—CO2CH3
25°
CH3
H
C=C
H
CO2CH3
+
CH3
CO2CH3
C=C
H
H

Reaction medium	% of *trans* product	% of *cis* product
$(CH_3)_2N$—CHO	97	3
$(CH_3)_2N$—CHO, $Li^{\oplus}X^{\ominus}$(X=Br, Cl, ClO_4, NO_3)	78–82	18–22
CH_3OH	62	38

(Ref. 104d)

In reactions of alkylidene phosphoranes that do not contain stabilizing alpha substituents, the highest proportions of *cis*-olefin have been obtained by the use

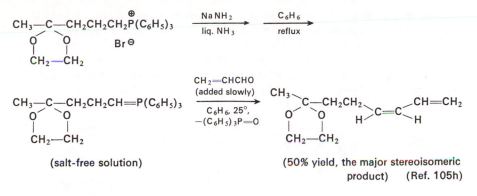

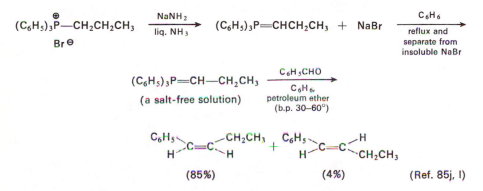

(50% yield, the major stereoisomeric product)   (Ref. 105h)

of salt-free ylids in non-polar, aprotic media.[83i,85j,k,l] In the following example the percentage of *trans*-isomer in the product is significantly increased by the addition of relatively soluble lithium salts such as lithium iodide (17% *trans*-

isomer in the product) and especially lithium tetraphenylboride (48% *trans*-isomer in the product).[85j,l] As noted earlier and also shown in the following example, even

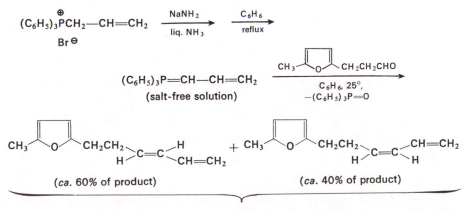

(67% yield)   (Ref. 105h)

ylids stabilized by α-substituents such as phenyl or vinyl groups normally yield mainly the *trans*-isomer even when salt-free solutions of these stabilized ylids are employed.[83i,85l,105h]

In all of the previously discussed cases the proportions of stereoisomers are believed to be determined by the relative rates of formation, dissociation, and decomposition of the intermediate betaines [68] and [69].[83i,85,95l,104,105] Although

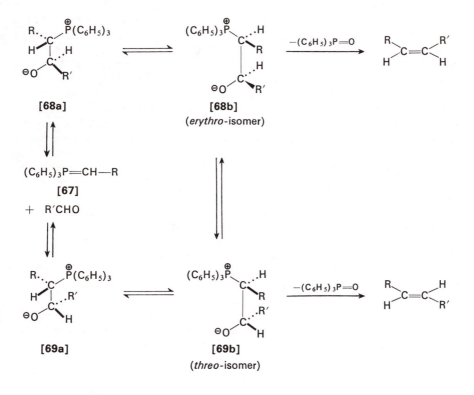

these betaine intermediates may be equilibrated without dissociation by removal of the proton alpha to phosphorus, this type of equilibration appears to be important only[85j] in subsequently discussed cases when an additional equivalent of a strong base (e.g., phenyllithium)[85i,l,105i] is added to the reaction mixture; equilibration of the betaines by any procedure normally enhances the proportion of the *trans*-olefin in the product.[83i,85i,k,l] The rates of decomposition of the polar betaine intermediates (e.g., [68] and [69]) to form olefins are accelerated by the presence of polar solvents which favor the dissociation of the ion pairs or ion aggregates present in non-polar media and by the absence of dissolved metal cations (especially lithium cation) which form tightly associated ion pairs (or covalent bonds) with the betaine. The decomposition of the betaines [68] and [69] to olefins is also favored by the presence of substituents (e.g., carbonyl, vinyl, or aryl groups) which can stabilize the forming double bond by orbital overlap as illustrated in structure [66]. As discussed earlier, the coplanarity requirement for such overlap is part of the reason

why stabilized ylids usually form primarily that olefin isomer in which the stabilizing substituent is *trans* to the larger substituent of the carbonyl reactant.

The stereochemical results obtained for all the previously discussed Wittig reactions are best explained by the hypothesis that in the initial reaction of the ylid [67] with the carbonyl compound there is a decided kinetic preference for the formation of the *erythro*-betaine [68] rather than the *threo*-isomer [69]. The reason(s) for this kinetic preference are unclear; although it has been argued that electrostatic interactions should favor the formation of betaines in the eclipsed conformations [68b] and [69b],[83i,104,105] this assumption is by no means secure since a combination of torsional strain (see Chapter 2) and dissipation of charge through aggregation of charged species may well favor conformations [68a] and [69a] for the intermediate betaines. In any event, the kinetic preference for forming the *erythro*-isomer [68] followed by a decomposition to the olefin which is faster than dissociation (i.e. [68]→[67]) of the betaines readily accounts for the production of mainly *cis*-olefins from unstabilized ylids in salt-free reaction media. When metal cations (especially Li⊕) are present in the reaction solution, decomposition of the betaines [68] and [69] is retarded, presumably by association of the metal cation with the alkoxide portion of the betaine. Under these circumstances, some equilibration of the betaines occurs to form the more stable *threo*-isomer [69] at a rate which is competitive with olefin formation.[85i,l] Consequently, an increased amount of the *trans*-olefin is produced. It might be noted in passing that Wittig reactions involving relatively unstable reactants and products such as those derived from allylphosphonium salts are best conducted under conditions where metal salts formed during the generation of the ylids will be insoluble.[94c,105h] Such conditions maximize the rate of betaine decomposition and so permit the use of short reaction times and mild reaction conditions.

When stabilizing α-substituents are present in the ylid (e.g., [67] where R = acyl, aryl, or vinyl), the rates of dissociation of the betaines [68] and [69] to the starting ylid [67] become either faster than or competitive with the decomposition of the betaines to olefins. For the ylids [67], the rate of dissociation of the corresponding betaines lies in the order: R = alkyl < R = phenyl < R = acyl or carboalkoxyl.[104g] It would appear that the rate of this dissociation with stabilized ylids is retarded by the presence of protic solvents or dissolved lithium salts.[104d] The enhanced dissociation rate for the betaines [68] or [69] from stabilized ylids (in salt-free aprotic media) allows the betaines to equilibrate relatively rapidly. A combination of this propensity for rapid equilibration and the previously discussed steric factors favoring the decomposition of the *threo*-betaine ([69] where R is a stabilizing substituent) appear to account for the predominant formation of *trans*-olefins from stabilized ylids and phosphonate anions.[95l]

An especially interesting modification of the Wittig reaction consists of treating the initially formed betaine-lithium salt complex with an equivalent of an organolithium reagent to form a new ylid reagent.[85i,l] Reaction of the ylid with various electrophilic reagents such as proton donors, halogen donors, and carbonyl compounds occurs with remarkable stereospecificity to form new betaines which decompose to olefins.[85i,k,l,105i] The following equations illustrate the application of this method to the preparation of *trans*-olefins and to the stereospecific synthesis of trisubstituted olefins. The reasons for the stereospecificity observed in the

reaction of these ylid alkoxides with electrophiles and for the subsequent selectivity in the direction of decomposition of the betaine are uncertain at the present time.

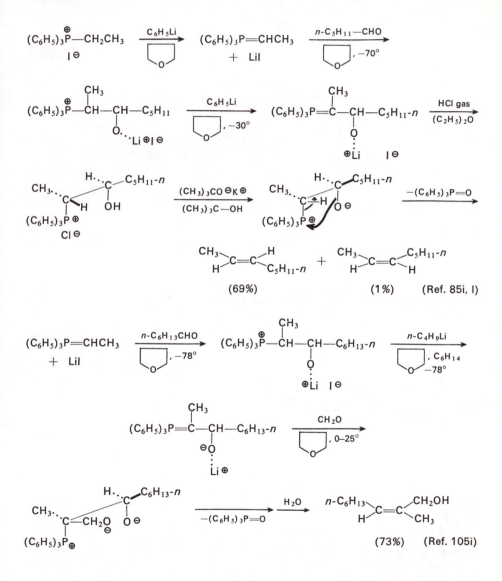

REACTIONS OF SULFUR YLIDS

Like phosphonium salts, sulfonium salts and sulfoxonium salts have alpha C—H bonds which are sufficiently acidic that they may be converted to the indicated sulfur ylids [70] and [71] by treatment with strong bases.[82a,107] It is appropriate to consider, along with the reactions of these sulfur ylids, the reactions of the

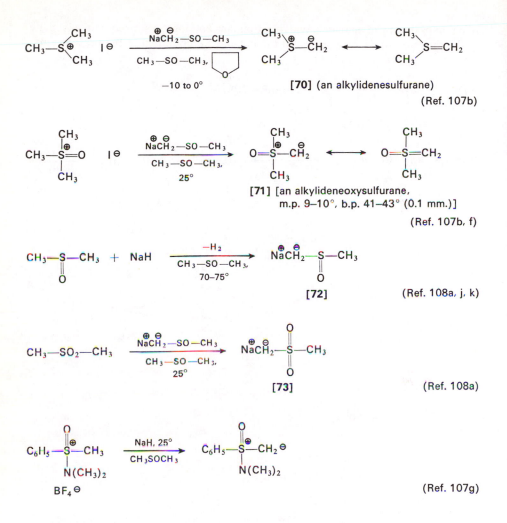

[70] (an alkylidenesulfurane)
(Ref. 107b)

[71] [an alkylideneoxysulfurane,
m.p. 9–10°, b.p. 41–43° (0.1 mm.)]
(Ref. 107b, f)

[72]   (Ref. 108a, j, k)

[73]   (Ref. 108a)

(Ref. 107g)

107. (a) A. W. Johnson, *Ylid Chemistry*, Academic Press, New York, 1966, pp. 304–366.
(b) E. J. Corey and M. Chaykovsky, *J. Am. Chem. Soc.*, **87**, 1353 (1965); *Org. Syn.*, **49**,
78 (1969). For a discussion of the spectra of sulfonium and sulfoxonium salts, see N. J.
Leonard and C. R. Johnson, *J. Am. Chem. Soc.*, **84**, 3701 (1962). (d) T. Durst, *Adv. Org.
Chem.*, **6**, 285 (1969). (e) D. Martin, A. Weise, and H. J. Niclas, *Angew. Chem., Intern.
Ed. Engl.*, **6**, 318 (1967). (f) H. Schmidbaur and W. Tronich, *Tetrahedron Letters*, **No. 51**,
5335 (1968). (g) C. R. Johnson, E. R. Janiga, and M. Haake, *J. Am. Chem. Soc.*, **90**, 3890
(1968); C. R. Johnson and C. W. Schroek, *ibid.*, **90**, 6852 (1968); C. R. Johnson and G. F.
Katekar, *ibid.*, **92**, 5753 (1970); C. R. Johnson, M. Haake, and C. W. Schroeck, *ibid.*,
92, 6594 (1970). This ylid can be obtained optically active and used for asymmetric
synthesis. (h) For the generation of alkylidenesulfuranes as transient intermediates in
aqueous solution, see M. J. Hatch, *J. Org. Chem.*, **34**, 2133 (1969); M. Yoshimine and
M. J. Hatch, *J. Am. Chem. Soc.*, **89**, 5831 (1967). (i) For the preparation and reactions of
ylids of the type, $R_2S^{\oplus}$—$N^{\ominus}$—$CO_2C_2H_5$, see G. F. Whitfield, H. S. Beilan, D. Saika, and
D. Swern, *Tetrahedron Letters*, **No. 41**, 3543 (1970).

anions [72] and [73] derived from dimethyl sulfoxide and dimethyl sulfone and other similar compounds since many of these reactions are related mechanistically.[107,108] The degree of reactivity of the most common of these methylides (and the basicity) increases in the following order:

$$CH_3—SO_2—\overset{\ominus}{C}H_2 < (CH_3)_2\overset{+}{\underset{O}{S}}—\overset{\ominus}{C}H_2 < (CH_3)_2\overset{\oplus}{S}—\overset{\ominus}{C}H_2 < CH_3—SO—\overset{\ominus}{C}H_2$$

The alkylidenesulfuranes (e.g., the methylide [70]) are sufficiently unstable that they must be produced at temperatures below 0° and used promptly before they decompose. This is especially true for the ylids derived from sulfonium salts containing alkyl groups larger than a methyl group such as [74][109] where decomposition of the ylid by a dimerization (which may involve a carbene intermediate)

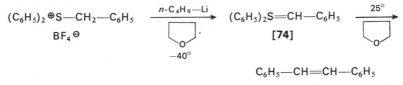

$$C_6H_5—CH{=}CH—C_6H_5$$

(61%, mixture of stereoisomers)   (Ref. 109a)

---

108. (a) E. J. Corey and M. Chaykovsky, *J. Am. Chem. Soc.,* **87**, 1345 (1965). (b) H. D. Becker and G. A. Russell, *J. Org. Chem.,* **28**, 1896 (1963). (c) G. A. Russell and H. D. Becker, *J. Am. Chem. Soc.,* **85**, 3406 (1963). (d) H. D. Becker, G. J. Mikol, and G. A. Russell, *ibid.,* **85**, 3410 (1963). (e) G. A. Russell, E. Sabourin, and G. J. Mikol, *J. Org. Chem.,* **31**, 2854 (1966). (f) G. A. Russell and G. J. Mikol, *J. Am. Chem. Soc.,* **88**, 5498 (1966). (g) G. A. Russell and E. T. Sabourin, *J. Org. Chem.,* **34**, 2336 (1969). (h) G. A. Russell, E. T. Sabourin, and G. Hamprecht, *ibid.,* **34**, 2339 (1969). (i) G. A. Russell and L. A. Ochrymowycz, *ibid.,* **34**, 3618, 3624 (1969); G. A. Russell and G. Hamprecht, *ibid.,* **35**, 3007 (1970). (j) The thermal decomposition of dimethyl sulfoxide solutions of the sodium salt [72] of dimethyl sulfoxide occurs slowly at 80° and becomes very rapid at temperatures above 100°. For a study of the various products formed in this decomposition, see C. C. Price and T. Yukuta, *ibid.,* **34**, 2503 (1969). (k) For the formation of relatively stable complexes of metal derivatives of dimethylsulfoxide with amines and ethers, see K. R. Martin, *J. Organometal. Chem.,* **24**, 7 (1970).
109. (a) A. W. Johnson, V. J. Hruby, and J. L. Williams, *J. Am. Chem. Soc.,* **86**, 918 (1964). (b) E. J. Corey and W. Oppolzer, *ibid.,* **86**, 1899 (1964). (c) J. Kiji and M. Iwamoto, *Tetrahedron Letters,* **No. 24**, 2749 (1966). (d) E. J. Corey, M. Jautelat, and W. Oppolzer, *ibid.,* **No. 24**, 2325 (1967). (e) E. J. Corey and M. Jautelat, *J. Am. Chem. Soc.,* **89**, 3912 (1967). (f) R. B. Bates and D. Feld, *Tetrahedron Letters,* **No. 4**, 417 (1968). (g) B. M. Trost and R. LaRochelle, *ibid.,* **No. 29**, 3327 (1968). (h) J. E. Baldwin and R. E. Peavy, *ibid.,* **No. 48**, 5029 (1968). (i) Y. Hayaski and R. Oda, *ibid.,* **No. 51**, 5381 (1968). (j) R. H. Mitchell and V. Boekelheide, *ibid.,* **No. 14**, 1197 (1970); V. Boekelheide and P. H. Anderson, *ibid.,* **No. 14**, 1207 (1970). (k) U. Schöllkopf, G. Ostermann, and J. Schossig, *ibid.,* **No. 31**, 2619 (1969). (l) J. E. Baldwin, R. E. Hackler, and D. P. Kelley, *Chem. Commun.,* **No. 10**, 537, 538 (1968). (m) R. W. C. Cose, A. M. Davies, W. D. Ollis, C. Smith, and I. O. Sutherland, *ibid.,* **No. 6**, 293 (1969). (n) J. E. Baldwin, W. F. Erickson, R. E. Hackler, and R. M. Scott, *ibid.,* **No. 10**, 576 (1970). (o) K. W. Ratts and A. N. Yao, *J. Org. Chem.,* **33**, 70 (1968). (p) B. M. Trost, R. LaRochelle, and M. J. Bogdanowicz, *Tetrahedron Letters,* **No. 39**, 3449 (1970).

or an elimination reaction involving the starting sulfonium salt could be serious side reactions. Alkylidenesulfuranes with two alkyl substituents at the carbon atom adjacent to sulfur (e.g., **[75]**) are very unstable and are best obtained by the

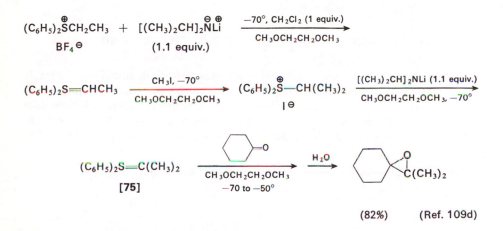

(82%) (Ref. 109d)

alkylation procedure indicated. Because of their instability, these dialkylylids should be used immediately after they have been prepared. Sulfur ylids formed from sulfonium salts which have allyl, benzyl, or acylmethyl groups bonded to the sulfur undergo a ready rearrangement[109f-o] which may complicate the use of such ylids in synthesis. These ylids may either rearrange by the concerted electrocyclic path illustrated in the following equations or they may undergo a radical dissociation–recombination reaction. This latter process, resulting in the shift of an alkyl group from sulfur to an adjacent atom is known as the Stevens rearrangement. The Stevens rearrangement has a higher activation energy than the electrocyclic rearrangement and is favored by heating the ylid in a non-polar solvent where proton transfer between isomeric ylids is unfavorable.

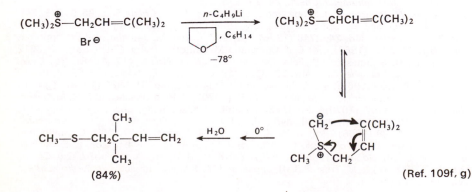

(84%) (Ref. 109f, g)

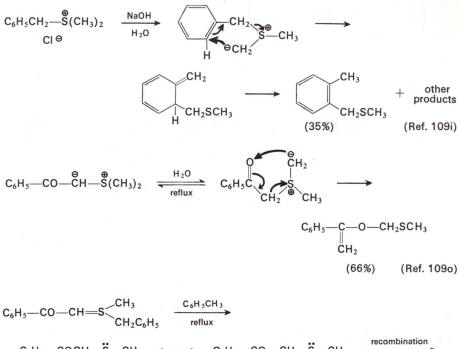

(35%) (Ref. 109i)

(66%) (Ref. 109o)

C₆H₅—CO—CH=S⟨CH₃ / CH₂C₆H₅ →(C₆H₅CH₃, reflux)

C_6H_5—COCH=$\ddot{S}$—CH_3 + ·$CH_2C_6H_5$ ⟷ C_6H_5—CO—$\overset{|}{C}H$—$\ddot{S}$—CH_3 →(recombination of radicals)

C_6H_5CO—CH—S—CH_3 $\overset{|}{C}H_2C_6H_5$ + other products

(45% of product) (Ref. 109k, n)

If substituents that can delocalize negative charge are present at the alpha carbon atom then the sulfur ylids, like the previously discussed phosphorus ylids, become more stable and can often be isolated (e.g., [76] and [77]).[82g,110,111] The sulfur ylids stabilized by α-acyl substituents also undergo thermal or photochemically induced decomposition to form 1,2,3-triacylcyclopropanes;[110b,h,111a] at least the photolytic reactions are thought to involve carbene intermediates.[101b,h,111l]

110. (a) A. W. Johnson and R. B. LaCount, *J. Am. Chem. Soc.*, **83**, 417 (1961). (b) A. W. Johnson and R. T. Amel, *J. Org. Chem.*, **34**, 1240 (1968). (c) K. W. Ratts and A. N. Yao, *ibid.*, **31**, 1185, 1689 (1966). (d) K. W. Ratts, *Tetrahedron Letters*, **No. 39**, 4707 (1966). (e) W. E. Truce and G. D. Madding, *ibid.*, **No. 31**, 3681 (1966). (f) P. Robson, P. R. H. Speakman, and D. G. Stewart, *J. Chem. Soc.*, C, 2180 (1968). (g) A. J. Speziale, C. C. Tung, K. W. Ratts, and A. Yao, *J. Am. Chem. Soc.*, **87**, 3460 (1965). (h) B. M. Trost, *ibid.*, **89**, 138 (1967). (i) G. B. Payne, *J. Org. Chem.*, **32**, 3351 (1967); **33**, 1284 (1968); G. B. Payne and M. R. Johnson, *ibid.*, **33**, 1285 (1968). (j) J. Adams, L. Hoffman, Jr., and B. M. Trost, *ibid.*, **35**, 1600 (1970). (k) W. J. Middleton, E. L. Buhle, J. G. McNally, Jr., and M. Zanger, *ibid.*, **30**, 2384 (1965).

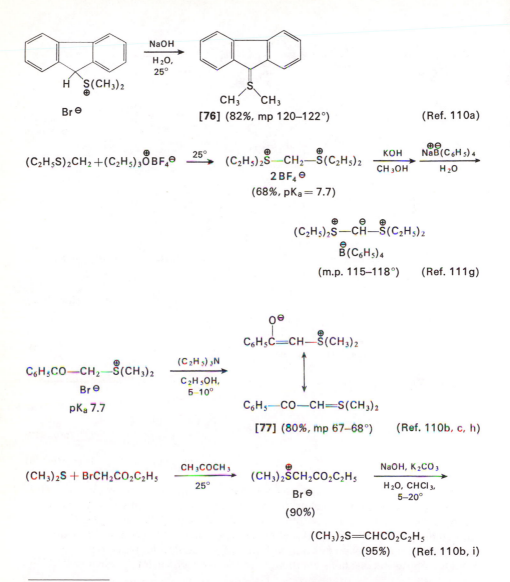

[76] (82%, mp 120–122°)   (Ref. 110a)

$(C_2H_5S)_2CH_2 + (C_2H_5)_3\overset{\oplus}{O}BF_4^{\ominus} \xrightarrow{25°} (C_2H_5)_2\overset{\oplus}{S}-CH_2-\overset{\oplus}{S}(C_2H_5)_2 \xrightarrow[CH_3OH]{KOH} \xrightarrow[H_2O]{Na\overset{\oplus\ominus}{B}(C_6H_5)_4}$

2 BF$_4^{\ominus}$

(68%, pK$_a$ = 7.7)

$(C_2H_5)_2\overset{\oplus}{S}-\overset{\ominus}{C}H-\overset{\oplus}{S}(C_2H_5)_2$

$\overset{\ominus}{B}(C_6H_5)_4$

(m.p. 115–118°)   (Ref. 111g)

$C_6H_5CO-CH_2-\overset{\oplus}{S}(CH_3)_2 \xrightarrow[C_2H_5OH,\ 5-10°]{(C_2H_5)_3N}$

Br$^{\ominus}$

pK$_a$ 7.7

$C_6H_5\overset{O^{\ominus}}{\overset{|}{C}}=CH-\overset{\oplus}{S}(CH_3)_2$

$\updownarrow$

$C_6H_5-CO-CH=S(CH_3)_2$

[77] (80%, mp 67–68°)   (Ref. 110b, c, h)

$(CH_3)_2S + BrCH_2CO_2C_2H_5 \xrightarrow[25°]{CH_3COCH_3} (CH_3)_2\overset{\oplus}{S}CH_2CO_2C_2H_5 \xrightarrow[H_2O,\ CHCl_3,\ 5-20°]{NaOH,\ K_2CO_3}$

Br$^{\ominus}$

(90%)

$(CH_3)_2S=CHCO_2C_2H_5$

(95%)   (Ref. 110b, i)

111. (a) H. Nozaki, M. Takaku, and K. Kondo, *Tetrahedron,* **22**, 2145 (1966). (b) H. Nozaki, D. Tunemoto, S. Matubara, and K. Kondo, *ibid.,* **23**, 545 (1967). (c) H. Nozaki, D. Tunemoto, Z. Morita, K. Nakamura, K. Watanabe, M. Takaku, and K. Kondo, *ibid.,* **23**, 4279 (1967). (d) M. Takaku, Y. Hayasi, and H. Nozaki, *ibid.,* **26**, 1243 (1970). (e) *Tetrahedron Letters,* **No. 24,** 2303 (1967); **No. 37,** 3179 (1969). (f) A. F. Cook and J. G. Moffatt, *J. Am. Chem. Soc.,* **90,** 740 (1968). (g) C. P. Lillya and P. Miller, *ibid.,* **88,** 1559, 1560 (1966); C. P. Lillya and E. F. Miller, *Tetrahedron Letters,* **No. 11,** 1281 (1968). (h) J. J. Tufariello, L. T. C. Lee, and P. Wojtkowski, *ibid.,* **89,** 6804 (1967). (i) J. T. Lumb, *Tetrahedron Letters,* **No. 8,** 579 (1970). (j) J. I. DeGraw and M. Cory, *ibid.,* **No. 20,** 2501 (1968). (k) S. H. Smallcombe, R. J. Holland, R. H. Fish, and M. C. Caserio, *ibid.,* **No. 57,** 5987 (1968). (l) R. H. Fish, L. C. Chow, and M. C. Caserio, *ibid.,* **No. 16,** 1259 (1969).

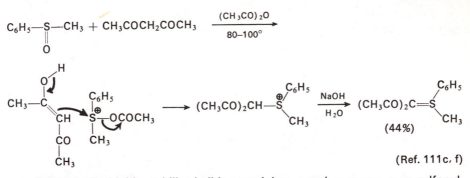

(Ref. 111c, f)

Although the highly stabilized ylids containing α-acyl, α-cyano, or α-sulfonyl substituents may fail to react even with the carbonyl function of aldehydes,[101b-f,h,k] the more reactive stabilized ylids (with α-aryl, α-carboxylate ion, or α-carboxamide substituents) react with aldehydes and the alkylidenesulfuranes react with both aldehydes and ketones to form epoxides. As shown in the following equations, this reaction is believed to proceed by the initial reversible formation of a betaine intermediate (e.g., [78]) as in the Wittig reaction. Subsequent intramolecular displacement is thought to produce the epoxide product. The predominant formation of the *trans*-epoxide in such cases is thought to result from the reversibility of the betaine formation. Closure of the betaine [78b] to form the *trans*-epoxide does not require eclipsing of the large aryl groups on adjacent carbons as would be the case in the closure of betaine [78a] to the *cis*-epoxide. It should be noted that the alkylidenesulfuranes also react readily with imines and nitroso compounds[107a,b] as well as trialkylboranes.[111h]

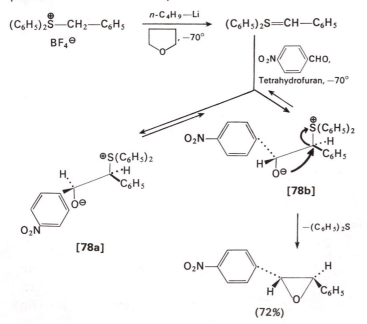

(72%)

(Ref. 109a)

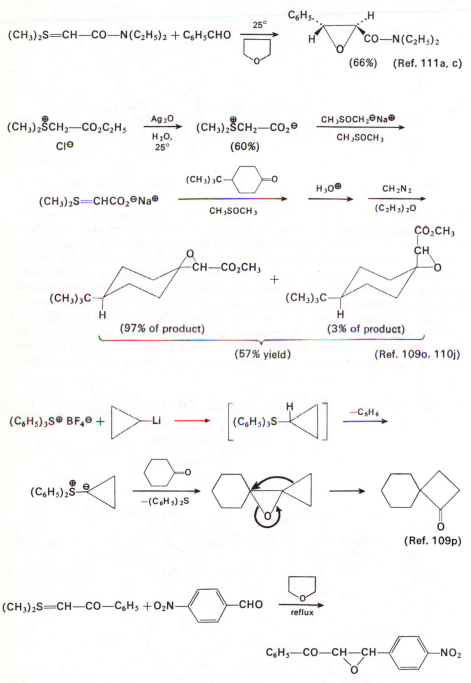

$(CH_3)_2S{=}CH{-}CO{-}N(C_2H_5)_2 + C_6H_5CHO \xrightarrow[\ ]{25°}$

(66%)   (Ref. 111a, c)

$(CH_3)_2\overset{\oplus}{S}CH_2{-}CO_2C_2H_5 \xrightarrow[H_2O,\ 25°]{Ag_2O} (CH_3)_2\overset{\oplus}{S}CH_2{-}CO_2^{\ominus} \xrightarrow[CH_3SOCH_3]{CH_3SOCH_2{}^{\ominus}Na^{\oplus}}$
$Cl^{\ominus}$ (60%)

$(CH_3)_2S{=}CHCO_2^{\ominus}Na^{\oplus} \xrightarrow[CH_3SOCH_3]{(CH_3)_3C{-}\bigcirc{=}O} \xrightarrow{H_3O^{\oplus}} \xrightarrow[(C_2H_5)_2O]{CH_2N_2}$

(97% of product)   +   (3% of product)

(57% yield)   (Ref. 109o, 110j)

$(C_6H_5)_3S^{\oplus} BF_4^{\ominus} + \triangleright{-}Li \longrightarrow \left[(C_6H_5)_3S{-}\triangleright\right] \xrightarrow{-C_6H_6}$

$(C_6H_5)_2\overset{\oplus}{S}{-}\overset{\ominus}{\triangleleft} \xrightarrow[-(C_6H_5)_2S]{} \longrightarrow$

(Ref. 109p)

$(CH_3)_2S{=}CH{-}CO{-}C_6H_5 + O_2N{-}\bigcirc{-}CHO \xrightarrow[reflux]{}$

$C_6H_5{-}CO{-}CH{-}CH{-}\bigcirc{-}NO_2$

(10%)   (Ref. 110b)

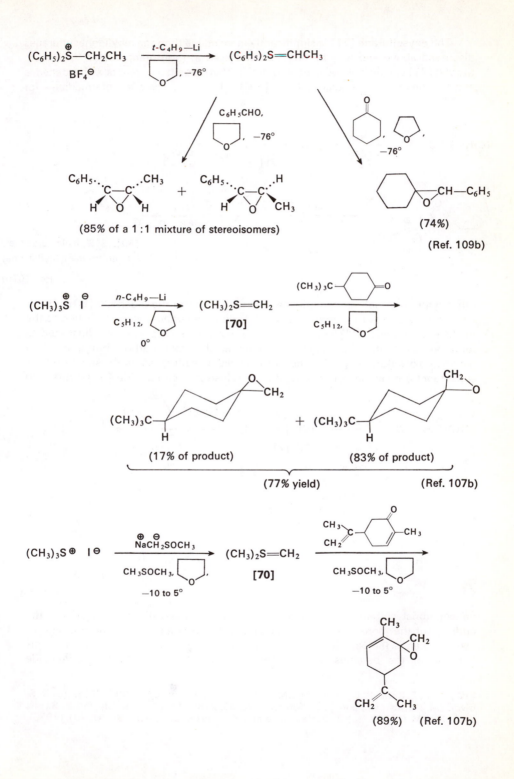

(85% of a 1:1 mixture of stereoisomers)

(74%)

(Ref. 109b)

(17% of product)

(83% of product)

(77% yield)

(Ref. 107b)

[70]

[70]

(89%)   (Ref. 107b)

The oxysulfurane [71] is significantly more stable than the alkylidenesulfuranes discussed above and may be prepared and used at room temperature; if desired, the ylid [71] can be isolated in pure form.[107f] This reagent also reacts with aldehydes and ketones to form epoxides (e.g., [79]) which may serve as intermediates for

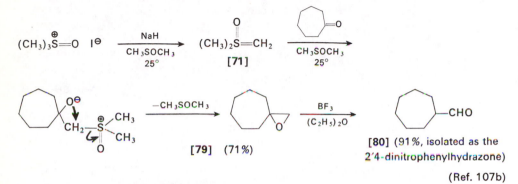

[80] (91%, isolated as the
2′4-dinitrophenylhydrazone)

(Ref. 107b)

other syntheses such as the indicated rearrangement to the aldehyde [80]. However, the reactions of the two types of ylids differ in certain respects. Whereas the methylenesulfurane [70] reacts with cyclohexanones as previously illustrated to form predominantly the epoxide with an axial carbon-carbon bond, the less reactive oxysulfurane [71] (and also alkylidenesulfuranes with stabilizing α-substituents) gives predominantly, if not exclusively, the epoxide [81] containing

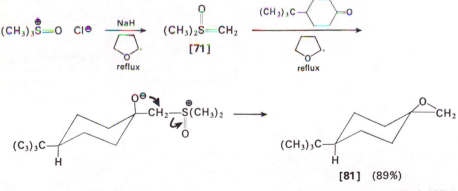

[81]   (89%)

(Ref. 107b)

an equatorial carbon-carbon bond.[107b,112] These observations suggest that the addition of the more stable ylid [71] to a carbonyl compound is more easily reversible than is the case for the ylid [70]. In this event, partial or complete equilibration of the betaines derived from the oxysulfurane [71] before epoxide

112. (a) C. E. Cook, R. C. Corley, and M. E. Wall, *J. Org. Chem.*, **33**, 2789 (1968). (b) R. S. Bly, C. M. DuBose, Jr., and G. B. Konizer, *ibid.*, **33**, 2188 (1968); R. K. Bly and R. S. Bly, *ibid.*, **34**, 304 (1969). (c) J. D. Ballantine and P. J. Sykes, *J. Chem. Soc.*, C, 731 (1970).

formation would favor the betaine indicated below with the larger substituent on an equatorial bond unless this arrangement is opposed by hindrance from neighboring substituents.[112c] Although the highly reactive methylenesulfurane sometimes reacts even with α,β-unsaturated ketones to form epoxides as illustrated in an earlier equation, reaction of the oxysulfurane [71] with α,β-unsaturated ketones, (e.g., [82]) and other conjugated olefins usually yields the corresponding cyclopropyl ketone as the major product. This reaction, which is believed to involve the indicated initial conjugate addition of the ylid to the unsaturated system followed by ring closure, is the more general reaction for all types of sulfur ylids with α,β-unsaturated carbonyl compounds.[113,114] Thus even the highly reactive isopropylidene ylid [75] reacts with α,β-unsaturated ketones to form cyclopropanes[109e]

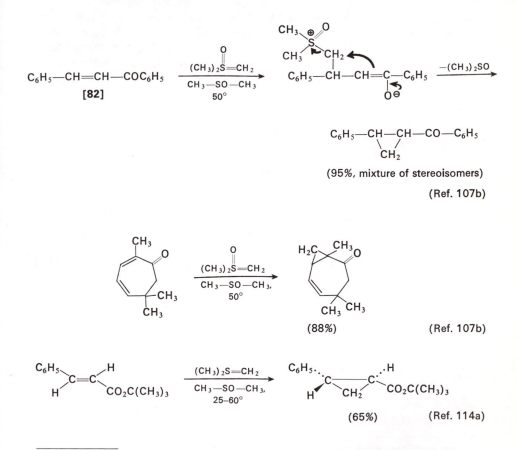

(95%, mixture of stereoisomers)

(Ref. 107b)

(88%)   (Ref. 107b)

(65%)   (Ref. 114a)

113. (a) N. H. Dyson, J. A. Edwards, and J. H. Fried, *Tetrahedron Letters,* **No. 17,** 1841 (1966). (b) H. van Kemp, P. Nissen, and E. van Vliet, *ibid.,* **No. 16,** 1457 (1967). (c) G. W. Krakower and H. A. Van Dine, *J. Org. Chem.,* **31,** 3467 (1966). (d) H. Nozaki, H. Ito, D. Tunemoto, and K. Kondo, *Tetrahedron,* **22,** 441 (1966). (e) S. R. Landor and N. Punja, *J. Chem. Soc.,* C, 2495 (1967). (f) C. Agami, C. Prevost, and J. Aubouet, *Bull. Soc. Chim. France,* 2299 (1967). (g) T. Kunieda and B. Witkop, *J. Am. Chem. Soc.,* **91,** 7751 (1969).

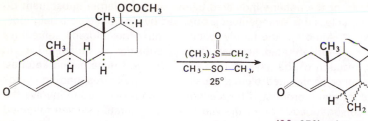

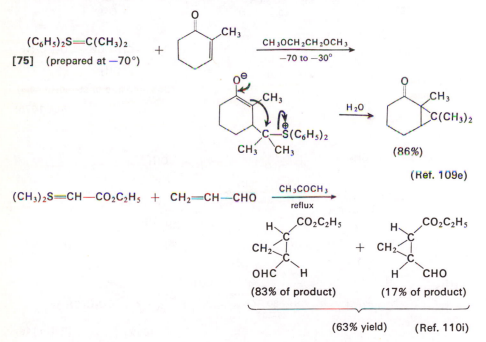

(32–65%, mixture of
approximately equal
amounts of two
stereoisomers)   (Ref. 113a)

and other alkylidenesulfuranes with stabilizing substituents react with α,β-unsaturated ketones and related compounds to form the cyclopropane derivatives.[110h,i,111b,e,114d,115] These reactions are not stereospecific and usually favor the production of the cyclopropane diastereoisomer which has less steric interference between substituents at adjacent carbons.[110i,114a]

(83% of product)   (17% of product)

                    (63% yield)   (Ref. 110i)

114. (a) C. Kaiser, B. M. Trost, J. Beeson, and J. Weinstock, *J. Org. Chem.,* **30,** 3972 (1965). (b) J. Ide and Y. Kishida, *Tetrahedron Letters,* **No. 16,** 1787 (1966). (c) P. T. Izzo, *J. Org. Chem.,* **28,** 1713 (1963). (d) J. Casanova, Jr., and D. A. Rutalo, Jr., *Chem. Commun.,* **No. 23,** 1224 (1967).
115. (a) V. Franzen and H. E. Driesen, *Chem. Ber.,* **96,** 1881 (1963). (b) W. E. Truce and V. V. Badiger, *J. Org. Chem.,* **29,** 3277 (1964). (c) W. E. Truce and C. T. Goralski, *ibid.,* **33,** 3849 (1968).

Cyclopropane derivatives have also been synthesized by the indicated reaction of vinylsulfonium salts with active methylene compounds.[116] This process involves a Michael addition of nucleophiles to the double bond of the sulfonium salts in a process which is analogous to the previously discussed additions of nucleophiles to vinylphosphonium salts.

$$C_6H_5CH{=}CH{-}SCH_3 \xrightarrow[90-100°]{(CH_3O)_2SO_2} C_6H_5{-}CH{=}CH{-}\overset{\oplus}{S}(CH_3)_2 \xrightarrow[C_2H_5OH,\ 0°]{NaCH(CO_2C_2H_5)_2}$$

$$CH_3OSO_3^{\ominus}$$
$$(60{-}80\%)$$

$$\underset{(C_2H_5O_2C)_2\overset{|}{C}H}{C_6H_5\overset{|}{C}H{-}\overset{\ominus}{C}H{-}\overset{\oplus}{S}(CH_3)_2} \underset{transfer}{\overset{proton}{\rightleftarrows}}$$

$$\underset{(C_2H_5O_2C)_2\overset{|}{C}{\ominus}}{C_6H_5{-}\overset{|}{C}H{-}CH_2{-}\overset{\oplus}{S}(CH_3)_2} \longrightarrow \underset{\overset{|}{C}H_2}{C_6H_5{-}\overset{|}{C}H{-}C(CO_2C_2H_5)_2}$$
$$(60\%)$$

(Ref. 116b)

It should be noted that the methyleneoxysulfurane [71] may react with ketones having relatively acidic C—H bonds (e.g., desoxybenzoin and $\varDelta^4$-cholesten-3-one) to form enolate anions from which the starting ketone is recovered after hydrolysis.[107b] This oxygenated ylid has also been found to react with imines, nitroso compounds, nitro compounds, nitriles, and esters.[107] Although the reaction of the oxysulfurane [71] with saturated esters is a slow process,[114a,117a] this ylid reacts with the more reactive phenyl esters or with acid chlorides to form β-ketosulfonium ylids, (e.g., [83]).[117a,c]

It will be noted that the C-acylation with phenyl esters (e.g., [84]), is more rapid than conjugate addition of the ylid [71] to form a cyclopropane; however, with alkyl esters such as [85] conjugate addition is more rapid[114b,117a] and is followed by an intramolecular C-acylation to form a cyclic ylid [86]. Subsequent cleavage of these β-keto ylids [86] by a dissolving metal reduction (see Chapter 3)

116. (a) M. C. Caserio, R. E. Pratt, and R. J. Holland, *J. Am. Chem. Soc.,* **88**, 5747 (1966). (b) J. Gosselck, L. Beress, H. Schenk, and G. Schmidt, *Angew. Chem., Intern. Ed. Engl.,* **4**, 1080 (1965); J. Gosselck, L. Beress, and H. Schenk, *ibid.,* **5**, 596 (1966). (c) J. Gosselck, G. Schmidt, L. Beress, and H. Schenk, *Tetrahedron Letters,* **No. 3**, 331 (1968). (d) G. Schmidt and J. Gosselck, *ibid.,* **No. 39**, 3445 (1969). (e) For a related synthesis of cyclopropyl derivatives by the reaction of α-halo carbonyl compounds with methyleneoxysulfurane [71], see P. Bravo, G. Gaudiano, C. Ticozzi, and A. Umani-Ronchi, *ibid.,* **No. 43**, 4481 (1968).
117. (a) E. J. Corey and M. Chaykovsky, *J. Am. Chem. Soc.,* **86**, 1640 (1964). (b) A. Winkler and J. Gosselck, *Tetrahedron Letters,* **No. 48**, 4229 (1969). (c) These β-keto sulfur ylids like the previously discussed stabilized phosphorus ylids, have an appreciable energy barrier to rotation about the bond between the alpha and beta carbon atoms; K. Kondo and D. Tunemoto, *Chem. Commun.,* **No. 20**, 1361 (1970).

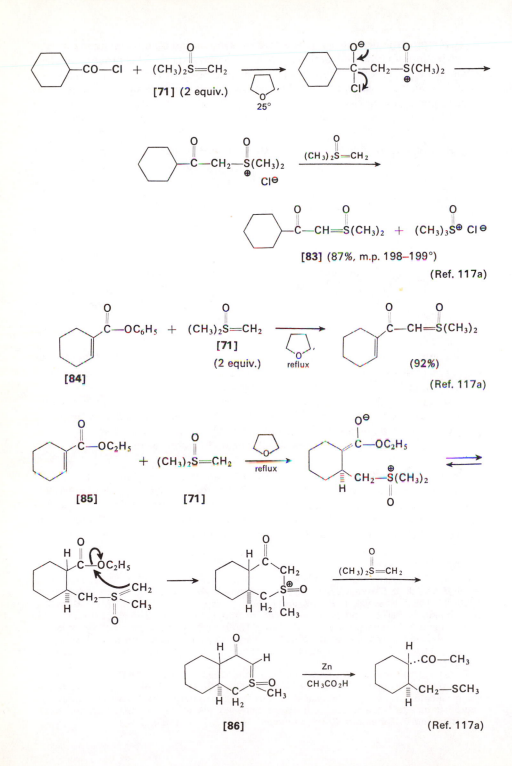

[71] (2 equiv.)

25°

[83] (87%, m.p. 198–199°)

(Ref. 117a)

[84]

[71]

(2 equiv.)   reflux

(92%)

(Ref. 117a)

[85]   [71]

(Ref. 117a)

[86]

(Ref. 117a)

is worthy of note as is the photochemically induced rearrangement of these β-keto sulfoxonium ylids to form ketones.[117a]

Stabilized sulfur ylids may also be alkylated and acylated.[110b,c,111a,d,117b] As in the case of phosphorus ylids stabilized by α-acyl substituents,[96e] the α-acyl-sulfur ylids react with acid chlorides in a kinetically controlled reaction to yield O-acyl products. The slower reaction of these sulfur ylids with acid anhydrides leads to the formation of C-acylated products, possibly by way of the intermediate enol esters as illustrated. Only C-acylated products have been found from the acylation of α-carboalkoxy ylids.[110b]

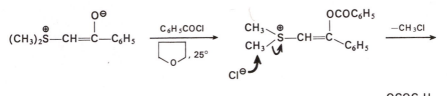

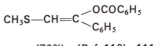

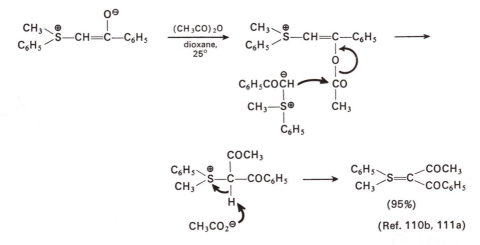

Esters undergo a C-acylation reaction with the anions [72], [73], and [87] derived from dimethyl sulfoxide,[108a,b,d,h,118a] from dimethyl sulfone,[108a,b,h,118b] from N,N-dimethylmethanesulfonamide,[108a] and from certain sulfinamides.[118c] These β-keto sulfur compounds (e.g., [88]), may be either hydrolyzed[118c] or

118. (a) E. J. Corey and M. Chaykovsky, *J. Am. Chem. Soc.,* **86**, 1639 (1964). (b) H. O. House and J. K. Larson, *J. Org. Chem.,* **33**, 61 (1968). (c) E. J. Corey and T. Durst, *J. Am. Chem. Soc.,* **88**, 5656 (1966); **90**, 5548, 5553 (1968). (d) G. J. Mikol and G. A. Russell, *Org. Syn.,* **48**, 109 (1968). (e) For an example of a stereospecific Pummerer rearrangement, see S. Glue, I. T. Kay, and M. R. Kipps, *Chem. Commun.,* **No. 18**, 1158 (1970).

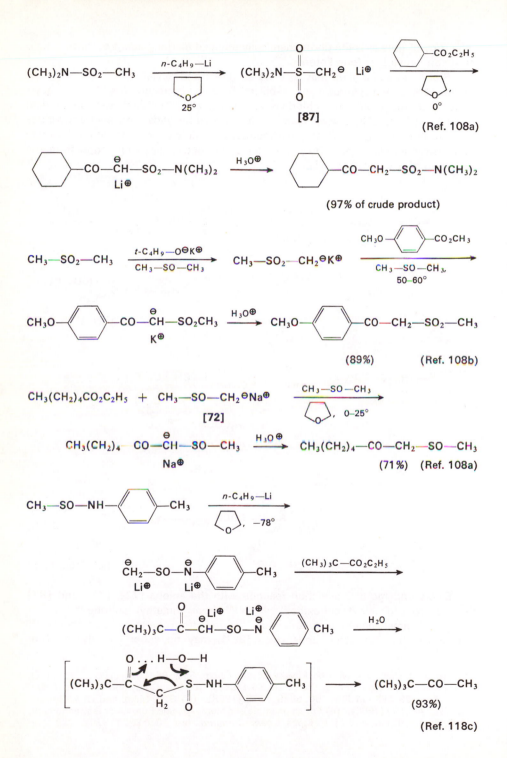

$(CH_3)_2N—SO_2—CH_3$ $\xrightarrow[\substack{\text{O}\\25°}]{n\text{-}C_4H_9—Li}$ $(CH_3)_2N—\overset{\text{O}}{\underset{\text{O}}{\overset{\|}{\underset{\|}{S}}}}—CH_2^{\ominus}$ $Li^{\oplus}$

[87]

(Ref. 108a)

(97% of crude product)

$CH_3—SO_2—CH_3$ $\xrightarrow[CH_3—SO—CH_3]{t\text{-}C_4H_9—O^{\ominus}K^{\oplus}}$ $CH_3—SO_2—CH_2^{\ominus}K^{\oplus}$

(89%)   (Ref. 108b)

$CH_3(CH_2)_4CO_2C_2H_5 + CH_3—SO—CH_2^{\ominus}Na^{\oplus}$

[72]

$CH_3(CH_2)_4—\overset{\ominus}{CO}—CH—SO—CH_3$ $\xrightarrow{H_3O^{\oplus}}$ $CH_3(CH_2)_4—CO—CH_2—SO—CH_3$
   $Na^{\oplus}$

(71%)   (Ref. 108a)

$CH_3—SO—NH—\langle\text{ring}\rangle—CH_3$ $\xrightarrow[\substack{\text{O},\ -78°}]{n\text{-}C_4H_9—Li}$

$\overset{\ominus}{CH_2}—SO—\overset{\ominus}{N}—\langle\text{ring}\rangle—CH_3$
 $Li^{\oplus}$ $Li^{\oplus}$

$\xrightarrow{(CH_3)_3C—CO_2C_2H_5}$

$(CH_3)_3C—\overset{O}{\overset{\|}{C}}—\overset{\ominus}{CH}—SO—\overset{\ominus}{N}—\langle\text{ring}\rangle—CH_3$ $^{\ominus}Li^{\oplus}$ $Li^{\oplus}$ $\xrightarrow{H_2O}$

$\left[(CH_3)_3C—\overset{O\ldots H—O—H}{\overset{\|}{C}}\underset{H_2}{\overset{C}{\cdots}}S—NH—\langle\text{ring}\rangle—CH_3\right]$ $\longrightarrow$ $(CH_3)_3C—CO—CH_3$

(93%)

(Ref. 118c)

reductively cleaved by reaction with aluminum amalgam or zinc dust to form the corresponding methyl ketone.[108a,f,i,118,119]

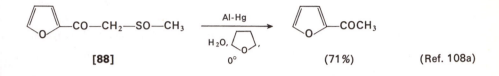

[88]   (71%)   (Ref. 108a)

$$CH_3(CH_2)_{16}-CO-CH_2-SO_2-CH_3 \xrightarrow[\substack{H_2O, \\ 65°}]{Al-Hg} CH_3(CH_2)_{16}-CO-CH_3$$

(81%)

(Ref. 118b)

Although the β-keto sulfoxides offer the advantage of apparently being cleaved more rapidly in the reduction step, the β-keto sulfones possess the advantages that they are more stable to oxidation and reduction[108d,e,118b] and that the introduction of a methanesulfonyl function does not introduce a center of asymmetry into the intermediate β-keto compound.[118b]

Being relatively acidic active methylene compounds the β-keto sulfoxides[108f,i,119] and the β-keto sulfones[118b,120] are readily alkylated at the central carbon atom to produce alkylated β-keto sulfoxide or sulfone derivatives which can be cleaved to ketones.[108f,118b,119] The overall result, illustrated in the following equation, is a ketone synthesis comparable to the preparation of ketones from β-keto esters (see Chapter 9).

$$C_6H_5-CO-CH_2-SO-CH_3 \xrightarrow[\substack{CH_3-SO-CH_3 \\ 25°}]{NaH}$$

$$C_6H_5-CO-\overset{\ominus}{C}H-SO-CH_3 \xrightarrow[CH_3SO-CH_3]{C_2H_5I}$$

Na$^{\oplus}$

$$C_6H_5-CO-\overset{\overset{\displaystyle C_2H_5}{|}}{C}H-SO-CH_3 \xrightarrow[\substack{H_2O, \\ 25°}]{Al-Hg} C_6H_5-COCH_2CH_2CH_3$$

(70%)

(Ref. 119a)

119. (a) P. G. Gassman and G. D. Richmond, *J. Org. Chem.*, **31**, 2355 (1966). (b) J. B. Lee, *Tetrahedron Letters*, **No. 46**, 5669 (1966). (c) J. B. Siddall, A. D. Cross, and J. H. Fried, *J. Am. Chem. Soc.*, **88**, 862 (1966).
120. (a) W. I. O'Sullivan, D. F. Tavares, and C. R. Hauser, *J. Am. Chem. Soc.*, **83**, 3453 (1961). (b) M. L. Miles and C. R. Hauser, *J. Org. Chem.*, **29**, 2329 (1964). (c) N. M. Carroll and W. I. O'Sullivan, *ibid.*, **30**, 2830 (1965).

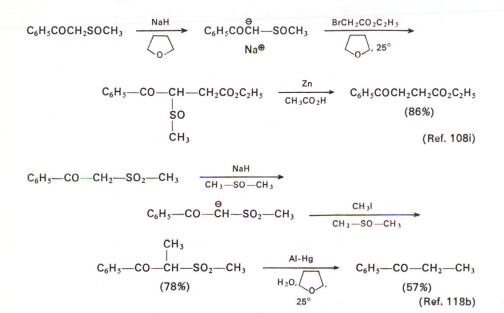

The β-keto sulfoxides may also serve as synthetic intermediates for other transformations.[108d,e,g,i,118d,e] For example, reaction of β-keto sulfoxides (e.g., [89]), with aqueous acids or other acidic reagents results in isomerization (the Pummerer rearrangement) of the starting sulfoxide to form the hemimercaptal (or its derivative) of an α-keto aldehyde.[108d,f,i,118d,e] A possible mechanism for this transformation is indicated in the accompanying equations.

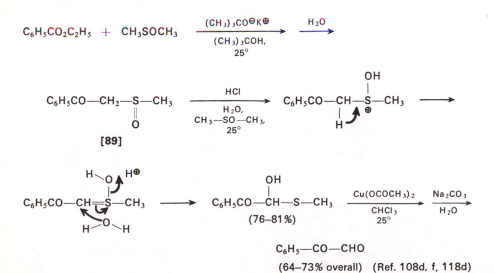

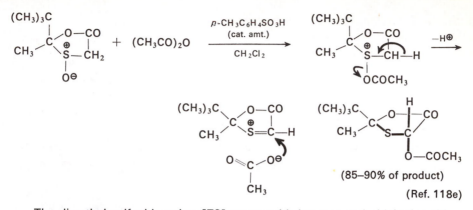

(85–90% of product)

(Ref. 118e)

The dimethyl sulfoxide anion [72] reacts with ketones and aldehydes either to abstract an alpha proton forming an enolate anion or to add to the carbonyl function in an aldol condensation.[108a,c,118c,121] The β-hydroxy sulfoxide (e.g., [90]), formed by this latter process is capable of undergoing a variety of further changes.[108c,121] Other examples of aldol condensations with carbanions stabilized

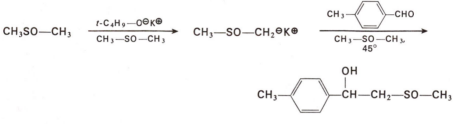

[90] (40%)   (Ref. 108c)

by sulfoxide and sulfinamide functions are presented below. The use of the aldol product from the sulfinamide for the preparation of olefins is worthy of note; however, in most cases this procedure appears to offer no advantage over the Wittig reaction.

C_6H_5—SO—CH_2—Cl $\xrightarrow[\text{, } -78°]{n\text{-}C_4H_9Li}$

C_6H_5—SO—$\overset{\ominus}{\underset{Li^{\oplus}}{CH}}$—Cl $\xrightarrow[\text{, } -78 \text{ to } -20°]{CH_3COCH_3}$ $\xrightarrow{H_2O}$

C_6H_5SO—CH—$\underset{Cl}{\overset{OH}{C}}(CH_3)_2$ $\xrightarrow[CH_3OH,\ 25°]{KOH}$ C_6H_5SO—CH—C$(CH_3)_2$ with O bridge

(75%, only one diastereoisomer)

(90%)   (Ref. 121c)

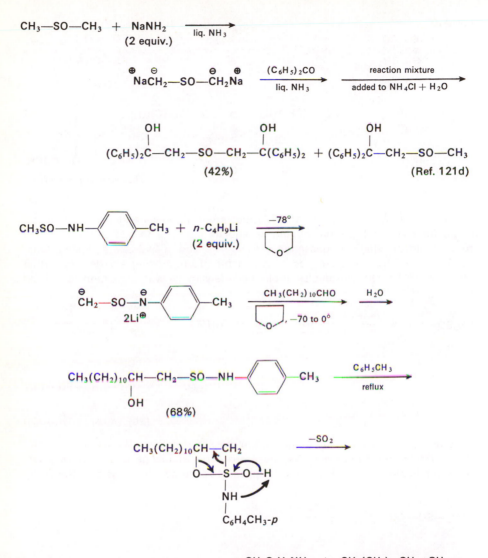

p-CH$_3$C$_6$H$_4$NH$_2$ + CH$_3$(CH$_2$)$_{10}$CH=CH$_2$

(83%) (Ref. 118c)

121. (a) M. Chaykovsky and E. J. Corey, *J. Org. Chem.*, **28**, 254 (1963). (b) C. Walling and L. Bollyky, *ibid.*, **28**, 256 (1963). (c) T. Durst, *J. Am. Chem. Soc.*, **91**, 1034 (1969). (d) E. M. Kaiser and R. D. Beard, *Tetrahedron Letters*, **No. 21**, 2583 (1968). (e) J. A. Gautier, M. Miocque, M. Plat, H. Moskowitz, and J. Blanc-Guenee, *ibid.*, **No. 21**, 895 (1970). (f) Since the unsymmetrically substituted sulfoxides are asymmetric, the two hydrogens at an α-methylene group are in different environments and their rates of exchange with deuterium are found to differ. S. Wolfe and A. Rauk, *Chem. Commun.*, **No. 21**, 778 (1966).

The carbanions stabilized by an adjacent sulfone function (e.g., **[73]**) are much weaker (and more stable) bases than those derived from the corresponding sulfoxides. These anions are useful both for aldol condensations and for alkylation reactions as the following equations indicate.[122]

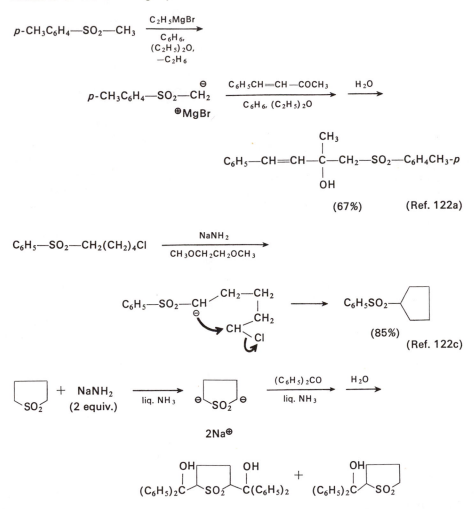

p-CH$_3$C$_6$H$_4$—SO$_2$—CH$_3$ $\xrightarrow[\substack{C_6H_6, \\ (C_2H_5)_2O, \\ -C_2H_6}]{C_2H_5MgBr}$

p-CH$_3$C$_6$H$_4$—SO$_2$—CH$_2^{\ominus}$ $\oplus$MgBr $\xrightarrow[C_6H_6, (C_2H_5)_2O]{C_6H_5CH=CH—COCH_3}$ $\xrightarrow{H_2O}$

$$C_6H_5—CH=CH—\overset{\overset{\displaystyle CH_3}{|}}{\underset{\underset{\displaystyle OH}{|}}{C}}—CH_2—SO_2—C_6H_4CH_3\text{-}p$$

(67%)   (Ref. 122a)

C_6H_5—SO$_2$—CH$_2$(CH$_2$)$_4$Cl $\xrightarrow[CH_3OCH_2CH_2OCH_3]{NaNH_2}$

C_6H_5—SO$_2$—CH → $C_6H_5SO_2$—⬠

(85%)

(Ref. 122c)

⬠SO$_2^-$ + NaNH$_2$ (2 equiv.) $\xrightarrow[liq. NH_3]{}$ ⬠ $^\ominus$SO$_2^\ominus$ 2Na$^\oplus$ $\xrightarrow[liq. NH_3]{(C_6H_5)_2CO}$ $\xrightarrow{H_2O}$

$$(C_6H_5)_2\overset{\overset{\displaystyle OH}{|}}{C}\text{⬠}SO_2\overset{\overset{\displaystyle OH}{|}}{C}(C_6H_5)_2 \quad + \quad (C_6H_5)_2\overset{\overset{\displaystyle OH}{|}}{C}\text{⬠}SO_2$$

(41%)   (48%)   (Ref. 122f)

122. (a) J. W. McFarland and D. N. Buchanan, *J. Org. Chem.,* **30,** 2003 (1965); J. W. McFarland and G. N. Coleman, *ibid.,* **35,** 1194 (1970). (b) W. E. Truce and T. C. Klingler, *ibid.,* **35,** 1834 (1970) and references therein. (c) W. E. Truce, K. R. Hollister, L. B. Lindy, and J. E. Parr, *ibid.,* **33,** 43 (1968). (d) W. E. Truce and D. J. Vrencur, *ibid.,* **35,** 1226 (1970). (e) Y. Shirota, T. Nagai, and N. Tokura, *Tetrahedron,* **25,** 3193 (1969). (f) E. M. Kaiser and C. R. Hauser, *Tetrahedron Letters,* **No. 34,** 3341 (1967). (g) W. E. Truce and L. W. Christensen, *ibid.,* **No. 36,** 3075 (1969).

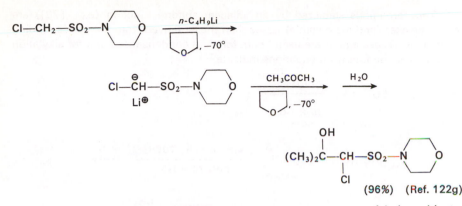

$$Cl-CH_2-SO_2-N \overset{n-C_4H_9Li}{\underset{\overset{\displaystyle O}{\displaystyle\quad}, -70°}{\longrightarrow}}$$

$$Cl-\overset{\ominus}{\underset{Li^{\oplus}}{CH}}-SO_2-N \overset{CH_3COCH_3}{\underset{\overset{\displaystyle O}{\displaystyle\quad}, -70°}{\longrightarrow}} \overset{H_2O}{\longrightarrow}$$

$$(CH_3)_2\overset{OH}{\underset{}{C}}-\underset{\overset{\displaystyle |}{\displaystyle Cl}}{CH}-SO_2-N \qquad O$$

(96%)   (Ref. 122g)

The α-sulfonyl carbanions exhibit the interesting property of being able to maintain asymmetry when they are generated from a sulfone with an asymmetric α-carbon atom.[123] The following equation illustrates a typical example in which hydrogen-deuterium exchange at the alpha carbon atom has occurred with retention of configuration. These results are explained in terms of the intermediate carbanion [91] either being pyramidal with a substantial energy barrier to inversion or being planar with a substantial energy barrier to rotation of the α-C—S bond.

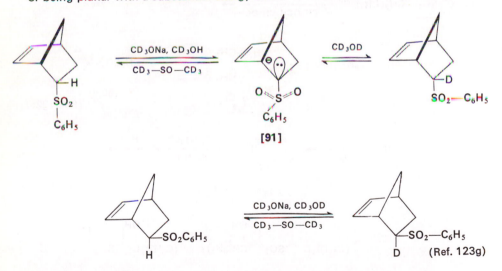

[91]

(Ref. 123g)

123. (a) D. J. Cram, *Fundamentals of Carbanion Chemistry*, Academic Press, New York, 1965, pp. 105–113. (b) R. R. Fraser and F. J. Schuber, *Chem. Commun.*, **No. 24**, 1474 (1969). (c) E. J. Corey, H. König, and T. H. Lowry, *Tetrahedron Letters*, **No. 12**, 515 (1962). (d) E. J. Corey and T. H. Lowry, *ibid.*, **No. 13**, 793, 803 (1965). (e) D. J. Cram, R. D. Trepka, and P. St. Janiak, *J. Am. Chem. Soc.*, **88**, 2749 (1966). (f) D. J. Cram and T. A. Whitney, *ibid.*, **89**, 4651 (1967). (g) G. Maccagnani, F. Montanari, and F. Taddei, *J. Chem. Soc.*, B, 453 (1968). (h) S. Wolfe, A. Rauk, and I. G. Csizmadia, *J. Am. Chem. Soc.*, **91**, 1567 (1969). (i) F. G. Bordwell, E. Doomes, and P. W. R. Corfield, *ibid.*, **92**, 2581 (1970).

Reactions related to the foregoing syntheses with sulfur ylids occur with the lithium salts of 1,3-dithiane derivatives.[124,125] Reaction of these salts (e.g., [92]), with primary and secondary alkyl halides leads to the indicated formation of dithio acetals or dithioketals of aldehydes or ketones which can be hydrolyzed to the corresponding aldehyde or ketone. These anions have been found to react with

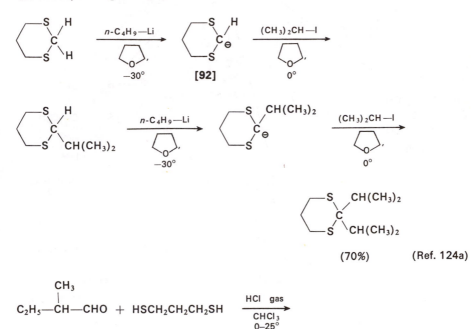

[92]

(70%)   (Ref. 124a)

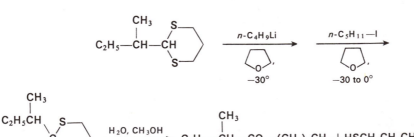

(51% overall)   [precipitated as the Hg (II) salt]

(Ref. 124d)

124. (a) E. J. Corey and D. Seebach, *Angew. Chem., Intern. Ed. Engl.,* **4**, 1075, 1077 (1965). (b) D. Seebach, *ibid.,* **8**, 639 (1969). (c) *ibid.,* **6**, 442, 443 (1967). (d) D. Seebach and D. Steinmüller, *ibid.,* **7**, 619 (1968); D. Seebach, D. Steinmüller, and F. Demuth, *ibid.,* **7**, 620 (1968). (e) E. J. Corey, D. Seebach, and R. Freedman, *J. Am. Chem. Soc.,* **89**, 434 (1967). (f) E. J. Corey and D. Crouse, *J. Org. Chem.,* **33**, 298 (1968). (g) D. Seebach, N. R. Jones, and E. J. Corey, *ibid.,* **33**, 300 (1968).

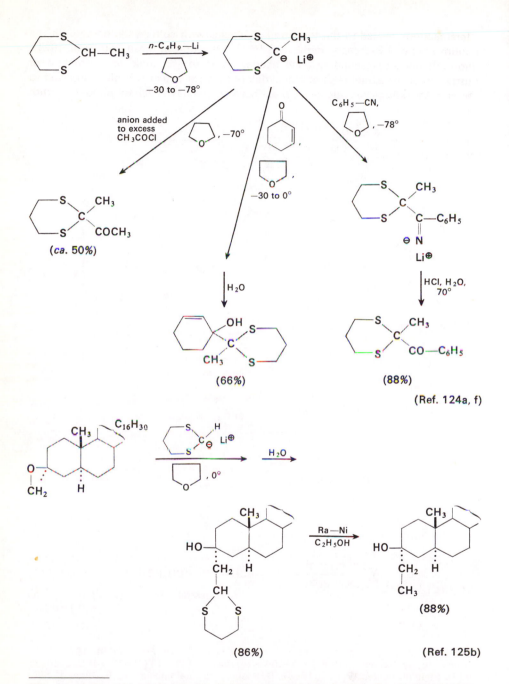

(*ca.* 50%)

(66%)

(88%)

(Ref. 124a, f)

(86%)

(88%)

(Ref. 125b)

125. (a) R. M. Carlson and P. M. Helquist, *Tetrahedron Letters*, **No. 3**, 173 (1969).
(b) J. B. Jones and R. Grayshan, *Chem. Commun.*, **No. 3**, 141 (1970) ; **No. 12**, 741 (1970).
(c) D. L. Coffen, T. E. McEntee, Jr., and D. R. Williams, *ibid.*, **No. 15**, 913 (1970).

esters and ketones to form either 2-acyl-1,3-dithianes or the corresponding tertiary alcohols. The acylated derivatives are more readily prepared by addition of the anion to an excess of an acyl halide or addition of the anion to a nitrile. The illustrated desulfurization of a 1,3-dithiane derivative with Raney nickel (see Chapter 1) constitutes another synthetically useful application of these intermediates.[125b]

Carbanions can also be stabilized by the presence of three or even one adjacent thioether grouping.[126] Two examples of the preparation and use of thiomethyllithium reagents are provided below.

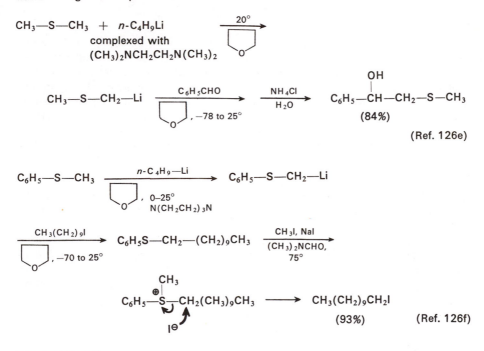

126. (a) S. Oae, W. Tagaki, and A. Ohno, *Tetrahedron*, **20**, 417, 427, 437, 443 (1964). (b) A. I. Shatenshtein and H. A. Gvozdeva, *ibid.*, **25**, 2749 (1969). (c) K. C. Bank and D. L. Coffen, *Chem. Commun.*, **No. 1**, 8 (1969). (d) S. Wolfe, A. Rauk, L. M. Tel, and I. G. Csizmadia, *ibid.*, **No. 2**, 96 (1970). (e) D. J. Peterson, *J. Org. Chem.*, **32**, 1717 (1967). (f) E. J. Corey and D. Seebach, *ibid.*, **31**, 4097 (1966); E. J. Corey and M. Jautelat, *Tetrahedron Letters*, **No. 55**, 5787 (1968).

11

ACYLATION AT CARBON

Reactions that result in the introduction of acyl groups at carbon include the base-catalyzed acylation of active methylene compounds; the acid-catalyzed acylation of active methylene compounds, olefins, and aromatic systems;[1a] and the acylation of certain organometallic compounds.[2] The commonly employed acylating agents are the following carboxylic acid derivatives, listed in order of increasing reactivity: $R-CO-NR_2 < R-CO-OH < R-CO-OC_2H_5 < R-CO-OC_6H_5 < R-CO-O-CO-R < R-CO-Cl$ and $R-CH=C=O$. Other acylating agents which may be employed include mixed anhydrides with carbonic acid ($R-CO-O-CO-OC_2H_5$) and phosphoric acid [$R-CO-O-\overset{\overset{O}{\parallel}}{P}(OC_2H_5)_2$],[1c] and the very powerful acylating agents such as the complexes of acid chlorides with Lewis acids (e.g., $Cl_2C=O^{\oplus}-Al^{\ominus}Cl_3$),[1d] and the acylium ion salts (e.g., $CH_3-C\equiv O^{\oplus} \leftrightarrow CH_3C^{\oplus}=O \; SbF_6^{\ominus}$).[1e,f,h]

Nitriles may also serve as acylating agents since the intermediate imines formed by the addition of a nucleophile to the nitrile function may be hydrolyzed to carbonyl functions. The nitrilium salts ($R-C\equiv N^{\oplus}-C_2H_5 \; BF_4^{\ominus}$) are much more reactive than nitriles toward attacking nucleophiles[1g] and may be useful for acylation at carbon. The discussion in this chapter will be restricted to the acylation of active methylene compounds, olefins and aromatic systems.

ACYLATION OF ACTIVE METHYLENE COMPOUNDS UNDER BASIC CONDITIONS

The principles involved in the acylation of an active methylene compound with an ester may be illustrated by the self-condensation of ethyl acetate [1], a reaction usually called the acetoacetic ester condensation.[3] When the condensation is

---

1. (a) D. P. N. Satchell, *Quart. Rev.,* **17**, 160 (1963). (b) E. Lindner, *Angew. Chem., Intern. Ed. Engl.,* **9**, 114 (1970). (c) D. S. Tarbell, *Accts. Chem. Res.,* **2**, 296 (1969); A. W. Friederang, D. S. Tarbell, and S. Ebine, *J. Org. Chem.,* **34**, 3825 (1969); M. A. Insalaco and D. S. Tarbell, *Org. Syn.,* **50**, 9 (1970). (d) Z. Iqbal and T. C. Waddington, *J. Chem. Soc.,* A, 1745 (1968). (e) F. P. Boer, *J. Am. Chem. Soc.,* **90**, 6706 (1968). (f) G. A. Olah and M. B. Comisarow, *ibid.,* **88**, 4442 (1966); **89**, 2694 (1967). (g) R. F. Borch, *Chem. Commun.,* No. 8, 442 (1968). (h) E. Lindner, *Angew. Chem., Intern. Ed. Engl.,* **9**, 114 (1970).
2. (a) D. A. Shirley, *Org. Reactions,* **8**, 28 (1954). (b) J. Cason, *Chem. Rev.,* **40**, 15 (1947). (c) M. S. Kharasch and O. Reinmuth, *Grignard Reactions of Nonmetallic Substances,* Prentice-Hall, Englewood Cliffs, N.J., 1954.
3. C. R. Hauser and B. E. Hudson, Jr., *Org. Reactions,* **1**, 266 (1942).

catalyzed by relatively weak bases such as sodium ethoxide or sodium methoxide,[5] each of the steps leading to the enolate anion [2] of the β-keto ester is reversible and the initial equilibrium to form the ester enolate anion [3] is not favored. In

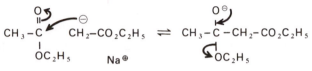

$$CH_3-CO_2C_2H_5 \xrightarrow[\substack{CH_3CO_2C_2H_5 \\ reflux}]{\substack{Na \\ cat.\ amt.\ C_2H_5OH}} CH_3-CO_2C_2H_5 + NaOC_2H_5 \rightleftharpoons$$

[1]

$$Na^{\oplus}$$

$$CH_2{}^{\ominus}-CO_2C_2H_5 + C_2H_5OH$$

[3]

$$CH_3-\overset{\overset{\displaystyle O}{\|}}{\underset{\underset{\displaystyle OC_2H_5}{|}}{C}}\quad CH_2^{\ominus}-CO_2C_2H_5 \rightleftharpoons CH_3-\overset{\overset{\displaystyle O^{\ominus}}{|}}{\underset{\underset{\displaystyle OC_2H_5}{|}}{C}}-CH_2-CO_2C_2H_5$$

$$Na^{\oplus}$$

$$\rightleftharpoons CH_3-CO-CH_2-CO_2C_2H_5 + C_2H_5O^{\ominus} \rightleftharpoons$$

$$CH_3-\overset{\overset{\displaystyle}{C}}{\underset{\underset{\displaystyle O}{\|}}{}}-CH^{\ominus}-CO_2C_2H_5 + C_2H_5OH \xrightarrow[\substack{H_2O}]{\substack{CH_3CO_2H}}$$

[2]

$$CH_3-CO-CH_2-CO_2C_2H_5 \qquad\qquad (Ref.\ 4)$$

(28–29%)

such cases the success of the condensation reaction is attributable to the final formation of the very stable enolate anion [2], which displaces previous un-favorable equilibria. In cases (e.g., [4]) where a very stable β-keto ester enolate anion cannot be formed, the condensation does not occur with bases such as sodium ethoxide but can be effected with strong bases (see Chapter 9) such as triphenylmethyl sodium. The success of this procedure results from the quantitative conversion of the starting ester [4] to its enolate anion [5]. Condensation then leads to the formation of ethoxide ion, which is a weaker base than the enolate anion [5] derived from the ester. Although the β-keto ester formed here is converted to the enolate anion [6] the latter step is not essential to the success of the con-densation.[6b]

4. J. K. H. Inglis and K. C. Roberts, *Org. Syn., Coll. Vol. 1*, 235 (1944).
5. (a) E. E. Royals, *J. Am. Chem. Soc.,* **70**, 489 (1948). (b) E. E. Royals, J. C. Hoppe, A. D. Jordan, Jr., and A. G. Robinson, III, *ibid.,* **73**, 5857 (1951). (c) E. E. Royals and D. G. Turpin, *ibid.,* **76**, 5452 (1954).
6. (a) C. R. Hauser and W. B. Renfrow, Jr., *J. Am. Chem. Soc.,* **59**, 1823 (1937). (b) B. E. Hudson, Jr., and C. R. Hauser, *ibid.,* **63**, 3156 (1941). (c) R. H. Hasek, R. D. Clark, E. U. Elam, and R. G. Nations, *J. Org. Chem.,* **27**, 3106 (1962).

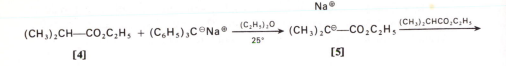

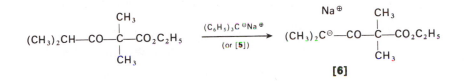

$$\xrightarrow[\text{H}_2\text{O}]{\text{CH}_3\text{CO}_2\text{H}} \quad (CH_3)_2CH-CO-\underset{\underset{CH_3}{|}}{\overset{\overset{CH_3}{|}}{C}}-CO_2C_2H_5 \qquad \text{(Ref. 6)}$$

(35%)

Even in cases where sodium ethoxide is successful as a catalyst for acetoacetic ester condensations, better yields have been obtained under forcing conditions, the ethanol being continuously removed from the reaction mixture to force the reaction to completion.[3,5] Alternatively, the yields obtained from acetoacetic ester condensations have been improved by the use of strong bases such as sodium amide,[7] diisopropylaminomagnesium bromide,[5,8] or sodium hydride.[5,9] Sodium hydride accompanied by a catalytic amount of alcohol, appears to be the catalyst of choice because it reacts irreversibly with the alcohol liberated during the reaction to form sodium alkoxide and hydrogen and force the reaction to completion. It is probable that the catalyst actually effecting condensation in most of these cases is not sodium hydride but rather the sodium alkoxide it generates.

For the synthesis of simple β-keto acid derivatives, a method often superior to the acetoacetic ester condensation of esters is the preparation and subsequent dimerization of ketenes.[10] The following equations illustrate the formation and hydrolysis of ketene dimers such as [7] to produce symmetrical ketones.

7. J. C. Shivers, M. L. Dillon, and C. R. Hauser, *J. Am. Chem. Soc.,* **69**, 119 (1947).

8. F. C. Frostick, Jr., and C. R. Hauser, *J. Am. Chem. Soc.,* **71**, 1350 (1949).

9. (a) F. W. Swamer and C. R. Hauser, *J. Am. Chem. Soc.,* **72**, 1352 (1950). (b) N. Green and F. B. LaForge, *ibid.,* **70**, 2287 (1948).

10. (a) W. E. Hanford and J. C. Sauer, *Org. Reactions,* **3**, 108 (1964). (b) R. N. Lacey, *Adv. Org. Chem.,* **2**, 213 (1960). (c) G. Quadbeck in W. Foerst, ed., *Newer Methods of Preparative Organic Chemistry,* Vol. 2, Academic, New York, 1963, pp. 133–161.

11. (a) J. C. Sauer, *Org. Syn.,* **Coll. Vol. 4,** 560 (1963). (b) L. J. Durham, D. J. McLeod, and J. Cason, *ibid.,* **Coll. Vol. 4,** 555 (1963). (c) N. J. Turro, P. A. Leermakers, and G. F. Vesley, *ibid.,* **47,** 34 (1967).

$$n\text{-}C_{10}H_{21}\text{---}CH_2\text{---}CO\text{---}Cl \xrightarrow[\;(C_2H_5)_2O\;]{(C_2H_5)_3N} [n\text{-}C_{10}H_{21}\text{---}CH\!=\!C\!=\!O] \rightarrow n\text{-}C_{10}H_{21}\text{---}CH\!=\!C\text{---}O$$

$$n\text{-}C_{10}H_{21}\text{---}CH\text{---}C\!=\!O$$

[7]

$$\xrightarrow[\substack{H_2O \\ \text{reflux}}]{H_2SO_4} \left[\begin{array}{c} n\text{-}C_{10}H_{21}\text{---}CH_2\text{---}CO\text{---}CH\text{---}CO_2H \\ | \\ n\text{-}C_{10}H_{21} \end{array} \right]$$

$$\xrightarrow{-CO_2} \quad n\text{-}C_{10}H_{21}\text{---}CH_2\text{---}CO\text{---}CH_2\text{---}C_{10}H_{21}\text{-}n \qquad \text{(Ref. 11a)}$$

(45–55%)

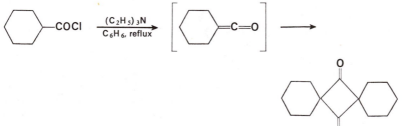

$$CH_3O_2C(CH_2)_4COCl \xrightarrow[C_6H_6,\,5\text{--}35°]{(C_2H_5)_2N} \begin{array}{c} CH_3O_2C(CH_2)_3CH\!=\!C\text{---}O \\ | \qquad\qquad | \\ CH_3O_2C(CH_2)_3\text{---}CH\text{---}C\!=\!O \end{array} \xrightarrow[\substack{H_2O, \\ \text{reflux}}]{KOH}$$

$$\xrightarrow[H_2O]{HCl} \quad HO_2C(CH_2)_4CO(CH_2)_4CO_2H$$

(60–64%)   (Ref. 11b)

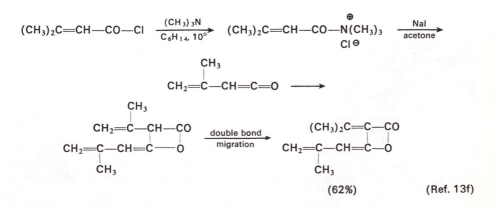

(49–58%)

[8]   (Ref. 11c)

$$(CH_3)_2C\!=\!CH\text{---}CO\text{---}Cl \xrightarrow[C_6H_{14},\,10°]{(CH_3)_3N} (CH_3)_2C\!=\!CH\text{---}CO\text{---}\overset{\oplus}{N}(CH_3)_3 \xrightarrow[\text{acetone}]{NaI}$$

$$Cl^{\ominus}$$

$$\begin{array}{c} CH_3 \\ | \\ CH_2\!=\!C\text{---}CH\!=\!C\!=\!O \end{array} \longrightarrow$$

(62%)   (Ref. 13f)

As the accompanying examples suggest, two structurally isomeric ketene dimers (e.g., [7] and [8]) may be formed.[10,13] Under appropriate conditions with basic catalysts, these dimers may be converted to trimers;[6c,13d,e,i] certain phenyl substituted ketene dimers may be also converted to the cyclobutane dimers by reaction with bases.[13h,i] The cyclobutanedione dimers are generally isomerized to the lactone dimers by treatment with acids such as amine hydrochlorides, zinc chloride, and aluminum chloride[13a,i] and the conversion of the ketenes directly to lactone dimers is catalyzed by the same acids[13a,i] as well as phosphite esters.[13b]

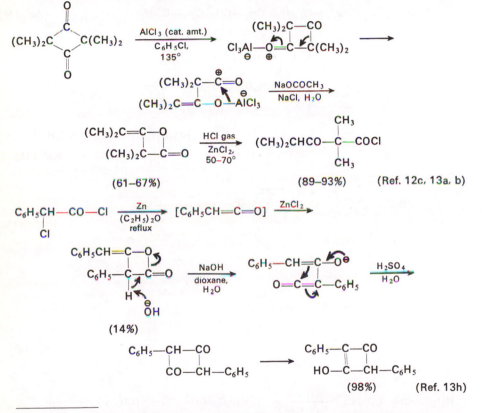

12. (a) S. O. Lawesson, S. Gronwall, and R. Sandberg, *Org. Syn.,* **42**, 28 (1962). (b) J. W. Williams and J. A. Krynitsky, *ibid.,* **Coll. Vol. 3**, 10 (1955). (c) R. H. Hasek, R. D. Clark, and G. L. Mayberry, *ibid.,* **48**, 72 (1968); E. U. Elam, P. G. Gott, and R. H. Hasek, *ibid.,* **48**, 126 (1968).

13. (a) R. H. Hasek, R. D. Clark, E. U. Elam, and J. C. Martin, *J. Org. Chem.,* **27**, 60 (1962). (b) E. U. Elam, *ibid.,* **32**, 215 (1967). (c) J. C. Martin, R. D. Burpitt, and H. U. Hostettler, *ibid.,* **32**, 210 (1967). (d) J. L. E. Erickson, F. E. Collins, Jr., and B. L. Owen, *ibid.,* **31**, 480 (1966). (e) E. Marcus, J. K. Chan, and C. B. Strow, *ibid.,* **31**, 1369 (1966). (f) G. B. Payne, *ibid.,* **31**, 718 (1966). (g) For a discussion of the products from α-halo ketenes, see W. T. Brady, F. H. Parry, R. Roe, Jr., E. F. Hoff, and L. Smith, *ibid.,* **35**, 1515 (1970) and references therein. (h) J. E. Baldwin and J. D. Roberts, *J. Am. Chem. Soc.,* **85**, 2444 (1963). (i) D. G. Farnum, J. R. Johnson, R. E. Hess, T. B. Marshall, and B. Webster, *ibid.,* **87**, 5191 (1965).

It is believed that the uncatalyzed dimerization of all substituted ketenes favors the formation of the cyclobutane dimers (e.g., [8]); in the presence of weak acid catalysts such as amine hydrochlorides, the sterically hindered ketoketenes $(R_2C=C=O)$ still give the same cyclobutane dimers but the aldoketenes $(RCH=C=O)$ frequently undergo an acid-catalyzed dimerization to form the lactone dimers (e.g., [7]). Although both types of dimers are susceptible to attack by various nucleophiles such as water, alcohols, amines, and enamines,[13c] the lactone dimers (e.g., [7]) are usually more reactive than the corresponding cyclobutane isomers.[13a,b] Examples of the formation of β-keto esters and β-keto amides are shown below.

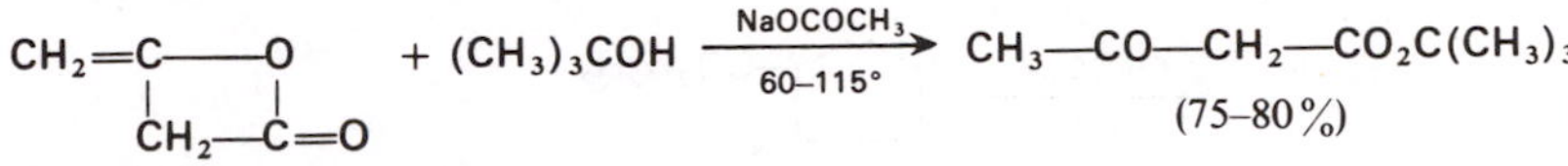

$$CH_2=C{-}O \quad + (CH_3)_3COH \xrightarrow[60-115°]{NaOCOCH_3} CH_3{-}CO{-}CH_2{-}CO_2C(CH_3)_3$$
$$CH_2{-}C=O$$

(75–80%)   (Ref. 12a)

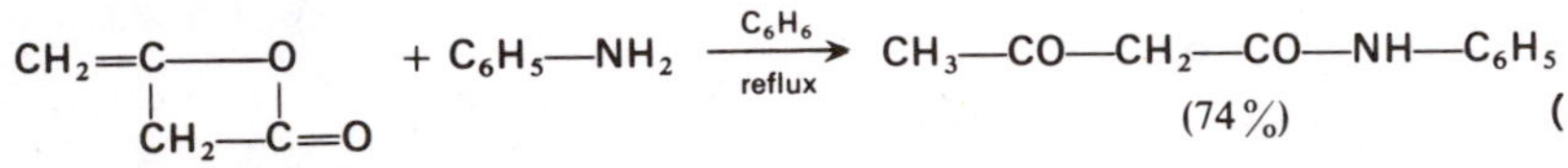

$$CH_2=C{-}O \quad + C_6H_5{-}NH_2 \xrightarrow[reflux]{C_6H_6} CH_3{-}CO{-}CH_2{-}CO{-}NH{-}C_6H_5$$
$$CH_2{-}C=O$$

(74%)   (Ref. 12b)

Another procedure for the preparation of symmetrical ketones, again often superior to acetoacetic ester condensations and subsequent hydrolysis, involves pyrolysis of the salt (e.g., [9]) of a carboxylic acid.[14,15c] Although the formation of an intermediate β-keto acid derivative, has been postulated, the mechanism of this reaction is uncertain at the present time.

$$n\text{-}C_{17}H_{35}{-}CO_2H + MgO \xrightarrow{335-340°} (n\text{-}C_{17}H_{35}{-}CO{-}O)_2Mg \xrightarrow{-CO_2}$$
$$[9]$$

$$n\text{-}C_{17}H_{35}{-}CO{-}C_{17}H_{35}\text{-}n$$

(81–87%)   (Ref. 14a)

$$CH_3CH_2CH_2CO_2H + Fe \xrightarrow{reflux}$$

$$(CH_3CH_2CH_2CO_2)_2Fe \xrightarrow[distil]{heat\ and} CH_3CH_2CH_2COCH_2CH_2CH_3$$

(69–75%)   (Ref. 14d)

14. (a) A. G. Dobson and H. H. Hatt, *Org. Syn.*, **Coll. Vol. 4**, 854 (1963). (b) J. F. Thorpe and G. A. R. Kon, *ibid.*, **Coll. Vol. 1**, 192 (1944). (c) R. M. Herbst and R. H. Manske, *ibid.*, **Coll. Vol. 2**, 389 (1943). (d) R. Davis, C. Granito, and H. P. Schultz, *ibid.*, **47**, 75 (1967).
15. (a) P. S. Pinkney, *Org. Syn.*, **Coll. Vol. 2**, 116 (1943). (b) R. Mayer in W. Foerst, ed., *Newer Methods of Preparative Organic Chemistry*, Vol. 2, Academic, New York, 1963, pp. 101–131. (c) J. P. Schaefer and J. J. Bloomfield, *Org. Reactions*, **15**, 1 (1967). (d) A. T. Nielsen and W. R. Carpenter, *Org. Syn.*, **45**, 25 (1965).

Esters of dicarboxylic acids may undergo an intramolecular acetoacetic ester condensation, called the Dieckmann condensation,[3,15] leading to cyclic β-keto esters. The reversibility of this reaction normally restricts its use to diesters (e.g., [10]) that can yield five- and six-membered ring products; however, it has occasion-

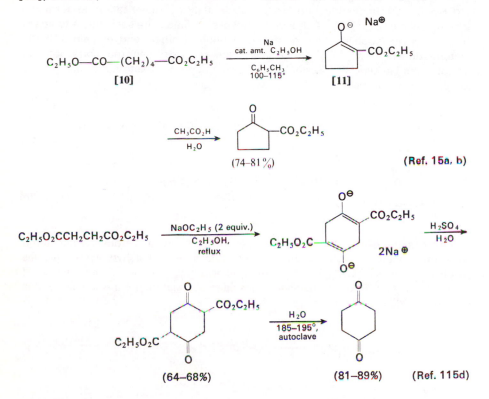

$$C_2H_5O—CO—(CH_2)_4—CO_2C_2H_5$$

[10]

Na
cat. amt. C_2H_5OH
——————————→
$C_6H_5CH_3$
100–115°

[11]

$$\xrightarrow[H_2O]{CH_3CO_2H}$$

(74–81%)   (Ref. 15a, b)

$$C_2H_5O_2CCH_2CH_2CO_2C_2H_5$$

$$\xrightarrow[\substack{C_2H_5OH, \\ reflux}]{NaOC_2H_5 \ (2 \ equiv.)}$$

$$\xrightarrow[H_2O]{H_2SO_4}$$

2Na$^{\oplus}$

$$\xrightarrow[\substack{185–195°, \\ autoclave}]{H_2O}$$

(64–68%)   (81–89%)   (Ref. 115d)

ally been successful for the preparation of larger rings.[15c,16a,b] Metallic sodium, accompanied by a catalytic amount of ethanol or methanol is often employed to effect both the Dieckmann condensations[15a,17] and intermolecular acetoacetic ester condensations.[4] Although the use of these materials serves to displace the various equilibria in favor of the final β-keto ester enolate anion (e.g., [2] and [11]), it may lead to complications from a competing acyloin condensation (see Chapter

16. (a) N. J. Leonard and R. C. Sentz, *J. Am. Chem. Soc.*, **74**, 1704 (1952). (b) W. S. Johnson, A. R. Jones, and W. P. Schneider, *ibid.*, **72**, 2395 (1950). (c) J. L. Baas, A. Davies-Fidder, and H. O. Huisman, *Tetrahedron*, **22**, 285 (1966). (d) C. G. Chin, H. W. Cuts, and S. Masamune, *Chem. Commun.*, **No. 23**, 880 (1966). (e) K. Sisido, K. Utimoto, and T. Isida, *J. Org. Chem.*, **29**, 2781 (1964). (f) W. L. Meyer, A. P. Lobo, and E. T. Marquis, *ibid.*, **30**, 181 (1965). (g) R. L. Augustine, Z. S. Zelawski, and D. H. Malarek, *ibid.*, **32**, 2257 (1967). (h) G. Büchi and E. C. Roberts, *ibid.*, **33**, 460 (1968). (i) For an example in which the direction of Dieckmann closure is controlled by bonding one end of a dicarboxylic acid to a polystyrene resin, see J. I. Crowley and H. Rapoport, *J. Am. Chem. Soc.*, **92**, 6363 (1970).
17. J. Dutta and R. N. Biswas, *J. Chem. Soc.*, 2387 (1963).

3),[5c,15c] especially if an insufficient amount of alcohol is present in the original reaction mixture. For this reason sodium hydride, accompanied by a catalytic amount of alcohol, as a condensing agent appears to be more desirable particularly in view of its previously discussed advantages.[15c,16f,18]

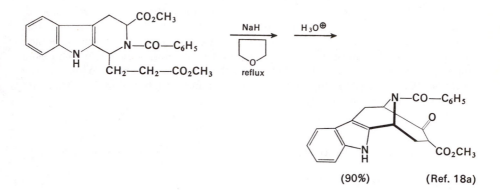

(90%)   (Ref. 18a)

Sodium hydride has been used both as a suspension in inert solvents such as tetrahydrofuran, benzene, and xylene[16f,18a] and to form a solution of the sodium salt of dimethyl sulfoxide.[16d,18b] Although the latter reaction conditions are some-times advantageous, the various side reactions possible with dimethyl sulfoxide (see Chapter 10) usually make the choice of an inert solvent preferable.[15c] Like the intermolecular acetoacetic ester condensation, the Dieckmann condensation usually fails if the product can not form a stable β-keto ester enolate.[15a,16c] An exception to this generality is the case of the diester [**12**] where cyclization is

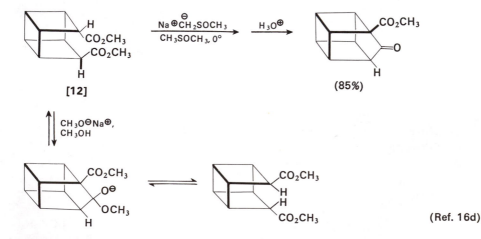

(Ref. 16d)

18. (a) J. D. Hobson, J. Raines, and R. J. Whiteoak, *J. Chem. Soc.*, 3495 (1963). (b) J. J. Bloomfield and P. V. Fennessey, *Tetrahedron Letters*, **No. 33**, 2273 (1964).

exceptionally favorable sterically; however, an acylic product is formed when the reaction is performed in alcohol solution. The ready reversibility of the Dieckmann condensation along with the tendency to form the least hindered metal enolate usually determine the structure of the cyclized product from an unsymmetrical diester.[15,16] The following examples are illustrative.

A reaction analogous to the Dieckmann condensation is the cyclization of dinitriles, known as the Thorpe-Ziegler condensation.[15c] By conducting this reaction under high-dilution conditions in an inert solvent so that the metal enolate formed is insoluble, the condensation becomes sufficiently irreversible to be used for the synthesis of large rings. However, this method is inferior to the acyloin reaction (Chapter 3) for this purpose because high-dilution conditions are not required in the acyloin reaction.

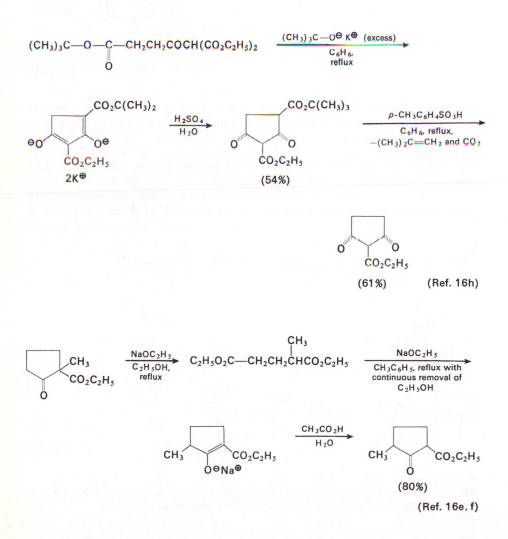

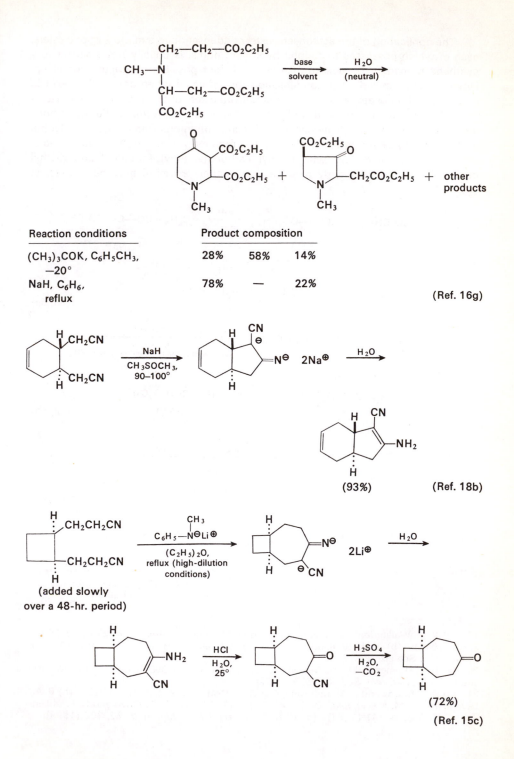

Reaction conditions	Product composition		
$(CH_3)_3COK$, $C_6H_5CH_3$, $-20°$	28%	58%	14%
NaH, C_6H_6, reflux	78%	—	22%

(Ref. 16g)

(93%)   (Ref. 18b)

(added slowly over a 48-hr. period)

(72%)

(Ref. 15c)

The application of the acetoacetic ester condensation to a mixture of two esters, each of which possesses two alpha hydrogen atoms, is usually not a satisfactory synthetic procedure because a mixture of all four possible products is obtained. However, such mixed-ester condensations are of value when one of the two has no alpha hydrogen atoms; commonly employed esters of this type include those of aromatic acids such as benzoic and furoic, as well as ethyl formate, diethyl carbonate, and diethyl oxalate. Whereas either forcing conditions (removal of the alcohol as it is formed) or strong bases are normally required for successful condensations with aromatic esters and with diethyl carbonate, ethyl formate[20] and diethyl oxalate[21,22a,b] are sufficiently reactive that neither provision may be necessary.

$$C_6H_5-CO_2CH_3 + CH_3CH_2-CO_2CH_3 \xrightarrow[\substack{C_6H_6 \\ reflux}]{NaH} C_6H_5-CO-\overset{\overset{\displaystyle CH_3}{|}}{\underset{\overset{\displaystyle \ominus}{}}{C}}-CO_2CH_3$$
$$Na^{\oplus}$$

$$\xrightarrow{H_3O^{\oplus}} C_6H_5-CO-\overset{\overset{\displaystyle CH_3}{|}}{CH}-CO_2CH_3$$
$$(56\%) \qquad\qquad\qquad\qquad\qquad \text{(Ref. 5c)}$$

$$C_6H_5-CH_2-CO_2C_2H_5 + (C_2H_5O)_2CO \xrightarrow[\substack{(C_2H_5O)_2CO \\ \text{reflux with continuous removal of } C_2H_5OH}]{NaOC_2H_5}$$

$$\overset{Na^{\oplus}}{C_6H_5-C^{\ominus}(CO_2C_2H_5)_2} \xrightarrow{H_3O^{\oplus}} C_6H_5-CH(CO_2C_2H_5)_2$$
$$(86\%) \qquad\qquad \text{(Ref. 19)}$$

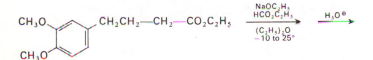

$$\xrightarrow[\substack{(C_2H_5)_2O \\ -10 \text{ to } 25°}]{\substack{NaOC_2H_5 \\ HCO_2C_2H_5}} \xrightarrow{H_3O^{\oplus}}$$

(Ref. 20a)

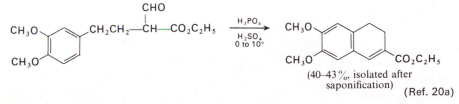

(40–43%, isolated after saponification)   (Ref. 20a)

19. (a) V. H. Wallingford, A. H. Homeyer, and D. M. Jones, *J. Am. Chem. Soc.,* **63**, 2056, 2252 (1941). (b) V. H. Wallingford, D. M. Jones, and A. H. Homeyer, *ibid.,* **64**, 576 (1942). (c) See also Ref. 9.
20. (a) H. L. Holmes and L. W. Trevoy, *Org. Syn.,* **Coll. Vol. 3**, 300 (1955). (b) H. Minato and I. Horibe, *J. Chem. Soc.,* C, 1575 (1967).
21. (a) R. F. B. Cox and S. M. McElvain, *Org. Syn.,* **Coll. Vol. 2**, 272, 279 (1943). (b) P. A. Levene and G. M. Meyer, *ibid.,* **Coll. Vol. 2**, 288 (1943). (c) D. E. Floyd and S. E. Miller, *ibid.,* **Coll. Vol. 4**, 141 (1963). (d) H. R. Schweizer, *Helv. Chim. Acta,* **52**, 322 (1969).

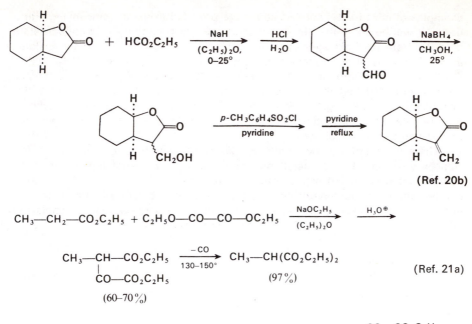

(Ref. 20b)

$$CH_3—CH_2—CO_2C_2H_5 + C_2H_5O—CO—CO—OC_2H_5 \xrightarrow[\text{(C}_2\text{H}_5)_2\text{O}]{\text{NaOC}_2\text{H}_5} \xrightarrow{\text{H}_3\text{O}^\oplus}$$

$$\underset{\substack{| \\ \text{CO—CO}_2\text{C}_2\text{H}_5 \\ (60–70\%)}}{CH_3—CH—CO_2C_2H_5} \xrightarrow[130–150°]{-CO} CH_3—CH(CO_2C_2H_5)_2$$

(97%)   (Ref. 21a)

$$\underset{\substack{| \\ \text{CH}_2—\text{CO}_2\text{C}_2\text{H}_5}}{CH_2—CO_2C_2H_5} + C_2H_5O—CO—CO—OC_2H_5 \xrightarrow[\substack{C_6H_5CH_3, \\ (C_2H_5)_2O, \\ 25°}]{\text{NaOC}_2\text{H}_5} \xrightarrow[\text{H}_2\text{O}]{\text{HCl}} \underset{\substack{| \\ \text{CH—CO}_2\text{C}_2\text{H}_5 \\ | \\ \text{CH}_2—\text{CO}_2\text{C}_2\text{H}_5}}{CO—CO_2C_2H_5}$$

(86–91%)

$$\xrightarrow[\substack{\text{H}_2\text{O} \\ \text{reflux}}]{\text{HCl}} HO_2C—CH_2—CH_2—CO—CO_2H \qquad \text{(Ref. 22a, b)}$$

(73–83%)

$$C_2H_5O_2C—CO_2C_2H_5 + C_2H_5O_2C—(CH_2)_4—CO_2C_2H_5 \xrightarrow[\substack{C_6H_5CH_3, \\ (C_2H_5)_2O, \\ 25–55°}]{\text{NaOC}_2\text{H}_5} \xrightarrow[\text{H}_2\text{O}]{\text{H}_2\text{SO}_4}$$

(added slowly)

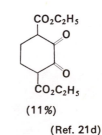

(11%)

(Ref. 21d)

22. (a) L. Friedman and E. Kosower, *Org. Syn.*, **Coll. Vol. 3**, 510 (1955). (b) E. M. Bottorff and L. L. Moore, *ibid.*, **44**, 67 (1964). (c) K. L. Rinehart, *ibid.*, **Coll. Vol. 4**, 120 (1963). (d) J. Cason, K. L. Rinehart, Jr., and S. D. Thornton, Jr., *J. Org. Chem.*, **18**, 1594 (1953).

Examples of mixed-ester condensations are provided in the accompanying equations. It will be noted that these examples also illustrate the use of condensation products from ethyl oxalate as precursors for the synthesis of α-keto and malonic acids. The thermal decarbonylation of systems of the type $-CO-CH_2-CO-CO_2C_2H_5$ to form products of the type $-CO-CH_2-CO_2C_2H_5$ is a general reaction provided that the beta carbonyl function is acyclic or is part of a six-membered or larger ring. Although the mechanisms of certain decarbonylation reactions have been investigated,[23] the reaction path followed in those described here remains to be established.

The acylation of esters with nitriles has been successful under special circumstances where an α-bromo ester is converted to its bromozinc enolate (a Reformatsky reagent, see Chapter 10) before reaction with a nitrile. The success of this mixed condensation is attributable to the relatively slow equilibration of zinc enolates in non-polar solvents accompanied by the formation of a very stable metal enolate in which the metal is coordinated with a bidentate ligand.

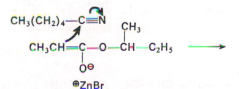

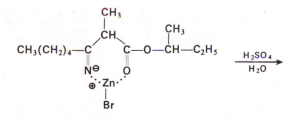

$$CH_3(CH_2)_4-CO-\overset{\overset{\displaystyle CH_3}{|}}{CH}-CO_2-\overset{\overset{\displaystyle CH_3}{|}}{CH}-C_2H_5$$

(50–58%)   (Ref. 22c, d)

23. (a) M. Calvin and R. M. Lemmon, *J. Am. Chem. Soc.,* **69**, 1232 (1947). (b) J. D. Roberts, D. R. Smith, and C. C. Lee, *ibid.,* **73**, 618 (1951).

The foregoing principles are also applicable to the acylation of ketones,[9,19a,24] nitriles,[19b,25] and other active methylene compounds by condensation with esters (sometimes called the Claisen reaction). As the accompanying examples show, sodium ethoxide is an adequate base to promote condensations with the very reactive esters, ethyl formate and diethyl oxalate. However, strong bases such as

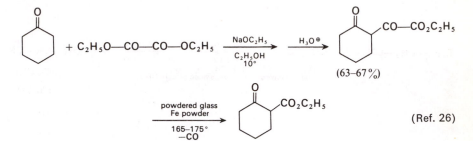

(63–67%)

(59–62% overall)

(Ref. 26)

C_6H_5—CH_2—CN + C_2H_5O—CO—CO—OC_2H_5 $\xrightarrow[\substack{C_2H_5OH \\ 25°}]{NaOC_2H_5}$ $\xrightarrow{H_3O^\oplus}$ C_6H_5—CH—CN
  |
  CO—$CO_2C_2H_5$

(69–75%)

(Ref. 27)

CH_3—CO—CH_3 + H—$CO_2C_2H_5$ $\xrightarrow[\substack{(C_2H_5)_2O \\ 25°}]{NaOC_2H_5}$ CH_3—CO—CH=CH—$O^\ominus$ $Na^\oplus$

$\xrightarrow[50°]{H_3O^\oplus}$ [CH_3—CO—CH_2—CHO] $\longrightarrow$ CH_3—CO—

(30–38%)

(Ref. 28)

24. (a) C. R. Hauser, F. W. Swamer, and J. T. Adams, *Org. Reactions,* **8**, 59 (1954). (b) R. Levine, J. A. Conroy, J. T. Adams, and C. R. Hauser, *J. Am. Chem. Soc.,* **67**, 1510 (1945). (c) J. J. Bloomfield, *J. Org. Chem.,* **27**, 2742 (1962). (d) S. D. Work and C. R. Hauser, *ibid.,* **28**, 725 (1963). (e) A. P. Krapcho, J. Diamanti, C. Cayen, and R. Bingham, *Org. Syn.,* **47**, 20 (1967). (f) J. P. John, S. Swaminathan, and P. S. Venkataramani, *ibid.,* **47**, 83 (1967).
25. (a) B. Abramovitch and C. R. Hauser, *J. Am. Chem. Soc.,* **64**, 2720 (1942). (b) R. Levine and C. R. Hauser, *ibid.,* **68**, 760 (1946). (c) R. S. Long, *ibid.,* **69**, 990 (1947). (d) C. J. Eby and C. R. Hauser, *ibid.,* **79**, 723 (1957).
26. (a) H. R. Snyder, L. A. Brooks, and S. H. Shapiro, *Org. Syn.,* **Coll. Vol. 2**, 531 (1943). (b) C. S. Marvel and E. E. Dreger, *ibid.,* **Coll. Vol. 1**, 238 (1944). (c) E. R. Riegel and F. Zwilgmeyer, *ibid.,* **Coll. Vol. 2**, 126 (1943). (d) For an example involving the condensation of ethyl oxalate at the methyl group of o-nitrotoluene, see W. E. Noland and F. J. Baude, *ibid.,* **43**, 40 (1963).
27. R. Adams and H. O. Calvery, *Org. Syn.,* **Coll. Vol. 2**, 287 (1943).
28. (a) R. L. Frank and R. H. Varland, *Org. Syn.,* **Coll. Vol. 3**, 829 (1955). (b) R. P. Mariella, *ibid.,* **Coll. Vol. 4**, 210 (1963). (c) W. Franke and R. Kraft in W. Foerst, ed., *Newer Methods of Preparative Organic Chemistry,* Vol. 2, Academic, New York, 1963, pp. 1–30.

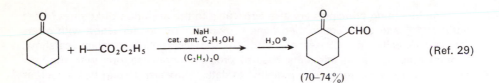

(Ref. 29)

(70–74%)

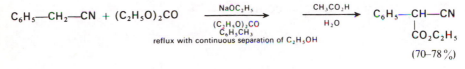

C_6H_5—CH_2—CN + $(C_2H_5O)_2CO$ $\xrightarrow[\substack{(C_2H_4O)_2CO\\C_6H_5CH_3}]{NaOC_2H_5}$ reflux with continuous separation of C_2H_5OH $\xrightarrow[H_2O]{CH_3CO_2H}$ C_6H_5—CH—CN

$CO_2C_2H_5$

(70–78%)

(Ref. 30)

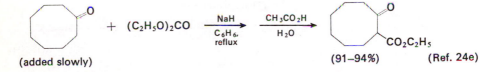

(added slowly)   (91–94%)   (Ref. 24e)

CH_3—CH=CH—CH_2CH_2—CO—CH_3 + $(C_2H_5O)_2CO$ $\xrightarrow[\substack{(C_2H_5)_2O\\25°}]{NaH}$ $\xrightarrow[H_2O]{CH_3CO_2H}$

CH_3—CH=CH—CH_2CH_2—CO—CH_2—$CO_2C_2H_5$ (Ref. 9b)

(85%)

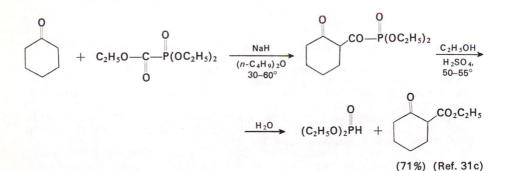

(71%) (Ref. 31c)

29. (a) C. Ainsworth, *Org. Syn.*, **Coll. Vol. 4**, 536 (1963). (b) S. Boatman, T. M. Harris and C. R. Hauser, *ibid.*, **48**, 40 (1968).
30. E. C. Horning and A. F. Finelli, *Org. Syn.*, **Coll. Vol. 4**, 461 (1963).
31. (a) A Magnani and S. M. McElvain, *Org. Syn.*, **Coll. Vol. 3**, 251 (1955). (b) H. Schiefer and G. Henseke, *Angew Chem., Intern. Ed. Engl.*, **4**, 527 (1965). (c) I. Shahak, *Tetrahedron Letters*, **No. 20**, 2201 (1966).

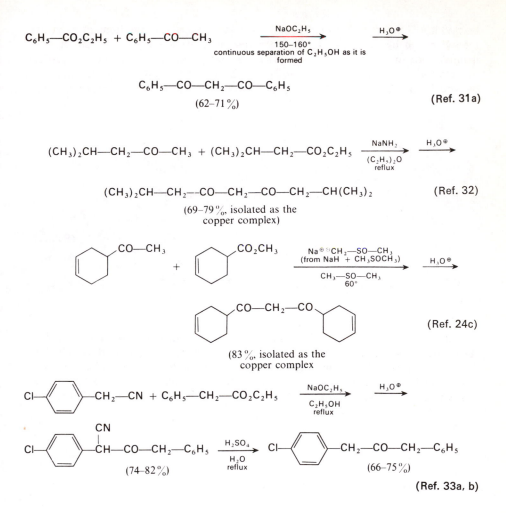

$$C_6H_5\!-\!CO_2C_2H_5 + C_6H_5\!-\!CO\!-\!CH_3 \xrightarrow[\substack{150-160° \\ \text{continuous separation of } C_2H_5OH \text{ as it is} \\ \text{formed}}]{NaOC_2H_5} \xrightarrow{H_3O^\oplus}$$

$$C_6H_5\!-\!CO\!-\!CH_2\!-\!CO\!-\!C_6H_5$$

(62–71%)   (Ref. 31a)

$$(CH_3)_2CH\!-\!CH_2\!-\!CO\!-\!CH_3 + (CH_3)_2CH\!-\!CH_2\!-\!CO_2C_2H_5 \xrightarrow[\substack{(C_2H_5)_2O \\ \text{reflux}}]{NaNH_2} \xrightarrow{H_3O^\oplus}$$

$$(CH_3)_2CH\!-\!CH_2\!-\!CO\!-\!CH_2\!-\!CO\!-\!CH_2\!-\!CH(CH_3)_2 \qquad (Ref.\ 32)$$

(69–79%, isolated as the
copper complex)

Na⊕ ⊖CH₂—SO—CH₃
(from NaH + CH₃SOCH₃)   →   H₃O⊕ →

CH₃—SO—CH₃
60°

(Ref. 24c)

(83%, isolated as the
copper complex

$$NaOC_2H_5 \xrightarrow{} H_3O^\oplus$$
C₂H₅OH
reflux

(74–82%)   H₂SO₄ / H₂O reflux →   (66–75%)

(Ref. 33a, b)

sodium amide, the sodium salt of dimethyl sulfoxide, or sodium hydride are often more satisfactory for condensations involving less reactive esters. From several of the foregoing examples, as well as from the acylation of the ketone **[13]** shown

32. C. R. Hauser, J. T. Adams, and R. Levine, *Org. Syn.*, **Coll. Vol. 3**, 291 (1955).
33. (a) S. B. Coan and E. I. Becker, *Org. Syn.*, **Coll. Vol. 4**, 174, 176 (1963). (b) P. L. Julian, J. J. Oliver, R. H. Kimball, A. B. Pike, and G. D. Jefferson, *ibid.*, **Coll. Vol. 2**, 487 (1943). (c) M. L. Miles, T. M. Harris, and C. R. Hauser, *ibid.*, **46**, 57 (1966); M. L. Miles and C. R. Hauser, *ibid.*, **46**, 60 (1966). (d) T. M. Harris and C. M. Harris, *Org. Reactions*, **17**, 155 (1969). (e) T. M. Harris, S. Boatman, and C. R. Hauser, *J. Am. Chem. Soc.*, **87**, 3186 (1965). (f) M. L. Miles, T. M. Harris, and C. R. Hauser, *J. Org. Chem.*, **30**, 1007 (1965); M. L. Miles and C. R. Hauser, *ibid.*, **29**, 2329 (1964). (g) D. M. von Schriltz, M. L. Miles, and C. R. Hauser, *ibid.*, **32**, 1774 (1967). (h) S. D. Work, D. R. Bryant, and C. R. Hauser, *ibid.*, **29**, 722 (1964). (i) J. F. Wolfe and G. B. Trimitsis, *ibid.*, **33**, 894 (1968). (j) T. M. Harris, C. M. Harris, and M. P. Wachter, *Tetrahedron*, **24**, 6897 (1968).

below, it will be seen that methyl alkyl ketones are acylated predominantly at the methyl group by this procedure. Since the reactions leading to the structurally isomeric enolate anions [14] and [15] are reversible, this preferential acylation reflects the previously noted (Chapter 9) fact that the stability of enolate anions derived from 1,3-dicarbonyl compounds is diminished by alkyl substitution (i.e

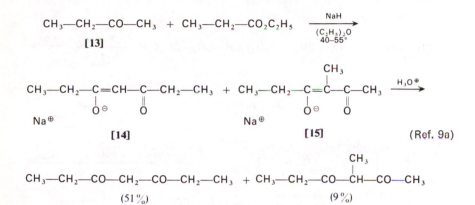

CH₃—CH₂—CO—CH₃ + CH₃—CH₂—CO₂C₂H₅ $\xrightarrow[\substack{(C_2H_5)_2O \\ 40-55°}]{NaH}$

[13]

CH₃—CH₂—C=CH—C—CH₂—CH₃ + CH₃—CH₂—C=C—C—CH₃ $\xrightarrow{H_3O^\oplus}$

Na⊕   [14]   Na⊕   [15]   (Ref. 9a)

CH₃—CH₂—CO—CH₂—CO—CH₂—CH₃ + CH₃—CH₂—CO—CH—CO—CH₃

(51%)   (9%)

[15] is less stable than [14]). This is especially true if acylation at the more highly substituted alpha position would form a 1,3-dicarbonyl compound which can not yield a very stable enolate ion. In accordance with this generalization, un-symmetrical cyclic ketones, such as the one shown, are acylated by esters at the less highly substituted α-carbon. As would be expected from previous discussions

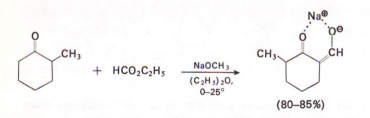

+ HCO₂C₂H₅ $\xrightarrow[\substack{(C_2H_5)_2O, \\ 0-25°}]{NaOCH_3}$

(80–85%)   (Ref. 29b)

(see Chapters 9 and 10), the dicarbanions derived from dicarbonyl compounds and a strong base condense with esters at the carbon corresponding to the more reactive (less highly stabilized) enolate.[33c–g,j] Similar products may be obtained by

CH₃C—CH=CH—O⊖ $\xrightarrow[liq. NH_3]{KNH_2}$ CH₂=C—CH=CH—O⊖ $\xrightarrow[liq. NH_3]{C_6H_5CO_2C_2H_5 \text{ (0.5 equiv.)}}$

‖ |
O   Na⊕   O⊖   Na⊕, K⊕

C₆H₅C=CH—C—CH=CH—O⊖ $\xrightarrow{H_2O}$ C₆H₅COCH₂COCH₂CHO

| ‖   (52%)
O⊖   O

Na⊕, K⊕   (Ref. 33e)

reaction of the dicarbonyl compound with sodium hydride and the ester of an aromatic acid. In these cases, prior formation of a dicarbanion does not appear to be involved.[33f] Similar methods have been used to effect the C-acylation of amides.[33h,i]

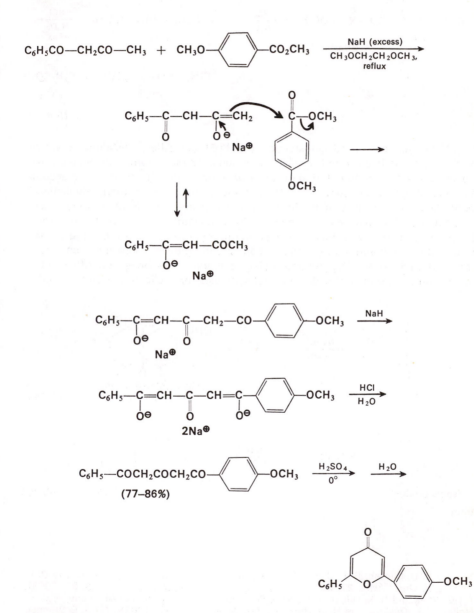

(Ref. 33c, f)

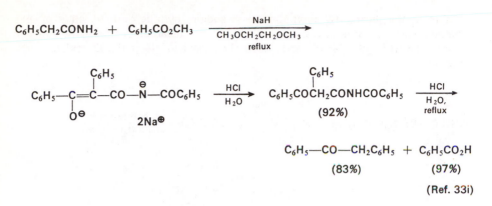

$$C_6H_5-CO-CH_2C_6H_5 \ + \ C_6H_5CO_2H$$
$$(83\%) \qquad\qquad (97\%)$$

(Ref. 33i)

The condensation of 3-ketosteroids (e.g., [16]) and other β-decalone derivatives with esters has been found to yield predominantly the 2-formyl derivatives, irrespective of whether the A-B ring fusion is *cis* or *trans* (cf. Chapter 8).[34] To account for this, it has been suggested[34a,c] that the acylation is a thermodynamically controlled reaction leading to the more stable 2-acyl derivative in which steric interactions are minimized. The behavior of a related α,β-(or β,γ-) unsaturated ketone is in accord with this idea since the formyl group is introduced at the less hindered position and not the position corresponding to the more stable enolate of the starting ketone. The intramolecular acylation of ketones and other active methylene compounds provides a useful synthesis for cyclic compounds provided

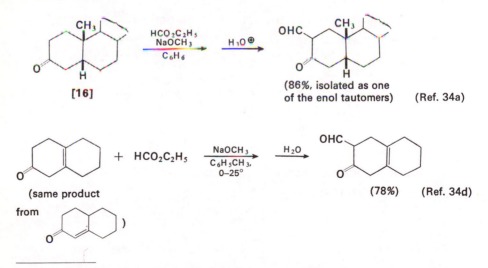

34. (a) R. O. Clinton, R. L. Clarke, F. W. Stonner, A. J. Manson, K. F. Jennings, and D. K. Phillips, *J. Org. Chem.*, **27**, 2800 (1962). (b) P. J. Palmer, *J. Chem. Soc.*, 3901 (1963). (c) G. Stork and R. K. Hill, *J. Am. Chem. Soc.*, **79**, 495 (1957). (d) K. Wiedhaup, A. J. H. Nollet, J. G. Korsloot, and H. O. Huisman, *Tetrahedron*, **24**, 771 (1968). (e) J. E. Brenner, *J. Org. Chem.*, **26**, 22 (1961).

the condensation can lead to five- or six-membered rings as in the equations below.[35]

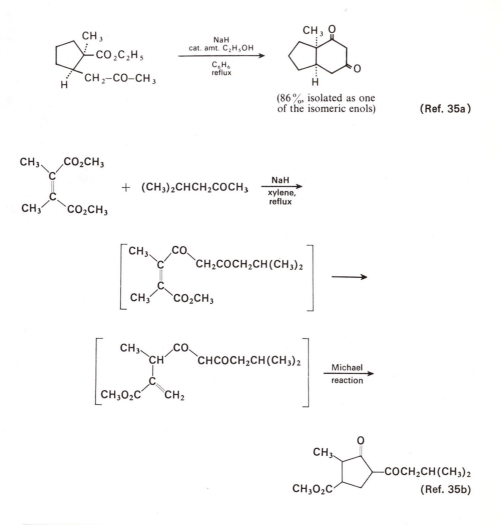

(86%, isolated as one of the isomeric enols)   (Ref. 35a)

(Ref. 35b)

35. (a) H. Conroy, *J. Am. Chem. Soc.*, **74**, 3046 (1952). (b) M. Elliott, N. F. Janes, and K. A. Jeffs, *J. Chem. Soc.*, C, 1845 (1969). (c) O. Touster, *Org. Reactions*, **7**, 327 (1953). (d) N. Kornblum, *ibid.*, **12**, 101 (1962). (e) A. P. Black and F. H. Babers, *Org. Syn.*, **Coll. Vol. 2**, 512 (1943). (f) H. Feuer and P. M. Pivawer, *J. Org. Chem.*, **31**, 3152 (1966) and references therein. (g) H. Feuer, A. M. Hall, S. Golden, and R. L. Reitz, *ibid.*, **33**, 3622 (1968). (h) H. Feuer and R. P. Monter, *ibid.*, **34**, 991 (1969). (i) H. Feuer and M. Auerbach, *ibid.*, **35**, 2551 (1970). (j) H. Feuer and J. P. Lawrence, *J. Am. Chem. Soc.*, **91**, 1856 (1969). (k) P. E. Pfeffer and L. S. Silbert, *Tetrahedron Letters*, **No. 10**, 699 (1970). (l) W. A. Mosher and W. E. Meier, *J. Org. Chem.*, **35**, 2924 (1970); H. H. Baer and S. R. Naik, *ibid.*, **35**, 2927 (1970).

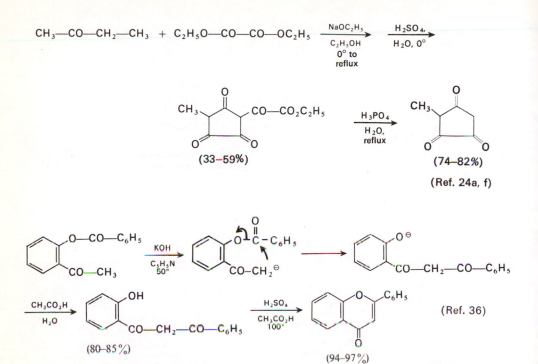

$$CH_3\!-\!CO\!-\!CH_2\!-\!CH_3 \; + \; C_2H_5O\!-\!CO\!-\!CO\!-\!OC_2H_5 \xrightarrow[\substack{C_2H_5OH \\ 0° \text{ to} \\ \text{reflux}}]{NaOC_2H_5} \xrightarrow[H_2O,\ 0°]{H_2SO_4,}$$

(33–59%)

(74–82%)

(Ref. 24a, f)

(Ref. 36)

(80–85%)

(94–97%)

Reactions related to the C-acylation of enolates with the esters of carboxylic acids include the reactions of enolates with esters of inorganic acids such as alkyl nitrites[35c] and alkyl nitrates.[35d–k] The condensation of alkyl nitrites with active methylene compounds is more frequently effected with acid catalysis;[35e] this process is related to the subsequently discussed acylation of enols. Examples of the base-catalyzed reactions follow.

$$n\text{-}C_4H_9\!-\!CH(CO_2C_2H_5)_2 \; + \; C_2H_5\!-\!O\!-\!NO \xrightarrow[\substack{C_2H_5OH, \\ -10°}]{NaOC_2H_5} \; n\text{-}C_4H_9\!-\!\overset{\ominus}{C}(CO_2C_2H_5)_2 \longrightarrow$$

$$C_2H_5O\!-\!N\!\!=\!\!O$$

$$n\text{-}C_4H_9\!-\!\underset{\substack{| \\ O=C}}{\overset{\substack{N=O \\ |}}{C}}\!-\!CO_2C_2H_5 \xrightarrow{-OC(OC_2H_5)_2} n\text{-}C_4H_9\!-\!\underset{\substack{\| \\ N\!-\!O^\ominus \\ Na^\oplus}}{C}\!-\!CO_2C_2H_5 \xrightarrow[H_2O]{HCl}$$

$$n\text{-}C_4H_9\!-\!\underset{\substack{\| \\ NOH}}{C}\!-\!CO_2C_2H_5$$

(80%)

(Ref. 35c)

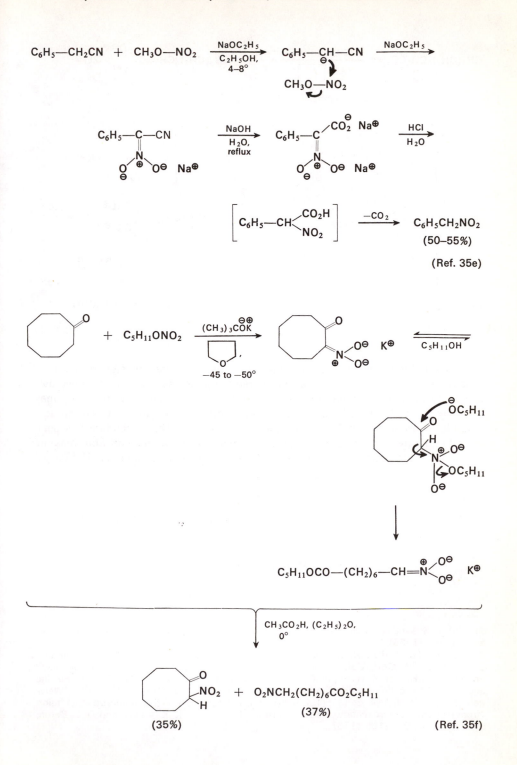

$$C_6H_5-CH_2CN \ + \ CH_3O-NO_2 \xrightarrow[\substack{C_2H_5OH, \\ 4-8°}]{NaOC_2H_5} \quad C_6H_5-\overset{\ominus}{CH}-CN \xrightarrow{NaOC_2H_5}$$

(50–55%)

(Ref. 35e)

$$C_5H_{11}OCO-(CH_2)_6-CH=\overset{\oplus}{N} \quad K^{\oplus}$$

CH₃CO₂H, (C₂H₅)₂O,
0°

(35%)

$+ \ O_2NCH_2(CH_2)_6CO_2C_5H_{11}$

(37%)

(Ref. 35f)

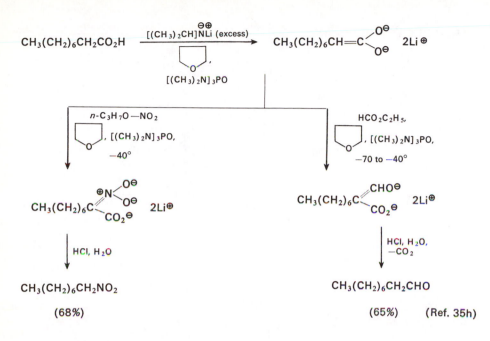

All the foregoing acylations suffer from the fact that the step resulting in introduction of the acyl group is reversible, requiring the use of one of the previously discussed methods to force the reaction to completion. The acylation step may normally be made irreversible by allowing an enolate anion to react with an acid chloride or an acid anhydride in an inert solvent.[24a,37] This procedure has been most frequently applied to the acylations of enolate anions derived from malonic esters or β-keto esters, as illustrated in the accompanying equations. The use of

$$CH_2(CO_2C_2H_5)_2 \xrightarrow[\substack{C_2H_5OH}]{\substack{Mg \\ cat.\ amt.\ CCl_4}} C_2H_5O-Mg^{\oplus \ominus}CH(CO_2C_2H_5)_2 \xrightarrow[(C_2H_5)_2O]{Cl-CO_2C_2H_5}$$

$$Cl-Mg^{\oplus \ominus}C(CO_2C_2H_5)_3 \xrightarrow[H_2O]{CH_3CO_2H} CH(CO_2C_2H_5)_3$$

$$(88-93\%) \qquad \text{(Ref. 37a)}$$

36. (a) T. S. Wheeler, *Org. Syn.*, **Coll. Vol. 4**, 478 (1963). (b) R. Mozingo, *ibid.*, **Coll. Vol. 3**, 387 (1955).
37. (a) H. Lund and A. Voigt, *Org. Syn.*, **Coll. Vol. 2**, 594 (1943). (b) B. B. Corson and J. L. Sayre, *ibid.*, **Coll. Vol. 2**, 596 (1943). (c) For the acylation of the cyclopentadienyl anion, see K. Hafner, K. H. Vöpel, G. Ploss, and C. König, *ibid.*, **47**, 52 (1967). (d) For the acylation of the cyanide anion from copper(I) cyanide, see T. S. Oakwood and C. A. Weisgerber, *ibid.*, **Coll. Vol. 3**, 112 (1955). (e) For the acylation of the trichloromethyl anion (from sodium trichloroacetate), see A. Winston, J. C. Sharp, K. E. Atkins, and D. E. Battin, *J. Org. Chem.*, **32**, 2166 (1967).

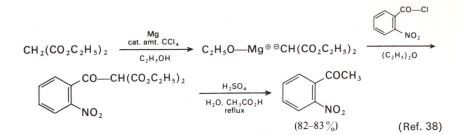

$$CH_2(CO_2C_2H_5)_2 \xrightarrow[C_2H_5OH]{\substack{Mg \\ cat.\ amt.\ CCl_4}} C_2H_5O\text{—}Mg^{\oplus\,\ominus}CH(CO_2C_2H_5)_2 \xrightarrow[(C_2H_5)_2O]{}$$

(aryl)—CO—CH(CO$_2$C$_2$H$_5$)$_2$ with NO$_2$ $\xrightarrow[\substack{H_2O,\ CH_3CO_2H \\ reflux}]{H_2SO_4}$ (aryl)—COCH$_3$ with NO$_2$

(82–83%)   (Ref. 38)

$$CH_3\text{—}CO\text{—}CH_2\text{—}CO_2C_2H_5 + CH_3\text{—}CO\text{—}Cl \xrightarrow[\substack{C_6H_6 \\ reflux}]{Mg} \xrightarrow[]{H_2O} \xrightarrow[]{Cu(OCOCH_3)_2}$$

copper complex $\xrightarrow[H_2O]{H_2SO_4}$ (CH$_3$—CO)$_2$CH—CO$_2$C$_2$H$_5$ (Ref. 39)

(46–52%)

$$C_6H_5\text{—}CO_2H + Cl\text{—}CO_2C_2H_5 \xrightarrow[\substack{C_6H_5CH_3 \\ 0°}]{(C_2H_5)_3N} C_6H_5\text{—}CO\text{—}O\text{—}CO_2C_2H_5$$

$$\xrightarrow[(C_2H_5)_2O,\ C_6H_5CH_3]{C_2H_5O\text{—}Mg^{\oplus}CH^{\ominus}(CO_2C_2H_5)_2} C_6H_5\text{—}CO\text{—}CH(CO_2C_2H_5)_2$$ (Ref. 40)

(68–75%)

$$C_6H_5\text{—}CH_2\text{—}CH(CO_2C_4H_9\text{-}t)_2 \xrightarrow[C_6H_6]{NaH} C_6H_5\text{—}CH_2\text{—}\overset{Na^{\oplus}}{C^{\ominus}}(CO_2C_4H_9\text{-}t)_2 \xrightarrow[C_6H_6]{C_6H_5\text{—}CO\text{—}Cl}$$

$$\underset{\underset{C_6H_5\text{—}CH_2}{|}}{C_6H_5\text{—}CO\text{—}C}(CO_2C_4H_9\text{-}t)_2 \xrightarrow[\substack{CH_3CO_2H \\ reflux}]{p\text{-}CH_3\text{—}C_6H_4\text{—}SO_3H} C_6H_5\text{—}CO\text{—}CH_2\text{—}CH_2\text{—}C_6H_5$$

(80%) + (CH$_3$)$_2$C=CH$_2$ + CO$_2$

(Ref. 41a)

38. (a) G. A. Reynolds and C. R. Hauser, *Org. Syn.,* **Coll. Vol. 4**, 708 (1963). (b) K. Meyer and H. S. Bloch, *ibid.,* **Coll. Vol. 3**, 637 (1955).
39. A. Spassow, *Org. Syn., ***Coll. Vol. 3**, 390 (1955).
40. (a) J. A. Price and D. S. Tarbell, *Org. Syn.,* **Coll. Vol. 4**, 285 (1963). (b) D. S. Tarbell and J. A. Price, *J. Org. Chem.,* **22**, 245 (1957).
41. (a) G. S. Fonken and W. S. Johnson, *J. Am. Chem. Soc.,* **74**, 831 (1952). (b) E. C. Taylor and A. McKillop, *Tetrahedron,* **23**, 897 (1967). (c) For related procedures involving the cleavage of trimethylsilyl esters rather than t-butyl esters, see U. Schmidt and M. Schwochau, *Tetrahedron Letters,* **No. 10**, 875 (1967). (d) H. Muxfeldt, G. Grethe, and W. Rogalski, *J. Org. Chem.,* **31**, 2429 (1966). (e) For the use of N-carbamoyl aziridines as acylating agents for enolates, see H. Stamm and G. Führling, *Tetrahedron Letters,* **No. 22**, 1937 (1970).

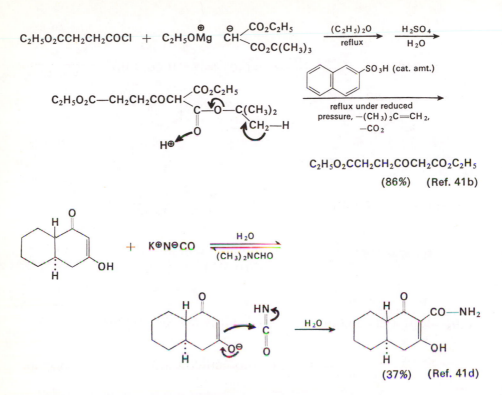

$$C_2H_5O_2CCH_2CH_2COCH_2CO_2C_2H_5$$

(86%)   (Ref. 41b)

(37%)   (Ref. 41d)

an inert solvent rather than an alcohol is necessary to prevent reaction of the acid chloride or anhydride with the solvent. The ethoxymagnesium cation employed in several of the examples offers the advantage of forming enolate salts that are soluble in inert solvents such as benzene or ether. However, the same solubility advantage may be gained by reaction of malonic esters or β-keto esters with sodium hydride in 1,2-dimethoxyethane to produce solutions of the sodium enolates.

An alternative procedure for the acylation of active methylene compounds utilizes the ability of the magnesium cation to form stable chelate structures with the enolate anions of β-keto acids and related compounds.[42] The introduction of

$$Mg(OCH_3)_2 + CO_2 \xrightleftharpoons[]{(CH_3)_2N-CHO} CH_3-O-Mg-O-CO-O-CH_3 \qquad (Ref. 42c)$$

[17]

42. (a) M. Stiles and H. L. Finkbeiner, *J. Am. Chem. Soc., 81*, 505 (1959). (b) M. Stiles, *ibid., 81*, 2598 (1959). (c) H. L. Finkbeiner and M. Stiles, *ibid., 85*, 616 (1963). (d) R. E. Ireland and J. A. Marshall, *ibid., 81*, 2907 (1959). (e) H. Finkbeiner, *ibid., 87*, 4588 (1965); *J. Org. Chem., 30*, 3414 (1965). (f) S. W. Pelletier, R. L. Chappell, P. C. Parthasarathy, and N. Lewin, *ibid., 31*, 1747 (1966). (g) S. N. Balasubrahmanyam and M. Balasubramanian, *Org. Syn., 49*, 56 (1969). (h) T. M. Harris and C. M. Harris, *J. Org. Chem., 31*, 1032 (1966). (i) G. Bottaccio and G. P. Chiusoli, *Chem. Commun., No. 17*, 618 (1966).

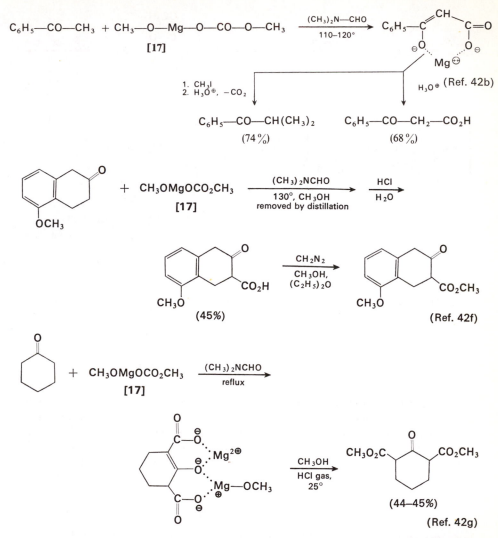

a carboxyl function at a methyl or methylene group alpha to a carbonyl or nitro group has been achieved by reaction of the active methylene compound with the reagent [17], called by the authors methyl magnesium carbonate, in dimethylformamide solution. As indicated in the scheme outlined above, the resulting solution of the magnesium chelate can be either hydrolyzed or alkylated (cf. Chapter 9) and then hydrolyzed and decarboxylated. A related procedure, employed for the synthesis of β-keto esters, utilizes the acylation of an intermediate magnesium chelate as shown. Enolate anions have also been found to react with carbon dioxide under various other conditions to form β-keto acids.[42h,i] This reaction with a dicarbanion is illustrated. However, the scope and possible utility of the reaction of enolate anions with carbon dioxide remains to be explored.

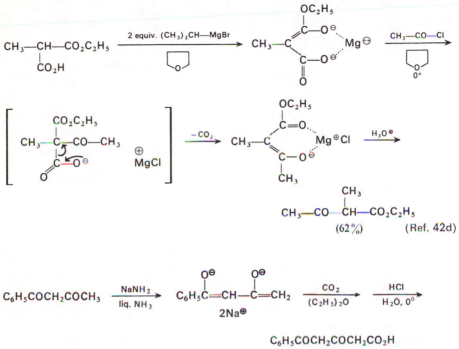

$$C_6H_5COCH_2COCH_3 \xrightarrow[\text{liq. NH}_3]{\text{NaNH}_2} \overset{\ominus O}{C_6H_5C}=CH-\overset{\ominus O}{C}=CH_2 \xrightarrow[\text{(C}_2\text{H}_5)_2\text{O}]{\text{CO}_2} \xrightarrow[\text{H}_2\text{O, 0}°]{\text{HCl}}$$

$$2Na^{\oplus}$$

$$C_6H_5COCH_2COCH_2CO_2H$$

$$(74\%) \qquad \text{(Ref. 42h)}$$

The products of β-keto ester acylations are readily cleaved by reaction with base, especially if there is no acidic proton at the alpha carbon atom. As illustrated in the accompanying equations, the cleavage normally occurs preferentially at the most reactive carbonyl function to produce the anion of the strongest acid (i.e. the most stable anion). Therefore, a corresponding cleavage of α-acylmalonic esters removes the acyl group; nonhydrolytic conditions (see Chapter 9 and

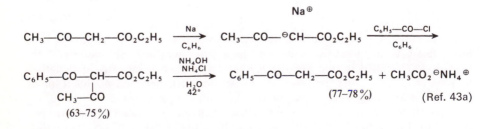

43. (a) R. L. Shriner, A. G. Schmidt, and L. J. Roll, *Org. Syn.*, **Coll. Vol. 2**, 266 (1943). See also J. M. Straley and A. C. Adams, *ibid.*, **Coll. Vol. 4**, 415 (1963). (b) M. Guha and D. Nasipuri, *ibid.*, **42**, 41 (1962). (c) J. J. Looker, *J. Org. Chem.*, **31**, 2714 (1966).

$$CH_3-CO-CH_2-CO_2C_2H_5 \xrightarrow[\text{(C}_2\text{H}_5)_2\text{O}]{\text{Na}} CH_3-CO-{}^{\ominus}CH-CO_2C_2H_5 \xrightarrow[\text{(C}_2\text{H}_5)_2\text{O}]{\overset{\text{Na}^{\oplus}}{\text{Cl}-CO-(CH_2)_3-CO_2C_2H_5}}$$

$$CH_3-CO-\underset{\underset{CO_2C_2H_5}{|}}{CH}-CO-(CH_2)_3-CO_2C_2H_5 \xrightarrow[\text{(C}_2\text{H}_5)_2\text{O}]{\text{NH}_3}$$

(61–66%)   (Ref. 43b)

$$C_2H_5-O-CO-CH_2-CO-(CH_2)_3-CO_2C_2H_5 \ + \ CH_3-CO-NH_2$$

[18]   (50–59%)

$$C_6H_5CH_2CO-\underset{\underset{C_6H_5}{|}}{CH}-SO_2C_6H_4Cl\text{-}p \ + \ \underset{\underset{H}{|}}{\overset{}{\boxed{}}} \xrightarrow[\text{reflux}]{C_6H_6}$$

$$C_6H_5CH_2-\overset{\overset{O^{\ominus}}{\nearrow}}{\underset{\underset{N}{|}}{C}}-CH\overset{SO_2C_6H_4Cl\text{-}p}{\underset{C_6H_5}{}} \xrightarrow{H^{\oplus}} C_6H_5CH_2CO-N\boxed{} \ + \ C_6H_5CH_2SO_2C_6H_4Cl\text{-}p$$

(62%)   (98%)

(Ref. 43e)

subsequent discussion) are required for the synthetically useful cleavage of these substances to give ketones. The cleavage of α-acyl-β-keto esters outlined above provides a good synthetic route to β-keto esters (e.g., **[18]**) of the type that would be obtained as one of a mixture of products from an acetoacetic ester condensation involving two different esters. The acid-catalyzed cleavage of α-acylmalonic esters provides a second such synthetic route, as the following equations and previous examples demonstrate.

$$CH_2(CO_2C_2H_5)_2 \xrightarrow[\underset{25°}{C_2H_5OH}]{KOH} \underset{\underset{CO_2C_2H_5}{|}}{CH_2}-CO_2^{\ominus} K^{\oplus} \xrightarrow[100\text{-}110°]{H_3O^{\oplus} \quad \overset{\overset{CO-Cl}{|}}{\boxed{}}\underset{CO-Cl}{}} \underset{\underset{CO_2C_2H_5}{|}}{CH_2}-CO-Cl$$

$$\xrightarrow[\underset{\text{reflux}}{\underset{(C_2H_5)_2O}{C_6H_5N(CH_3)_2}}]{(CH_3)_3COH} \underset{\underset{CO_2C_2H_5}{|}}{CH_2}-CO_2C(CH_3)_3 \xrightarrow[\underset{\text{reflux}}{\underset{(C_2H_5)_2O}{C_2H_5OH}}]{\overset{Mg}{\text{cat. amt. } CCl_4}} C_2H_5OMg^{\oplus} \ {}^{\ominus}\underset{\underset{CO_2C_2H_5}{|}}{CH}-CO_2-C(CH_3)_3$$

$$\xrightarrow[(C_2H_5)_2O]{C_6H_5-CH_2-CO-Cl} C_6H_5-CH_2-CO-\underset{\underset{CO_2C_2H_5}{|}}{CH}-CO_2C(CH_3)_3 \xrightarrow[\underset{\text{reflux}}{C_6H_6}]{p\text{-}CH_3-C_6H_4-SO_3H}$$

$$C_6H_5-CH_2-CO-CH_2-CO_2C_2H_5 \ + \ CO_2 \ + \ (CH_3)_2C{=}CH_2 \qquad \text{(Ref. 44)}$$

(47%)

44. (a) D. S. Breslow, E. Baumgarten, and C. R. Hauser, *J. Am. Chem. Soc.*, **66**, 1286 (1944). (b) W. H. Miller, A. M. Dessert, and G. W. Anderson, *ibid.*, **70**, 500 (1948)

$$C_2H_5OMg^{\oplus} \ ^{\ominus}CH(CO_2C_2H_5)_2 \xrightarrow[\substack{(C_2H_5)_2O \\ \text{reflux}}]{CH_3-CH_2-CO-Cl} \xrightarrow{H_3O^{\oplus}} CH_3-CH_2-CO-CH(CO_2C_2H_5)_2$$

$$\xrightarrow[200°]{\text{(naphthalene)}-SO_3H} CH_3-CH_2-CO-CH_2-CO_2C_2H_5 \qquad \text{(Ref. 45)}$$

$$(57\%)$$

Acid chlorides and anhydrides may also be used to acylate enolate anions derived from simple esters (e.g., [19]) and from ketones (e.g., [20] and [21]).[24a,47]

$$(CH_3)_2CH-CO_2C_2H_5 \xrightarrow[(C_2H_5)_2O]{(C_6H_5)_3C^{\ominus}Na^{\oplus}} (CH_3)_2C^{\ominus}\overset{Na^{\oplus}}{-CO_2C_2H_5} \xrightarrow[(C_2H_5)_2O]{C_6H_5-CO-Cl}$$

$$[19]$$

$$C_6H_5-CO-\overset{\overset{\displaystyle CH_3}{|}}{\underset{\underset{\displaystyle CH_3}{|}}{C}}-CO_2C_2H_5 \qquad \text{(Ref. 46)}$$

$$(50-55\%)$$

$$(CH_3)_3C-CO-CH_3 \xrightarrow[(C_2H_5)_2O]{NaNH_2} (CH_3)_3C-CO-CH_2^{\ominus} \overset{Na^{\oplus}}{} \xrightarrow[\substack{(C_2H_5)_2O \\ 0°}]{1 \text{ equiv. } (CH_3)_3C-CO-Cl}$$

$$[20] \qquad\qquad (3 \text{ equiv.})$$

$$(CH_3)_3C-CO-\overset{Na^{\oplus}}{CH^{\ominus}}-CO-C(CH_3)_3 \xrightarrow{H_3O^{\oplus}} (CH_3)_3C-CO-CH_2-CO-C(CH_3)_3$$

$$(58\%) \qquad \text{(Ref. 47a)}$$

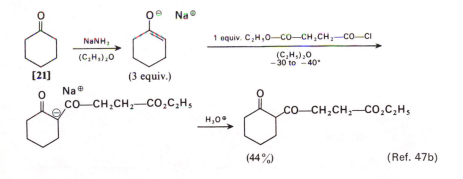

$$(44\%) \qquad\qquad\qquad \text{(Ref. 47b)}$$

45. (a) B. Riegel and W. M. Lilienfeld, *J. Am. Chem. Soc.*, **67**, 1273 (1945).   (b) Acyl-malonates have also been cleaved in boiling water to form β-keto esters. See B. R. Baker, R. E. Schaub, and J. H. Williams, *J. Org. Chem.*, **17**, 116 (1952).
46. (a) C. R. Hauser and W. B. Renfrew, Jr., *Org. Syn.*, **Coll. Vol. 2**, 268 (1943). (b) D. F. Thompson, P. L. Bayless, and C. R. Hauser, *J. Org. Chem.*, **19**, 1490 (1954).
47. (a) B. O. Linn and C. R. Hauser, *J. Am. Chem. Soc.*, **78**, 6066 (1956). (b) C. R. Hauser and B. O. Linn, *ibid.*, **79**, 731 (1957).

In order to obtain good yields of C-acylated ketones by this procedure, it is normally necessary to employ an excess of the enolate anion (usually two or three equivalents per equivalent of the acid chloride or anhydride). Note that in cases where the product is an enolizable 1,3-dicarbonyl compound, one equivalent of the starting enolate will be consumed in converting the initial, relatively acidic product to its metal enolate. If a different reaction procedure is followed in which the enolate from an aldehyde, ketone, or β-keto ester is added to an excess of the acid chloride or acid anhydride, the O-acylated derivative (an enol ester such as [22]) is often the major product.[48] Two of the accompanying examples also illustrate the fact that the proportion of O-acylated product may be enhanced by conducting the acylation reaction in a relatively polar solvent with an enolate metal cation (e.g., Na⊕ or Li⊕) which does not favor the formation of tightly associated ion pairs.[48b,d,e,49a] The proportions of C- and O-acylation are also influenced by the presence of electron-donating or withdrawing substituents in the reactants,[48d] and by the degree of steric hindrance in the reactant.[48b,d,e,f] These various factors are similar to those which influence the previously discussed C- and O-alkylation reactions (see Chapter 9). The principal difference seems to be that O-acylation is usually the most serious side reaction with alkali metal enolates of simple aldehydes and ketones, the compounds which offer the least problem with regard to O-alkylation.

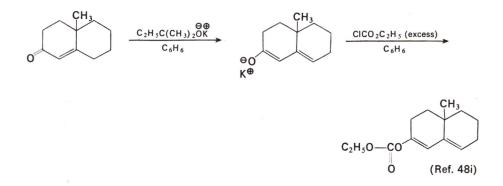

(Ref. 48i)

48. (a) H. O. House and V. Kramar, *J. Org. Chem.,* **28**, 3362 (1963). (b) J. P. Ferris, C. E. Sullivan, and B. G. Wright, *ibid.,* **29**, 87 (1964); J. P. Ferris, B. G. Wright, and C. C. Crawford, *ibid.,* **30**, 2367 (1965). (c) H. O. House and D. J. Reif, *J. Am. Chem. Soc.,* **77**, 6525 (1955). (d) For studies of the C- and O-acylation of various β-diketone metal enolates, see D. C. Nonhebel and J. Smith, *J. Chem. Soc.,* C, 1919 (1967); H. D. Murdock and D. C. Nonhebel, *ibid.,* 2153 (1962); C, 2298 (1968). (e) H. O. House, W. L. Respess, and G. M. Whitesides, *J. Org. Chem.,* **31**, 3128 (1966). (f) H. O. House, L. J. Czuba, M. Gall, and H. D. Olmstead, *ibid.,* **34**, 2324 (1969); H. O. House, M. Gall, and H. D. Olmstead, To be published. (g) W. M. Muir, P. D. Ritchie, and D. J. Lyman, *J. Org. Chem.,* **31**, 3790 (1966). (h) K. Yoshida and Y. Yamashita, *Tetrahedron Letters,* **No. 7**, 693 (1966). (i) N. K. Basu, U. R. Ghatak, G. Sengupta, and P. C. Dutta, *Tetrahedron,* **21**, 2641 (1965).
49. (a) R. Gompper, *Angew Chem., Intern. Ed. Engl.,* **3**, 560 (1964). (b) R. E. Davis, *Tetrahedron Letters,* **No. 41**, 5021 (1966). (c) R. G. Pearson and J. Songstad, *J. Am. Chem. Soc.,* **89**, 1827 (1967).

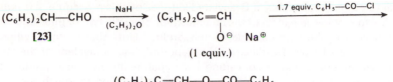

$(C_6H_5)_2CH\text{—}CHO \xrightarrow[(C_2H_5)_2O]{NaH} (C_6H_5)_2C\text{=}CH \xrightarrow{\text{1.7 equiv. } C_6H_5\text{—}CO\text{—}Cl}$

[23] $\underset{O^\ominus \quad Na^\oplus}{|}$

(1 equiv.)

$(C_6H_5)_2C\text{=}CH\text{—}O\text{—}CO\text{—}C_6H_5$

(39%)   (Ref. 48c)

$(CH_3)_2CH\text{—}CH\text{=}\underset{\underset{M^\oplus}{\overset{|}{\underset{O^\oplus}{}}}}{C}\text{—}CH_3 \xrightarrow[\text{solvent}]{(CH_3CO)_2O} \xrightarrow{H_2O}$

$(CH_3)_2CHCH\text{=}\underset{\underset{OCOCH_3}{|}}{C}\text{—}CH_3 \quad + \quad (CH_3)_2CH\text{—}CH(COCH_3)_2$

[22]

$M^\oplus$	Solvent	Product yields	
Li	$CH_3OCH_2CH_2OCH_3$	75%	<1%
Li	$(C_2H_5)_2O$	68%	4%
BrMg	$(C_2H_5)_2O$	36%	34%

(Ref. 48e)

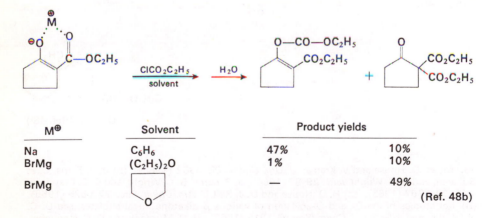

$M^\oplus$	Solvent	Product yields	
Na	C_6H_6	47%	10%
BrMg	$(C_2H_5)_2O$	1%	10%
BrMg		—	49%

(Ref. 48b)

The successful C-acylations which have been accomplished by reaction of an acid chloride or anhydride with excess (2–3 equivalents) ketone enolate[24a,47] are now known to be the result of the further reaction of the O-acylated product with more enolate ion as in structure **[23]**.[48a,f,g,h] Thus, the product from the kinetically controlled acylation of an enolate is often the O-acylated product which is capable of reacting with an additional equivalent of the enolate ion to form the more stable anion of the C-acylated product.[48]

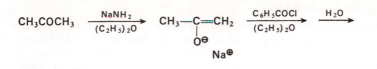

$$C_6H_5COO-C=CH_2 \quad + \quad C_6H_5COCH_2COCH_3$$
$$\qquad\qquad | $$
$$\qquad\quad CH_3$$

Reaction conditions	Product yields	
1 equiv. of acid chloride added to 2 equiv. of enolate	9%	33%
1 equiv. of enolate added to 2 equiv. of acid chloride	41%	6%

(Ref. 48g)

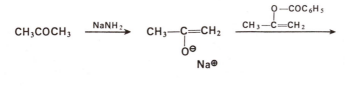

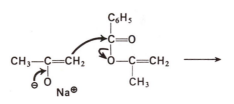

[23]

$$CH_3-C=CH_2 \quad + \quad CH_3COCH_2COC_6H_5 \longrightarrow$$
$$\quad\ \ |$$
$$\quad O^\ominus$$

$$CH_3COCH_3 \ + \ CH_3C-CH=C-C_6H_5 \ \xrightarrow{\ H_2O\ } \ C_6H_5COCH_2COCH_3$$
$$\qquad\qquad\quad \| \qquad\quad |$$
$$\qquad\qquad\quad O \qquad\quad O^\ominus \qquad\qquad\qquad (71\%)$$

(Ref. 48g)

In certain cases involving hindered ketones[47] the O-acylated compound has been isolated either as a by-product or as a major product from the reaction of an acid chloride with excess enolate anion. This result presumably reflects the fact that the reaction of the excess enolate anion with the O-acylated product is relatively slow.

THE ACYLATION OF ENAMINES

An alternative procedure for the acylation of ketones utilizes the reaction of enamines (see Chapters 9 and 10) with acid chlorides or acid anhydrides.[50,51] Unlike that with alkylating agents, the reaction of enamines with acylating reagents at nitrogen forms an N-acylammonium salt (e.g., [24]) which is still a good acylating agent. Consequently, good yields of C-acylated products may be formed. It will be noted in the accompanying examples that the C-acylated product (e.g., [25]), a weakly basic enamine, does not absorb the acid formed during the acylation. Consequently, it is necessary to employ either two equivalents of the enamine or a

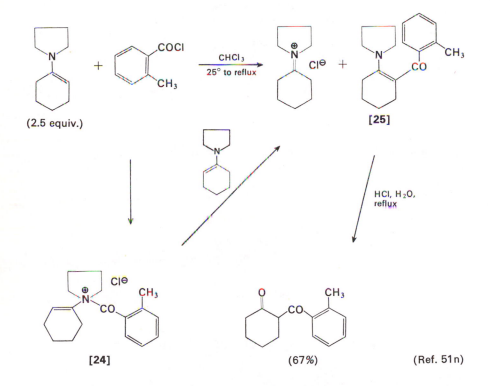

50. (a) G. Stork, A. Brizzolara, H. Landesman, J. Szmuszkovicz, and R. Terrell, *J. Am. Chem. Soc.*, **85**, 207 (1963). (b) J. Szmuszkovicz, *Adv. Org. Chem.*, **4**, 1 (1963).
51. (a) S. Hünig and E. Lücke, *Chem. Ber.*, **92**, 652 (1959). (b) S. Hünig, E. Lücke, and E. Benzing, *ibid.*, **91**, 129 (1958). (c) S. Hünig and W. Lendle, *ibid.*, **93**, 909, 913 (1960). (d) S. Hünig and M. Salzwedel, *ibid.*, **99**, 823 (1966). (e) S. Hünig and H. J. Buysch, *ibid.*, **100**, 4010, 4017 (1967); S. Hünig, H. J. Buysch, H. Hoch, and W. Lendle, *ibid.*, **100**, 3996 (1967). (f) S. Hünig, E. Lücke, and W. Brenninger, *Org. Syn.*, **43**, 34 (1963). (g) G. Opitz and E. Tempel, *Justus Liebigs Ann. Chem.*, **699**, 74 (1966). (h) G. Opitz and M. Kleemann, *ibid.*, **665**, 114 (1963); C. Wakselman, *Bull. Soc. Chim. France*, 3763 (1967); A. Kirrmann and C. Wakselman, *ibid.*, 3766 (1967). (i) S. Hünig and H. Hoch, *Tetrahedron Letters*,

mixture of one equivalent of the enamine and one equivalent of a second tertiary amine (usually triethylamine) for each equivalent of the acid chloride in order that the hydrogen chloride liberated may be absorbed. If the initially formed β-acyl

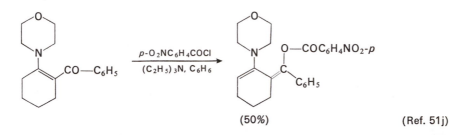

(50%)   (Ref. 51j)

enamine is treated with excess of the acid chloride and tertiary amine, an O-acylated product usually results.[51j−l] Although the reaction path just discussed appears to be applicable to some reactions of enamines with *aliphatic acid chlorides*, in many cases an alternative path is followed in which the first step is the previously described reaction of the aliphatic acid chloride with one equivalent of the enamine (or other added base) to form a ketene.[10] The ketene then undergoes a cycloaddition reaction with the enamine to form a cyclobutanone intermediate.[51d,g−i,k,m] It will be noted

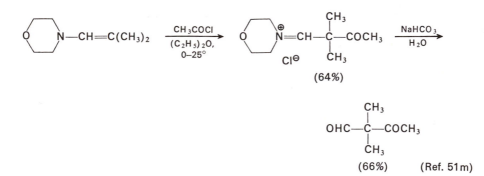

(64%)

(66%)   (Ref. 51m)

from the accompanying examples that the cyclobutanone intermediates (e.g., [26] and [27]) may undergo ring opening in either of two directions during the hydrolysis step. The intermediate cyclobutanones obtained from the enamines of five-, six-, seven-, and eight-membered ketones usually are opened to form 2-acylcycloalkanones whereas the other mode of ring opening (e.g., [27a]) to form cycloalkane-1,3-diones is favored with larger sized rings.[51h,i]

No. 42, 5215 (1966). (j) R. Helmers, *ibid.,* No. 18, 1905 (1966). (k) G. H. Alt, *J. Org. Chem.,* 31, 2384 (1966); G. H. Alt and A. J. Speziale, *ibid.,* 29, 794, 798 (1964). (l) G. A. Berchtold, G. R. Harvey, and G. E. Wilson, Jr., *ibid.,* 30, 2642 (1965). (m) T. Inukai and R. Yoshizawa, *ibid.,* 32, 404 (1967). (n) H. O. House, R. G. Carlson, H. Müller, A. W. Noltes, and C. D. Slater, *J. Am. Chem. Soc.,* 84, 2614 (1962).

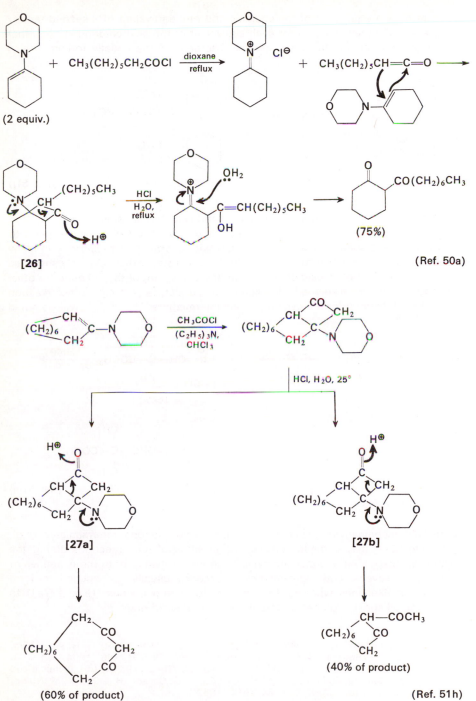

(2 equiv.)

[26]

(75%)

(Ref. 50a)

[27a]   **[27b]**

(60% of product)

(40% of product)

(Ref. 51h)

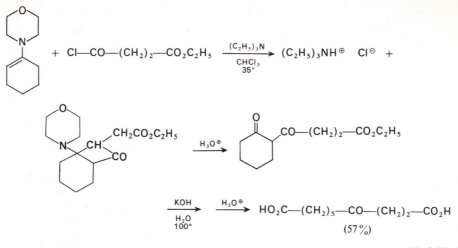

(Ref. 51a)

Enamines may also be acylated with reagents such as acid anhydrides,[50] ethyl chloroformate,[50] cyanogen chloride,[52a,c] nitrosyl chloride (to form an α-oximino-ketone),[52b] phosgene,[52d,e] and imino chlorides.[52f,g] Examples of these reactions are provided in the following equations.

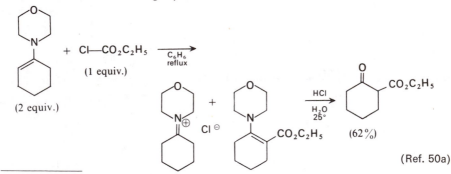

(Ref. 50a)

52. (a) M. E. Kuehne, *J. Am. Chem. Soc.*, **81**, 5400 (1959). (b) H. Metzger, *Tetrahedron Letters*, **No. 4**, 203 (1964). (c) R. T. Parfitt, *J. Chem. Soc.*, C, 140 (1967). (d) A. Halleux and H. G. Viehe, *ibid.*, C, 881 (1970). (e) R. Buyle and H. G. Viehe, *Tetrahedron*, **24**, 3987, 4217 (1968). (f) Y. Ito, S. Katsuragawa, M. Okano, and R. Oda, *ibid.*, **23**, 2159 (1967). (g) W. Ziegenbein, *Angew. Chem., Intern. Ed. Engl.*, **4**, 358 (1965). (h) W. Sobotka, W. N. Beverung, G. G. Munoz, J. C. Sircar, and A. I. Meyers, *J. Org. Chem.*, **30**, 3667 (1965); A. I. Meyers and J. C. Sircar, *ibid.*, **32**, 1250 (1967). (i) R. J. Friary, R. W. Franck, and J. F. Tobin, *Chem. Commun.*, **No. 5**, 283 (1970). (j) R. Fuks and H. G. Viehe, *Tetrahedron*, **25**, 5721 (1969). (k) P. W. Hickmott and J. R. Hargreaves, *ibid.*, **23**, 3151 (1967). (l) N. F. Firrell and P. W. Hickmott, *J. Chem. Soc.*, C, 2320 (1968). (m) J. R. Hargreaves, P. W. Hickmott, and B. J. Hopkins, *ibid.*, C, 2599 (1968); *ibid.*, C, 592 (1969). (n) P. W. Hickmott and B. J. Hopkins, *ibid.*, C, 2918 (1968). (o) N. F. Firrell, P. W. Hickmott, and B. J. Hopkins, *ibid.*, C, 1477 (1970). (p) P. W. Hickmott, B. J. Hopkins, and C. T. Yoxall, *Tetrahedron Letters*, **No. 29**, 2519 (1970); R. Gelin, S. Gelin, and R. Dolmazon, *ibid.*, **No. 42**, 3657 (1970). (q) W. Steglich and G. Höfle, *ibid.*, **No. 13**, 1619 (1968); G. Singh and S. Singh, *ibid.*, **No. 50**, 3789 (1964) and references therein. (r) R. H. Wiley and O. H. Borum, *Org. Syn.*, **Coll. Vol. 4**, 5 (1963).

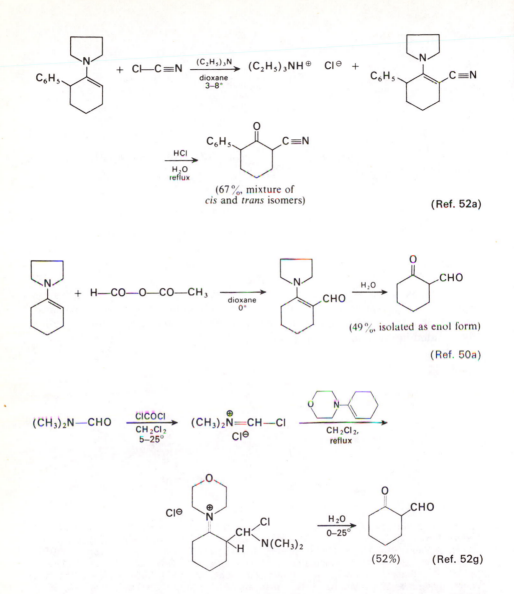

(67%, mixture of
cis and trans isomers)

(Ref. 52a)

(49%, isolated as enol form)

(Ref. 50a)

(52%)   (Ref. 52g)

Various ring-closure reactions which involve enamine acylations have been reported.[52h-p] The accompanying equations illustrate the intramolecular acylation of an enamine by an ester and the formation of cyclic products from reaction of enamines with α,β-unsaturated acid chlorides.[52k-p] The latter reactions are believed to involve the initial formation of N-acyl derivatives (e.g., [28]) which subsequently undergo the sigmatropic rearrangement and cyclization indicated.

An acylation reaction related to those previously discussed is the Dakin-West reaction which is used to convert an α-amino acid to an α-acylamido ketone.[52q,r]

$$C_6H_5-NH-CH_2CO_2C_2H_5 \; + \; (CH_3)_2N-C\equiv C-C_6H_5 \quad \xrightarrow[\;(C_2H_5)_2O\;]{\text{HCl (cat. amt.)}}$$

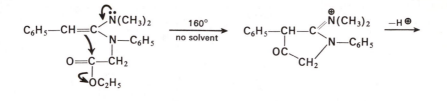

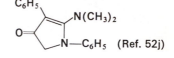

(Ref. 52j)

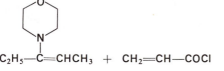

$$C_2H_5-C{=}CHCH_3 \; + \; CH_2{=}CH-COCl \quad \xrightarrow[\text{reflux}]{C_6H_6}$$

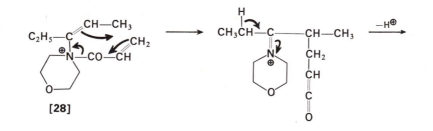

[28]

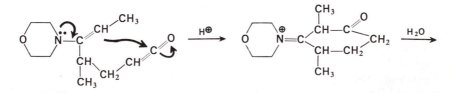

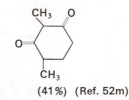

(41%) (Ref. 52m)

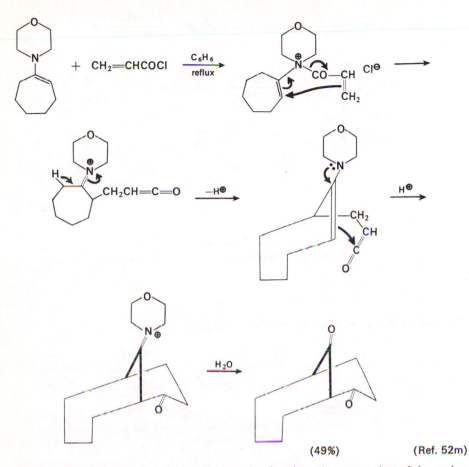

(49%)   (Ref. 52m)

As the following equation shows, the reaction involves the conversion of the amino acid to an oxazolone (also called an azlactone) which is acylated first at oxygen and then rearranged to a C-acylated oxazolone. Subsequent acetolysis yields the β-keto acid which decarboxylates.

THE ACYLATION OF ACTIVE METHYLENE COMPOUNDS UNDER ACIDIC CONDITIONS

The acylation of active methylene compounds with acid anhydrides or acid chlorides to form 1,3-dicarbonyl compounds may also be accomplished in the presence of acid catalysts such as boron trifluoride,[24a,53,54] aluminum chloride,[55a-e] or various

53. (a) D. Kästner in *Newer Methods of Preparative Organic Chemistry*, Wiley-Interscience, New York, 1948, pp. 249–313. (b) H. S. Booth and D. R. Martin, *Boron Trifluoride and its Derivatives*, Wiley, New York, 1949. (c) 1,3-dicarbonyl compounds have also been prepared by reaction of the borofluoride complexes of β-keto acids with an acid anhydride at elevated temperatures: see H. Musso and K. Figge, *Justus Liebigs Ann. Chem.*, **668**, 1, 15 (1963). (d) For a study of the reactions of the borofluoride complexes of 1,3-dicarbonyl compounds, see R. A. J. Smith and T. A. Spencer, *J. Org. Chem.*, **35**, 3220 (1970).

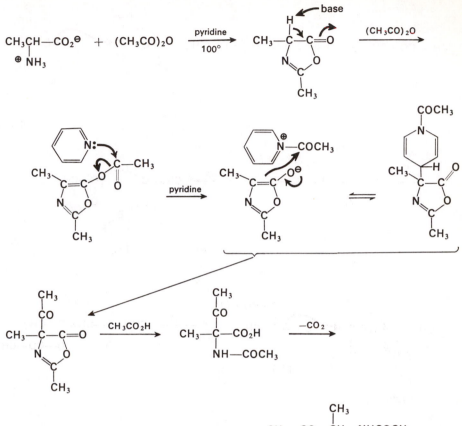

$$CH_3-CO-\overset{\overset{\displaystyle CH_3}{|}}{CH}-NHCOCH_3$$

(81–88%)   (Ref. 52r)

proton acids. The boron trifluoride-catalyzed reaction of ketones with acid anhydrides is thought to involve electrophilic attack by the complex of the anhydride and the Lewis acid (e.g., [29]) on the enol derivative of the ketone (e.g., [30]). In the presence of a proton acid, the enol of the ketone is apparently first acylated

54. (a) C. R. Hauser and J. T. Adams, *J. Am. Chem. Soc.*, **66**, 345 (1944). (b) J. T. Adams and C. R. Hauser, *ibid.*, **67**, 284 (1945). (c) C. R. Hauser, F. C. Frostick, Jr., and E. H. Man, *ibid.*, **74**, 3231 (1952). (d) R. M. Manyik, F. C. Frostick, Jr., J. J. Sanderson, and C. R. Hauser, *ibid.*, **75**, 5030 (1953). (e) C. E. Denoon, Jr., *Org. Syn.*, **Coll. Vol. 3**, 16 (1955). (f) J. F. Wolfe, C. J. Eby, and C. R. Hauser, *J. Org. Chem.*, **30**, 55 (1965). (g) E. M. Kaiser, S. D. Work, J. F. Wolfe, and C. R. Hauser, *ibid.*, **32**, 1483 (1967). (h) C. L. Mao, F. C. Frostick, Jr., E. H. Man, R. M. Manyik, R. L. Wells, and C. R. Hauser, *ibid.*, **34**, 1425 (1969). (i) M. Gorodetsky, E. Levy, R. D. Youssefyeh, and Y. Mazur, *Tetrahedron*, **22**, 2039 (1966). (j) B. C. Elmes, M. P. Hartshorn, and D. N. Kirk, *J. Chem. Soc.*, 2285 (1964). (k) R. D. Youssefyeh, *J. Am. Chem. Soc.*, **85**, 3901 (1963). (l) J. A. Durden and D. G. Crosby, *J. Org. Chem.*, **30**, 1684 (1965). (m) Enol esters react rapidly with hydrogen fluoride to form acid fluorides and ketones. E. S. Rothman, G. G. Moore, and S. Serota, *ibid.*, **34**, 2486 (1969).

$$CH_3\!-\!CO\!-\!CH_3 \ + \ (CH_3\!-\!CO)_2O \xrightarrow[0°]{BF_3}$$

(excess)

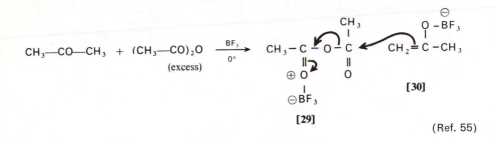

[29]   [30]

(Ref. 55)

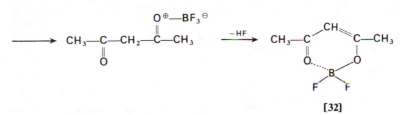

[32]

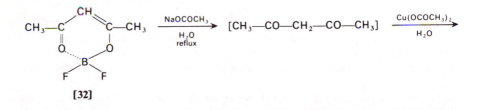

[29]   [31]

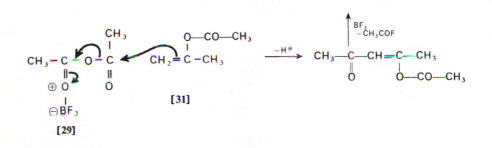

[32] $[CH_3\!-\!CO\!-\!CH_2\!-\!CO\!-\!CH_3]$ $\xrightarrow[H_2O]{Cu(OCOCH_3)_2}$

NaOCOCH₃, H₂O reflux

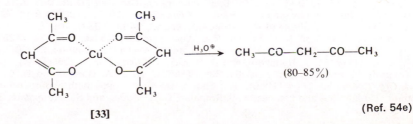

$\xrightarrow{H_3O^{\oplus}}$ $CH_3\!-\!CO\!-\!CH_2\!-\!CO\!-\!CH_3$

(80–85%)

[33]

(Ref. 54e)

at oxygen and the resulting enol ester (e.g., [31]) is then attacked by the anhydride-Lewis acid complex. In either case, the resulting 1,3-dicarbonyl compound is subsequently converted to its borofluoride complex (e.g., [32]), which may be either isolated from the reaction mixture or, more commonly, directly hydrolyzed to form the β-diketone. The conversion of the 1,3-dicarbonyl compound to its copper complex (e.g., [33]) is often employed to facilitate separation of the product from other components in the reaction mixture.

The various conditions which have been used in this acylation of ketones with anhydrides include saturation of the reaction mixture with gaseous boron trifluoride,[24a,54a−d] reaction of the ketone with the anhydride in the presence of p-toluenesulfonic acid followed by the addition of boron trifluoride as its solid or liquid complex with acetic acid,[54f,h] or addition of the boron trifluoride as its liquid complex with diethyl ether.[54i−k] The reaction path is believed to change with variations in reaction conditions; the use of gaseous boron trifluoride in the absence of proton acids favors acylation of the initially formed enol derivative (e.g., [30]) and usually leads to substitution at the less highly substituted α-carbon. The reactions performed in the presence of a proton acid and with one of the boron trifluoride complexes favor the initial formation and equilibration of the enol

$$CH_3COCH(CH_3)_2 \xrightarrow[\text{(CH}_3\text{CO)}_2\text{O}]{\text{BF}_3 \text{ catalyst}} \xrightarrow[\text{H}_2\text{O, reflux}]{\text{NaOCOCH}_3}$$

$$CH_3COCH_2COCH(CH_3)_2 \;+\; CH_3CO-\overset{\overset{\displaystyle CH_3}{|}}{\underset{\underset{\displaystyle CH_3}{|}}{C}}-COCH_3$$

[34]

Reaction conditions	Product composition	
BF$_3$ gas, fast saturation, 0–10°	70%	30%
BF$_3$ gas, slow saturation, 0–10°	37%	63%
BF$_3$(CH$_3$CO$_2$H)$_2$ complex, 25°	2%	98%
BF$_3$(CH$_3$CO$_2$H)$_2$ complex + p-CH$_3$C$_6$H$_4$SO$_3$H, 25°	—	100%
		(Ref. 54h)

55. (a) P. C. Doolan and P. H. Gore, J. Chem. Soc., C, 211 (1967). (b) H. Schick, G. Lehmann, and G. Hilgetag, Angew. Chem., Intern. Ed. Engl., 6, 80, 371 (1967); J. Prakt. Chem., 35, 28 (1967); H. Schick and G. Lehmann, ibid., 38, 391 (1968). The reversibility of the aluminum chloride-catalyzed acylations of active methylene compounds is suggested by the work of M. Frangopol, A. Genunche, N. Negoita, P. T. Frangopol, and A. T. Balaban, Tetrahedron, 23, 841 (1967). (c) V. J. Grenda, G. W. Lindberg, N. L. Wendler, and S. H. Pines, J. Org. Chem., 32, 1236 (1967). (d) F. Merenyi and M. Nilsson, Acta Chem. Scand., 21, 1755 (1967). (e) E. S. Rothman and G. G. Moore, J. Org. Chem., 35, 2351 (1970); E. S. Rothman, G. G. Moore, and A. N. Speca, Tetrahedron Letters, No. 59, 5205 (1969). (f) R. Sciaky and U. Pallini, ibid., No. 28, 1839 (1964). (g) K. Ikawa, F. Takami, Y. Fukui, and K. Tokuyama, ibid., No. 38, 3279 (1969). (h) D. Burn and co-workers, Tetrahedron, 20, 597 (1964); 21, 569 (1965); D. Burn, J. P. Yardley, and V. Petrow, ibid., 25, 1155 (1969). (i) L. A. Paquette, B. A. Johnson, and F. M. Hinga, Org. Syn., 46, 18 (1966). (j) J. P. Dusza, J. P. Jospeh, and S. Bernstein, J. Am. Chem. Soc., 86, 3908 (1964). (k) W. H. Hartung and F. Crossley, Org. Syn., Coll. Vol. 2, 363 (1943). See also N. Levin and W. H. Hartung, ibid., Coll. Vol. 3, 191 (1955).

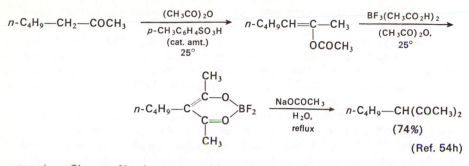

(Ref. 54h)

esters (see Chapter 9); the more stable enol ester is then acylated (see structure [31]) and often leads to substitution at the more highly substituted α-carbon. The accompanying examples illustrate the differences in product composition which may be observed and the possibility of preferential acylation at the more highly substituted alpha position. The successful formation of the β-diketone [34] with no hydrogen atoms at the center carbon indicates that the formation of a borofluoride complex such as [32] is not essential to the success of this acylation reaction.

Other examples of boron trifluoride-catalyzed acylation reactions are provided below. The reactions using boron trifluoride etherate as a catalyst are less vigorous than other acylating conditions; under these conditions reactions at an α-carbon with two alkyl substituents often do not occur.[54i] It should be noted in the first two of the following examples that the position of acylation in this acid catalyzed procedure is determined by the stereochemistry of the ring fusion. This result is in agreement with the previous generalization (see Chapter 8) that formation of the $\Delta^{3,4}$-enol is kinetically favored with a *cis*-ring fusion whereas the $\Delta^{2,3}$-enol is formed most rapidly in the corresponding *trans*-fused system. The previously described base-catalyzed reactions were controlled by equilibration among the

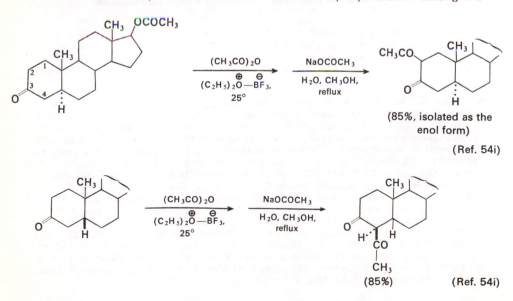

(85%, isolated as the enol form)

(Ref. 54i)

(85%)

(Ref. 54i)

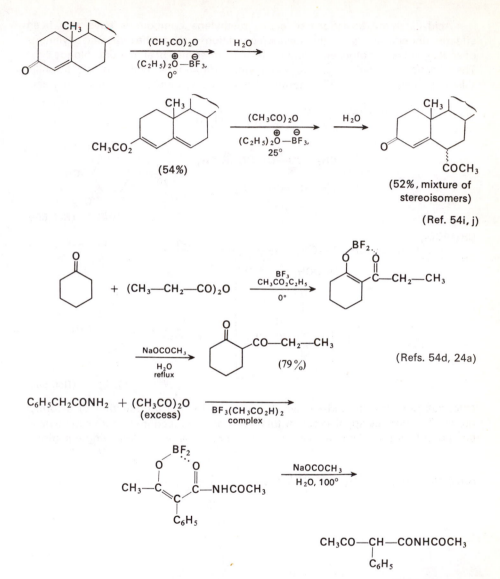

possible products, the enolates of 1,3-dicarbonyl compounds. As a result the base-catalyzed acylations invariably lead to reaction at the less hindered 2-position of a 3-keto steroid irrespective of the stereochemistry of fusion of the A and B rings. The illustrated acylation of an α,β-unsaturated ketone at the γ-position is also worthy of note; this position of attack corresponds to that observed with other electrophilic reagents and dienols or their derivatives (see Chapter 9).

Acid-catalyzed acylations of active methylene compounds have also been effected under a variety of other reaction conditions such as the indicated acylations of enol acetates and other enol derivatives in the presence of aluminum chloride.[55a-e] The reaction of various active methylene compounds with the electrophilic Vilsmeier reagent (e.g., [35]), formed from dimethylformamide and various acid

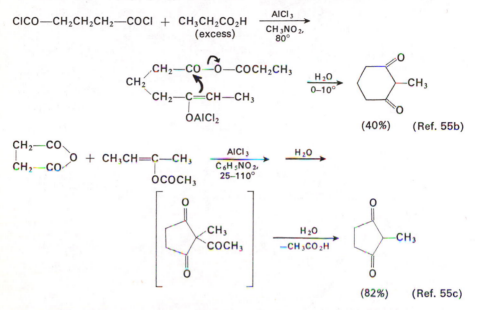

$$ClCO-CH_2CH_2CH_2-COCl + CH_3CH_2CO_2H \xrightarrow[\substack{CH_3NO_2, \\ 80°}]{AlCl_3}$$
(excess)

(40%)   (Ref. 55b)

(82%)   (Ref. 55c)

chlorides provides an acid-catalyzed procedure for the introduction of a formyl group.[55f-i] Various applications of this formylation procedure to active methylene compounds or the related enol derivatives are provided in the following equations.

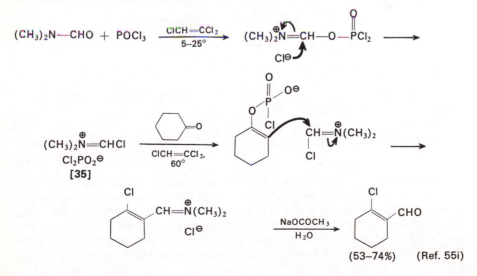

$$(CH_3)_2N-CHO + POCl_3 \xrightarrow[5-25°]{ClCH=CCl_2}$$

(53–74%)   (Ref. 55i)

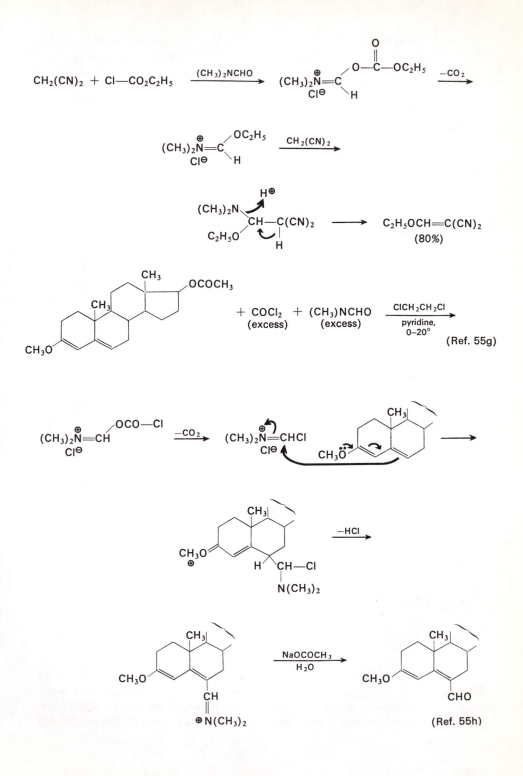

(80%)

+ COCl₂ + (CH₃)NCHO   (Ref. 55g)
(excess)   (excess)

(Ref. 55h)

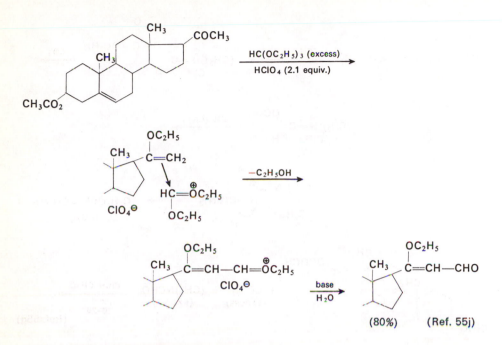

It was mentioned earlier in this chapter that the use of alkyl nitrites for the nitrosation of active methylene compounds is most often performed with acid catalysis.[35c] The application of this reaction to the preparation of the monoxime of an α-diketone is shown below.

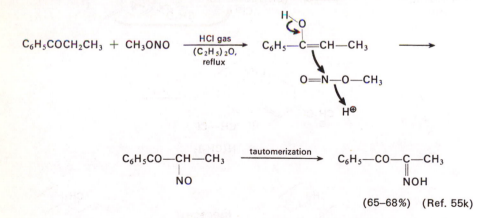

Intramolecular acylation may also be effected with acid catalysts,[55b-d,56] as illustrated by the cyclization of the keto acid [36]. The β-diketone produced in this reaction cannot form a stable enol, enolate anion, or borofluoride complex; it is presumably for this reason that attempts to effect intramolecular base-catalyzed

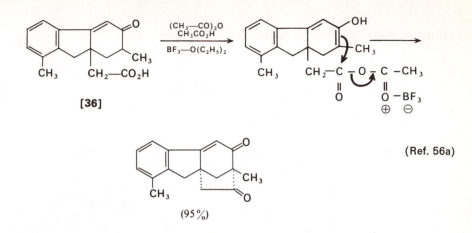

[36]

(Ref. 56a)

(95%)

cyclizations with the ester corresponding to the acid [36] failed.[56a] Cyclizations of this type may also be effected by treatment of the keto acid with a proton acid such as an arenesulfonic acid in a refluxing hydrocarbon solvent with the continuous

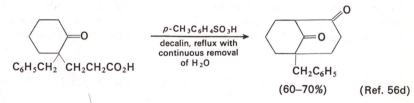

(60–70%)   (Ref. 56d)

removal of water.[56c,d]   A number of workers[54g,l,56e,f] have applied comparable acylation reactions to the preparation of pyrones and pyrylium salts such as those indicated in the following examples.

$C_6H_5CH_2COCH_2C_6H_5$ + CH_3CO_2H $\xrightarrow[130-135°]{\text{polyphosphoric acid}}$
(excess)

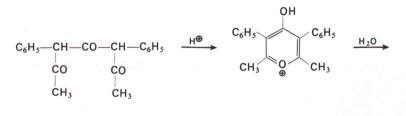

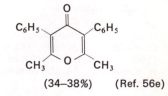

(34–38%)   (Ref. 56e)

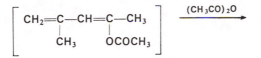

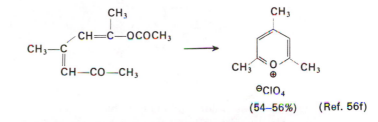

(54–56%)   (Ref. 56f)

Enol esters (e.g., [37]) of ketones may be isomerized thermally, photochemically, or by treatment with various acids such as boron trifluoride (as well as by treatment with bases) to form 1,3-dicarbonyl compounds.[24a,57] The acid-catalyzed reaction is believed to involve electrophilic attack of a Lewis acid–enol ester complex (e.g., [38]) on a second molecule of the enol ester, as shown below.

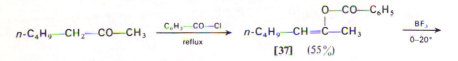

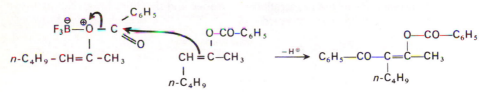

[38]

(Ref. 57a)

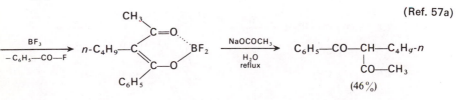

(46%)

56. (a) Y. Kos and H. J. E. Leowenthal, *J. Chem. Soc.*, 605 (1963). (b) D. Becker and H. J. E. Loewenthal, *ibid.*, 1338 (1965). (c) H. J. E. Loewenthal and Z. Neuwirth, *J. Org. Chem.*, **32**, 517 (1967). (d) R. Fusco and F. Tenconi, *Tetrahedron Letters*, **No. 18**, 1313 (1965). (e) T. L. Emmick and R. L. Letsinger, *Org. Syn.*, **47**, 54 (1967). (f) K. Hafner and H. Kaiser, *ibid.*, **44**, 101 (1964). See also A. T. Balaban and C. D. Nenitzescu, *ibid.*, **44**, 98 (1964).

This procedure appears to be particularly useful for the introduction of an aroyl group under acidic conditions. Other examples of the rearrangement of enol esters are given in the following equations. The photochemical and thermal

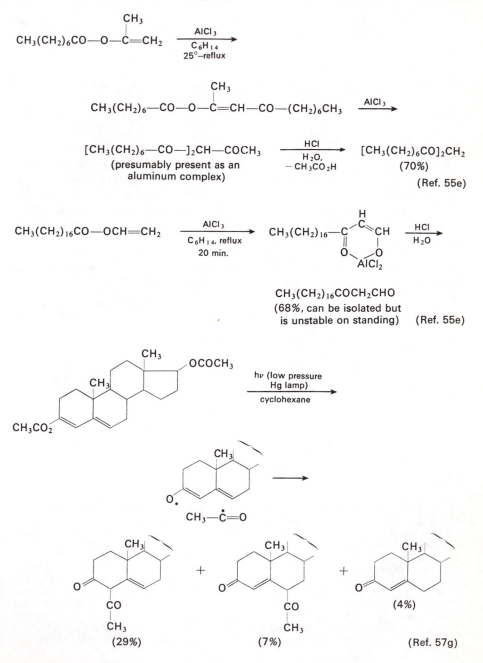

$$CH_3(CH_2)_6CO—O—\underset{\underset{CH_3}{|}}{C}{=}CH_2 \xrightarrow[\substack{C_6H_{14} \\ 25°–reflux}]{AlCl_3}$$

$$CH_3(CH_2)_6—CO—O—\underset{\underset{CH_3}{|}}{C}{=}CH—CO—(CH_2)_6CH_3 \xrightarrow{AlCl_3}$$

$$[CH_3(CH_2)_6—CO—]_2CH—COCH_3 \xrightarrow[\substack{H_2O, \\ -CH_3CO_2H}]{HCl} [CH_3(CH_2)_6CO]_2CH_2$$
(presumably present as an aluminum complex) (70%)

(Ref. 55e)

$$CH_3(CH_2)_{16}CO—OCH{=}CH_2 \xrightarrow[\substack{C_6H_{14}, \text{ reflux} \\ 20 \text{ min.}}]{AlCl_3} CH_3(CH_2)_{16}—\underset{\underset{O}{\diagdown}\underset{\underset{AlCl_2}{|}}{}}{C}\overset{\overset{H}{|}}{\underset{}{\diagup}{C}}\diagdown\underset{O}{\diagdown}CH \xrightarrow[H_2O]{HCl}$$

$$CH_3(CH_2)_{16}COCH_2CHO$$
(68%, can be isolated but is unstable on standing)   (Ref. 55e)

CH₃CO₂ — [steroid structure with OCOCH₃, CH₃, CH₃ substituents] $\xrightarrow[\text{cyclohexane}]{hν \text{ (low pressure Hg lamp)}}$

[bicyclic structure with CH₃, O•, and CH₃—Ċ=O] →

[bicyclic structure with CH₃, O=, CO, CH₃]
(29%)

+

[bicyclic structure with CH₃, O=, CO, CH₃]
(7%)

+

[bicyclic structure with CH₃, O=]
(4%)

(Ref. 57g)

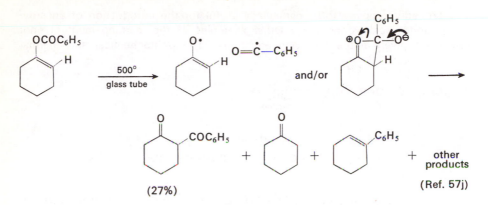

(27%)   (Ref. 57j)

processes appear to be intramolecular in nature; they would appear to involve either the indicated cleavage of the acyl carbon-oxygen bond to form a pair of radicals or the formation of a four-membered ring intermediate. Because the yields of 1,3-dicarbonyl products are relatively low in these thermal and photochemical reactions, the acid-catalyzed (or base-catalyzed) rearrangement of enol esters is usually the method of choice for synthetic work.

The α-acylcyclohexanones and α-acylcyclopentanones obtainable from the previously described acylations (particularly from the acylation of enamines) may be cleaved by reaction with aqueous alkali to form keto acids.[50a,51,58] As illustrated in the accompanying equations, this cleavage, involving attack by hydroxide ion (e.g., [39]) at the cycloalkanone carbonyl function, is the predominant

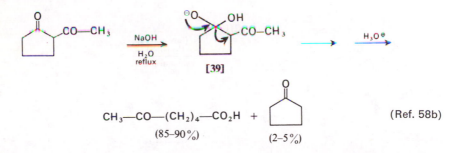

CH_3—CO—$(CH_2)_4$—CO_2H +   (Ref. 58b)

(85–90%)   (2–5%)

57. (a) F. G. Young, F. C. Frostick, Jr., J. J. Sanderson, and C. R. Hauser, *J. Am. Chem. Soc.,* **72**, 3635 (1950). (b) H. H. Wasserman and S. H. Wentland, *Chem. Commun.,* **No. 1,** 1 (1970). (c) V. I. Denes and G. Ciurdarn, *ibid.,* **No. 8,** 510 (1970). (d) D. Cohen and G. E. Pattenden, *J. Chem. Soc.,* C, 2314 (1967). (e) J. Correa and R. M. Mainero, *J. Org. Chem.,* **34**, 2192 (1969). (f) P. N. Rao, J. E. Burdett, Jr., and B. E. Edwards, *ibid.,* **34**, 2767 (1969); A. J. Liston and P. Toft, *ibid.,* **33**, 3109 (1968). (g) M. Gorodetsky and Y. Mazur, *Tetrahedron,* **22**, 3607 (1966); *J. Am. Chem. Soc.,* **86**, 5213 (1964). (h) A. Yogev, M. Gorodetsky, and Y. Mazur, *ibid.,* **86**, 5208 (1964). (i) J. Libman, M. Sprecher, and Y. Mazur, *ibid.,* **91**, 2062 (1969). (j) R. J. P. Allan, R. L. Forman, and P. D. Ritchie, *J. Chem. Soc.,* 2717 (1955); R. J. P. Allan, J. McGee, and P. D. Ritchie, *ibid.,* 4700 (1957).
58. (a) C. R. Hauser, F. W. Swamer, and B. I. Ringler, *J. Am. Chem. Soc.,* **70**, 4023 (1948). (b) P. J. Hamrick, Jr., C. F. Hauser, and C. R. Hauser, *J. Org. Chem.,* **24**, 583 (1959).

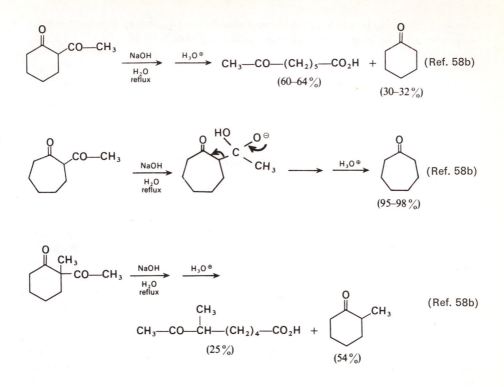

reaction only for α-acylcycloalkanones that have no other alpha substituent and that are derived from cyclopentanones or cyclohexanones.   An analogous cleavage reaction may be performed with 2-alkylcycloalkane-1,3-diones;[51,59] since these intermediates are symmetrical, only one mode of cleavage is observed irrespective of ring size in the cycloalkane derivative.   Subsequent reduction of the keto acids

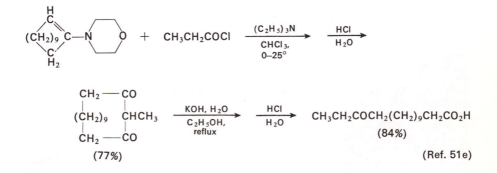

59. H. Stetter in W. Foerst, ed., *Newer Methods of Preparative Organic Chemistry*, Vol. 2, Academic Press, New York, 1963, pp. 51–99.

formed in this way by the Wolff-Kishner method (see Chapter 4) provides a useful synthetic route to long-chain carboxylic acids (e.g., [40]).[50a,51]

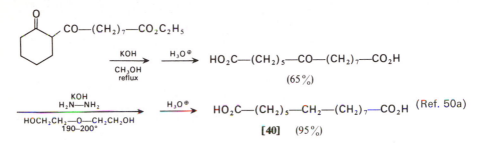

$$HO_2C—(CH_2)_5—CO—(CH_2)_7—CO_2H$$
$$(65\%)$$

$$HO_2C—(CH_2)_5—CH_2—(CH_2)_7—CO_2H \quad (Ref.\ 50a)$$
$$[40] \quad (95\%)$$

ACYLATION OF OLEFINS AND ACETYLENES

Acylations of both olefins and aromatic systems, which may be considered examples of the Friedel-Crafts reaction,[1,60] are usually effected either with an acid chloride or anhydride in the presence of one of the Lewis acids—aluminum chloride,[61] boron trifluoride,[53] stannic chloride, or zinc chloride—or with a carboxylic acid in the presence of a protonic acid such as hydrogen fluoride,[62] sulfuric acid, or polyphosphoric acid.[63] The carboxylic acids are employed most frequently for intramolecular reactions leading to cyclic ketones.[60c,63] The Lewis acids, whose activities follow the order $BCl_3 > AlCl_3 > TiCl_4 > BF_3 > SnCl_4 > ZnCl_2$[60,61,b,g] require anhydrous conditions since the presence of appreciable amounts of water reduces the activity of these catalysts. However, it has been noted[60f] that truly anhydrous conditions are probably never achieved and, in fact, traces of water seem to be essential to the success of many Friedel-Crafts reactions. The Lewis acid catalysts are normally used in solvents such as carbon disulfide, methylene chloride, nitromethane, 1,2-dichloroethane, or nitrobenzene or an excess of the hydrocarbon being acylated. [Anhydrous aluminum chloride is soluble in methylene chloride, 1,2-dichloroethane, nitromethane, and nitrobenzene, but is insoluble in carbon disulfide.] The reaction of an acid halide with a Lewis acid may form either a

60. (a) P. H. Gore, *Chem. Rev.,* **55**, 229 (1955). (b) G. Baddeley, *Quart. Rev.,* **8**, 355 (1954). (c) W. S. Johnson, *Org. Reactions,* **2**, 114 (1944). (d) E. Berliner, *ibid.,* **5**, 229 (1949). (e) G. A. Olah, ed., *Friedel-Crafts and Related Reactions,* 4 volumes, Wiley-Interscience, New York, 1963–1965. (f) R. Corriu, M. Dore, and R. Thomassin, *Tetrahedron Letters,* **No. 23**, 2759 (1968).
61. (a) C. A. Thomas, *Anhydrous Aluminum Chloride in Organic Chemistry,* Reinhold, New York, 1941. (b) J. F. Deters, P. A. McCusker, and R. C. Pilger, Jr., *J. Am. Chem. Soc.,* **90**, 4583 (1968). (c) G. A. Olah, J. Lukas, and E. Lukas, *ibid.,* **91**, 5319 (1969). (d) S. E. Rasmussen and N. C. Broch, *Chem. Commun.,* **No. 13**, 289 (1965). (e) D. E. H. Jones and J. L. Wood, *J. Chem. Soc.,* A, 1140 (1967). (f) G. Oulevey and B. P. Susz, *Helv. Chim. Acta,* **47**, 1828 (1964); **48**, 1965 (1965). (g) D. P. N. Satchell and R. S. Satchell, *Chem. Rev.,* **69**, 251 (1969).
62. K. Wiechert in *Newer Methods of Preparative Organic Chemistry,* Wiley-Interscience, New York, 1948, pp. 315–368.
63. (a) F. Uhlig and H. R. Snyder, *Adv. Org. Chem.,* **1**, 35 (1960). (b) F. D. Popp and W. E. McEwen, *Chem. Rev.,* **58**, 321 (1958). (c) J. P. Marthe and S. Munavalli, *Bull. Soc. Chim. France,* 2679 (1963).

dipolar complex or the salt of an acylium ion depending on the structures of the reactants.[1e,f,61b−g] These very reactive species, illustrated below, are believed to be the actual reactants in the Friedel-Crafts acylations of olefins and aromatic systems.

$$CH_3CH_2-CO-F + SbF_5 \xrightarrow[0°]{\substack{F_2ClC-CFCl_2 \\ (Freon-113)}} CH_3CH_2-\overset{\oplus}{C}{=}O \quad SbF_6^{\ominus} \qquad (Ref.\ 61c)$$

$$C_6H_5-CO-Cl + AlCl_3 \xrightarrow{CS_2} C_6H_5-\underset{Cl}{\overset{\overset{\oplus}{O}-\overset{\ominus}{AlCl_3}}{C}} \qquad (Ref.\ 61d)$$

Hydrogen fluoride is most commonly employed as an anhydrous liquid (b.p. 19.5°) in apparatus constructed entirely of copper or polyethylene. Intramolecular acylations are usually accomplished by dissolving the acid to be cyclized in excess hydrogen fluoride and allowing the solution to stand at room temperature until the hydrogen fluoride has evaporated.[60c,62,64] Reactions catalyzed by sulfuric acid are normally performed by adding the reactant(s) to excess concentrated sulfuric acid, which then also serves as the solvent. Such reactions are run at temperatures ranging from 0 to 100°; at elevated temperatures sulfonation of aromatic systems and acid-catalyzed aldol condensation (see Chapter 10) of the ketonic products may be serious side reactions. The crude product is isolated after pouring the reaction mixture onto ice. Polyphosphoric acid,[63,65]

$$(HO)_2\overset{O}{\overset{\|}{P}}-O-(-\overset{O}{\overset{\|}{\underset{OH}{P}}}-O-)_n-\overset{O}{\overset{\|}{P}}(OH)_2,$$

which may be prepared by dissolving phosphorus pentoxide in 85% aqueous phosphoric acid, is a very viscous liquid, sufficiently fluid to permit stirring, only when warmed to 40–50°. Reactions with this reagent are commonly run by adding the reactant(s) to warm polyphosphoric acid with stirring and then heating the resulting solution to 50–90° on a steam bath. The reaction mixture is poured onto ice prior to isolation of the crude product. Since aromatic hydrocarbons which have no polar substituents are often not soluble in the protonic acids (HF, H_2SO_4, or polyphosphoric acid), the use of these reagents for intermolecular acylation is frequently not a satisfactory procedure because the reactants (the hydrocarbon and the acylating agent) are not both in solution.

The acylation of olefins and acetylenes with acid chlorides in the presence of aluminum chloride is illustrated by the following equations. As noted earlier, these

---

64. L. F. Fieser and E. B. Hershberg, *J. Am. Chem. Soc.*, **61**, 1272 (1939); **62**, 49 (1940).
65. Commercially available from Matheson, Coleman, and Bell.

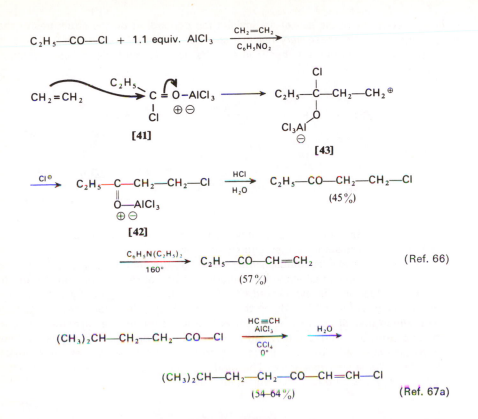

reactions are believed to proceed via the indicated electrophilic attack on the olefin by a complex of the Lewis acid with the acid chloride (e.g., [41]); alternatively, the acylating agent may be formulated as an acyl cation.[60,61] The ketone formed in the reaction mixture is converted to its conjugate acid (e.g., [42]) by the aluminum chloride present.[61e,g] For this reason it is necessary to use at least one full equivalent of the aluminum chloride catalyst to ensure complete reaction. The stereochemistry of the acylation of acetylenes has been suggested to change as the reaction conditions are varied; conditions favoring the acid chloride complex [43a] are believed to yield mainly the *cis*-chloro ketone whereas reactions involving the ionic intermediate [43b] are thought to form mainly the *trans*-isomer.[67c,d]

The conditions employed in these acylations are sufficiently vigorous that the intermediate carbonium ion or the ion produced by reaction of the initially formed

66. E. M. McMahon, J. N. Roper, Jr., W. P. Utermohlen, Jr., R. H. Hasek, R. C. Harris, and J. H. Brant, *J. Am. Chem. Soc.*, **70**, 2971 (1948).
67. (a) C. C. Price and J. A. Pappalardo, *Org. Syn.*, **Coll. Vol. 4**, 186 (1963). (b) W. R. Benson and A. E. Pohland, *J. Org. Chem.*, **29**, 385 (1964). (c) H. Martens and G. Hoornaert, *Tetrahedron Letters*, **No. 21**, 1821 (1970). (d) A related result has been observed in the intramolecular cyclization of a 4-cyclooctenecarboxylic acid chloride. W. F. Erman and H. C. Kretschmar, *J. Org. Chem.*, **33**, 1545 (1968).

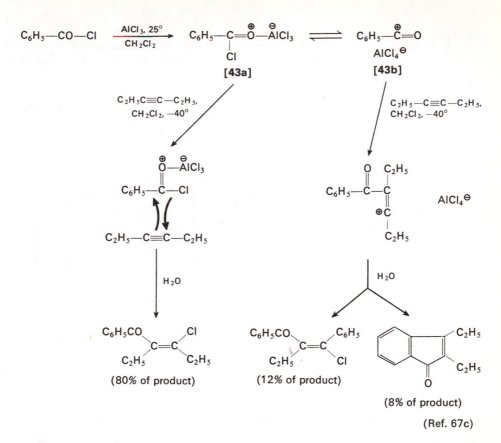

chloro compound with aluminum chloride may undergo further reaction, especially in cases involving more complex olefins.[68,69] This possibility is demonstrated in the equations that follow. In at least some of the cases when rearrangement of the carbon skeleton has accompanied the acylation of olefins, it is probable that careful attention to reaction conditions (no excess Lewis acid, use of minimum reaction times and temperatures) would either suppress or eliminate the secondary reactions leading to rearranged products.

68. (a) R. E. Christ and R. C. Fuson, *J. Am. Chem. Soc.,* **59**, 893 (1937). (b) C. L. Stevens and E. Farkas, *ibid.,* **75**, 3306 (1953). (c) H. O. House, V. Paragamian, R. S. Ro, and D. J. Wluka, *ibid.,* **82**, 1457 (1960). (d) For other examples, see C. D. Gutsche and W. S. Johnson, *ibid.,* **68**, 2239 (1946).

69. (a) J. H. Burckhalter and J. R. Campbell, *J. Org. Chem.,* **26**, 4232 (1961). (b) J. A. Blair, G. P. McLaughlin, and J. Paslawski, *Chem. Commun., No. 1,* 12 (1967); J. A. Blair and C. J. Tate, *ibid., No. 24,* 1506 (1969). (c) I. Tabushi, K. Fujita, and R. Oda, *Tetrahedron Letters, No. 40,* 4247 (1968); *No. 52,* 5455 (1968). (d) D. D. Phillips and A. W. Johnson, *J. Org. Chem.,* **21**, 587 (1956). (e) T. S. Cantrell, *ibid.,* **32**, 1669 (1967). T. S. Cantrell and B. L. Strasser, *ibid.,* **36**, 670 (1971). (f) W. S. Trahanovsky, M. P. Doyle, P. W. Mullen, and C. C. Ong, *ibid.,* **34**, 3679 (1969). (g) G. J. Martin, C. Rabiller, and G. Mabon, *Tetrahedron Letters, No. 36,* 3131 (1970).

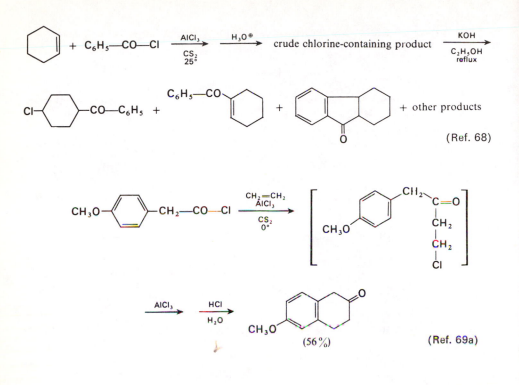

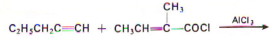

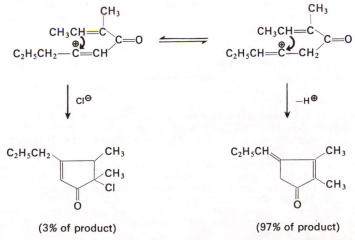

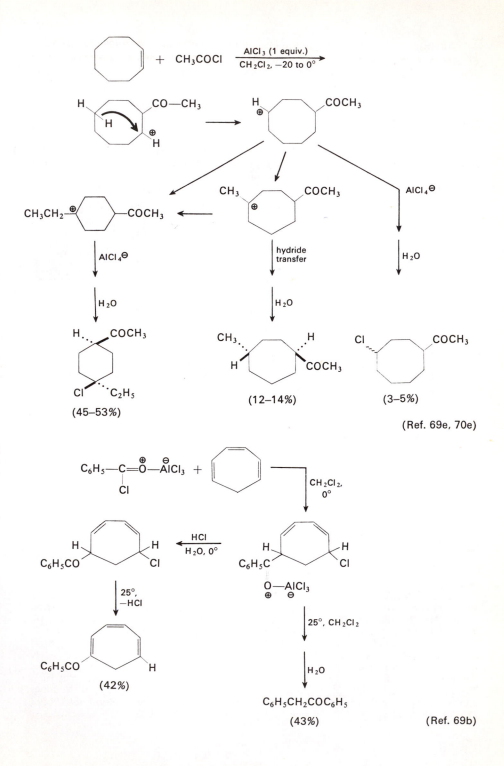

(12–14%)

(3–5%)

(45–53%)

(Ref. 69e, 70e)

(42%)

(43%)   (Ref. 69b)

Alternative procedures requiring less vigorous conditions include the acylation of olefins (e.g., [44]) with acid anhydrides or acid chlorides in the presence of stannic chloride[68c],[70] or zinc chloride[70] and the preparation of solutions of acid chloride-aluminum chloride complexes in methylene chloride followed by treatment of these complexes with the olefin (e.g., [45]) to be acylated under conditions where no excess aluminum chloride is present (sometimes called the Perrier procedure).[71] It will be noted that the position of acylation of the olefin [44] corresponds to the production of the more stable tertiary carbonium ion intermediate [46]; also, the acylating agent attacks from the less hindered side of the

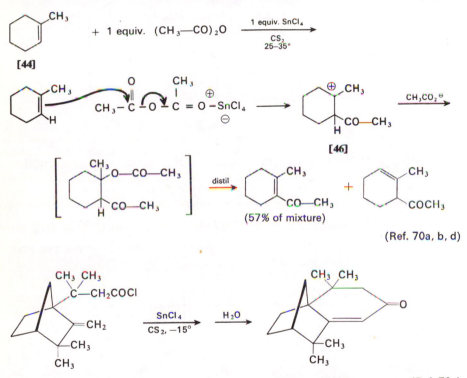

(57% of mixture)

(Ref. 70a, b, d)

(84%)   (Ref. 70c)

70. (a) E. E. Royals and C. M. Hendry, *J. Org. Chem.,* **15**, 1147 (1950). (b) H. O. House and W. F. Gilmore, *J. Am. Chem. Soc.,* **83**, 3980 (1961). (c) R. R. Sobti and S. Dev, *Tetrahedron Letters,* **No. 30,** 2893 (1967). (d) J. K. Groves and N. Jones, *Tetrahedron Letters,* **No. 14,** 1161 (1970); *J. Chem. Soc.,* C, 608 (1969); *ibid.,* C, 2215, 2898 (1968). (e) *ibid.,* C, 1718, 2350 (1969). (f) J. Allard and N. Dufort, *Can. J. Chem.,* **47,** 2403 (1969). (g) P. J. Kropp, D. C. Heckert, and T. J. Flautt, *Tetrahedron,* **24,** 1385 (1968). (h) J. A. Marshall, N. H. Andersen, and J. W. Schlicher, *J. Org. Chem.,* **35,** 858 (1970).
71. (a) G. Baddeley, H. T. Taylor, and W. Pickles, *J. Chem. Soc.,* 124 (1953). (b) H. T. Taylor, *ibid.,* 3922 (1958). (c) N. Jones and H. T. Taylor, *ibid.,* 4017 (1959). (d) N. Jones and H. T. Taylor, *ibid.,* 1345 (1961). (e) N. Jones, H. T. Taylor, and E. Rudd, *ibid.,* 1342 (1961). (f) N. Jones, E. J. Rudd, and H. T. Taylor, *ibid.,* 2354 (1963).

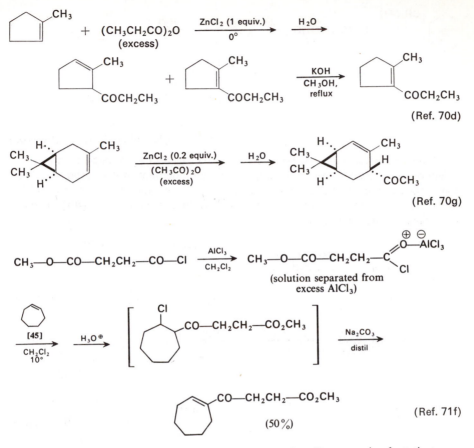

(Ref. 70d)

(Ref. 70g)

(Ref. 71f)

(50%)

olefin. Several of the accompanying examples also illustrate the fact that even under mild acylating conditions, isomerization of the carbon-carbon double bond may precede or follow the introduction of the acyl group.

Reaction conditions		Product yields	
CH_3COCl, CH_2Cl_2, deactivated $AlCl_3$ (exposed to air), $-15°$	—	68%	—
CH_3COCl, CS_2, $SnCl_4$, $25°$	38%	25%	—
$(CH_3CO)_2O$, $ZnCl_2$, $20°$	—	—	45%
$(CH_3CO)_2O$, $(C_2H_5)_2O$, BF_3, $20°$	—	—	45%

(Ref. 70e)

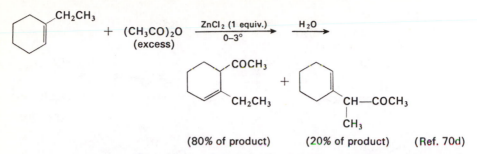

(80% of product)   (20% of product)   (Ref. 70d)

Olefin acylation has also been achieved by reaction with carboxylic acids in polyphosphoric acid.[72] It will be noted that rearrangement of the carbon skeleton may also accompany these acylations. When α,β-unsaturated acids (e.g., [47]) are employed, cyclization of the intermediate acylation product (e.g., [48]) leads to a cyclopentenone derivative.

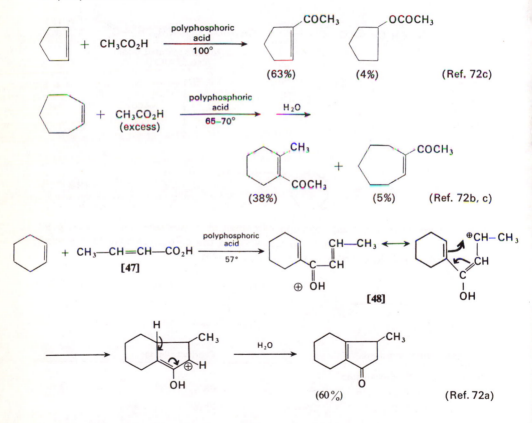

72. (a) S. Dev, *J. Indian Chem. Soc.*, **32**, 255, 403 (1955); **33**, 703 (1956); **34**, 169 (1957). (b) S. B. Kulkarni and S. Dev, *Tetrahedron*, **24**, 545, 553, 561 (1968). (c) L. Rand and R. J. Dolinski, *J. Org. Chem.*, **31**, 3063, 4061 (1966).

Reactions of certain unsaturated esters, hydroxy acids, γ-lactones (e.g., [49]), and δ-lactones with phosphoric acid or phosphorus pentoxide have similarly been found to yield cyclic unsaturated ketones; the cyclopentenone and cyclo-hexenone derivatives presumably arise via intramolecular acylation of the inter-

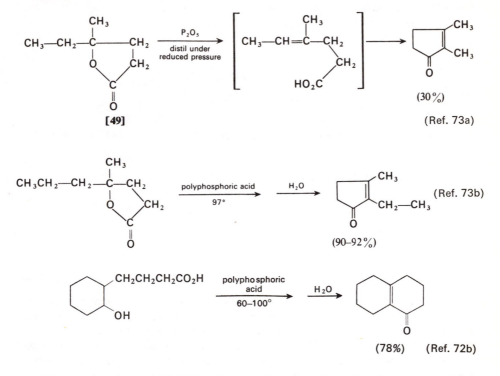

[49]

(30%)

(Ref. 73a)

(Ref. 73b)

(90–92%)

(78%)   (Ref. 72b)

mediate unsaturated acids.[72b,73] Saturated esters may also be converted to unsaturated ketones as the following example indicates.

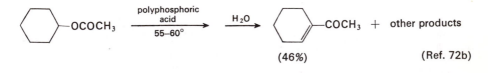

(46%)

(Ref. 72b)

Unsaturated carboxylic acids may also be used for cyclizations with poly-phosphoric acid. At a reaction temperature of 100°, the isomerization of the carbon-carbon double bond in the starting acid is apparently faster than the acylation

73. (a) R. L. Frank, R. Armstrong, J. Kwiatek, and H. A. Price, *J. Am. Chem. Soc.,* **70,** 1379 (1948). (b) S. Dev and C. Rai, *J. Indian Chem. Soc.,* **34,** 178, 266 (1957). (c) M. F. Ansell and M. H. Palmer, *Quart. Rev.,* **18,** 211 (1964). (d) M. F. Ansell, J. C. Emmett, and R. V. Coombs, *J. Chem. Soc.,* C, 217 (1968). (e) M. F. Ansell, J. E. Emmett, and B. E. Grimwood, *ibid.,* C, 141 (1969). (f) M. F. Ansell and T. M. Kafka, *Tetrahedron,* **25,** 6025 (1969).

reaction so that comparable mixtures of cyclopentenone and cyclohexenone derivatives are formed irrespective of the initial position of the double bond in the starting acid. One of the following examples shows that rearrangement of the carbon skeleton may also occur during the cyclization of unsaturated acids.

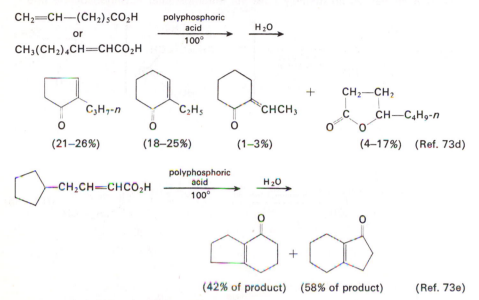

$$CH_2\!\!=\!\!CH\!\!-\!\!(CH_2)_5CO_2H$$

or

$$CH_3(CH_2)_4CH\!\!=\!\!CHCO_2H$$

polyphosphoric acid 100° → H₂O →

(21–26%)   (18–25%)   (1–3%)   (4–17%)   (Ref. 73d)

$$\text{—}CH_2CH\!\!=\!\!CHCO_2H$$

polyphosphoric acid 100° → H₂O →

(42% of product)   (58% of product)   (Ref. 73e)

The acylation of olefins as well as of aromatic systems by reaction with carboxylic acids is also promoted by trifluoroacetic anhydride.[70h,73d,74] As illustrated in the accompanying equations, this reaction is believed to involve acylation by the mixed anhydride (e.g., [50]), catalyzed by trifluoroacetic acid. Double bond migration seems to occur less readily during this acylation procedure than is the case for the other methods discussed.[73d] Application of the same reaction to acetylenes (e.g., [51]) offers a synthetic route to 1,3-dicarbonyl compounds.

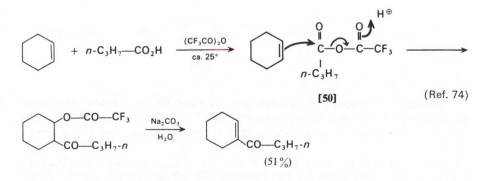

$$+ \; n\text{-}C_3H_7\text{—}CO_2H$$

(CF₃CO)₂O ca. 25° →

[50]   (Ref. 74)

Na₂CO₃ H₂O →

(51%)

74. A. L. Henne and J. M. Tedder, *J. Chem. Soc.*, 3628 (1953); R. J. Ferrier and J. M. Tedder, *ibid.*, 1435 (1957).

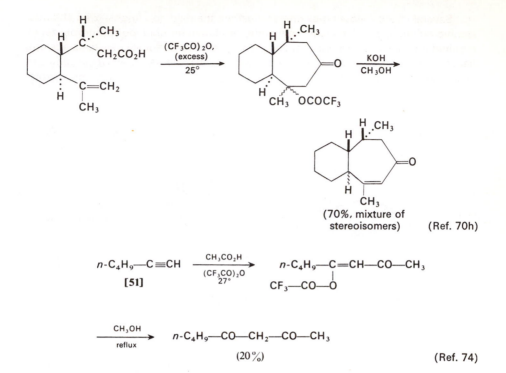

(70%, mixture of
stereoisomers)   (Ref. 70h)

$$n\text{-}C_4H_9\text{---}C\equiv CH \quad \xrightarrow[\substack{(CF_3CO)_2O \\ 27°}]{CH_3CO_2H} \quad n\text{-}C_4H_9\text{---}C\text{=}CH\text{---}CO\text{---}CH_3$$

[51] $CF_3\text{---}CO\text{---}O$

$$\xrightarrow[\text{reflux}]{CH_3OH} \quad n\text{-}C_4H_9\text{---}CO\text{---}CH_2\text{---}CO\text{---}CH_3$$

(20%)   (Ref. 74)

ACYLATION OF AROMATIC SYSTEMS

Some typical Friedel-Crafts acylations of aromatic systems are described in the equations below. As indicated in the first of these, reaction of an acid anhydride with aluminum chloride may yield an acid chloride-aluminum chloride complex which is the actual acylating agent. Subsequent electrophilic attack of the Lewis acid-acid chloride complex (e.g., [52]) on the aromatic system, followed by loss of a proton from the intermediate ion (e.g., [53]) or sigma complex (see Chapter 8), leads to formation of the ketone, present in the reaction mixture as its conjugate acid.[61e] It will be noted that two equivalents of aluminum chloride are required in this sequence with acid anhydrides. In some acylations with anhydrides, it seems likely that a complex of the acid anhydride with the acid catalyst is the acylating agent[75c] and the prior formation of an acid chloride is not required. The relative rates for the acylation of benzene with solutions of some typical acid chlorides and aluminum chloride in nitromethane are: CH_3COCl (1.00); C_2H_5COCl (0.92); $n\text{-}C_3H_7COCl$ (0.64); $(CH_3)_2CHCOCl$ (0.20); C_6H_5COCl (0.06).[75d] It is apparent that both steric hindrance and, especially, conjugation reduce the reactivity of these acylating agents.

75. (a) R. Adams and C. R. Noller, *Org. Syn.,* **Coll. Vol. 1,** 109 (1944). (b) F. E. Ray and G. Rieveschl, Jr., *ibid.,* **Coll. Vol. 3,** 23 (1955). (c) W. R. Edwards, Jr., and R. J. Eckert, Jr., *J. Org. Chem.,* **31,** 1283 (1966). (d) P. H. Gore, J. A. Hoskins, and S. Thorburn, *J. Chem. Soc.,* B, 1343 (1970).

Several of the following examples illustrate the mild conditions (zinc chloride, stannic chloride, trifluoroacetic anhydride, or aluminum chloride in nitrobenzene) required for reactive aromatic nuclei. These reactive aromatic systems include benzene derivatives with electron-donating substituents, polycyclic aromatic hydrocarbons, and the five-membered heterocycles, thiophene, furan, and pyrrole.

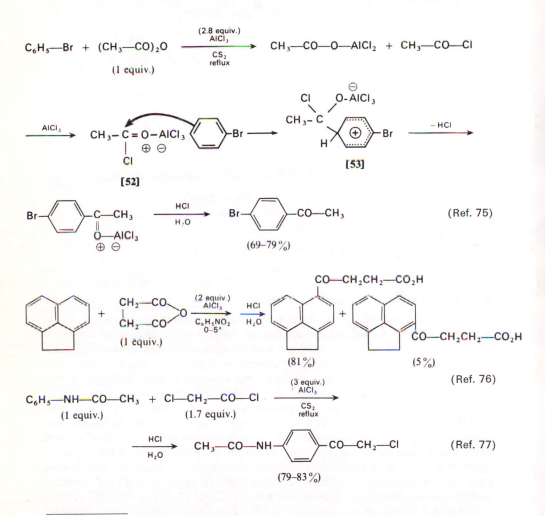

(Ref. 75)

(Ref. 76)

(Ref. 77)

76. (a) L. F. Fieser, *Org. Syn.*, **Coll. Vol. 3**, 6 (1955). (b) O. Grummitt, E. I. Becker, and C. Miesse, *ibid.*, **Coll. Vol. 3**, 109 (1955). (c) L. F. Fieser, *ibid.*, **Coll. Vol. 1**, 517 (1944). (d) L. F. Somerville and C. F. H. Allen, *ibid.*, **Coll. Vol. 2**, 81 (1943). (e) C. Merritt, Jr., and C. E. Braun, *ibid.*, **Coll. Vol. 4**, 8 (1963). (f) K. R. Tatta and J. C. Bardhan, *J. Chem. Soc.*, C, 893 (1968).
77. (a) J. L. Leiserson and A. Weissberger, *Org. Syn.*, **Coll. Vol. 3**, 183 (1955). (b) C. F. H. Allen, *ibid.*, **Coll. Vol. 2**, 3 (1943). (c) W. Minnis, *ibid.*, **Coll. Vol. 2**, 520 (1943). (d) L. Friedman and R. Koca, *J. Org. Chem.*, **33**, 1255 (1968).

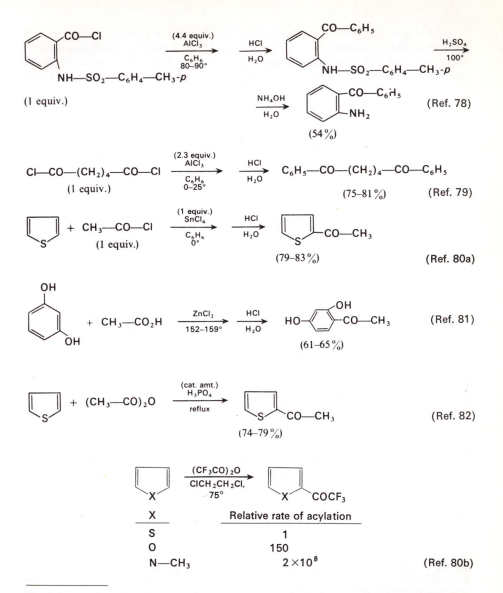

78. (a) H. J. Scheifele, Jr., and D. F. DeTar, *Org. Syn.*, **Coll. Vol. 4**, 34 (1963). (b) C. F. H. Allen and W. E. Barker, *ibid.*, **Coll. Vol. 2**, 156 (1943). (c) F. J. Villani and M. S. King, *ibid.*, **Coll. Vol. 4**, 88 (1963).
79. (a) R. C. Fuson and J. T. Walker, *Org. Syn.*, **Coll. Vol. 2**, 169 (1943). (b) R. E. Lutz, *ibid.*, **Coll. Vol. 3**, 248 (1955).
80. (a) J. R. Johnson and G. E. May, *Org. Syn.*, **Coll. Vol. 2**, 8 (1943). (b) S. Clementi, F. Genel, and G. Marino, *Chem. Commun.*, **No. 10**, 498 (1967).
81. (a) S. R. Cooper, *Org. Syn.*, **Coll. Vol. 3**, 761 (1955). (b) I. C. Badhwar and K. Venkataraman, *ibid.*, **Coll. Vol. 2**, 304 (1943).
82. A. I. Kosak and H. D. Hartough, *Org. Syn.*, **Coll. Vol. 3**, 14 (1955).

Other procedures that result in acylation of reactive aromatic nuclei include acid-catalyzed condensation with nitriles (e.g., [54]), called the Hoesch reaction,[83] and acid-catalyzed condensation with amides (e.g., [55]). The previously mentioned[55f–i] reaction of N,N-disubstituted amides with aromatic systems in the

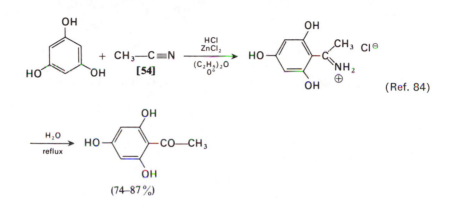

(Ref. 84)

(74–87%)

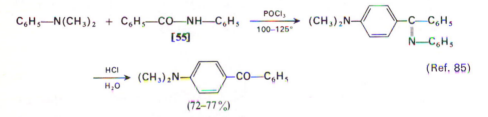

[55]

(Ref. 85)

(72–77%)

presence of phosphorus oxychloride, thionyl chloride, or phosgene, known as the Vilsmeier reaction, has proven especially valuable.[86] Its use to introduce the formyl group into reactants containing electron-rich olefinic bonds or reactive aromatic rings (e.g., [57]) is illustrated in the following equations. The reaction is thought[86c,d] to involve a salt such as [56] as an electrophilic intermediate. Introduction of formyl groups has also been effected by reaction of aromatic systems (e.g., [58])

83. (a) P. E. Spoerri and A. S. DuBois, *Org. Reactions,* **5**, 387 (1949). (b) E. A. Jeffery and D. P. N. Satchell, *J. Chem. Soc.,* B, 579 (1966). (c) For examples of the intramolecular acylation of olefins with nitriles in the presence of polyphosphoric acid, see R. K. Hill and R. T. Conley, *J. Am. Chem. Soc.,* **82**, 645 (1960).
84. K. C. Gulati, S. R. Seth, and K. Venkataraman, *Org. Syn.,* **Coll. Vol. 2**, 522 (1943).
85. C. D. Hurd and C. N. Webb, *Org. Syn.,* **Coll. Vol. 1**, 217 (1944).
86. (a) M. R. de Maheas, *Bull. Soc. Chim. France,* 1989 (1962). (b) K. Hafner and co-workers, *Angew. Chem., Intern. Ed. Engl.,* **2**, 123 (1963). (c) Z. Arnold and A. Holy, *Collection Czech. Chem., Commun.,* **27**, 2886 (1962). (d) G. Martin and M. Martin, *Bull. Soc. Chim. France,* 1637 (1963). (e) T. D. Smith, *J. Chem. Soc.,* A, 841 (1966). (f) M. D. Scott and H. Spedding, *ibid.,* C, 1603 (1968). (g) J. G. Dingwall, D. H. Reid, and K. Wade, *ibid.,* C, 913 (1969). (h) M. L. Filleux-Blanchard, M. T. Quemeneur, and G. J. Martin, *Chem. Commun.,* **No. 15**, 837 (1968).

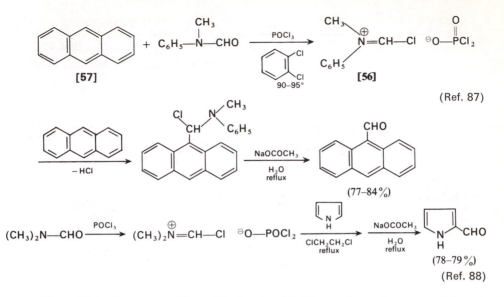

(Ref. 87)

(77–84%)

(78–79%)

(Ref. 88)

either with hydrogen cyanide and hydrogen chloride or with carbon monoxide and hydrogen chloride, procedures developed by Gattermann.[89] Aromatic aldehydes have also been synthesized by reaction of aromatic systems with the ion obtained from dichloromethyl methyl ether and titanium tetrachloride.[91b,c]

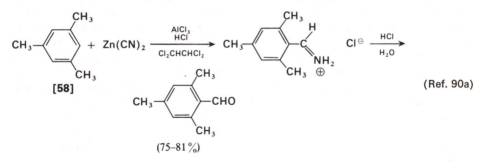

(75–81%)

(Ref. 90a)

87. L. F. Fieser, J. L. Hartwell, and J. E. Jones, *Org. Syn.,* **Coll. Vol. 3**, 98 (1955).
88. (a) R. M. Silverstein, E. E. Ryskiewicz, and C. Willard, *Org. Syn.,* **Coll. Vol. 4**, 831 (1963); see also (b) P. N. James and H. R. Snyder, *ibid.,* **Coll. Vol. 4**, 539 (1963). (c) A. W. Weston and R. J. Michaels, Jr., *ibid.,* **Coll. Vol. 4**, 915 (1963). (d) E. Campaigne and W. L. Archer, *ibid.,* **Coll. Vol. 4**, 331 (1963).
89. W. E. Truce, *Org. Reactions,* **9**, 37 (1957).
90. (a) R. C. Fuson, E. C. Horning, S. P. Rowland, and M. L. Ward, *Org. Syn.,* **Coll. Vol. 3**, 549 (1955). (b) For the use of s-triazine rather than hydrogen cyanide in the Gatterman aldehyde synthesis, see A. Kreutzberger, *Angew. Chem., Intern. Ed. Engl.,* **6**, 940 (1967).
91. (a) G. H. Coleman and D. Craig, *Org. Syn.,* **Coll. Vol. 2**, 583 (1943). (b) A. Rieche, H. Gross, and E. Höft, *ibid.,* **47**, 1 (1967); H. Gross, A. Rieche, E. Höft, and E. Beyer, *ibid.,* **47**, 47 (1967). (c) The introduction of a formyl group *ortho* to the hydroxyl group of a phenol may be accomplished by reaction with hexamethylenetetramine and an acid catalyst (the Duff reaction). Y. Ogata, A. Kawasaki, and F. Sugiura, *Tetrahedron,* **24**, 5001 (1968) and references therein.

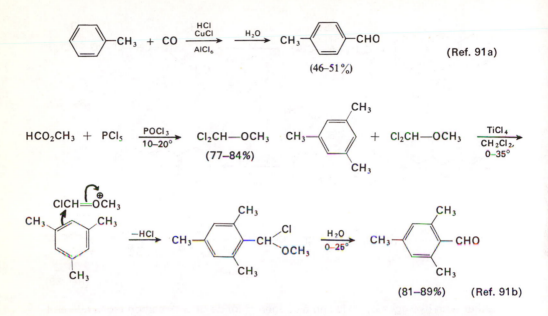

(46–51%)

(Ref. 91a)

(77–84%)

(81–89%)   (Ref. 91b)

An alternative method for the preparation of acylated phenols is based on the rearrangement of phenyl esters in the presence of aluminum chloride. This reaction, known as the Fries rearrangement,[92,93] may involve both intermolecular and intramolecular transfer of the acyl group.[92c,d] The intermolecular transfer presumably occurs by electrophilic attack of an acid chloride or an acyl cation on either the phenyl ester or the aluminum alkoxide derived from the phenol whereas the intra-molecular process is believed to proceed via a pi-complex (e.g., [59]) which yields primarily, if not exclusively the *ortho*-substituted product.[92d] Since the intermolecular process is favored by the use of excess Lewis acid (or by the presence of a proton acid), the proportion of the *ortho* isomer in the product is found to decrease as the amount of the aluminum halide is increased from one to two equivalents. Since the first step in the reaction of an acid halide and an aluminum halide with a phenol appears to be the formation of a phenyl ester, all such reactions can be considered as modifications of the Fries rearrangement.

From a number of the preceding examples, it is apparent that the position taken by an entering acyl group corresponds to the usual orientation observed in electrophilic substitutions of aromatic systems. The reactivity of the system is enhanced by electron-donating substituents and reduced by electron-withdrawing substituents. In fact, Friedel-Crafts acylations of benzene derivatives containing powerful electron-withdrawing substituents (e.g., nitro or carbonyl functions)

92. (a) A. H. Blatt, *Org. Reactions,* **1,** 342 (1942). (b) C. R. Hauser and E. H. Man, *J. Org. Chem.,* **17,** 390 (1952). (c) Y. Ogata and H. Tabuchi, *Tetrahedron,* **20,** 1661 (1964). (d) M. J. S. Dewar and L. S. Hart, *ibid.,* **26,** 973, 1001 (1970). (e) F. G. Baddar, I. Enayat, and S. M. Abdel-Wahab, *J. Chem. Soc.,* C, 343 (1967). (f) M. H. Palmer and G. J. McVie, *ibid.,* B, 742, 745, 856 (1968).

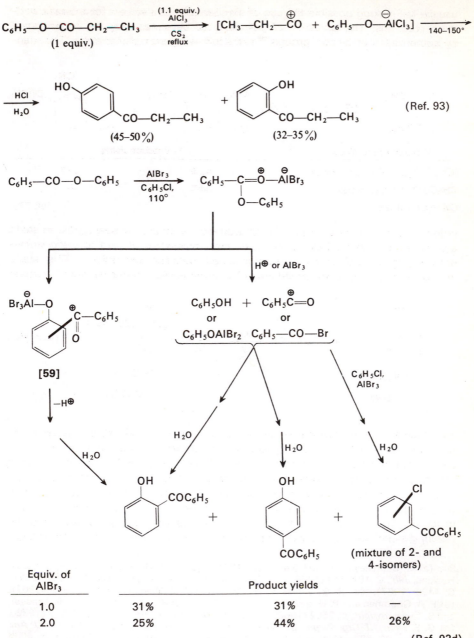

(Ref. 93)

Equiv. of AlBr$_3$	Product yields		
1.0	31%	31%	—
2.0	25%	44%	26%

(Ref. 92d)

93. (a) E. Miller and W. H. Hartung, *Org. Syn.*, **Coll. Vol. 2**, 543 (1943). (b) G. C. Amin and N. M. Shah, *ibid.*, **Coll. Vol. 3**, 280 (1955). (c) A. Russel and J. R. Frye, *ibid.*, **Coll. Vol. 3**, 281 (1955).

usually fail making possible the use of nitrobenzene as a solvent for aromatic acylation reactions. The acylation of polyalkylated benzenes may be accompanied
by isomerization of the alkyl groups[77d] as the following example indicates. The orien-

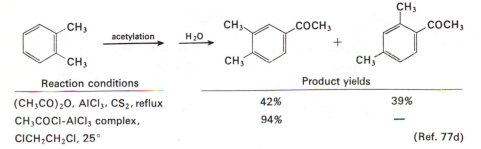

Reaction conditions	Product yields	
(CH$_3$CO)$_2$O, AlCl$_3$, CS$_2$, reflux	42%	39%
CH$_3$COCl-AlCl$_3$ complex,	94%	—
ClCH$_2$CH$_2$Cl, 25°		(Ref. 77d)

tation resulting from a Friedel-Crafts acylation is unusually susceptible to steric
effects, as is evidenced by the relatively high proportion of *para* to *ortho* isomer
obtained in acylations of monosubstituted benzene derivatives. This steric
influence is obvious in the acetylation of *p*-cymene [**60**], where the acetyl group is

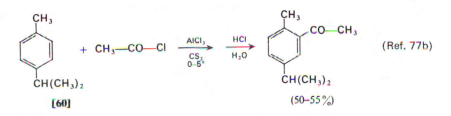

[**60**]   (50–55%)

introduced *ortho* to the smaller alkyl group. The use of nitrobenzene (or nitromethane) as a solvent not only reduces the activity of aluminum chloride as a
Lewis acid but also appears to increase the effective steric bulk of the acylating
agent, presumably because the nitrobenzene becomes associated with the
aluminum chloride–acid chloride complex. The change in product orientation
that results when naphthalene acylation is carried out in its presence rather than
in its absence (see accompanying equations) has been suggested to reflect the
differing steric sizes of the acylating agents in the two cases.[60a,94b–i] Alternatively,

94. (a) G. Baddeley, *J. Chem. Soc.*, S99 (1949). (b) R. B. Girdler, P. H. Gore, and J. A.
Hoskins, *ibid.*, C, 181, 518 (1966). (c) P. H. Gore and C. K. Thadani, *ibid.*, C, 1729 (1966);
C, 1498 (1967). (d) R. B. Girdler, P. H. Gore, and C. K. Thadani, *ibid.*, C, 2619 (1967).
(e) P. A. Goodman and P. H. Gore, *ibid.*, C, 2502 (1968). (f) P. H. Gore, C. K. Thadani,
and S. Thorburn, *ibid.*, C, 2502 (1968). (g) P. H. Gore and J. A. Hoskins, *ibid.*, C, 517 (1970).
(h) G. E. Lewis, *J. Org. Chem.*, **31**, 749 (1966). (i) L. Friedman and R. J. Honour, *J. Am.
Chem. Soc.*, **91**, 6344 (1969). (j) E. Rothstein and W. G. Schofield, *J. Chem. Soc.*, 4566
(1965). (k) H. A. Bruson and H. L. Plant, *J. Org. Chem.*, **32**, 3356 (1967). (l) H. Hopff
and R. von Rütte, *Helv. Chim. Acta*, **49**, 1416 (1966). (m) D. G. Pratt and E. Rothstein,
J. Chem. Soc., C, 2548 (1968). (n) M. H. Palmer and N. M. Scollick, *ibid.*, C, 2833, 2836
(1968).

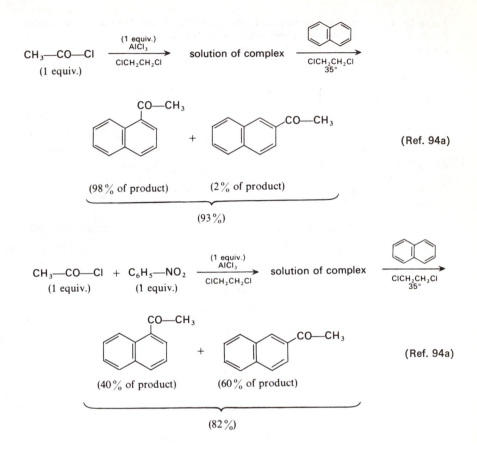

(Ref. 94a)

(Ref. 94a)

it has been suggested that in the acylation of reactive aromatic systems (where the acylation is reversible) formation of the thermodynamically more stable β-acyl derivative results from equilibration of the initially formed products when a less reactive acylating agent is used.[60a,94b-g] Similar solvent effects have been observed in the acylation of other polycyclic aromatic compounds; the results obtained with several aromatic systems are summarized in Table 11–1. It has been noted repeatedly that the proportions of isomeric acylated products may also vary with the conditions used to mix the reactants with one another. A striking example of this variation is seen in the following equation; this variation is suggested to arise from the reversible formation of the initial sigma complexes (e.g., [53]) unless excess aluminum chloride is present in the reaction mixture to trap these intermediates as they are formed.[94i] Steric effects are also observed in acylation reactions employing α-substituted succinic and glutaric anhydrides.[60d,76f] In such cases, the aromatic system is usually acylated predominantly by the less sterically hindered carbonyl function.

Table 11–1 Yields of monoacetylated products obtained from the reaction of various polycyclic aromatic compounds with AlCl$_3$ and CH$_3$COCl

(The reaction solvent is indicated in parentheses)

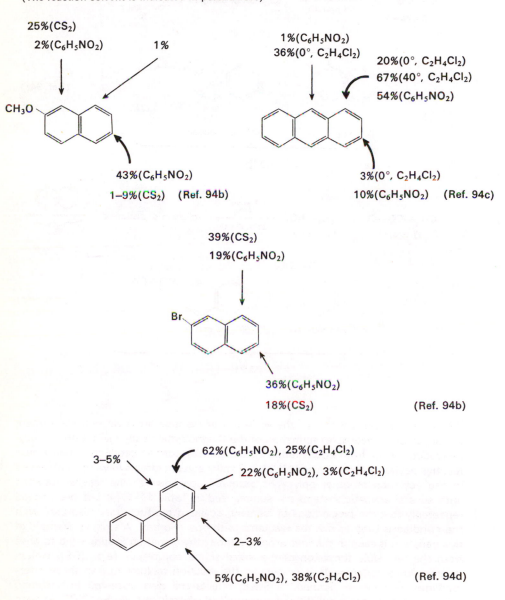

25%(CS$_2$)

2%(C$_6$H$_5$NO$_2$)   1%

1%(C$_6$H$_5$NO$_2$)
36%(0°, C$_2$H$_4$Cl$_2$)

20%(0°, C$_2$H$_4$Cl$_2$)
67%(40°, C$_2$H$_4$Cl$_2$)
54%(C$_6$H$_5$NO$_2$)

CH$_3$O

43%(C$_6$H$_5$NO$_2$)

1–9%(CS$_2$)   (Ref. 94b)

3%(0°, C$_2$H$_4$Cl$_2$)

10%(C$_6$H$_5$NO$_2$)   (Ref. 94c)

39%(CS$_2$)
19%(C$_6$H$_5$NO$_2$)

Br

36%(C$_6$H$_5$NO$_2$)

18%(CS$_2$)   (Ref. 94b)

62%(C$_6$H$_5$NO$_2$), 25%(C$_2$H$_4$Cl$_2$)

3–5%

22%(C$_6$H$_5$NO$_2$), 3%(C$_2$H$_4$Cl$_2$)

2–3%

5%(C$_6$H$_5$NO$_2$), 38%(C$_2$H$_4$Cl$_2$)   (Ref. 94d)

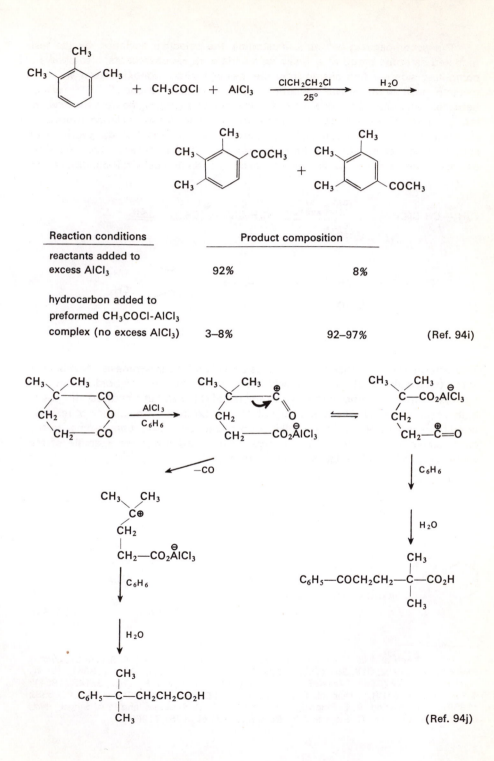

Reaction conditions	Product composition	
reactants added to excess AlCl$_3$	92%	8%
hydrocarbon added to preformed CH$_3$COCl-AlCl$_3$ complex (no excess AlCl$_3$)	3–8%	92–97% (Ref. 94i)

(Ref. 94j)

The accompanying example, illustrating the selective acylation by the less hindered carbonyl group of a cyclic anhydride with α-substituents, also shows a competing side reaction arising from the loss of carbon monoxide from an intermediate acylium ion. This decarbonylation is a side reaction[92f,94j,m,n] which becomes particularly favorable when the loss of carbon monoxide from the acylium ion will form a relatively stable carbonium ion. The decarbonylation process is reversible so that a relatively stable carbonium ion, generated in the presence of excess carbon monoxide, may be converted to an acyl derivative. The following example appears to involve this reverse reaction under Friedel-Craft conditions.[94k,l]

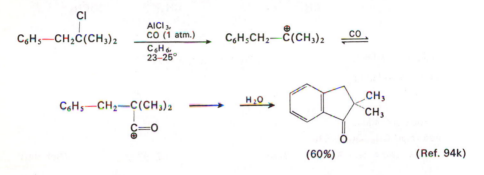

(60%)   (Ref. 94k)

This carbonylation reaction also serves as a method for the synthesis of carboxylic acids (called the Koch-Haaf reaction) from olefins or other carbonium ion precursors.[95] In the second of the two examples of this synthetic method, the possibility of forming either diastereoisomer of the acid product by the choice of reaction conditions is illustrated. This equilibration of the less stable *trans*-isomer to the more stable *cis*-acid in fuming sulfuric acid provides another example of the ready reversibility of the decarbonylation reaction.[95d,e]

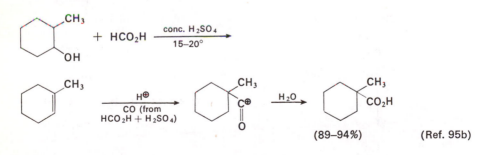

(89–94%)   (Ref. 95b)

---

95. (a) H. Koch and W. Haaf, *Angew. Chem.*, **70**, 311 (1958); **72**, 628 (1960); *Justus Liebigs Ann. Chem.*, **618**, 251 (1958); **638**, 122 (1960); *Org. Syn.*, **44**, 1 (1964). (b) W. Haaf, *ibid.*, **46**, 72 (1966). (c) K. E. Möller, *Angew. Chem., Intern. Ed. Engl.*, **3**, 148 (1964); **4**, 535 (1965). (d) R. E. Pincock, E. Grigat, and P. D. Bartlett, *J. Am. Chem. Soc.*, **81**, 6332 (1959); P. D. Bartlett, R. E. Pincock, J. H. Rolston, W. G. Schindel, and L. A. Singer, *ibid.*, **87**, 2590 (1965). (e) G. Stork and M. Bersohn, *ibid.*, **82**, 1261 (1960).

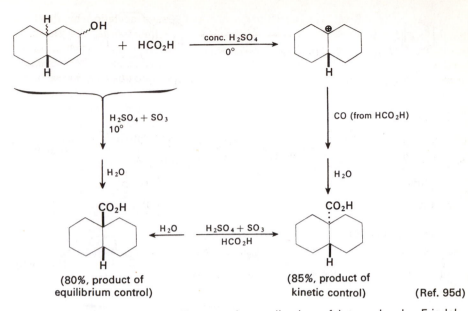

(80%, product of
equilibrium control)

(85%, product of
kinetic control)   (Ref. 95d)

The following equations illustrate the application of intramolecular Friedel-Crafts acylations to the preparation of cyclic ketones.[60c] The synthesis of the eight-membered ring ketone [62] from the acid chloride [61] requires high-dilution

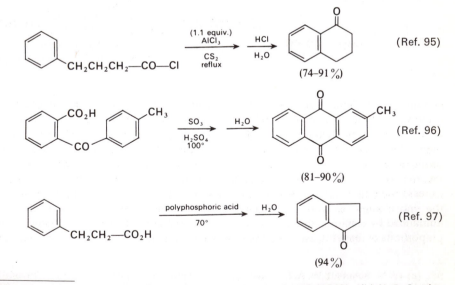

(Ref. 95)

(74–91%)

(Ref. 96)

(81–90%)

(Ref. 97)

(94%)

96. (a) E. L. Martin and L. F. Fieser, *Org. Syn.,* **Coll. Vol. 2,** 569 (1943). (b) H. R. Snyder and F. X. Werber, *ibid.,* **Coll. Vol. 3,** 798 (1955). (c) G. D. Johnson, *ibid.,* **Coll. Vol. 4,** 900 (1963). (d) L. F. Fieser, *ibid.,* **Coll. Vol. 1,** 353 (1944). (e) A. U. Rahman and O. L. Tombesi, *Tetrahedron Letters,* **No. 36,** 3925 (1968).
97. J. Koo, *J. Am. Chem. Soc.,* **75,** 1891 (1953).

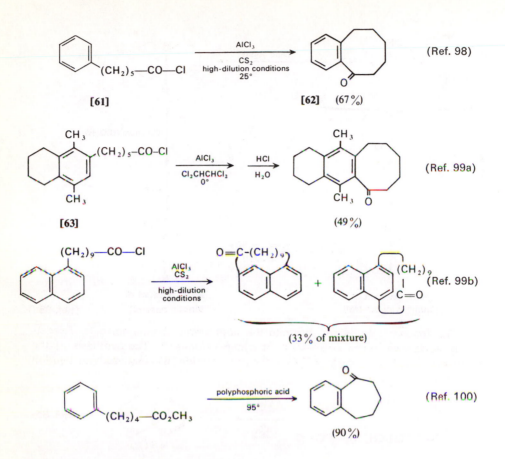

conditions. A similar cyclic ketone has been prepared without high dilution by cyclization of the acid chloride [63]; however, closure of the next higher homolog to form a nine-membered ring failed.[99a] With longer side chains, cyclization under high-dilution conditions has occurred at positions other than the *ortho* position to produce cyclic ketones.[99b] Although intramolecular acylations are generally favored, with relative reactive aromatic compounds as solvents (present in large excess) the intermolecular reaction may predominate.[92f,94n] As would be expected, the ring-closure reactions are facilitated and the direction of cyclization may be influenced by electron-donating substituents in the aromatic ring.[60c] However, the proportions of reagents used and the order of mixing may alter the effect of electron

98. (a) W. M. Schubert, W. A. Sweeney, and H. K. Latourette, *J. Am. Chem. Soc.*, **76**, 5462 (1954). (b) G. D. Hedden and W. G. Brown, *ibid.*, **75**, 3744 (1953).
99. (a) H. Stetter, B. Schäfer, and H. Spangenberger, *Chem. Ber.*, **89**, 1620 (1956). (b) R. Huisgen and U. Rietz, *Tetrahedron*, **2**, 271 (1958) and references therein.
100. (a) R. C. Gilmore, Jr., *J. Am. Chem. Soc.*, **73**, 5879 (1951). (b) R. C. Gilmore, Jr., and W. J. Horton, *ibid.*, **73**, 1411 (1951).

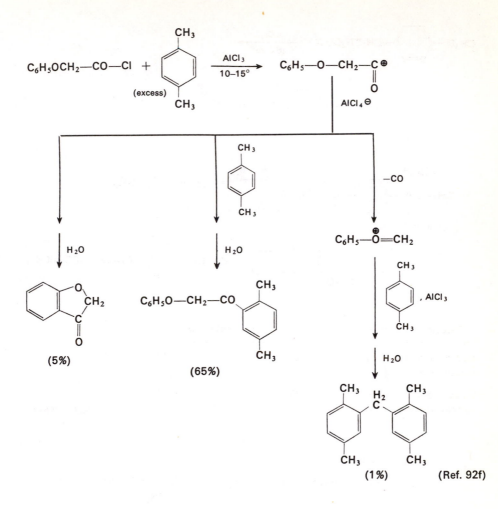

donating substituents. For example, in the following cyclization the presence of excess aluminum chloride throughout the reaction period served to convert the activating methoxy substituent into a positively charged complex which deactivated the aryl ring toward intramolecular acylation.

In the absence of opposing steric interactions, the ease of ring closure in intramolecular acylations follows the order 6-membered > 5-membered > 7-membered.[60c] This generalization is exemplified by the cyclizations of the diacids [64] and [65]. However, the preference for six-membered-ring formation is sufficiently small that it may be altered by opposing steric interactions. For example, the acid [66] produced predominantly the indanone [68] under conditions where the diastereoisomeric acid [67] formed predominantly the tetralone [69]. In each case that ring closure is favored which avoids eclipsing the two large substituents in the transition state. Configuration about a double bond will also control the direction of cyclization, as shown by the closure of the unsaturated

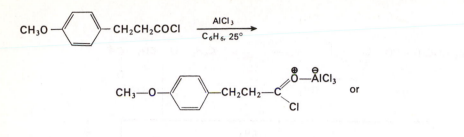

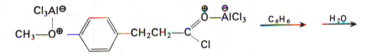

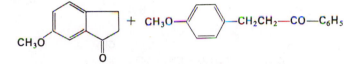

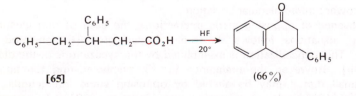

Reaction conditions	Product yields	
acid chloride added to excess AlCl₃	21%	40%
1 equiv. of AlCl₃ added to acid chloride	90%	8%

(Ref. 101b–d)

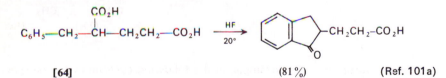

[64]   (81%)   (Ref. 101a)

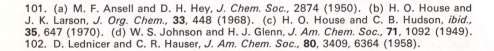

[65]   (66%)

101. (a) M. F. Ansell and D. H. Hey, *J. Chem. Soc.*, 2874 (1950). (b) H. O. House and J. K. Larson, *J. Org. Chem.*, **33**, 448 (1968). (c) H. O. House and C. B. Hudson, *ibid.*, **35**, 647 (1970). (d) W. S. Johnson and H. J. Glenn, *J. Am. Chem. Soc.*, **71**, 1092 (1949). 102. D. Lednicer and C. R. Hauser, *J. Am. Chem. Soc.*, **80**, 3409, 6364 (1958).

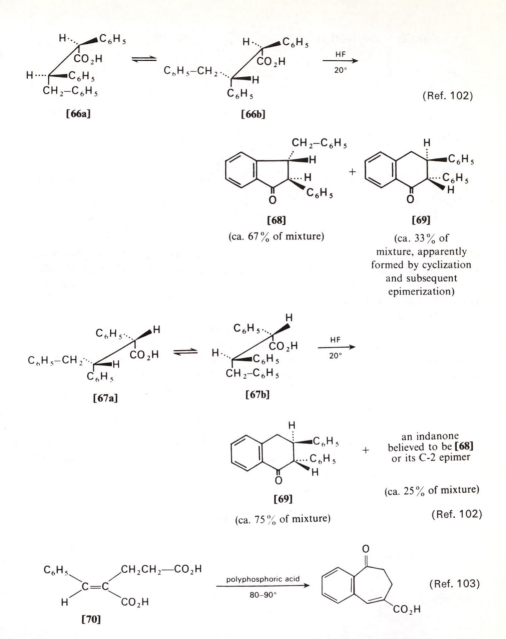

[66a]   [66b]   (Ref. 102)

[68]   [69]

(ca. 67% of mixture)   (ca. 33% of mixture, apparently formed by cyclization and subsequent epimerization)

[67a]   [67b]

[69]   an indanone believed to be [68] or its C-2 epimer

(ca. 75% of mixture)   (ca. 25% of mixture)

(Ref. 102)

[70]   polyphosphoric acid   (Ref. 103)
80–90°

acid [70] to form a seven-membered rather than a five-membered ring. This behavior has been used to assign configurations to the isomeric α-arylidenesuccinic acids (e.g., [71] and [72]) obtained from the Stobbe condensation (see Chapter

103. H. O. House, V. Paragamian, R. S. Ro, and D. J. Wluka, *J. Am. Chem. Soc.*, **82**, 1452 (1960).

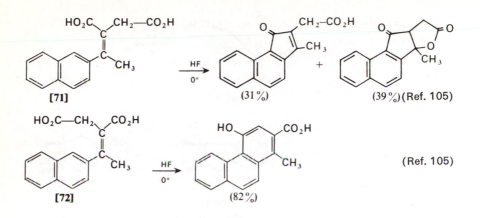

[71]   (31%)   (39%)(Ref. 105)

[72]   (82%)   (Ref. 105)

10)[104,105] It should be noted that cyclization of the unsaturated acid [72] leads to isolation of a phenol derivative rather than of the thermodynamically less stable unsaturated ketone. This procedure for making stereochemical assignments fails whenever acid-catalyzed interconversion of the geometrical isomers is faster than cyclization of each isomer.[101b,105b]

Normally, the configuration at an asymmetric center not adjacent to the carbonyl function is unaffected by the conditions of the Friedel-Crafts acylation, as illustrated by the cyclization of the diastereoisomeric acids [73] and [74]. However, if one of the centers of asymmetry is adjacent to the carbonyl function,

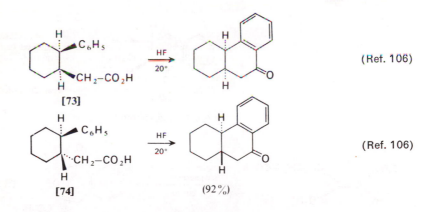

[73]   (Ref. 106)

[74]   (92%)   (Ref. 106)

104. W. S. Johnson and G. H. Daub, *Org. Reactions,* **6**, 1 (1951).
105. (a) W. S. Johnson and A. Goldman, *J. Am. Chem. Soc.,* **66**, 1030 (1944). (b) F. G. Baddar, M. F. El-Neweihy, and R. O. Loutfy, *J. Chem. Soc.,* C, 620 (1970).
106. (a) C. D. Gutsche, *J. Am. Chem. Soc.,* **73**, 786 (1951). (b) C. D. Gutsche and W. S. Johnson, *ibid.,* **68**, 2239 (1946). (c) A similar result appears to have been observed with a cyclization performed in anhydrous hydrogen fluoride. See S. Bien, L. Cohen, and K. Scheinmann, *J. Chem. Soc.,* 1495 (1965).

epimerization of the initial product to a more stable stereoisomer may be observed. For example, addition of aluminum chloride to the *trans*-acid chloride [75] leads to formation of the more stable *cis*-ketone [76]. The same type of epimerization is

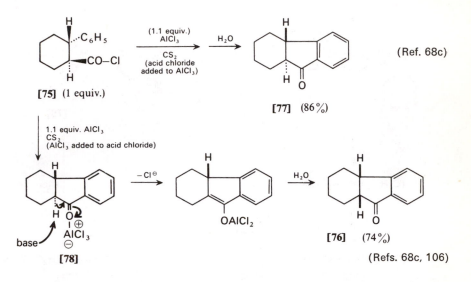

[75] (1 equiv.)

[77] (86%)

[78]

[76] (74%)

(Refs. 68c, 106)

(Ref. 68c)

seen in the previously mentioned cyclization of the acid [66] to the ketone [69]. Interestingly, the reverse procedure, in which the acid chloride [75] is added to excess aluminum chloride, produces the less stable *trans*-isomer [77]. In both cases, the initial product is believed to be the conjugate acid [78] of the *trans*-ketone. The formation of the conjugate acid [78] in the presence of excess acid chloride (which can serve as a base: see Chapter 8) allows enolization to occur and, consequently, permits epimerization of the asymmetric center adjacent to the carbonyl function. However, under conditions where aluminum chloride is always in excess, no base is present to abstract a proton from the conjugate acid [78] and epimerization is not observed.

Acylation of aromatic systems with the derivatives of α,β-unsaturated acids frequently yields mixtures of ketones and substituted acids.[107] These products are thought to arise by the pathways indicated in the following equation in which reaction occurs at the β-position of either the acid chloride or the initially formed α,β-unsaturated ketone (e.g., [79]). Further attack at the β-carbon of the initially formed α,β-unsaturated ketone (e.g., [79]) can be minimized by the use of low reaction temperatures with no excess aluminum chloride.[107b]

107. (a) K. M. Johnston and R. G. Shotter, *J. Chem. Soc.*, C, 1703 (1966); *ibid.*, C, 2476 (1967). (b) K. M. Johnston and J. F. Jones, *ibid.*, C, 814 (1969). (c) M. F. Ansell and G. F. Whitfield, *Tetrahedron Letters*, No. 26, 3075 (1968). (d) K. M. Johnston, *Tetrahedron*, 24, 5595 (1968).

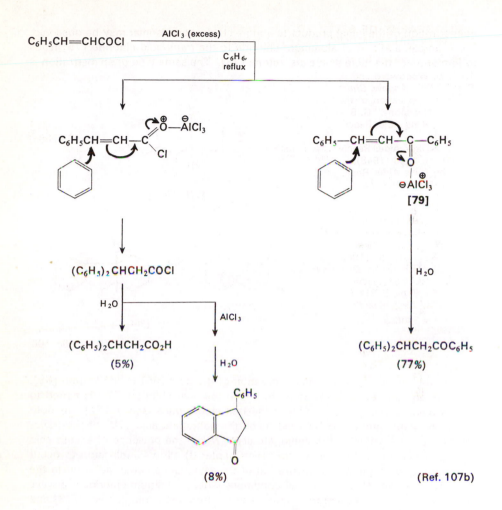

BIBLIOGRAPHY OF SELECTED REFERENCES TO SOME OTHER SYNTHETICALLY IMPORTANT REACTIONS

The Diels-Alder Reaction and Related Electrocyclic Processes

1. R. B. Woodward and R. Hoffmann, *The Conservation of Orbital Symmetry,* Academic Press, 1970; *Angew. Chem., Intern. Ed. Engl.,* **8**, 781 (1969).
2. "The Application of the Woodward-Hoffmann Orbital Symmetry Rules to Concerted Organic Reactions," G. B. Gill, *Quart. Rev.,* **22**, 338 (1968).
3. "The Diels-Alder Reaction with Maleic Anhydride," M. C. Kloetzel, *Org. Reactions,* **4**, 1 (1948).
4. "The Diels-Alder Reaction: Ethylenic and Acetylenic Dienophiles," H. L. Holmes, *Org. Reactions,* **4**, 60 (1948).
5. "The Diels-Alder Reaction: Quinones and Other Cyclenones," L. W. Butz and A. W. Rytina, *Org. Reactions,* **5**, 136 (1949).
6. "The Diene Synthesis," K. Alder in *Newer Methods of Preparative Organic Chemistry,* Vol. 1, Wiley-Interscience, New York, 1948, pp. 381–511.
7. "Stereochemistry of the Diels-Alder Reaction," J. G. Martin and R. K. Hill, *Chem. Rev.,* **61**, 537 (1961).
8. "Orientation in Diene-Synthesis and Its Dependence on Structure," Y. A. Titov, *Russ. Chem. Rev.,* **31**, 267 (1962).
9. A. S. Onishchenko, *Diene Synthesis,* Daniel Davey, New York, 1964.
10. "Diels-Alder Reactions. Part I: New Preparative Aspects," J. Sauer, *Angew. Chem., Intern. Ed. Engl.,* **5**, 211 (1966).
11. "Diels-Alder Reactions. Part II: The Reaction Mechanism," J. Sauer, *Angew. Chem., Intern. Ed. Engl.,* **6**, 16 (1967).
12. "The Reverse Diels-Alder or Retrodiene Reaction," H. Kwart and K. King, *Chem. Rev.,* **68**, 415 (1968).
13. "Diels-Alder Syntheses with Heteroatomic Compounds," S. B. Needleman and M. C. C. Kuo, *Chem. Rev.,* **62**, 405 (1962).
14. "The Chemistry of Perchlorocyclopentenes and Cyclopentadienes," H. E. Ungnade and E. T. McBee, *Chem. Rev.,* **58**, 249 (1958).
15. J. Hamer, ed., *1,4-Cycloaddition Reactions,* Academic Press, New York, 1967.
16. "The Ene Reaction," H. M. R. Hoffmann, *Angew. Chem., Intern. Ed. Engl.,* **8**, 556 (1969).
17. "Cyclobutane Derivatives from Thermal Cycloaddition Reactions," J. D. Roberts and C. M. Sharts, *Org. Reactions,* **12**, 1 (1962).
18. "Cycloadditions with Polar Intermediates," R. Gompper, *Angew. Chem., Intern. Ed. Engl.,* **8**, 312 (1969).
19. "Cycloadditions—Definition, Classification, and Characterization," R. Huisgen, *Angew. Chem., Intern. Ed. Engl.,* **7**, 321 (1968).
20. "The Claisen Rearrangement," D. S. Tarbell, *Org. Reactions,* **2**, 1 (1944).
21. "Rearrangements Proceeding Through 'No-Mechanism' Pathways: The Claisen, Cope, and Related Rearrangements," S. J. Rhoads in P. de Mayo, ed., *Molecular Rearrangements,* Vol. 1, Wiley-Interscience, New York, 1963, pp. 655–706.
22. "Molecular Rearrangements Related to the Claisen Rearrangement," A. Jefferson and F. Scheinmann, *Quart. Rev.,* **22**, 391 (1968).
23. "1,3-Dipolar Cycloadditions, Past and Future," R. Huisgen, *Angew. Chem., Intern. Ed. Engl.,* **2**, 565 (1963).
24. "Kinetics and Mechanism of 1,3-Dipolar Addition," R. Huisgen, *Angew. Chem., Intern. Ed. Engl.,* **2**, 633 (1963).
25. "Neues über 1,3-Cycloadditionen," R. Huisgen, *Helv. Chim. Acta,* **50**, 2421 (1967).
26. "Decomposition and Addition Reactions of Organic Azides," G. L'Abbe, *Chem. Rev.,* **69**, 345 (1969).
27. "Mechanisms of Cycloaddition," P. D. Bartlett, *Quart. Rev.,* **24**, 473 (1970).

The Use of Organomagnesium and Organolithium Reagents in Synthesis

1. G. E. Coates, M. L. H. Green, and K. Wade, *Organometallic Compounds,* Vols. 1 and 2, Methuen, London, 3rd ed., 1967–1968.

2. "Methods for the Preparation of Organometallic Compounds," R. G. Jones and H. Gilman, *Chem. Rev.,* **54**, 835 (1954).

3. M. S. Kharasch and O. Reinmuth, *Grignard Reactions of Nonmetallic Substances,* Prentice-Hall, New York, 1954.

4. "Recent Advances in the Chemistry of Organomagnesium Compounds," B. J. Wakefield, *Organometal. Chem. Rev.,* **1**, 131 (1966).

5. "Organometallic Reaction Mechanisms," R. E. Dessy and W. Kitching, *Adv. Organometal. Chem.,* **4**, 267 (1966).

6. "Grignard Reagents. Composition and Mechanisms of Reaction," E. C. Ashby, *Quart. Rev.,* **21**, 259 (1967).

7. "Alkenylmagnesium Halides," H. Normant, *Adv. Org. Chem.,* **2**, 1 (1960).

8. "The Synthesis of Ketones from Acid Halides and Organometallic Compounds of Magnesium, Zinc, and Cadminium," D. A. Shirley, *Org. Reactions,* **8**, 28 (1954).

9. "The Reaction Between Grignard Reagents and The Oxirane Ring," N. G. Gaylord and E. I. Becker, *Chem. Rev.,* **49**, 413 (1951).

10. "Ethynyl Ethers and Thioethers as Synthetic Intermediates," J. F. Arens, *Adv. Org. Chem.,* **2**, 117 (1960).

11. "The Coupling of Acetylenic Compounds," G. Eglinton and W. McCrae, *Adv. Org. Chem.,* **4**, 225 (1963).

12. "Syntheses with Organolithium Compounds," G. Wittig in *Newer Methods of Preparative Organic Chemistry,* Vol. 1, Wiley-Interscience, New York, 1948, pp. 571–591.

13. "The Halogen-Metal Interconversion with Organolithium Compounds," R. G. Jones and H. Gilman, *Org. Reactions,* **6**, 339 (1951).

14. "The Metalation Reaction with Organolithium Compounds," H. Gilman and J. W. Morton, Jr., *Org. Reactions,* **8**, 258 (1954).

15. "The Structures of Organolithium Compounds," T. L. Brown, *Adv. Organometal. Chem.,* **3**, 365 (1965).

16. "The Preparation of Ketones from the Reaction of Organolithium Reagents and Carboxylic Acids," M. J. Jorgenson, *Org. Reactions,* **18**, 1 (1970).

17. "Chemistry of Stable α-Halogenoorganolithium Compounds and the Mechanism of Carbenoid Reactions," G. Köbrich and co-workers, *Angew. Chem., Intern. Ed. Engl.,* **6**, 41 (1967).

18. "The Role of Ate Complexes as Reaction-Determining Intermediates," G. Wittig *Quart. Rev.,* **20**, 191 (1966).

19. "Structures and Reactions of Organic Ate-Complexes," W. Tochtermann, *Angew. Chem., Intern. Ed. Engl.,* **5**, 351 (1966).

Elimination Reactions to Form Carbon-Carbon Multiple Bonds

1. D. V. Banthorpe, *Elimination Reactions,* Elsevier, New York, 1963.

2. "The Preparation of Substituted Styrenes by Methods Not Involving Hydrocarbon Cracking," W. S. Emerson, *Chem. Rev.,* **45**, 347 (1949).

3. "The Mechanism of Bimolecular β-Elimination Reactions," J. F. Bunnett, *Angew. Chem., Intern. Ed. Engl.,* **1**, 225 (1962).

4. "Olefins from Amines : The Hoffmann Elimination Reaction and Amine Oxide Pyrolysis," A. C. Cope and E. R. Trumbull, *Org. Reactions,* **11**, 317 (1960).

5. "Pyrolytic Cis Eliminations," C. H. DePuy and R. W. King, *Chem. Rev.,* **60**, 431 (1960).

6. "The Preparation of Olefins by the Pyrolysis of Xanthates. The Chugaev Reaction," H. R. Nace, *Org. Reactions,* **12**, 57 (1962).

7. "Dehydration of Alcohols on Aluminum Oxide," H. Knözinger, *Angew. Chem., Intern. Ed. Engl.,* **7**, 791 (1968).

8. "Eliminations in Cyclic *cis-trans*-Isomers," W. Hückel and M. Hanack, *Angew. Chem., Intern. Ed. Engl.,* **6**, 534 (1967).

9. "The Carbanion Mechanism of Olefin-forming Elimination," D. J. McLennan, *Quart. Rev.,* **21**, 490 (1967).

10. "The Synthesis of Acetylenes," T. L. Jacobs, *Org. Reactions,* **5**, 1 (1949).

11. R. A. Raphael, *Acetylenic Compounds in Organic Synthesis,* Academic Press, New York, 1955.

12. "The Formation of the Acetylenic Bond," W. Franke, W. Ziegenbein, and H. Meister in W. Foerst, ed., *Newer Methods of Preparative Organic Chemistry,* Vol. 3, Academic Press, New York, 1964, pp. 425–450.

13. "Eliminations from Olefins," G. Köbrich, *Angew. Chem., Intern., Ed. Engl.* **4**, 49 (1965).

14. "Heterolytic Fragmentation. A Class of Organic Reactions," C. A. Grob and P. W. Schiess, *Angew. Chem., Intern. Ed. Engl.,* **6**, 1 (1967).

15. "Mechanisms and Stereochemistry of Heterolytic Fragmentation," C. A. Grob, *Angew. Chem., Intern. Ed. Engl.,* **8** 535 (1969).

Acylation at Oxygen and Nitrogen

1. P. F. G. Praill, *Acylation Reactions,* MacMillan, New York, 1963.

2. "An Outline of Acylation," D. P. N. Satchell, *Quart. Rev.,* **17**, 160 (1963).

3. "The Reactions of Aliphatic Acid Chlorides," N. O. V. Sonntag, *Chem. Rev.,* **52**, 237 (1953).

4. "Protective Groups," J. F. W. McOmie, *Adv. Org. Chem.,* **3**, 191 (1963).

5. "Advances in the Chemistry of Carbodiimides," F. Kurzer and F. Douraghi-Zadeh, *Chem. Rev.,* **67**, 107 (1967).

6. "Syntheses Using Heterocyclic Amides (Azolides)," H. A. Staab and W. Rohr in W. Foerst, ed., *Newer Methods of Preparative Organic Chemistry,* Vol. 5, Academic Press, New York, 1968, pp. 61–108.

7. "Cyclic Carboxylic Monoimides," M. K. Hargreaves, J. G. Pritchard, and H. R. Dave, *Chem. Rev.,* **70**, 439 (1970).

8. "Esters of Carbamic Acid," P. Adams and F. A. Baron, *Chem. Rev.,* **65**, 567 (1965).

9. "The Synthesis of β-Lactams," J. C. Sheehan and E. J. Corey, *Org. Reactions,* **9**, 388 (1957).

10. "Azlactones," H. E. Carter, *Org. Reactions,* **3**, 198 (1946).

11. "Methods of Peptide Synthesis," T. Wieland and H. Determann, *Angew. Chem., Intern. Ed. Engl.,* **2**, 358 (1963).

12. "Preparation of Peptides and Ureas Using Reactive Amides or Imides," S. Goldschmidt and H. L. Krauss in W. Foerst, ed., *Newer Methods of Preparative Organic Chemistry,* Vol. 2, Academic Press, New York, 1963, pp. 31–50.

13. "Synthesis of Peptides with Mixed Anhydrides," N. F. Albertson, *Org. Reactions,* **12**, 157 (1962).

14. "Selectively Removable Amino Protective Groups Used in the Synthesis of Peptides," R. A. Boissonnas, *Adv. Org. Chem.,* **3**, 159 (1963).

15. "Acidic and Basic Amide Hydrolysis," C. O'Connor, *Quart. Rev.,* **24**, 553 (1970).

INDEX

Acetals, reduction, 85
Acetoacetic ester condensation, 734–736, 740–746
  forcing conditions, 735, 736, 744
  with ester mixtures, 744–746
α-Acetoxy ketones, synthesis, 380–384, 398–400
Acetoxylation with lead tetraacetate, 376–384, 399, 400
Acetoxymercurium ion intermediates, 387–398
Acetylchromic anion, 258
Acetylenes, 818, 819
  acylation, 788, 789, 796, 797
  epoxidation, 313
  halogenation, 430, 431
  hydrogenation to olefins, 19, 20
  isomerization to allenes, 208, 211
  oxidation, 378, 397, 398
  reaction with borane, 119, 121, 125, 126
  reaction with diborane, 119, 121, 125, 126
  reaction with hydrogen halides, 446, 448, 449
  reaction with iodine azide, 445
  rearrangement during acylation, 788, 789
  reduction, 19, 20, 91, 92, 119, 121, 125, 126, 172, 205–211, 252
  reduction with metal hydrides, 91, 92
  reductive cyclization, 172
  stereochemistry of acylation, 788, 789
  synthesis, 246, 247, 385, 406
Acetylenic esters, cycloaddition to enamines, 616, 618
  as Michael acceptors, 608, 618
Acetylenic ethers, 818
Acid catalysts
  aluminum chloride, 533, 666, 738, 772, 778, 783, 786–793, 797–799, 801–812, 815, 816
  antimony pentafluoride, 787

Acid catalysts—*Continued*
  boric acid, 640
  boron trichloride, 786
  boron trifluoride, 533, 654, 718, 772–777, 781, 782, 786
  boron trifluoride-acetic acid complex, 775, 776
  boron trifluoride etherate, 654, 718, 773, 775–777, 781
  hydrogen chloride, 640, 641
  hydrogen fluoride, 665, 773, 786, 787, 812–814
  Lewis acids, 734, 786
  perchloric acid, 533, 780, 782
  phosphoric acid, 799
  phosphorus pentoxide, 795
  polyphosphoric acid, 781, 786, 787, 794–796, 809, 810, 813
  stannic chloride, 786, 792, 799
  sulfuric acid, 640, 642, 643, 666, 786, 787, 808, 809
  titanium tetrachloride, 786, 801
  p-toluenesulfonic acid, 642, 781
  trifluoroacetic acid, 655, 796, 797
  zinc chloride, 533, 534, 632, 665, 738, 786, 792–794, 799, 800
Acidities
  of active methylene compounds, 492–494
  effect of alkyl substituents, 492–494, 557–560
  effect of halogen substituents, 492, 494
  of organic compounds, 492–494
Activated aryl halides, as alkylating agents, 536
Active methylene compounds
  acylation, 734–765, 772–786
  addition to vinylsulfonium salts, 721
  alkylation, 510–623
  equilibrium acidities, 492–494, 501, 502, 557–560

821